T0254182

Mechanisms and Machine Science

Volume 109

This book series establishes a well-defined forum for monographs, edited Books, and proceedings on mechanical engineering with particular emphasis on MMS (Mechanism and Machine Science). The final goal is the publication of research that shows the development of mechanical engineering and particularly MMS in all technical aspects, even in very recent assessments. Published works share an approach by which technical details and formulation are discussed, and discuss modern formalisms with the aim to circulate research and technical achievements for use in professional, research, academic, and teaching activities.

This technical approach is an essential characteristic of the series. By discussing technical details and formulations in terms of modern formalisms, the possibility is created not only to show technical developments but also to explain achievements for technical teaching and research activity today and for the future.

The book series is intended to collect technical views on developments of the broad field of MMS in a unique frame that can be seen in its totality as an Encyclopaedia of MMS but with the additional purpose of archiving and teaching MMS achievements. Therefore, the book series will be of use not only for researchers and teachers in Mechanical Engineering but also for professionals and students for their formation and future work.

The series is promoted under the auspices of International Federation for the Promotion of Mechanism and Machine Science (IFToMM).

Prospective authors and editors can contact Mr. Pierpaolo Riva (publishing editor, Springer) at: pierpaolo.riva@springer.com

Indexed by SCOPUS and Google Scholar.

More information about this series at https://link.springer.com/bookseries/8779

Milan Rackov · Radivoje Mitrović · Maja Čavić
Editors

Machine and Industrial Design in Mechanical Engineering

Proceedings of KOD 2021

Editors
Milan Rackov
Faculty of Technical Sciences
University of Novi Sad
Novi Sad, Serbia

Radivoje Mitrović
Faculty of Mechanical Engineering
University of Belgrade
Belgrade, Serbia

Maja Čavić
Faculty of Technical Sciences
University of Novi Sad
Novi Sad, Serbia

ISSN 2211-0984 ISSN 2211-0992 (electronic)
Mechanisms and Machine Science
ISBN 978-3-030-88467-3 ISBN 978-3-030-88465-9 (eBook)
https://doi.org/10.1007/978-3-030-88465-9

This Springer imprint is published by the registered company Springer Nature Switzerland AG
The registered company address is: Gewerbestrasse 11, 6330 Cham, Switzerland

Committees

Conference Chairs

Milan RACKOV, Novi Sad
Radivoje MITROVIĆ, Belgrade
Maja ČAVIĆ, Novi Sad

Honorary Committee

Carmen ALIC, Hunedoara
Kyrill ARNAUDOW, Sofia
Radoš BULATOVIĆ, Podgorica
Danica JOSIFOVIĆ, Kragujevac
Siniša KUZMANOVIĆ, Novi Sad
Vojislav MILTENOVIĆ, Niš
Slobodan NAVALUŠIĆ, Novi Sad
Milosav OGNJANOVIĆ, Belgrade
Victor E. STARZHINSKY, Gomel
Miroslav VEREŠ, Bratislava
Simon VILMOS, Budapest
Aleksandar VULIĆ, Niš
Miodrag ZLOKOLICA, Novi Sad

With Support of

ADEKO—Association for Design, Elements and Constructions
IFToMM—International Federation for the Promotion of Mechanism and Machine Science

Conference Proceedings Published by

International Scientific Committee

Sandro DOBOVIČEK
Rade DOROSLOVAČKI
Mircea-Viorel DRAGOI
Ágota DRÉGELYI-KISS
Zorica ĐORĐEVIĆ
Aleksija ĐURIĆ
Igor FÜRSTNER
Jozsef GAL
Tale GERAMITCIOSKI
Gilbert Rainer GILLICH
Ladislav GULAN
Jana GULANOVA
Fuad HADŽIKADUNIĆ
Miodrag HADŽISTEVIĆ
Gorazd HLEBANJA
Sava IANICI
Milan IKONIĆ
Lozica IVANOVIĆ
Kristina JAKIMOVSKA
Juliana JAVOROVA
Zorana JELI
Dejan JEREMIĆ
Mitar JOCANOVIĆ
Jasmin KALJUN
Velibor KARANOVIĆ
Imre KISS
Darko KNEŽEVIĆ
Peter KOSTAL
Nenad KOSTIĆ
Dražan KOZAK
Milija KRAIŠNIK
Ružena KRÁLIKOVÁ
Remigiusz LABUDZKI
Sergey A. LAGUTIN
Tihomir LATINOVIĆ
Stanisław LEGUTKO
Erwin-Christian LOVASZ
Dejan LUKIĆ
Tamas MANKOVITS
Aleksandar MARINKOVIĆ
Dragan MARINKOVIĆ
Nenad MARJANOVIĆ
Biljana MARKOVIĆ
Miloš MATEJIĆ
Waldemar MATYSIAK

Štefan MEDVECKY
Athanassios MIHAILIDIS
Cristina Carmen MIKLOS
Imre Zsolt MIKLOS
Dragan MILČIĆ
Miloš MILOVANČEVIĆ
Dobrica MILOVANOVIĆ
Aleksandar MILTENOVIĆ
Žarko MIŠKOVIĆ
Natasa NAPRSTKOVA
Dorian NEDELCU
Peter NENOV
Vlastimir NIKOLIĆ
Dragoljub NOVAKOVIĆ
Milenko OBAD
Sándor PÁLINKÁS
Zoran PANDILOV
Nenad T. PAVLOVIĆ
José I. PEDRERO
Alexandru-Viorel PELE
László POKORÁDI
Vanessa PRAJOVA
Sanel PURGIĆ
Zoltán RAJNAI
Nebojša RAŠOVIĆ
Sorin RAŢIU
Mileta RISTIVOJEVIĆ
Božidar ROSIĆ
Anna RUDAWSKA
Roman RUZAROVSKY
József SÁROSI
Dan SĂVESCU
Mile SAVKOVIĆ
Frank Helmut SCHAEFER
Serge V. SHIL'KO
Hasan SMAJIĆ
Dušan STAMENKOVIĆ
Sergiu-Dan STAN
Stevan STANKOVSKI
Blaža STOJANOVIĆ
Stefan STEFANOV
Aleksandar SUBIĆ
Senka ŠEKULARAC-IVOŠEVIĆ
Dávid TAKACS
Amra TALIĆ-ČIKMIŠ

Milan TICA
Mykola TKACHUK
Radoslav TOMOVIĆ
Adam TOROK
Lucian TUDOSE
Krasimir TUJAROV
Attila TURI
Nicolae UNGUREANU
Karol VELISEK
Adisa VUČINA
Krešimir VUČKOVIĆ
Đorđe VUKELIĆ
Igor VUŠANOVIĆ
Wojciech WIELEBA
Milan ZELJKOVIĆ
Samir ŽIC
Aleksandar ŽIVKOVIĆ

Organizing Committee

Mirjana BOJANIĆ ŠEJAT
Sanja BOJIĆ
Dijana ČAVIĆ
Jovan DORIĆ
Radomir ĐOKIĆ
Ivan KNEŽEVIĆ
Srđan NIKAČEVIĆ
Nebojša NIKOLIĆ
Marko PENČIĆ
Nenad POZNANOVIĆ
Dragan RUŽIĆ
Boris STOJIĆ
Atila ZELIĆ
Ninoslav ZUBER
Dragan ŽIVANIĆ

Preface

Dear Ladies and Gentlemen, respectable Colleagues and Friends of KOD,

The 11th International Conference on Machine and Industrial Design in Mechanical Engineering (KOD 2021) is being organized by the Faculty of Technical Sciences, University of Novi Sad, from June 10 till 12, 2021.

The Conference KOD 2021 is being held under very specific conditions due to the COVID-19 pandemic, for the first time. However, we hope that the pandemic stays behind us and that every day brings us all safety. With that in mind, the Conference KOD is organized as a hybrid event. We will do our best to connect all participants who stay in their countries with all the participants who attend the conference in person.

The conference chair would like to extend special gratitude to the editor of the Springer Book Series Mechanisms and Machine Science, Prof. Marco Ceccarelli, for supporting the conference and giving the chance to all authors to publish a paper in an edition titled Machine and Industrial Design in Mechanical Engineering—Proceedings of KOD 2021.

The basic goal of this conference is to assemble experienced researchers and practitioners from universities, scientific institutes and different enterprises and organizations from within this field. Also, it should instigate more intensive cooperation and exchange of practical professional experiences in the field of shaping, forming and design in mechanical and graphical engineering, industrial design and shaping, product development and management. As there is a pressing need, under the cover of Industry 4.0, for more effective, simpler, smaller, cheaper, noiseless and more esthetically pleasing products that can easily be recycled and are not harmful to the environment, the cooperation between specialists in these fields should certainly be well developed and intricate.

Finally, we would like to thank all the people who have supported the conference and have helped and encouraged us in all the activities.

We are very grateful to our reviewers whose enormous work of assessing the papers is gratefully appreciated.

We would also like to express our appreciation to our keynote speakers for their invaluable contribution.

We would like to thank the authors themselves for contributing research papers, without whose expert input there would have been no conference.

And last but not least, we are pleased to acknowledge the assistance provided by the members of the International Scientific Committee and Technical Program Committee of the Conference KOD 2021.

We wish You to have a safe and stimulating conference that will bring about numerous fruitful discussions and lay the foundations for future collaboration.

Have an unforgettable stay in the city of Novi Sad, catch up with old friends, and make some new ones.

We wish You good healthish and success in Your further research and great fortune and happiness in Your personal life.

Novi Sad, Serbia
10th June 2021

Prof. Milan Rackov, Ph.D. Eng.
Prof. Radivoje Mitrović, Ph.D. Eng.
Prof. Maja Čavić, Ph.D. Eng.
Conference Chairs of KOD 2021

Contents

Part I
Invited Keynote Lectures

Chapter 1
Analytical Model for Spur Gears with Profile Modification: Simulation of the Meshing Stiffness, Load Sharing, and Transmission Error

José I. Pedrero

Abstract The determination of the load sharing, meshing stiffness, and transmission error is essential for evaluating the behavior of the gear pair in such important aspects as the strength, dynamic load, load carrying capacity, and noise and vibrations levels. The use of computational techniques as the Finite Element Method provides accurate results, but it is time consuming and requires high computational cost, which involve great difficulties for performing repetitive calculations as required in the initial design steps. In this paper, a simple, analytical model for spur gears with profile modification is presented. The load sharing, transmission error, and time varying mesh stiffness are expressed as simple functions of the gear geometry and profile modification, which provide accurate enough results to take preliminary decisions for design.

Keywords Spur gears · Meshing stiffness · Load sharing · Transmission error · Profile modification

1.1 Introduction

Spur gear drives are extensively used in the aeronautical, marine, and automotive industries, as well as many other mechanical applications. But all these applications are currently subject to increasingly demanding requirements from mechanical, environmental, and sustainability point of view. It is clear that not only classic criteria—as power density, efficiency or reliability—should be govern the calculations, but also modern demands, as noise emissions, vibration levels, or maintenance requirements, should be also kept in mind during the design process.

The load sharing and transmission error have decisive influence on the gear behavior. The load sharing affects the critical stresses and load conditions, which determines the load carrying capacity; the efficiency and wear are also affected by the load and teeth geometry; the transmission error is the source of noise, vibrations, and dynamic load. To consider all these aspects in the design process, a model

J. I. Pedrero (✉)
Dep. Mecánica, UNED, Juan del Rosal 12, 28040 Madrid, Spain
e-mail: jpedrero@ind.uned.es

M. Rackov et al. (eds.), *Machine and Industrial Design in Mechanical Engineering*, Mechanisms and Machine Science 109,
https://doi.org/10.1007/978-3-030-88465-9_1

relating the meshing stiffness, load sharing, transmission error, and teeth geometry is required. Finite Element models provide accurate and reliable results but require a lot of time for the model preparation and high computational cost. This makes it a useful tool for the validation or final design steps but involve great difficulties for performing repetitive calculations at the initial steps. Calculation methods based on analytic equations are therefore suitable for design, even if less accurate than FEM models.

In this paper, a simple analytical model for the meshing stiffness (MS), load sharing, and quasi static transmission error (QSTE) for spur gears, including the influence of profile modification, is presented. The load sharing ratio (LSR) is calculated from the hypothesis of minimum elastic potential energy, which results in equal deflection of all the tooth pairs in simultaneous contact. From this result, the gear MS and the QSTE are calculated from the equations of the linear theory of elasticity. The profile modification is considered as a gap between meshing teeth which reduces the tooth pair deflection, and consequently the tooth pair stiffness, of the specific pair.

1.2 Single Mesh Stiffness

The single mesh stiffness (SMS), or mesh stiffness of the tooth pair, can be computed from [1]:

$$K_M = \left(\frac{1}{k_{x1}} + \frac{1}{k_{s1}} + \frac{1}{k_{n1}} + \frac{1}{k_{x2}} + \frac{1}{k_{s2}} + \frac{1}{k_{n2}} + \frac{1}{k_H} + \frac{1}{k_{B1}} + \frac{1}{k_{B2}} \right)^{-1} \tag{1.1}$$

where k_x is the bending stiffness, k_s the shear stiffness, k_n the compressive stiffness, k_H the contact stiffness, k_B the gear body stiffness, and subscripts 1 and 2 denote the driving and driven gear, respectively. k_x, k_s, and k_n can be computed from the equations of the elastic potential energy [2], although the equations of the tooth profile are complicated, specifically at the root trochoid, and the integrals should be solved by numerical methods. For k_H and k_B several approaches can be considered. Usually, k_H is calculated from the Hertz [3] or Weber and Banaschek [4] equations; k_B is often evaluated from the Weber and Banaschek [4] or Sainsot et al. [5] equations. The most suitable approach will probably depend on the specific gear set to study (material, ring body thickness, etc.). All of them can be considered by replacing the proper equation in Eq. (1.1).

Obviously, the SMS depends on the contact point, which can be described by the following parameter:

$$\xi = \frac{z_1}{2\pi} \sqrt{\frac{r_{C1}^2}{r_{b1}^2} - 1} \tag{1.2}$$

in which z is the number of teeth, r_C the radius of the contact point, and r_b the base radius. It can be proved that ξ is also a linear coordinate along the line of action, which takes the following values at the tangency points with driving and driven gear base circumference, T_1 and T_2:

$$\xi_{T1} = 0$$
$$\xi_{T2} = \frac{z_1 + z_2}{2\pi} \tan \alpha'_t \tag{1.3}$$

being α'_t the operating transverse pressure angle. This means that the SMS of the tooth pair –as well as any other parameter depending on the contact point– can be expressed as a function of the ξ coordinate along the pressure line, $K_M = K_M$ (ξ). It can be proved that the difference between the contact point parameter of the theoretical outer and inner points of contact, ξ_o and ξ_{inn}, is equal to the contact ratio ε_α:

$$\xi_o - \xi_{inn} = \varepsilon_\alpha \tag{1.4}$$

and the difference between the contact point parameter of two pairs is:

$$\xi_{(i+j)} = \xi_{(i)} + j \tag{1.5}$$

The curve of SMS $K_M(\xi)$ will depend on the considered approaches for the contact stiffness and gear body stiffness; however, the shape of the curve is always quite similar. Figure 1.1 shows the normalized SMS curve $K_M(\xi)/K_{M\max}$ for six different combinations of the approaches above. In all the cases the curve has a parabolic shape, symmetric respect the midpoint of the domain—i.e., the theoretical contact

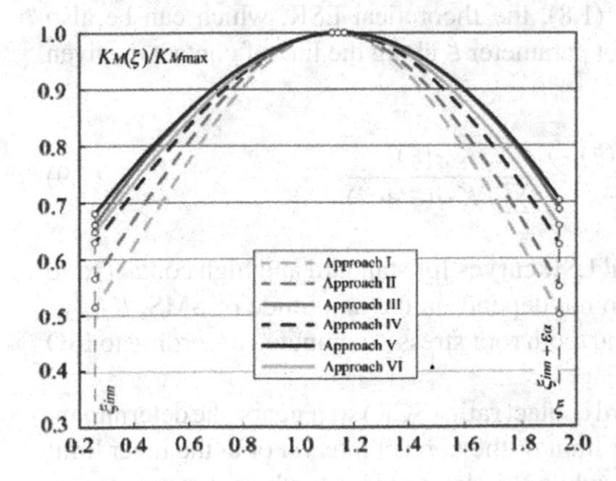

	BSC	Contact	GB	K_1	K_2
I	C	(-)	(-)	0.86	0.80
II	C	H	(-)	1.11	1.17
III	C	W	(-)	2.40	4.07
IV	C	H	W	1.56	2.00
V	C	W	W	2.50	4.38
VI	C	50% W	W	4.38	2.81

BSC: bending, shear, compression.
C / (-): considered / neglected.
H / W: Hertzian / Weber approach.

Fig. 1.1 Curves of SMS from different approaches

interval—, with a maximum at this point. The same shape is obtained from other numerical or computational techniques, as FEM [6].

These curves of SMS can be accurately approximated by the Eq. [1]:

$$K_M(\xi) = K_{M\max} \cos b_0(\xi - \xi_m) \tag{1.6}$$

with:

$$b_0 = \left[\frac{1}{2}\left(\kappa_1 + \frac{\varepsilon_\alpha}{2}\right)^2 - \kappa_2\right]^{-\frac{1}{2}}; \ \xi_m = \xi_{inn} - \frac{\varepsilon_\alpha}{2}; \ \xi_{inn} \le \xi \le \xi_o \tag{1.7}$$

and $K_{M\max}$ can be computed as described in [1]. To adjust the Eq. (1.6) to each specific approach, appropriate values of coefficients κ_1 and κ_2 should be chosen. Table in Fig. 1.1 shows the values of the coefficients for six approaches represented in the diagram. New values can be easily calculated for other approaches.

1.3 Theoretical Load Sharing Ratio and Determinant Stresses

From the minimum elastic potential energy hypothesis, the load at the tooth pair (*i*) can be computed from [2]:

$$F_{(i)} = \frac{K_{M(i)}}{\sum_j K_{M(j)}} F_T \tag{1.8}$$

where F_T is the total transmitted load and the sum is extended to all the tooth pairs in simultaneous contact. From Eq. (1.8), the theoretical LSR, which can be also expressed as a function of the contact parameter ξ along the line of contact, is given by:

$$R(\xi) = \frac{F(\xi)}{F_T} = \frac{K_M(\xi)}{\sum_j K_M(\xi + j)} \tag{1.9}$$

Figure 1.2 presents the theoretical LSR curves for standard and high contact ratio (HCR) spur gears. Note that they do not depend on the amplitude of SMS, $K_{M\max}$. The evolution of the contact stress and tooth root stress—computed according to ISO 6336 [7], are also shown in Fig. 1.2.

It can be observed that, for standard contact ratio (SCR) spur gears, the determinant contact stress may occur at the inner limit of the contact interval or at the inner limit of the single tooth contact interval, while the determinant tooth root stress occurs for the load acting at the outer limit of the single tooth contact interval. For HCR

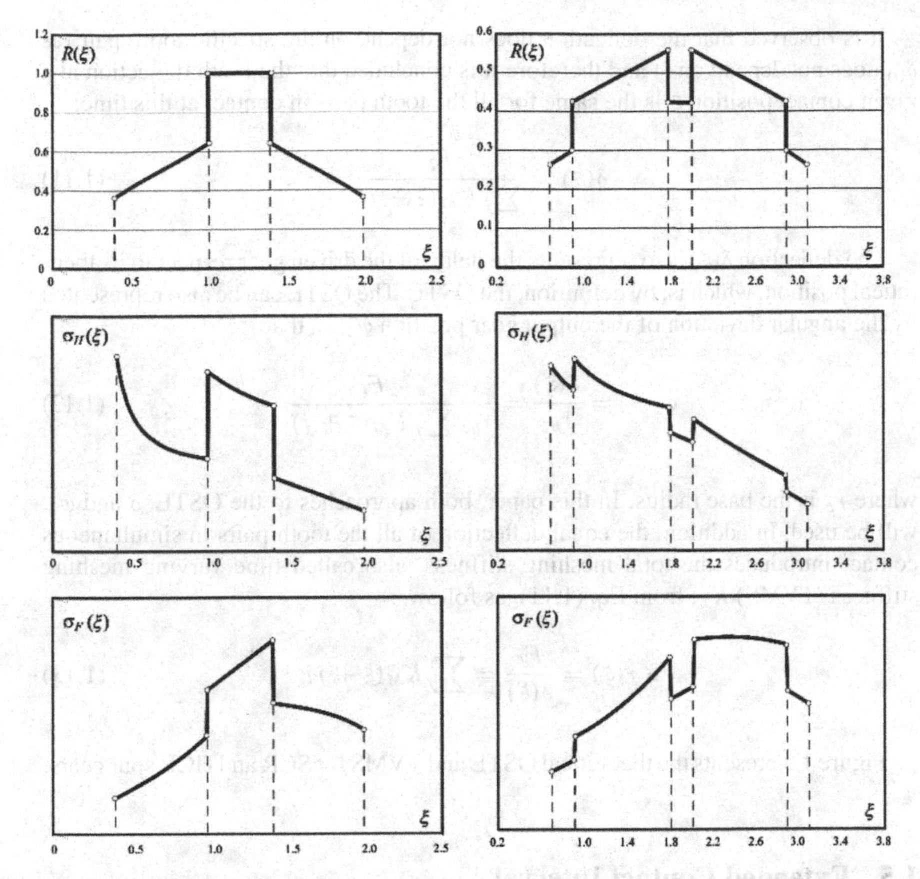

Fig. 1.2 Curves of theoretical LSR (up), contact stress (middle), and tooth root stress (down) for SCR (left) and HCR (right) spur gears

spur gears, critical contact stress occurs at the inner point of two pair tooth contact (though in some cases may occur at the inner point of contact), while the critical tooth root stress arises for contact at some point of the outer interval of two pair tooth contact, along which the tooth root stress is quite uniform.

1.4 Tooth Pair Deflection: Time Varying Meshing Stiffness and Quasi Static Transmission Error

The tooth pair deflection δ can be computed from the SMS and the LSR as follows:

$$\left(\begin{array}{l} F_{(i)} = K_{M(i)}\delta_{(i)} \\ F_{(i)} = \frac{K_{M(i)}}{\sum_j K_{M(j)}} F_T \end{array} \right) \Rightarrow \delta_{(i)} = \frac{F_T}{\sum_j K_{M(j)}} \tag{1.10}$$

It is observed that the deflection does not depend on the specific tooth pair (as $\delta_{(i)}$ does not depend on i) and therefore it is concluded that the tooth deflection at a given contact position ξ is the same for all the tooth pairs in contact at this time:

$$\delta(\xi) = \frac{F_T}{\sum_j K_M(\xi + j)} \tag{1.11}$$

The deflection $\delta(\xi)$ also represents the delay of the driven gear respect to its theoretical position, which is, by definition, the QSTE. The QSTE can be also represented by the angular deviation of the output gear position φ_2, so that:

$$\varphi_2(\xi) = \frac{\delta(\xi)}{r_{b2}} = \frac{1}{r_{b2}} \frac{F_T}{\sum_j K_M(\xi + j)} \tag{1.12}$$

where r_b is the base radius. In this paper, both approaches to the QSTE, δ and φ_2, will be used. In addition, the equal deflection of all the tooth pairs in simultaneous contact introduces the total meshing stiffness, also called time varying meshing stiffness (TVMS) K_T, from Eq. (1.11), as follows:

$$K_T(\xi) = \frac{F_T}{\delta(\xi)} = \sum_j K_M(\xi + j) \tag{1.13}$$

Figure 1.3 presents the theoretical QSTE and TVMS for SCR and HCR spur gears.

1.5 Extended Contact Interval

Due to the output gear delay, contact between the driving tooth root and the driven tooth tip starts earlier than expected and outside the line of action, which induces the so-called mesh-in impact. Figure 1.4 shows the delayed position of the output gear tooth, corresponding to point a at the line of action, when the driving gear tooth position is depicted by point b. The segment $\overline{ab}$ represents the delay of the driven gear and its length is equal to that of segment $\overline{cd}$, which represents the tooth pair deflection. Due to this delay/deflection, the root of the incoming driving tooth hits the tip of the driven one at point I in Fig. 1.4, which induces the mesh-in impact.

The theoretical inner point of contact is point e in Fig. 1.4 and corresponds to $\xi = \xi_{inn}$. The actual inner point of contact is point b and will correspond to $\xi = \xi_{\min}$. The same occurs at the end of contact with the theoretical outer point of contact ξ_o and the actual outer point of contact $\xi_{\max}$. The extended contact interval will be $\xi_{\min} \le \xi \le \xi_{\max}$. The limits $\xi_{\min}$ and $\xi_{\max}$ can be approximately computed from [8]:

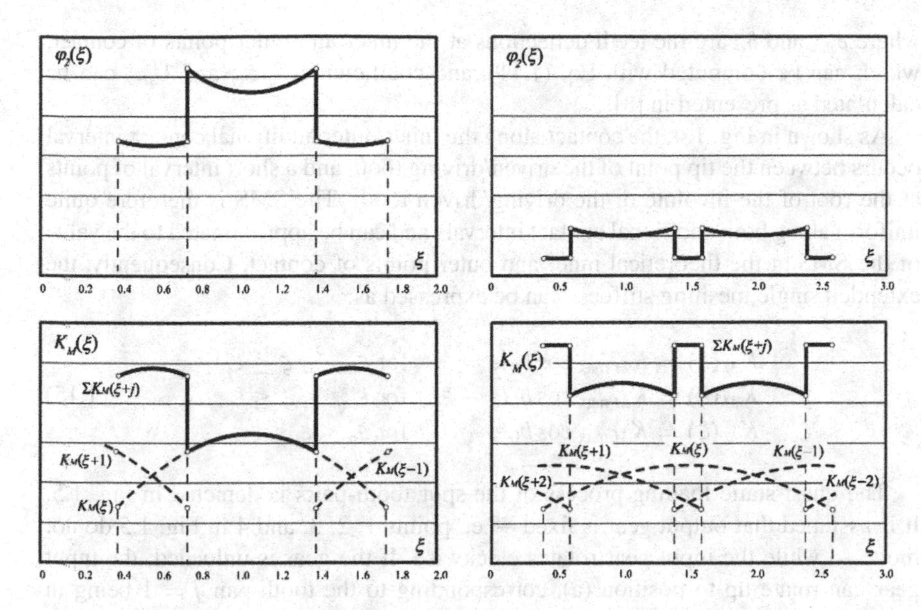

Fig. 1.3 Curves of theoretical QSTE (up) and TVMS (down) for SCR (left) and HCR (right) spur gears

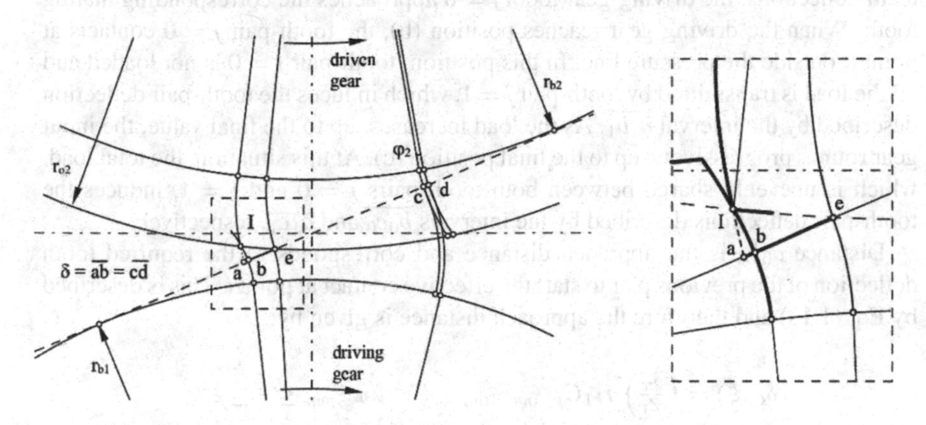

Fig. 1.4 Mesh-in impact due to load induced tooth deflections

$$
\xi_{\min} \approx \xi_{inn} - \frac{z_1}{2\pi}\sqrt{\frac{1}{C_{p-inn}}\left(\frac{\delta_{inn}}{r_{b1}}\right)}
$$

$$
\xi_{\min} \approx \xi_o + \frac{z_1}{2\pi}\sqrt{\frac{1}{C_{p-o}}\left(\frac{\delta_o}{r_{b1}}\right)} \tag{1.14}
$$

where δ_{inn} and δ_o are the teeth deflections at the inner and outer points of contact, which can be computed with Eq. (1.11), and coefficients $C_{p\text{-}inn}$ and $C_{p\text{-}o}$ can be calculated as presented in [8].

As shown in Fig. 1.4, the contact along the inner/outer additional contact interval occurs between the tip point of the driven/driving tooth and a short interval of points at the root of the involute of the driving/driven tooth. The SMS is therefore quite uniform along both additional contact intervals and can be approximated to the value of the SMS at the theoretical inner and outer points of contact. Consequently, the extended single meshing stiffness can be expressed as:

$$\begin{array}{ll} K_M(\xi) = K_{M\max} \cos b_0 \frac{\varepsilon_\alpha}{2} & \text{for } \xi_{\min} \le \xi \le \xi_{inn} \\ K_M(\xi) = K_{M\max} \cos b_0 (\xi - \xi_m) & \text{for } \xi_{inn} \le \xi \le \xi_o \\ K_M(\xi) = K_{M\max} \cos b_0 \frac{\varepsilon_\alpha}{2} & \text{for } \xi_o \le \xi \le \xi_{\max} \end{array} \tag{1.15}$$

The quasi-static loading process of the spur tooth-pairs is depicted in Fig. 1.5. It is assumed that output gear is fixed—i.e., points 1, 2, 3, and 4 in Fig. 1.5 do not move—, while the input gear rotates clockwise. If the gear is unloaded, the input gear can rotate up to position (a), corresponding to the tooth-pair $j = 1$ being in contact at point a_1 on the pressure line. At this position, the incoming tooth-pair $j = 0$ is not in contact. As the load increases, the driving gear rotates and, due to the teeth deflections, the driving-gear tooth $j = 0$ approaches the corresponding mating tooth. When the driving gear reaches position (b), the tooth-pair $j = 0$ contacts at point i, outside the pressure line. In this position, tooth-pair $j = 0$ is not loaded and all the load is transmitted by tooth-pair $j = 1$, which induces the tooth-pair deflection described by the interval $\overline{a_1 b_1}$. As the load increases, up to the final value, the input gear rotates progressively, up to the final position (c). At this situation, the total load, which is unevenly shared between both tooth-pairs $j = 0$ and $j = 1$, induces the tooth-pair deflections described by the intervals $\overline{b_0 c_0}$ and $\overline{a_1 c_1}$, respectively.

Distance $\overline{a_0 b_0}$ is the approach distance and corresponds to the required tooth deflection of the previous pair to start the effective contact at point i. This is described by Eq. (1.13) and therefore the approach distance is given by:

$$\begin{array}{ll} \delta_G(\xi) = \left(\frac{2\pi}{Z_1}\right)^2 r_{b1} C_{p-inn} (\xi_{inn} - \xi)^2 & \text{for } \xi_{\min} \le \xi \le \xi_{inn} \\ \delta_G(\xi) = 0 & \text{for } \xi_{inn} \le \xi \le \xi_o \\ \delta_G(\xi) = \left(\frac{2\pi}{Z_1}\right)^2 r_{b1} C_{p-o} (\xi - \xi_o)^2 & \text{for } \xi_o \le \xi \le \xi_{\max} \end{array} \tag{1.16}$$

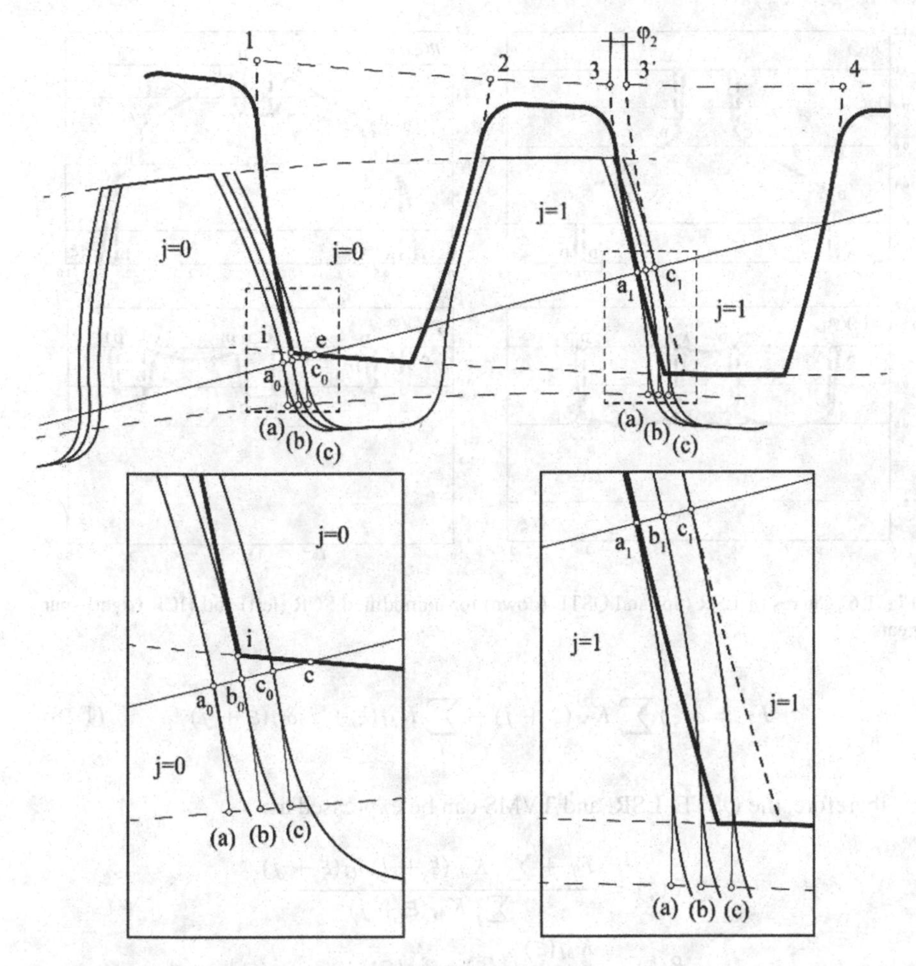

Fig. 1.5 Quasi static loading process of spur gear pair

1.6 Load Sharing Ratio, Quasi Static Transmission Error, and Time Varying Meshing Stiffness for Unmodified Teeth

It is very clear that, inside the additional contact intervals, the tooth pair deflection is equal to $(\delta - \delta_G)$. Since $\delta_G = 0$ inside the theoretical contact interval, the load at any contact point of the extended contact interval can be expressed as follows:

$$F(\xi) = K_M(\xi)(\delta(\xi) - \delta_G(\xi)) \tag{1.17}$$

Accordingly, the total load will be:

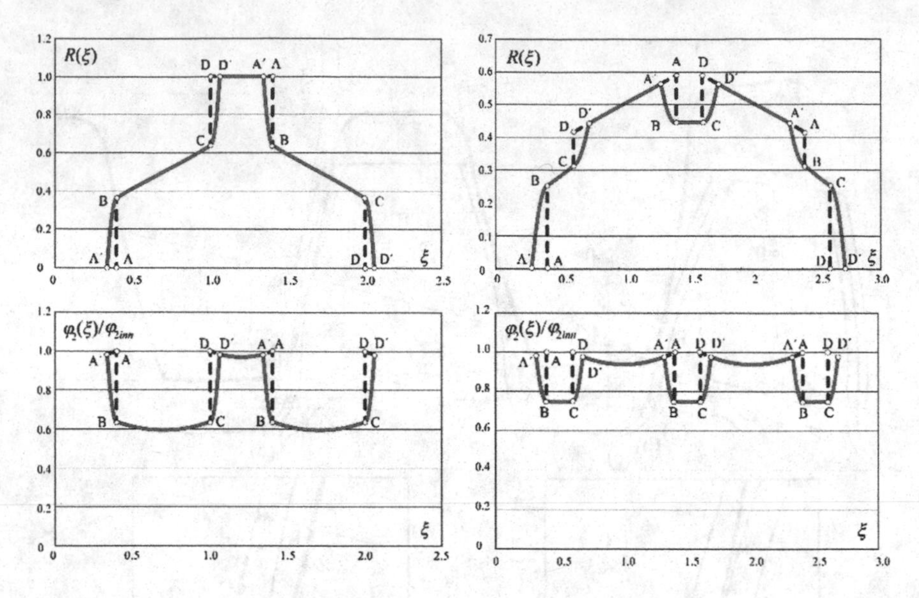

Fig. 1.6 Curves of LSR (up) and QSTE (down) for unmodified SCR (left) and HCR (right) spur gears

$$F_T = \delta(\xi) \sum_j K_M(\xi + j) - \sum_j K_M(\xi + j)\delta_G(\xi + j) \tag{1.18}$$

therefore, the QSTE, LSR, and TVMS can be expressed as:

$$\begin{aligned}
\delta(\xi) &= \frac{F_T + \sum_j K_M(\xi + j)\delta_G(\xi + j)}{\sum_j K_M(\xi + j)} \\
R(\xi) &= \frac{K_M(\xi)}{F_T}(\delta(\xi) - \delta_G(\xi)) \\
K_T(\xi) &= \frac{F_T}{\delta(\xi)}
\end{aligned} \tag{1.19}$$

Figure 1.6 presents the LSR and QSTE curves of spur gear pair of SCR and HCR, obtained from Eq. (1.19). The dashed lines represent the corresponding theoretical curves of LSR and QSTE.

1.7 Load Sharing Ratio, Quasi Static Transmission Error, and Time Varying Meshing Stiffness for Modified Teeth

Figure 1.7-left shows the effective start of contact of a spur tooth pair, in which mesh-in impact occurs at point I. Figure 1.7-right shows how a tip relief at driven tooth tip

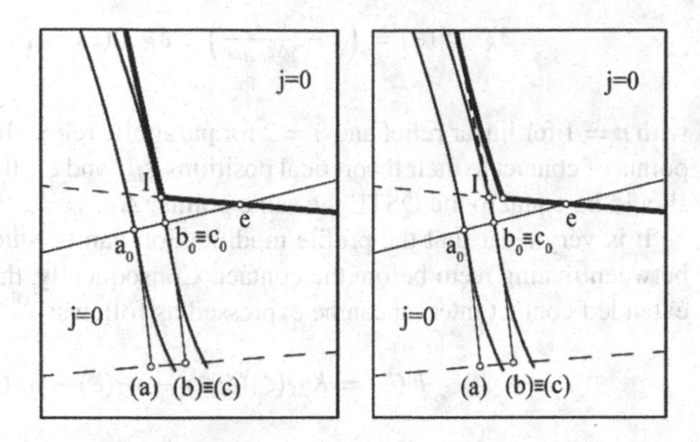

Fig. 1.7 Quasi static loading process of spur gear pair

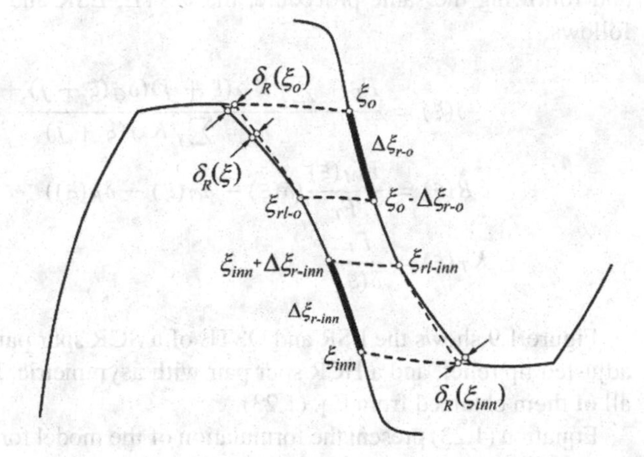

Fig. 1.8 Geometry of the profile modification

delays the start of contact. If the amount of relief is appropriate, the start of contact is delayed to the theoretical inner point of contact, namely point e in Fig. 1.7. Figure 1.8 shows the geometry of profile modification. As other geometrical parameters, the profile modification δ_R will be expressed as a function of the linear parameter along the line of action ξ, as follows:

$$\begin{aligned}
&\delta_R(\xi) = \Delta_{R-inn} && \text{for } \xi_{\min} \le \xi \le \xi_{inn} \\
&\delta_R(\xi) = \Delta_{R-inn}\delta_{R-inn}(\xi) && \text{for } \xi_{inn} \le \xi \le \xi_{inn} + \Delta\xi_{r-inn} \\
&\delta_R(\xi) = 0 && \text{for } \xi_{inn} + \Delta\xi_{r-inn} \le \xi \le \xi_o - \Delta\xi_{r-o} \\
&\delta_R(\xi) = \Delta_{R-o}\delta_{R-o}(\xi) && \text{for } \xi_o - \Delta\xi_{r-o} \le \xi \le \xi_o \\
&\delta_R(\xi) = \Delta_{R-o} && \text{for } \xi_o \le \xi \le \xi_{\max}
\end{aligned} \tag{1.20}$$

where Δ_R is the amount of modification, $\Delta\xi_r$ is the length of modification, and $\delta_{R\text{-}o}(\xi)$ and $\delta_{R\text{-}inn}(\xi)$ represent the shape of modification. For example, for linear or parabolic tip relief, these functions are:

$$\delta_{R-inn}(\xi) = \left(1 - \frac{\xi - \xi_{inn}}{\Delta\xi_{r-inn}}\right)^n ; \ \delta_{R-o}(\xi) = \left(1 - \frac{\xi_o - \xi}{\Delta\xi_{r-o}}\right)^n \quad (1.21)$$

with $n = 1$ for linear relief and $n = 2$ for parabolic relief. To shift the inner and outer points of contact to their theoretical positions ξ_{inn} and ξ_o, the amount of modification should be equal to the QSTE the same points, $\Delta_{R\text{-}inn} = \delta(\xi_{inn})$ and $\Delta_{R\text{-}o} = \delta(\xi_o)$.

It is very clear that the profile modification can be studied as an additional gap between mating teeth before the contact. Consequently, the load at any point of the extended contact interval can be expressed as follows:

$$F(\xi) = K_M(\xi)(\delta(\xi) - \delta_G(\xi) - \delta_R(\xi)) \quad (1.22)$$

and following the same procedure, the QSTE, LSR and TVMS are expressed as follows:

$$\delta(\xi) = \frac{F_T + \sum_j K_M(\xi + j)(\delta_G(\xi + j) + \delta_R(\xi + j))}{\sum_j K_M(\xi + j)}$$
$$R(\xi) = \frac{K_M(\xi)}{F_T}(\delta(\xi) - \delta_G(\xi) - \delta_R(\xi))$$
$$K_T(\xi) = \frac{F_T}{\delta(\xi)} \quad (1.23)$$

Figure 1.9 shows the LSR and QSTE of a SCR spur pair with symmetric, linear, adjusted tip relief, and a HCR spur pair with asymmetric, linear, adjusted tip relief, all of them obtained from Eq. (1.23).

Equation (1.23) present the formulation of the model for QSTE, LSR, and TVMS for spur gears with profile modification. All the results presented in the following sections have been obtained from these equations.

1.7.1 Influence of the Shape of Modification

The shape of modification influences the shape of the QSTE and LSR curves. Figure 1.10 shows the curves of LSR and QSTE of a SCR gear pair with linear and parabolic, symmetric profile modification. The diagrams correspond to the same spur pair of Fig. 1.9-left, but considering a specific length of modification which verifies the condition $\Delta\xi_{r\text{-}inn} = \Delta\xi_{r\text{-}o} = (\varepsilon_a - 1)/2$, in such a way that the interval of modification coincides with the whole interval of two pair tooth contact, and therefore points B_1 and C_1 are coincident.

It is remarkable that the shape of the branches of the curves of QSTE and LSR are governed by the equation of the profile modification through Eq. (1.23). Thereby, for a given equation of the QSTE or LSR, the required profile modification can be computed from:

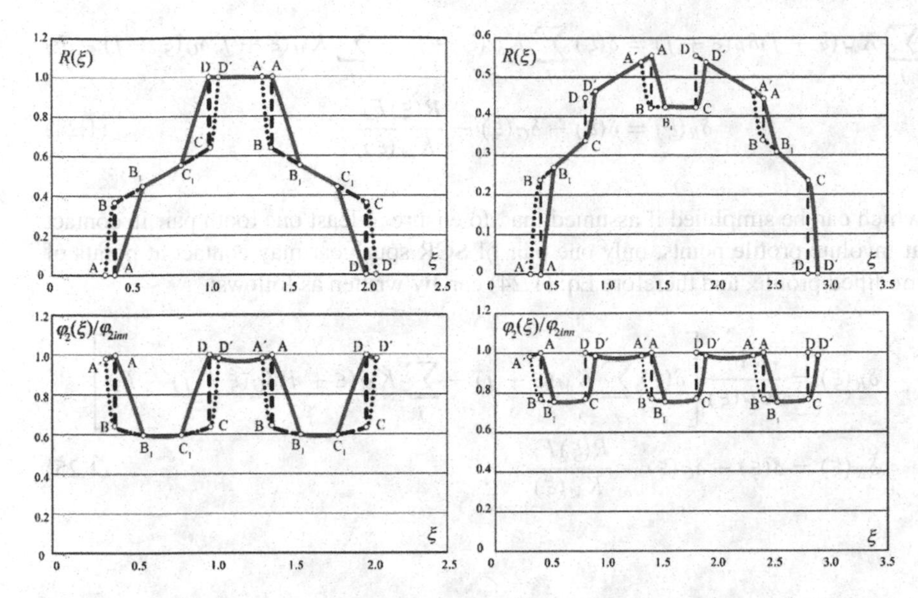

Fig. 1.9 Curves of LSR (up) and QSTE (down) for SCR with symmetric, linear, adjusted profile modification (left) and HCR with asymmetric, linear, adjusted profile modification (right)

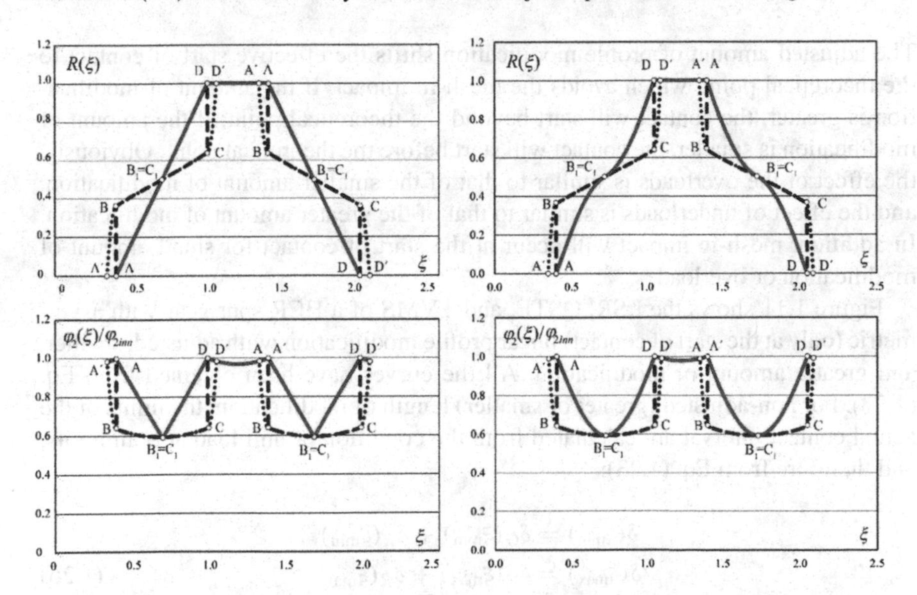

Fig. 1.10 Curves of LSR (up) and QSTE (down) for SCR spur gear with symmetric, adjusted, linear (left) and parabolic (right) profile modification

$$\sum_j K_M(\xi + j)\delta_R(\xi + j) = \delta(\xi)\sum_j K_M(\xi + j) - \sum_j K_M(\xi + j)\delta_G(\xi + j) - F_T$$

$$\delta_R(\xi) = \delta(\xi) - \delta_G(\xi) - \frac{R(\xi)F_T}{K_M(\xi)} \tag{1.24}$$

which can be simplified if assumed that, to ensure at least one tooth pair in contact at involute profile points, only one pair of SCR spur gear may contact at points of modified profile, and therefore Eq. (1.24) can by written as follows:

$$\delta_R(\xi) = \frac{1}{K_M(\xi)}\left[\delta(\xi)\sum_j K_M(\xi + j) - \sum_j K_M(\xi + j)\delta_G(\xi + j) - F_T\right]$$

$$\delta_R(\xi) = \delta(\xi) - \delta_G(\xi) - \frac{R(\xi)F_T}{K_M(\xi)} \tag{1.25}$$

1.7.2 *Influence of the Amount of Modification*

The adjusted amount of profile modification shifts the effective start of contact to the theoretical point, which avoids the mesh-in impact. If the amount of modification is greater, the contact will start beyond the theoretical point; if the amount of modification is smaller, the contact will start before the theoretical point. Obviously, the effect of the overloads is similar to that of the smaller amount of modification, and the effect of underloads is similar to that of the greater amount of modification. In addition, mesh-in impact will occur at the start of contact for small amount of modification or overload.

Figure 1.11 shows the LSR, QSTE, and TVMS of a HCR spur gear with asymmetric (only at the start of contact) linear profile modification, with adjusted, smaller, and greater amount of modification. All the curves have been obtained from Eq. (1.23). For non-adjusted (greater or smaller) length of modification, the limits of the actual contact interval are calculated from the condition of null load at both limits, and therefore from Eq. (1.23):

$$\begin{aligned} \delta(\xi_{\min}) &= \delta_G(\xi_{\min}) + \delta_R(\xi_{\min}) \\ \delta(\xi_{\max}) &= \delta_G(\xi_{\max}) + \delta_R(\xi_{\max}) \end{aligned} \tag{1.26}$$

Equation (1.26) can be simplified by considering that for overloads (or smaller amount of modification) $\delta_R(\xi_{\min/\max}) = \delta_R(\xi_{inn/o})$ and $\delta(\xi_{\min/\max}) \approx \delta(\xi_{inn/o})$; while for underloads (or greater amount of modification) $\delta_G(\xi_{\min/\max}) = 0$.

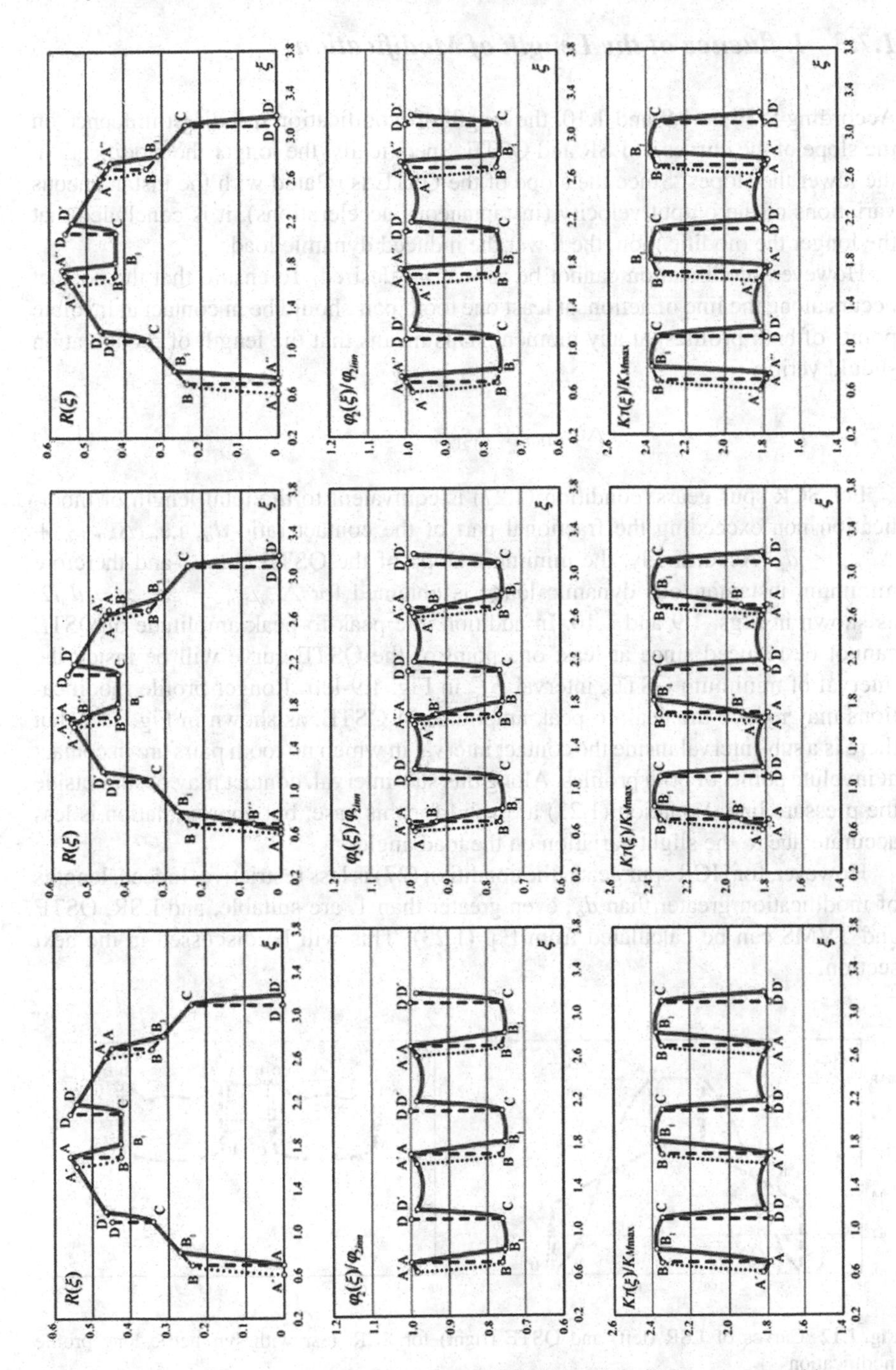

Fig. 1.11 Curves of LSR (up), QSTE (middle), and TVMS (down) for HCR spur gear with asymmetric linear profile modification, with adjusted (left), smaller (center), and greater (right) amount of modification

1.7.3 Influence of the Length of Modification

According to Figs. 1.9 and 1.10, the length of modification has direct influence on the slope of the curves of LSR and QSTE. Specifically, the longer the modification, the lower the slopes. Since the slope of the QSTE is related with the instantaneous variations of the output velocity (instantaneous accelerations), it is concluded that the longer the modification, the lower the induced dynamic load.

However, modification cannot be as long as desired. To ensure that the contact occurs along the line of action, at least one tooth pair should be in contact at involute points of both profiles, at any moment. This means that the length of modification should verify:

$$\Delta\xi_{r-inn} + \Delta\xi_{r-o} \leq \varepsilon_\alpha - 1 \tag{1.27}$$

For SCR spur gears, condition (1.27) is equivalent to the total length of modification non exceeding the fractional part of the contact ratio d_α, i.e., $\Delta\xi_{r\text{–}inn} + \Delta\xi_{r\text{–}o} \leq d_\alpha$. Accordingly, the minimum slope of the QSTE curve –and therefore minimum instantaneous dynamic load– is obtained for $\Delta\xi_{r\text{–}inn} = \Delta\xi_{r\text{–}o} = d_a/2$, as shown in Figs. 1.9 and 1.10. In addition, the peak-to-peak amplitude of QSTE cannot be reduced since at least one point of the QSTE curve will be inside the interval of minimum QSTE, interval $\overline{BC}$ in Fig. 1.9-left. Longer profile modifications may reduce the peak-to-peak amplitude of QSTE, as shown in Fig. 1.12, but there is a sub-interval inside the contact interval in which no tooth pairs are in contact at involute points of both profiles. Along this sub-interval, contact may occur outside the pressure line. Equation (1.23) are valid for this case, but approximation is less accurate due to the slight variation on the load angle.

However, for HCR spur gears, the condition (27) is less restrictive. Indeed, lengths of modification greater than d_α, even greater than 1, are suitable, and LSR, QSTE and TVMS can be calculated from Eq. (1.23). This will be discussed in the next section.

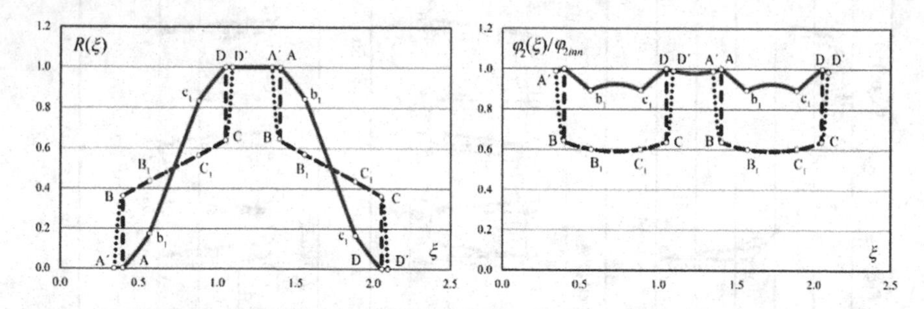

Fig. 1.12 Curves of LSR (left) and QSTE (right) for SCR gear with symmetric long profile modification

1.8 High Contact Ratio Spur Gears with Long Profile Modifications

Figure 1.13 shows the LSR and QSTE of a HCR spur gear with short-symmetric, long-symmetric, and long-asymmetric linear profile modification. The amount of modification is adjusted. All of them have been computed from Eq. (1.23).

The effect of symmetric modification on the LSR and QSTE is quite different from that of asymmetric modification. Although both provides interesting design tools.

1.8.1 Symmetric Long Profile Modification

As seen in Fig. 1.13-center, symmetric profile modification drastically reduces the peak-to-peak amplitude of QSTE. This is due to as the length of modification increases, point B_1 moves to the right and point C_1 moves to the left. Once they cross, the curve of QSTE does not reach the interval of minimum QSTE $\overline{BC}$, and therefore the peak-to-peak amplitude is reduced. Note that at point b_1, the curve of QSTE leaves the trajectory $\overline{AB_1}$ because at this point the outgoing tooth pair begins to contact inside the interval of modification, and therefore the mesh stiffness decreases and the QSTE increases.

The peak-to-peak amplitude of QSTE will be minimum when QSTE at points A and M are equal. Is has been studied in [9], where the following approximate equation were found for the length of symmetric modification for minimum peak-to-peak amplitude of QSTE:

$$\Delta\xi_{r-inn} = \Delta\xi_{r-o} = 0.8672 d_\alpha \tag{1.28}$$

Figure 1.14-left shows the LSR and QSTE for the same HCR spur gear of Fig. 1.13, with profile modification for minimum pear-to-peak amplitude of QSTE.

Other interesting possibility is the control of the maximum instantaneous induced dynamic load with the length of modification. As mentioned above, the time variation of the QSTE is related with the induced dynamic load. Although minimum slope of QSTE may not ensure minimum dynamic load (because the dynamic load depends on some other factors as shaft stiffness, bearing stiffness, inertia, etc.), minimizing the slope of the QSTE curve will minimize the contribution of the QSTE to the dynamic load.

The mathematical study can be found in [9], but from Fig. 1.13-center it is intuitive that the maximum slope of the QSTE curve will be minimum if segment $\overline{Ab_1}$ is eliminated. This occurs for point C_1 coincident with point B (and point B_1 coincident with point C), which is expressed as follows:

$$\Delta\xi_{r-inn} = \Delta\xi_{r-o} = d_\alpha \tag{1.29}$$

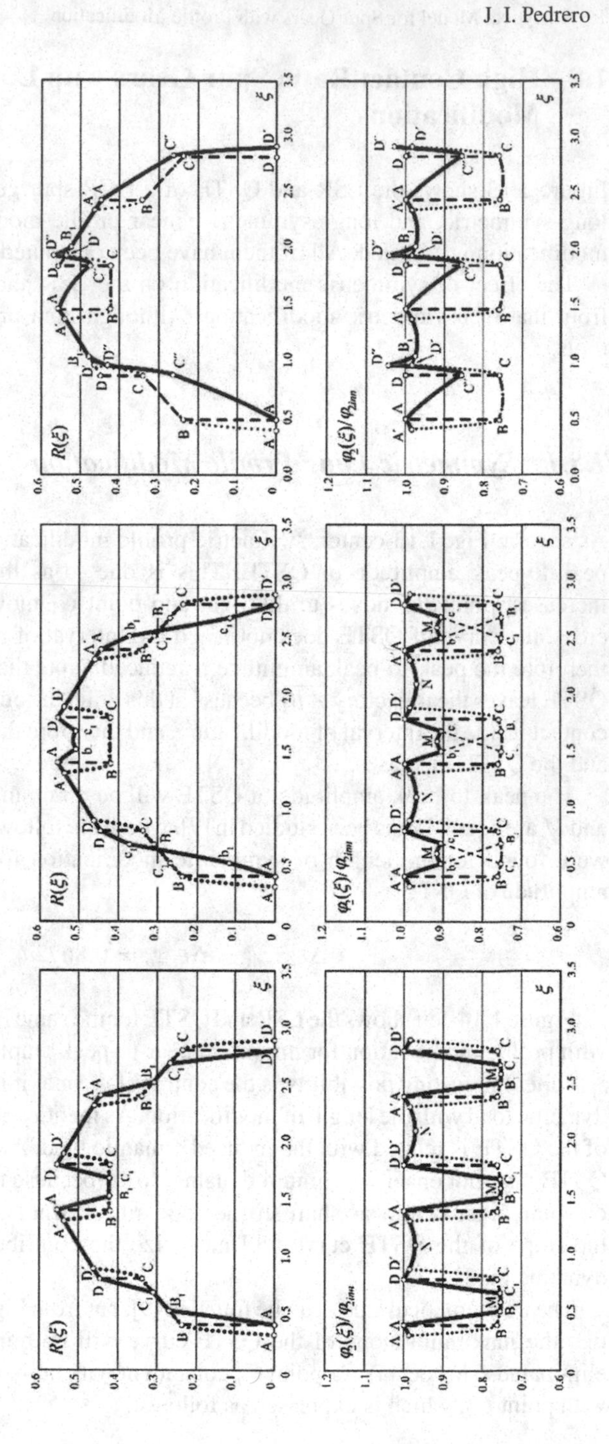

Fig. 1.13 Curves of LSR (up) and QSTE (down) for HCR spur gear with short-symmetric (left), long-symmetric (center), and long-asymmetric (right) adjusted linear profile modification

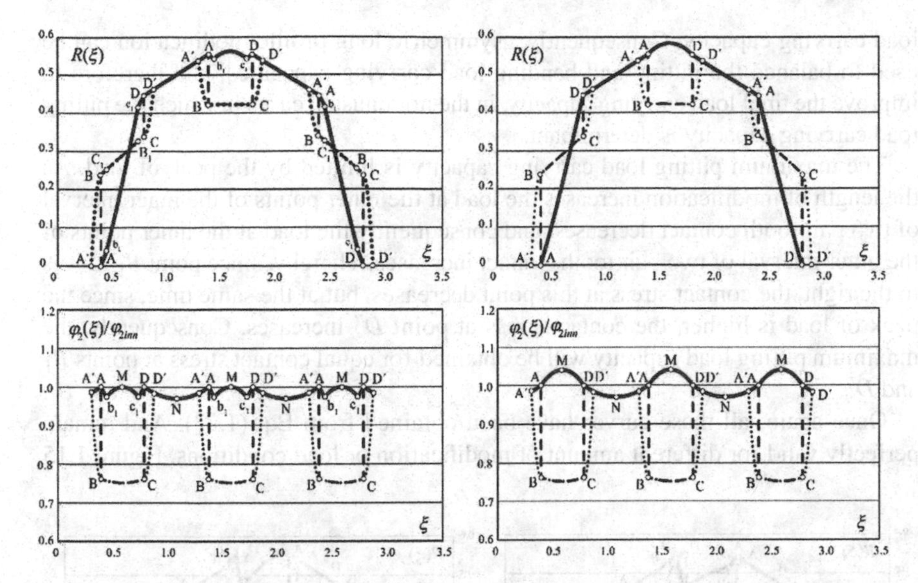

Fig. 1.14 Curves of LSR (up) and QSTE (down) for HCR gear with optimal long profile modification for minimum peak-to-peak amplitude of QSTE (left) and minimum instantaneous dynamic load (right)

Figure 1.14-right shows the LSR and QSTE for the same HCR spur gear of Fig. 1.13, with profile modification for minimum instantaneous induced dynamic load.

1.8.2 Asymmetric Long Profile Modification

Figure 1.13-right shows the LSR and QSTE curves of HCR spur gear with asymmetric long profile modification. The peak of load arising at the inner limit of the outer interval of two pair tooth contact is due to the modification interval contains the beginning of the inner interval of two pair tooth contact, and therefore the load is lower. Since the load is shared between two tooth pairs, if it is lower in one of them, it should necessarily be greater in the other.

This provides a new interesting design tool. As shown in Fig. 1.2, for HCR spur gears, the critical contact stress occurs for contact at the inner point of the inner interval of two pair tooth contact (near the point with $\xi = 1.0$ in the figure). The critical tooth root stress occurs for contact at some point of the outer interval of two pair tooth contact, along which the tooth root stress is quite uniform. According to Fig. 1.13-right, if due to the asymmetric long profile modification, the load decreases at the point of maximum contact stress and increases at the interval of maximum tooth root stress, the critical contact stress will decrease and the critical tooth root stress will increase, resulting in higher pitting load carrying capacity and lower bending load carrying capacity. Consequently, asymmetric long profile modification can be used to balance the pitting and bending load carrying capacities, and therefore to improve the final load carrying capacity, in the not unusual cases in which the pitting load carrying capacity is determinant.

The maximum pitting load carrying capacity is limited by the peak of load. As the length of modification increases the load at the inner points of the inner interval of two pair tooth contact decreases, and consequently the load at the inner points of the outer interval of two pair tooth contact increases. Thereby, since point B_1 moves to the right, the contact stress at this point decreases, but at the same time, since the peak of load is higher, the contact stress at point D'' increases. Consequently, the maximum pitting load capacity will be obtained for equal contact stress at points B_1 and D''.

Once again, all these curves have been obtained from Eq. (1.23). And remain perfectly valid for different amount of modification or load conditions. Figure 1.15

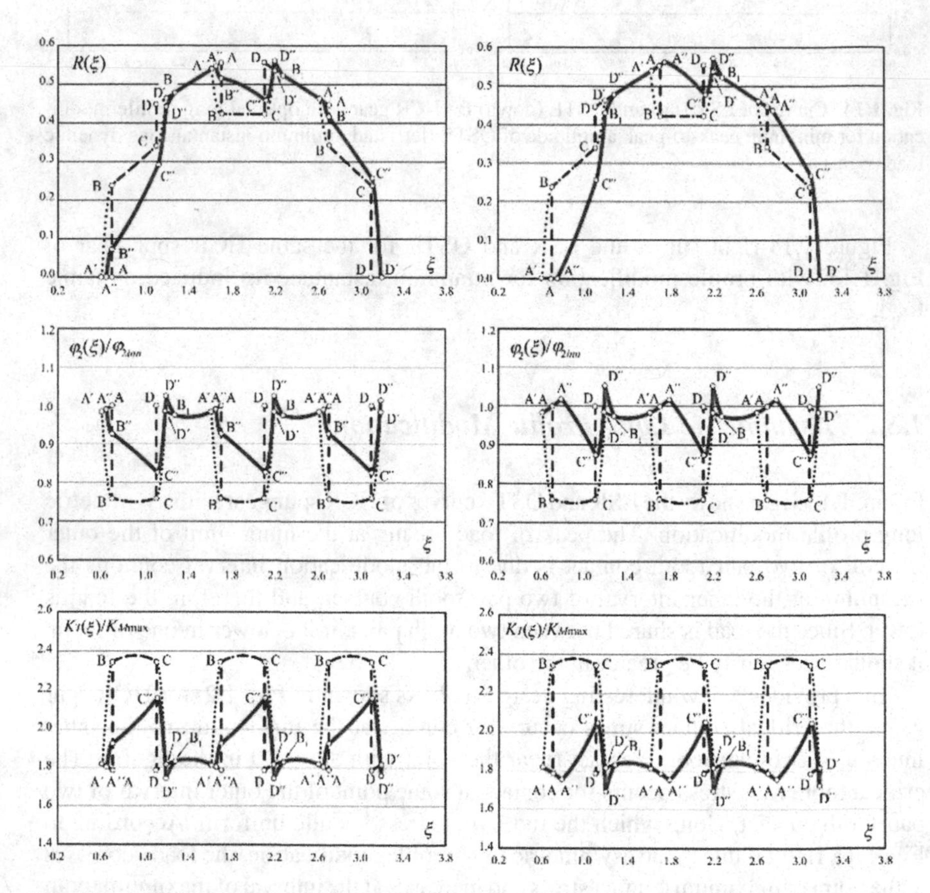

Fig. 1.15 Curves of LSR (up), QSTE (middle), and TVMS (down) for HCR gear with asymmetric long profile modification under overload (left) and underload (right) conditions

shows the effect of asymmetric long profile modification on the LSR, QSTE, and TVMS of HCR spur gear under overload and underload conditions, obtained from Eq. (1.23).

1.9 Non-Standard Tooth Dimensions

The expression for the SMS presented in Eqs. (1.6) and (1.15) and drawn in Fig. 1.1, is valid for spur gears with addendum factor h_a equal to 1 and standard center distance.

For shorter or longer addenda, the SMS curve should be truncated or prolonged, but Eqs. (1.6) and (1.15) remain valid, although the coefficient b_0 in Eq. (1.7) should be computed from the fictitious contact ratio, namely the contact ratio of the spur gear with $h_a = 1$, standard center distance, and identical values of all the other geometrical parameters [10]. Similarly, shorter or longer operating center distance results in longer or shorter contact interval, respectively, decreasing or increasing the contact interval at both limits in the same amount. Obviously, in these cases the point of maximum SMS may not coincide with the midpoint of the contact interval.

Figure 1.16 shows the SMS, LSR, and QSTE of an unmodified SCR spur gear with standard tooth addendum, and with reduced tooth addendum on the driven gear and enlarged tooth addendum on the driving gear. It can be observed the asymmetric

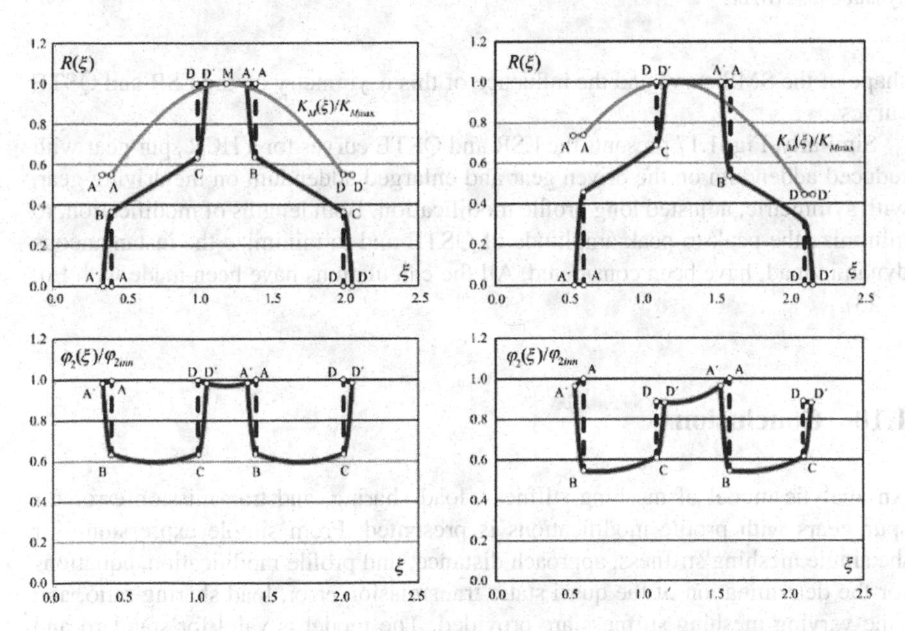

Fig. 1.16 Curves of SMS and LSR (up), and QSTE (down) for SCR unmodified gear with standard (left) and non-standard (right) tooth addendum

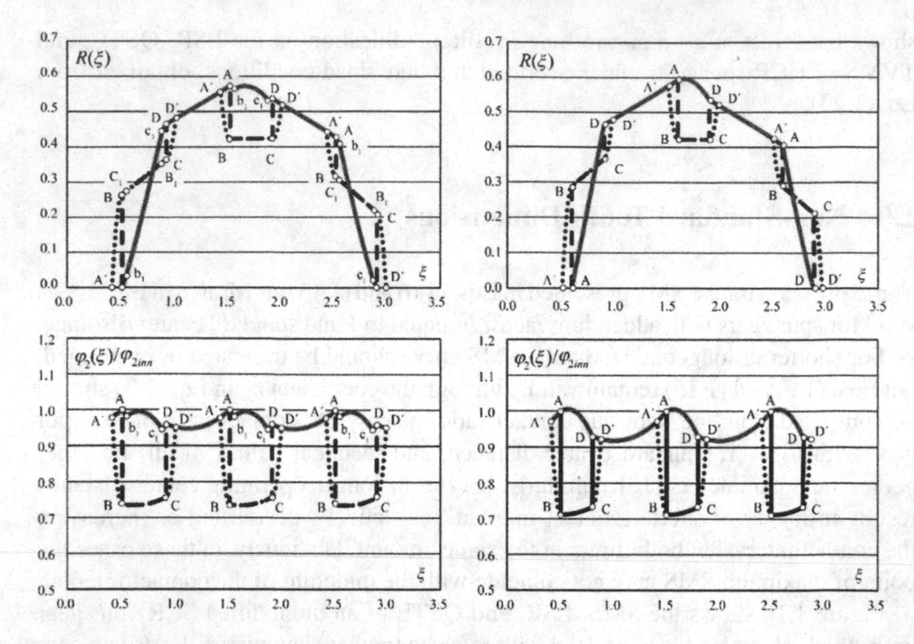

Fig. 1.17 Curves of LSR (up) and QSTE (down) for HCR gear with non-standard tooth addendum and long profile modification for minimum peak-to-peak amplitude of QSTE (left) and minimum dynamic load (right)

shape of the SMS curve and the influence of this asymmetry on the LSR and QSTE curves.

Similarly, Fig. 1.17 presents the LSR and QSTE curves for a HCR spur gear with reduced addendum on the driven gear and enlarged addendum on the driving gear, with symmetric, adjusted long profile modification. Both lengths of modification, to minimize the peak-to-peak amplitude of QSTE and to minimize the instantaneous dynamic load, have been considered. All the calculations have been made with Eq. (1.23).

1.10 Conclusion

An analytic model of meshing stiffness, load sharing, and transmission error for spur gears with profile modifications is presented. From simple expressions for the single meshing stiffness, approach distance, and profile modification, equations for the determination of the quasi static transmission error, load sharing ratio, and time varying meshing stiffness are provided. The model is valid for standard and non-standard tooth height and center distance, standard and high contact ratio gear pairs, and arbitrary values of the amount, length, and shape of profile modification, including:

- Adjusted and non-adjusted amount of modification (corresponding to nominal and non-nominal load conditions),
- Linear and parabolic modifications,
- Short and long length of modification, and
- Symmetric and asymmetric modifications.

The presented model provides an interesting design tool allowing to control the load sharing and transmission error by selecting the profile modification parameters. Specifically, for high contact ratio spur gears, optimal profile modifications for minimum average dynamic load, minimum instantaneous dynamic load, and optimal load carrying capacity have been obtained. In addition, the simplicity of the equations allows to study not only the load sharing ratio and quasi static transmission error for a given profile modification, but also the profile modification required for preestablished functions of load sharing ratio or quasi static transmission error.

The model will be completed in the future with the consideration of the influence of new parameters which can be expressed as new additional gaps between contacting surfaces, as for example the machining errors, surfaces temperature, and wear. The study of the influence of wear has a singularity owing to the wear depth depends on the number of meshing cycles. Since the load sharing at each cycle depends on the accumulated wear, and the instantaneous wear depends on the instantaneous load sharing, calculations should be made sequentially.

Acknowledgements The author expresses his gratitude to the Spanish Council for Scientific and Technological Research for the support of the project PID2019-110996RB-I00 "Simulation and control of transmission error of cylindric gears", as well as the School of Engineering of UNED for the support of the action 2021-MEC23, "Control of transmission error in cylindric gears with profile modification". The author also acknowledges the contributions to this research from the members of the UNED's Gear Research Group, Profs. Miguel Pleguezuelos and Miryam Sánchez.

References

1. Sánchez, M.B., Pleguezuelos, M., Pedrero, J.I.: Approximate equations for the meshing stiffness and the load sharing ratio of spur gears including Hertzian effects. Mech. Mach. Theory **109**, 231–249 (2017)
2. Pedrero, J.I., Pleguezuelos, M., Artés, M., Antona, J.A.: Load distribution model along the line of contact for involute external gears. Mech. Mach. Theory **45**, 780–794 (2010)
3. Hertz, J.: On the Contact of Elastic Solids. Macmillan, Miscellaneous Papers (1896)
4. Weber, C., Banaschek, K.: Formänderung und profilrücknahme bei gerad und schrägverzahnten rädern. Schriftenreihe Antriebstechnik. Vieweg Verlag, Braunschweig (1955)
5. Sainsot, P., Velex, PH., Duverger, O.: Contribution of gear body to tooth deflections—a new bidimensional analytical formula. J. Mech. Des. **126**, 748–752 (2014)
6. Fernández del Rincón, A., Viadero, F., Iglesias, M., García, P., de Juan, A., Sancibrián, R.: A model for the study of meshing stiffness in spur gear transmissions. Mech. Mach. Theory **61**, 30–58 (2013)
7. International Organization for Standardization—ISO: Calculation of load capacity of spur and helical gears, Standard 6336, Parts 1, 2, and 3. ISO, Geneva (2019)

8. Sánchez, M.B., Pleguezuelos, M., Pedrero, J.I.: Influence of profile modifications on meshing stiffness, load sharing, and transmission error of involute spur gears. Mech. Mach. Theory **139**, 506–525 (2019)
9. Pleguezuelos, M., Sánchez, M.B., Pedrero, J.I.: Control of transmission error of high contact ratio spur gears with symmetric profile modifications. Mech. Mach. Theory **149**, 103839-1–103839-16 (2020)
10. Sánchez, M.B., Pleguezuelos, M., Pedrero, J.I.: Enhanced model of load distribution along the line of contact for non-standard involute external gears. Meccanica **48**, 527–543 (2013)

Chapter 2
Lifetime Mechanics of Machines as Cluster of Classical Mechanics, Reliability Theory and Digital Twin Concept

Vladimir Algin

Abstract The model approach and digital twin (DT) are the main features of Industry 4.0, based on the rapidly growing digitalization capabilities, including real-time modeling of the behavior/state of objects, the fusion of virtual and real objects and their updating during the life cycle. Since 2015, there are an avalanche increase in publications on DT. They cover the understanding of the term “digital twin”, general approaches and practical applications, and standardization issues. However, these and other issues cannot be resolved without accumulating a critical mass of tools and cases of their application in practice. The aim of the work is to develop the architecture of the information model for a technically complicated item and its basic components: the characteristic representations, models and the tools/methods serving them. The examples used relate to vehicles and their drives. The main idea is that the meaningful models and representations of the design stage should be further applied and, if necessary, adjusted for individual parameters at the operational stage of the object. They are supplemented with diagnostic data and are used to build the DT for the manufactured exemplar. This realizes the digital thread of representations, models and data beginning from the design stage. In addition, new tools are added during the operation stage (diagnostic system, lifetime expense evaluation). Components of the information model that developed in the Lifetime Mechanics of Machines provide a comprehensive assessment of the functional and life properties of technically complicated items and can be applied to a wide range of mechanical engineering objects.

Keywords Digital twin · Architecture · Dynamics · Dependability · Functional properties · Diagnostics · Lifetime expense

V. Algin (✉)
The State Scientific Institution “The Joint Institute of Mechanical Engineering of the National Academy of Sciences of Belarus”, 220072 Minsk, Belarus

M. Rackov et al. (eds.), *Machine and Industrial Design in Mechanical Engineering*, Mechanisms and Machine Science 109,
https://doi.org/10.1007/978-3-030-88465-9_2

2.1 Introduction

The paper describes the way of creating an architecture of digital twins (information models) for predicting and assessing the state of complex products using meaningful (sense, physical) models, representations and tools. Architecture is considered as fundamental concepts or properties of a system in its environment embodied in its elements, relationships, and in the principles of its design and evolution [1]. Meaningful models are used at the design stage and then at the operational stage. At the same time, the uncertainty of a number of model parameters under design stage (operation conditions, driving style, technological parameters, etc.) is eliminated in operation due to sensors data. The set of models at operation is expanded with diagnostic models, as well as object state models, based entirely on current operational data. This approach is particularly effective when creating digital twins of complex products developed using heterogeneous meaningful models from different disciplines.

The evolution of the concept proposed by M. Greaves in 2002, which led to the emergence of the now accepted term "digital twin" (DT), is presented in the works of Greaves [2, 3].

The DT concept is related to the so-called "digital thread", also called "digital stream". The term was coined by Lockheed Martin Corporation and implies a continuous flow of production chain modeling from the design stage. Digital threads refer to the digitization and traceability of product "from cradle to grave" [4]. See also some inventions related digital thread [5, 6].

It should be noted that the work [7] anticipated and combined the ideas of the digital twin and digital thread in the form of an information model of a machine that accompanies its life cycle. This work contains the following provisions: "The creation and use of high-tech products requires a new information technology for the construction and development at all stages of the machine's life cycle of its information model as the basis of this technology. This leads to the need to improve the technical documentation of machines by including an information model of the machine at various stages of the life cycle."

The following aspects are essential:

- each part of the machine is represented as a source of information signals,
- the machine units in which it is possible and appropriate to implement the principles of reflexive control are allocated (the control object and the controlling part change functions at some point), and
- procedures for identifying information sources, control objects and reflexive nodes are provided, their interrelationships are determined.

The information model should be designed in such a way as to allow the use of various sources:

- semantic, structural (logical), parametric (quantitative, mathematical) models,
- measurement results,
- expert evaluations, and

– means of simulating the elements and units of the machine (in slow, accelerated and real time scales in relation to the current, retrospective and predicted state) [7].

(The assumption that there are possible conflicts among the subsystems of the machine, endowed with intelligence, can be considered as a forecast for the forthcoming i4.0 technologies. These conflicts can be solved on the basis of the principles of reflexive control).

Recently, an increasing number of DT reviews have been published, covering an increasing number of publications. One of the most representative review for the beginning of 2021 is [8], which analyzes publications, standards and patents on the DT topic. Based on this review and other publications, the following conclusions can be drawn.

1. There are no general standards in the field of DT. The exception is [9], which refers only to the stage of production (making) of an object.

In [10], devoted to the development of a special standard, it is noted that nowadays a large number of different CAE tools are available for the design and analysis of a gear unit and its components, each of which has its own strengths. A major milestone for Industry 4.0 is the establishment of industry-wide standards.

The paper [10] presents an industry-wide standard under development for simple data exchange in transmission development under the name REXS (Reusable Engineering EXchange Standard) [11]. The REXS initiative pursues the goal of providing a digital twin in transmission development and calculation. REXS has the potential to establish itself on a large scale as a standard model for data exchange in the field of gear unit design and analysis. This would result in a number of advantages for the software manufacturers, for the companies using the tools and for the users: (1) Transmission designs and analyses at any level of detail, i.e. from the overall system view via the analysis of individual components to the individual physical phenomena, can always be carried out on the basis of a single data model, (2) A simple data exchange between the classical analytical gear design programs towards universal dynamic, FE and CAD systems would be possible. Expenses for additional, specific modeling could be greatly reduced, etc.

2. There are no DTs in which the functional and life properties of the product would be meaningfully and comprehensively displayed. Operational models mainly use big data.

The complication of technical objects, as well as the situation when the developer is one party, the manufacturer is the other, the consumer is a third party, and the service team is the other, leads to the fact that the consumer cannot use the models and tools of the design stage. And the consumer prefers to work with big data from operation. However, the use of big data becomes problematic when the operation conditions for using the object change. And special tools need to solve the problem. For example, the life prediction with considering the switching of mission profile, which is composed of different operating conditions is proposed in [12].

3. In addition, the processes of damage accumulation, especially fatigue ones, are poorly predicted.

In [13] it is indicated that when developing a mathematical model to simulate the fatigue damage of steel members, many sources of uncertainty are identified. First, the model input parameters have a wide range of variability. Furthermore, additional uncertainty arises by the simplification of mathematical models adopted, as well as, measurement errors and varying application conditions. The material response to applied stress is considered complicated since there are several factors that can alter the endurance limit. The proposed approach [13] is based on fatigue damage prediction models integrated with real-time damage records.

In [14], an approach, based on DT, is developed for performing mission optimization under uncertainty to ensure the safety of the system with respect to fatigue cracking. This is achieved by designing mission load profiles for the mechanical component such that the damage growth in the component is minimized, while the component performs the desired work. The proposed approach includes: damage diagnosis, damage prognosis, and mission optimization. All of them are affected by uncertainty regarding system properties, operational parameters, loading and environment, as well as uncertainties in sensor data and prediction models.

4. It should also be noted that the idea of using models developed at the R&D stage for the operation stage is gradually penetrating the scientific, technical, industrial and consumer community.

Thus, in [15], the concept of an Executable Digital Twin is proposed, which, from the point of view of Siemens specialists, will become a key aspect in any future Digital Twin driven application. The following definition is given [15]: "An Executable Digital Twin is a specific encapsulated realization of a Digital Twin with its execution engines. As such they enable the reuse of simulation models outside R&D. In order to do so, the Executable Digital Twin needs to be prepared suitably for a specific application out of existing data and models. In particular it must have the right accuracy and speed. The Executable Digital Twin can be instantiated on the edge, on premise, or in the cloud and used autonomously by a non-expert or a machine through a limited set of specific APIs".

Considering the above, the rest of this paper includes a description of the general architecture of the information model of a technically complicated item (TCI) (Sect. 2.2), the features of its implementation at the design and operation stages of a TCI using the concepts, representations, models and tools developed in the Lifetime Mechanics of Machines (Sect. 2.3) and final conclusions (Sect. 2.4).

2.2 Lifetime Mechanics of Machines and General Architecture of Information Model of Technically Complicated Item

The term “Lifetime mechanics of machines” (LMM) reflects the collaborative study of (1) movement and functioning, and (2) the life state (health) of a machine.

LMM uses the provisions and approaches of a number of disciplines (mechanics, reliability and system theories) and develops its own conceptual and computational apparatus (Fig. 2.1). *Mechanics* serves as the developer and “supplier” of the initial models of loading and damage to engineering components, with the help of which lifetimes are calculated (operating times to the limit state). *Reliability theory* provides a probabilistic approach, concepts for describing the properties of products and quantitative indicators for their evaluation. *System theory* adds the synergistic aspects inherent in any complex system, such as the dependent behavior of elements. *The new approach* is also to take into account the factors and problematic issues that appear during the transition from analog to digital models and methods for modeling individual product behavior.

The LMM considers the machine as a TCI. The proposed architecture of the TCI information model is shown in Fig. 2.2.

(TCI is a product with a hierarchical structure, the presence of an operator (driver) and a large number of components of different types. Such a product is used in various conditions and modes of operation. Typical technically complicated items are: mobile machines; agricultural machinery; main technological equipment [16]).

The architecture includes basic representations (views) and models of the TCI as well as methods that serving them. New basic representations are shown in rectangular frames as abbreviations. They are: LSC = Lifetime-strength curves; RMS =

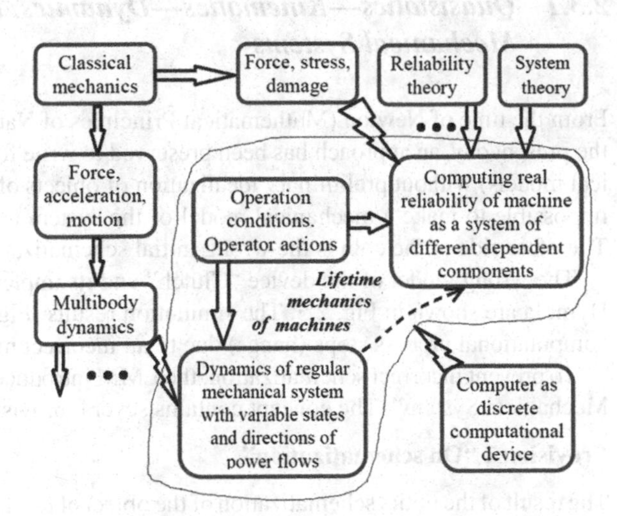

Fig. 2.1 LMM and its integrating role

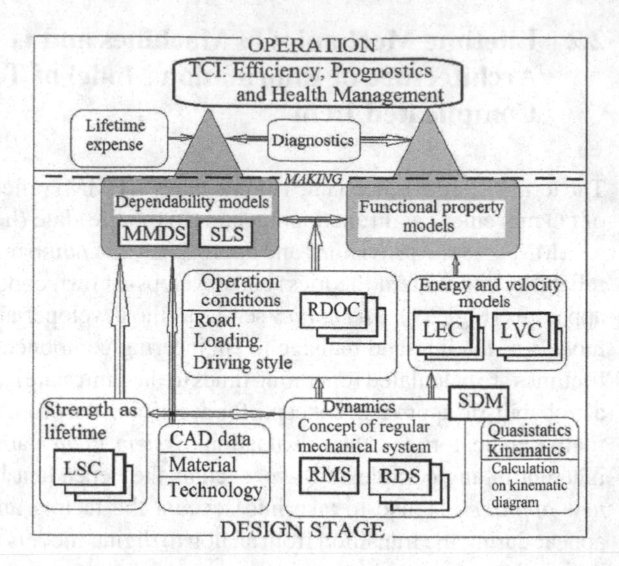

Fig. 2.2 Architecture of the TCI information model in LMM

Regular mechanical system; RDS = Regular dynamic scheme; SDM = Structural-distribution matrix; RDOC = Relative duration of operation conditions; LEC = Loading-energy/fuel curve; LVC = Loading-velocity curve; MMDS = Multilevel "Mechanics-Dependability" scheme; SLS = Scheme of limiting state.

2.3 Main Features of TCI Information Model Architecture

2.3.1 Quasistatics—Kinematics—Dynamics. Regular Mechanical Systems

From the time of Newton (Mathematical Principles of Natural Philosophy, 1687) to the present day, an approach has been preserved to write formulas (to get mathematical models) without preliminary idealization of objects of the real world. It is often impossible to make a mechanical model of the system using the written formulas. The other side of the coin is the wrong initial schematization.

The wrong model of the device "Clutch" and its implementation in the software Dymola are shown in Fig. 2.3. The simulation results (Fig. 2.4) are predictable: the computational process stops (hangs) due to an incorrect mechanical model.

To prevent incorrect schematization, the LMM introduces the concept of "Regular Mechanical System". The concept contains several provisions.

Provision 1 "On schematization".

The result of the initial schematization of the object of mechanics is a regular mechanical system (RMS) as a set of main parts (concentrated masses) and inertialess devices

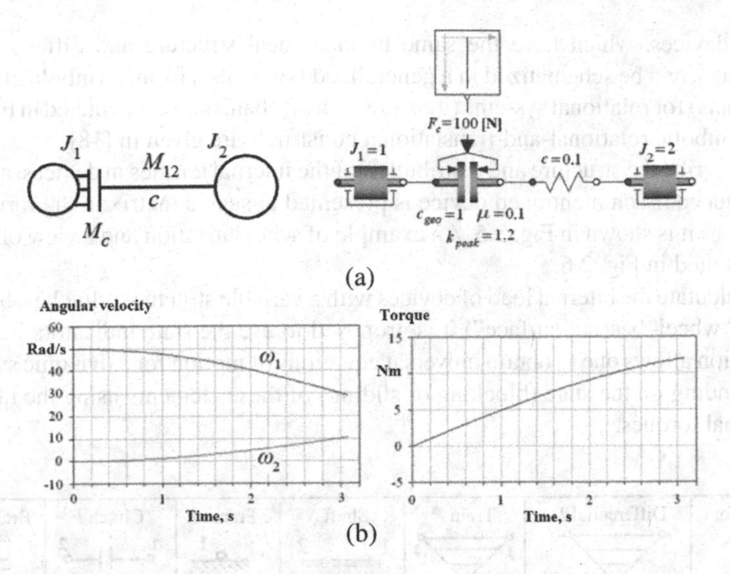

Fig. 2.3 **a** Wrong clutch model (left) and its realization in software Dymola (right) and **b** results of modeling: velocities (left) and torque (right)

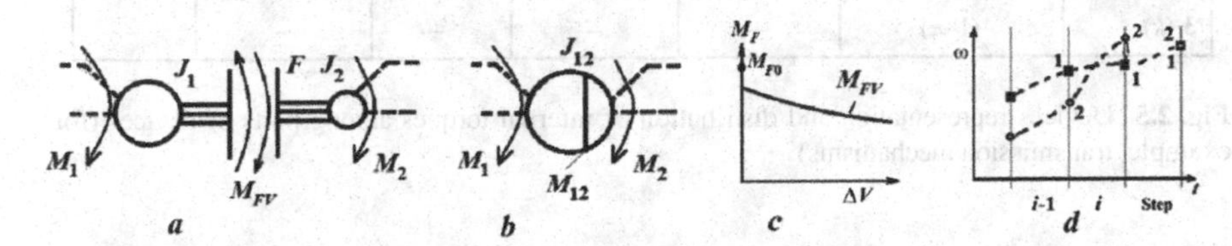

Fig. 2.4 States of the RMS with a clutch: **a** slipping, **b** locked state, **c** the friction torque MF, **d** change of sign in the difference of angular velocities ω of masses in a computational process

for their connection; direct connection of devices is prohibited; the masses can come into contact; forces act only on masses.

This provision reflects the principle of regularity of the systems under consideration. Other provisions concern the simulation of the process using a computer as a discrete device. One of them is presented below.

Provision 2 "About changing the state of the contact"

In the process of dynamic calculation of RMS, when conditions arise for the transition to a new state of the contacting masses (for example, closing or opening of a clutch), this state must be represented in the calculation at least one step of the computational process (Fig. 2.4).

Construction of universal mathematical models

The next point of the concept is the construction of universal mathematical models of quasi-static, kinematics and dynamics.

The devices, which have the same mathematical structure and differ only in parameters, can be schematized in a generalized (symbolic) form. Symbolical views (constructs) for rotational systems (transmission mechanisms) are depicted in Fig. 2.5 [17]. Symbolic rotational-and-translational constructs are given in [18].

To describe the structure and distribution of the internal torques in devices, a SDM is introduced. Each mentioned device is presented at such a matrix in the form of a column, as it is shown in Fig. 2.5. An example of schematization and a view of SDM are presented in Fig. 2.6.

To calculate the internal load of devices with a variable structure (clutches, brakes, system "wheel–bearing surface") it is proposed to use the state indicators λ. Their application allows one to obtain universal equations of motion for a dynamic system, no depending on the state (blocking or sliding) of these elements using the method of internal torques.

Index for device part	Differential D	Train P	Shaft S	Frame R	Clutch F	Brake T
1 (i)	1	1	1	1	1	1
2 (j)	$-u$	$-u$	-1 (u=1)	0 (u=0)	-1	0
3 (k)	$-(1-u)$	—	—	—	—	—

Fig. 2.5 Devices representation and distribution of internal torques among parts of devices (for example, transmission mechanisms)

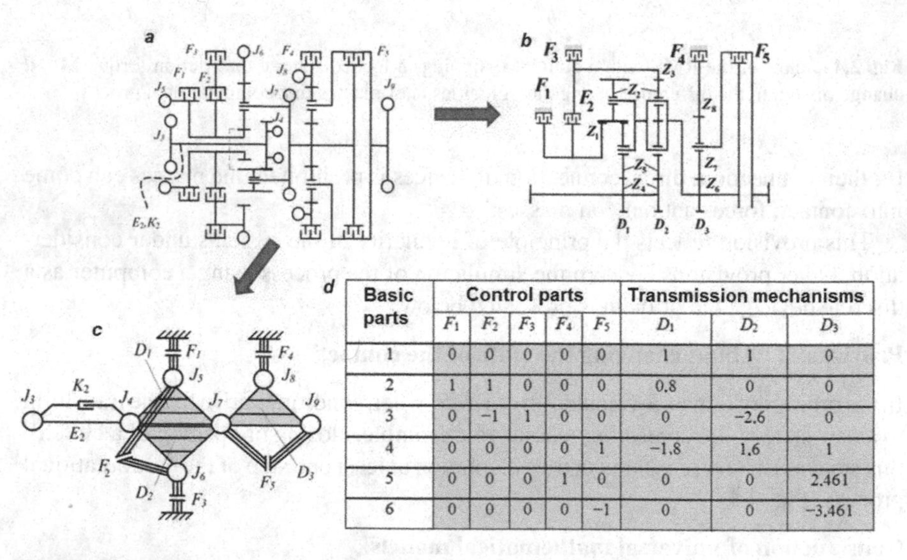

Basic parts	Control parts					Transmission mechanisms		
	F_1	F_2	F_3	F_4	F_5	D_1	D_2	D_3
1	0	0	0	0	0	1	1	0
2	1	1	0	0	0	0.8	0	0
3	0	−1	1	0	0	0	−2.6	0
4	0	0	0	0	1	−1.8	1.6	1
5	0	0	0	1	0	0	0	2.461
6	0	0	0	0	−1	0	0	−3.461

Fig. 2.6 Schematization of the MZKT-7922 transmission: **a** mechanical model, **b** equivalent kinematic diagram, **c** dynamic scheme and **d** structural-distribution matrix

The internal torque M_C takes on the values of the frictional torque M_F (when slipping) or the torque M_1, which holds the moving parts of the clutch/brake in a locked state. The alternative action of these torques is determined by the value of λ, which, depending on the process parameters. So the clutch or brake, can take on the values $\lambda = 1$ (slipping) or $\lambda = 0$ (locked state). (The reverse designation of states is also possible.) Using the λ indicator, the internal torque in a device with variable structure is presented as $M_C = (1 - \lambda)M_1 + \lambda M_F$.

Example of dynamic computation

The rotational-and-translational dynamic scheme of a mobile machine (Fig. 2.7a) has the rotational masses (inertial parts) with the moments of inertia J_i, elastic (E_i), and damping (K_i) constructs, and translational machine mass m_A. The engine torque M_E acts on the rotational mass J_E. Elements $F_{K1(2)}$ simulate the friction contact for coupling wheels with the surface.

Subsystem "Wheels-machine mass" is presented by a multibody model of the reasonable degree of complexity. It based on multibody mechanical model of a driving wheel (Fig. 2.7b). The model is suitable to reproduce the modes of elastic and non-elastic slipping of wheels.

The key components of the model are the ramifying joints "J_k–m_k–m_A" which transform the wheel torque M_W to the force couple ($F_W = M_W/r$; r is a wheel radius)

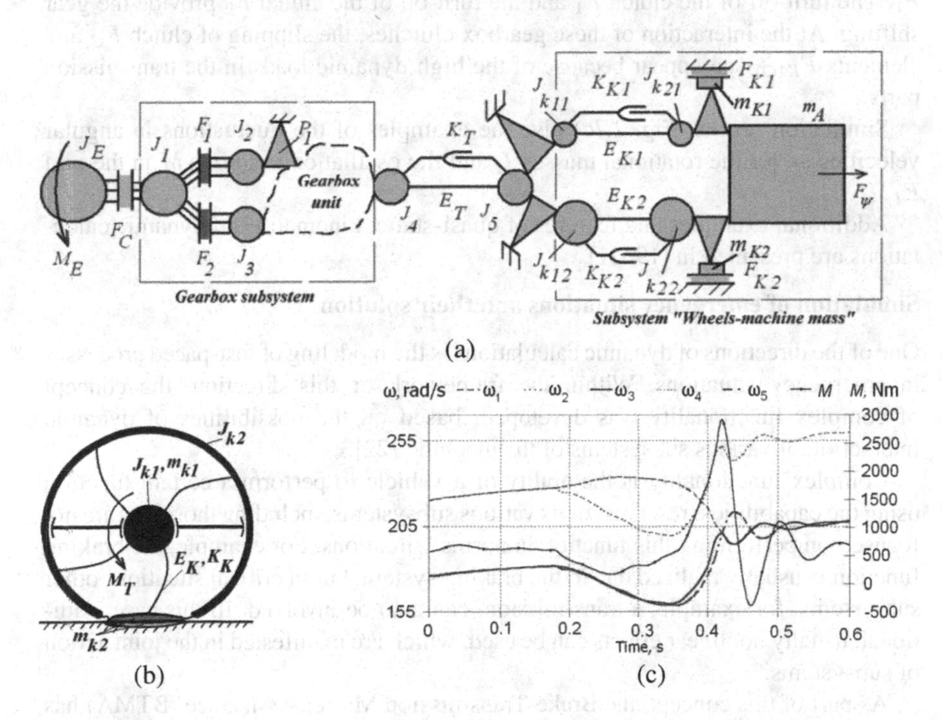

Fig. 2.7 **a** Dynamic scheme of a machine unit with five friction elements F, **b** multibody mechanical model of a driving wheel and **c** simulation results of gear shifting

actuating the masses m_k and m_A. The mass m_k includes the part of wheel mass, contacted with the surface, and the part of road surface mass "adhering" to the wheel and moving with the tire by the non-elastic slipping. The element m_k is connected with the surface through the frictional contact F_K and it can have the stationary or skidding state. Its behavior is described by the all-purpose dynamic equations which do not depend on the state of a contact [18]:

$$\begin{gathered} m_{k2}\dot{v}_{k2} = F_W - R_X \\ R_X = (1-\lambda)F_X - \lambda F_\varphi \end{gathered} \tag{2.1}$$

where $\dot{v}_{k2}$—derivative of velocity of mass m_{k2}, F_X—the force within the stationary tire/surface contact; F_φ = the friction force for the skidding wheel; λ—the indicator of the contact mode determined during the solving of dynamic equations ($\lambda = 0$ by stationary contact; $\lambda = 1$ by skidding). The force F_φ is described either analytically or via a plot depending on the parameters of wheel movement; the force F_X is defined through the common system of differential, algebraic and logical equations of machine movement.

Friction clutches F_1 and F_2 are controlled by the hydraulic pressure. At the initial point of time the clutch F_1 is closed. The torque is transferred through the train P_1. The turn-off of the clutch F_1 and the turn-on of the clutch F_2 provide the gear shifting. At the interaction of these gearbox clutches, the slipping of clutch F_C and elements $F_{K1(2)}$ can appear because of the high dynamic loads in the transmission parts.

Simulation results (Fig. 2.7c) give the examples of the fluctuations in angular velocities ω_i for the rotational masses J_i and the oscillation in torque M in the part E_T.

Additional examples and features of quasi-static, kinematic and dynamic calculations are presented in [19–21].

Simulation of emergency situations and their solution

One of the directions of dynamic calculations is the modeling of fast-paced processes in emergency situations. Within the framework of this direction, the concept of complex functionality was developed, based on the possibilities of dynamic interaction of various subsystems of the machine [22].

Complex functionality is the ability of a vehicle to perform a certain function using the capabilities (reserves) of its various subsystems, including those that are not focused on performing this function in normal situations. For example, the braking function is usually realized due to the braking system, but in critical situations other subsystems, for example, a trans-mission, can also be involved. In this case, additional, usually nonlinear effects can be used, which are manifested in the joint action of subsystems.

As part of this concept, the Brake-Transmission Master Assistance (BTMA) has been developed, which uses a transmission and braking system during emergency braking [23]. This assistant is based on the transmission gears switching "down", which allows unloading the basic brake system. Temperature heating, brake pressure levels, hysteresis losses value are decreased. As a result, the higher brake efficiency is secured. Offered BTMA is intended for the driver's using in critical situations, it also can be put into operation when the driver loses the control, for example, when he falling asleep or loosing of consciousness and in other statuses qualified as critical and demanding most effective job of a brake system.

The proposed method of emergency braking is protected by a Eurasian patent [24].

2.3.2 Dependability Calculation of Technically Complicated Items [25]

General procedure

The fundamental limitation of the known methods and software for predicting reliability is their scope, which is based on approaches of structural reliability. The starting point for such calculations is the data on the reliability of the elements included in the system. This is not acceptable for mechanical elements, the lifetimes of which are always individual. In addition, the lifetimes of mechanical components are dependent and correlated within the system in which they are located.

LMM uses the Multilevel Mechanics-Dependability Scheme (MMDS) to predict reliability of mechanical and combine systems (Fig. 2.8).

The MMDS has 6 levels in its standard form. In special cases, an arbitrary number of levels can be used, as well as combinations with a different number of levels for calculation according to different calculation chains.

The general procedure reproduces the probabilistic nature of the operating conditions and the properties of components, as well as the effects of their dependent behavior in the system. Monte Carlo simulation are used, in which local procedures are embedded that provide dependent behavior of elements and schemes of limiting states.

The lower physical levels (6, 5, and in some cases 4) reflect mechanical/physical processes and rest are structural ones using SLSs.

The preliminary procedures are: (1) selection of the distributions of the characteristics of the load carrying ability of the components (if their behavior is reproduced from the lower physical levels, see Fig. 2.8, bottom left) or the failure characteristics of the elements (when the behavior of individual components is reproduced from the structural levels); (2) construction of LSCs (see Fig. 2.8, bottom middle); (3) obtaining a spectrum of the RDOC (see Fig. 2.8, bottom right).

Features of predicting reliability

The calculation of the lifetime (operation life, service life) taking into account the physical processes (physics of failures) leading to the limiting states of machine parts is a key factor for combining the models of mechanics and the reliability theory. To

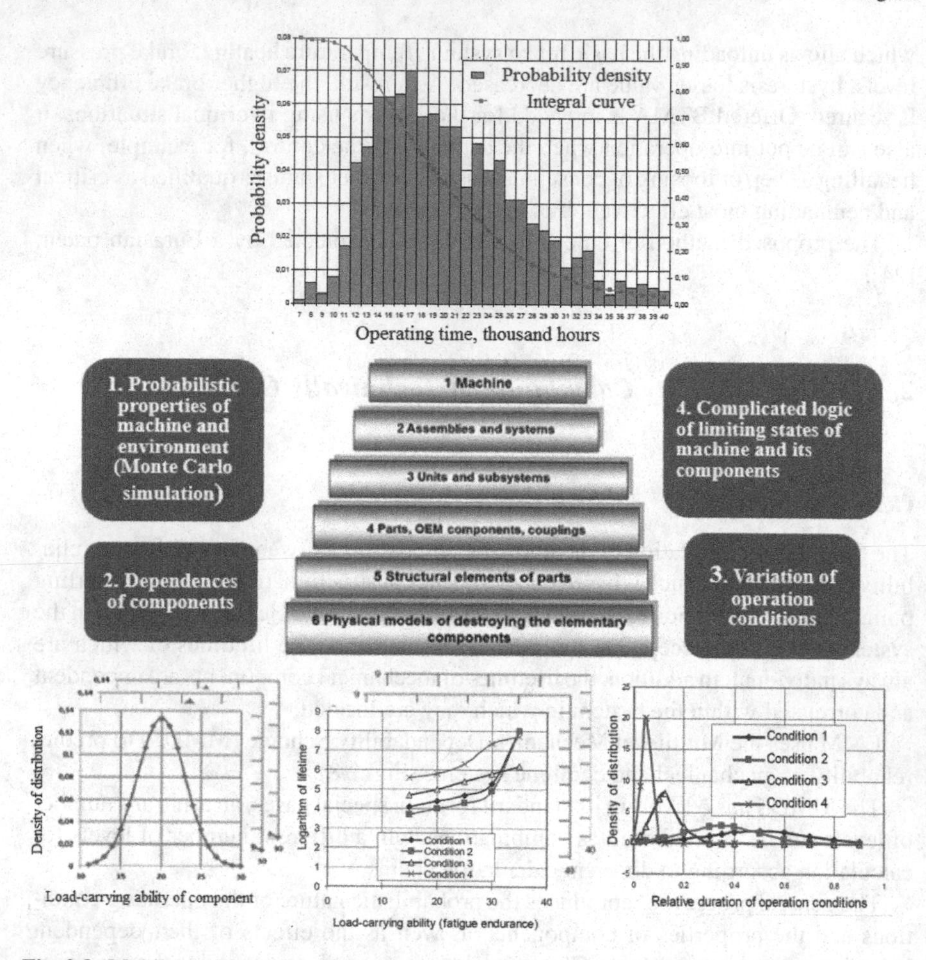

Fig. 2.8 Multilevel "mechanics-dependability" scheme

perform lifetime calculation of any mechanical component, meaningful models are needed to realize the way: {load-carrying ability & operating loads} → lifetime.

The strength calculation results of the various mechanical components take different forms. The results can be presented in the form of operating and permissible stresses, safety factors, gamma-percentage lifetimes, etc. (see ISO standards for the calculation of gears, bearings, etc. [26, 27]). Such heterogeneous results cannot be: (1) compared and (2) used in reliability calculations. Therefore, the first step is the presentation of strength calculations according to the "Strength as lifetime" scheme, when the result is the lifetime:

$$L = R/q \tag{2.2}$$

where R—measure of load-carrying ability (resistance to load) of a component, q—measure of damage per unity of operating time/life (damage extent for one km or hour), which is determined by the load mode of the component.

The load modes for the calculation of different limiting states for the machine parts can differ significantly. But all load modes are determined by the operation conditions (OC). Therefore, the lifetime of each component can be calculated as a result of modeling its loading under certain OC of the machine and their use to determine the loading parameters in lifetime calculations. For a set of parameters describing the OC and characteristics of the bearing capacity, it is advisable to form the lifetime space in the form of "life-strength curves (LSC)". This makes it possible not to carry out laborious calculations anew during statistical modeling.

In addition, in this case, the principle of *life-dependent behavior* of the elements of a loaded mechanical system is implemented, since the generated operation conditions act as a common factor for all calculated elements.

OC are categorized by road type and load, including traffic situation and load. The driving style can be assessed separately as Quiet, Active and Sporting (see, for example, [19, 22]). As a result, the *variation in operation conditions* is described by a probabilistic spectrum of the *relative duration of operation conditions* (RDOC).

OC vary from machine to machine. This inevitable variation (as a reflection of uncertainty) should be taken into account at the design stage when predicting the functional and life properties of machines. In the next stages, this uncertainty can be significantly reduced when considering a specific exemplar (instance) of the machine.

Next specificity is *using SLSs* instead of known Reliability block diagrams (RBD) and Fault trees from Fault Tree Analysis (FTA) to describe complex failures (limiting states). SLSs are simpler and present usual multilevel structure of the machine from top (first) level (machine) to any lower levels (2 = assemblies, … 6 = physical models of elementary components) [21, 22]. This structure is added by description of conditions for limiting states for the elements of higher and intermediate ranks.

An example of probabilistic modeling the lifetime of the system using the SLS

Modeling the lifetime failures of the driving axle (DA) by the method of Monte Carlo simulation in the Excel environment is shown in Fig. 2.9. This is a fragment of the simulation of life failures of the vehicle. The DA with identification number 9 (ID9) is represented by two limiting components: crankcase (ID12) and pinion gear (ID13). The Y parameter describes the types of these constituent parts (CPs).

The SLS of DA has the form (1, 1). This means that its limiting state is reached when the limiting state is reached simultaneously or sequentially by its CPs of the first and second types.

(In the general case, the schematic record has the form (X1, X2, etc.), which means that the limiting state of the object occurs if the limiting states are reached with its X1 parts of the first type (here X1 is the number standing in the first position), its X2 parts of the second type (here X2 is the number standing in the second position), etc. The object (unit, assembly, machine) can have some the SLS).

For the crankcase, the normal law (N) was adopted, and for the drive gear, the lognormal law (LN) was adopted with an average value of 300 unities and Var =

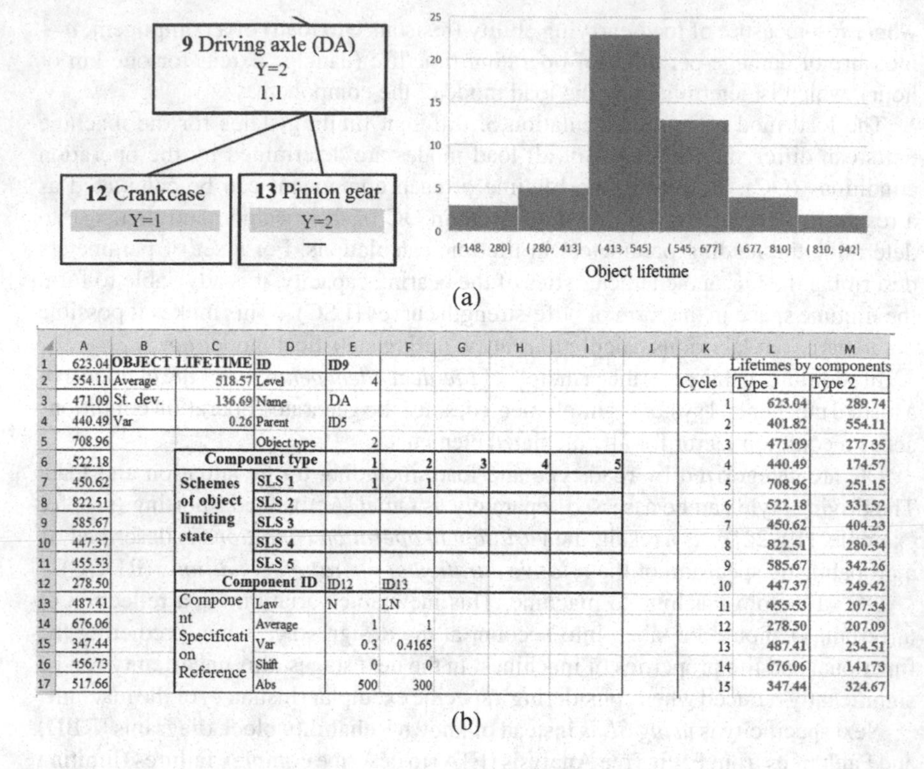

	A	B	C	D	E	F	G	H	I	J	K	L	M
1	623.04	**OBJECT LIFETIME**		ID	ID9							Lifetimes by components	
2	554.11	Average	518.57	Level	4						Cycle	Type 1	Type 2
3	471.09	St. dev.	136.69	Name	DA						1	623.04	289.74
4	440.49	Var	0.26	Parent	ID5						2	401.82	554.11
5	708.96			Object type	2						3	471.09	277.35
6	522.18		**Component type**		**1**	**2**	**3**	**4**	**5**		4	440.49	174.57
7	450.62		**Schemes of object limiting state**	**SLS 1**	1	1					5	708.96	251.15
8	822.51			**SLS 2**							6	522.18	331.52
9	585.67			**SLS 3**							7	450.62	404.23
10	447.37			**SLS 4**							8	822.51	280.34
11	455.53			**SLS 5**							9	585.67	342.26
12	278.50		**Component ID**		ID12	ID13					10	447.37	249.42
13	487.41		Component Specification Reference	Law	N	LN					11	455.53	207.34
14	676.06			Average	1	1					12	278.50	207.09
15	347.44			Var	0.3	0.4165					13	487.41	234.51
16	456.73			Shift	0	0					14	676.06	141.73
17	517.66			Abs	500	300					15	347.44	324.67

Fig. 2.9 **a** Two-level SLS of the drive axle (left) and histogram of its lifetime (right) and **b** statistical modeling fragment

0.4165. Column A (Fig. 10b) shows the array of life obtained from the life of the CPs that are generated in columns L and N. It can be seen that the choice of the DA life in each test cycle goes according to the largest of the two values, which corresponds to the given SLS (1, 1) and confirms the correctness of the calculation.

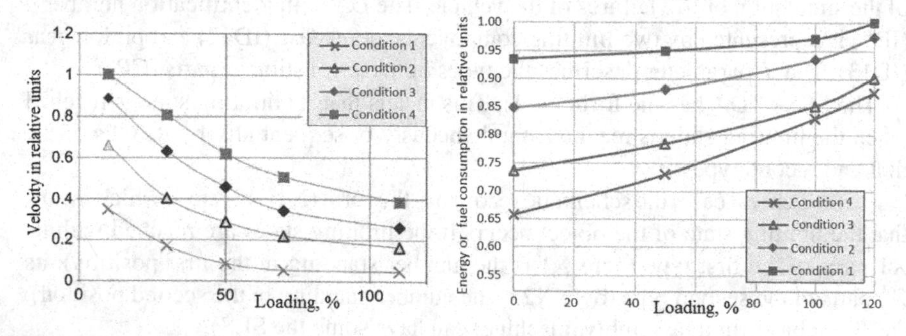

Fig. 2.10 VLC (left) and ELC (right)

2.3.3 Functional Properties Calculation

Energy consumption and performance are the main functional properties of mobile machines. The procedure for their assessment at the design stage in terms of modeling is in many ways similar to the assessment of reliability properties. But instead of LSC, LEC and VLC are used (Fig. 2.10) [28]. When constructing the LEC, energy consumption is determined for electric vehicles, and fuel consumption is found for vehicles with internal combustion engines.

One of the key problem is to determine the energy consumption of an electric vehicle, taking into account its design parameters, charging configuration and route features. This and other problems were considered and solved in the PLATON project [29, 30].

The impacts on energy consumption of driving style, passenger loading, length of segment "from-stop-to-stop", rolling resistance, route obstacles/interferences have been determined [31]. A fragment of the results of the experimental study is presented in Table 2.1.

Analysis of the data on the operation and testing of electric buses shows that the energy consumption even on one route has a significant variation. Finding and taking the calculated (design) value is a problem. A radical probabilistic approach was proposed in [33, 34] and then developed in [32]. This approach is based on two provisions. Firstly, it is necessary to create a space of possible solutions in the form of a distribution of an indicator that describes the property under study (for example, energy consumption for the route cycle considered). Secondly, it is necessary to introduce a stakeholder into the assessment procedure. This person must accept the

Table 2.1 Energy consumption versus driving style and gross bus weight (based on [32])

Gross bus weight m, t	Route length L, km	Driving style	Expended energy S_1, Wh	Recovered energy S_2, Wh	Energy consumption per km $(S_1 - S_2)/L$, kWh/km	TCI = Transport costs indicator = $(S_1 - S_2)/(m{\cdot}L)$, kWh/(t·km)	Driver factor (for the same m)
12	9.58	Calm (Light)	12,282	3860	0.88	73	1
12	9.58	Aggressive	14,377	5225	0.96	79	1.09
12	9.58	Super aggressive	18,270	6686	1.21	101	1.38
18	9.18	Calm (Light)	15,306	5150	1.11	61	1
18	9.58	Aggressive	18,431	7512	1.14	63	1.03
18	9.61	Super aggressive	20,217	8168	1.25	70	1.13

probability with which the choice of the calculated (design) indicator of energy consumption is made.

In a frame of the PLATON project the PLATON Toolkit was developed [35]. One of the tool is ECBus+ . This tool consists of ECBus v4.0 software and two procedures implemented in an Excel environment.

The software ECBus v4.0 is designed to calculate the energy, fuel and environmental performance of electric and diesel buses on the route. Even if the user does not set the corresponding data for the electric or diesel bus, the calculation is performed with the data specified in the program by default.

ECBus v 4.0 calculates the following basic parameters:

1. Energy consumption of the electric and diesel bus on the route (parameters are in Fig. 2.11; graphs of energy changes are in Fig. 2.12, left); for diesel buses additionally:
2. Fuel consumption (Fig. 2.11);
3. Pollutant emissions: ecology (Fig. 2.11).

The software also calculates some additional parameters: graph of changes in traction force (Fig. 2.12, right), distributions of velocities and accelerations of the bus (Fig. 2.13).

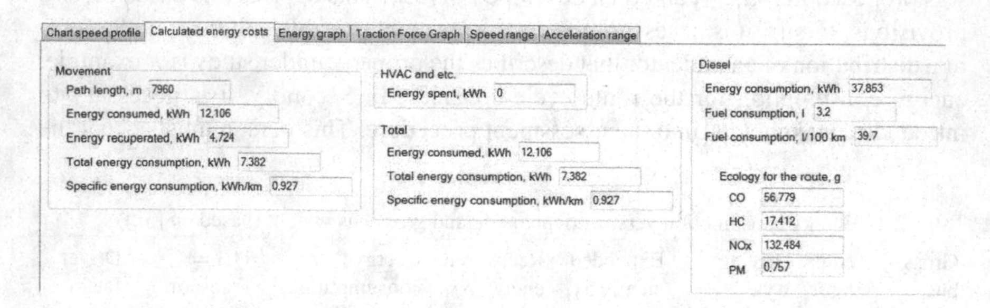

Fig. 2.11 Results of energy, fuel and ecology calculations

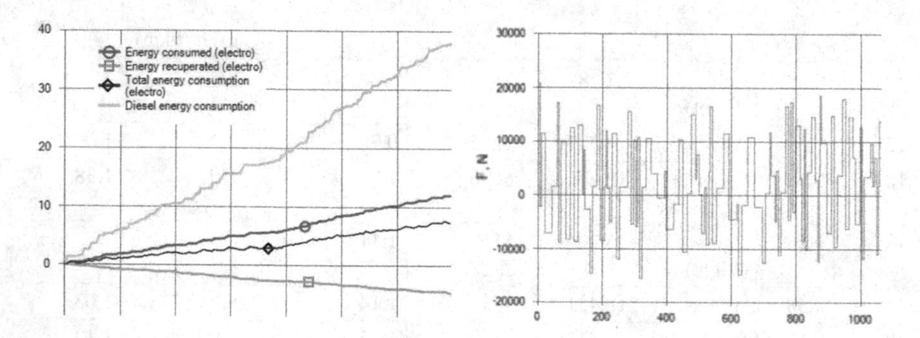

Fig. 2.12 Graphs of changes in energy (left) and tractive force (right)

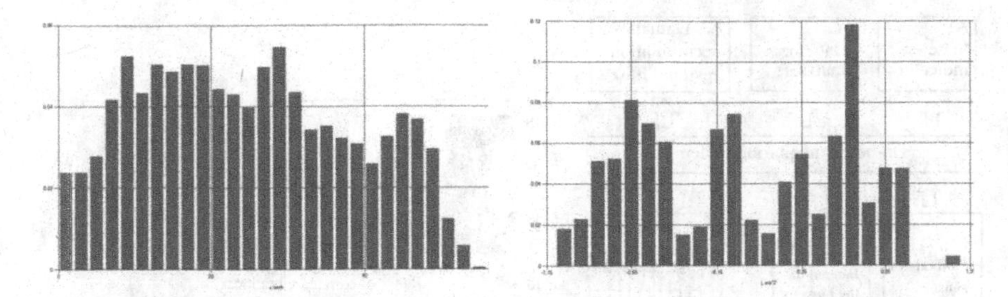

Fig. 2.13 Distributions of velocities (left) and accelerations (right) of the bus on the route

In addition to being posted on the PLANON project site [35], ECBus software is registered with the National Center for Intellectual Property of the Republic of Belarus [36].

The EC-Compare procedure is designed to calculate the energy consumption of an electric bus on a route taking into account fuel consumption of a diesel bus similar in weight. The result gives energy consumption by the electric bus under the same conditions as of the diesel bus-analog. The EC-Compare procedure is implemented in the Excel file: EC-Compare.xlsx.

The developed ECPro procedure provides the opportunity to select the calculated (design) energy consumption value of an electric bus on the route, taking into account the probability accepted by the interested party, reflecting the risk of not exceeding this value during operation. The ECPro procedure is implemented in the Excel file: ECPro.xlsx.

2.3.4 Diagnostics

The main individual component of the load mode of any mechanical unit is the level of its internal dynamic loading. Changes in this level are due to the peculiarities of making the unit and the operation conditions. Components such as gears and bearings are highly susceptible to changes in the internal dynamic loading of the transmission units. The greatest damage to the couplings is associated with short-term transient processes, which requires individual control over these processes.

A complex problem in the diagnosis of machines and their assemblies is determining loads on moving parts: gear wheels, bearings, couplings, etc. One approach is tracking the reactions of these loads on stationary parts (housings), and using these reactions for evaluating the loads of the moving components.

The feature of the developed diagnostic method is using conceptual modeling the oscillating process for the gear drive and the propagation of vibrations in the transmission [20, 21]. It is advisable to applicate together integral diagnostic models and predictive ones based on damage accumulation. Such a "two-coordinate" approach

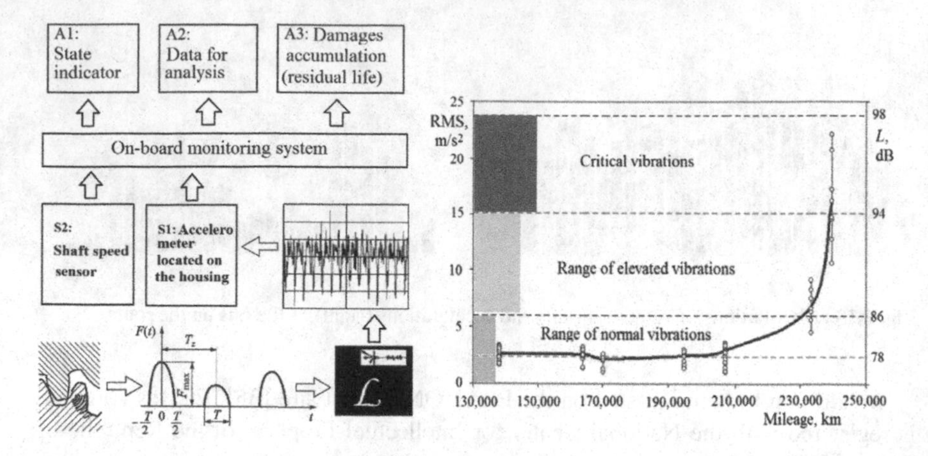

Fig. 2.14 Processes in the MWR and its monitoring system (left) and values of RMS (right)

(from two points of view) ensures a higher veracity of the individual lifetime forecasts.

An example of such an approach is presented below by diagnostics of a motor-wheel reducer (MWR) of a mining dump truck. The main processes for the emergence, transformation and processing of signals in the MWR and its diagnostic system are shown in Fig. 2.14.

The system periodically interrogates the sensors, processes the diagnostic information and evaluates the technical state of the reducer using root mean square (RMS) of vibration acceleration as main integral diagnostic index. The system compares the RMS with the maximum allowable values for each of the reducer states and constantly informs the driver using the corresponding light signal on the instrument panel in the truck cab.

At the same time, lifetime forecasting is performed based on a predicative model. The method of forecasting the lifetime includes the determination of the shock pulse in a meshing according to the results of vibration monitoring. From these data, the actual circumferential force and contact stresses in the meshing are calculated. It is taken into account that a linear relationship exists between the amplitude of the shock pulse and the peak value of the vibration acceleration. The growth of peak values means an increase in the dynamic factor K_{Hv}, which is used in the calculation of contact stresses σ_H.

The measure of fatigue damage to the gear is determined for each fixed i-th interval of the running time by the formula:

$$\Delta Q_{Hi} = \sigma_{Hi}^{w_H} N_i \quad (2.3)$$

where σ_{Hi}—contact stress; w_H—the exponent of the S–N (Stress σ–Number of cycles N) curve under calculating the gear; N_i—the number of loading cycles of the gear tooth.

2.3.5 Concept and Method for Assessment of TCI Lifetime Expense [37]

The term "life consumption" for an individual component is self-explanatory, as is its complementary term "residual life". The assessment of the life state of a complex object consisting of several components is ambiguous. In addition, some components can be replaced or repaired. To eliminate this uncertainty, the term "lifetime expense" for a TCI and a corresponding indicator are introduced. This indicator reflects the life potential of the product as a whole, taking into account the indicators of its main parts.

The lifetime expense KP of the main part is determined taking into account two factors: mileage and time (age), and is calculated according to the formula:

$$K_P = 1 - (1 - K_L)(1 - K_T) \tag{2.4}$$

where K_L—lifetime expense by mileage (operating time); K_T—lifetime expense by time (age); K_L relates to damage processes under loads during operation; K_T relates to damage processes under the influence of time (aging processes).

Usually, the processes that determine K_L and K_T can be considered as independent. Then K_L can be interpreted as the probability of failure under the influence of loads in the duty cycle, and K_T as the probability of failure under the influence of time (age). In this case, K_P is the probability of failure of the main part of the TCI under the combined action of loads and time (age) on the considered date.

The lifetime expense of the TCI is determined as follows:

$$K_A = \xi_1 K_{P1} + \xi_2 K_{P2} + \ldots + \xi_n K_{Pn} \tag{2.5}$$

where K_{Pi}—lifetime expense of the ith main part of the TCI (presented as relative value); ξ_i—the weight factor that determines the contribution of the main part to the total lifetime expense of the TCI. It is proposed to consider ξ_i as the relative mass of the ith main part (the fraction of this part mass in the total mass of n parts that determine the TCI lifetime).

Figure 2.15 shows the graphs K_L, K_T, and K_P of the main part for two typical cases of using the item. In the second case, aging processes prevail. Table 2.2 presents the calculation of the lifetime expense K_A for a vehicle on duty and making small runs. For such cases, traditional calculations based on operating time data are not suitable.

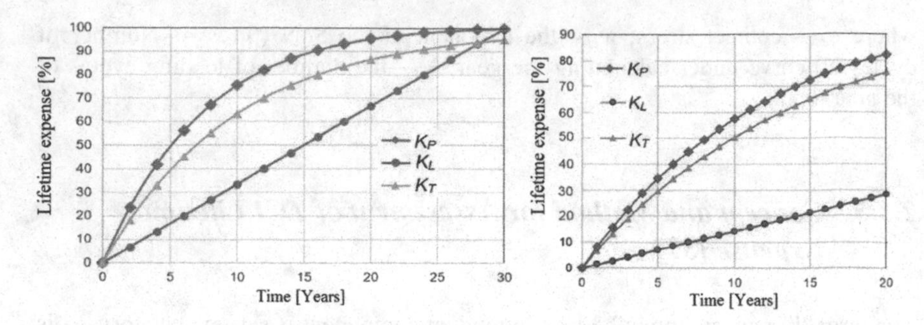

Fig. 2.15 Lifetime expense K_P for linear law K_L and exponential K_T for the item with regular use (left) and little use (right)

Table 2.2 Calculation of vehicle lifetime expense K_A

Vehicle main part	K_L [%]	K_T [%]	K_P [%]	m [kg]	K_{PY} [%]	K_A [%]
Frame	23.10	68.12	75.48	456	12.09	61.68
Engine (after overhaul)	6.17	31.95	43.39*	490	7.47	
Gearbox (after overhaul)	6.70	32.86	44.83*	100	1.57	
Transfer case (used)	27.70	47.96	62.38	120	2.63	
Front axle	23.10	59.27	68.68	495	11.94	
Middle axle	23.10	59.27	68.68	432	10.42	
Rear axle (replacement)	7.80	42.31	46.81	432	7.10	
Cabin	38.12	59.27	74.80	322	8.46	

Notes * The K_P of the main parts after overhaul increases by twenty percent in comparison with those that are not repaired; $K_{PY} = K_P\ (m/m_0)$; $m_0 = \Sigma m_i = 2847$ kg

The calculation of TCI lifetime expense, taking into account the mileage and time, gives an estimate that corresponds to reality. This allows you to assess the individual condition of the item during its life cycle, including periods of long downtime. In such cases, conventional Digital Twins using sensor readings are meaningless.

The developed technique is implemented in the State standard of the Republic of Belarus [37].

2.4 Conclusions

The developed **architecture** contains a number of **new representations**, models and tools that must be used to assess the functional and life properties of technically complicated items and it can be applied to the **wide range of engineering objects**.

The **lifetime expense model** should be applied in PHM (Prognostics and Health Management) of engineering systems **and gives comprehensive individual assessment of the life potential** of an item during its operation under the combined action of **loads** (mechanical, thermal, etc.) and **time (age)**.

Basic **meaningful representations** of TCI (kinematic diagrams, set of dynamic schemes for different types of dynamic calculations, schemes of limiting states, etc.) appear in the development process and go through all stages of the item life cycle. These components are complemented by TCI diagnostics and evaluation of lifetimes expense and serve as **accumulators of new knowledge and data** on the item and its operation conditions.

Lifetime Mechanics of Machines provides the interconnection of models from various disciplines and methodologies, the creation of an **information model as an integrator of digital twin and digital thread.**

Acknowledgements The part of this paper dealing with the functional properties of electric buses was prepared using some of the results of the PLATON project (Planning Process and Tool for Step-by-Step Conversion of the Conventional or Mixed Bus Fleet to a 100% Electric Bus Fleet, runtime: 01.2018–06.2020), which received funding from the ERANET COFUND Electric Mobility Europe (Horizon 2020 program).

References

1. ISO/IEC/IEEE 42010 Systems and software engineering—architecture description
2. Grieves, M.: Origins of the digital twin concept. https://www.researchgate.net/publication/307509727_Origins_of_the_Digital_Twin_Concept. Last Accessed on 22 March 2021
3. Grieves, M.: Virtually intelligent product systems: digital and physical twins. In: Flumerfelt, S., Schwartz, K.G., Mavris, D., Briceno, S. (eds.) Complex Systems Engineering: Theory and Practice, pp. 175–200. American Institute of Aeronautics and Astronautics (2019)
4. Dontha, R.: Data and Trending Technologies: Role of Data in Digital Thread. https://tdan.com/data-and-trending-technologies-role-of-data-in-digital-thread/24055. Last Accessed on 22 Mach 2021
5. EP 3 208 757 A1. Tracking production in a production facility, using searchable digital threads (2017)
6. US 8 991 692 B2. Managing component information during component lifecycle (2015)
7. Grinberg, A., Algin, V.: Information models and resources of machines. In: Proceedings of the International Conference Mechanics of Machines on the Threshold of the Third Millennium, pp. 272–281. IMM NAS Belarus, Minsk (2001) (in Russian)
8. Rathore, M.M., Shah, S.A., Shukla, D., Bentafat, E., Bakiras, S.: The role of AI, machine learning, and big data in digital twinning: a systematic literature review, challenges, and opportunities. IEEE Access **9**, 32030–32052 (2021)
9. ISO/DIS 23247-1(en) Automation systems and integration—digital twin framework for manufacturing—Part 1: overview and general principles
10. Keuthen, M.: REXS—standardized gear unit model. In: Proceedings of the International Conference International Conference on Gears, pp. 701–712. Garching/Munich (2019)
11. REXS (Reusable Engineering EXchange Standard). https://www.rexs.info/rexs_en.html. Last Accessed on 22 March 2021
12. Yin, J., Zheng, H., Yang, Y., Xu, M.: A new life prediction scheme for mechanical system with considering the mission profile switching. Appl. Sci. **10**, 673-1–673-16 (2020)

13. Wehbi, N.A., Slika W.G.: A health monitoring framework for optimal service life predictions of steel structures under fatigue loading. In: Papadrakakis, M., Papadopoulos, V., Stefanou, G. (eds.) Uncertainty Quantification in Computational Sciences and Engineering: UNCECOMP 2019, pp. 653–662. Greece (2019)
14. Karve, P.M., Guo, Y., Kapusuzoglu, B., Mahadevan, S., Haile, M.A.: Digital twin approach for damage-tolerant mission planning under uncertainty. Eng. Fract. Mech. **225**, 1–28 (2020)
15. Hartmann, D., Van der Auweraer, H.: Digital twins. https://www.researchgate.net/publication/338853051. Last Accessed on 26 March 2021
16. State standard of Republic of Belarus STB 2465–2016 Dependability in technics. Dependability management of technically complicated items
17. Algin, V., Ivanov, V.: Kinematic and dynamic computation of vehicle transmission based on regular constructs. In: Merlet J.-P., Dahan, M. (eds.) Proceeding of 12th IFToMM World Congress, pp. A14-1–A14-6. Besançon (2007)
18. Algin, V., Ivanov, V.: Application of regular rotational and translational constructs to vehicle dynamics problems. In: Brennan, M.J. (ed.) Proceeding of 7th European Conference on Structural Dynamics, pp. E16-1–E16-12. University of Southampton, Southampton (2008)
19. Algin, V.: From Newton's mechanics to dynamics of regular mechanical systems with variable states and power flows. In: Proceedings of 14th IFToMM World Congress, pp. 1–10. Taipei (2015)
20. Algin V., Ishin, M., Paddubka, S.: Models and approaches in design and diagnostics of vehicles planetary transmissions. IOP Conf. Ser.: Mater. Sci. Eng. **393**, 012042-1–012042-10 (2018)
21. Algin V., Ishin, M., Paddubka, S., Starzhinsky, V., Shil'ko, S., Rackov, M., Čavić, M.: Development of information model of power transmissions in the light of industry 4.0. Int. Sci. J. Math. Model. **4**, 54–63 (2020)
22. Algin, V., Paddubka, S.: Lifetime Mechanics of Vehicle Transmissions. Belaruskaya Navuka, Minsk (2019) (in Russian)
23. Algin, V., Tretsiak, D., Drobyshevskaya, O.: Investigations in advanced brake assistant systems. In: Proceedings of 12th EAEC European Automotive Congress Europe in the Second Century of Auto-Mobility, pp. 08_008-1–08_008-14. Bravislava (2009)
24. EA022858 B1. Method of vehicle's emergency braking (2016)
25. State standard of the Republic of Belarus STB 2466-2016. Dependability in Technics. Dependability Calculation of Technically Complicated Items
26. ISO 6336-6: 2019 Calculation of load capacity of spur and helical gears—Part 6: calculation of service life under variable load
27. ISO 281: 2007 Rolling bearings—dynamic load ratings and rating life
28. Algin, V.: Calculation of Mobile Technics: Kinematics, Dynamics, Life. Belaruskaya Navuka, Minsk (2014) (in Russian)
29. Algin, V.: Electrification of urban transport. Basic stages in creating electric buses fleet. Mechanics of Machine, Mechanisms and Materials **3**(44), 5–17 (2018)
30. Algin, V., Czogalla, O., Kovalyov, M., Krawiec, K., Chistov, S.: Essential functionalities of ERA-NET electric mobility Europe PLATON project. Mech. Mach., Mech. Mater. **4**(45), 24–35 (2018)
31. Algin, V., Goman, A., Skorokhodov, A.: Main operational factors determining the energy consumption of the urban electric bus: Schematization and modelling. In: Topical Issues of Mechanical Engineering: Collection of Scientific Papers 8, pp. 185–194. Minsk (2019)
32. Algin, V., Goman, A. Skorokhodov, A., Bytsko, O., Chistov, S., Fedasenka, S.: Methodology for probabilistic assessment of energy consumption by electric buses on routes. In: Krawiec, K., Markusik, S., Sierpiński, G. (eds.) Electric Mobility in Public Transport—Driving Towards Cleaner Air, pp. 83–105. Springer, Cham (2021)
33. Algin, V.: Calculated modes for assessing operation properties and dependability of vehicles. In: Uhl, T. (ed.) Advances in Mechanism and Machine Science: IFToMM WC2019. MMS, vol. 73, pp. 3749–3758. Springer, Cham (2019)
34. Algin, V.: Justification of calculated cases for the assessment of the basic properties of land vehicles. In: Beer, M., Zio, E. (eds.) Proceedings of the 29th European Safety and Reliability Conference, ESREL 2019, pp. 3510–3517. Research Publishing, Singapore (2019)

35. PLATON. http://service.ifak.eu/PLATON-Web/downloads.html. Last Accessed on 29 March 2021
36. Algin, V.B., Lahvinets, T.S., Goman, A.M. Shportko, V.V.: Certificate of registration of the computer program No. 1342. Name of the computer program: ECBus. In: Register of Registered Computer Programs of the National Center for Intellectual Property of the Republic of Belarus (2020)
37. State standard of the Republic of Belarus STB 2578-2020. Dependability in technics. Estimation of lifetime expense for technically complicated items

Chapter 3
Mechanism Applications in Robotics

Erwin-Christian Lovasz

Abstract The paper focuses on different applications of linkages and geared linkages in robotic applications. The applications show the use of geared linkages with linear actuators in development of a knee prosthesis, an active elbow orthosis and as legs for a planar parallel manipulator. The main characteristics of the geared linkage with linear actuation are the large rotation angle with proper transmission angle and the approximately linear transmission function in a large range.

Keywords Geared linkages · Knee prosthesis · Elbow orthosis · Exoskeleton · Planar parallel manipulators · Gripper · Axial centering precision

3.1 Introduction

Geared linkages is defined as a combination of a linkage and a gear mechanism [1]. The geared linkages with rotating input motion were studied by many researches, considering two structure types of geared linkages: with parallel connected gear train and with serial connected gear train (see Fig. 3.1a, b).

The first type of geared linkages with serial connected gear train were studied by Reuleaux [2] for some applications as sewing, stamping, steering and straight-line generator mechanisms. Freudenstein et al. in [3] studied the kinematical behaviour of the coupler point. A numerical synthesis method of the geared five-bar linkage for path and function generating tasks was proposed by Roth et al. in [4], respectively by Oleksa et al. in [5]. Mundo et al. in [6, 7] studied the geared linkages with serial connected gear train in the frame by using non-circular gear train. Later, Parlaktas et al. in [8] developed a novel analysis method based on the relationships of transmission angle.

The second type of geared linkages with parallel connected gear train were studied by Neumann [9] as step mechanisms with non-uniform continuous motion with high transmission ratio, with high swing angle [10], respectively with instantaneous dwell

E.-C. Lovasz (✉)
Politehnica University of Timişoara, Bv.Mihai Viteazul, 1, Timişoara, Romania
e-mail: erwin.lovasz@upt.ro

M. Rackov et al. (eds.), *Machine and Industrial Design in Mechanical Engineering*, Mechanisms and Machine Science 109,
https://doi.org/10.1007/978-3-030-88465-9_3

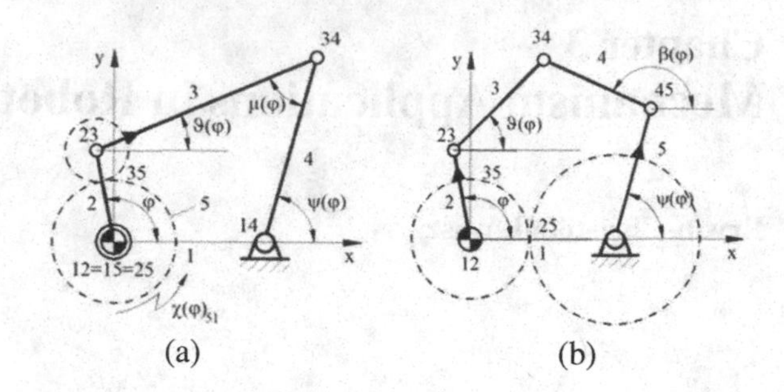

Fig. 3.1 **a** Geared linkages with serial and **b** parallel connected gear train

or pilgrim step [11, 12] and Hain [13]. Horani [14] and Rankers [15] developed analysis and synthesis methods for geared linkages with oscillating motion.

The kinematical characteristics of this type of geared linkages recommend it for generating a large oscillating motion (e.g. windscreen wiper transmissions) or pilgrim-step motions [16, 17]. In order to obtain a dwell or a constant transmission ratio within a range, the higher derivatives of the transmission function have to be zero. Modler et al. in [18] shows that such transmission functions can be generated with five-links geared linkages and a non-circular gear pair (Fig. 3.2).

Litvin in [19, 20] presented the non-circular gears generation employing the enveloping method, using tools similar to the circular gears generation. This method fulfils the rolling condition without sliding. The proposed non-circular gears applications would be: reducing torque and speed fluctuations in rotating shaft by Dooner in [21]; actively balancing shaking moments and torque fluctuations in planar linkages by Kochev in [22] and Yao et al. in [23]; maximising the human output during low speed pedalling on the power drive mechanism for the high performance bicycle by planetary gear train by Mundo in [24] or by band or tape drives by Freudenstein et al. in [25]; steering mechanism by Emura in [26]; modulation of the blood flow in external circulations machines—generating a function with quick forward

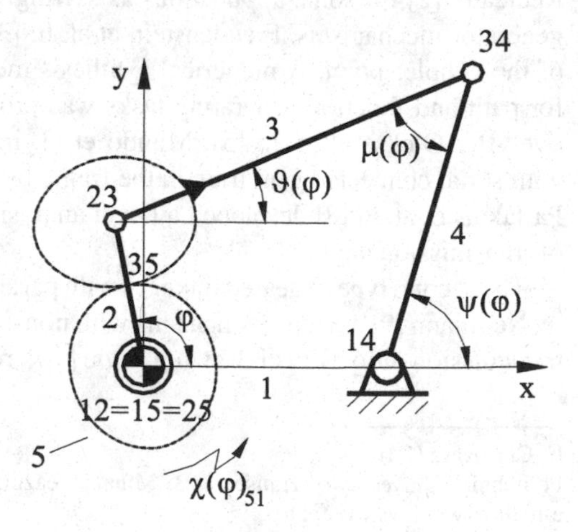

Fig. 3.2 Geared linkage with non-circular gear pair

stroke corresponding to the systolic phase and slow return stroke corresponding to the diastolic phase—through a slider-crank linkage driven by non-circular gears by Ottaviano et al. in [27, 28]; kinematic optimization of ball-screw transmissions in order to reduce the peak acceleration of the screw—through a serial connected non-circular gear with a ball-screw transmission by Khatait et al. in [29].

3.2 Geared Linkages with Linear Actuation

About the geared linkages with parallel connected geared train and linear actuation there are a few theoretical and experimental researches. Gnasa in [30] and [31] studied linkages and more link geared linkages with linear hydraulic actuation to be used as acting mechanism between the manipulator links. In [32–34] Lovasz et al. presented a unitary developed type synthesis, kinematic analysis and dimensional synthesis methods for the geared linkages with linear actuation.

3.2.1 Type Synthesis of Geared Linkages with Linear Actuation

The type synthesis pursues to find out the possible kinematic chains and the mechanism structures of the geared linkages with linear actuator. The basic equation for a mechanism with a constrained motion, according to Alt, is:

$$2 \cdot (e_1 + e_2/2) - 3 \cdot n + 3 + F = 0 \tag{3.1}$$

where: e_1 is the number of kinematic pairs with $DoF = 1$, e_2—the number of kinematic pairs with $DoF = 2$, n—the number of links and M—the degree of freedom of the mechanism DoF.

In the case of geared linkages the mechanism contains at least one kinematic pair with $DoF = 2$, i.e. $e_2 = 1$. Considering the mechanism degree of freedom $M = 1$, from the Eq. (3.1) follows the correlation between n and e_1, in the form:

$$n = (2 \cdot e_1 + 5)/3 \quad e_1 = 2 + 3 \cdot k \quad (k = 0, 1, ...) \tag{3.2}$$

The solution for the geared linkages chain with minimal structure is $e_1 = n = 5$. The chain contains at least a prismatic kinematic pair, which is the drive kinematic pair. The number of closed loops for the planar kinematic chain is to be computed with the relationship:

$$N = \sum_{i=1}^{2} e_i - n + 1 \quad (3.3)$$

The number of contours for the geared linkages with linear displacement actuator is $N = 2$, where the degree of freedom $M_j > 0$ for each loop must be positive. The numbers of the elements of different ranks satisfy the diophantine equations system:

$$n = n_2 + n_3 + n_4 + n_5$$

$$2 \cdot \sum_{i=1}^{2} e_i = 2 \cdot n_2 + 3 \cdot n_3 + 4 \cdot n_4 + 5 \cdot n_5 \quad (3.4)$$

The elements n_4 and n_5 are useless, so that the equations system (3.4) has the solutions $n_2 = 3$ and $n_3 = 2$. By a systematic development of all possible kinematic chains results a number of 11 possible chains, but only 6 kinematic chains fulfil the conditions $Mj > 0$ and the gear contact condition Using the Reuleaux method to develop the 6 kinematic chains and the criteria:

- the drive is of slider type,
- the mechanism should not consist of basical mechanisms in serial order, and
- all the links must be included in the motion transmission,

then 4 structure of geared linkages mechanisms remain to be considered. These will be classified and noted as:

- Planetary geared linkages, type A_Si (Fig. 3.3), and
- Cycloidal geared linkages, type Z_Si (Fig. 3.3).

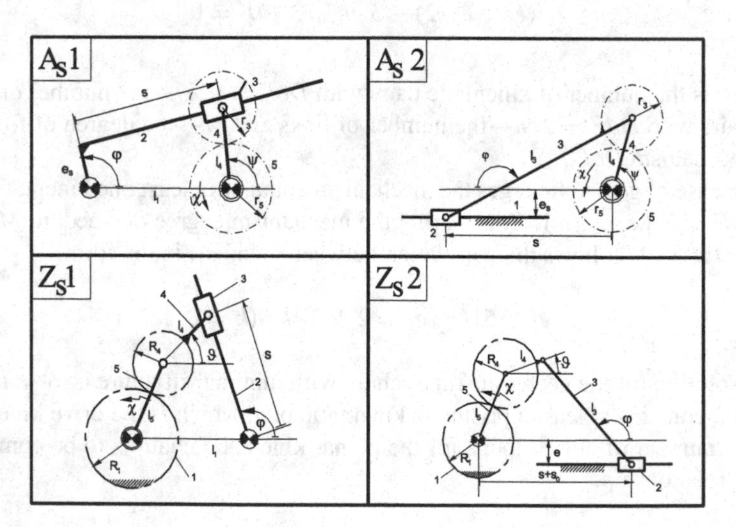

Fig. 3.3 Planetary geared linkages of type A_Si and cycloidal geared linkages of type Z_Si

3.2.2 Kinematic-Positional Analysis of the Geared Linkages with Linear Actuation

The kinematic analysis of planetary geared linkage with linear actuator considers the mechanism consisting of two basic structures: a 4-bar linkage and a planetary gear train (Fig. 3.4).

The transmission functions $\phi(s)$ and $\psi(s)$ of the 4-bar linkage depend on the input parameter, the stoke s, and it can be determined by considering the closure loop vector equation as follow:

– for geared linkage with inverted slider-crank as base structure (Fig. 3.3 A_s1)

$$\varphi(s) = 2 \cdot \arctan\Big((B_1(s) \pm \sqrt{A_1(s)^2 + B_1(s)^2 - C_1(s)^2})/(A_1(s) - C_1(s)) \Big) \tag{3.5}$$

with:

$$A_1(s) = 2 \cdot l_1 \cdot e_S, \quad B_1(s) = 2 \cdot l_1 \cdot (s_0 + s), \quad C_1(s) = -l_1^2 - e_S^2 - (s_0 + s)^2 + l_4^2 \tag{3.6}$$

and:

$$\psi(s) = \arccos\big((e_S^2 + (s_0 + s) - l_1^2 - l_4^2)/2l_1 l_4 \big) \tag{3.7}$$

– for geared linkage with slider-crank as base structure (Fig. 3.3 A_s2)

$$\varphi(s) = 2 \cdot \arctan\Big((B_2(s) \pm \sqrt{A_2(s)^2 + B_2(s)^2 - C_2(s)^2})/(A_2(s) - C_2(s)) \Big) \tag{3.8}$$

with:

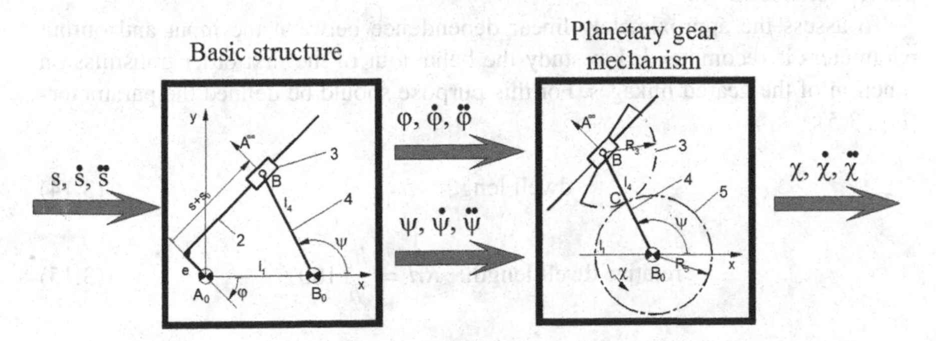

Fig. 3.4 Kinematic analysis of the geared linkage with linear actuation

$$A_2(s) = 2 \cdot l_3 \cdot (s_O + s), \quad B_2(s) = -2 \cdot l_3 \cdot e_S, \quad C_2(s) = e_S^2 + (s_O + s)^2 + l_3^2 - l_4^2 \tag{3.9}$$

and:

$$\psi(s) = \arcsin((l_3 \sin\varphi - e_S)/l_4) \tag{3.10}$$

The transmission function of the geared linkage is obtained in the form:

$$\chi(s) = (1 - \rho) \cdot \psi(s) + \rho \cdot \varphi(s) \tag{3.11}$$

where: $\chi(s)$ is the output parameter—transmission function of the geared linkage, $\rho = \pm r_3/r_4$—gear ratio and $\phi(s)$, $\psi(s)$—transmission functions of the 4-bar linkage.

The first order transmission function $\chi'(s)$ and the second order transmission function $\chi''(s)$ can be computed in the form:

$$\chi'(s) = (1 - \rho) \cdot \psi'(s) + \rho \cdot \varphi'(s) \tag{3.12}$$

$$\chi''(s) = (1 - \rho) \cdot \psi''(s) + \rho \cdot \varphi''(s) \tag{3.13}$$

3.2.3 Quality Parameters of the Geared Linkages with Linear Actuation

The geared linkages with linear displacement actuator reproduce an approximately linear dependence between the input and output movement for a very large swivel angle range, i.e. an approximate constant first order transmission function or transmission ratio.

To assess the approximately linear dependence between the input and output parameters is recommended to study the behaviour of the first order transmission function of the geared linkages. For this purpose should be defined the parameters (Fig. 3.5):

$$\text{dwell length:} \; h_d \tag{3.14}$$

$$\text{relative dwell length:} \; Rd = \frac{h_d}{h} 100\% \tag{3.15}$$

$$\text{dwell clearance :} \; \Delta\chi_d' \tag{3.16}$$

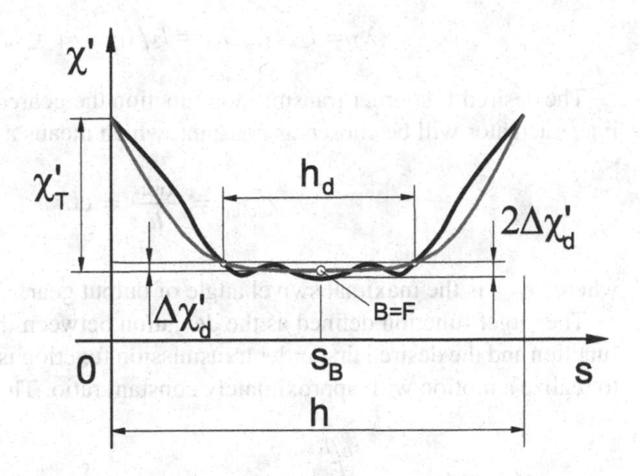

Fig. 3.5 Dwell quality parameters

$$\text{relative dwell deviation: } Rdd = \frac{\Delta \chi_d'}{\chi_T'} 100\% \tag{3.17}$$

where: h is the stroke of the linear actuator, χ_F'—the maximum value of the first order transmission function.

3.2.4 *Optimum Synthesis of the Geared Linkages with Linear Actuation*

This function described above is a flat-point function (Fig. 3.5). The mathematical conditions to become a flat-point function are:

$$\chi'' = 0, \quad \chi''' = 0, \quad \chi'''' = 0 \tag{3.18}$$

The conditions (3.18) lead to an equations system, which cannot be solved with analytical methods. An alternative method to realize the dimensional synthesis of the geared linkages using the condition (3.18) is provided by the optimum synthesis.

The input parameters for the optimum synthesis problem are the desired function χ_{desire}', the selected structure of the geared linkage (Fig. 3.3), the given stroke h and the gear ratio ρ. The first order transmission function will be determined for the selected structure of geared linkage $\chi'(s)$. As vector of the variable will be considered:

$$\mathbf{x} = (\lambda_2, \lambda_3, \lambda_4)^T \tag{3.19}$$

where the normalised variables are:

$$\lambda_2 = l_2/l_3, \quad \lambda_3 = l_3/l_1, \quad \lambda_4 = l_4/l_1 \tag{3.20}$$

The desired first order transmission function the geared linkage mechanism with linear actuator will be chosen as constant, which means a constant ratio:

$$\chi'_{\text{desire}} = \frac{\chi_{\max}}{h} = \text{ct.} \tag{3.21}$$

where: $\chi_{\max}$ is the maximal swivel angle of output gear.

The target function defined as the deviation between the first order transmission function and the desired first order transmission function is to be minimized, in order to realize a motion with approximately constant ratio. The target function is:

$$F(\mathbf{x}) = \int_{o}^{s_H/l_1} \left|\chi'(s, \lambda_2, \lambda_3, \lambda_4) - \chi_{\max}/h\right| \mathrm{ds} := Min! \tag{3.22}$$

The restrictions are given as geometrical boundary and transmission angle of the base mechanism condition:

$$g_k(\mathbf{x}) = g_k(\lambda_2, \lambda_3, \lambda_4) \geq b_k \tag{3.23}$$

The start values of the variables vector:

$$\mathbf{x}^{(0)} = \left(\lambda_2^{(0)}, \lambda_3^{(0)}, \lambda_4^{(0)}\right)^T \tag{3.24}$$

will be conveniently chosen. Generally, the local optimum values give the optimum values for the links length.

3.3 Applications of the Geared Linkages with Linear Actuation in Medical Robotics

3.3.1 *Active Knee Prosthesis with Geared Linkage with Linear Actuation*

The development in the field of knee prostheses pursues a biological static and dynamic behaviour of the lower human limb, a uniform distribution of the body weight both on the prosthetic limb and the healthy limb, ensuring the walking stability and identical movement for both limbs. All these requirements improve the life quality and the work capacity of the amputees. In order to insure the walking stability, usually a large series of mechanical, pneumatic and hydraulic solutions is used for the knee prosthesis. These prostheses are designed either as passive or active systems. The mechanical solutions can use a simple joint or an expensive kinematic joint with dampening system or polycentric (physiological) hinge [35].

The human knee joint allows a rotation motion with an angle of 130°. Usually, the active prostheses with 1 DOF uses an inverted slider-crank or a slider crank. Such mechanisms provide a non-linear transmission function and allow a limited rotation angle between the thigh and the shank.

The disadvantages of the inverted slider-crank can be avoided by using a geared linkage with a linear actuation. The geared linkage contains an inverted slider-crank as basic structure connected in parallel with a gear train.

In order to obtain an approximately constant transmission ratio for a large rotation angle, an optimization synthesis was previously recommended. The imposed parameters for the optimization problem are given in the Table 3.1.

Through the optimization synthesis results one local minimum value for the non-dimensional links lengths from the contour line diagram for $\lambda_2 = 0$ and $\lambda_4 = 0.9$. Thus, the links lengths are $l_2 = 0$ mm and $l_4 = 36$ mm for the frame length of $l_1 = 400$ mm. This mechanism allows a maximum rotation angle of $\chi^\circ_{\max} \cong 144°$ for a start position $s_0 = 364$ mm and a stroke $h = 36$ mm. The rotation angle is limited to 120° (Fig. 3.6) with the displacements $s_1 = 0.5$ mm and $s_2 = 32$ mm, which means that for the knee-joint, the start position and the stroke are $s_0 = 364.5$ mm and $h = 396$ mm, respectively.

The developed CAD model of the proposed geared linkage with the assembly of the actuator used for a knee prosthesis is shown in Fig. 3.7a, b [36, 37]. A double acting cylinder was chosen to drive the prosthesis in order to fulfill the requirement of stability in stepping.

Table 3.1 Input parameters for the optimal synthesis

Nr	Parameter	Value
1	Gear ratio	$\rho = 0.62$
2	Minimum transmission angle	$\mu_{\min} = 10°$
3	Maximum unitary stroke	$h/l_1 = 0.08$

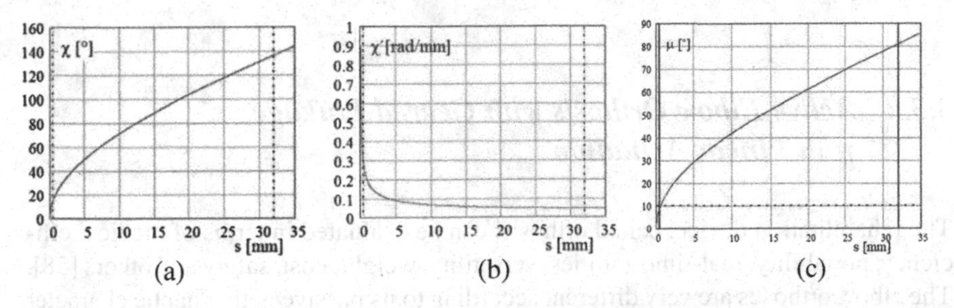

Fig. 3.6 Transmission function (**a**) instantaneous transmission ratio (**b**) and transmission angle (**c**) of the geared linkage with inverted slider-crank used for the knee joint

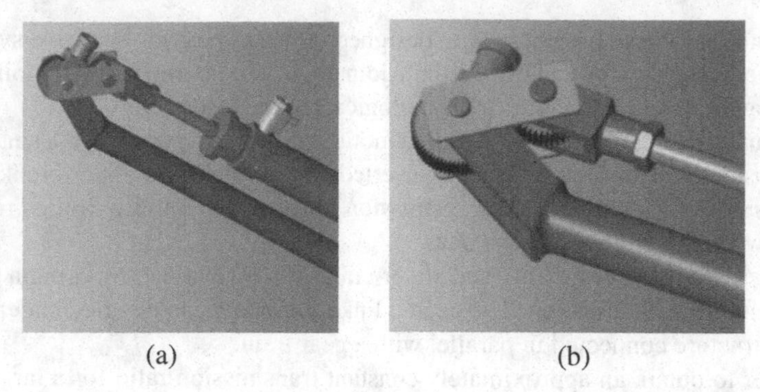
(a) (b)

Fig. 3.7 CAD model of the knee joint prosthesis

Fig. 3.8 Active knee prosthesis demonstrative model

The manufactured active prosthesis as a demonstrative model (Fig. 3.8) was developed with myoelectric sensors for controlling the valves and monitoring the velocity of the parts on different loads. In order to mimic the human gait in the best way it is necessary to control the braking action based on adjustment of throttles.

3.3.2 Active Elbow Orthosis with Geared Linkage with Linear Actuation

The rehabilitation device, called orthosis, can be evaluated in terms of medical efficiency, portability, real-time abilities, versatility, weight, cost, safety and others [38]. The elbow orthoses are very different according to its passive/active/haptic character and main task.

Table 3.2 Geometrical parameters of the geared linkage with linear actuation

Nr	Parameter	Dimension	Nr	Parameter	Parameters
1	Frame length	$l_1 = 212$ mm	4	Gear ratio	$\rho = 1.3$
2	Crank length	$l_4 = 34.5$ mm	5	Stroke	$s = 56.5$ mm
3	Initial stroke	$s_0 = 177.5$ mm	6	Minimum transmission angle	$\mu_{min} = 50°$

The elbow joint is composed of three joints (humeroulnar, humeroradial and proximal radioulnar) connected inside the same single articular capsule that defines a single articular cavity. From kinematical point of view the elbow joint ensures 1 DoF, the flexion–extension of the forearm with a maximum swing angle of 150°–160°.

By developing an active elbow orthosis for rehabilitation of the upper limb mobility with a large swing angle, the geared linkage with linear actuation is suitable. This mechanism reproduces an approximately linear transmission function for a large rotation angle [34], shows a very compact design and a singularity free extended position of the forearm. The geometical parameters of the geared linkage used for actuating the orthosis are shown in the Table 3.2 [39].

In order to have a better approximated linear transmission function, which is useful for the control of the motion, is to be satisfied the condition in the flat domain:

$$\chi'(s_1) = \chi'(s_2) \tag{3.25}$$

which sets the initial stroke to $s_0 = 195$ mm and the stroke to s = 38.5 mm ($s_1 =$ 17.5 mm and $s_2 = 56$ mm). With this condition, the angular displacement of the forearm is limited to about $\chi_{max} = 120°$ (Fig. 3.9).

The computed geared linkage with linear actuation was CAD modeled in Pro Engineer (Creo 5) together with the design of a commercial passive orthosis. Figure 3.10 provides a general view of the orthosis.

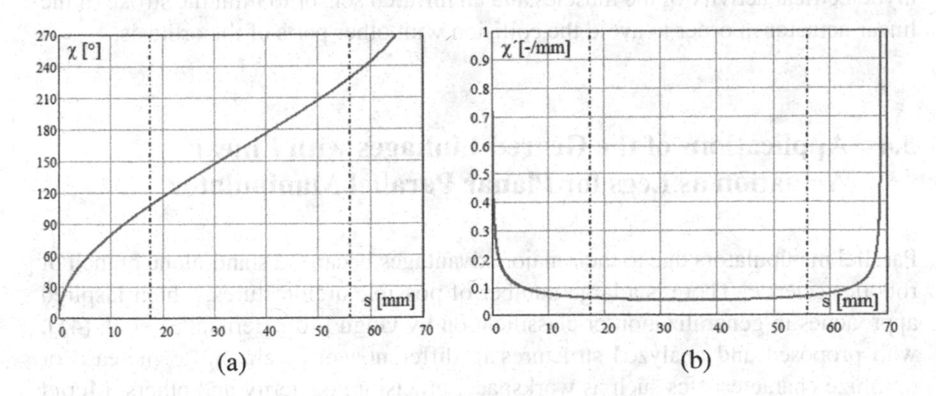

Fig. 3.9 **a** Transmission functions 0-ord and **b** 1-ord of the geared linkage

Fig. 3.10 General view of the elbow orthosis—CAD model

Fig. 3.11 Experimental prototype of the elbow orthosis

The practical achievement of the orthosis is illustrated in Fig. 3.11. The control of the linear actuator is achieved by using a microcontroller board, a H-bridge for reversing the motion of the actuator, a myosensor type Myo Ware for detecting the myoelectrical activity of the muscles and an infrared sensor to limit the stroke of the linear actuator in order to avoid the collision with other parts of the orthosis.

3.4 Applications of the Geared Linkages with Linear Actuation as Legs for Planar Parallel Manipulator

Parallel manipulators due to their major advantages became a stand-alone branch of robotics science. There is a large number of possible architectures, which inspired approaches in generalization or classification by Gogu [40], Hernandez et al. [41], who proposed and analyzed structures in different works, aiming to increase or optimize characteristics such as workspace, precision, dexterity and others. Merlet in [42] presents the problems of parallel robots regarding the structural synthesis, the direct and inverse kinematic analysis, singularities, workspace, static and dynamic analysis, calibration and design. Figures 3.12 and 3.13 show the development of the specific class of planar parallel manipulators using geared linkages with linear actuation as kinematic chain [43–45].

The simplified computing model (Fig. 3.14) considers the Chasles vector equation. The matrix equation for solving the forward and inverse kinematics of the planar parallel manipulator 3-R(R̲PRGR)RR are:

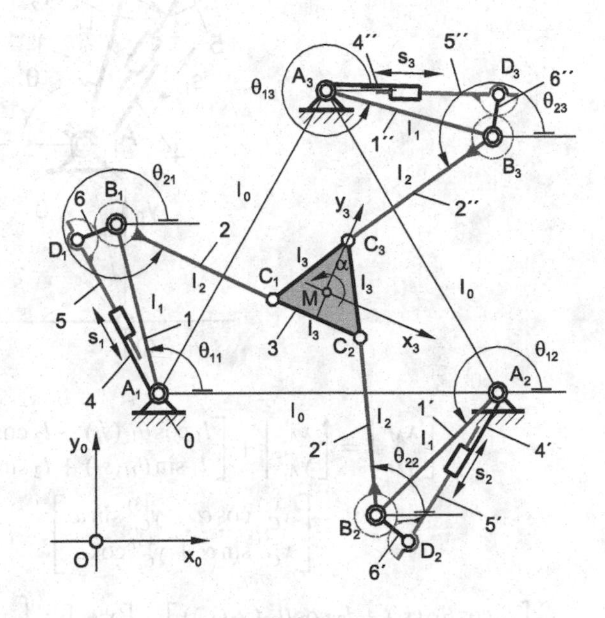

Fig. 3.12 Kinematic schema of the planar parallel manipulator 3-R(R̲PRGR)RR [43] using geared inverted slider crank with linear actuation

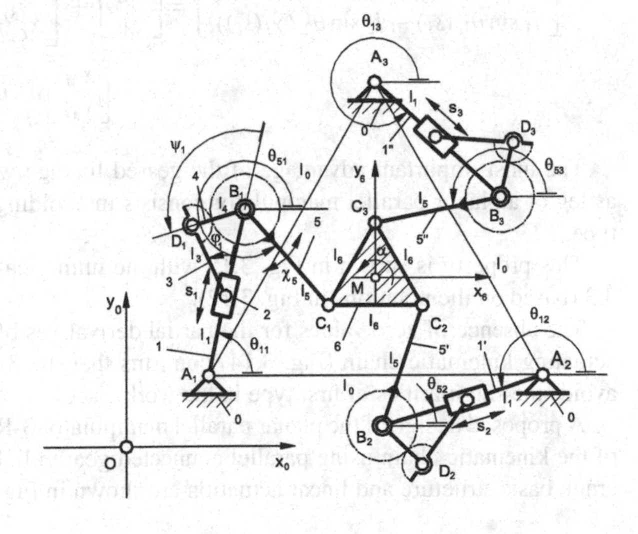

Fig. 3.13 Kinematic schema of the planar parallel manipulator 3-R(P̲PRGR)RR [46] using geared slider crank with linear actuation

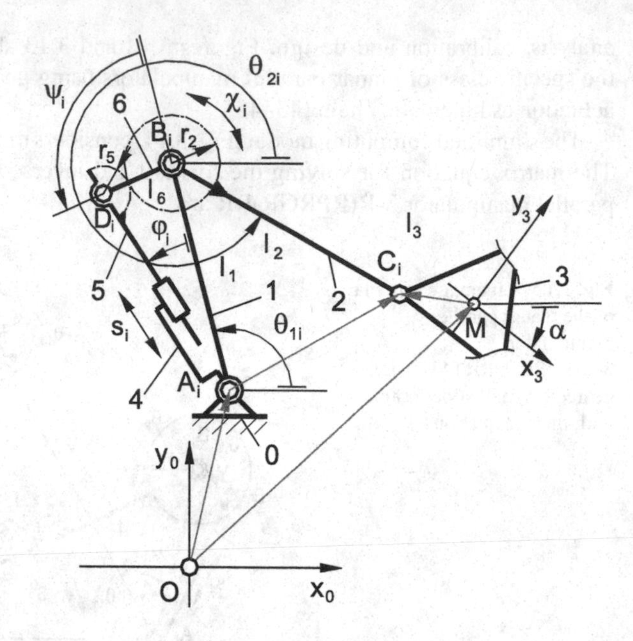

Fig. 3.14 Simplified computing model for the kinematic analysis 3-R(R̲PRGR)RR [43]

$$\begin{bmatrix} x_M \\ y_M \end{bmatrix} = \begin{bmatrix} x_{A_i} \\ y_{A_i} \end{bmatrix} + \begin{bmatrix} l_1 \cos\theta_{1i}(s_i) + l_2 \cos\theta_{2i}(\chi_i(s_i)) \\ l_1 \sin\theta_{1i}(s_i) + l_2 \sin\theta_{2i}(\chi_i(s_i)) \end{bmatrix} - \begin{bmatrix} x_{C_i}^{(3)} \cos\alpha - y_{C_i}^{(3)} \sin\alpha \\ x_{C_i}^{(3)} \sin\alpha + y_{C_i}^{(3)} \cos\alpha \end{bmatrix}, \quad i = \overline{1,3} \tag{3.26}$$

$$\begin{bmatrix} l_1 \cos\theta_{1i}(s_i) + l_2 \cos\theta_{2i}(\chi_i(s_i)) \\ l_1 \sin\theta_{1i}(s_i) + l_2 \sin\theta_{2i}(\chi_i(s_i)) \end{bmatrix} = \begin{bmatrix} x_{A_i} \\ y_{A_i} \end{bmatrix} + \begin{bmatrix} x_{C_i}^{(3)} \cos\alpha - y_{C_i}^{(3)} \sin\alpha \\ x_{C_i}^{(3)} \sin\alpha + y_{C_i}^{(3)} \cos\alpha \end{bmatrix} - \begin{bmatrix} x_M \\ y_M \end{bmatrix}, \quad i = \overline{1,3} \tag{3.27}$$

The most important advantage of the geared linkage with linear actuation used as leg of a planar parallel manipulator consists in avoiding the singularities of first type.

This property is shown in Fig. 3.15 with the numerical example given in Table 3.3 (based on the notations in Fig. 3.12).

The absence of zero values for the partial derivatives of each parallel connected actuating kinematic chain (Fig. 3.14) confirms that the 3-R(R̲PRGR)RR structure avoids the singularities of first type in the workspace.

A proposed design of the planar parallel manipulator 3-R(R̲PRGR)RR and a view of the kinematic chain using parallel connected geared linkage with inverted slider crank basic structure and linear actuation are shown in Fig. 3.16a, b.

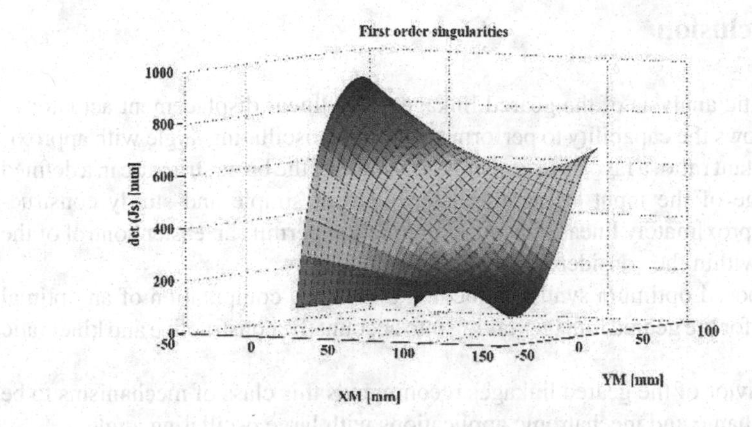

Fig. 3.15 Analysis of the singularities of first type for the planar 3-R(RPRGR)RR manipulator

Table 3.3 Geometrical parameters of the planar 3-R(RPRGR)RR manipulator

Frame platform length (0)	$l_0 = 120$ mm	Carrier length (6)	$l_6 = 10$ mm
Mobile platform length (3)	$l_3 = 25$ mm	Gear ratio	$\rho = 1.5$
Chain link length (1,1′,1″)	$l_1 = 50$ mm	Initial stroke	$s_0 = 42.5$ mm
Chain link length (2,2′,2″)	$l_2 = 50$ mm	Stroke	$h = 15$ mm

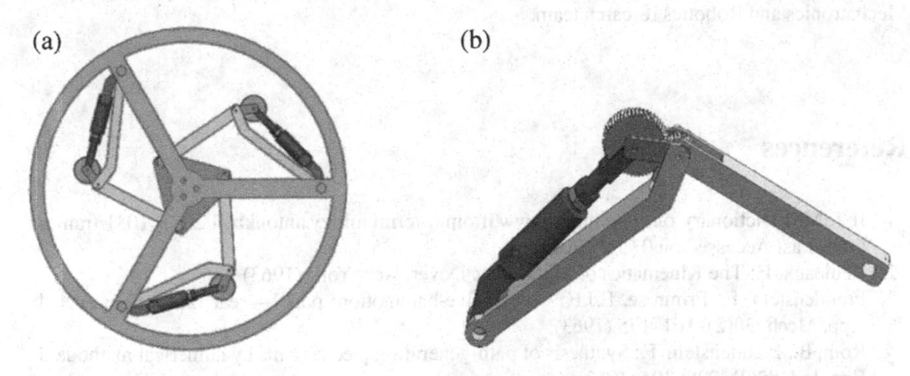

Fig. 3.16 a Design of the planar parallel manipulator 3-R(RPRGR)RR and b one kinematic chain with geared linkage and linear actuation

3.5 Conclusion

The kinematic analysis of the geared linkages with linear displacement actuator of first type shows the capability to perform a very large oscillating angle with approximately constant ratio, a favorable transmission angle of the basic structure in a defined motion range of the input element, small weight at simple and sturdy construction. The approximately linear transmission function permits an easier control of the movement within the considered range.

The proposed optimum synthesis method allows the computation of an optimal links length for the geared linkages taking into account the constructive and kinematic constrains.

The behavior of the geared linkages recommends this class of mechanisms to be used in mechanic and mechatronic applications with large oscillating angle.

The proposed drive mechanism with geared linkage is an original solution for the knee prosthesis and orthosis, which were designed and manufactured in the laboratories of the Politehnica University of Timisoara. The prosthesis implements a geared linkage in order to mimic the movement of the human knee during the flexure/extension and to improve the similar design solutions using inverted slider crank or slider crank as driving mechanism for the knee joint movement. The optimal dimensioning of the links was performed on the criterion of getting natural values of movement extension.

The analysis of the singularities of the manipulator shows that the geared linkages with linear actuation allow the avoiding of singularities of the first type by choosing of optimal initial stroke and working stroke.

Acknowledgements This work was developed at the Politehinca University of Timişoara at Department of Mechatronics through several PhD thesis, master and bachelor theses, respectively by the Mechatronics and Robotics research team.

References

1. IFToMM Dictionary on-line. http://www.iftomm-terminology.antonkb.nl/2057_1031/frames.html. Last Accessed on 07 July 2021
2. Reuleaux, F.: The Kinematics of Machinery. Dover, New York (1963)
3. Freudenstein, F., Primrose, E.J.F.: Geared five-bar motion: part I—gear ratio minus one. J. App. Mech. **30**(2), 161–175 (1963)
4. Roth, B., Freudenstein, F.: Synthesis of path-generating mechanisms by numerical methods. J. Eng. Ind. **85**(3), 298–304 (1963)
5. Oleksa, S.A., Tesar, D.: Multiply separated position design of the geared five-bar function generator. J. Eng. Ind **93**(1), 74–84 (1971)
6. Mundo, D., Gatti, G.: A graphical-analytical technique for the synthesis of noncircular gears in path-generating geared five-bar mechanisms. T. Can. Soc. Mech. Eng. **32**(3–4), 487–495 (2008)
7. Mundo, D., Gatti, G., Dooner, D.B.: Optimized five-bar linkages and non-circular gears for exact path generation. Mech. Mach. Theory **44**(4), 751–760 (2009)

8. Parlaktas, V., Soylemez, E., Tanik, E.: On the synthesis of a geared four-bar mechanism. Mech. Mach. Theory **45**(8), 1142–1152 (2010)
9. Neumann, R.: Hochübersetzende Getriebe. Maschinenbautechnik **26**(7), 297–305 (1977)
10. Neumann, R.: Fünfgliedrige Räderkoppelgetriebe für große Schwingwinkel. In: Proceedings of the Conference Dynamic und Getriebetechnik. TU Dresden (1985)
11. Neumann, R.: Fünfgliedrige Räderkoppel-Schrittgetriebe. TU Dresden (1986)
12. Neumann, R.: Fünfgliedrigen Räderkoppel-Schrittgetriebe, Aufbau, Synthese. Eigenschaften. Maschinenbautechnik **36**(10), 456–459 (1987)
13. Hain, K.: Zweiräder-Punktrasstgetriebe. Ind. Anz. **11**, 1–23 (1987)
14. Horani, M.: Untersuchungen zur Analyse und Synthese zykloidengesteuerter Zweischläge. Dissertation. TU Dresden (1977)
15. Rankers, H.: Rückkehrender Koppelrädermechanismus mit Doppelschleife. Konstruktion **39**(9), 335–357 (1987)
16. Volmer, J.: Getriebetechnik. Koppelgetriebe. VEB Verlag Technik, Berlin (1979)
17. Luck, K., Modler, K.-H.: Getriebetechnik: Analyse, Synthese. Optimierung. Springer, Berlin (1990)
18. Modler, K.-H., Lovasz, E.-C., Bähr, G., Neumann, R., Perju, D., Perner, M., Margineanu, D.: General method for the synthesis of geared linkages with non-circular gears. Mech. Mach. Theory **44**, 726–738 (2009)
19. Litvin, F.: Theory of Gearing. NASA Publication, Washington DC (1989)
20. Litvin, F.: Gear Geometry and Applied Theory. Cambridge University Press (2004)
21. Dooner, D.B.: Use of noncircular gears to reduce torque and speed fluctuations in rotating shafts. J. Mech. Design **119**, 299–306 (1997)
22. Kochev, I.S.: General method for active balancing of combined shaking moment and torque fluctuations in planar linkages. Mech. Mach. Theory **25**, 679–687 (1990)
23. Yao, Y.A., Yan, H.S.: A new method for torque balancing of planar linkages using non-circular gears. Proc. Inst. Mech. Eng., Part C **217**, 495–503 (2003)
24. Mundo, D.: Geometric design of a planetary gear train with non-circular gears. Mech. Mach. Theory **41**(4), 456–472 (2006)
25. Freudenstein, F., Chen, C.K.: Variable-ratio chain drives with noncircular sprockets and minimum Slack-theory and application. J. Mech. Design **113**, 253–262 (1991)
26. Emura, T., Arakawa, A.: A new steering mechanism using non circular gears. JSME Int. J., Series III **35**(4), 604–610 (1992)
27. Ottaviano, E., Mundo, D., Danieli, G.A., Ceccarelli, M.: Numerical and experimental analysis of non-circular gears and cam-follower systems as function generators. Mech. Mach. Theory **43**(8), 996–1008 (2007)
28. Ottaviano, E., Ceccarelli, M., Danieli, G.A., Mundo, D.: Analysis of non-circular gears and cam-follower systems as function generators. In: Proceedings of the International Symposium on Multibody Systems and Mechatronics, MUSME 2005, pp. 344–353. Uberlandia (2005)
29. Khatait, J.P., Mukherjee, S., Seth, B.: Compliant design for flapping mechanism: a minimum torque approach. Mech. Mach. Theory **41**, 3–16 (2006)
30. Gnasa, U., Modler, K.-H., Richter, E.-R.: Mechanismen zur Kraftübertragung bei hydraulisch angetriebenen Manipulatoren. In: 44 Internationales Kolloquium. TU Ilmenau (1999)
31. Gnasa, U.: Virtuelle Entwicklung von Gelenkarmmechanismen. Dissertation. TU Dresden (2001)
32. Lovasz, E.-C., Modler, K.-H., Hollman, C.: Auslegung der Räderkoppelgetriebe mit linearem Antrieb. In: 47. Internationales Kolloquium. TU Ilmenau (2002)
33. Modler, K.-H., Hollmann, C., Lovasz, E.-C., Perju, D.: Geared linkages with linear displacement actuator used as function generating mechanisms. In: Proceedings of the 11th IFToMM World Congress, pp. 1254–1259. Tianjin (2004)
34. Lovasz, E.-C., Modler, K.-H., Pop, C., Pop, F., Mărgineanu, D.T., Maniu, I.: Type synthesis and analysis of geared linkages with linear actuation. Mechanika **24**(1), 108–114 (2018)
35. Hutten, H.: Biomedizinische Technik. Springer, Graz (1991)

36. Popescu, I.A., Lovasz, E.-C., Ciupe, V.: Active prosthesis for lower limb amputated above knee. Robot. Manag. **15**(1), 59–60 (2010)
37. Lovasz, E.-C., Ciupe, V., Modler, K.-H., Gruescu, C.M., Hanke, U., Maniu, I., Mărgineanu, D.: Experimental design and control approach of an active knee prosthesis with geared linkage. In: Petuya, V., Pinto, C., Lovasz, E.-C. (eds.) New Advances in Mechanisms, Transmissions and Applications. MMS, vol. 17, pp. 149–156. Springer, Dordrecht (2014)
38. Mavroidis, C., Nikitczuk, J., Weinberg, B., Danaher, G., Jensen, K., Pelletier, P., Prugnaro-la, J., Stuart, R., Arango, R., Leahey, M., Pavone, R., Provo, A., Yasevac, D.: Smart portable rehabilitation devices. J. Neuroeng. Rehabil. **12**(2), 1–18 (2005)
39. Lovasz, E.-C., Sticlaru, C., Suciu, C., Gruescu, C.M., Ceccarelli, M., Maniu, I., Moldovan, C.E.: Novel actuation design of an active elbow orthosis. In: Uhl, T. (ed.) Advances in Mechanism and Machine Science: IFToMM WC 2019. MMS, vol. 73, pp. 1527–1534. Springer, Cham (2019)
40. Gogu, G.: Structural Synthesis of Parallel Robots. Part 3: Topologies with Planar Motion of the Moving Platform. Springer, Netherlands (2010)
41. Hernandez, A., Ibarreche, J.I., Petuya, V., Altuzarra, O.: Structural synthesis of 3-DoF spatial fully parallel manipulators. Int. J. Adv. Robot. Syst. **11**(101), 1–8 (2014)
42. Merlet, J.-P.: Parallel Robots. Springer (2006)
43. Lovasz, E.-C., Grigorescu, S., Margineanu, D., Gruescu, C.M., Pop, C., Ciupe, V., Maniu, I.: Geared linkages with linear actuation used as kinematic chains of a planar parallel manipulator. In: Corves, B., Lovasz, E.-C., Hüsing, M. (eds.) Mechanisms, Transmissions and Applications. MMS, vol. 31, pp. 21–31. Springer, Cham (2015)
44. Grigorescu, S.M., Lovasz, E.-C., Margineanu, D.T., Pop, C., Pop, F.: Dimensional synthesis of planar parallel manipulator using geared linkages with linear actuation as kinematic chains. In: Corves, B., Lovasz, E.-C., Hüsing, M., Maniu, I., Gruescu, C. (eds.) New Advances in Mechanisms, Mechanical Transmissions and Robotics. MMS, vol. 46, pp. 77–85. Springer, Cham (2016)
45. Lovasz, E.-C., Grigorescu, S., Margineanu, D., Gruescu, C.M., Pop, C., Maniu, I.: Kinemat-ics of the planar parallel manipulator using geared linkages with linear actuation as kinematic chains 3-R($R\underline{P}RGR$)RR. In: Proceedings of 14th IFToMM World Congress, pp 1–6. Taipei (2015)
46. Grigorescu, S.M., Lupuţi, A.-M.-F., Maniu, I., Lovasz, E.-C.: Novel planar parallel manipulator using geared slider-crank with linear actuation as connection kinematic chain. In: Pisla, D., Corves, B., Vaida, C. (eds.) New Trends in Mechanism and Machine Science: EuCoMeS 2020. MMS, vol. 89, pp. 496–505. Springer, Cham (2020)

Chapter 4
Robotics as Assistive Technology for Treatment of Children with Developmental Disorders—Example of Robot MARKO

Branislav Borovac, Mirko Raković, Milutin Nikolić, Vlado Delić, Srđan Savić, Marko Penčić, and Dragiša Mišković

Abstract In this paper we present the development of a robot designed as an assistive technology in therapy of children suffering from developmental disorders, tested on example of cerebral palsy. To achieve desired results the therapy should start as early as possible and it should be as intensive as possible. However, the therapeutic exercises are tiresome, boring and sometimes painful for children, and therefore the children are usually not motivated to undergo such therapy. This problem is particularly hard if patients are very young children. The earlier a treatment begins and the more intensive it is, the results are better. The anticipated benefit of such robot is to support the therapist efforts to conduct specific therapeutic exercises by establishing affective attachment of the child to the robot. Our aim was to develop robot to motivate the children to perform the therapist's instructions, and to perform more exercises than they usually do.

Keywords Cerebral palsy · Three-party therapy · Human-robot interaction · Affective attachment · Motivation of patient

4.1 Introduction

Developmental disorders primarily refer to conditions of abnormal gross and fine motor functioning and organization that occurred early in human childhood. Up to the age of 8, the brain development is intensive, and has abilities to reroute signals affected by an initial trauma. Therefore, the therapy should start as early as possible and it should be as intensive as possible. Nevertheless, the therapy practicing may also be tiresome, uncomfortable or even painful for children, and they often try to avoid it. In addition, it is not easy to explain the importance of therapeutic practice to

---

B. Borovac (✉) · M. Raković · M. Nikolić · V. Delić · S. Savić · M. Penčić · D. Mišković
Faculty of Technical Sciences, University of Novi Sad, Trg Dositeja Obradovića 6, 21000 Novi Sad, Serbia
e-mail: borovac@uns.ac.rs

M. Rackov et al. (eds.), *Machine and Industrial Design in Mechanical Engineering*, Mechanisms and Machine Science 109,
https://doi.org/10.1007/978-3-030-88465-9_4

children, and therefore it is essential that they are additionally motivated to undergo the therapy.

Recent research in the field of robot-assisted therapy for children suggests that a robotic system may increase motivation and trigger social interactions between the children and the therapist [1–4]. The approach reported in this paper supports these findings.

This paper reports our work [5] on prototypical robotic system intended to be used as an assistive tool (basically as a motivator) in therapy for children with cerebral palsy. A therapeutic scenario includes speech-based interaction between the child, the therapist and the robotic system. The anticipated benefit of such robot is to support the therapist efforts to conduct specific therapeutic exercises by establishing affective attachment of the child to the robot. We are convinced that if such an attachment is established, the child will be more motivated to accept and perform the robot instructions. We developed robot for the habilitation of gross motor functions and acquiring spatial relationships. The role of the robot was to motivate the children to perform the therapist's instructions, and to perform more exercises than they usually do.

The question that instantly arises is: what are the basic requirements that a robot must fulfill to achieve this. We believe that the requests are following:

- A robot must be of appealing appearance to children (i.e. children should like it),
- A robot must be able to attract and preserve the child's attention. To achieve this, the robot's ability to engage in verbal and non-verbal communication is extremely important. Thus, a robot should be of such a capacity to be engaged in a natural language dialogue, and must be able to perform, to certain extent, non-verbal actions,
- A robot must be able to demonstrate therapeutic exercises, which implies that its structure should include, at least, arms and legs.

Therapy (the child, the therapist and the robotic system are participating) may follow one of the following scenarios. In the first scenario, the therapist verbally instructs the child to perform exercise, for example, to perform gross motor movement. If the child does not understand what have to be done, the therapist asks the robot to perform the requested exercise. When robot perform exercise the therapists asks the child to repeat the movements of the robot. In addition, the robot encourages the child with appropriate verbal comments. In the second scenario the entire therapy is led by robot. The therapist is just supervisor, while the robot instructs the child to perform exercises, encourages the child, etc. It is important to underline that therapist is present in both scenarios, supervising therapy and having the control over the robot (e.g., the therapist can override or switch-off the robot at any given moment).

4.2 Robot Design

First question we should answer was overall structure of the robot (sketched in Fig. 4.1a). Robot should hear what humans (patient or therapist) are speaking. Then it should understand what has been said (comment or instruction i.e. command) and decide what response should be—answer or action. If it is expected that robot needs to say something, an reply should be composed and spoken (answering a question, encouraging the patient to try again to do exercise or to praise the patient for a well-done exercise). If demonstration of exercise is required appropriate exercise should be performed. It is clear that robot must have system for speech recognition and synthesis and sort of cognitive system. Also, active vision system should not be avoided. We decided that during therapy contact between robot and patient should not be allowed. Thus, robot should monitor where patient is and, for security reason, should stop any movement if patient is too close.

Based in such considerations our first guideline in the robot development was our understanding that robot must be able to demonstrate exercises by hands and legs. It is clear that our first decision was that robot should be of anthropomorphic structure i.e. to have arms and legs, to be able to demonstrate exercises. However, it was also very clear that we couldn't build a replica of human that normally stays "on his feet" and performs exercises. Our decision was to place the robot on a special support (cart), such that the arms and legs remain free for exercise demonstration. Two artists from our team designed alternative solutions. Finally, we decided that the robot is sitting on a horse-like cart (as sketched in Fig. 4.1b) by styrofoam model.

The movements expected to be performed by arms are more complex than those performed by legs. The robot's arm should have mobility close to the mobility of human arm, so we designed a robotic arm having 7 DOFs. The movements of legs required for exercise demonstration (and not for walking) are much simpler, including just hip abduction and adduction, knee bending, and foot swinging in ankle (4 DOFs). It is important to emphasize that this robot is not designed to walk, but is placed on a platform, enabling it to demonstrate most of the expected therapeutic exercises.

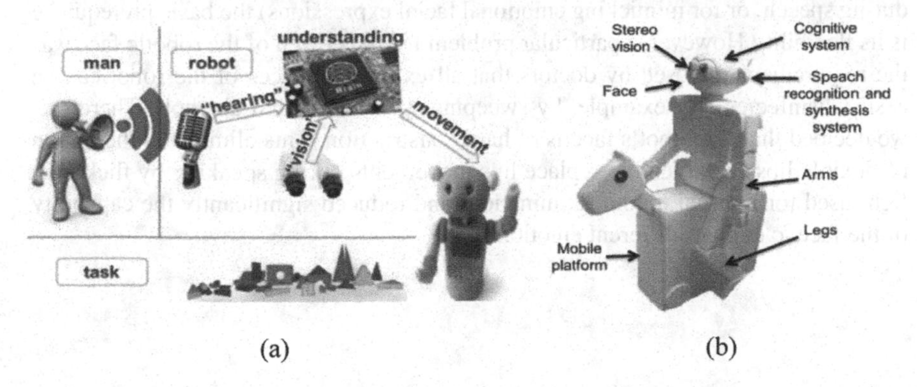

Fig. 4.1 Requirements for robot structure

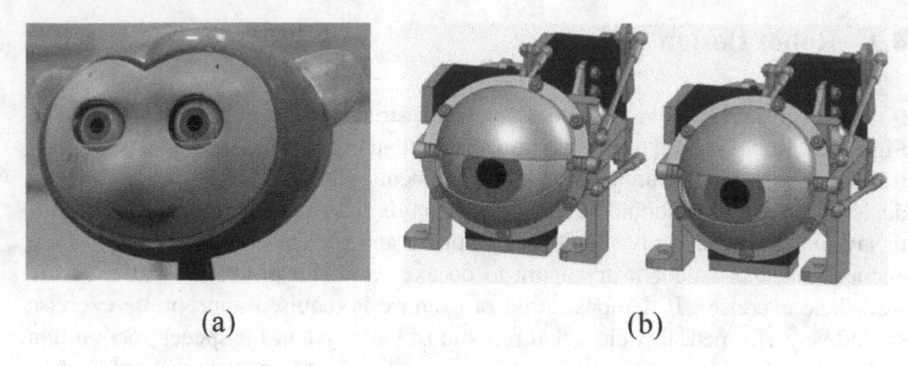

Fig. 4.2 **a** The actual robot's head mimicking the facial emotions; **b** eyes mechanical design

The next important question to be answered was about robot overall appearance. The general requirement we imposed was that children should like it. The psychologists from our team conducted a testing of healthy children in a kinder-garden with respect how they perceive a set of images of different robotic designs. The findings [5, 6] were that children prefer robots of a round shape (fatty than slim appearance) and of blue color. We included these finding in the final design of the robot.

Another important decision was related to the robotic head. This decision was important from two aspects: it contributes to the robot's general appearance, and it is able to express emotions. At one hand, we considered the option that the head (and particularly the face) should be as close as possible to a face of a human, and wondering about should it have male or female appearance. On the other hand, the alternative option was to adopt a non-human appearance of the head, and we were considering whether it should resemble an alien-like, animal-like, or something else. Based on the testing conducted on children in a kinder-garden, the psychologists' professional opinion and the experience of medical doctors, we decided to adopt a monkey-like face, appropriately designed and with no visible gender characteristics. The robotic head is shown in Fig. 4.2a).

To achieve a high degree of facial expressiveness (for example, proper lips motion during speech, or for mimicking emotional facial expressions) the basic prerequisite is its flexibility However, a particular problem for the design of the robotic face was the requirement imposed by doctors that all exterior surfaces of the robot can be easily disinfected (for example, by sweeping with cotton and alcohol). Therefore, we decided that the robot's face is of hard construction. This eliminated the option of flexible lips (we intend to replace lips movements during speaking by flickering light used for internal mouth illumination) and reduced significantly the capability of the face to express different emotions.

To compensate for this, we decided to make an additional effort to design vivid eyes [7], with motion capabilities as close as possible to those of humans. In addition, the robot is able to detect and track[1] the interlocutor's head, which makes the impression that it establishes eye-contact with the interlocutor (i.e. the robot is addressing the interlocutor or paying attention to him). To achieve this functionality, we integrated cameras in the robot's eyeballs. Besides the establishing eye-contact, the functionality of the robot's eyes significantly contributes to the robot's capability of mimicking emotional expressions. (please see Fig. 4.2a, c for an example of mimicking the emotion of surprise). As additional possibilities to reinforce emotional expressions, we intend also to engage the eyebrows, the internal lighting of the transparent ears (an illustration can also be seen in Fig. 4.2a) and the transparent oval at the robot stomach. The eyebrows were already implemented, but we are still not satisfied with their effect and try to improve them. The internal lighting of transparent oval at the robot stomach is still not implemented.

A particularly interesting question relates to the design of an appropriate cognitive system. We developed a initial dialogue system based on the focus tree model [8, 9] that enables the robot to autonomously engage in verbal interaction. An important feature of this system is that it does not request the user to follow a preset system's grammar when speaking. Instead, the user can spontaneously generate his dialogue.

Also, a special dialogue strategy was designed and implemented for the purpose of therapeutic interaction [10, 11]. However, although the robot may autonomously engage in interaction, in experimental robot-child interactions, the robot is controlled by a trained human operator, due to the sensitivity of the target group of children, and difficulties in the pronunciation of some patients.[2]

4.3 Experimental Use of Robot in Therapy

The use of the robot MARKO in therapy was experimentally assessed in a realistic therapeutic context. The experiments took place in the kinesitherapeutic room at the Clinic of Paediatric Rehabilitation in Novi Sad, Serbia (see Fig. 4.3).

Two different kinds of sessions were organized. First type of session was aimed to confirm children's acceptance of robot. Session consisted of three parts. In the first part, the robot tries to engage the child in verbal interaction. In the second part, the robot verbally presents a simple situational context, and tries to motivate

[1] The robot follows the interlocutor only with eye movements while the neck does not move and the head is immobile. To achieve more natural appearance of the robot we should divide needed movement between eyes and neck. However, this has not been realized yet.

[2] We have learned that dialogue is much more complex issue than we initially expected. First problem we faced was that children asked robot unexpected and unusual questions (like, Marko, what's new?). Also, they consider it normal for the current session to continue on the previous one and expect the robot to remember their previous conversation. This is a new requirement that should be considered when creating a cognitive system.

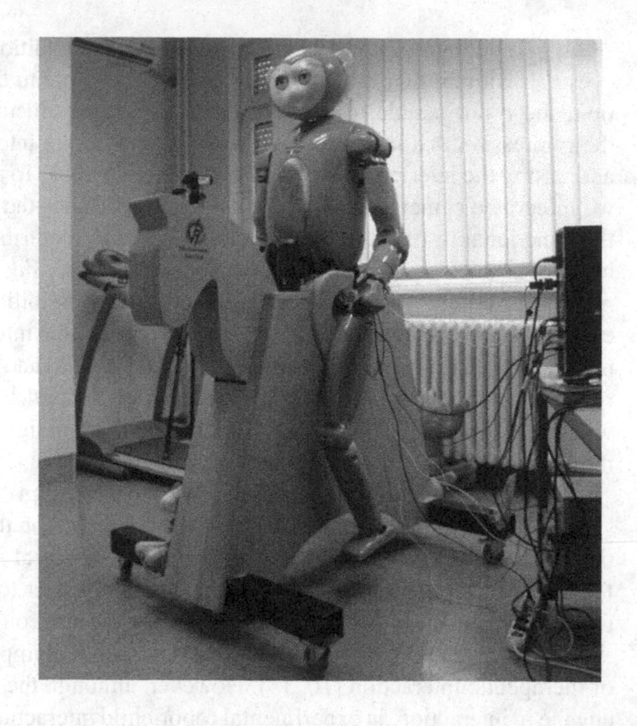

Fig. 4.3 Robot Marko in the kinesitherapeutic room at the Clinic of Pediatric Rehabilitation in Novi Sad, Serbia

the child to perform nonverbal actions. In the third part, the robot performs therapy-relevant nonverbal acts, and appropriately asks the child to reproduce them. In second experiment, the robot led (in the presence of a therapist) a complete session and therapist just supervise it. All experimental sessions were recorded by two cameras.

The preliminary results are promising: the positive motivation was observed in all subjects. Some details on these experiments are available in [12].

4.4 Conclusion

On the basis of the conducted experiments, we can conclude that, in general, the patients accepted the robot well. It should be underlined that some of them developed extremely strong emotional attachment, which was also validated by the therapists. We also have noticed that positive effects are particularly visible if cognitive functions of the patient are preserved to a larger extent. However, the therapists observed that even patients having disabilities of a large extent showed an increased interest to interact with the robot. However, this issue requires additional research and testing.

However, the kind and severity of disability, and the communication and intellectual capabilities may vary significantly among patients. In order to tailor the therapy specifically for each of them, a therapist should be able adjust the system according to the needs of a each particular patient by themselves i.e. without help of a programmer. An insight into this aspect of research is given in [8].

It is also particularly interesting that the patients often remembered what they were talking about with the robot during previous sessions, and expected that robot remembered this as well. Thus, the history of previous interaction has to be taken into account as an important part of current session.

Our opinion is that the therapist MUST be present and supervise all therapeutic sessions, at least to ensure the safety, to validate the children performance, and to perform on-line modification of therapy due to the actual patient condition. However, we believe that a robot can still play an important role in therapy, particularly related to the motivation of children.

Thus, we can conclude that there is an obvious potential of the robot used as assistive technology to contribute to the therapy of children with disabilities.

Acknowledgements The work presented was performed within project "Design of Robots as Assistive Technology for the Treatment of Children with Developmental Disorders" (III44008) and in part within project "Development of Dialogue Systems for Serbian and Other South Slavic Languages" (TR32035), funded by the Ministry of Education, Science and Technological Development of the Republic of Serbia.

References

1. Colton, M.B., Ricks, D.J., Goodrich, M.A., Dariush, B., Fujimura, K., Fujiki, M.: Toward therapist-in-the-loop assistive robotics for children with autism and specific language impairment. In: Proceedings of the 1st Symposium on New Frontiers in Human-Robot Interaction, AISB 2009, pp. 1–5. Edinburgh (2009)
2. Blázquez, M.P.: Clinical application of robotics in children with cerebral palsy. In: Pons, J., Torricelli, D., Pajaro, M. (eds.) Converging Clinical and Engineering Research on Neurorehabilitation: Biosystems and Biorobotics, vol. 1, pp. 1097–1102. Springer, Berlin, Heidelberg (2013)
3. Krebs, H.I., Ladenheim, B., Hippolyte, C., Monterroso, L., Mast, J.: Robot-assisted task-specific training in cerebral palsy. Dev. Med. Child. Neurol. **51**(Suppl. 4), 140–145 (2009)
4. Belokopytov, M., Fridin, M.: Motivation of children with cerebral palsy during motor involvement by RAC-CP Fun. In: Proceedings of the Workshops on Motivational Aspects of Robotics in Physical Therapy within IEEE/RSJ International Conference on Intelligent Robots and Systems, IROS 2012, pp. 40–45. Vilamoura, Algarve (2012)
5. Borovac, B., Gnjatović, M., Savić, S., Raković, M., Nikolić, M.: Human-like robot MARKO in the rehabilitation of children with cerebral palsy. In: Bleuler, H., Bouri, M., Mondada, F., Pisla, D., Rodic, A., Helmer, P. (eds.) New Trends in Medical and Service Robots. MMS, vol. 38, pp. 191–203. Springer, Cham (2016)
6. Oros, M., Nikolić, M., Borovac, B., Jerković, I.: Children's preference of appearance and parents' attitudes towards assistive robots. In: Proceedings of the IEEE-RAS International Conference on Humanoid Robots, Humanoids 2014, pp. 360–365. IEEE Press (2015)
7. Sivčev, S, Raković, M, Borovac, B, Nikolić, M.: Anthropomorphic robot eyes with realistic movements for non-verbal communication and emotion expressions. In: Proceedings of the 2nd Regional Conference Mechatronics in Practice and Education, MECHEDU 2013, pp. 84–88. Subotica (2013)

8. Gnjatović, M.: Therapist-centered design of a robot's dialogue behavior. Cogn. Comput. **6**, 775–788 (2014)
9. Gnjatović, M., Tasevski, J., Nikolić, M., Mišković, D., Borovac, B., Delić, V.: Adaptive multimodal interaction with industrial robot. In: Proceedings of the IEEE 10th Jubilee International Symposium on Intelligent Systems and Informatics, SISY 2012, pp. 329–333. IEEE Press (2012)
10. Gnjatović, M., Tasevski, J., Mišković, D., Savić, S., Borovac, B., Mikov, A., Krasnik, R.: Pilot corpus of child-robot interaction in therapeutic settings. In: Proceedings of the 8th IEEE International Conference on Cognitive Infocommunications, CogInfoCom 2017, pp. 253–257. IEEE Press (2018)
11. Tasevski, J., Gnjatović, M., Borovac, B.: Assessing the children's receptivity to the robot MARKO. Acta Polytech. Hung. **15**(5), 47–66 (2018)
12. Delić, V., Borovac, B., Gnjatović, M., Tasevski, J., Mišković, D., Pekar, D., Sečujski, M.: Toward more expressive speech communication in human-robot interaction. In: Ronzhin, A., Rigoll, G., Meshcheryakov, R. (eds.) Interactive Collaborative Robotics: ICR 2018. LNCS, vol. 11097, pp. 44–51. Springer, Cham (2018)

Chapter 5
Development of Terminology for Gearing: IFToMM Notions, Reference-Dictionary Book, Tooth Failure Modes

Victor E. Starzhinsky

Abstract The problems of developing terminology in gearing and transmissions are considered in the paper. A wide range of informative sources (international and national standards, reference books, dictionaries on mechanics, mechanical engineering, theory of mechanisms and machines, international translators, collectors of recommended terms et al.) is studied to state and analyzed. The following approaches in this field are presented: (a) becoming and developing terminology in the frame of IFToMM Permanent Commission "Standardization of Terminology on TMM" (PC A) activity; (b) compilation of reference dictionary book on gearing; (c) identification, classification and description of gear failure modes in the scope of developing interstate (for CiS-countries) standard on forms of gear failures. The paper describes the mechanism of preparation, correction, systematic update and editing of terminology texts by language editors. Process of evolutionary development of terminology by IFToMM PC A is analyzed—from the constitution of the IFToMM PC A in 1969 till the present time with accent to development of gearing terminology. Information about development and using of IFToMM electronic dictionary is given.

Keywords Gears · IFToMM terminology · Identification · Classification · Electronic dictionary · Reference-dictionary book · Gear failure modes

5.1 Introduction

In the paper the experience of author's participation in development of terminology for gearing is presented. The following directions are considered. Development of terminology for gearing in the frame of IFToMM Permanent Commission "Standardization of Terminology for MMS" (PC A) activity. Gathering information on gears from different sources and compiling it's in the vocabulary interpreting terms and notions. Identification of gear failure modes and elaboration of interstate standard with its description and classification. These directions are elaborated by corporate

V. E. Starzhinsky (✉)
V.A. Belyi Metal-Polymer Research Institute of National Academy of Sciences of Belarus, Kirov Street, 32-A, 246050 Gomel, Belarus

M. Rackov et al. (eds.), *Machine and Industrial Design in Mechanical Engineering*, Mechanisms and Machine Science 109,
https://doi.org/10.1007/978-3-030-88465-9_5

authors, whose names somebody can find in the references. Aforementioned directions are founded at analysis, summation and systematizing of terms, notions and definitions from the numerous engineering, reference and normative sources, listed below:

- International standards ISO, IEC.
- National standards: German—DIN, USA—AGMA, French—NF, British—BS, USSR—GOST, Bulgarian—BDS, Swiss—VSM, Ukrainian—DSTM.
- Illustrated dictionaries on Mechanical Engineering [1–3] with divisions relative to gearing: overall terms (148 ones), gear units (11 ones), gear manufacturing (44 ones).
- Gear terms from the MAAG Gear Book [4].
- Terminology from the reference-dictionary book on the metal working tool [5] with divisions: (1) disc type gear cutters (4 terms); (2) gear hobs (30 ones); (3) pinion type cutters for spur gears (31 ones); (4) showing tools (12 ones); (5) cutters for spiral bevel gears (14 ones); (6) special terms used Gleason Co. (99 ones).
- Fundamental glossary on the mechanics of machines [6] (above 130 terms, relative to elements of gears mechanisms and gear manufacturing technology).
- Terminological dictionary [7] (22 terms).
- Collection of recommended terms [8] (basic definitions—12 ones, gears—74 ones, gear meshing and gear trains—60 ones, figures to terms—116, lettering—34).
- Special terms of the "Gleason" Company [9] (487 ones).

In review [10] somebody can find detailed references on gearing terminology. Ibidem [10] the various rating of term designations in Russian—language according the GOST and English—language ones along ISO and AGMA terminology is discussed.

The results of activity in terminology analysis were reported continually at the sessions of committees and commissions ISO (Tune, 1988) and IFToMM (Izhevsk, Russia, 1998; Paris, France, 1999; Oulu, Finland, 1999). Generalized results of this activity is published in the proceedings of scientific seminar [11]. Further activity under the whole of above directions is fulfilled in parallel. Regular discussions of terminological problems online and at the Scheduled meetings of IFToMM Permanent Commission A, professional editing data by highly qualified editors, language speakers (English—Prof. Charles W. Stammers; French—Prof. J. P. Lallemand, Prof. Didier Remond; German—Prof. Gerhard Boegeljack; Russian—Prof. Victor E. Starzhinsky, Prof. Vladimir D. Plakhtin)—have been assured high quality presentation of enveloped terminological information.

5.2 Development of Gearing Terminology in the Frame of IFToMM Permanent Commission "Standardization of Terminology for the Mechanism and Machine Science"

The settled gearing terminology in the primary IFToMM Issues [7, 12] contains 22 terms and definitions, distributed on different chapters. In accordance with IFToMM decision to enlarge the Theory of Mechanisms and Machines (TMM) notion to it interpretation as Mechanism and Machine Science (MMS) one [13] the new divisions (7–13), including the Chap. 12 "Gearing" as a separate division, have been turned in IFToMM terminology, originally in English-language alphabetical [14]. At the 25th IFToMM PC A Working Meeting (2010, Minsk/Gomel, Belarus) the Chap. 12 with new thematic structure of 6 subdivisions (Gear and Tooth Geometry—94 terms, Gear Pair Basics—40 ones, Gear Pair with Parallel Axes-19 ones, Gear Pair with Intersting Axes—35 ones, Crossed-Axes Gear Pair—28 ones, Geared Mechanisms—10 ones) [15] has been confirmed. Hereinafter it was included in IFToMM electronic dictionary [16]).

5.3 Reference-Dictionary Book on Gearing

The start for the first issue of Reference-dictionary book on gearing was given by the paper [11]. In the subsequent issues, the volume of terminological information gradually extends: 68 pages (2002), 90 (2004), 112 (2005), 190 (2008), and 220 ones (2011). In the last issue, the reference-dictionary book contains about 900 terms. The following divisions in the form of tables, pictures and drawings are presented (terms quantity are in brackets): Geometrical and Kinematic Parameters and Elements of Gears and Gear Pairs. Terms and Symbols (115). Principal subscripts and signs (27); Extraction from Electronic dictionary "IFToMM Terminology for Mechanism and Machine Science" Chapt. 12. Gearing (226); Spiroid gears (79); Facial toothed joints and gearings (37); Accuracy and inspection of gearing (172); Calculation of bending strength, pitting resistance, scuffing and service life of gears (130); Calculation of scuffing load capacity of gears (118); Forms and location of tooth contact pattern (118); Gear failure modes (83); Classification of the gear drives, including gear pairs with parallel, crossing and skew axes (about 70 terms with definitions in shape of initial bodies and its relative arrangements, tooth and profile shape, etc.) [17].

5.4 Notion Identification on Gear Failure Modes. Development of Interstate Standard

Identification of the terms in gear failure modes was provided with exhaustive normative sources [18–22]. Above 200 gear failure modes have been analyzed [23]. A basic contain of the standard was published in [24, 25]. Official issue of the standard [26] was in 2009. The standard includes a list of terms for durability, interchangeability, metal corrosion, friction, wear and lubrication, calculation methods of gear drives on strength and scuffing, classification of metal fracture modes. There is a corresponding table of gear failure mode classification with different levels (classes), general failure modes, sub-modes, degree of failure. All together, the description of 76 gear failure modes are given, and 104 standard photo-examples are shown, as well as possible reasons of their appearance and recommendations for failure prevention. For details refer [27].

5.5 Conclusions

Results of author corporate activities in the field of gearing terminology have been published in numerous papers, in particular [28–30]. Detailed Russian—language version of given information somebody can find in the volumetric three—part paper [31–33].

Acknowledgements Author extend gratefulness to colleagues and coauthors of joint publications, proposed essential input in Mechanism and Machine Terminology—the departed Dr. Yuri L. Soliterman (mechanics, tooth failure modes), Prof. Veniamin I. Goldfarb (mechanics, spiroid gearing), and Prof. Efim I. Tesker (gear design and computation, tooth failure classification), as well as Prof. Vladimir B. Algin (mechanics of mobile machine transmissions, multibody dynamics), Prof. Mark M. Kane (mechanism and machine quality management), Dr. Eugeni V. Shalobaev (gear accuracy, mechatronics), Dr. Serge V. Shil'ko (mechanics, biomechanics, mechatronics). Also author would like to tell "Many thanks!" to my partners in IFToMM PC A, together with whom me be in successful joint operation, especially to PC a Post Chairmen Dr. Ing. Theodor Ionescu (Commission Presidency during 1998–2006) and Prof. Antonius J. Klein Breteler (Similar 2006–2013) for creative help and mutual understanding during the Working Meeting sessions and on-line communication.

References

1. Shvarts, V.V.: The Concile Illustrate Russian—English Dictionary on Mechanical Engineering (3759 terms). Russkii yazyk, Moscow (1983)
2. Shwarts, V.V.: Illustrated Dictionary on Mechanical Engineering. English, French, Spanish, Russian (3614 terms). Russkii yazyk, Moscow (1986)
3. Shwarts, V.V.: Dictionario Illustrato Russo-Italiano Delle Construsioni Meccanoege (3614 termini). Russkii yazyk, Mosca; Libreria Italia-URSS, Roma (1987)

4. MAAG Gear Book: Calculation and Manufacture of Gears and Gear Drives for Designers and Works Engineers. MAAG Gear-Wheel Company, LTD. Zurich (1963)
5. Kershenbaum, V.J.: Dictionary—Reference Book on Metal—Working Tools, Russian, English, German, French. Science and Technique, Moscow (1993)
6. Krainev, A.F.: Mechanics of Machines: Fundamental Glossary. Mashinostroenie, Moscow (2000)
7. IFToMM Commission standards for terminology: Terminology for the theory of machines and mechanisms. Mech. Mach. Theory **20**(5), 435–539 (1991)
8. Gears, Gear Meshing and Gear Trains with Constant Gear Ratio/Collection of Recommended Terms. Is. 57. Academy of Science of USSR, Moscow (1962)
9. Gleason Terminology. Copyright Gleason Work (1967–1971)
10. Starzhinsky, V.E., Soliterman, Y.L., Goman, A.M. et al.: Compiling a terminological reference dictionary on gearing. In: Proceedings of International Conference Power Transmissions, BAPT 2003, pp. 180–186. Sofia (2003)
11. Berestnev, O.V., Starzhinsky, V.E., Goman, A.M., Shalobaev, E.V.: Compilation of glossary of international terms in gear design, manufacture and serviceability: Concepts and contents. In: Proceeding of the Scientific Seminar Terminology of the Theory of Machines and Mechanisms, pp. 21–27. Technologija, Kaunas (2000)
12. Boegelsack, G., Gierse, F.J., Oravsky, V., Prentis, J.M., Rossi, A.: Terminology for the theory of machines and mechanisms (744 definitions). Mech. Mach. Theory **18**(6), 379–408 (1983)
13. Ceccarelli, M.: From TMM to MMS: A Vision of IFToMM. Bulletin IFToMM Newsletter (2001)
14. Ionescu, T.: Special issue: standardization of terminology (771 definitions in French, German and Russian, 1594 definitions in English). Mech. Mach. Theory **38**(7–10), 597–1111 (2003)
15. Starzhinsky, V., Algin, V.: Terminology for the mechanism and machine science. In: Proceedings of the 23rd Working Meeting of the IFToMM PC. Belarusian State Institute of Standardization and certification, Minsk (2010)
16. www.iftomm.net, www.iftomm.3me.tudelft.nl
17. Starzhinsky, V.E., Antonyuk, V.E., Goldfarb, V.I., Kane, M.M., Shil'ko, S.V., Goman, A.M., Raichman, G.N., Soliterman, Y.L., Shalobaev, E.V.: Reference Dictionary Book on Gearing (Russian-English-German-French). MPRI NASB, Gomel (2011)
18. ISO 10825:1995. Gears-Wear and Damage to Gear Teeth—Terminology—Engrenages-usure et défauts dentures—Terminologie. ISO (1995)
19. ANSI/AGMA 1010-E 95. Appearance of Gear Teeth—Terminology of Wear and Failure
20. DIN 3979:1979. Zahnschäden an Zahnradgetriebe. Bezeichnungen, Merkmale, Ursachen (1979)
21. Errichello, R.: Gear Failure Analysis. Gear tech, Montana (2000)
22. ZFN 201:1990. Zahnradschaeden Begriffs bestimmung. Bereichnungen und Ursachen
23. Starzhinsky, V.E., Soliterman, Yu.L., Goman, A.M., Ossipenko, S.A., Arnaudov, K.B.: Analysis of gear damage modes and preparation standard on their classification and description. Bulletin Vestnik NTU KhPI **22**, 70–77 (2006)
24. Starzhinsky, V.E., Soliterman, Yu.L., Goman, A.M., Ossipenko, S.A.: Forms of damage to gear wheels: typology and recommendation on prevention. J. Frict. Wear **29**(5), 340–353 (2008)
25. Starzhinsky, V.E., Soliterman, Yu.L., Tesker, E.I., Goman, A.M., Ossipenko, S.A.: Modes of gear failures: typology and recommendations on failure prevention. Trenie i Iznos/Russian J. Friction Wear **29**(5), 465–482 (2008)
26. GOST 31381-2009. Gear Wheels. Modes of Damages. Classification and Description (76 notions in Russian, English, French and German). Gosstandart. Minsk (2009)
27. Algin, V.B., Starzhinsky, V.E.: Participation of Belarussian scientists in IFToMM Activity. In: Algin, V.B., Starzhisky, V.E. (eds.) Gears and Gear Transmissions in Belarus: Design, Technology Estimation of Properties, pp. 384–392. Belaruskaya Navuka, Minsk (2017)
28. Shalobaev, E.V., Shil'ko, S.V., Tolocka, R.T., Starzhinsky, V.E., Iurkova, G.N., Surikov, D.G.: State of art in separate sections of MMS terminology and some proposals. In: Corves, B., Lovasz, E.C., Hüsing, M., Maniu, I., Gruescu, C. (eds.) New Advances in Mechanisms, Mechanical Transmissions and Robotics. MMS, vol. 46, pp. 217–225. Springer, Cham (2017)

29. Starzhinsky, V.E., Shalobaev, E.V., Kane, M.M., Goldfarb, V.I.: Activities of Russian—speaking scientists in development of MMS terminology. In: Corves, B., Lovasz, E.C., Hüsing, M., Maniu, I., Gruescu, C. (eds.) New Advances in Mechanisms, Mechanical Transmissions and Robotics. MMS, vol. 46, pp. 209–217. Springer, Cham (2017)
30. Starzhinsky, V.E., Goldfarb, V.I, Algin, V.B., Shalobaev, E.V., Kane, M.M.: Participation of Scientists from former Soviet Republics and CIS Countries in IFToMM activity. In: Processing of the Scientific Seminar Terminology for the Mechanism and Machine Science Victor, pp. 13–42. MPRI NASB, Gomel-Saint-Petersburg (2016)
31. Starzhinsky, V.E, Goldfarb, V.I., Shil'ko, S.V., Shalobaev, E.V., Tesker, E.I.: Development of terminology in the field of gears and gear transmissions. Intellect. Syst. Manuf. **15**(Part 1), pp 30–36 (2017) (in Russian)
32. Starzhinsky, V.E, Goldfarb, V.I., Shil'ko, S.V., Shalobaev, E.V., Tesker, E.I.: Development of terminology in the field of gears and gear transmissions. Intellect. Syst. Manuf. **15**(Part 2), 51–61 (2017) (in Russian)
33. Starzhinsky, V.E, Goldfarb, V.I., Shil'ko, S.V., Shalobaev, E.V., Tesker, E.I.: Development of terminology in the field of gears and gear transmissions. Intellect. Syst. Manuf. **15**(Part 3), 60–66 (2017) (in Russian)

Part II
Design, Manufacturing and Management of Machine Elements, Equipment and Systems

Chapter 6
The Self-Compensation Approach for Backlash on Gear Train

Bahadır Karba, Nihat Yıldırım, Mert Vardar, Fatih Karpat, and Milan Rackov

Abstract Even if rotates unidirectional or bidirectional, a certain amount of space (free angular movement usually called "backlash") between meshing teeth flanks are allowed for functional reasons. The actual value and effects of the mechanical clearance that is critical in the mesh condition of gears have been a subject to be investigated by gear engineers. The actual backlash stems from time-varying backlash and constant backlash. User friendly software has been developed in MATLAB, to guide assembly personnel/staff to create a combination of different gearbox components from the measurement data bank to obtain a specified backlash range for three gear train. Provided that, different gearboxes with different backlash values/ranges will be assembled by using manufactured and measured components. Unless out of design limits, almost no component will be wasted. This condition will also avoid any slowing of manufacturing and helping to increase production capacity. In the present paper, operating behaviour of system is evaluated in terms of backlash also the automatically self-compensation of sources of backlash methodology is applied to three gear train for backlash parameter.

Keywords Manufacturing errors · Installation errors · Backlash prediction · Backlash optimization · Low backlash gear train

6.1 Introduction

In literature, there is no resemble methodology which can directly predict and find best optimized configurations considering tooth-to-tooth contacts characteristics to

B. Karba (✉) · N. Yıldırım
Faculty of Engineering, Gaziantep University, Gaziantep 27410, Turkey

B. Karba · M. Vardar
TR Transmisyon Power Transmission Systems Inc., Ankara 06378, Turkey

F. Karpat
Faculty of Engineering, Uludağ University, Bursa 16059, Turkey

M. Rackov
Faculty of Technical Sciences, Novi Sad University, 21102 Novi Sad, Serbia

M. Rackov et al. (eds.), *Machine and Industrial Design in Mechanical Engineering*, Mechanisms and Machine Science 109,
https://doi.org/10.1007/978-3-030-88465-9_6

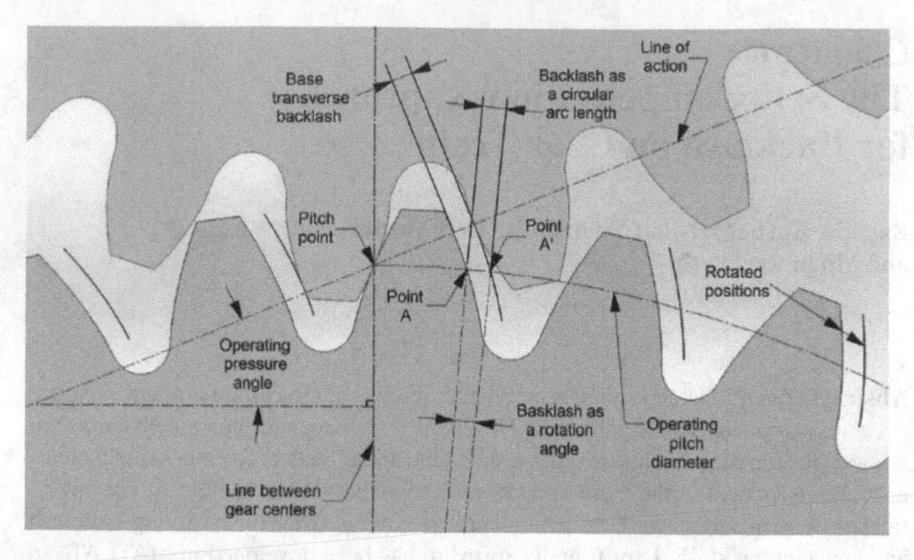

Fig. 6.1 Backlash in a gear mesh [2]

evaluate actual backlash on every mesh between gears. According to Michalec, backlash is not exist until a gear is mated with pinion in an assembly. But, part of the backlash which is then revealed for the gear pair is attributable to sources identifiable in the individual gears before they are mated. Hence, it is possible and proper to consider the inherent mechanical clearance of a single gear before assembly [1]. The definition of backlash, which is depicted as a schematic in Fig. 6.1, is very simple, but accurately predicting backlash is often difficult. However, optimum/correct value of the required backlash is not well defined and calculated before the actual gear pairs are put into operation. Therefore, usually greater backlash than required value is used to prevent any likely back surface rubbing. This additional backlash causes unwanted "lost motion" phenomenon when motion reverses.

To eliminate or reduce backlash (lost motion), there are 3 fundamental ways such as:

- Very precise gear and components manufacturing,
- Special designed or cluster gears, and
- Special design and assembly which use components except for gears.

6.1.1 Motivation of Study

This study has been conducted in master thesis research [3]. In first paper of series of papers, range of backlash prediction have been investigated both their own effects and their integration effects via combinations [4]. In the second paper, backlash of gear pair is predicted by the in-house software including manufacturing and installation errors and the results have been validated both numerically and experimentally. Two much-used commercial products were used for numerical validation of the predictions whereas the special purpose test rig was designed and manufactured for experimental measurements [5]. In this article, standalone program computes range of backlash which is applicable under sort of gear meshing configuration for spur and helical three gears train by using the novel methodology. In this proposed method, manufacturing tolerances are not as tight as the ones in previous method but there are certainly the upper and lower limits for manufacturing tolerances. The components manufactured are individually measured for different individual parameters like tooth thickness, runout, profile error, center distance etc. Referring to manufactured gears reports, all data is applicable to software as inputs, so range of backlash can be known and also it can be minimized considering the only absolute condition, which assembly stays within design limits, and abide by the gear fundamental law by using proper assembly configuration.

These assembly configurations contain such as:

- Possible center distance allowance to remove backlash within non-interference region,
- Prepared an algorithm what applying manual phasing methodology to find minimum and maximum backlash options, and
- Another algorithm is prepared to be capable of auto-phasing for each gear pair under self-compensating methodology.

The software has capabilities to carry out:

- Determining actual pertinent macro gear dimensions,
- Predicting assembly backlash under manufacturing and installation parameters,
- Estimating time-varying (one cycle rotation) operating backlash,
- Determining possible center distance variations under non-interference region,
- Minimizing backlash by using self-compensation under phasing methodology
- Determining manual phasing permission to operate numbered gears, and
- Determining output backlash according to direction of rotation.

The purpose of study, backlash of system can be minimized under novel method that is called self-compensating methodology to stable system backlash variation. Software gives feedback to user about how gears of gear train will be mounted according to 3 type of configurations such as 1st finding minimum backlash within all mesh combinations that contain smallest minimum value, 2nd finding minimum backlash within all mesh combinations that contain smallest maximum value, and 3rd finding minimum backlash within all mesh combinations that contain smallest peak-to-peak value for full cycle rotation.

6.2 Minimizing Methodology

In this novel method, backlash can be minimized by using ideal phasing compensating methodology for gears with marked numbers. If selected number of teeth of gears are relatively prime hunting tooth meshing is completed under each tooth contacts via each tooth. In this approach, probabilities of minimum backlash configurations have been performed creating adjacent pitch deviation document which is effectful parameter based on manufacturing reports approximately. Thus, how backlash is minimum could via certain configurations between gears have been evaluate using phasing methodology under defined contact teeth which are provides minimum backlash chance. Standalone program works like mounting simulator tool by using aforementioned methodology to predict actual operating backlash for three gear trains considering manufacturing reports of gears also mounting conditions.

6.3 Case Study of Three Gear Train

In this case, designed helical three gear trains are examined in terms of backlash via manufacturing reports by using software standalone program (see Fig. 6.2). Besides that, nominal tooth thicknesses of Three meshed gears (see Fig. 6.3) are calculated by span/over pin measurement methodology considering given macro properties of gears.

Elemental sources of manufacturing parameters which are taken from measurement reports were added to software to evaluate backlash precisely. In this case study, influence of adjacent pitch error is also examined under manufacturing reports according to direction of rotation of gears considering from first measured flank to last one of gears respectively. Angular backlash is explained as a formula to evaluate backlash in controls precision units.

Best configurations which provide to minimize backlash through full cycle rotations are given considering type of rotation (see Figs. 6.4 and 6.5). Backlash graphs of Gear (1–2) were given under mounting configuration considering clockwise rotation at following Figs. 6.6, 6.7 and 6.8 for pair (1–2) and Figs. 6.9, 6.10 and 6.11 for pair (2–3) respectively.

At the end of evaluation of best configurations for Gear pair 1–2 and 2–3. Entire Torsional Angle of gear train can be founded by Software under integrated best configurations methodologies. According to selected driving gear, so, first gear of chain or last gear of chain Z1–Z2–Z3, respectively considering at the following Figs. 6.12, 6.13 and 6.14 under clockwise rotation for first gear driving.

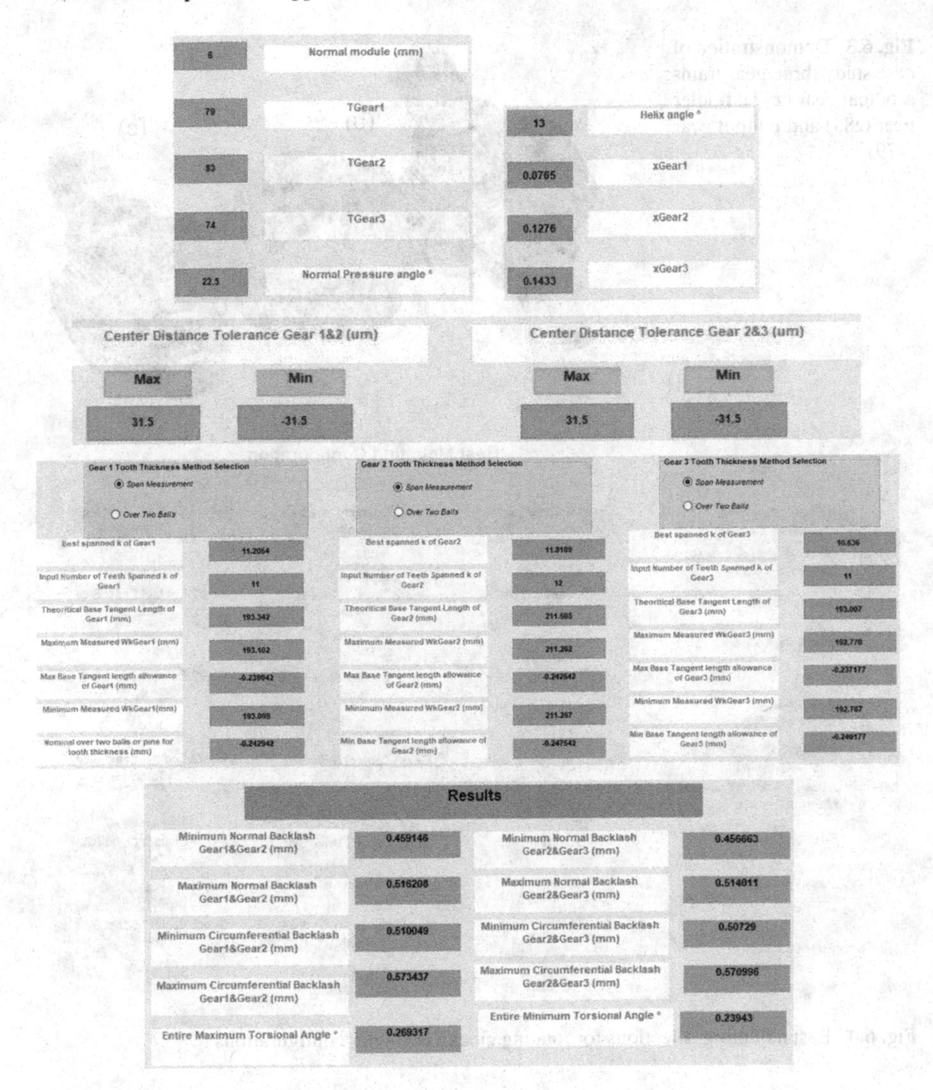

Fig. 6.2 Self-compensated backlash gear train program under case study

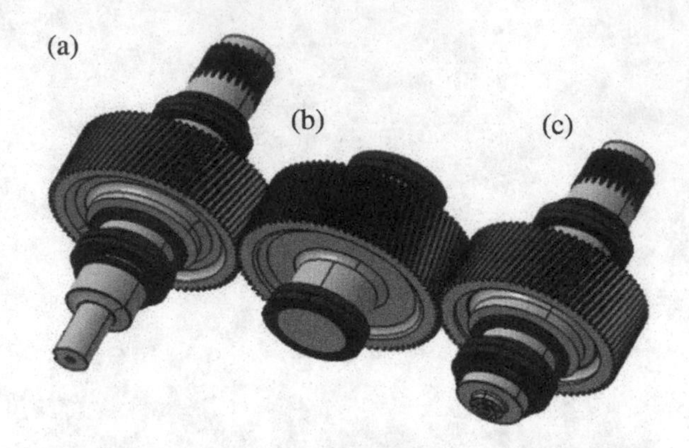

Fig. 6.3 Demonstration of case study three gear trains: **a** output gear (z74), **b** idler gear (z83) and **c** input gear (z79)

	Best Mounting Configuration		
Gear1-Gear2	Gear1 Starting Tooth	Gear2 Contact Drive Side Tooth	Gear2 Contact Coast Side Tooth
1st Configuration Based Smallest Minimum Value	9	83	1
	Gear1 Starting Tooth	Gear2 Contact Drive Side Tooth	Gear2 Contact Coast Side Tooth
2nd Configuration Based Smallest Maximum Value	44	83	1
	Gear1 Starting Tooth	Gear2 Contact Drive Side Tooth	Gear2 Contact Coast Side Tooth
3rd Configuration Based Smallest Peak-Peak Value	44	83	1
Gear2-Gear3	Gear3 Starting Tooth	Gear2 Contact Drive Side Tooth	Gear2 Contact Coast Side Tooth
1st Configuration Based Smallest Minimum Value	1	13	14
	Gear3 Starting Tooth	Gear2 Contact Drive Side Tooth	Gear2 Contact Coast Side Tooth
2nd Configuration Based Smallest Maximum Value	1	35	36
	Gear3 Starting Tooth	Gear2 Contact Drive Side Tooth	Gear2 Contact Coast Side Tooth
3rd Configuration Based Smallest Peak-Peak Value	1	56	57

Fig. 6.4 Best mounting selections for rotating clockwise under configurations

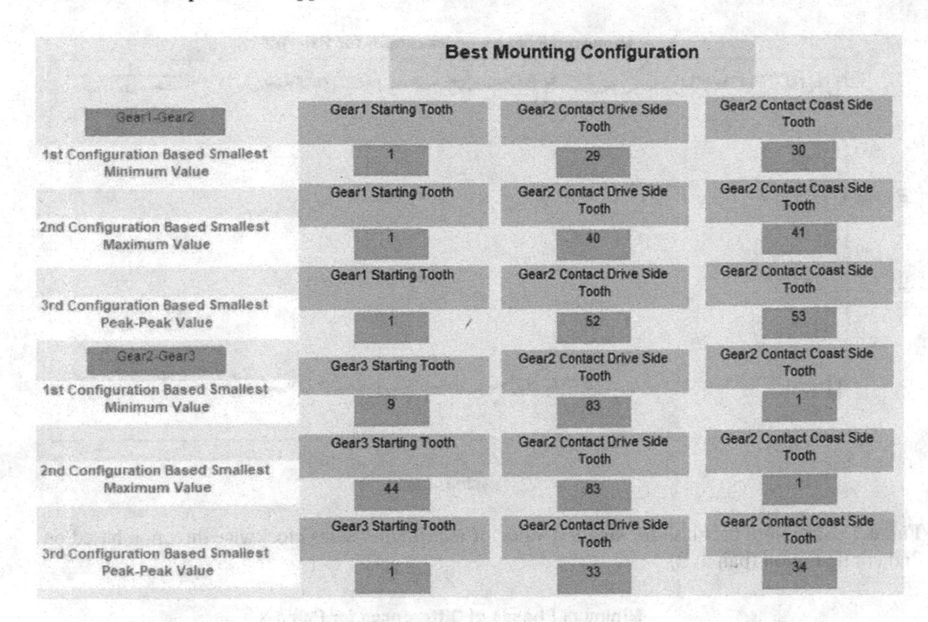

Fig. 6.5 Best mounting selections for counterclockwise rotating under configurations

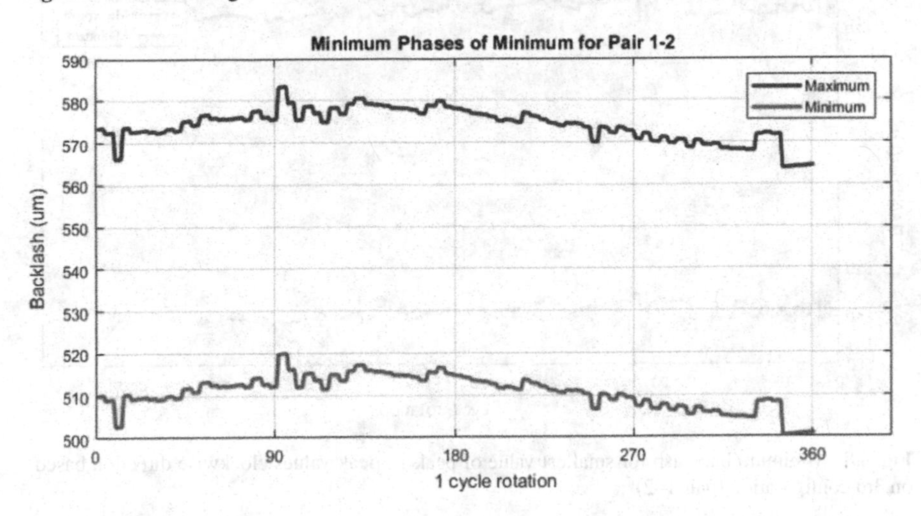

Fig. 6.6 Minimum backlash for smallest value of minimum values clockwise direction based on 1st configuration (pair 1–2)

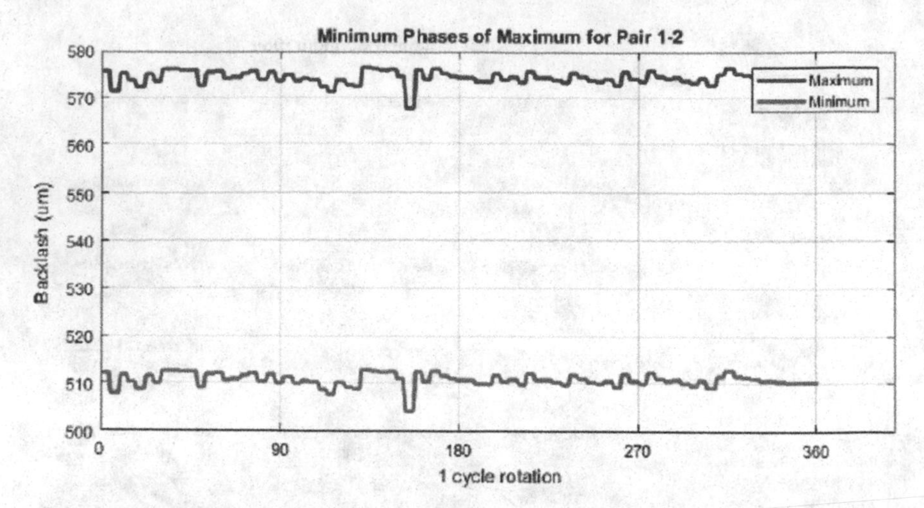

Fig. 6.7 Minimum backlash for smallest value of maximum values clockwise direction based on 2nd configuration (pair 1–2)

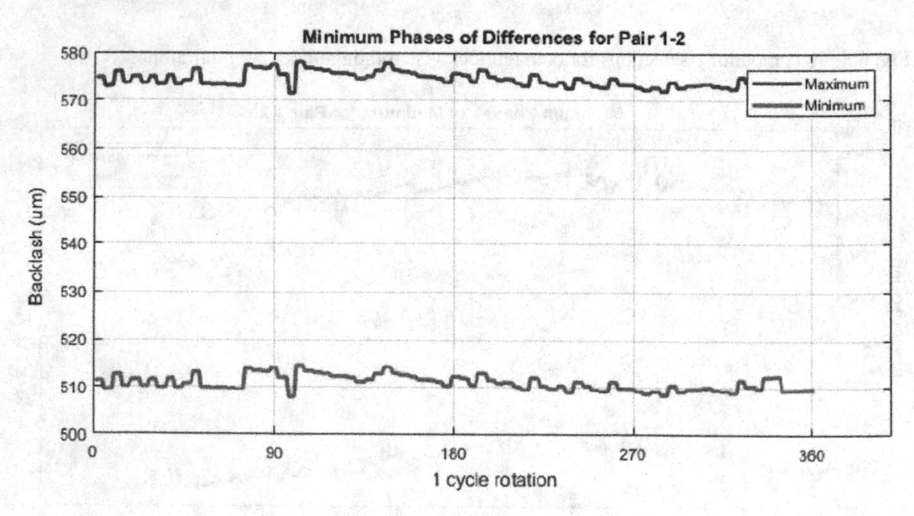

Fig. 6.8 Minimum backlash for smallest value of peak-to-peak values clockwise direction based on 3rd configuration (pair 1–2)

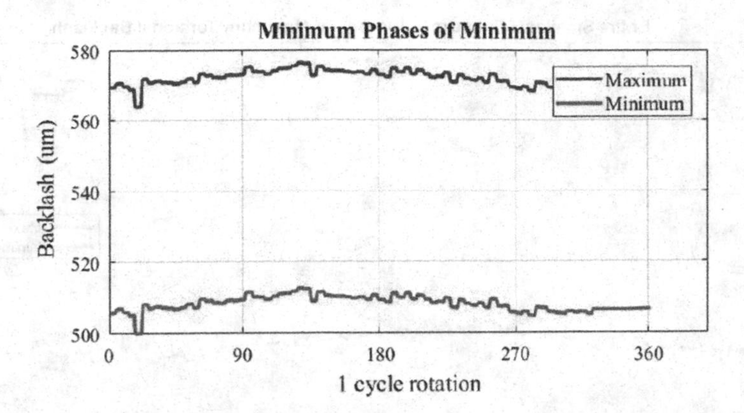

Fig. 6.9 Minimum backlash for smallest value of minimum values clockwise direction based on 1st configuration (pair 2–3)

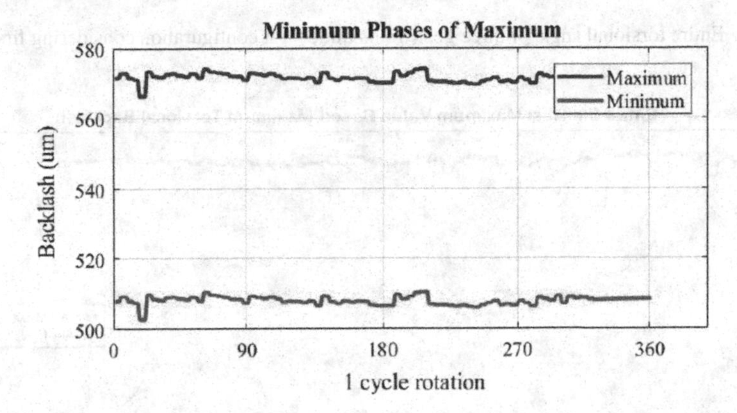

Fig. 6.10 Minimum backlash for smallest value of maximum values clockwise direction based on 2nd configuration (pair 2–3)

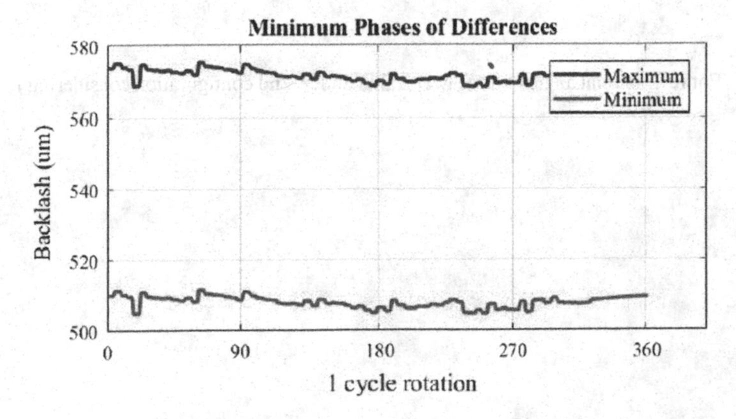

Fig. 6.11 Minimum backlash for smallest value of peak-to-peak values clockwise direction based on 3rd configuration (pair 2–3)

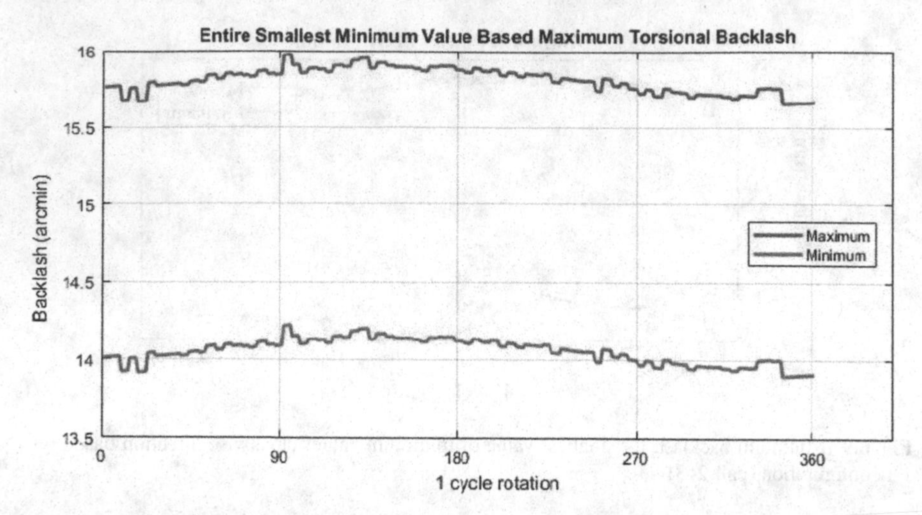

Fig. 6.12 Entire torsional angle of three gear trains under 1st configuration considering first gear driving

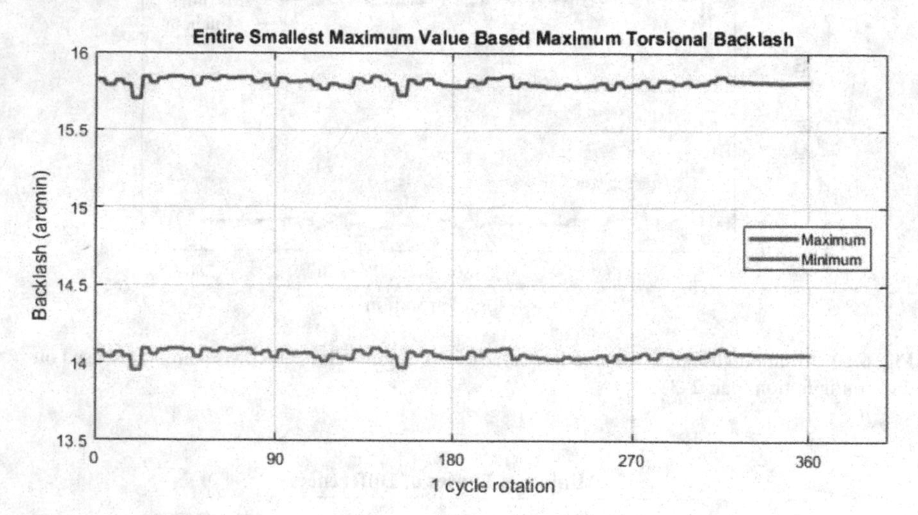

Fig. 6.13 Entire Torsional angle of three gear trains under 2nd configuration considering first gear driving

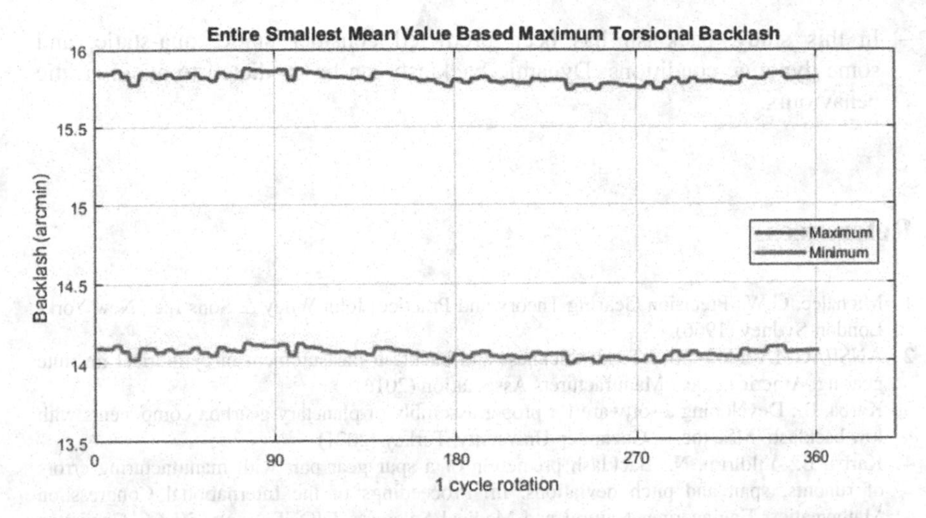

Fig. 6.14 Entire torsional angle of three gear trains under 3rd configuration considering first gear driving

6.4 Conclusion

Mostly, manufacturing errors of individual gear is distributed uniformly or as maximum and minimum. However, each tooth has own individual errors or deviations. For that reason, unless gear engineer considers that or mounting personnel obey to mounting procedure, actual behavior cannot be predicted while gears are operating. To avoid that circumstance, if there is any consider about that, it must be applied properly to stay within tolerance range of behavior of system. If gears with marked according to numbers of teeth respectively based on manufacturing reports. Assembly procedure for low backlash can be obtained with high probability with taken from software outputs according to founded best configurations. Case study for backlash prediction have been performed in that way. Hence, backlash range of train was clarified to understand considering each rolling degree during meshing. An algorithm allow to interpretation and find optimum mounting configurations for lowest backlash using self-compensation automatically. If backlash graph is taken as a reference, there are 3 important possibilities which are provided to interpret minimum backlashes. The software outputs whose best configurations give to user showed that backlash can be reduced comparing to randomly mounted gearbox. As a result of these examinations, backlashes of related systems are reduced by using this methodology. The followings may be considered as future works:

- All efforts which have been performed unloaded system during this study. Loaded gear systems can be investigated for further backlash prediction,
- Statistical analysis can be added to backlash prediction to find nearest real backlash using literature theories e.g. [6], and perfectible theories gear pair, gear trains, and planetary gearbox, and

– In this study, backlash has been predicted consider static, qua-static, and some dynamic conditions. Dynamic backlash can be predicted to avoid erratic behaviours.

References

1. Michalec, G.W.: Precision Gearing Theory and Practice. John Wiley & Sons Inc., New York-London-Sydney (1966)
2. ANSI/AGMA 2002-C16.: Tooth thickness and backlash measurement of cylindrical involute gearing. American Gear Manufacturers Association (2016)
3. Karba, B.: Developing a software for proper assembly of planetary gearbox components with low backlash. MSc thesis. Gaziantep University, Turkey (2021)
4. Karba, B., Yıldırım, N.: Backlash prediction of a spur gear pair with manufacturing errors of runouts, span and pitch deviations. In: Proceedings of the International Congress on Mathematics, Engineering, Natural and Medical Sciences, EJONS V, pp. 50–66. Gaziantep (2018)
5. Karba, B., Yıldırım, N., Erdoğan, F., Vardar, M.: A study on prediction and validation of meshing gear pair backlash under manufacturing and assembly errors. MATEC Web Conf. **287**, 07001-1–07001-10 (2019)
6. Michalec, G.W.: Correlation of probabilistic backlash with measurements. Mech. Mach. Theory **8**(2), 161–173 (1973)

Chapter 7
The Modification Effects of the Sum of Pitch Angles in Case of Bevel Gear Pairs Having Straight Teeth for the TCA

Sándor Bodzás and Gyöngyi Szanyi

Abstract The purpose of this work is the analysis of the effects of the modification of the sum of pitch angles for the TCA's (Tooth Contact Analysis) parameters. We design five different types of straight bevel gear pairs whose the sum of the pitch angles will be modified beside the constancy of other geometric parameters. After that the CAD (Computer Aided Designing) models could be created. Knowing of the exact geometries TCA could be done because of the analysis of the received normal stress', normal strain's and normal deformation's distributions on the connecting tooth areas. Finally, the diagrams could be determined in the function of the changing of the modified geometric parameter and the analysed TCA's parameters.

Keywords Bevel gear pairs having straight teeth · TCA · CAD · Sum of pitch angles

7.1 Introduction

The bevel gear pair having straight teeth is preferably used in many engineering constructions where the modification of the transmission ratio and the position of the shafts are needed (vehicle, working machine, etc.). Many parameters are needed for the correct geometric designing because they have fairly complex geometry [1–7]. One of the most important parameters is the sum of pitch angles (Fig. 7.1).

If the $\Sigma = 90°$ sum of pitch angles arrangement is not solvable but angular shaft position is needed the sum of the pitch angles will have to be modified. In these special cases the appropriate tooth geometry is significant because of the effect of given load. The tooth connection is along the pitch cones that is why the sum of pitch angles is (Fig. 7.2) [5, 8, 9]:

S. Bodzás (✉) · G. Szanyi
Faculty of Engineering, University of Debrecen, Ótemető str. 2-4, Debrecen, Hungary
e-mail: bodzassandor@eng.unideb.hu

G. Szanyi
e-mail: szanyi.gyongyi@eng.unideb.hu

M. Rackov et al. (eds.), *Machine and Industrial Design in Mechanical Engineering*, Mechanisms and Machine Science 109,
https://doi.org/10.1007/978-3-030-88465-9_7

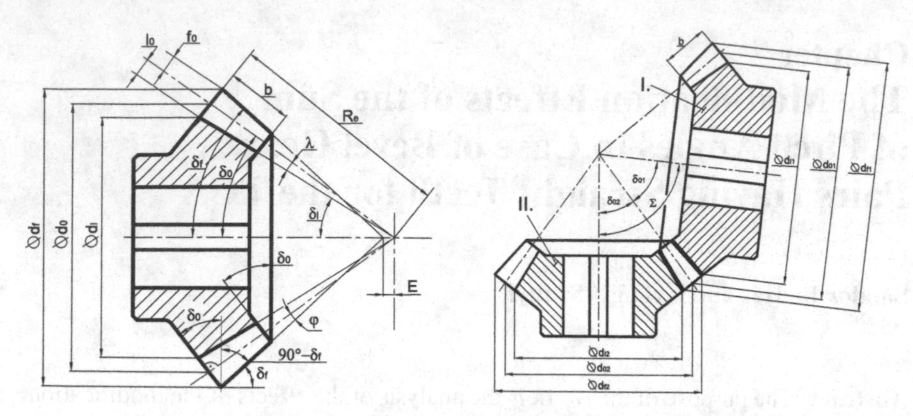

Fig. 7.1 The geometric establishment of the bevel gear pairs having straight teeth

Fig. 7.2 The coordinate system's arrangement for the mathematical analysis

$$\Sigma = \delta_{01} + \delta_{02} \tag{7.1}$$

7.2 The Mathematical Description of the Gear Connection

Given the two parametric vector—scalar function of the surface of the pinion in general case:

$$\vec{r}_{1R} = \vec{r}_{1R}(\eta, \vartheta) = \begin{bmatrix} x_{1R}(\eta, \vartheta) \\ y_{1R}(\eta, \vartheta) \\ z_{1R}(\eta, \vartheta) \end{bmatrix} \tag{7.2}$$

The necessary transformation matrices between the K_{1R} and K_{2R} coordinate systems are:

$$\mathbf{M_{2R,1R}} = \begin{bmatrix} \cos\varphi_2 \sin\Sigma & \begin{matrix}\cos\varphi_1 \sin\varphi_2 \\ +\cos\varphi_2 \cos\Sigma \sin\varphi_1\end{matrix} & \begin{matrix}\sin\varphi_2 \sin\varphi_1 \\ -\cos\varphi_2 \cos\Sigma \cos\varphi_1\end{matrix} & c\cos\varphi_2 \\ -sin\varphi_2 \sin\Sigma & \begin{matrix}\cos\varphi_2 \cos\varphi_1 \\ -\cos\Sigma \sin\varphi_2 \sin\varphi_1\end{matrix} & \begin{matrix}\cos\varphi_1 \cos\Sigma \sin\varphi_2 \\ +\cos\varphi_2 \sin\varphi_1\end{matrix} & -c\sin\varphi_2 \\ \cos\Sigma & -\sin\varphi_1 \sin\Sigma & \cos\varphi_1 \sin\Sigma & a \\ 0 & 0 & 0 & 1 \end{bmatrix} \tag{7.3}$$

$$\mathbf{M_{1R,2R}} = \begin{bmatrix} \cos\varphi_1 \sin\Sigma & -\sin\Sigma \sin\varphi_2 & 1 & -a+\cos\Sigma - c\sin\Sigma \\ \begin{matrix}\cos\Sigma \cos\varphi_2 \sin\varphi_1 \\ +\cos\varphi_1 \sin\varphi_2\end{matrix} & \begin{matrix}\cos\varphi_1 \cos\varphi_2 \\ -\cos\Sigma \sin\varphi_1 \sin\varphi_2\end{matrix} & -\sin\varphi_1 \sin\Sigma & \begin{matrix}a\sin\varphi_1 \sin\Sigma \\ -c\cos\Sigma \sin\varphi_1\end{matrix} \\ \begin{matrix}\sin\varphi_1 \sin\varphi_2 \\ -\cos\varphi_1 \cos\Sigma \cos\varphi_2\end{matrix} & \begin{matrix}\cos\varphi_2 \sin\varphi_1 \\ +\cos\varphi_1 \cos\Sigma \sin\varphi_2\end{matrix} & \cos\varphi_1 \sin\Sigma & \begin{matrix}c\cos\varphi_1 \cos\Sigma \\ -a\cos\varphi_1 \sin\Sigma\end{matrix} \\ 0 & 0 & 0 & 1 \end{bmatrix} \tag{7.4}$$

where:

$$\mathbf{M_{2R,1R}} = \mathbf{M_{2R,2S}} \cdot \mathbf{M_{2S,0}} \cdot \mathbf{M_{0,01}} \cdot \mathbf{M_{01,1S}} \cdot \mathbf{M_{1S,1R}} \tag{7.5}$$

$$\mathbf{M_{1R,2R}} = \mathbf{M_{1R,1S}} \cdot \mathbf{M_{1S,01}} \cdot \mathbf{M_{01,0}} \cdot \mathbf{M_{0,2S}} \cdot \mathbf{M_{2S,2R}} \tag{7.6}$$

The two tooth surfaces are wrapped each other during the connection that is why [5]:

$$\varphi_2 = i \cdot \varphi_1 \tag{7.7}$$

The relative velocity vector is between the two contact surfaces [5]:

$$\vec{v}_{2R}^{(1,2)} = \frac{d}{dt} \cdot \vec{r}_{2R} = \frac{d}{dt}\mathbf{M_{2R,1R}} \cdot \vec{r}_{1R} \tag{7.8}$$

Considering the correlation between the relative velocity vectors on the K_{1R} and K_{2R} coordinate systems [5]:

$$\vec{v}_{1R}^{(1,2)} = \mathbf{M_{1R,2R}} \cdot \vec{v}_{2R}^{(1,2)} = \mathbf{M_{1R,2R}} \cdot \frac{d}{dt}\mathbf{M_{2R,1R}} \cdot \vec{r}_{1R} = \mathbf{P} \cdot \vec{r}_{1R} \tag{7.9}$$

Knowing of the parametric surface equation of the pinion the equation of the contact points on the contact surfaces are the solution of the following equation [5]:

$$\left.\begin{array}{l} \vec{r}_{1R} = \vec{r}_{1R}(\eta, \vartheta) \\ \vec{n}_{1R} \cdot \vec{v}_{1R}^{(1,2)} = \begin{bmatrix} \vec{i} & \vec{j} & \vec{k} \\ \frac{\partial x_{1R}}{\partial \eta} & \frac{\partial y_{1R}}{\partial \eta} & \frac{\partial z_{1R}}{\partial \eta} \\ \frac{\partial x_{1R}}{\partial \vartheta} & \frac{\partial y_{1R}}{\partial \vartheta} & \frac{\partial z_{1R}}{\partial \vartheta} \end{bmatrix} \cdot \begin{bmatrix} \vec{v}_{1Rx}^{(1,2)} \\ \vec{v}_{1Ry}^{(1,2)} \\ \vec{v}_{1Rz}^{(1,2)} \end{bmatrix} = 0 \end{array}\right\} \tag{7.10}$$

The common solution could be calculated by numerical way that is why the exact position of the contact points could be calculated and analyzed [5].

7.3 Computer Aided Design of Different Bevel Gear Pairs

Considering the suggestion of the references five types of bevel gear pairs having straight teeth had to be designed. The number of step was 5° in case of each gear pairs. The difference between them is only the modification of the sum of pitch angle (Table 7.1).

For the facilitation of the designing process the GearTeq designing software was used. This is a quite complex software with which many types of gears could be designed (Fig. 7.3). The input parameters have to be given and after the calculation this pro-gram can built up the geometric models of the pinion. The computer aided model of the driven gear could be also built up by direct kinematical method in mathematical way. The received gear pair could be visualized on the display. After the geometric design the gear pair could be saved into the SolidWorks software where the assembly corrections and simulations could be done before the TCA (Fig. 7.3). The simulation analysis is important because of the inspection of the connection. During this analysis we can do beat examinations and check the correct teeth connection.

7.4 TCA Analysis in the Function of Selected Geometric Parameter

After the geometric correct designing the TCA is an important analysis where the mechanical parameters (stress, strain, deformation) could be calculated on the tooth contact zone for the effect of given load (force or torque) [6–8].

Dense meshing (element size 1.1 mm) was used on the teeth contact zone [1–5]. The driven gear was totally fixed. One rotation freedom degree was permitted on the pinion. 800 Nm torque was applied as a load on the pinion. We analyzed the effect of this load on the tooth connection. The selected material type is structural steel (Table 7.2). The friction coefficient was $\mu = 0.15$ on the teeth connection.

Table 7.1 The parameters of the designed gear pairs

Parameter	I	II	III	IV	V
Module [mm]	10				
Pressure angle [°]	20				
Spiral angle [°]	0				
Shaft angle [°]	70	75	80	85	90
Number of teeth I. z_I	20				
Number of teeth II. z_2	30				
Face width [mm]	100				
Working depth [mm]	20				
Whole depth [mm]	21.9				
Pitch diameter I. [mm] d_{01}	200				
Pitch diameter II. [mm] d_{02}	300				
Pitch angle I. [°] δ_{01}	27	28.8	30.5	32.1	33.7
Pitch angle II. [°] δ_{02}	43	46.2	49.5	52.9	56.3
Circular pitch [mm]	31.4				
Addendum [mm] I. f_{01}	12.6				
Dedendum [mm] I. l_{01}	9.3				
Addendum [mm] II. f_{02}	7.4				
Dedendum [mm] II. l_{02}	14.4				
Clearance [mm]	1.9				
Root angle [°] I. δ_{l1}	24.6	26.2	27.8	29.3	30.7
Root angle [°] II. δ_{l2}	39.2	42.2	45.3	48.5	51.7
Outside diameter [mm] I. d_{f1}	222.4	222	221.6	221.3	220.9
Outside diameter [mm] II. d_{f2}	310.9	310.3	309.7	309	308.3
Tooth thickness [mm] I	17.2				
Tooth thickness [mm] II	14.2				
Addendum modification I. [mm]	2.6				
Addendum modification II. [mm]	−2.6				
Addendum modification coefficient I	0.26				
Addendum modification coefficient II	−0.26				
Fillet radius [mm]	2.8				

7.4.1 The Analysis of the Normal Stresses

The formed normal stress distributions were analysed on the teeth surfaces. The normal stress distributions could be seen on Fig. 7.4 in case of $\Sigma = 70°$. We adopted a coordination system in the middle of the tooth surface. The x-axis is perpendicular for the tooth surface which is the normal direction. We analysed the received TCA results into the normal direction (Fig. 7.4).

Fig. 7.3 A CAD assembly models of a bevel gear pairs by SolidWorks software ($\Sigma = 70°$)

Table 7.2 The parameters of the selected structural steel

Density	7850 kg/m^3
Yield limit	250 MPa
Ultimate strength	460 MPa

Based on Fig. 7.5 the lowest stresses are in case of 75°. The highest stresses are in case of 90°. The stress values are significantly higher in case of 90° in absolute value.

7.4.2 *The Analysis of the Normal Elastic Strain*

The formed normal elastic strain distributions were analyzed on the teeth surfaces. The normal elastic strain distributions could be seen on Fig. 7.6 in case of $\Sigma = 70°$.

Based on Fig. 7.7 the lowest strains are in case of 75°. The highest strains are in case of 90° degree. The strain values are significantly higher in case of 90° degree in absolute value.

7.4.3 *The Analysis of the Normal Deformation*

The formed normal deformation distributions were analysed on the teeth surfaces. The normal deformation distributions could be seen on Fig. 7.8 in case of $\Sigma = 70°$.

Based on Fig. 7.9 the lowest deformations are in case of 85°. The highest deformations are in case of 90°. The deformation values are continuously being decreased from 70° to 85°. The deformation values are significantly higher in case of 90° in absolute value.

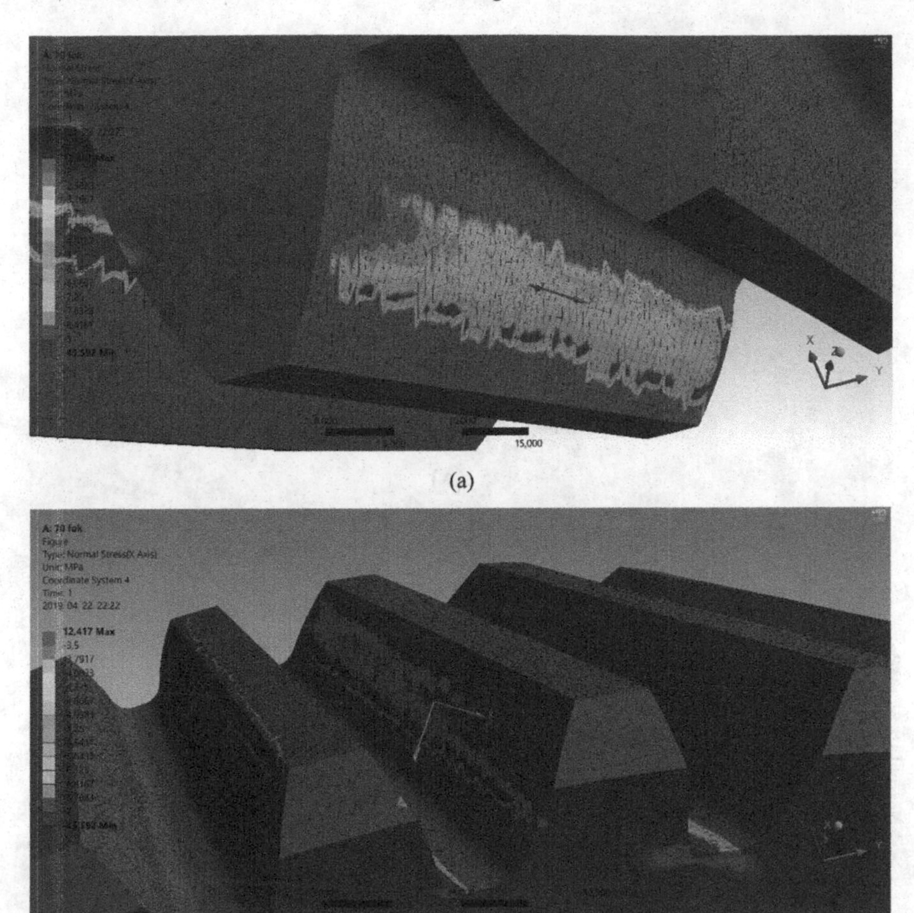

Fig. 7.4 The normal stress distribution in case of $\Sigma = 70°$ sum of pitch angles: **a** tooth side of the pinion and **b** tooth side of the driven gear

7.5 Conclusion

The aim of this publication is the analysis of the TCA results in the function of the modification of the sum of pitch angles in case of bevel gears having straight teeth. Using of the GearTeq geometric designing software five types of bevel gears are designed. All of the geometric parameters of them is the same except the sum of pitch angles were modified. Because of this modification some connection parameters were also changed. The CAD models were created by SolidWorks software. After the geometric modelling TCA was done by Ansys software. The normal stress, the normal elastic strain and the normal deformation were analysed in the function of the modified parameter based on the effects of given load. Many functions were drawn.

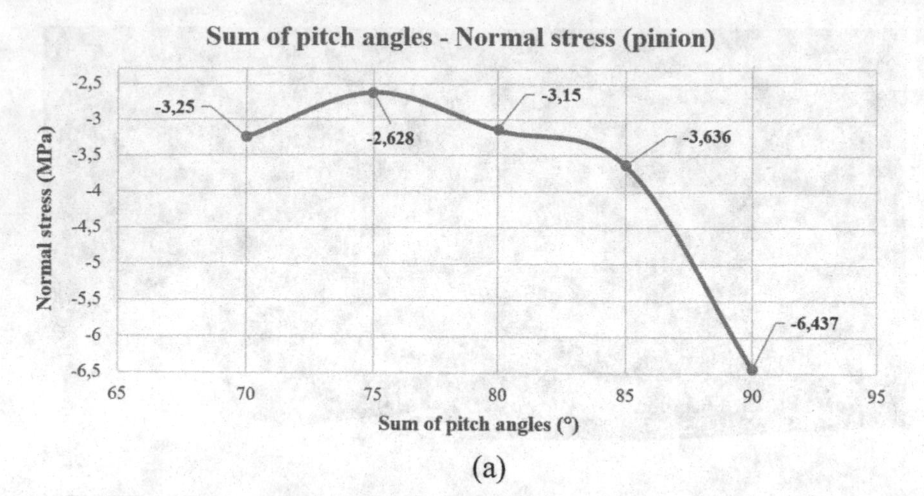

(a)

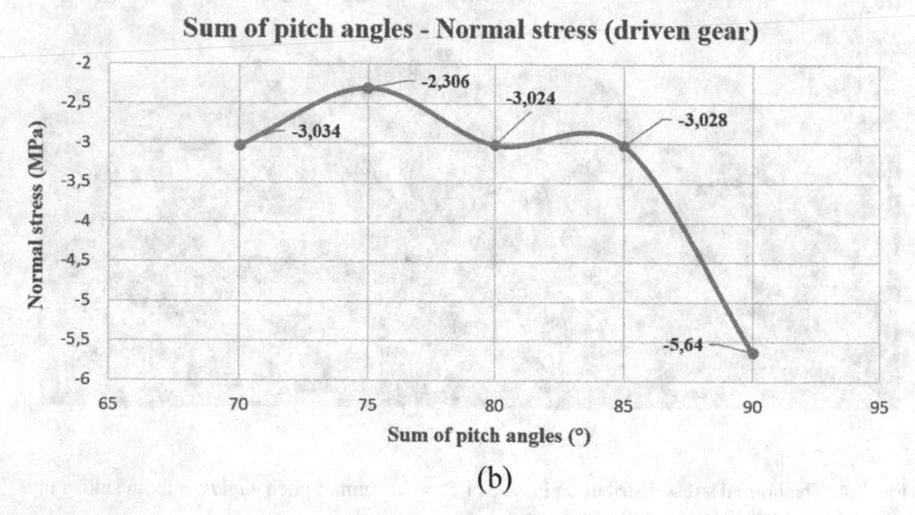

(b)

Fig. 7.5 The average normal stresses in the function of the sum of pitch angles: **a** pinion and **b** driven gear

As a result the highest values were received in case of 90° in absolute value that is why the 90° arrangement is the most sensitive in aspect of loads. Generally, the perpendicular shaft application is the most frequent in the designing of mechanical construction. Occasionally non-perpendicular shaft arrangement could be also used depending of the mechanical designing and the assembly possibilities.

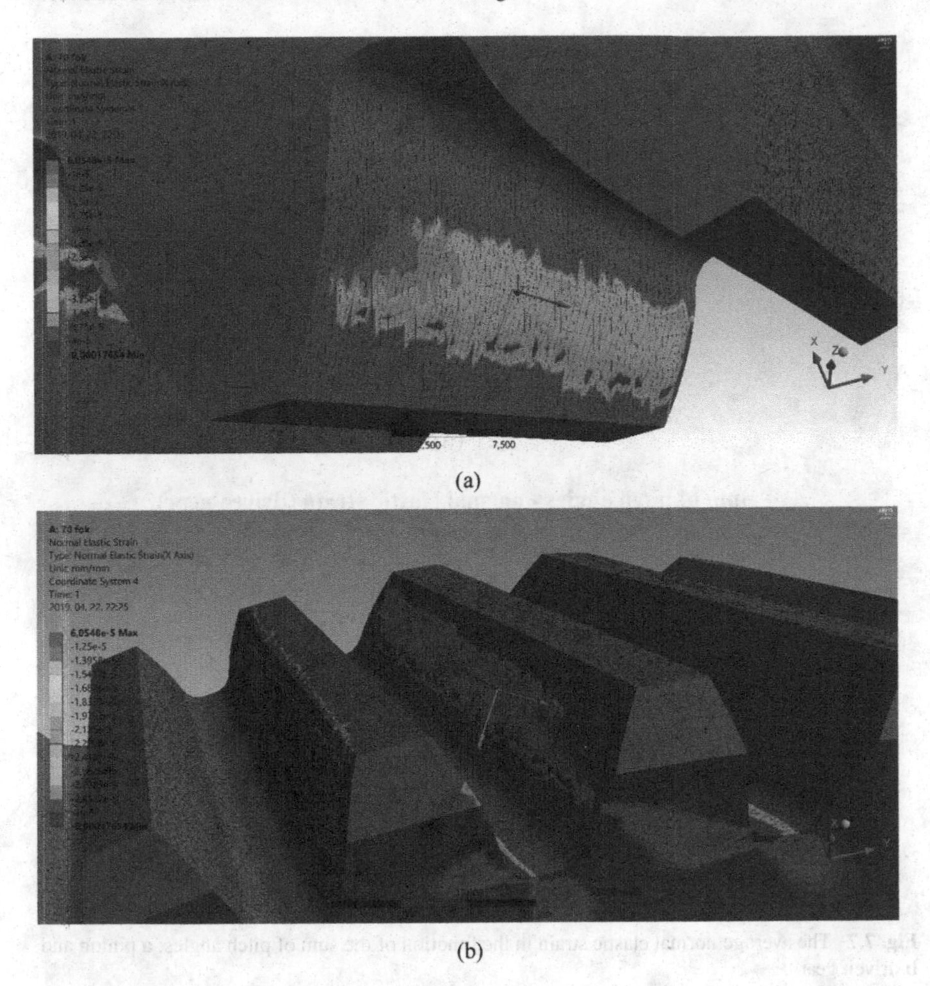

Fig. 7.6 The normal elastic strain distribution in case of $\Sigma = 70°$ sum of pitch angles: **a** tooth side of the pinion and **b** tooth side of the driven gear

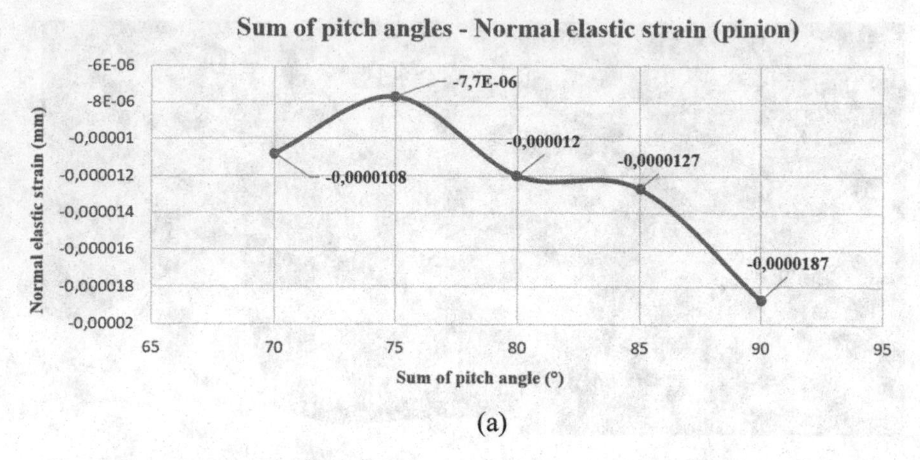

(a)

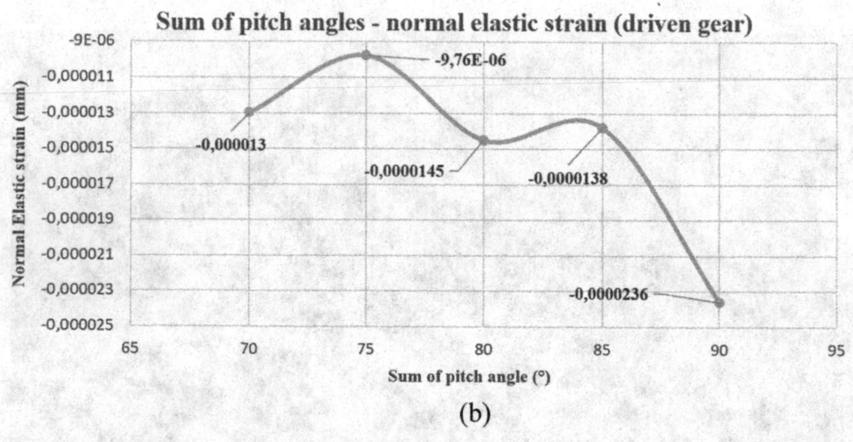

(b)

Fig. 7.7 The average normal elastic strain in the function of the sum of pitch angles: **a** pinion and **b** driven gear

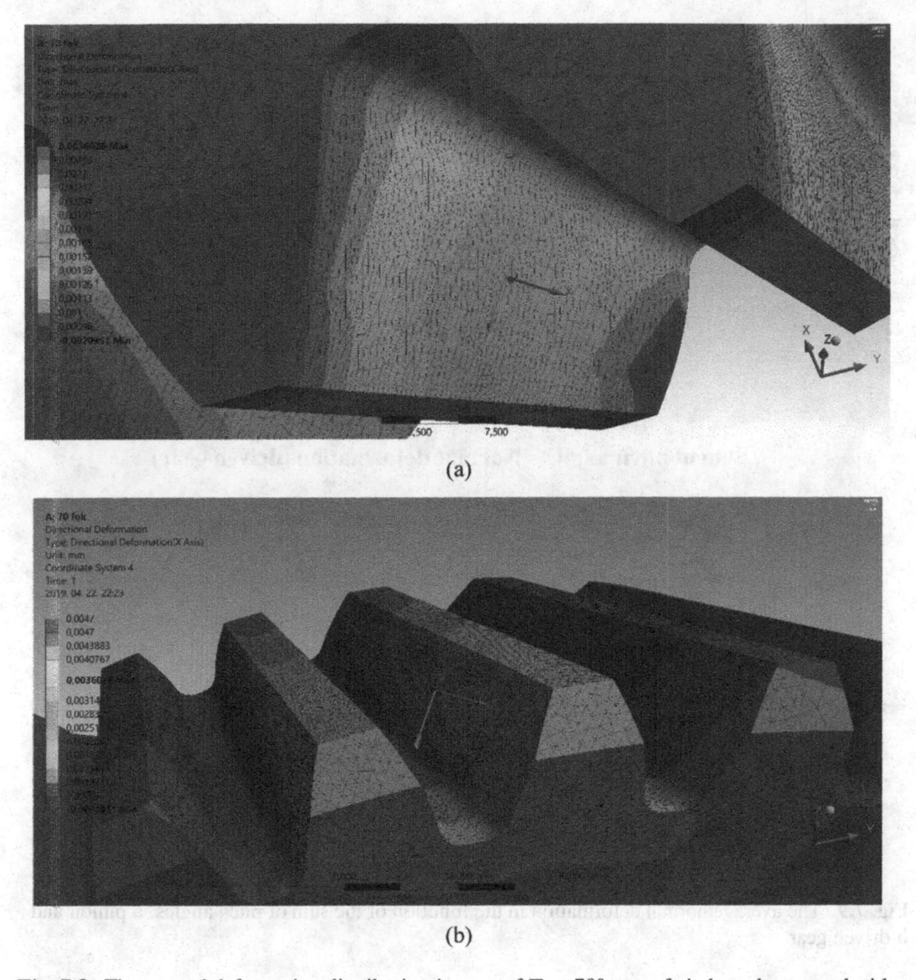

Fig. 7.8 The normal deformation distribution in case of $\Sigma = 70°$ sum of pitch angles: **a** tooth side of the pinion and **b** tooth side of the driven gear

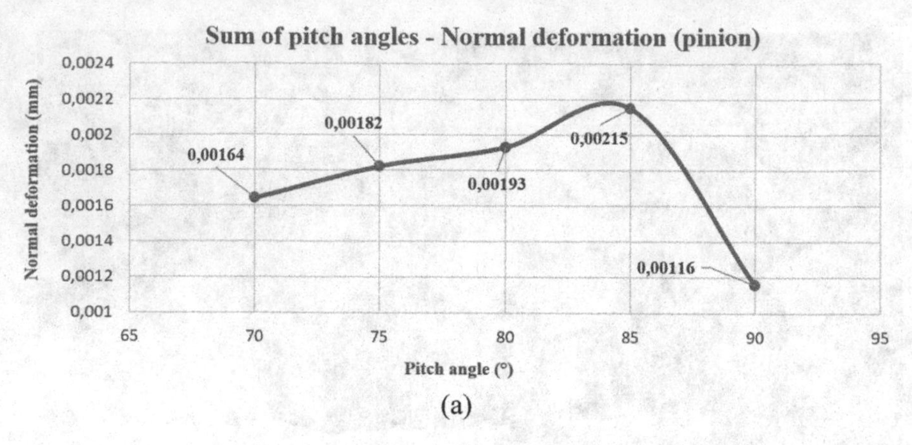

(a)

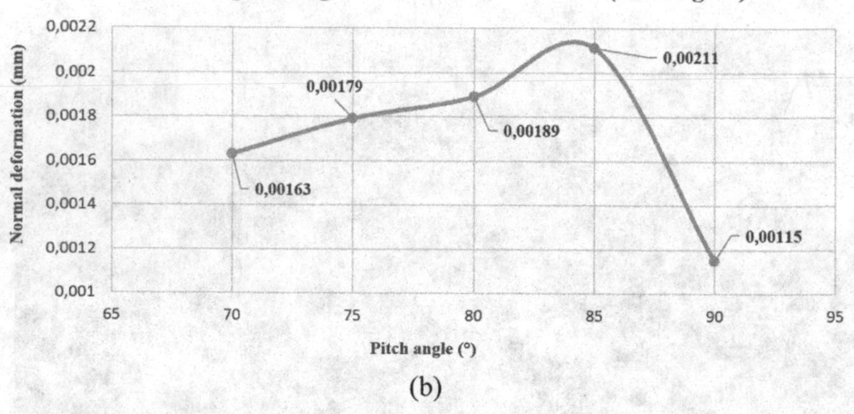

(b)

Fig. 7.9 The average normal deformation in the function of the sum of pitch angles: **a** pinion and **b** driven gear

Acknowledgements Project no. TKP2020-NKA-04 has been implemented with the support provided from the National Research, Development and Innovation Fund of Hungary, financed under the 2020-4.1.1-TKP2020 funding scheme. This research was partly supported by the János Bolyai Research Scholarship of the Hungarian Academy of Sciences.

References

1. Argyris, J., Fuentes, A., Litvin, F.L.: Computerized integrated approach for design and stress analysis of spiral bevel gears. Comput. Method Appl. M. **191**(11–12), 1057–1095 (2002)
2. Fuentes, A., Iserte, J.L., Gonzalez-Perez, I., Sanchez-Marin., F.T.: Computerized design of advanced straight and skew bevel gears produced by precision forging. Comput. Method Appl. M. **200**(29–32), 2363–2377 (2011)

3. Fuentes-Aznar, A., Yague-Martinez, E., Gonzalez-Perez, I.: Computerized generation and gear mesh simulation of straight bevel gears manufactured by dual interlocking circular cutters. Mech. Mach. Theory **122**, 160–176 (2018)
4. Gosselin, C.: Multi axis CnC manufacturing of straight and spiral bevel gears. In: Goldfarb, V., Trubachev, E., Barmina, N. (eds.) Advanced Gear Engineering. MMS, vol. 51, pp. 167–204. Springer, Cham (2018)
5. Litvin, F.L., Fuentes, A.A.: Gear Geometry and Applied Theory. Cambridge University Press, Cambridge (2004)
6. Zeng, Q.L., Wang, K., Wan, L.R.: Modelling and straight bevel gear transmission and simulation of its meshing performance. Int. J. Simul. Model **17**(3), 521–533 (2018)
7. Zolfaghari, A., Goharimanesh, M., Akbari, A.A: Optimum design of straight bevel gears pair using evolutionary algorithms. J. Braz. Soc. Mech. Sci. **39**, 2121–2129 (2017)
8. Dudás, L.: Kapcsolódó felületpárok gyártásgeometriai feladatainak megoldása az elérés modell alapján. TMB, Budapest, (1991)
9. Dudley, D.W.: Gear Handbook. McGraw Hill, New York (1962)

Chapter 8
Influence of Wear on Load Sharing and Transmission Error of Spur Gears with Profile Modifications

Miguel Pleguezuelos, José I. Pedrero, Miryam B. Sánchez, and Elena Vicente

Abstract The gear mesh stiffness, and therefore the load sharing, tooth deflections, and transmission error, are strongly influenced by the tooth dimensions and profile shape. But the profile shape and contact conditions are affected by the wear produced by the relative sliding of the teeth surfaces, so that load sharing and transmission error change as the wear depth increases. Consequently, the wear depth increase is different at any contact point—because the load and sliding velocity are different—and in any meshing cycle as well—because the load sharing changes with the accumulated wear—. In addition, the length, depth, and shape of profile modifications may sensibly affect the initial load sharing, and therefore the complete wear process. In this paper, a model of meshing stiffness, load sharing, and transmission error for spur gears with profile modifications including the influence of wear is presented.

Keywords Spur gears · Wear · Profile modifications

8.1 Introduction

The profile shape of gear teeth has strong influence on the load sharing between couples of teeth in simultaneous contact, and therefore on the teeth deflections and transmission error [1–5]. A small difference on the profile geometry, for example a few microns relief, induces a deflection on this tooth of this few microns less, which results in a much smaller load on this tooth pair than that on the other pairs in contact.

Due to the relative sliding of the tooth surfaces along the meshing contact, wear arises after a number of meshing cycles. This affects the contact conditions in the same way as a profile modification along the whole active profile, in such a manner that the acting load affects the instantaneous wear, and the accumulated wear affects the acting load [6, 7]. The main difference is that profile modification due to wear varies with the number of meshing cycles, but for a given number of cycles the wear can be studied as a long profile modification, which affects all the active profile.

M. Pleguezuelos · J. I. Pedrero (✉) · M. B. Sánchez · E. Vicente
Dep. Mecánica, UNED, Juan del Rosal 12, 28040 Madrid, Spain
e-mail: jpedrero@ind.uned.es

M. Rackov et al. (eds.), *Machine and Industrial Design in Mechanical Engineering*, Mechanisms and Machine Science 109,
https://doi.org/10.1007/978-3-030-88465-9_8

To describe the wear on contact surfaces the Archard's model [8] is frequently used. It has been successfully applied to the wear prediction in tooth gear surfaces [9–11], as well as to study its influence on the load and transmission error [12]. But all these studies were focused on involute spur gears, without profile modifications. Nowadays, profile modifications are often used to avoid the mesh-in impact induced by the earlier start of contact due to the teeth deflections and control the load sharing and transmission error [1–4, 13]. It is therefore necessary to know the evolution of the load sharing and transmission error with the wear depth, in spur gears with profile modifications. In this paper, the wear depth is introduced in a simple analytical model previously developed [2, 4] which considers profile modifications.

8.2 Background

It has been demonstrated that, if assumed the minimum elastic potential energy principle, the deflection of all the tooth pairs in simultaneous contact is the same [2, 4, 13]. This equal deflection represents the delay of the actual position of the driven gear respect to its theoretical conjugate position, which is called quasi-static transmission error (QSTE). From this assumption, the load at a specific tooth pair j which is in contact at the point of the line of action described by ξ can be expressed as follows:

$$F_j(\xi) = K_M(\xi)[\delta(\xi) - \delta_G(\xi) - \delta_R(\xi) - \delta_W(\xi)] \tag{8.1}$$

where F_j is the load on tooth pair j, K_M the single meshing stiffness of the tooth pair, $\delta(\xi)$ the driven gear delay measured on the line of action, $\delta_G(\xi)$ the approach distance to start the contact before the theoretical inner point of contact, $\delta_R(\xi)$ the profile modification, and $\delta_W(\xi)$ the sum of the wear depth at both contact surfaces. ξ is a linear coordinate along the line of action, which is equal to 0 at the tangency point with the driving gear base circumference, and equal to $[(Z_1 + Z_2)/2\pi] \cdot \tan\alpha'_t$ at the tangency point with the driven gear base circumference, being Z the number of teeth, α'_t the operating transverse pressure angle, and subscripts 1 and 2 denotes the driving and driven gear, respectively. It can be proved that the difference between the contact point parameter of the theoretical outer and inner points of contact, ξ_o and ξ_{inn}, is equal to the theoretical contact ratio ε_α:

$$\xi_o - \xi_{inn} = \varepsilon_\alpha \tag{8.2}$$

and the difference between the contact point parameter of two pairs is:

$$\xi_{i+j} = \xi_i + j \tag{8.3}$$

From Eq. (8.1), the total transmitted load F_T is:

$$F_T = \delta(\xi) \sum_j K_M(\xi + j) - \sum_j K_M(\xi + j)(\delta_G(\xi + j) + \delta_R(\xi + j) + \delta_W(\xi + j)) \tag{8.4}$$

from which:

$$\delta(\xi) = \frac{1}{\sum_j K_M(\xi + j)} \left[F_T + \sum_j K_M(\xi + j)(\delta_G(\xi + j) + \delta_R(\xi + j) + \delta_W(\xi + j)) \right] \tag{8.5}$$

Equation (8.5) provides the QSTE of the gear pair at contact position described by ξ. The load sharing ratio (LSR) can be obtained by replacing Eq. (8.5) in Eq. (8.1), resulting in:

$$\begin{aligned} R(\xi) &= \frac{F(\xi)}{F_T} \\ &= \frac{K_M(\xi)}{F_T} \left[\frac{F_T + \sum_j K_M(\xi + j)(\delta_G(\xi + j) + \delta_R(\xi + j) + \delta_W(\xi + j))}{\sum_j K_M(\xi + j)} \right. \\ &\quad \left. -[\delta_G(\xi) + \delta_R(\xi) + \delta_W(\xi)] \right] \end{aligned} \tag{8.6}$$

The single meshing stiffness is very approximately described by [14]:

$$\begin{aligned} K_M(\xi) &= K_{M\max} \cos b_0 \tfrac{\varepsilon_\alpha}{2} && \text{for } \xi_{\min} \le \xi \le \xi_{inn} \\ K_M(\xi) &= K_{M\max} \cos b_0 (\xi - \xi_m) && \text{for } \xi_{inn} \le \xi \le \xi_o \\ K_M(\xi) &= K_{M\max} \cos b_0 \tfrac{\varepsilon_\alpha}{2} && \text{for } \xi_o \le \xi \le \xi_{\max} \end{aligned} \tag{8.7}$$

where $K_{M\max}$ can be computed as described in [14], and:

$$\begin{aligned} b_0 &= \left[\frac{1}{2}\left(1.11 + \frac{\varepsilon_\alpha}{2}\right)^2 - 1.17 \right]^{-\frac{1}{2}} \\ \xi_m &= \xi_{inn} - \frac{\varepsilon_\alpha}{2} \end{aligned} \tag{8.8}$$

$\xi_{\min}$ and $\xi_{\max}$ are the limits of the actual contact interval, which is wider than the theoretical one due to the delay of the driven tooth respect to the driving tooth. The calculation of their values can be found in [2, 4]. Obviously, $K_M(\xi) = 0$ outside the interval $\xi_{\min} \le \xi \le \xi_{\max}$.

The approach distance depends on the load and geometry of the meshing teeth, and is very approximately described by [2, 4, 13]:

$$\begin{array}{ll} \delta_G(\xi) = \left(\frac{2\pi}{Z_1}\right)^2 r_{b1} C_{p-inn} (\xi_{inn} - \xi)^2 & \text{for } \xi_{\min} \le \xi \le \xi_{inn} \\ \delta_G(\xi) = 0 & \text{for } \xi_{inn} \le \xi \le \xi_o \\ \delta_G(\xi) = \left(\frac{2\pi}{Z_1}\right)^2 r_{b1} C_{p-o} (\xi - \xi_o)^2 & \text{for } \xi_o \le \xi \le \xi_{\max} \end{array} \quad (8.9)$$

where r_b is the base radius, and coefficients $C_{p\text{-}inn}$ and $C_{p\text{-}o}$ can be calculated as presented in [2, 4].

The equation of the profile modification will obviously depend on the designer's decision. A typical equation for a linear tip relief at both mating teeth is:

$$\begin{array}{ll} \delta_R(\xi) = \Delta_{R-inn}\left(1 - \frac{\xi - \xi_{inn}}{\Delta\xi_{r-inn}}\right) & \text{for } \xi_{inn} \le \xi \le \xi_{inn} + \Delta\xi_{r-inn} \\ \delta_R(\xi) = 0 & \text{for } \xi_{inn} + \Delta\xi_{r-inn} \le \xi \le \xi_o - \Delta\xi_{r-o} \\ \delta_R(\xi) = \Delta_{R-o}\left(1 - \frac{\xi_o - \xi}{\Delta\xi_{r-o}}\right) & \text{for } \xi_o - \Delta\xi_{r-o} \le \xi \le \xi_o \end{array} \quad (8.10)$$

being $\Delta_{R\text{-}inn}$ and $\Delta_{R\text{-}o}$ the amount of modification at the driven and driving tip, respectively, and $\Delta\xi_{R\text{-}inn}$ and $\Delta\xi_{R\text{-}o}$ the respective lengths of modification. To avoid the mesh-in impact, the amount of modification $\Delta_{R\text{-}inn}$ should be equal to the approach distance at the inner point of contact of the actual contact interval $\delta_G(\xi_{\min})$ [2, 4, 13].

Equations (8.5) and (8.6) provide the curves of LSR and QSTE for spur gears with profile modification. All the parameters taking part in Eqs. (8.5) and (8.6) can be computed with Eqs. (8.7)–(8.10), except the wear depth, which is studied in the next section. Figure 8.1 shows the curves of LSR and QSTE obtained from the above equations, for a specific spur gear pair, with and without tip relief at the driven tooth.

8.3 Wear Depth

From the Archard's equation [8], the increment of wear depth h_W at contact point ξ due to the relative sliding corresponding to one meshing cycle can be expressed as:

$$\Delta h_{W1/W2}(\xi) = K_W \frac{F(\xi)}{b} \frac{|\mathrm{d}l_1(\xi) - \mathrm{d}l_2(\xi)|}{\mathrm{d}l_{1/2}(\xi)} \quad (8.11)$$

where K_W is a constant depending on the material, b the face width, and $\mathrm{d}l$ the length of the profile arc corresponding to a small rotation of the driving gear. The distance between both mating profiles due to this increment of wear depth will obviously be the sum of both wear depths, and therefore:

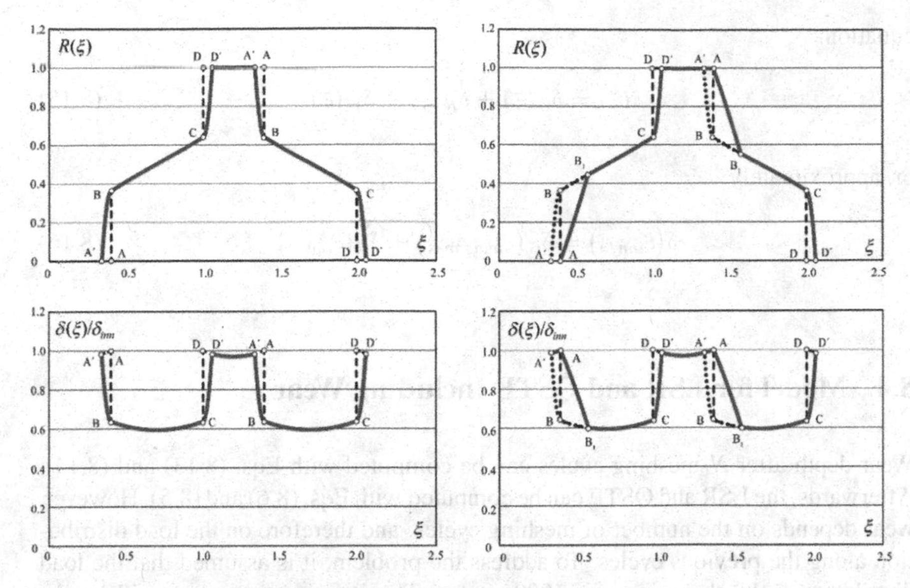

Fig. 8.1 Curves of LSR (up) and QSTE (down) for spur gear pair without profile modification (left) and with tip relief at the driven tooth (right)

$$\Delta\delta_W(\xi) = \Delta h_{W1}(\xi) + \Delta h_{W2}(\xi) = K_W \frac{F(\xi)}{b} |\mathrm{d}l_1(\xi) - \mathrm{d}l_2(\xi)| \left(\frac{1}{\mathrm{d}l_1(\xi)} + \frac{1}{\mathrm{d}l_2(\xi)} \right) \tag{8.12}$$

which, after some calculations, can be expressed as follows:

$$\Delta\delta_W(\xi) = K_W \frac{F(\xi)}{b} \left| \frac{2\pi}{Z_1}\xi - \tan\alpha_t' \right| \left(\frac{\frac{Z_1}{Z_2}}{\frac{2\pi}{Z_1+Z_2}\xi} + \frac{1}{\tan\alpha_t' - \frac{2\pi}{Z_1+Z_2}\xi} \right) \tag{8.13}$$

The distance δ_W after N meshing cycles should be calculated through an iterative procedure:

$$\delta_{W(i+1)}(\xi) = \delta_{W(i)}(\xi) + \Delta\delta_{W(i)}(\xi) \tag{8.14}$$

in which $\Delta\delta_{W(i)}(\xi)$ is computed from the values of $F(\xi)$ given by Eq. (8.6) for $\delta_W(\xi) = \delta_{W(i)}(\xi)$, starting from $\delta_{W(0)}(\xi) = 0$.

It should be noted that the actual contact interval $\xi_{\min} \le \xi \le \xi_{\max}$ becomes wider with wear. This is because the load at the limits of the contact interval is null, as presented in Fig. 8.1, and therefore the wear is also null at this point. But in these contact positions, the load and wear at the inner and outer points of single tooth contact are not null, so that the delay of the driven gear increases, and contact starts earlier and ends later. According to Eqs. (8.5) and (8.6), $\xi_{\min}$ and $\xi_{\max}$ are the solutions of the equation:

$$\delta(\xi) = \delta_G(\xi) + \delta_R(\xi) + \delta_W(\xi) \tag{8.15}$$

or, approximately:

$$\delta\left(\xi_{inn/o}\right) = \delta_G\left(\xi_{\min/\max}\right) + \delta_R\left(\xi_{inn/o}\right) \tag{8.16}$$

8.4 Model for LSR and QSTE Including Wear

Wear depth after N meshing cycles can be computed with Eqs. (8.13) and (8.14). Afterwards, the LSR and QSTE can be computed with Eqs. (8.6) and (8.5). However, wear depends on the number of meshing cycles, and therefore on the load distribution along the previous cycles. To address the problem, it is assumed that the load distribution only changes every 1000 cycles. The iterative procedure will be the following:

1. Take $\delta_{W(0)}(\xi) = 0$, and $i = 0$.
2. Compute $\xi_{\min(i)}$ and $\xi_{\max(i)}$ with Eqs. (8.5) and (8.15) or (8.16).
3. Compute $\delta_{(i)}(\xi)$ with Ec. (8.5), and $F_{(i)}(\xi)$ with Eq. (8.6).
4. Compute $\Delta\delta_{W(i)}(\xi)$ with Ec. (8.13), from values of $F_{(i)}(\xi)$ of step 3.
5. Compute $\delta_{W(i+1)}(\xi)$ with Eq. (8.14).
6. If $i < N$, do $i = i + 1$ and return to step 2.

Figure 8.2 presents the LSR, QSTE, wear depth, and accumulated wear depth of spur gear pair without profile modification considering wear, after 150,000 load cycles. Figure 8.3 presents the same curves for the same spur gear with tip relief at the driven tooth. The evolution of the curves every 50,000 cycles is also presented.

It is observed that the load decreases at the inner interval of two pair tooth contact and increases at the outer one. This is due to the sliding velocity at the pinion root is greater than that at the wheel root, and therefore wear is greater at the beginning of the contact, as seen in right diagrams of Figs. 8.2 and 8.3. This produces the less worn tooth pair to support higher deflections, and consequently higher loads. The QSTE increases (obviously, the gap is longer as wear depth increases); however, the peak-to-peak amplitude of QSTE decreases, and therefore the induced dynamic load decreases as well.

In addition, since the load increases around the outer point of single tooth contact, the teeth deflection increases, and the contact starts earlier, as shown in the detail of upper-right diagram of Fig. 8.3. This means that mesh-in impact occurs, even for suitable tip relief at the driven gear teeth.

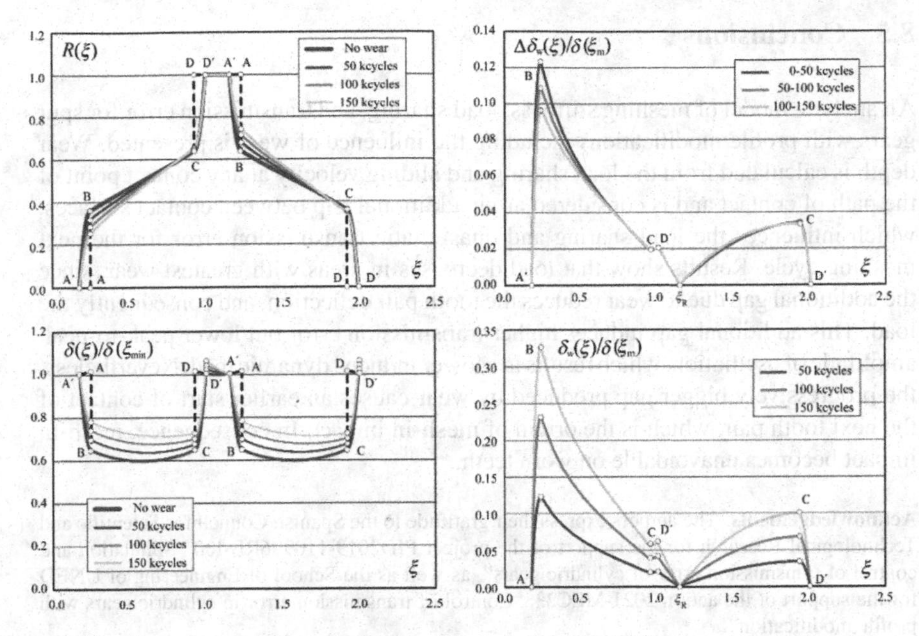

Fig. 8.2 Curves of LSR (up/left), QSTE (down/left), wear depth (up/right), and accumulated wear depth (down/right) for spur gear pair without profile modification, after 0, 50, 100, and 150 meshing kcycles

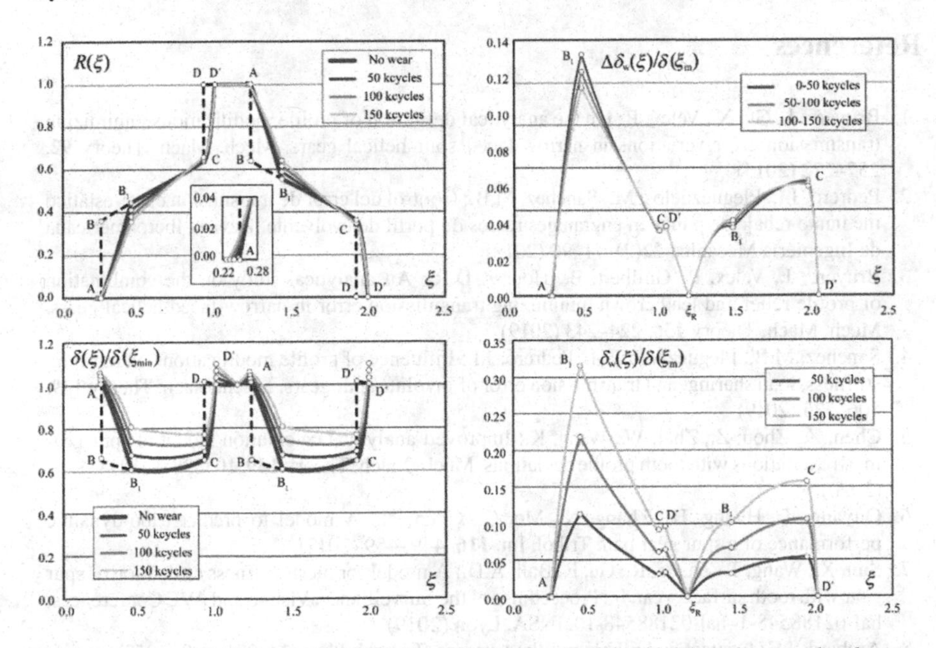

Fig. 8.3 Curves of LSR (up/left), QSTE (down/left), wear depth (up/right), and accumulated wear depth (down/right) for spur gear pair with tip relief at the driven tooth, after 0, 50, 100, and 150 meshing kcycles

8.5 Conclusions

An analytic model of meshing stiffness, load sharing, and transmission error for spur gears with profile modifications including the influence of wear is presented. Wear depth is calculated from the load sharing and sliding velocity at any contact point of the path of contact and is considered as an additional gap between contact surfaces, which influences the load sharing and quasi static transmission error for the next meshing cycle. Results show that load decreases in areas with greatest wear, since the additional gap due to wear reduces the tooth pair deflection, and consequently the load. This additional gap induces higher transmission error but lower peak-to-peak amplitude of oscillation, which results in a lower induced dynamic load. Nevertheless, the progressively bigger gap produced by wear causes an earlier start of contact of the next tooth pair, which is the origin of mesh-in impact. In consequence, mesh-in impact becomes unavoidable on worn teeth.

Acknowledgements The authors express their gratitude to the Spanish Council for Scientific and Technological Research for the support of the project PID2019-110996RB-I00 "Simulation and control of transmission error of cylindric gears", as well as the School of Engineering of UNED for the support of the action 2021-MEC23, "Control of transmission error in cylindric gears with profile modification".

References

1. Bruyere, J., Gu, X., Velex, P.: On the analytical definition of profile modifications minimizing transmission error variations in narrow-faced spur helical gears. Mech. Mach. Theory **92**, 257–272 (2015)
2. Pedrero, J.I., Pleguezuelos, M., Sánchez, M.B.: Control del error de transmisión cuasi-estático mediante rebaje de punta en engranajes rectos de perfil de evolvente. Revista Iberoamericana de Ingeniería Mecánica **22**(2), 71–90 (2018)
3. Bruyere, J., Velex, P., Guilbert, B., Houser, D.R.: An analytical study on the combination of profile relief and lead crown minimizing transmission error in narrow-faced helical gears. Mech. Mach. Theory **136**, 224–243 (2019)
4. Sánchez, M.B., Pleguezuelos, M., Pedrero, J.I.: Influence of profile modifications on meshing stiffness, load sharing, and transmission error of involute spur gears. Mech. Mach. Theory **139**, 506–525 (2019)
5. Chen, Z., Zhou, Z., Zhai, W., Wang, K.: Improved analytical calculation model of spur gear mesh excitations with tooth profile deviations. Mech. Mach. Theory **149**, 103838-1–103838-17 (2020)
6. Ouyang, T., Huang, H., Zhang, N., Mo, C., Chen, N.: A model to predict tribo-dynamic performance of a spur gear pair. Tribol. Int. **116**, 449–459 (2017)
7. Sun, X., Wang, T., Zhang, R., Gu, F., Ball, A.D.: A model for mesh stiffness evaluation of spur gear with tooth surface wear. In: Proceedings of the Surveillance, Vishno and AVE Conferences, hal-02188548-1–hal-02188548-10. INSA, Lyon (2019)
8. Archard, J.F.: Contact and rubbing of flat surfaces. J. Appl. Phys. **24**, 981–988 (1953)
9. Flodin, A., Andersson, S.: Simulation of mild wear in spur gears. Wear **207**, 16–23 (1997)
10. Flodin, A., Andersson, S.: Wear simulation of spur gears. Tribotest J. **5**(3), 225–249 (1999)

11. Xu, X., Lai, J., Lohmann, C., Tenberge, P., Weibring, M., Dong, P.: A model to predict initiation and propagation of micro-pitting on tooth flanks of spur gears. Int. J. Fatigue **122**, 106–115 (2019)
12. Chen, W., Lei, Y., Fu, Y., Hou, L.: A study of effects of tooth surface wear on time-varying mesh stiffness of external spur gear considering wear evolution process. Mech. Mach. Theory **155**, 104055-1–104055-19 (2021)
13. Pleguezuelos, M., Sánchez, M.B., Pedrero, J.I.: Control of transmission error of high contact ratio spur gears with symmetric profile modifications. Mech. Mach. Theory **149**, 103839-1–103839-16 (2020)
14. Sánchez, M.B., Pleguezuelos, M., Pedrero, J.I.: Approximate equations for the meshing stiffness and the load sharing ratio of spur gears including Hertzian effects. Mech. Mach. Theory **109**, 231–249 (2017)

Chapter 9
Service Life of Universally Loaded Deep Groove Ball Bearing Depending on Internal Clearance

Tatjana M. Lazović, Ivan M. Simonović, and Aleksandar B. Marinković

Abstract The service life of a rolling bearing is calculated using a standardized formula, based on the bearing's dynamic load rate, equivalent load, and the appropriate exponent, depending on the rolling elements' shape. Deep groove ball (DGB) bearings are supplied to the market with radial internal clearance. Clearance ranges are classified into several size groups, with values depending on the bearing bore diameter. Radial clearance causes the presence of axial clearance. DGB bearing is a bearing with initial radial contact between balls and raceways (zero contact angle). This type of bearing can be loaded with pure radial load or universally loaded with combined radial and axial load. When a radial bearing is loaded only with a radial load or a combined load with a negligibly small axial component, the contact angle retains the zero value. When the axial component of the load is significantly large, there is an axial relative displacement of the rings for the value of the axial clearance. Then the radial bearing becomes a bearing with angular contact. In the standard calculation method, the expression of the service life does not explicitly take into account the internal clearance. However, the influence of the clearance is calculated indirectly, through the corresponding values of the equivalent dynamic load's parameters. These parameters depend on the contact angle, and the contact angle depends on the internal clearance. Theoretical quantitative analysis of the internal clearance influence on the service life of a universally loaded DGB bearing is presented in this paper.

Keywords Ball bearing · Equivalent load · Internal clearance · Service life

9.1 Introduction

The last version of the international standard for the calculation of rolling bearings' dynamic load rating and rating life was published in 2007 [1]. The basic bearings' life formula has not been changed since 1962 when the standard was established.

T. M. Lazović (✉) · I. M. Simonović · A. B. Marinković
Faculty of Mechanical Engineering, University of Belgrade, Belgrade, Serbia
e-mail: tlazovic@mas.bg.ac.rs

M. Rackov et al. (eds.), *Machine and Industrial Design in Mechanical Engineering*, Mechanisms and Machine Science 109,
https://doi.org/10.1007/978-3-030-88465-9_9

Due to the identified effects of reliability, materials and technology of bearing parts, lubrication and lubricant contamination, the basic service life is adjusted by appropriate modification factors, so that the calculated modified service life is supposed to be equal to the actual service life with min. 90% reliability.

The practical experience of manufacturers and users of rolling bearings, as well as the results of scientific research, have shown that one of the most important design features of rolling bearings—internal clearance is not explicitly taken into account in calculating the bearings' service life. When a deep groove ball (DGB) bearing is subjected to a pure radial load, the balls are not equally loaded. Furthermore, with increasing internal radial clearance, the load distribution between balls becomes more unequal. The effect of internal clearance of radially loaded DGB bearing on load distribution and service life was investigated by many authors [2–6] using different approaches, but with the same conclusion that increasing clearance results in reduced bearing life.

When a DGB bearing is subjected to a universal (combined radial and axial) load, the load distribution between balls depends on the axial force. In the case of a large axial component related to the radial one, all balls are engaged in the load transfer. The load distribution is less unequal and the bearing service life is longer. So the increase in the internal clearance causes an increase in the service life. The question arises, how to take into account the influence of the internal clearance in the calculation of the bearing's service life and what is the level of this impact. The explanation is given through a change of the contact angle between balls and raceways under universal loading. The relation between the internal clearance, contact angle and equivalent bearing load, based on standard expressions, is obtained in this paper. Based on the derived function, the analysis of the internal clearance influence on the ball bearing's service life was carried out.

9.2 Internal Clearance and Initial Contact Angle

The DGB bearings of all series (with the same bore diameter) can have different internal radial clearances, depending on the user's needs. The clearance ranges with which the bearings are delivered to the market are classified into several size groups (CN, C2, C3, C4 and C5) [7]. These are the values of the initial internal radial clearance before mounting. Usually, the operating internal clearance in a bearing is smaller than its initial internal clearance, due to mounting (interference fits between inner ring and shaft or between outer ring and housing or both) and rising operating temperature (thermal expansion of the bearing components and other components of the bearing arrangement). The operating internal clearance is the internal clearance when the mounted bearing is in operation and has reached an operating temperature. It is a result of thermal effects in the fitted condition [8].

The radial clearance induces and determines the axial clearance. Due to the presence of an axial clearance and large axial component of operational load, the radial ball bearing can be considered as a ball bearing with angular contact (Fig. 9.1).

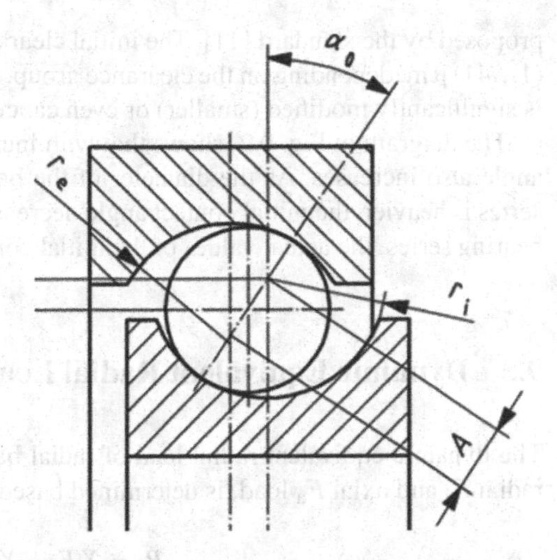

Fig. 9.1 Initial contact angle in DGB bearing caused by internal clearance [9]

The initial contact angle depends on internal clearance and dimensions of balls and raceways grooves [9]:

$$\cos \alpha_0 = 1 - \frac{s}{2(r_i + r_e - D_w)} = 1 - \frac{s}{0.08 D_w} \tag{9.1}$$

where s is radial operating clearance; D_w is ball diameter; $r_i = r_e = 0.52D_w$ [10] are cross-sectional raceway groove radii of the inner and outer ring, respectively (Fig. 9.1).

The diagram in Fig. 9.2 shows the change of the initial contact angle depending on the clearance and ball diameter of the DGB bearings of the same bore diameter ($d = 30$ mm) and different series (60, 62, 63 and 64). The ball diameters are

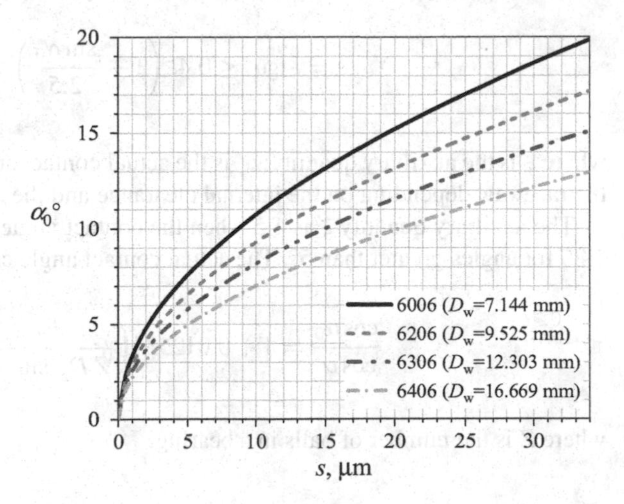

Fig. 9.2 Initial contact angle versus operating internal radial clearance

proposed by the standard [11]. The initial clearance of these bearings is in the range (1…41) μm, depending on the clearance group. However, actual operating clearance is significantly modified (smaller) or even cancelled due to fitting and heating [8].

The diagram in Fig. 9.2, shows that with increasing clearance, the initial contact angle also increases. As the diameter of the balls increases, which means that the series is heavier, the initial contact angle decreases. Depending on the clearance and bearing series, the actual values of the initial contact angle can be approx. up to 20°.

9.3 Dynamic Equivalent Radial Load

The dynamic equivalent radial load of radial ball bearing, loaded with a combined radial F_r and axial F_a load, is determined based on expression:

$$P_r = XF_r + YF_a \tag{9.2}$$

where X is radial load factor and Y is axial load factor.

The load distribution in the ball bearing depends on the F_a/F_r ratio. If the axial load is zero or negligibly small in relation to the radial force, not all balls participate in the load transfer. As the axial component of the load increases, the number of active balls, which come into contact with the raceways is increasing. Also, the contact angle is increased. With a further increase in the F_a/F_r ratio, all the balls participate in the load transfer, and the radial bearing became a bearing with an angular contact. The radial and axial force participation in the equivalent bearing load is determined by the load factors X and Y. The values of these factors are determined by F_a/F_r ratio and appropriate parameter e. This parameter is the limiting value of F_a/F_r for the applicability of different values of load factors and the standard expression for this parameter is [10]:

$$e = \xi \mathrm{tg}\alpha' \leq 0.4\xi \left(1 - \frac{\sin \alpha_0}{2.5}\right)^{-1} \tag{9.3}$$

where ξ is the auxiliary quantity; α' is the actual contact angle between the balls and the raceway, depending on the internal clearance and the axial load.

The auxiliary quantity ξ is 1.5 when the contact angle is less or equal to 5° and 1.25 for angles greater than 5°. The actual contact angle can be determined by [10]:

$$\frac{\cos \alpha_0}{\cos \alpha'} = 1 + 0.012534 \left(\frac{F_a}{ZD_w^2 \sin \alpha'}\right)^{\frac{2}{3}} \tag{9.4}$$

where Z is the number of balls in a bearing.

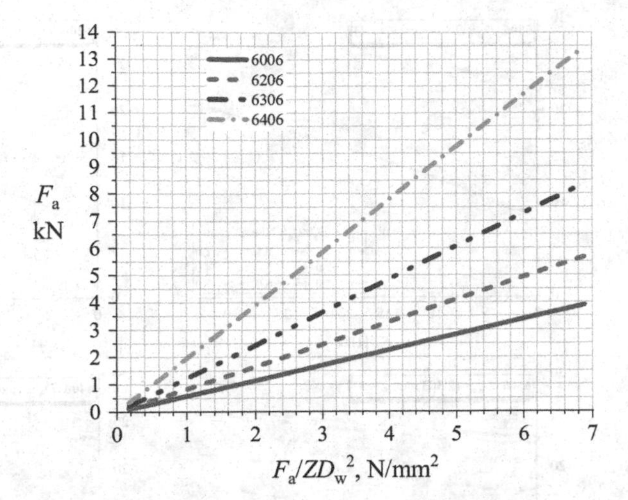

Fig. 9.3 Axial load versus load–geometry ratio for DGB bearings with bore diameter 30 mm

In Eqs. (9.3) and (9.4), if $\alpha^\circ \leq 5^\circ$, then in calculations should be used $\alpha^\circ = 5^\circ$ [10]. To have comparable results for different bearing series, one of the variables in the following analyzes is load–geometry ratio F_a/ZD_w^2 (Eq. 9.4). This ratio is in the range (0.1…7.0) for ball bearings [10]. The change of axial component of the bearing load depending on the load–geometry ratio, for different series of DGB bearings with a bore diameter of 30 mm, is given in the diagram in Fig. 9.3.

The diagrams of actual contact angle α', determined for different clearances and ratio F_a/ZD_w^2, are shown in Fig. 9.4. As the clearance and the axial component of the operating load increase, the actual contact angle also increases. Under conditions of extremely large clearances and axial loads, the actual contact angle can reach values of up to approx. 25°…30°. The influence of the clearance on the contact angle is greater in the case of lighter series of bearings. When $F_a/F_r \leq e$ (Eq. 9.3), then $X = 1$, $Y = 0$ and the dynamic equivalent radial load of the bearing is $P_r = F_r$ according to Eq. 9.2. When $F_a/F_r > e$, then factors X and Y are determined based on the following expressions [10]:

$$X = 1 - 0.4\xi\left(1 - \frac{\sin\alpha_0}{2.5}\right)^{-1}; \quad Y = 0.4\cot\alpha'\left(1 - \frac{\sin\alpha_0}{2.5}\right)^{-1} \geq 1 \qquad (9.5)$$

In Eqs. (9.5), if $\alpha^\circ \leq 5^\circ$, then in calculations should be used $\alpha^\circ = 5^\circ$ [10]. The dependence of the auxiliary parameter e, as well as the load factors X and Y on the radial clearance for DGB bearing 6206 used in this study as an example, is shown in Fig. 9.5. The load factor X does not show a significant dependence on the clearance. Its values are between 0.44 and 0.56. This means that in the case of a universally loaded radial bearing, the radial load component participates with approx. half of its value. The load factor Y cannot be less than 1, which corresponds to large clearances and large values of the axial load component. Under these conditions, almost half of the radial load component and the entire axial component are included in the

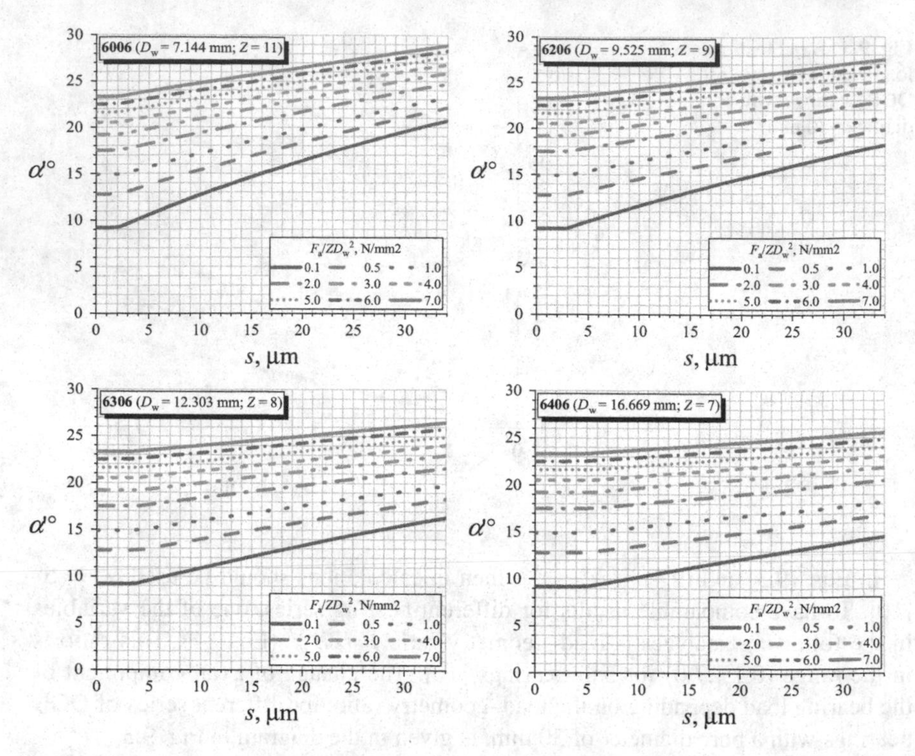

Fig. 9.4 Actual contact angle versus operating internal clearance

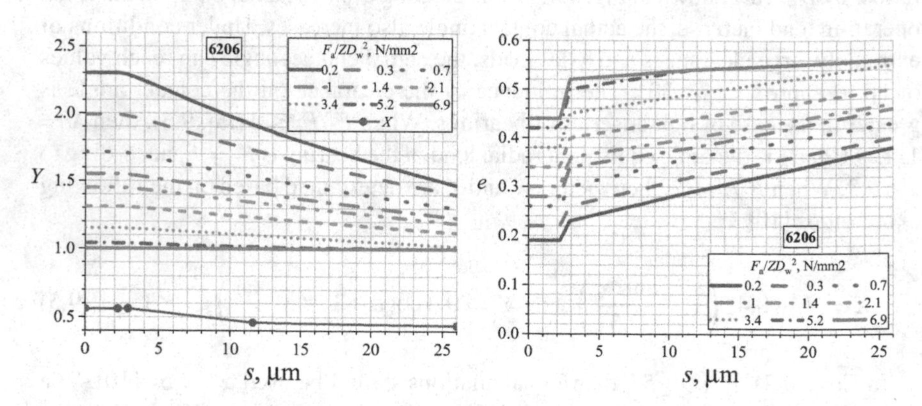

Fig. 9.5 Load factors X, Y and auxiliary parameter e versus operating internal clearance

dynamic equivalent load P_r. However, in cases of smaller clearances and axial load, the axial load factor can have values of $1 < Y < 2.3$ (Fig. 9.5), which means that the contribution of axial force in the equivalent load can be multiple.

9.4 Service Life

The standard formula for the basic life rating of a radial ball bearing is [1, 9, 10]:

$$L = \left(\frac{C_r}{P_r}\right)^3 = \left(\frac{1.3 f_c (1 - \frac{s}{0.08 D_w})^{0.7} Z^{\frac{2}{3}} D_w^{1.8}}{X F_r + Y F_a}\right)^3 10^6 \text{revolutions} \quad (9.6)$$

where C_r is dynamic load rating of the bearing, f_c is a factor that depends on bearing geometry parameter $D_w \cos\alpha_0 / D_{pw}$ [1], D_{pw} is pitch diameter of the bearing's ball set.

The influence of the clearance on the service life was analyzed on the example of the DGB bearing 6206 ($D_w = 9.525$ mm; $D_{pw} = 46$ mm; $Z = 9$). Four loading cases are considered, so that the values of load-geometry ratio are $F_a/ZD_w^2 = \{0.3, 1.0, 2.1, 6.9\}$. The radial and axial components of the load are chosen so that their ratio is $F_a/F_r = 0.6 > e$. The load factors X and Y are obtained from Eq. 9.5, i.e. from diagram in Fig. 9.5. The calculated dynamic equivalent load and service life are given in Table 9.1

The ratio of service life of the bearing with some positive internal clearance to the service life in the case of zero-clearance can be named as clearance life factor:

$$a_s = \frac{L_s}{L_{s=0}} \quad (9.7)$$

A diagram of the clearance life factor of DGB bearing 6206, depending on actual operational internal clearance and load–geometry ratio is shown in Fig. 9.6. The functions show, that the bearing's service life is increasing with an increase of internal clearance and a decrease in the axial load–geometry ratio. The clearance life factor a_s can reach the maximum value of approx 2.5, under conditions of large clearance (s

Table 9.1 Bearing load and service life depending on internal clearance and load-geometry ratio

6206		F_a/ZD_w^2							
		0.3 N/mm²		1.0 N/mm²		2.1 N/mm²		6.9 N/mm²	
F_a, kN		0.25		0.82		1.71		5.63	
F_r, kN		0.41		1.36		2.85		9.38	
F_a/F_r		0.6		0.6		0.6		0.6	
α_0	s µm	P_r kN	L 10^6 rev	P_r kN	L 10^6 rev	P_r kN	L 10^6 rev	P_r kN	L 10^6 rev
0°	2.2	0.73	18,768.8	2.03	872.8	3.84	128.9	10.88	5.7
5°	2.9	0.73	18,768.8	2.03	872.8	3.84	128.9	10.88	5.7
10°	11.6	0.62	30,164.6	1.78	1274.7	3.41	181.3	9.94	7.0
15°	26	0.53	46,071.6	1.61	1643,6	3.17	215.3	9.76	7.4

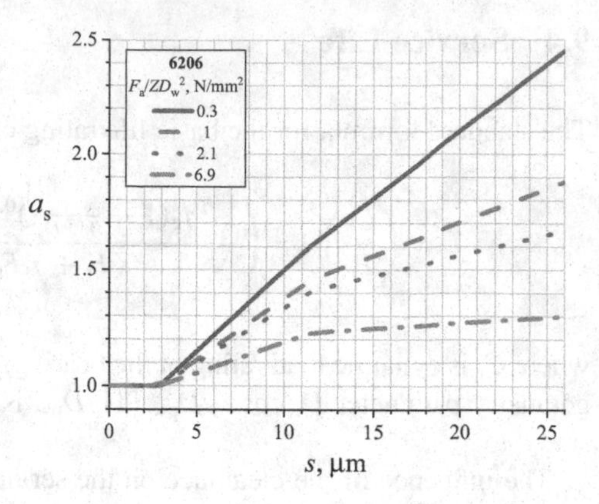

Fig. 9.6 Clearance life factor depending on actual operating clearance and load–geometry ratio

= 26 μm) and light axial load ($F_a/ZD_w^2 = 0.3$). Under the small values of the axial component of load, the bearing is more sensitive to change of service life depending on the internal clearance.

9.5 Conclusion

In the universally loaded DGB bearing, there is an axial relative displacement of the rings due to the axial load component and the presence of an operational internal clearance (after fitting and heating under actual operational temperature). Consequently, the radial ball bearing becomes a ball bearing with angular contact. Because of that, the load distribution between rolling elements is more favourable, load carrying capacity is larger, and the service life is longer. A numerical example carried out in this paper has shown that depending on the combination of the clearance and the intensity of the axial load, the service life of the universally loaded DGB bearing with internal radial clearance can be increased by more than two times in relation to the service life of the same bearing with zero clearance. In further theoretical research, a general model of the clearance life factor should be developed, including the influence of operational internal radial clearance, axial load-geometry ratio and the ratio between axial and radial load on the bearing's service life.

Acknowledgements This work was supported by the Ministry of Education, Science and Technological Development of the Republic of Serbia (Contract 451-03-9/2021-14/200105).

References

1. International Standard ISO 281: Rolling bearings—dynamic load ratings and rating life (2007)

2. Oswald, F., Zaretsky, E., Poplawski, J.: Effect of internal clearance on load distribution and life of radially loaded ball and roller bearings. Tribol. T. **55**(2), 245–265 (2012)
3. Ambrożkiewicz, B., Arkadiusz, S., Nicolas, M., Grzegorz, L., Anthimos, G.: Radial internal clearance analysis in ball bearings. Eksploat. Niezawodn. **23**(1), 42–54 (2021)
4. Sahoo, V., Sinha, R.: Effect of relative movement between bearing races on load distribution on ball bearings. SN Appl. Sci. **2**, 2100-1–2100-12 (2020)
5. Lazović, T.: Influence of internal radial clearance of rolling bearing on load distribution between rolling elements. J. Mech. Eng. Des. **4**(1), 25–32 (2001)
6. Lazović, T., Mitrović, R., Ristivojević, M.: Influence of internal radial clearance on the ball bearing service life. J. Balk. Tribol. Assoc. **16**(1), 1–8 (2010)
7. International Standard ISO 5753-1: Rolling bearings—internal clearance—Part 1: radial internal clearance for radial bearings (2009)
8. Ristivojevic, M., Mitrovic, R., Lazovic, T.: Investigation of causes of fan shaft failure. Eng. Fail. Anal. **17**, 1188–1194 (2010)
9. Technical Specification ISO/TS 16281: Rolling bearings—Methods for calculating the modified reference rating life for universally loaded bearings (2008)
10. Technical Report ISO/TR 1281-1: Rolling bearings—explanatory notes on ISO 281—Part1: basic dynamic load rating and basic rating life (2008)
11. International Standard ISO 3290-1: Rolling bearings—balls—Part 1: steel balls (2014)

Chapter 10
Investigation of Low to Medium Load Operating Regimes for Roller Bearings in Heavy Industrial Equipment

Radomir Karakolev and Lubomir Dimitrov

Abstract Large spherical roller bearings are typically used in heavy low speed industrial applications. This sort of sophisticated bearings has achieved nowadays too high dynamic capacity, thanks to the latest modern bearing production technologies and the usage of highest quality bearing steels. Heavy industrial machines from chemical, mining or cement plants are using such type of bearings. Modern industrial plants require higher reliability of their rotating equipment thus to avoid costly downtimes and expensive repairs. However, some heavily loaded applications still struggle of outages. For the bearing's selection process is typically used C [kN] basic dynamic load rating, but one additional catalogue parameter looks very important—P_u [kN] fatigue load limit. The ISO 281 standard does not recommend one bearing simply to be selected just on that P_u value, rather advices more investigations to be done how P_u could correlate to specified working conditions. In this case study will be considered relevant raceway contact friction from rolling and micro-sliding, and lubrication regimes, for low to medium bearing loadings of a rotary tube furnace (kiln) support roller.

Keywords Rolling bearings · Friction · Support roller

10.1 Introduction

The fatigue load limit value P_u [kN] is relevant to bearing material. It is defined by the bearing producers in order to be able to modify the calculated bearings lifetime L_{10} according to the ISO 281:2007 by an introduced stress life modification factor a_{iso} [1]. ISO/TS 16281:2008 describes more advanced calculation method using also factor a_{iso} [2]. In this way the basic rating life methodology from R281:1947 created by Lundberg-Palmgren considering sub-surface bearing steel fatigue [3], is adjusted

R. Karakolev (✉) · L. Dimitrov
Technical University of Sofia, Sofia, Bulgaria

L. Dimitrov
e-mail: lubomir_dimitrov@tu-sofia.bg

M. Rackov et al. (eds.), *Machine and Industrial Design in Mechanical Engineering*, Mechanisms and Machine Science 109,
https://doi.org/10.1007/978-3-030-88465-9_10

much closer to the real working conditions: bearing rings tilting, clearance, rolling contact surface events—lubrication film thickness, and particles contamination.

The P_u [kN] is specified from maximum Hertzian contact pressure for the most loaded roller element $\sigma_u = 1500$ MPa, and for orthogonal sub-surface shear stress τ_u around 350 MPa. When a bearing operates under these stresses, in theory, the fatigue can be avoided if the lubricant film completely separates the rolling elements from the raceways, and surface distresses are excluded [1].

It is under question the bearing SKF 24168 that has a ratio $C/P_u = 8.7$ (Table 10.1), usually used for heavy loaded and low speed applications, could rotate satisfactory on a working condition between *medium* ($C/P = 8$) to *low* ($C/P = 15$) modes as shown on [4, see pp. 112–113], thus to function below fatigue load limit P_u [kN].

For some projects, the load P [kN] can be modified and based on the preferred bearing with specific C [kN], a good balance can be found between the load regime, raceway contact stresses, fatigue life, friction losses and lubrication conditions, which for heavy industrial machines such as in chemical, mining or cement plants will result in reliable and smooth continuous rotation. The practice shows that in some cases, spherical roller bearings operate for no more than 4–7 years, despite their significant design capacity and calculated life. Inspections of dismantled bearings observe significantly more contact surface wear like pitting or corrosion than spalling from normal sub-surface fatigue. However, are there opportunities to achieve a significant resource of production time, reduced: bearing wear; lubrication consumption, and energy losses through low friction in the bearings? This can be discussed based on the computational results with SKF SimPro Expert 4.4 software [5].

In the (Fig. 10.1) with a circle is shown the area of operation of roller bearings in medium to low mode, which considers a larger factor a_{skf} (a_{iso}), leading to 5–20 times theoretically increased bearing life, but still as a disadvantage should be noted its significant combined sensitivity to changes in lubrication regimes, operating cleanliness, and load (acc. to ratio P_u/P) seen from the steep curves of the graph. For the durability of the bearings, in such medium to low operating regimes, the phenomena in the surface contact (raceways and rolling elements) layers will be of interest, and not so much in the sub-surface.

10.2 Friction in Rolling Bearings and Its Sources

Most often, contact surface wear in rolling bearings is a result of an adhesion, a relative sliding motion or through the intervention of a third body—abrasive particles. All of these have a contribution to the rolling friction. Therefore, the relationship of the *bearing friction moment* and the *load* will be considered to find their optimal acceptable values. The rolling bearings frictional *moment* cannot be eliminated, at least it can be evaluated by one of its destructive components—friction by *micro-slip/slide/*. In practice, the friction of rolling elements and micro-sliding motion cannot be completely avoided even in a pure rolling condition [6]. The sliding M_{sl} friction moment, in fact, arises from the osculation (the ratio of the ball/rolls/ radius

Table 10.1 Dynamic load capacity C [kN] and Fatigue load limit values Pu [kN] for 24168 [4]

Principal dimensions			Basic load ratings		Fatique load limit	Speed ratings		Mass	Designations
			Dynamic	Static		Reference speed	Limiting speed		Bearing with cylindrical bore
d	D	B	C	C_0	P_u				
mm			kN			r/min		kg	–
340	460	90	1490	2800	216	1300	1400	45.5	23968 CC/W33
	520	133	2812	4550	335	1000	1300	105	23068 CC/W33
	520	180	3621	6200	475	750	1100	140	24068 CC/W33
	580	190	4445	6800	480	800	1000	210	23168 CC/W33
	580	190	4452	6800	490	–	240	210	23168-2CS5/VT143
	580	243	5487	8650	630	430	630	280	24168 ECCJ/W33

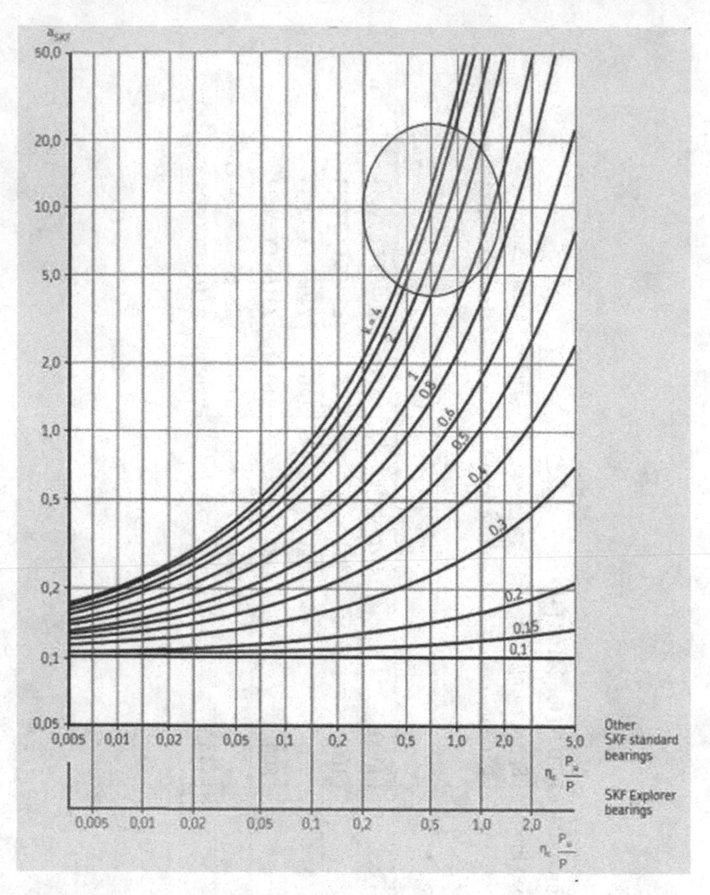

Fig. 10.1 Stress factor a_{skf} (a_{iso}) for roller bearings considering, equivalent load P, fatigue load limit P_u, cleanliness η_c, and lubrication regimes κ [4]

to the raceway radius) and elastic contact deformations between the rolling elements and the raceways. An estimate of the applied load will be made against rating life and friction losses. In this case, the *micro-slip/slide/* M_{sl} [Nm], the Slide [Nm]/Roll [Nm] ratio, as well as the power losses created by friction Power loss [W] are considered. A similar study was conducted and described in the publication [7], but with a focus on high-speed rotating machines like electrical motors in contrast to the specific case here, with components in slow-running units.

10.3 Experimental Data

The studied industrial application is a support roller (Fig. 2a–c) from a rotating tube furnace (kiln) located in a chemical plant.

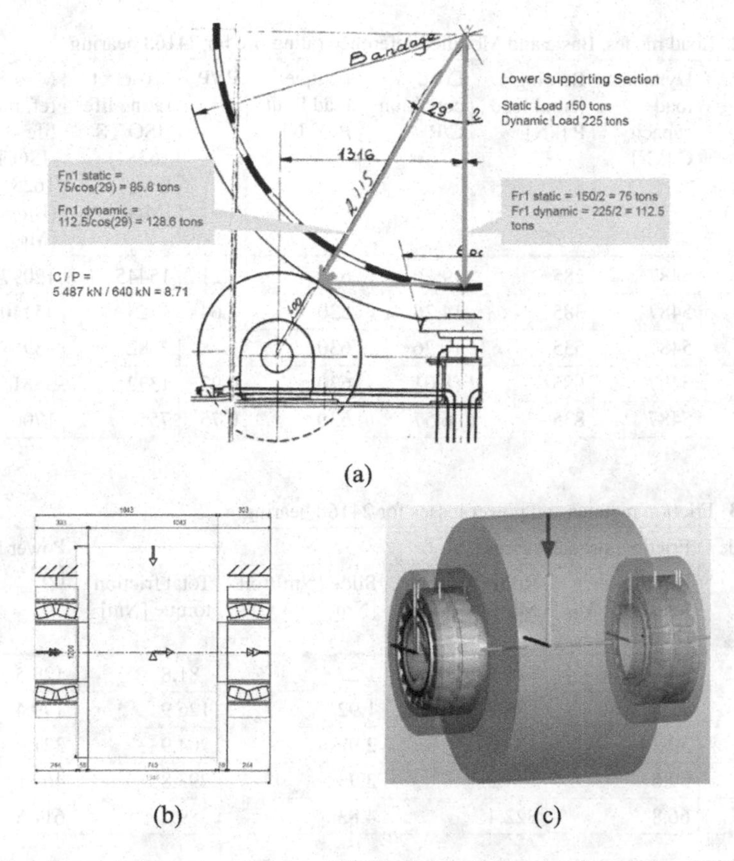

Fig. 10.2 **a** Support roller—general view with local normal dynamic force F_{n1} = 128.6 tons, **b** Support roller dimensions and **c** Support roller model with 2 × 24168 bearings

Only one of the two bearings of the same type are considered and the presence of a uniform radial load (w/o axial) of the two bearings is assumed. The operating temperature 50 °C (estimated) and speed of 15 rpm are constant for all 5 load scenarios /modes/. (Static_3) from (Tables 10.2 and 10.3) is close to the existing designed load of the support roller, which can be said being compact designed based on the ratio C/P = 8. This load falls into the category *moderate* [4, page 116, Table 5]. From the study, optimal results are obtained at (Static_2) case (Table 10.2) with a *low to medium* mode of an operation and ratio C/P = 10.26, and P_u/P = 1.18 where the resulting friction from local *micro-slip /slide/*M_{sl} (Table 10.3) has lower value than original design. For the friction and life calculations, the grease base oil viscosity is the same due to the constant temperature for all modes. The rolling resistance M_{rr} [Nm] is relatively independent of the load, but rather is influenced by the lubrication regime and the rotational speed, that are kept stable for the case study (Table 10.2).

(Static_0 and _1) are *low* load modes of the bearing that are also acceptable by delivering lower friction and power loses. From the results in (Table 10.3) it can

Table 10.2 Load modes, Basic and Modified reference rating life for 24168 bearing

Load mode	Dynamic load capacity C [kN]	Bearing radial load P [kN]	Load condition C/P	Fatique load limit P_u [kN]	P_u/P	Basic ref. rating life (ISO/TS 16281) (L_{10r}) [M_{Rev}]	Modified ref. rating life (ISO/TS 16281) (L_{10mr}) [M_{Rev}]
Static_0	5487	285	19.26	630	2.21	15445	120970
Static_1	5487	385	14.26	630	1.64	6723	35440
Static_2	5487	535	10.26	630	1.18	2682	9596
Static_3	5487	685	8.01	630	0.92	1332	3661
Static_4	5487	835	6.57	630	0.75	755	1706

Table 10.3 Friction regimes and power losses for 24168 bearing

Load mode	Friction torque sources				Power loss [W]
	Rolling resistance M_{rr} [Nm]	Rolling sliding M_{sl} [Nm]	Slide [Nm]/roll [Nm]	Total friction torque [Nm]	
Static_0	36	45.8	1.27	81.8	128.5
Static_1	43.5	83.5	1.92	126.9	199.4
Static_2	52.3	152.6	2.92	204.9	321.9
Static_3	59.6	234.2	3.93	293.8	461.5
Static_4	66.8	322.4	4.83	389.2	611.3

be seen that change in the load has a significant nonlinear effect on the sliding M_{sl} friction moment, while M_{rr} changes insignificantly. When comparing the sliding friction moments for different load modes, it is noticeable that M_{sl} dominates within the total friction moment M, and hence the frictional power Power loss [W]. (Static_4) is a much heavier load mode and the long-term operation of the bearings within it would not be stable due to the need for an additional cooling. It is allowed for short-term overloads or transient modes.

Comparatively, in (Table 10.2) are shown, the fatigue load limit P_u [kN] relative to the working loads P—low modes are (Static_0, _1) with $P_u/P = 2.21$, and resp. 1.64; design load of (Static_3) with $P_u/P = 0.92$, and heavier load mode (Static_4) with $P_u/P = 0.75$.

Slide/Roll friction ratio (Table 10.3) are with optimal values for (Static_2) with value 2.92, and (Static_1) with 1.92 reflecting lower potential for surface wear, than the heavier modes with 3.93, and 4.83. One of the evaluation criteria is undoubtedly the obtained increased Modified reference rating life (ISO_TS16281) (L_{10mr}) [M_{Rev}]. ISO 16281 [2] is used as a more advanced than ISO 281 [1]. (Static_2) mode achieves 2.6 times higher value of rating life than (Static_3). This is due to the power factor

10/3 from the life formula from ISO 281. For a comparison, a column with values for (ISO_TS16281) (L_{10r}) [M_{Rev}] has been added to the table, which shows bearing life without considering the positive effect of lubrication, but only the subsurface fatigue of the metal rings. The contact pressures in (Static_2) from (Fig. 10.3) have a completely acceptable level for the most loaded rolling elements—rollers numbers 14, and 15 from each set, which may lead to long sub-surface fatigue resistance: stresses are $\tau = 350$ MPa, and ($\sigma = 1400$ MPa) < ($\sigma_u = 1500$ MPa).

On (Fig. 10.4) the most loaded rollers are those with numbers 13–17, which fall into the narrow strip zones with a relative IR *velocity slip* of (±0.01) Rolling in these local areas can cause severe damage to the contact surfaces if a sufficiently stable oil layer with $\lambda > 1.2$ (mixed lubrication) is not provided [8]. In this case,

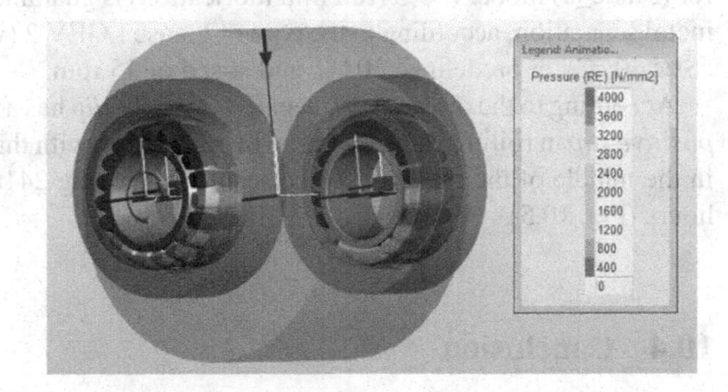

Fig. 10.3 Contact pressures for the low-to-moderate mode (static_2)

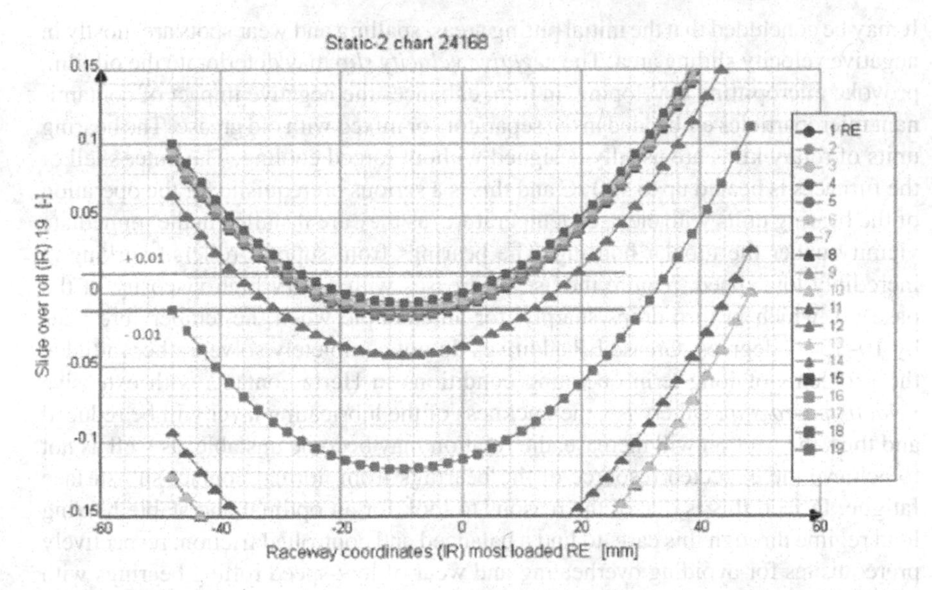

Fig. 10.4 Sliding over roll velocity conditions for all bearing rollers 1–19 on the inner ring (IR)

Fig. 10.5 Bearing 24168 with predominant wear within the negative velocity slip zone

for (Static_2) mode, $\lambda = 3$ (full film lubrication) is guaranteeing complete metal-to-metal separation, according to the type of grease LGEV 2 (viscosity of base oil 1020 cSt@40 °C), at predefined 50 °C, and speed of 15 rpm.

According to the research [9], negative *velocity slip* has a more negative effect than *positive slip* in rolling Hertz contacts. In connection with this, a specific contact zone in the middle of the raceway can be seen on a bearing 24168 after 37500 operating hours (Fig. 10.5).

10.4 Conclusion

It may be concluded that the initial pitting areas, spalling and wear spots are mostly in negative velocity sliding area. The *negative velocity slip* may deteriorate the oil film, provoke micropitting developing, in turn, enhances the negative impact of contaminant microparticles embedded in the separators or mixed with old grease. The bearing units of rotary kilns are usually designed without forced cooling. The outer shell of the furnaces is heated up to 400 °C and this is a serious prerequisite for the operation of the bearing units with elevated temperature, as they are installed in the immediate vicinity under the tubular housing. The bearings from support rollers, working at incredibly low speed, require the use of greases with a very high viscosity of the base oil, which in turn drops sharply, for an example when the temperature rises by 10–15 °C degrees. Grease EP additives do not completely solve, rather mitigate the problems of long-term operating conditions in Hertz contacts with extensive *velocity micro-slip*. Otherwise, the thickness of the lubricating layer will be reduced and thus, the friction will increase, the rotation may become unstable, as well as not to achieve the expected resource of the bearings from normal contact sub-surface fatigue. In fact, this is one of the reasons to look for an optimal and stable bearing load regime through this case, to find a balanced and controlled friction, respectively prerequisites for avoiding overheating and wear of low-speed rolling bearings with large dimensions.

To facilitate the design work, it could be recommended, for a given kinematic and dimensional parameters of a shaft system for heavy slow-moving equipment, to optimize the structure and supporting roller locations with a choice of rolling bearing loads P [kN], below the fatigue limit P_u [kN], e.g. $P_u/P = 1.1–1.6$ times.

Than for the specified size of rolling bearing, needs to be selected in details the lubrication regime and special lubricant type, thus reducing the wear of the contact surfaces and making a significant contribution to the long-term, reliable, profitable, and energy-efficient operation of heavy industrial equipment.

References

1. ISO 281:2007: Rolling bearings—dynamic load ratings and rating life (2007)
2. ISO/TS16281:2008: Rolling bearings—methods for calculating the modified reference rating life for universally loaded bearings (2008)
3. Lundberg, G., Palmgren, A.: Dynamic capacity of rolling bearings. J. Appl. Mech. **16**(2), 165–172 (1949)
4. SKF Bearing Catalogue PUB17000. Sweden (2018)
5. SKF SimPro Expert 4.4 software user manual (2020)
6. Morales-Espejel, G.E.: Wear and surface fatigue in rolling bearings. SKF (2019).
7. Karakolev, R., Dimitrov, L.: Analysis of electrical motor mechanical failures due to bearings. IOP Conf. Ser.: Mater. Sci. Eng. **393**, 012064–1–012064–6 (2018)
8. Morales, G.: Micro-pitting modelling in rolling–sliding contacts: application to rolling bearings. Tribol. T. **54**(4), 625–643 (2011)
9. Everitt, C.: The influence of gear surface roughness on rolling contact fatigue under thermal elasto-hydrodynamic lubrication with slip. Tribol. Int. **151**, 106394-1–106394-19 (2020)

Chapter 11
Performance of Water Lubricated Journal Bearings Under Elastic Contact Conditions

Juliana Javorova and Alexandru Radulescu

Abstract Hydrodynamic journal bearings are one of the most popular bearings used in many rotating machines. It is known that in some applications including underwater machinery and environmentally friendly applications, water lubricated bearings have become increasingly used. The purpose of this study is to carry out theoretical performance analysis of hydrodynamic journal bearings lubrication using water as a lubricating fluid, taking into account the elastic deformation of the bearing stationary surface. The bearing bush is covered with a very thin resilient liner whose radial distortions are of the same order of magnitude as the fluid film thickness. The elastic deformation of the bush layer is estimated using the Winkler model of an elastic foundation. The finite difference method is adopted for the simulation of the characteristics of water lubricated bearing. The lubrication performance is studied with respect to materials parameters and structural parameters. The comparison results between bearings with and without consideration of the effects of the elastic deformations, at various eccentricity ratios and aspect ratios, are presented.

Keywords Journal bearings · Water lubrication · Hydrodynamic characteristics · Elastic deformations

11.1 Introduction

Journal bearings are used widely and successfully for hundreds of years and probably they are one of the most used machine element in our civilization.

Along with good performance bearings lubricated with oil and grease, nowadays an increase in the use of water-lubricated bearings is observed due to continuous

J. Javorova (✉)
University of Chemical Technology and Metallurgy, 1756 Sofia, Bulgaria
e-mail: july@uctm.edu

A. Radulescu
University Politehnica, 060042 Bucharest, Romania

M. Rackov et al. (eds.), *Machine and Industrial Design in Mechanical Engineering*, Mechanisms and Machine Science 109,
https://doi.org/10.1007/978-3-030-88465-9_11

environmental concerns. This kind of bearings has found an application in shipbuilding, industrial machinery and equipment, transportation industry and others [1–4]. The researches in the field of water-lubricated bearings are focused on different directions [5–7], as only few of them have addressed the influence of the elastic deformations on bearing lubrication performance. In order to show this effect, in the paper is presented the performance analysis of water lubricated journal bearings in consideration of deformability of the elastic bearing layer.

11.2 Analysis

The configuration of the considered water lubricated plain bearing under steady operation performance is shown in Fig. 11.1. The bearing is fully submerged in water. It is assumed that the journal and bearing are circular and their surfaces are smooth, the constant external load is applied in a vertical direction, the groove is filled with a water lubricant with constant properties, and the journal rotates with a constant angular velocity about its axis.

Since the viscosity of water is much less than typical oils and the order of the lubricating film thickness is of micrometer, the Reynolds number of journal bearings is commonly small when the velocity is not extremely high [8–10]. On that base, in the current paper the laminar flow for the numerical analysis will be considered.

As a whole the process of hydrodynamic (HD) lubrication in the journal bearing can be presented shortly by following manner [1, 8]: At a normal HD lubrication in the bearing the journal rotates with an angular velocity under the vertical load and finally arrives at an equilibrium position. Hydrodynamic lubrication process

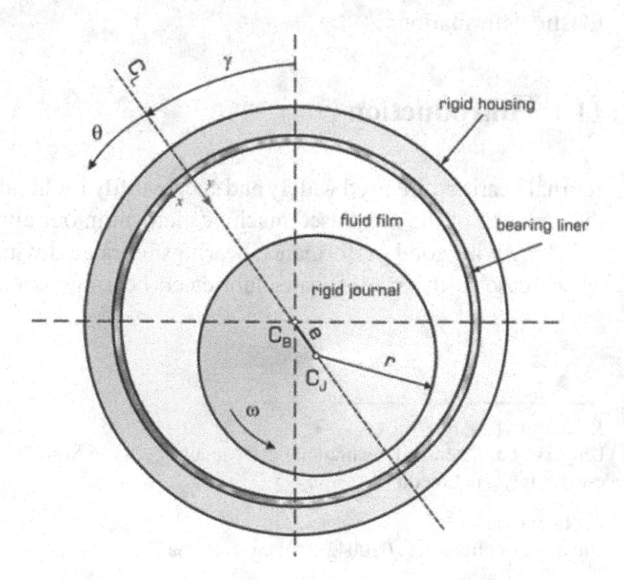

Fig. 11.1 Configuration of the journal bearing

within a thin film of water fluid leads to generation of dynamic pressure, mainly in the convergent zone of the bearing gap, in order to acts against the load there via separation both surfaces (of the journal and of the bearing). In the divergent zone of the gap the generated HD pressure terminates, because there the pressure may drop below the value of water-cavitation pressure. The eccentricity ratio, based on the journal eccentricity at the equilibrium position, has an important influence on hydrodynamic performances of journal bearings with water as lubricant.

11.2.1 Reynolds Equation

The pressure generation in water-lubricated finite length journal bearing is assumed to be governed by the Reynolds equation, presented in Cartesian coordinates, as follows:

$$\frac{\partial}{\partial x}\left(\frac{h^3}{\eta}\frac{\partial p}{\partial x}\right)+\frac{\partial}{\partial z}\left(\frac{h^3}{\eta}\frac{\partial p}{\partial z}\right)=6\omega r\frac{\partial h}{\partial x} \tag{11.1}$$

Here p is hydrodynamic pressure, h – fluid film thickness, η – fluid dynamic viscosity, ω – angular velocity, r – journal radius, x, z – Cartesian rectangular coordinates.

11.2.2 Fluid Film Thickness and Elasticity Equation

The required hydrodynamic pressure of the bearing generates in the lubricant film between the journal and the bush. By reason of this, along with the hydrodynamic pressure, the fluid film thickness is one of the most important parameters in the Reynolds equation. Therefore, the fluid film thickness equation is needed to be carefully considered in the analysis.

It is assumed that the pressure generated in the fluid film acts normally on the bearing surfaces. All tangential forces to these surfaces are neglected. The present analysis will take into consideration the elastic deflection of the bush liner. The other components of the bearing and the journal will be treated as rigid.

The approach in a present study aims to superimpose the deformation of bush layer (caused by hydrodynamic pressure generated) onto the fluid film thickness. The gap thickness is then modified to account for the estimated elastic deformation /represented by the last term of equation/ as follows:

$$h(x,z)=c+e\cos\theta+\delta \tag{11.2}$$

Determination of the radial displacements of the liner's surface points is carried out in accordance to the plain strain hypothesis (column or Winkler model) [11–14, etc.]:

$$\delta = \frac{(1+\mu)(1-2\mu)}{(1-\mu)} \frac{d}{E} p \tag{11.3}$$

where μ is Poisson's ratio, E – Young's modulus, d – bearing liner thickness.

In conformity with [11, 12] the column or Winkler model is applicable for the cases when lining is thin compared to the dimensions of the bearing. The applicability of this formula for mechanical deformations is also verified by comparison with the results obtained using a full deformation model.

11.2.3 Non-dimensional Form of Equations

For the numerical solution the above Eqs. (11.1)–(11.3) need to be represented in dimensionless form by introducing the appropriate coordinate transformations and the corresponding non-dimensional variables according to the following substitutions:

$$\theta = \frac{x}{R} \approx \frac{x}{r}, \quad z_1 = \frac{z}{L/2}, \quad \alpha = \frac{2r}{L},$$
$$\beta = \frac{c}{r}, \quad \varepsilon = \frac{e}{c}, \quad H = \frac{h}{c}, \quad \Pi = \frac{c^2}{6\eta U r} p \tag{11.4}$$

where: θ – bearing circumferential coordinate, z_1 – dimensionless axial coordinate, L – bearing axial length, α – diameter to length ratio, β – clearance ratio ($c = R - r$ – radial clearance), ε – eccentricity ratio; H – dimensionless film thickness; Π – non-dimensional HD pressure, $U = \omega r$ – journal circumferential velocity.

After substitution of (11.4) in (11.1) the 2D Reynolds equation for pressure distribution is obtained as:

$$\frac{\partial}{\partial \theta}\left(H^3 \frac{\partial \Pi}{\partial \theta}\right) + \alpha^2 \frac{\partial}{\partial z_1}\left(H^3 \frac{\partial \Pi}{\partial z_1}\right) = \frac{\partial H}{\partial \theta} \tag{11.5}$$

Film thickness Eq. (11.2) can be rendered dimensionless through the relevant substitutions and written as:

$$H = 1 + \varepsilon \cos \theta + \overline{\delta} \tag{11.6}$$

On the other side, the radial distortions of the bearings liner surface points, given by Eq. (11.3), are presented in a following non-dimensional form:

$$\overline{\delta} = \frac{6\eta\omega r^2}{c^3} \frac{(1+\mu)(1-2\mu)}{(1-\mu)} \frac{d}{E} \Pi \tag{11.7}$$

11.2.4 Bearing Characteristics

Integrating the steady film pressure over the film region gives the steady load-carrying capacity of the bearing, calculated by:

$$\overline{W} = \sqrt{\overline{W}_1^2 + \overline{W}_2^2} = \frac{\beta^2}{6\eta\omega r L} W \tag{11.8}$$

Here $\overline{W}_1$ and $\overline{W}_2$ represent the components along and perpendicular to the line of centres and they are given respectively as:

$$\overline{W}_1 = -\int_{-1}^{1}\int_{0}^{2\pi} \Pi \cos\theta \, d\theta dz_1, \quad \overline{W}_2 = \int_{-1}^{1}\int_{0}^{2\pi} \Pi \sin\theta \, d\theta dz_1 \tag{11.9}$$

And the Sommerfeld number is defined as:

$$S = \frac{W\beta^2}{\eta\omega r L} \tag{11.10}$$

11.3 Boundary Conditions and Numerical Method

For pressure field in water lubricated bearings, several boundary conditions can be used including the Reynolds, Jakobsson-Floberg-Olsson (JFO), half-Sommerfeld and Sommerfeld boundary conditions. Generally speaking, JFO conditions take into account the cavitation in the best way but [7] Reynolds boundary conditions are closer to the practical engineering operations. Moreover, several numerical solutions for pressure under the Reynolds boundary conditions, for example [7], show better correspondence with the published experimental results in the literature. In the current analysis the Reynolds boundary conditions are used: – pressure at the journal bearing axial edges is equal to zero; – pressure is equal to zero also in those zones where the gradient of pressure in the circumferential direction becomes zero.

The presented elastohydrodynamic (EHD) problem presupposes simultaneous solution of the Reynolds equation for the pressure distribution (11.1), film thickness equation (11.2) and elasticity equation (11.3). The dimensionless modified

Reynolds equation is solved numerically using the finite difference method. Furthermore, in order to improve the convergence rate the iterative scheme with application of a successive over-relaxation procedure [15–18] is employed. For EHD analysis, an iterative process is repeated until the required convergence is achieved. The converged nodal pressures are then used to calculate the nodal displacements. By taking into account the radial component of the nodal displacements, the film thickness is changed to obtain the solution for the nodal pressures. Iterations are also required to get performance characteristics for different values of the parameters included in the elasticity equation to consider the flexibility of the bush liner. For greater accuracy, a convergence tolerance of 10^{-6} is applied for all residual terms.

11.4 Results and Discussions

The presented results correspond to the steady-state HD lubrication of water journal bearing at constant values of angular velocity and external load. The lubricant is water with dynamic viscosity $\eta = 1.002 \times 10^{-3}$ [Pa s]. The current analysis showed that the effect of deformability of the bushing' layer can be presented by the properties of the material μ, E and the liner thickness d (the last one will be kept constant in this study). Then, considering the mathematical model, the governing parameters are eccentricity ratio ε, diameter to length ratio α, and elastic layer parameters μ, E. The results are obtained for α equal to 0.5; 1.0; 1.5, whereas ε was varied from 0.1 to 0.9. The elasticity parameters are set to two different cases: $E = 2.1 \times 10^{11}$ [Pa], $\mu = 0.25$ (rigid case); $E = 1.63 \times 10^{8}$ [Pa], $\mu = 0.38$ (soft/elastic case).

The effect of eccentricity ratio on the pressure distribution in circumferential direction is shown in Fig. 2a. It has been observed that the pressure profiles are significantly different at different ε, as the maximum pressure values increase with the eccentricity ratio increases. Moreover, the peak of pressure is visible narrower at high eccentricity ratios.

The same effect of epsilon on the pressure values and profiles is presented in Fig. 2b as in addition the effect of deformability of the bush layer is taken into account. It is evident that if the radial elastic displacements are considered (soft case), the pressure values decrease compared with rigid case, as it was normally to expected the same tendency for both eccentricity ratio (as well as for all range of possible values of ε). The difference between the rigid and elastic case gradually increases with the eccentricity ratio.

The profile of fluid film thickness with variation of ε is shown in the next two figures – Fig. 3a, b. From the first of them can be concluded that the incresing eccentricity ratio leads to decreasing the minimum film thickness, as the film thickness profile as whole is similar for all ε and differs only by values.

In Fig. 3b along with the effect of epsilon (rigid case), the influence of the deformability of the bush layer (soft case) is also shown. It can be observed a visible change of the film thickness geometry at rendering into account the elastic deformations, as it is more pronounced for the high eccentrity ratios.

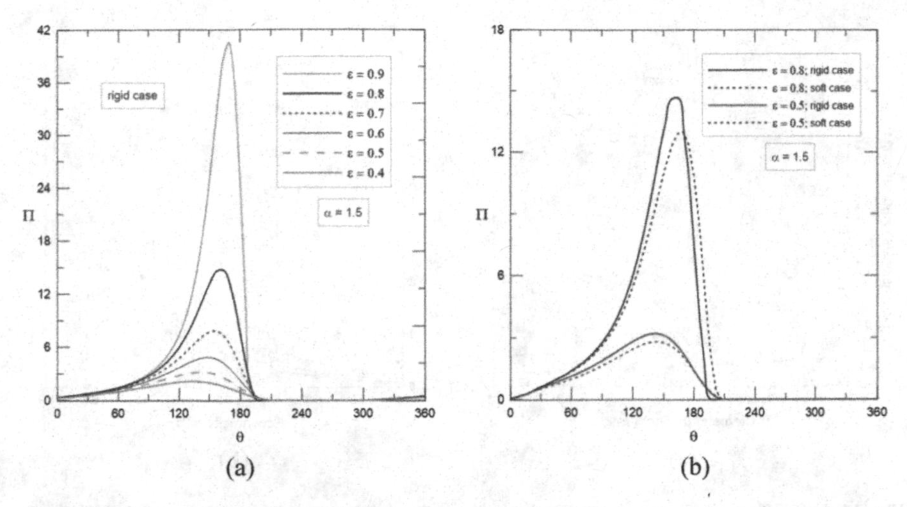

Fig. 11.2 **a** HD pressure distribution at different eccentricity ratios, and **b** Effect of layer deformability on the pressure distribution

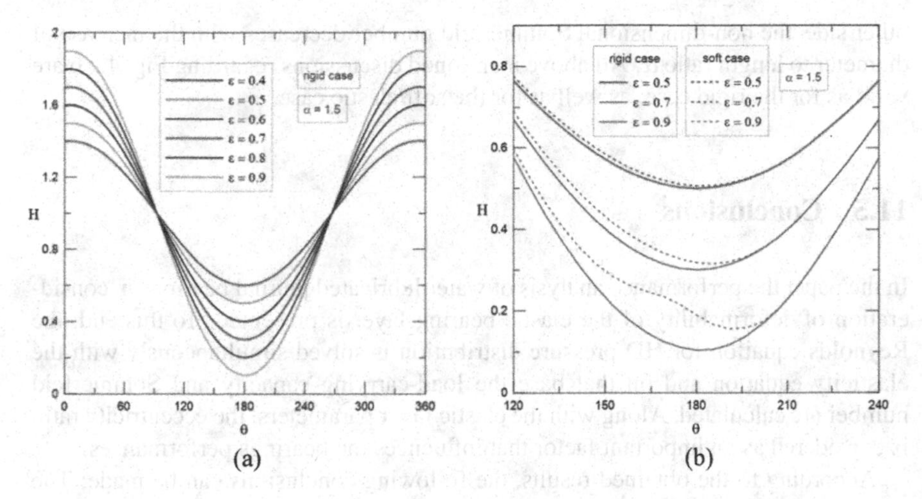

Fig. 11.3 **a** Film thickness geometry at different eccentricity ratios, and **b** Effect of layer deformability on the fluid film thickness profile

Figure 4a shows the effect of diameter to length ratio α on the dimensionless load-carrying capacity $\overline{W}$ of the bearing at several eccentricity ratios ε for the so-called in the paper rigid case. From the results here it is clear that the load-carrying capacity decreases with diameter to length ratio, but increases with eccentricity ratio.

The dependence of Sommerfeld number S on the eccentricity ratio ε for the soft (elastic) case is given in Fig. 4b. It is found that the values of load-carrying capacity coefficient S increase with increasing the eccentricity ratio, as this effect is more pronounced for the middle and especially for the high eccentrity ratios. On the

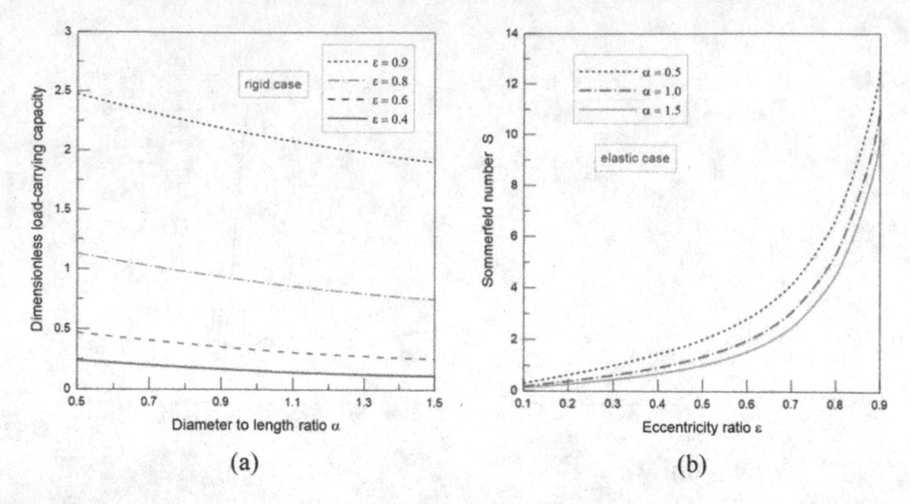

Fig. 11.4 **a** Effect of diameter to length ratio on the load-carrying capacity, and **b** Effect of eccentricity ratio on the Sommerfeld number

other side, the non-dimensional Sommerfeld number decreases with the increase of diameter to length ratio α. All above mentioned discussions regarding Fig. 4a, b are valid as for the rigid case, as well as for the soft/elastic case.

11.5 Conclusions

In the paper the performance analysis of water lubricated journal bearings in consideration of deformability of the elastic bearing layer is presented. To this end, the Reynolds equation for HD pressure distribution is solved simultaneously with the elasticity equation and on that base the load-carrying capacity and Sommerfeld number are calculated. Along with the elastic layer parameters, the eccentricity ratio is considered as an important factor that influences the bearings performances.

According to the obtained results, the following conclusions can be made: The deformability of the bearing elastic liner leads to a visible decrease in the pressure peak and an increase the minimum film thickness, as these effects are more significant for the softer liner materials. In result of this, the load-carrying capacity decreases under elastic contact conditions (soft cases). On the other side, with the increase in the eccentricity ratio, the maximum pressure increases whereas minimum film thickness decreases. Dimensionless load-carrying capacity and Sommerfeld number increase with the increasing in eccentricity ratio. The conclusions drown can be useful for the design, and optimization of journal bearings.

Acknowledgements This paper has been supported by the Bulgarian National Science Fund under the project КП-06-Н27-8/2018, which is gratefully acknowledged by the authors. Collaboration through the CEEPUS network CIII-BG-0703 is also acknowledged.

References

1. Gao, G., Yin, Z., Jiang, D.: Numerical analysis of plain journal bearing under hydrodynamic lubrication by water. Tribol. Int. **75**, 31–38 (2014)
2. Wang, Y.Q, Chao, L.: Numerical analysis of hydrodynamic lubrication on water-lubricated rubber bearings. Adv. Mater. Res. **299–300**, 12–6 (2011)
3. Pai, R., Hargreaves, D.: Water lubricated bearings. In: Nosonovsky, M., Bhushan, B. (eds.) Green Tribology: Green Energy and Technology, pp. 347–391. Springer, Berlin (2012)
4. Zhang, X., Yin, Z., Gao, G., Li, Z.: Determination of stiffness coefficients of hydrodynamic water-lubricated plain journal bearings. Tribol. Int. **85**, 37–47 (2015)
5. Elsayed, K., El-Butch, A.: A study on hydrodynamic water lubricated journal bearing. Eng. Res. J. **153**, M1–M15 (2017)
6. Shi, Y., Li, M., Zhu, G., Yu, Y.: Dynamics of a rotor system coupled with water-lubricated rubber bearings. P. I. Mech. Eng. C.-J. Mec. **232**(23), 4263–4277 (2018)
7. Xie, Z., Rao, Z., Liu, H.: Effect of surface topography and structural parameters on the lubrication performance of a water-lubricated bearing: theoretical and experimental study. Coatings **9**(1), 23-1–23-20 (2019)
8. Gao, G., Yin, Z., Jiang, D., Zhang, X., Wang, Y.: Analysis on design parameters of water-lubricated journal bearings under hydrodynamic lubrication. J. Eng. Trib. **230**, 1019–1029 (2016)
9. Hamrock, B., Schmid, S., Jacobson, B.: Fundamentals of Fluid Film Lubrication. Marcel Dekker Inc., NY (2004)
10. Szeri, A.: Fluid Film Lubrication. Cambridge University Press, Cambridge (2005)
11. Glavatskih, S., Fillon, M.: TEHD analysis of thrust bearings with PTFE-faced pads. J. Tribol. **128**, 49–58 (2006)
12. Kuznetsov, E., Glavatskih, S., Fillon, M.: THD analysis of compliant journal bearings considering liner deformation. Tribol. Intern. **44**, 1629–1641 (2011)
13. Javorova, J., Alexandrov, V.: Effects of fluid inertia and bearing flexibility on the performance of finite length journal bearing. IOP Conf. Ser.: Mater. Sci. Eng. **174**, 012039-1–012039-8 (2017)
14. Usov, P.: Elastohydrodynamic problem for journal sliding bearing under reciprocating motion. J. Frict. Wear **37**, 204–212 (2016)
15. Guha, S.: On the steady-state performance of hydrodynamic flexible journal bearings of finite width lubricated by ferro fluids with micro-polar effect. Int. J. Mech. Eng. Robot. Res. **1**(2), 32–49 (2012)
16. Meng, F., Yuanpei, C.: Analysis of elasto-hydrodynamic lubrication of journal bearing based on different numerical methods. Ind. Lubr. Tribol. **67**(5), 486–497 (2015)
17. Javorova, J., Mazdrakova, A., Andonov, I., Radulescu, A.: Analysis of HD journal bearings considering elastic deformation and non-Newtonian Rabinowitsch fluid model. Tribol. Ind. **38**(2), 186–196 (2016)
18. Mahdi, M., Hussein, A.: Investigation the combined effects of wear and turbulent on the performance of hydrodynamic journal bearing operating with couple stress fluids. Int. J. Struct. Integr. **10**(6), 825–837 (2019)

7. [illegible] (2016)
8. [illegible] (2009)
9. [illegible]
10. [illegible] (2005)
11. [illegible] (2002)

Chapter 12
Influence of Low Temperature on the Pumping Rate of Radial Lip Seals

Sumbat Bekgulyan, Simon Feldmeth, and Frank Bauer

Abstract Elastomeric radial lip seals are complex tribological systems. Their sealing performance is essentially determined by the back pumping capability of the system. During the dynamic operation, radial lip seals can actively pump leaking fluid back to the fluid side. For the assessment of the pumping capability and the leak-tightness of the sealing system, the pumping rate can be used. The pumping rate measures the pumped fluid volume per time or distance. In this paper, the influence of low operating temperatures on the pumping rate of elastomeric radial lip seals is analysed experimentally. For this purpose, the pumping rate was measured at different oil sump temperatures between +40 and –20 °C. The experimental analyses were performed with radial lip seals made of nitrile rubber (NBR) and fluoro rubber (FKM) in combination with mineral and synthetic (polyglycol and polyalphaolefin) oils.

Keywords Radial lip seal · Pumping rate · Low temperature

12.1 Introduction

Radial lip seals are complex tribological dynamic sealing systems. In several research works at the University of Stuttgart, e.g. [1–5], radial lip seals were evaluated using the pumping rate. In this work, the influence of low temperatures on the pumping rate of elastomeric radial lip seals is analysed experimentally and presented.

12.1.1 Pumping Rate of Radial Lip Seal

During the dynamic operation, the lubricant enters the contact area between the sealing edge and the shaft. Thereby, a thin sealing gap, which is filled with lubricant,

S. Bekgulyan (✉) · S. Feldmeth · F. Bauer
Institute of Machine Components (IMA), University of Stuttgart, Pfaffenwaldring 9, 70569 Stuttgart, Germany
e-mail: sumbat.bekgulyan@ima.uni-stuttgart.de

M. Rackov et al. (eds.), *Machine and Industrial Design in Mechanical Engineering*, Mechanisms and Machine Science 109,
https://doi.org/10.1007/978-3-030-88465-9_12

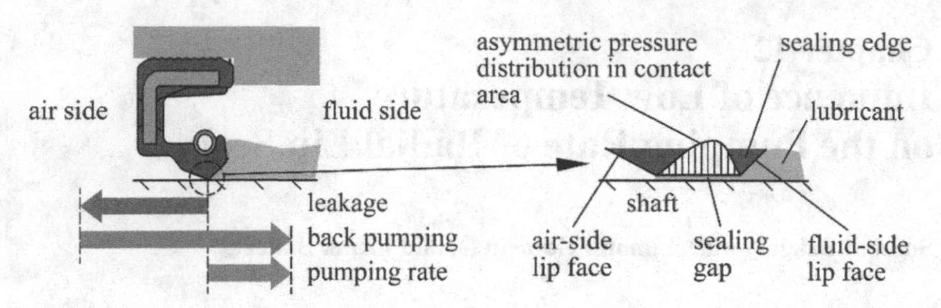

Fig. 12.1 Sealing mechanism of radial lip seals

is formed between the sealing edge and the shaft surface, Fig. 12.1. Due to the different lip face angles, an asymmetric pressure distribution results in the contact area. Additionally, the elastomer material of the sealing edge deforms elastically in circumferential direction while sliding on the shaft surface. Mainly the combination of these processes in the contact area results in an active back pumping capability, which prevents the lubricant from leaking out of the sealing gap. Furthermore, there are different operating principles explaining the back pumping capability of radial lip seals [6, 7].

The active back pumping capability can be observed by offering oil to the radial lip seal on the air side. The offered oil is pumped from the air side to the oil side. The quotient of the pumped lubricant quantity and the elapsed time equals the pumping rate, which describes the back pumping capability quantitatively. Pumping rate measurements can show the influence of individual system parameters on the leak tightness of radial lip seals [2–4]. There are several test methods to measure the pumping rate of a sealing system, which are based on two measuring principles [1–3]. Either the amount of pumped lubricant (mass or volume) for a certain period of time (principle 1) or the time required to pump a certain amount of lubricant (principle 2) is measured. In this work, the pumping rate is given in [μl/m] as the pumped lubricant volume per sliding distance on the shaft surface according to Eq. (12.1), where Δm represents the mass of the pumped lubricant, D the diameter of the shaft, Δt the pumping time, ρ the density of the lubricant and n the shaft speed in rpm.

$$pumpingrate = \frac{\Delta m}{\pi \cdot D \cdot \Delta t \cdot \rho \cdot n} \tag{12.1}$$

12.1.2 *Effects of Low Temperature*

At low temperatures, the elasticity of elastomer materials decreases as the glass transition temperature is approached or is even passed below. Lower elasticity can influence the back pumping capability of the radial lip seal and reduce the ability of the sealing lip to follow the shaft surface. As a result, even a small dynamic eccentricity (dynamic run out) can cause leakage. The exact range of the glass transition temperature depends on the actual elastomer compound material and also on the load frequency [8]. Standards like [9] and [10] and manufacturer catalogues (e.g. [11]) provide users information on the minimum specified operating temperature for radial lip seal materials.

Furthermore, the viscosity of the lubricant increases as operating temperature decreases. Depending on the oil-elastomer combination, the viscosity of the lubricant influences the pumping rate strongly [2–4]. For the minimum application temperature of liquid lubricants, the pour point can be used. The pour point is the lowest temperature at which a sample of an oil product barely flows under certain standard conditions during a cooling process. The pour point is determined according to ISO 3016 [12] and specified in oil manufacturers' catalogues and data collection catalogues (e.g. [13]).

12.2 Experimental Approach

The pumping rate measurements were performed with an inversely mounted radial lip seal by completely flooding the seal air side with lubricant, according to [3]. The inverse assembly results in a continuous pumping of lubricant to the observed seal oil side. The pumped lubricant is collected in a beaker and weighed. Afterwards the pumping rate is determined by the pumped lubricant quantity and the pumping duration according to Eq. (12.1).

Figure 12.2 shows the test rig chamber prepared for a pumping rate measurement at low temperature. Via a control computer, different test conditions can be set. The test shaft is driven by a motor via a drive belt. The aluminium housing walls of the oil chamber contain heating cartridges and circulating channels filled with cooling fluid. The oil is thus heated or cooled indirectly via a large surface without

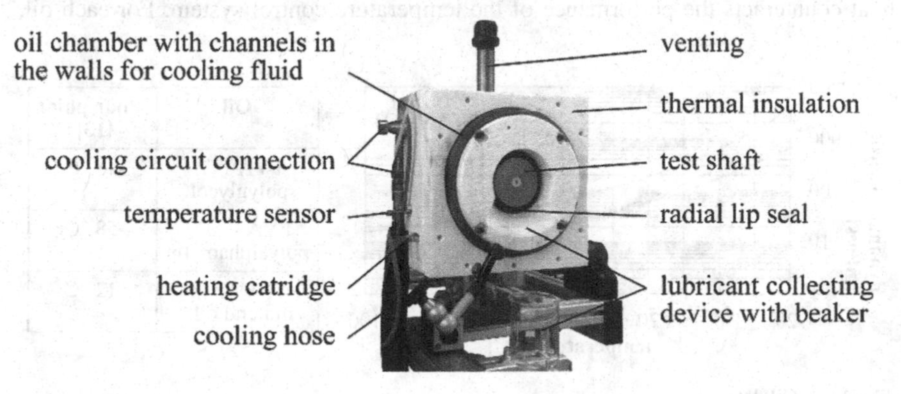

Fig. 12.2 Test rig chamber

being thermally damaged, even at high heating or cooling power. The oil sump temperature is measured by a temperature sensor inside the chamber. For tests in the low temperature range, the oil chamber was equipped with a thermal insulation to improve the thermal efficiency. For cooling, an external temperature control system Julabo Magnum 91 was connected to the cooling circuit of the oil chamber.

The test runs were conducted at five different oil sump temperatures. The rotating speed and direction as well as the oil sump temperature were constant during each test. Each test duration was 10 h. The whole test duration was used as the pumping time for calculating the pumping rate according to Eq. (12.1). The oil chamber was completely filled with oil. The nominal diameter of the sealing system was 80 mm. The test shaft was rotating with 1552 revolutions per minute, which corresponds to a circumferential speed of 6.5 m/s. Shaft sleeves made of 100Cr6 with a lead free, plunge ground surface with a roughness Rz of 1.96–2.33 μm were used as test shafts. Standard radial lip seals (dimensions 80 × 100 × 10 mm) with trimmed sealing edges made of fluoro rubber 75 FKM 585 (BAUMX7) and nitrile rubber 72 NBR 902 (BAUX2) from the manufacturer Freudenberg Sealing Technologies were used. The minimum operating temperature is –25 °C for the FKM seals and –40 °C for the NBR seals [11].

One mineral and two synthetic reference oils were used, all of which are standardised by the FVA (German Research Association for Drive Technology). The pour points of the oils are given in Fig. 12.3. The viscosity-temperature behaviour of the oils was measured using a rheometer Anton Paar MRC 302 with a parallel-plate measuring system. Figure 12.3 shows the measured viscosity-temperature dependence. The oil viscosity decreases with increasing temperature. The dynamic viscosity of the polyglycol oil FVA PG1 is higher than the viscosity of the polyalphaolefin oil FVA PAO1 and the mineral oil FVA2 at each temperature. The viscosity of FVA2 was not measured at temperatures below –10 °C because of its pour point at –12 °C.

The oil in the test rig chamber can be cooled to temperatures below −50 °C if the shaft is not rotating. However, during the operation, frictional heat is generated in the fluid, at the sealing contact and on the drive side of the shaft. The frictional heat counteracts the performance of the temperature control system. For each oil,

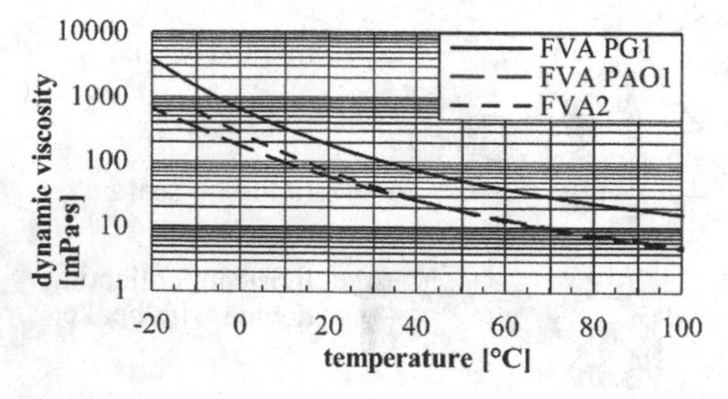

Oil	Pour point [13]
FVA PG1 polyglycol	-48 °C
FVA PAO1 polyalphaolefin	-58 °C
FVA2 mineral oil	-12 °C

Fig. 12.3 Oil data

Table 12.1 Overview of the oil sump temperatures

Oil name	Oil sump temperature (°C)					Explanation
	−20	−10	0	10	40	
FVA2	– (*)	+	+	+	+	(*)—below pour point (**)—fluid friction to high
FVA PG1	– (**)	+	+	+	+	
FVA PAO1	+	+	+	+	+	

preceding tests were conducted to determine the minimum oil sump temperature, which can be maintained at the shaft speed of 1552 rpm for the entire duration of the pumping rate measurement. For FVA2 and FVA PG1, the minimum oil sump temperature with rotating shaft was 10 °C. For FVA PAO1 a minimum oil sump temperature of 20 °C was achieved. The oil sump temperature was varied between the possible minimum temperature and 40 °C. Table 12.1 shows an overview of the oil sump temperatures for each oil. Two pumping rate measurements were performed per oil-elastomer combination for each temperature. A total of 52 pumping rate measurements were carried out.

The most critical operating conditions are in the test with FKM seal and –20 °C, as the application limit of the FKM material is at –25 °C, [11]. To prove the functionality, a function test was carried out with FKM seal and FVA PAO1 oil at –20 °C. Thereby, the sealing system showed no leakage.

For each measurement a new sealing system with a new radial lip seal and a new running track on the shaft surface was used. The seals used did not pass a running-in period before the pumping rate was measured. Before each measurement the sealing system was tempered to the aspired temperature while the shaft did not rotate. Then the shaft drive was switched on and the measurement was performed.

During the tests, the temperature at the sealing edge was measured at the oil side of the radial lip seal with a thermal imager Fluke Ti32. The emission coefficient ε was set to 0.95. The emission coefficient might change during test runs because of the wetting of the surface by the pumped lubricant. Therefore, the measured temperatures are only qualitative and should only be used as a rough indicator.

12.3 Results

The results presented in this work are partly based on the analyses conducted in the research project [3]. Figure 12.4 shows the pumping rate as an arithmetic mean of all measurements for each oil-elastomer combination in the oil sump temperature range between –10 and 40 °C. This mean pumping rate indicates the level of measured pumping rates. For both seal materials, FKM and NBR, the maximal level of the pumping rate is achieved with FVA PG1 oil. For the FVA PAO1 and FVA2 oils, the level of the pumping rate is similar and is significantly lower than for FVA PG1. The higher pumping rate level for measurements with FVA PG1 can be explained

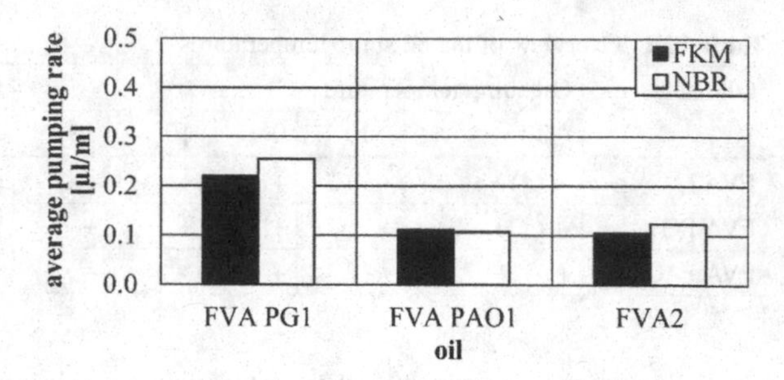

Fig. 12.4 Influence of the oil-elastomer combination on the pumping rate in the oil sump temperature range between –10 and 40 °C

by the higher viscosity of FVA PG1. The viscosity of FVA PG1 is by a factor of 3 higher compared to FVA PAO1 and FVA2, Fig. 12.3. For NBR seals the level of the pumping rate is slightly higher than for FKM seals with FVA PG1 and FVA2 oils. This behaviour is within the expectations for low viscosity oils according to previous research works [2] and [3].

12.3.1 Pumping Rates at Low Temperatures

Figure 12.5 shows the influence of the oil sump temperature on the pumping rate for each oil-elastomer combination. For the lubricant FVA PG1 the pumping rates with the NBR seals are slightly higher than those with the FKM seals at 0 °C and above. Both seal types show a slight increase in the pumping rate when the temperature falls from 40 to 0 °C. The pumping rate with the NBR seals decreases significantly with decreasing temperature from 0 to –10 °C.

For FVA PAO1 at temperatures at –10 °C and above, the pumping rates of both seal types are at a similar level for respective temperatures. In the case of FKM seals, the pumping rates decreases slightly with decreasing temperature to –10 °C. From –10 to –20 °C, an increase of pumping rate with FKM seals can be observed. For

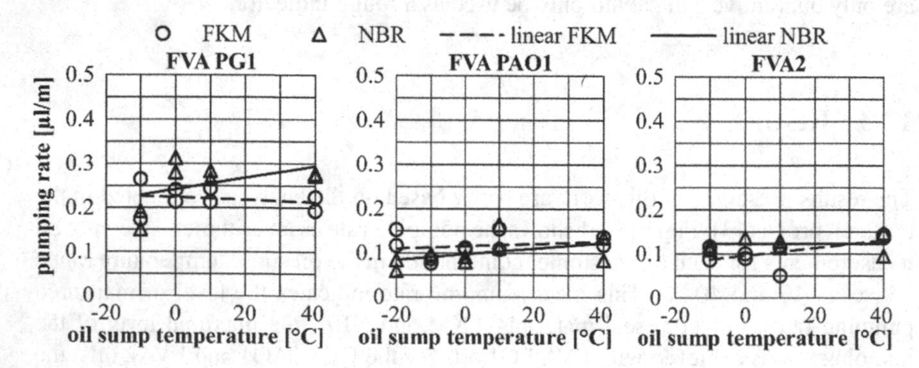

Fig. 12.5 Influence of oil sump temperature

NBR seals, there is a tendency to decreasing pumping rates with decreasing oil sump temperature for the whole temperature range.

For FVA2 oil at temperatures at 10 °C and below, the pumping rates measured with FKM seals are slightly lower than with NBR seals. The FKM seals show a slightly decreasing pumping rate over the whole temperature range with decreasing oil sump temperature. The pumping rates for NBR seals are approximately at the same level for each temperature.

12.3.2 Temperature at the Sealing Edge

Figure 12.6 shows the highest temperatures measured at the sealing edge for each oil depending on the oil sump temperature. The temperatures at the sealing edge are at the same level for both sealing types with the corresponding oils and at the corresponding oil sump temperatures. The temperature at the sealing edge is much higher as in the oil sump and determines the viscosity of the fluid during the pumping. The fluid viscosity influences the pumping rate strongly [2–4]. Therefore, the temperature at the sealing edge is an important parameter for fluid pumping.

The viscosity of the oils decreases considerably by increasing the oil sump temperature by 50 K between –10 and 40 °C, Fig. 12.3. For FVA PG1 and FVA2, a decrease in viscosity by a factor of about 20 can be observed. With FVA PAO1 the viscosity decreases only by a factor of about 13. Assuming a large change in viscosity due to the variation of the oil sump temperature, large changes in pumping rates would be expected. However, actual measured pumping rates show rather small changes depending on the oil sump temperature, Fig. 12.5.

As soon as the shaft starts rotating, frictional heat between the shaft and the sealing edge results in a higher temperature at the seal contact area compared to the oil sump temperature. The highest temperature at the sealing edge varies in the range between 40 and 70 °C, Fig. 12.6. For all test runs with the same oil, the temperature difference at the sealing edge is less than 23 K and thus less than half of the difference in the oil sump temperature. Therefore, the viscosity change at the sealing edge is much lower than in the oil sump. Within the measured temperature range at the sealing edge, the

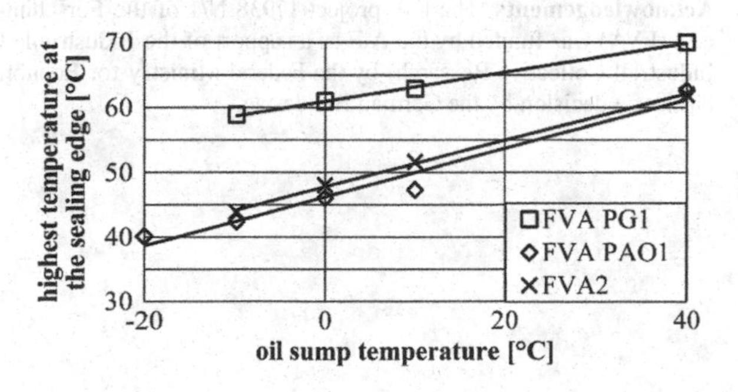

Fig. 12.6 Temperature at the sealing edge

viscosity only changes by a factor of approximately 2, Fig. 12.3. Thus, the viscosity change at the sealing edge and its influence on the pumping rate is much lower than indicated by the oil sump temperature.

12.4 Conclusion

The pumping behaviour of radial lip seals at low temperatures was analysed. For this purpose, pumping rate measurements were carried out at oil sump temperatures between 40 and –20 °C. The experimental analyses were conducted for six oil-elastomer combinations. Depending on the oil sump temperature, slight changes in the pumping rate were observed. The pumping rates with mineral oil FVA2 and polyalphaolefin oil FVA PAO1 are at a similar level as well as their viscosities. The viscosity of the polyglycol oil FVA PG1 and the pumping rates measured with FVA PG1 are higher than for FVA2 and FVA PAO1.

During a pumping rate measurement, the contact area between the rotating shaft and the sealing edge heats up due to frictional heat. This results in a higher temperature in the seal contact area compared to the oil sump. At the sealing edge, the temperature variation and thus the change in viscosity is much lower compared to the oil sump. This might explain, why the influence of the oil sump temperature on the pumping rate is lower than expected. In addition, the thermo-viscoelastic behaviour of the elastomer might influence the pumping rate. The elastomer elasticity decreases with decreasing temperature. According to the common operating principles [7], this reduces the pumping rate as well. Therefore, the influences of fluid viscosity and elastomer elasticity on the pumping rate counteracts each other and might be responsible for nearly constant pumping rates in the analysed range of the oil sump temperature.

Further measurements could be performed to analyse the pumping behaviour of radial lip seals at low temperatures. For example, even lower temperatures in the sealing contact might be achieved by using oils with lower viscosity and a lower viscosity-temperature dependency as well as a more efficient temperature control system in combination with a cooled shaft.

Acknowledgements The IGF project 17938 N/1 of the Forschungsvereinigung Antriebstechnik e.V. (FVA) was funded by the AiF as a support of the Industrielle Gemeinschaftsforschung (IGF, Industrial Collective Research) by the Federal Ministry for Economics Affairs and Energy on the basis of a decision by the German Bundestag.

References

1. Remppis, M., Bauer, F., Haas, W.: Measurement of the pump rate of radial lip seals during long term tests. In: Proceedings of the VII Iberian Conference on Tribology, IBERTRIB 2013, Porto, pp. 94–95 (2013)
2. Remppis, M.: Untersuchungen zum Förderverhalten von Dichtsystemen mit Radial-Wellendichtringen aus Elastomer. PhD Thesis, University of Stuttgart (2016)
3. Bekgulyan, S., Bauer, F., Haas, W.: Rechnerische Abschätzung der Dichtgüte von Radial-Wellendichtringen durch Kenntnis der Systemparametereinflüsse II. Abschlussbericht, FVA, Forschungsvorhaben Nr. 617 II, FVA-Heft 1259, IGF-Nr. 17938 N/1, Frankfurt/Main (2017)
4. Bekgulyan, S., Feldmeth, S., Bauer, F.: Influence of static and dynamic eccentricity on the pumping rate of radial lip seals. In: Proceedings of the 25th International Conference on Fluid Sealing, BHR Group, Manchester, pp. 79–91 (2020)
5. Sommer, M., Haas, W.: A new approach on grease tribology in sealing technology: influence of the thickener particles. Tribol. Int. **103**, 574–583 (2016)
6. Müller, H.K., Nau, B.S.: Fluid Sealing Technology. Principles and Applictions. M. Dekker, New York (1998)
7. Bauer, F.: Federvorgespannte-Elastomer-Radial-Wellendichtungen. Grundlagen der Tribologie & Dichtungstechnik, Funktion und Schadensanalyse. Springer Vieweg (2021)
8. Jaunich, M.: Tieftemperaturverhalten von Elastomeren im Dichtungseinsatz. PhD Thesis, Technische Universität, Berlin (2012)
9. DIN 3760: Rotary shaft lip type seals, September 1996
10. DIN 3761-2: Rotary shaft lip seals for automobiles; applications, November 1983
11. Freudenberg Sealing Technologies GmbH & Co. KG: Simmerring and Rotary Seals—Catalog (FNST)—Vol. 11. Weinheim (2015)
12. ISO 3016:2019: Petroleum and related products from natural or synthetic sources—Determination of pour point
13. FVA-Heft 660: Referenzölkatalog. Datensammlung, May 2007

Chapter 13
Design of an Automated Handling System

Pavol Varga and Andrej Smelík

Abstract This work deals with the design of an automated handling system for a project from the company Microstep. The first part of the work is a theoretical basis for the design of a machine vision system and manipulation effector, the basic components of an automated manipulation system. First, it acquaints the reader with basic information about machine vision and the advantages and disadvantages of its use. Secondly, it describes the possibilities of using SV in industry, the basic components of machine vision and their principles of their operation and an overview of available technologies in this area. It also describes various types of manipulation effectors, realized projects using machine vision for manipulation. Subsequently, the design of suction cup handling effector, design of stands for cameras and accessories, design of intermediate storage and sorting, selection of machine vision components and the principle of operation and control of the handling system are described.

Keywords Handling system · Automation system · Laser machine · Sorting

13.1 Introduction

The paper deals the design of an automated line (Fig. 13.1) for the production of segments of steel structures. The production line consists of several modules or functional units to which various technological and handling activities belong. The whole system according to Fig. 13.2 consists of a machine for the production of flange parts 1, a robotic handling unit 2 for removing and sorting parts produced by this machine, a robotic system 3 which provides automatic assembly with welding and an automatic machine which ensures insertion of the profile blank into the drilling

---

P. Varga (✉)
Faculty of Mechanical Engineering, Slovak University of Technology, Nám. Slobody 17, 81231 Bratislava, Slovakia
e-mail: palo@microstep.sk

A. Smelík
Microstep a.S, Vajnorská 158, 83104 Bratislava, Slovakia

M. Rackov et al. (eds.), *Machine and Industrial Design in Mechanical Engineering*, Mechanisms and Machine Science 109,
https://doi.org/10.1007/978-3-030-88465-9_13

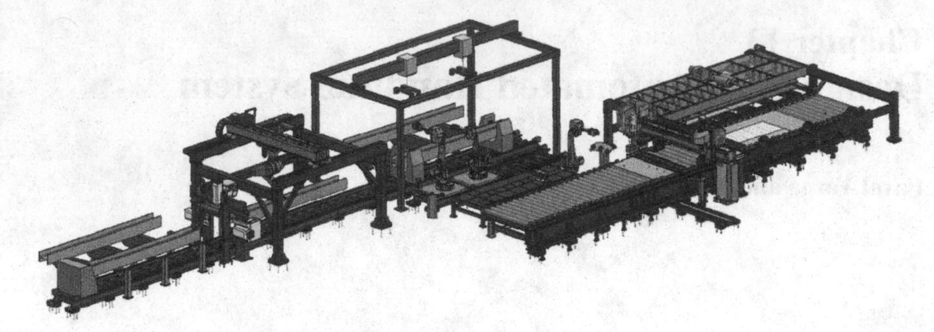

Fig. 13.1 Input—3D model of an automated line

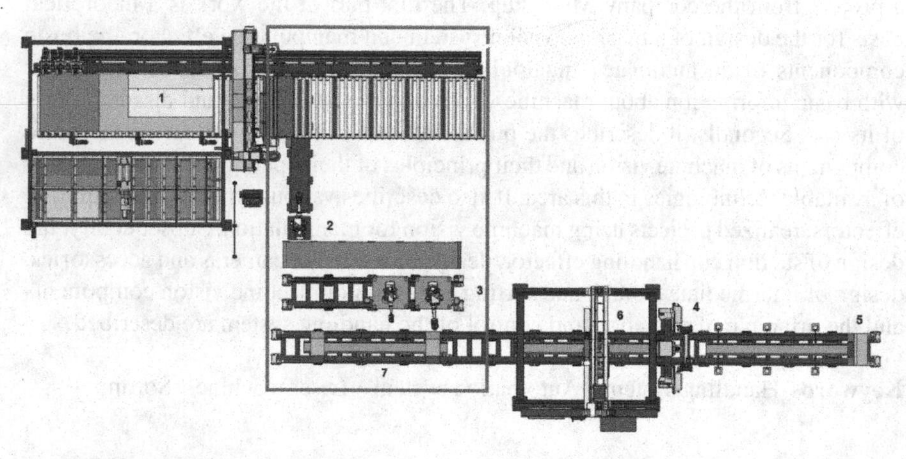

Fig. 13.2 The automated production line

part by means of a sliding unit 5. 4, subsequently to the cutting part 6 and from there after cutting to the handling part 7 [1].

The result of the project should be products representing separate machines and in various assemblies representing a comprehensively automated production cell for the production of segments of steel structures. Therefore, it is necessary to design a flexible handling/sorting a system that easily adapts to different configurations [2].

The part of the line which is the subject of the solution is shown in Fig. 13.2 bordered by a red rectangle. It is a part in which it is necessary to solve handling tasks—removal and sorting of flanged parts of different shapes and sizes from the conveyor to a possible intermediate storage and their exact location in the desired position with respect to the profile to which they will be by welding robot welded [3].

13.2 Handling Head with Vacuum Cups

For the handling of flanges, an effector was designed using a vacuum cup gripping technique (Fig. 13.3), which is a proven tool for handling steel sheets. The parameters for the design of the suction head from the given project were the dimensions of steel flanges in a combination of flange sizes (500 × 500 × 30) mm to (200 × 200 × 15) mm. The problem of holding such a large dimensional range of objects, using only one effector, has been solved by designing a mechanism that allows the suction cups to be positioned with respect to the dimensions of the object being manipulated. Using vacuum cups of two different diameters. The positioning of the suction cups is based on measuring the position and dimensions of the sheets by means of machine vision.

The head contains a movable mechanism for positioning the suction cups. The mechanism consists of an electric motor driving a moving ball screw mounted in bearings, along which the ball nut moves. On the motor side—Rexroth, there are double-row angular contact ball bearings, where on one side the inner ring of the bearing is fitted with a ball screw and on the other side a nut is screwed on the thread of the ball screw creating a preload. On the other side is a clamping sleeve with a ball bearing. A circular flange with holes for guide rods is attached to the ball nut, around the perimeter of which 6 hinges are evenly attached. A lock nut is attached to the ball nut from the bottom. In the event that the balls come loose and fall out, the lock nut will jam and prevent the ball nut from falling and damaging the mechanism. Suction cups with a diameter of ϕ80 mm (Fig. 13.4) are each mounted separately in a hinge holder.

The hinges on the holders are individually connected to the hinges on the ball nut by rods to form two joints. The suction cup holders are housed in the radial grooves of the lower support ring plate, between two sliding washers fixed by a nut which is screwed onto the outer thread of the holder so that the holders can move along the groove. Suction cups with a diameter of ϕ60 mm are mounted in separate holders, which are placed in the grooves in a similar way as larger suction cups. The

Fig. 13.3 Handling head with vacuum cups

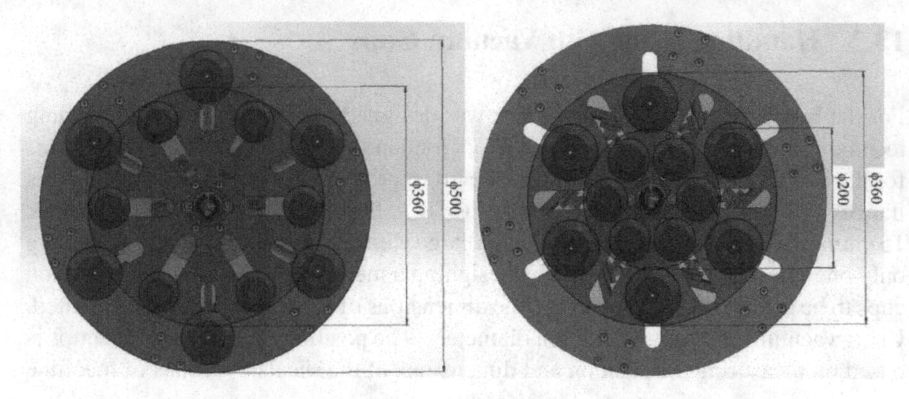

Fig. 13.4 Maximum span of vacuum cup (left) and minimum span of vacuum cup (right)

movement of the smaller suction cups is derived from the movement of the larger suction cups, to which they are connected by means of a ball rail guide. The bevel gear allows the motor to be placed perpendicular to the moving screw, four guide rods in which the ball nut is mounted, 6 support rods made of rectangular profiles, which reinforce the structure and vacuum components are attached to it and there is a flange on top for fastening to the ABB robot IRB 4600 via screws. The head also contains a holder for the motor, 4 support columns, a bearing housing and holders for holding vacuum components [4–6].

13.2.1 *Design of Structural Parts of the Device*

The parameters of the motor, bolts and washers interact. The selected stepper motor type is commonly used in positioning mechanisms. This type of motor is controlled by means of electrical pulses, where the position of the motor shaft is given the number of electrical pulses and the speed depend on the pulse frequency. The disadvantage of the stepper motor is the relatively low torque, which decreases rapidly from a certain increasing value of the pulse frequency. For stepper motors, it exists worldwide extended standard NEMA (National Electrical Manufacturers Association), which determines the size of the motor resp. the size of the motor mounting flange and the motor torque. This motor is also supplied with an electromagnetic brake that holds the load in vertical applications when the motor is at rest. The size of the motor torque Mk depends on the weight of the load and the external force. The weight of the load consists of the weight of the bolt, flange, hinges and bolts. Any external force that must be overcome by the motor torque to move the mechanism is the frictional force between the suction cup holders and the support plate.

To calculate the friction force, it is necessary to make a static decomposition of forces in the mechanism. In the calculation, we do not consider the gravitational force from the weight of the rod (Fig. 13.5). Hinges are represented by rotating bonds [7].

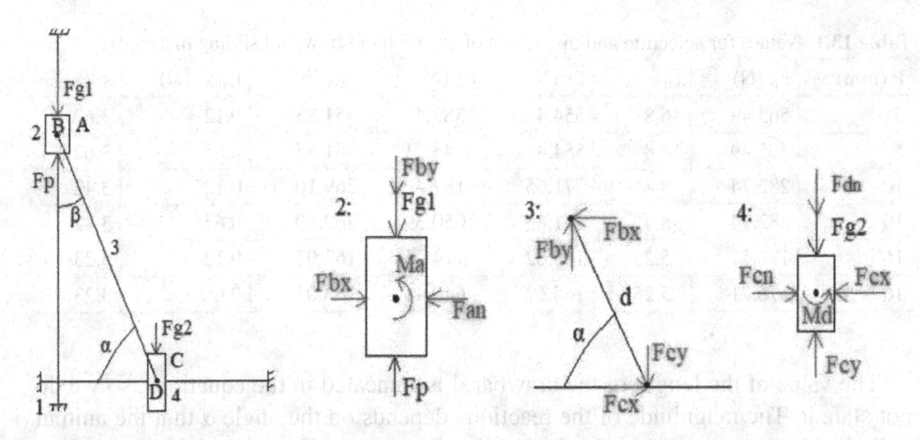

Fig. 13.5 Kinematic scheme of the mechanism, Fp—power from the drive, Fg1— gravitational force from the weight of the screw, flange …, Fg2—gravitational force from the weight of the vacuum cup, holder …, 1—base body, 2—ball nut, 3—drawbar, 4—suction cup, A, C—sliding bonds, B, D—rotational bonds

The movement of the nut on the screw and the movement of the vacuum cup in the groove is represented by sliding links.

Equilibrium equation:

$$x : \sum Fx = O = F_{bx} - F_{an} \gg F_{bx} = F_{an} \tag{13.1}$$

$$y : \sum F_y = 0 = F_p - F_{g1} - F_{by} \gg F_{by} = F_p - F_{g1} \tag{13.2}$$

$$z : \sum M = 0 = M_a = 0 \tag{13.3}$$

$$x : \sum Fx = O = -F_{bx} + F_{cx} \gg F_{bx} = F_{cx} \tag{13.4}$$

$$y : \sum F_y = 0 = F_{by} - F_{cy} = 0 \gg F_{by} = F_{cy} \tag{13.5}$$

$$z : \sum F_z = 0 = -F_{by} \cdot d \cdot cos\alpha + F_{bx} \cdot d \cdot sin\alpha \gg F_{bx} = \frac{F_{by} \cdot cos\alpha}{sin\alpha} \tag{13.6}$$

$$x : \sum Fx = O = F_{cn} - F_{cx} \gg F_{cx} = F_{cx} \tag{13.7}$$

$$y : \sum F_y = 0 = F_{cy} - F_{g2} - F_{dn} \gg F_{dn} = F_{cy} - F_{g2} \tag{13.8}$$

$$z : \sum M = 0 = M_d = 0 \tag{13.9}$$

Table 13.1 Values for selection and inspection of motor, ball screw and sliding materials

P (mm)	F_p (N)	t_m (s)	F_1 (N)	F_2 (N)	F_{dn} (N)	f_{0max} (N)	s
5	565.49	16.8	554.4	382.45	551.85	0.12	3.62
5	565.49	16.8	554.4	2143.71	551.85	0.65	3.62
10	282.74	8.4	271.65	187.4	269.10	0.12	3.42
10	282.74	8.4	271.65	1050.39	269.10	0.65	3.42
16	176.71	5.25	165.62	114.25	163.07	0.12	3.23
16	176.71	5.25	165.62	640.41	163.07	0.65	3.23

The value of the length of the drawbar d is truncated in the equations, so we do not state it. The magnitude of the reactions depends on the angle α that the animal pulled with the support plate. I calculate the reactions in the extreme positions of the mechanism, where αmax = 55.4° and αmin = 14.5°. It follows from Table 13.1 that αmax is decisive for the calculation of the motor, because in this position a smaller force Fcx = F2 arises, which moves the mechanism.

The direction of the force Fp depends on the direction of rotation of the ball screw. The magnitude of the force Fp arises from the transformation of the engine torque into a ball screw-ball nut pair:

$$Fp = \frac{Mk.2.\pi.\eta, 1000}{P} \tag{13.10}$$

The path between the extreme positions of the mechanism is 140 mm. In addition to the path and pitch P, this time tm also depends on the engine speed. For the NEMA 17 engine, I choose a speed value n = 100 rpm, which has the corresponding torque Mk = 0.5 Nm.

13.3 Strength Analysis

Strength analysis was performed using FEM simulation in Solidworks. The size of the load was chosen with respect to the maximum weight of the steel flange (500 × 500 × 30) mm, which weighs 58.5 kg. This mass creates a loading force F = 573.89 N. For the connection of the contact surfaces of the suction cup holders and the support plate, we have defined the contact set as a bonded connection.

Fig. 13.6 Displacements at horizontal load in the direction of force 2

Table 13.2 Results of simulation under load in the horizontal direction

Direction of force	Stress (MPa)	Displacement (mm)	Deformation	Safety
1	78.7296	0.142382	0.000245	3.8861
2		0.150246	0.000235	
3		0.138867	0.000226	
4		0.119856	0.000226	

13.3.1 Load in Horizontal Direction

The head can enter this type of load while holding the flange in the welding process. In the horizontal direction (Fig. 13.6), I do the calculation at four angular rotations of force, because the construction of the head is not axially symmetrical.

In Table 13.2, the maximum values are recorded individual simulation results. The largest displacement occurs when the load is in direction 2. The highest stresses under horizontal loading occur in the support columns and bolts due to the bending moment. The displacement is greatest at the bottom of the plate and gradually decreases upwards. The offset value at the upper end of the ball screw at the connection point of the bevel gear for connecting the drive is 0.023 mm.

13.3.2 Load in the Vertical Direction

Loading in the vertical direction is the most common way of loading the head. The load in the vertical direction is checked in the extreme positions and in the middle position of the mechanism. The largest displacement occurs in the extreme position when the suction cups are closest to each other (Fig. 13.7). When loaded, the plate bends with a maximum value of 0.06 mm (Table 13.3). There must be sufficient

Fig. 13.7 Displacements under vertical load when the suction cups are loaded in the position closest to each other

Table 13.3 Results of simulation under loading in the vertical direction

Direction of force	Stress (MPa)	Displacement (mm)	Deformation	Safety
Midpoint	78.730	0.0578	0.00023	3.886
Vacuum cups closest to each other		0.0615	0.000227	
Vacuum cups furthest apart		0.0443	0.000232	

clearance in the hinges, given the value of the displacement, to allow the suction cup holders to rest on the plate, thus eliminating the stress on the positioning mechanism.

Based on the results of the strength analysis, it can be stated that the proposed structure is satisfactory. A pneumatic circuit and electrical connection were designed. The next step is the creation of drawing documentation for the production of the robot's positioning head.

13.4 Conclusion

A structural design of the handling head was made using a suction cup gripping technique, which is flexible for the dimensional range of manipulated objects—steel flanges. The effector contains a kinematic mechanism for positioning the suction cups, based on a ball screw. The operation of the control subsystem, which controls the position of the mechanism and the active branch of the suction cups, is based on obtaining information through machine vision.

Acknowledgements The research presented in this paper is an outcome of the project LACUT Excellent team for research into diagnostics and classification of quality and dimensional tolerances of energy beam cutting machines stimuli for research and development funded by the Rector of Slovak University of technology in Bratislava.

References

1. Onderová, I., Kolláth, L., Ploskuňáková, L.: Experimental verification of the structural and technological parameters of the PKS. Acta Polytech. **54**(1), 59–62 (2014)
2. Onderová, I., Šooš, Ľ: Analysis of parameters affecting the quality of a cutting machine. Acta Polytech. **54**(1), 63–67 (2014)
3. Shivakoti, I.; Kibria, G.; Cep, R.; Pradhan, B.B.; Sharma, A.: Laser surface texturing for biomedical applications: a review. Coatings **11**(2), 124-1–124-15 (2021)
4. Čilliková, M., Mičietová, A., Čep, R., Mičieta, B., Neslušan, M., Kejzlar, P.: Asymmetrical barkhausen noise of a hard milled surface. Materials **14**(5), 1293-1–1293-11 (2021)
5. Šooš, Ľ., Onderová, I.: Parameters influencing the components of quality. In: Proceedings of the 2nd International Quality Conference, Quality Festival 2008, Kragujevac, pp. 96–108 (2008)
6. Onderová, I., Šooš, Ľ.: Innovation of gantry design. In: Proceedings of the International Congress MATAR PRAHA 2008: Part 1: Drives & Control, Design, Models & Simulation, pp. 105–107. Society for Machine Tools, Prague (2008)
7. Wang, Z., Wang, D., Wu, Y., Dong, H., Yu, S.: An invariant approach replacing Abbe principle for motion accuracy test and motion error identification of linear axes. Int. J. Mach. Tools Manuf. **166**, 103746 (2021)

Chapter 14
Analysis of Two-Stage Cycloid Speed Reducers Dimensions and Efficiency

Milos Matejic, Vladimir Goluza, Milan Vasic, and Mirko Blagojevic

Abstract Cycloid speed reducers are characterized by good working characteristics such as very compact design, low level of noise and vibration, wide range of transmission ratios, very reliable operation in conditions of dynamic loads, high level of efficiency, etc. All these listed operating characteristics give them an advantage over conventional speed reducers. Of these good features, the extremely compact design and high efficiency stands out. In this paper, a comparative analysis of the dimensions of a conventional and new concept of two-stage cycloid speed reducer is performed. A comparison was made of the total: length, width, height and volume. A comparative analysis of the efficiency according to the Kudrijavcev method was performed as well. The method of determining the efficiency was adapted so that it could be used in determination in the new concept of a two-stage cycloid speed reducer. At the end of the paper, conclusions are drawn and further directions of research are suggested.

Keywords Two-stage cycloid speed reducer · Overall dimensions · Efficiency

14.1 Introduction

Cycloid power transmission gearboxes can be used both as reducers and as multipliers. Their possibility of application as a reducer has been much researched through various scientific and practical papers, while their possibility of application as a multiplier is still insufficiently investigated. During the last decades, cycloid speed reducers have found a very wide application in engineering practice. Cycloid speed reducer have a wide range of very good performance characteristics. The most prominent good working characteristics of cycloid speed reducers are very compact design, low level of noise and vibration, wide range of transmission rations, very reliable operation in conditions of dynamic loads, high level efficiency, etc. All these good

M. Matejic (✉) · V. Goluza · M. Vasic · M. Blagojevic
Faculty of Engineering, University of Kragujevac, Sestre Janjic 6, 34000 Kragujevac, Serbia
e-mail: mmatejic@kg.ac.rs

M. Rackov et al. (eds.), *Machine and Industrial Design in Mechanical Engineering*, Mechanisms and Machine Science 109,
https://doi.org/10.1007/978-3-030-88465-9_14

performance characteristics give them a slight advantage over conventional gearboxes. Cycloid speed reducers are used in robots, satellites, manipulation devices, crane machines, conveyors, mixers of high viscosity materials, renewable energy sources (wind generators, mini hydro power plants…), etc. In addition to the fact that these gearboxes belong to the new generation of mechanical power transmissions, their price is in range with conventional mechanical power transmissions. The price aspect is very important when choosing gearboxes in the power source-transmission-machine chain, so it has enabled cycloid speed reducers to become very competitive with conventional gearboxes.

Dimensions and efficiency of cycloid speed reducers are certainly one of the most important aspects in their design. The first mathematical model for determining the efficiency, in several different ways and for several different conceptions, was defined by Kudrijavcev [1]. Malhotra and Parameswaran [2] made significant improvements to Kudrijavcev's model for determining the efficiency in cycloid speed reducers, but only for one concept. Gorla et al. [3] performed a comparative analysis of experimental and theoretical determination of the cycloid speed reducer efficiency, and then made a new mathematical model for the concept presented in that paper. As part of his doctoral dissertation, Blagojevic [4] developed an algorithm for determining the forces, which are acting on the cycloid gear. Blagojevic et al. [5] conducted a study of the influence of friction that occurs in the contact of cycloid gears and corresponding rollers on the cycloid speed reducer efficiency. Mackic et al. [6] analyzed the design parameters influence on cycloid reducer efficiency. Blagojevic et al. [7] made a comparative analytical and experimental analysis of the cycloid speed reducer efficiency, which is mostly made of plastic elements. In his doctoral thesis, Matejic [8] theoretically elaborated on the three most well-known methods for determining cycloid reducer efficiency: according to Kudrijavcev, according to Malhotra and according to Gorla. Matejic et al. [9] performed a comparative analysis of obtaining the cycloid speed reducers efficiency according to the three previously mentioned methods. In [10], an analysis of the new concept cycloid speed reducer efficiency is conducted in dependence on different input parameters (power, angular speed and transmission ratio) was performed. Vasic et al. [11] performed a comparative analysis of the efficiency of classical cycloid speed reducer concept depending on the input parameters variation. In his bachelor thesis, Goluza [12] dealt with the design of a two-stage cycloid speed reducer of the classical concept. In almost all the mentioned sources, in addition to the theoretical efficiency, its experimental verification was also performed. The differences between theoretical models and experimental research are not significant, so all these theoretical models can be used with a high level of confidence.

In this paper, a comparative analysis of the dimensions of a conventional two-stage cycloid speed reducer and a two-stage cycloid speed reducer of a new concept is performed. A comparison was made of the total: length, width, height and volume. The graphical interpretation of results is very clearly presented. A comparative analysis of the efficiency according to the Kudrijavcev method was performed as well. The method of determining the efficiency was adapted so that it could be used for determination of the new two-stage cycloid speed reducer efficiency. At the end of the paper, conclusions are drawn and possible further directions of research are suggested.

14.2 Two-Stage Cycloid Reducer Concepts

In addition to conventional, planetary and wave reducers, cycloid speed reducers are increasingly used in modern machine systems. Their usage is especially pronounced when it is necessary to achieve large transmission ratios in combination with high efficiency, e.g. at the robot. The use of cycloid speed reducers has experienced expansion due to the excellent performance characteristics:

- They have a very long and reliable service life. This is a direct consequence of the quality materials selection for manufacturing vital elements, precise manufacturing, very strict quality control and careful installation. Another reason for a reliable and long service life is the absence of sliding friction in meshing of the cycloid gear and ring gear.
- They have a wide range of transmission ratios. The transmission ratio at one gear of the cycloid speed reducer is equal to the number of teeth of the cycloid gear. This feature opens up opportunities for achieving various values of transmission ratios. For single-stage cycloid speed reducers, the value of the transmission ratio ranges from $i = 3$ to $i = 119$. For two-stage cycloid speed reducers, the maximum value of the transmission ratio is $i = 119^2$.
- They are very reliable in conditions of pronounced dynamic loads. For classical concepts of cycloid speed reducers, two or more cycloid gears rotated by an angle of 180° are used per transmission stage. Due to this design solution, the centrifugal forces in the cycloid speed reducers are mutually reciprocating, so this gearbox is very suitable for operation where frequent starting and stopping of the system is required.
- They have a very high efficiency. In most cycloid speed reducer concepts, sliding friction is replaced by rolling friction, so these mechanical power transmissions achieve very high efficiency. In single-stage cycloid speed reducers, the efficiency is around 95%, while in two-stage cycloid speed reducers efficiency levels it is approximately 85%, [4, 8, 9].
- Cycloid speed reducers has a lower mass properties compared to other types of gearboxes. This good characteristic is achieved thanks to the very compact design. The mass of cycloid speed reducers is significantly smaller related to conventional types of multi-stage reducers. In addition, their overall dimensions are smaller, so it is possible to install them where there are significant spatial limitations.

In the continuation of this chapter, descriptions of the classical concept of a two-stage cycloid speed reducer will be given as well as the new concept of a two-stage cycloid speed reducer.

Fig. 14.1 Conventional concept of two-stage cycloid speed reducer

14.2.1 Conventional Concept of Two-Stage Cycloid Speed Reducer

The classical conception of a two-stage cycloid speed reducer consists of two coaxial single-stage cycloid speed reducers. The output shaft of the first stage is also the input shaft of the second stage. In practice, these reducers are made with transmission ratios from $i = 104$ to $i = 731$, [13]. As previously stated, it is theoretically possible to achieve a transmission ratio up to $i = 119^2$. In practice, such solutions are not performing in two reasons. The first reason is that the need for such transmission ratios is quite rare. The second reason is that the increase in transmission ratios according to the transmission stage, due to the complex and precise construction, has an great impact on the total price increase. If there were a need for such large gear ratios, it would be much more cost-effective to make a three-stage cycloid speed reducer whose stages would have lower transmission ratio values. Figure 14.1 shows the classical concept of a two-stage cycloid speed reducer.

14.2.2 New Concept of Two-Stage Cycloid Speed Reducer

With the new concept of cycloid speed reducers [4], a big step forward was made in the design of cycloid reducers. The biggest difference related to classic cycloid speed reducers is that in this new concept of a two-stage cycloid speed reducer uses one cycloid gear per transmission stage. For the dynamic balancing of this cycloid speed reducers concept, a similar principle was used as in the classical concepts. Namely, the cycloid gears of the first and second transmission stage are rotated in relation to the other by an angle of 180°. By using this principle, the compactness of the two-stage cycloid speed reducer was significantly increased, which directly affected

Fig. 14.2 New concept of two-stage cycloid speed reducer

the reduction of its mass and overall dimensions. Unlike the two-stage cycloid speed reducer of the classical concept, where the output shaft of the first transmission stage is also the input shaft of the second transmission stage, the new two-stage cycloid speed reducer concept has gears for both transmission stages on the same shaft. This shaft is the drive shaft for the first cycloid gear, while in the second transmission stage it only used gear nesting. The drive for the second transmission stage is obtained through a central mechanism, which is also the output mechanism of the first transmission stage. In this concept, the output torque is obtained on the ring gear of the second transmission stage, unlike the classic concept of a two-stage cycloid speed reducer, is movable and has an output shaft on it. Figure 14.2 shows a two-stage cycloid speed reducer of a new concept with an insight into its interior.

Since the new concept of a two-stage cycloid speed reducer uses one cycloid gear per transmission stage, the working principle described above is much more complex. A detailed description of the working principles of the presented new concept of two-stage cycloid speed reducer is presented in the literature [4].

14.3 Comparative Analysis of Cycloid Speed Reducers Overall Dimensions

In order to make an accurate comparison of the two-stage cycloid speed reducer of the classical and new concept, a complete design of both concepts with the same operating characteristics was performed. The operating characteristics of the two-stage cycloid speed reducers used in this research are given in Table 14.1.

One of the most important characteristics of reducers in general is certainly their overall dimensions. Figure 14.3 shows a comparative view of the length and height of a two-stage cycloid speed reducer of the classical and new concept. Figure 14.4 shows a comparative view of the height and width of the classical and new concepts

Table 14.1 Working characteristics of two-stage cycloid reducers

Name	Designation, unit	Classic concept	New concept
Input power	*P*, kW	2.53	2.53
Input RPM	*n*, min^{-1}	750	750
Total transmission ratio	*i*, ul	121	121
Partial transmission ratio-1	i_1, ul	11	11
Partial transmission ratio-2	i_2, ul	11	11
Dividing circle radius-1	r_1, mm	70	130
Dividing circle radius-2	r_2, mm	130	130

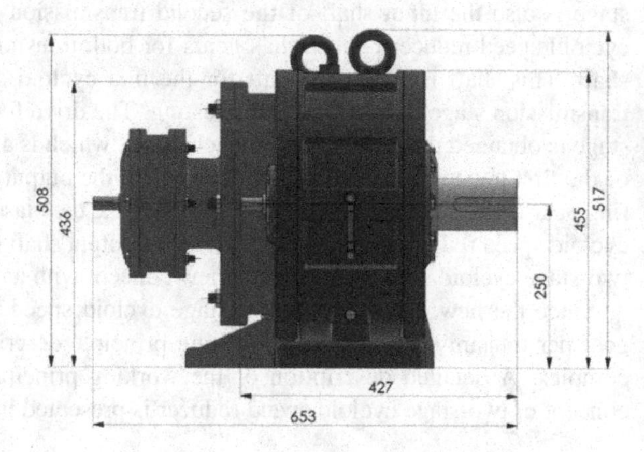

Fig. 14.3 Comparative characteristics of the length and height of two-stage cycloid speed reducer of the classical and new concept

of two-stage cycloid speed reducer. Figure 14.5 shows a comparative representation of the volume characteristics of the classical and new concept of a two-stage cycloid speed reducer.

According to the design of both two-stage cycloid speed reducers, their masses were determined. The mass of the classic concept of a two-stage cycloid speed reducer is about 230 kg, while the mass of the new concept is about 180 kg. Table 14.2 shows the comparative characteristics of dimensions, mass and volume of a two-stage cycloid speed reducer of the conventional and new concept.

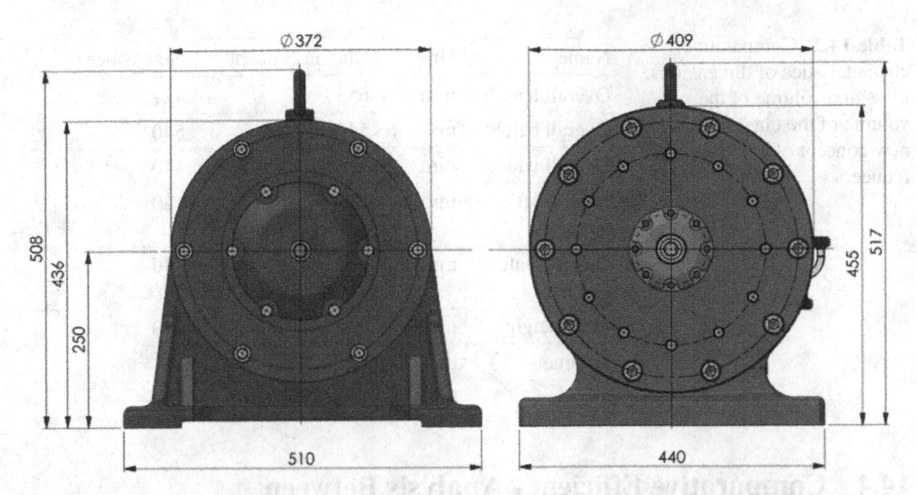

Fig. 14.4 Comparative characteristics of width and height of two-stage cycloid speed reducers of classical and new concept

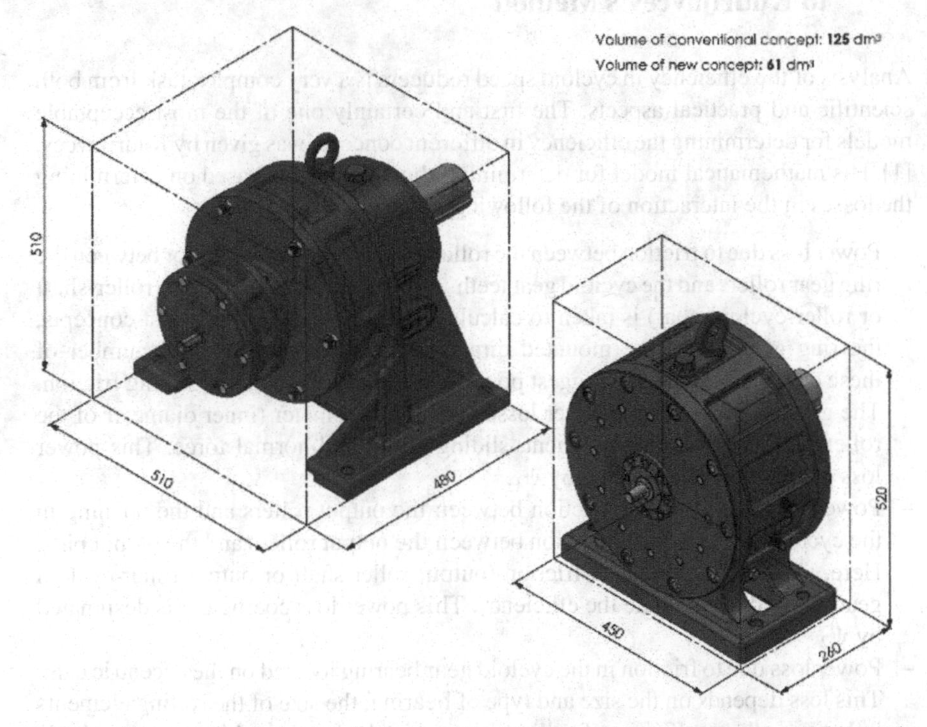

Fig. 14.5 Comparative view of the volume of the classical and new concept of cycloid speed reducers

Table 14.2 Comparative characteristics of dimensions, mass and volume of the volume of the classical and new concept of cycloid speed reducers

Name	Unit	Classic concept	New concept
Overall length	mm	653	427
Overall height	mm	540	540
Overall width	mm	510	440
Input shaft diameter	mm	20	20
Output shaft diameter	mm	80	80
Shaft height	mm	250	250
Volume	dm^3	125	61
Mass	kg	230	180

14.4 Comparative Efficiency Analysis Between Conventional and New Concept According to Kudrijavcev's Method

Analysis of the efficiency in cycloid speed reducers is a very complex task from both scientific and practical aspects. The first and certainly one of the most acceptable models for determining the efficiency in different concepts was given by Kudrijavcev, [1]. His mathematical model for determining the efficiency is based on determining the losses in the interaction of the following elements:

- Power loss due to friction between the rollers and the ring gear pins, or between the ring gear rollers and the cycloid gear teeth. A lower friction coefficient (roller-shaft or roller-cycloid-gear) is taken to calculate the efficiency. In the most concepts, the ring gear rollers are mounted directly on the shafts. Since the number of these contacts is large, the biggest power losses occur here due to sliding friction. The greatest influence on power losses has shaft diameter (inner diameter of the roller), sliding friction coefficient, sliding speed and normal force. This power loss coefficient is designated by ψ_1.
- Power loss due to rolling friction between the output rollers and the opening in the cycloid gear, or due to friction between the output rollers and the output pins. Here, a lower coefficient of friction (output roller-shaft or output roller-cycloid gear) is used to calculate the efficiency. This power loss coefficient is designated by ψ_2.
- Power loss due to friction in the cycloid gear bearing located on the eccentric cam. This loss depends on the size and type of bearing, the size of the rolling elements in bearing, the coefficient of rolling friction in the bearing, the force magnitude on the eccentric cam and the angular velocity. This power loss can be quite large depending on the size of the eccentric bearing. The power loss coefficient in the eccentric bearing is designated by ψ_3.

The total power loss coefficient ψ is calculated as the sum of the previously described losses and is represented by Eq. (14.1):

$$\psi = \psi_1 + \psi_2 + \psi_3 \tag{14.1}$$

Efficiency calculation per one transmission stage η_I is given by Eq. (14.2):

$$\eta_I = \frac{1 - \psi_I}{1 + z_1 \psi_I} \tag{14.2}$$

Total efficiency calculation η of conventional two-stage cycloid reducer is given by Eq. (14.3):

$$\eta = \eta_I \cdot \eta_{II} = \frac{1 - \psi_I}{1 + z_1 \psi_I} \cdot \frac{1 - \psi_{II}}{1 + z_2 \psi_{II}} \tag{14.3}$$

Total efficiency calculation η of the new concept of two-stage cycloid reducer is given by Eq. (14.4):

$$\eta = \eta_I \cdot \eta_{II} = \frac{1 - \psi_I}{1 + z_1 \psi_I} \cdot \frac{1 - \psi_{II}}{1 + z_1 \psi_{II}} \tag{14.4}$$

A detailed procedure for determining the efficiency according to this method is described in [7–9]. According to the previously described procedure, the efficiency was determined for both the classical and the new concept of a two-stage cycloid speed reducer according to the done design. Partial loss coefficients are shown in Table 14.3.

When the total efficiency levels are calculated, it is obtained that the two-stage cycloid speed reducer of the classical concept has a efficiency of $\eta = 0.79$, while the two-stage cycloid speed reducer of the new concept has $\eta = 0.782$.

Table 14.3 Partial power losses coefficients in the classical and in the new concept of two-stage cycloid speed reducer

Classic concept of two-stage cycloid speed reducer $\eta_{CR} = 0.79$					
	ψ_1	ψ_2	ψ_3	ψ	η
1. Transmission stage	0.0047	0.0042	0.0021	0.0110	0.882
2. Transmission stage	0.0067	0.0036	0.0024	0.0127	0.887
New concept of two-stage cycloid speed reducer $\eta_{CR} = 0.782$					
	ψ_1	ψ_2	ψ_3	ψ	η
1. Transmission stage	0.00405	0.004	0.002	0.01005	0.891
2. Transmission stage	0.00525	0.003	0.002	0.01025	0.889

14.5 Conclusions

This paper describes two concepts of two-stage cycloid speed reducers. The classical and new concept of cycloid speed reducers are described. The classical concept of a cycloid speed reducer consists of two single-stage coaxial cycloid speed reducers in which the output shaft of the first transmission stage is also the input shaft of the second transmission stage. The classic concept has two gears per transmission stage. The new two-stage cycloid speed reducer concept has one gear per transmission stage and the output torque is transmitted to the ring gear of the second transmissions stage.

The total length of the conventional two-stage cycloid speed reducer concept is about 34% longer than the new concept. The width of the classical concept is, as well about 14% larger than the new concept. The height of the classic concept is 4% lower than the new concept. If the volume of the body of the cycloid speed reducer is observed, the volume of the new concept is about 52% smaller related to the classical concept of the two-stage cycloid speed reducer. The mass of the new concept is about 21% less than the classical concept. From the aspect of geometric and mass characteristics, it is concluded that the usage of the new concept would lead to significant savings in the space required for the installation of a two-stage cycloid speed reducer, as well as significant savings in material.

The efficiency of the classical concept of a two-stage cycloid speed reducer is only about 1% higher related to the new concept, according to Kudrijavcev's method. According to the results presented in this paper, the new concept has significant advantages over the classical concept of a two-stage cycloid speed reducer. However, during the design, possible problems were noticed in the eventual usage of the new concept, such as more demanding, precise, and thus more complicated production, greater complexity of vital elements, which entails difficult installation, etc. All this would lead to a higher cost of manufacturing a new concept.

In further research, the plan is to make both cycloid speed reducers, the classic and new concepts, and to make a comparative analysis of the required time of production and installation. In addition, after the development, an experimental verification of the efficiency of analytically obtained results is planned.

Acknowledgements The Ministry of Education and Science of the Republic of Serbia under the contract TR33015 funded this research.

References

1. Kudrijavcev, V.N.: Planetary Transmissions. Moscow (1966)
2. Malhotra, S.K., Parameswaran, M.A.: Analysis of a cycloid speed reducer. Mech. Mach. Theory **18**(6), 491–499 (1983)
3. Gorla, C., Davoli, P., Rosa, F., Chiozzi, F., Samarani, A.: Theoretical and experimental anaysis of a cycloidal speed reducer. J. Mech. Des. **130**(11), 112604-1–112604-8 (2008)
4. Blagojevic, M.: Stress and strain state of cycloidal speed reducer elements during the dynamic loads. PhD Thesis, Kragujevac (2008)

5. Blagojevic, M., Kocic, M., Marjanovic, N., Stojanovic, B., Djordjevic, Z., Ivanovic, L., Marjanovic, V.: Influence of the friction on the cycloidal speed reducer efficiency. J. Balk. Tribol. Assoc. **18**(2), 217–227 (2012)
6. Mackic, T., Blagojevic, M., Babic, Z., Kostic. N.: Influence of design parameters on cyclo drive efficiency. J. Balk. Tribol. Assoc. **19**(4), 497–507 (2013)
7. Blagojevic, M., Matejic, M., Kostic, N., Petrovic, N., Marjanovic, N., Stojanovic, B.: Theoretical and experimental testing of plastic cycloid reducer efficiency in dry conditions. J. Balk. Tribol. Assoc. **23**(2), 367–375 (2017)
8. Matejic, M.: A new approach to design and optimization of cycloid speed reducers. PhD Thesis, Kragujevac (2019)
9. Matejic, M., Blagojevic, M., Cofaru, I. I., Kostic, N., Petrovic, N., Marjanovic, N.: Determining efficiency of cycloid reducers using different calculation methods. In: MATEC Web of Conference, Vol. 290, pp. 01008-1–01008-8 (2019)
10. Matejic, M., Blagojevic, M., Kostic, N., Petrovic, N., Marjanovic, N.: Efficiency analysis of new two-stage cycloid drive concept. Tribol. Ind. **42**(2), 337–344 (2020)
11. Vasic, M., Matejic, M., Blagojevic, M.: A comparative calculation of cycloid drive efficiency. In: Proceedings of the Conference on Mechanical Engineering Technologies and Applications, COMETa 2020, East Sarajevo, pp. 259–266 (2020)
12. Goluza, V.: Design of two-stage cycloid reducer. BSc Thesis, Kragujevac (2020)
13. Sumitomo Drive Technologies: Cyclo Drive 6000 Gearmotors & Speed Reducers (2012)

Chapter 15
Balancing Rotating Parts. A New Method and Device

Mircea-Viorel Dragoi, Marius Daniel Nasulea, Milos Matejic, and Gheorghe Oancea

Abstract Rotating parts affected by an eccentricity create problems in a proper working of the entire ensemble that contains such parts. Since eccentricity, even if a very small one exists, it has to be compensated in order to ensure a smooth working of the part and ensemble it is mounted in. The paper presents a device that attached to a rotating part can balance it, so it rotates smoothly, without any runout. The principle the device is based on is adding a mass at the proper place on the eccentric part. The novelty of the device consists of that the mass of added material does not depend on the eccentricity of the part. It seeks its correct position on the rotating part moving its center of gravity towards the rotation axis along a certain trajectory until the balance is achieved. The trajectory is placed in a perpendicular plane to the rotation axis. Being an adjustable one, the device can be used to balance parts having different eccentricity or mass. Having a mobile centre of gravity, that is an own adjustable eccentricity strictly kept under control, the device can compensate the entire ensemble's one. This concept removes the main disadvantage of the classical methods, which are much time consuming, caused of the repeating procedures. The device can be adjusted individually for each part, or for more parts in a batch, which are sorted in groups, to each group fitting a certain adjustment. Tuning and attaching the device to the eccentric part are very easy and safe.

Keywords Eccentricity balancing · Rotating parts · Adjustable device

15.1 Introduction

A good balancing of the rotating parts or subassemblies is essential for a smooth rotation, free of vibrations or runout in functioning. Because almost any part displays

M.-V. Dragoi (✉) · M. D. Nasulea · G. Oancea
Department of Manufacturing Engineering, Transilvania University of Brasov, B-dul Eroilor 29, 500036 Brașov, Romania
e-mail: dragoi.m@unitbv.ro

M. Matejic
Faculty of Engineering, University of Kragujevac, Sestre Janjić 6, 34000 Kragujevac, Serbia

M. Rackov et al. (eds.), *Machine and Industrial Design in Mechanical Engineering*, Mechanisms and Machine Science 109,
https://doi.org/10.1007/978-3-030-88465-9_15

an eccentricity due to manufacturing processes, and/or assembling imperfection, the single solution to get a well-balanced part is to compensate the existing eccentricity. To do that, several methods are available, as can be found in literature. An important work in these terms is [1] where a fundamental concept is defined: the influence coefficient, which is the basis for many balancing methods. It provides an extensive theory on the balancing techniques, which takes into account many of the previous achievements. Balancing the rotor of an electric motor of 15,000 rpm, by adding loads (weights) in two different planes displaced perpendicularly to the rotation axis is presented in [2]. An in situ balancing method, based on measuring eccentricity by means of an accelerometer, and remove material from appropriate place by laser ablation is described in [3]. A method to diagnose the unbalancing using an eddy current sensor, instead of a classic accelerometer is presented in [4], and another solution to compensate the eccentricity of the spindle and disk of the DVD drives in [5]. The influence of nonlinear forces on the balancing procedure is analyzed in [6]. The nonlinear response is substituted with a linear approximation that is precise enough to avoid affecting the accuracy of diagnose. In [7] a side issue is presented: a material suitable for vibration transducer used in balancing rotating parts. A method to statically balance a part in order to get its dynamic stability is described in [8]. This is a time and cost saving method. A specific problem is treated in [9]: a study devoted to large thin disks. Even if the unbalanced mass is not significant, the unbalancing vector might be an important one, which causes high vibration. Estimating the unbalancing vector is a challenge for authors' approach. An original way to balance rotors while they are in rotating movement, by removing material through laser ablation is presented in [10]. This is in fact a technology used in balancing large parts in situ. A method to balance dissymmetrical parts subjected to rotation movement is set up in [11]. This is an analytical method used to find the means of balancing rotating parts that do not benefit from symmetry.

Among the several available methods to balance rotating parts, one of the most used is that which adds material to the subject part, in the appropriate amount and position, with its different variants. Regardless the variant adopted, this method has the disadvantage that it is a repetitive one, time consuming, and affected by some lack of precision. This paper proposes an original approach; the added mass has a moveable gravity center. Moving continuously and in the appropriate direction the center of gravity of the added mas, it comes to the position where the aggregate body gets its balance. The right location of the added mass can be determined analytically, as well. This operation is automated by means of an original piece of software designed by the authors.

15.2 Materials and Methods

The principle the method is based on is that a device is attached to the unbalanced part, as an additional mass. The device has a symmetry axis, and by moving some of its parts, the gravity centre travels along the symmetry axis. This axis has to be

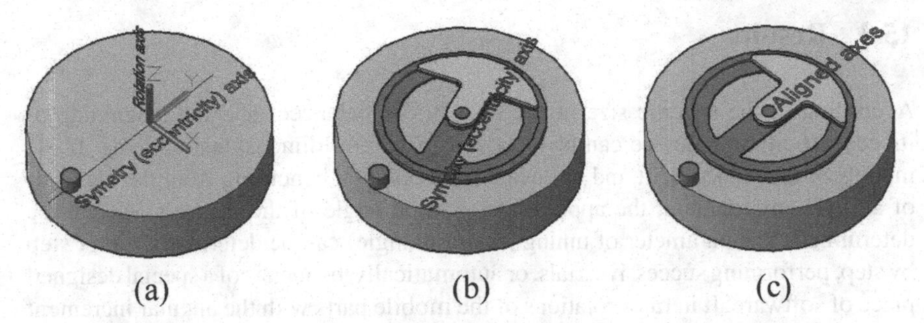

Fig. 15.1 Aligning the device to the part: **a** Part and its eccentricity axis, **b** Device and its eccentricity axis (not aligned to the part's one) and **c** The aligned axes of part and device

aligned to the axis determined by the rotating axis and the centre of gravity of the unbalanced part, as shown in Fig. 15.1, where the red coloured bolt is intended to materialize the eccentricity of the subject part. It emphasizes, as well, the eccentricity axis. After the alignment is done, so the centre of gravity of the device and the part's one are placed in opposition related to the rotation axis, the mobile elements (ME) of the device are rotated symmetrically, so the centre of the aggregate body migrates to the axis of rotation. When it is placed right on the rotation axis, the balance is achieved. The device benefits of adjustability, and is free of the disadvantage of a discrete tuning. Tuning the device can be done according to individual products, or for batches of products. In this case, the parts of the batch can be first sorted in smaller groups that have eccentricity in the same narrower range. The same tuning of the device might fit to an entire subset of parts.

Rotating symmetrically ME of the device, its gravity centre (and at the same time the aggregate gravity centre—AGC) migrates along the eccentricity axis to the rotation axis, achieving the balanced position, as shown in Fig. 15.2. The device is designed in two variants, so it is able to work on the front side of a rotating part, or surrounding a shaft of an assembly.

Fig. 15.2 Rotating the mobile parts of the device to get the balance of the aggregate body

15.3 Results

According to the specific size of the body to be balanced, and the magnitude of its eccentricity, the device can be designed in several dimensional variants. If 3D models of the subject part and of device are available, benefiting from the facilities of a CAD environment, the appropriate rotation angle of the device's ME can be determined. The parameter of tuning (rotation angle) can be determined either step by step, performing successive trials, or automatically, by means of a special designed piece of software. It iterates rotations of the mobile parts with the angular increment input by the user, and computes the position of AGC position. Iteration stops when this position is close enough to the rotation axis. User can control the balancing precision by means of a tolerance value he used as input.

A useful tool for the user is the chart of the gravity centre position as a function of the rotation angle. Such a diagram is easily drawn using the results of running a software tool, as well designed by the authors. Three such diagrams are presented in Fig. 15.3, for three different shapes and sizes of ME.

In Table 15.1 are presented the data that describe the position of AGC when device is mounted on an eccentric part.

The support disk and ME are marked with signs that help a precise control of their rotated position. Different solutions for fastening the device on the part or assembly, and to keep ME in their proper position can be applied, without affecting the symmetry of the construction. Manufacturing the parts of the device does not pose any difficulty in terms of technology or precision of execution.

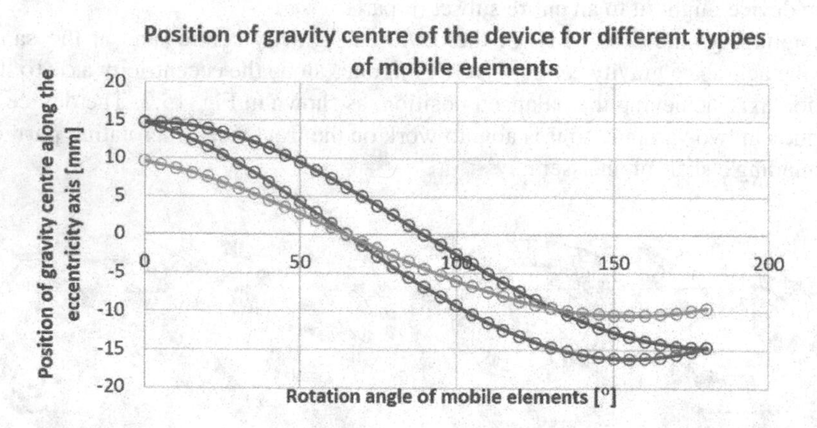

Fig. 15.3 Position of gravity centre of the device as ME rotate

Table 15.1 Dependence of AGC position on rotation angle of the tuning elements

Rotation angle of ME (°)	AGC distance from rotation axis (mm)	Rotation angle of ME (°)	AGC distance from rotation axis (mm)
0	0.5036	24	0.2870
3	0.4840	27	0.2512
6	0.4622	31	0.2138
9	0.4381	33	0.1749
12	0.4119	36	0.1345
15	0.3836	39	0.0928
18	0.3533	42	0.0499
21	0.3211	45	0.0059

15.4 Discussion

The design of the device can be easily adapted to the specific of the part or assembly subjected to balancing. Versions for blind (Figs. 15.1 and 15.2), or shafted parts (Fig. 15.4) are available.

The shape and size of ME can be determined according to the particular conditions of any unbalanced part. As can be seen in Table 15.2, an appropriate size of ME allows a very fine tuning of the device: a rotation of 3° produces an AGC displacement of 5–15 μm, depending on the magnitude of the total rotation angle. The chart in Fig. 15.3 shows, as well, that the intensity of influence of rotation angle on the tuning precision can be controlled by the shape and/or size of the ME: designing them appropriately, can be selected the side of the dependence graphic, with bigger or smaller slope. If needed, for special cases, when restrictions are posed on the size of ME, designer has the chance to choose for them a material with higher density. A special attention must be payed to rotate symmetrically the two ME. In this way the symmetry of the

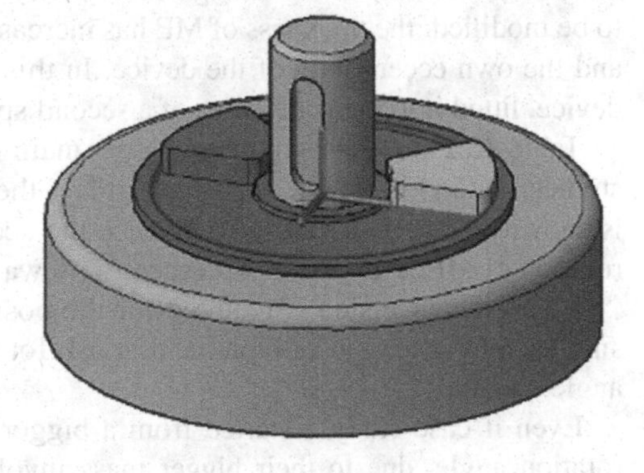

Fig. 15.4 Balancing device for shafted part

Table 15.2 Data that characterize the two case studies

Characteristic	Case study 1	Case study 2
Part mass (kg)	5.98	5.98
Part eccentricity (mm)	– 0.253	– 0.401
Device mass (kg)	0.158	0.191
Thickness of ME (mm)	2.5	3.5
Device eccentricity (mm)	14.72	17.05
Tolerance for AGC position	0.01	0.01
Increment of rotation angle (°)	3	3
AGC position before balancing (mm)	0.133	
AGC position before balancing with 2.5 mm elements thickness (mm)		– 0.013
AGC position before balancing with 3.5 mm elements thickness (mm)		0.14
AGC position after balancing	0.008	0.004

device is maintained, and AGC does not deviate from the eccentricity vector. The method has a weakness: the need to determine the direction (vector) of eccentricity and to align the axes of the part and of the device before starting the procedure of balancing.

15.5 Case Studies

To illustrate the validity of the balancing proposed method, two case studies are presented. They target the same part, having a mass of 5.98 kg, in two variants: with eccentricity of 0.253 mm, and 0.401 mm, respectively. The device that balanced the first sample had not a big enough eccentricity to balance the second one, so it had to be modified: the thickness of ME has increased. This increased, as well the mass, and the own eccentricity of the device. In this way, a minimal modification of the device, fitted it to the necessities of a second specific case.

Table 15.2 presents synthetically the main data that characterizes the two case studies, and in Fig. 15.5 a chart that displays the movement of AGC when ME rotate is shown for the two cases. To balance the second sample, was enough a smaller rotation of ME, even if the part eccentricity was bigger than that of the first sample.

If a better precision is required for the position of AGC, user may introduce a smaller tolerance as data input, and/or ask for a smaller increment of ME rotation angle.

Even if case study 2 started from a bigger eccentricity, it needed smaller ME rotation angle, due to their bigger mass involved in balancing. The precision of balancing was, as well, better for the second sample. Yet, precision of balancing

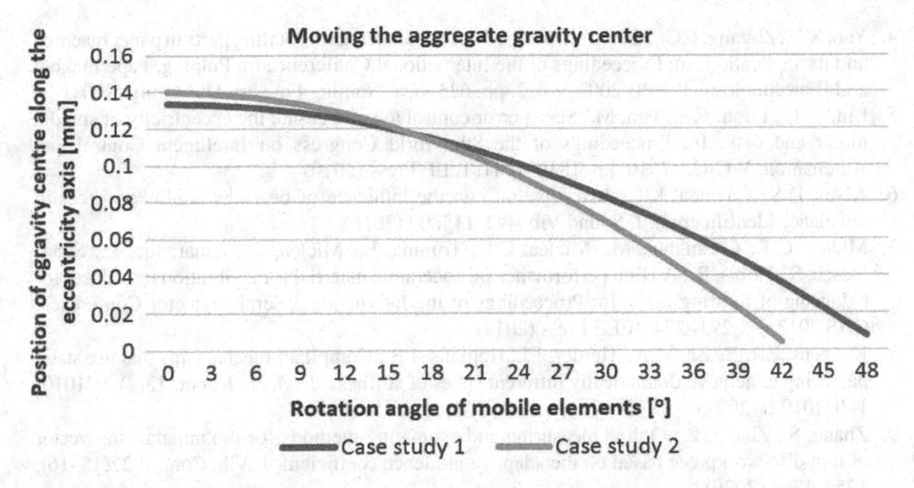

Fig. 15.5 Case studies. AGC position while ME rotate symmetrically to get the balance

can be improved, if keep rotating ME 1° or 2° after the tolerance of 0.01 mm was achieved.

15.6 Conclusion

The method and device for balancing eccentric part are new concepts. Their main novelty consists of the continuous adjustment of gravity center position of the device which is attached to the unbalanced part. The device is easy fastened to the part subjected to balancing. ME are, as well, easy to be fixed in the right position after the balanced position was achieved. The device can be easily adapted to fit in terms of shape, size, and own eccentricity to any specific case. The continuous adjustment of the device is an advantage, it makes balancing easier to be done than through other similar methods. The disadvantage of the method is the necessity of pre-setup that consists of finding the eccentricity direction and aligning the axes of the part and device.

References

1. Darlow, M.S.: Balancing of high-speed machinery: theory, methods and experimental results. Mech. Syst. Signal Pr. **1**(1), 105–134 (1987)
2. Korotkov, V.S., Sicora, E.A., Nadeina, L.V., Yongzheng, W.: Balancing fast-rotating parts of hand-held machine drive. In: IOP Conference Series: Materials Science and Engineering, vol. 327, pp. 042054-1–042054-5 (2018)
3. Stoesslein, M., Axinte, D.A., Guillerna, A.B.: Pulsed laser ablation as a tool for in-situ balancing of rotating parts. Mechatronics **38**, 54–67 (2016)

4. Yao, X.Y., Zhuang, F.C., Zhang, H.: Method of site balancing for rotating parts in paper machine and its application. In: Proceedings of the International Conference on Pulping, Papermaking and Biotechnology, ICPPB 2008, vol. 2, pp. 623–628. Nanjing Forestry University (2008)
5. Lin, Y.-H., Luoh, F.-B. Pan, M.: Servo-error control to compensate the eccentricity of spindle motor and disk. In: Proceedings of the 8th World Congress on Intelligent Control and Automation, WCICA 2010, pp. 1836–1841. IEEE Press (2010)
6. Alves, D.S., Cavalca, K.L.: Investigation into the influence of bearings nonlinear forces in unbalance identification. J. Sound Vib. **492**, 115807 (2016)
7. Miclea, C.T., Cioangher, M., Miclea, C.F., Trupina, L., Miclea, C., Amarande, L., Spanulescu, S., Faibis, R.: A High performance piezoceramic material for a vibration transducer for balancing of rotating parts. In: Proceedings of the International Semiconductor Conference, CAS 2012, pp. 291–294. IEEE Press (2012)
8. Kuppens, P.R, Bessa, M.A., Herder, J.J., Hopkins, J.B.: Compliant mechanisms that use static balancing to achieve dramatically different states of stiffness. J. Mech. Robot. **13**(2), 021010-1–021010-6 (2021)
9. Zhang, S., Zhang, Z.: Online measuring and estimating methods for the unbalancing vector of thin-disc workpiece based on the adaptive influence coefficient. J. Vib. Control **27**(15–16), 1753–1764 (2020)
10. Stoesslein, M., Axinte, D., Gilbert, D.: On-the-fly laser machining: A case study for in situ balancing of rotative parts. J. Manuf. Sci. E.-T. ASME **139**(3), 031002-1–031002-14 (2017)
11. He, Z.H., Rui, Y.N., Ma, Y.P.: Method of balancing-design of the non-symmetric rotary part with thin wall by 3D technology. J. Soochow Univ. Eng. Sci. Ed. **25**(4), 69–70 (2005)

Chapter 16
Machine Retrofitting for Tissue Paper Industry—INTERFOLDER Case

Milos Milivojevic, Bozica Bojovic, Vladimir Babic, and Djordje Djuric

Abstract This paper is a presentation of applied design knowledge as well as practical experience in mechanical engineering and industrial automation in order to retrofit machines. The main goal is to offer affordable solution for automated log transfer and at the same time extend the life of those tissue paper converting machines that are already in operation. We exhibit practically tested solutions for a module for sheet counting and log separation that can be mounted onto existing semi-automated machine. The digital prototype e.g. CAD model is generated for a preliminary module design. Rapid prototyping is used to refine delicate geometry of moving segment before tangible manufacturing. By machine retrofitting we attain a fully automated production that consequently increased efficiency of the line by elimination of the log manual transfer operation. Also, data collection from built-in sensors and analysis provide optimization of the overall equipment effectiveness. Additionally, retrofitting includes repairs, replacements and adjustments of electrical and pneumatic systems, implementation of software for controlling that adapts machine for conformity marking and for regulations.

Keywords Retrofitting · Fully automated · Modular design

M. Milivojevic · B. Bojovic (✉)
Department of Production Engineering, Faculty of Mechanical Engineering, Belgrade University, Belgrade, Serbia
e-mail: bbojovic@mas.bg.ac.rs

V. Babic
Consulting Agency, 25. Maj, Belgrade, Serbia

D. Djuric
Enerteh, Belgrade, Serbia

M. Rackov et al. (eds.), *Machine and Industrial Design in Mechanical Engineering*, Mechanisms and Machine Science 109,
https://doi.org/10.1007/978-3-030-88465-9_16

16.1 Introduction

Producers and consumers of tissue papers are turning to interfolded wipes, as an economically and environmentally friendly alternative, resulting in savings in transport and storage space and 40% lower consumption according to [1]. They are additionally supported by more demanding hygiene standards, especially recent after COVID-19. Such interfolded wipes are produced on INTERFOLDER machines. Currently, the wipes transfer from INTERFOLDER machine to the automatic packaging machine which requires increased workers for transferring wipes by hand, according to [2]. The producers in tissue paper converting industry are under pressure to optimize production, to reduce costs and to introduce automation technologies. At the same time, the benefits from I4.0 are encouraging enough to provoke the next stage of tissue paper industry [3].

If the current machines are still operational or purchasing of the new machines is costly, the retrofitting offers acceptable solutions. Retrofitting increases machine functionality and reliability, as well as enabling older machines to be compatible with new technologies [4]. Traditional retrofitting is prerequisites for upgrading machines suitable for smart retrofitting, which is discussed in [5]. Industry 4.0 requires smart retrofitting for taking full advantage of it. Although practitioners deal with retrofitting long before this topic gains attention in research domain, the retrofitting process still suffers variation in the implementation in different industries [6], which makes the identification of common points in retrofitting difficult. Useful generalization will provide the platform presented in [6] that assists in the process identification of requirements for retrofitting and implementation technologies for integration with Industry 4.0 and the methodology presented in [7] that provides efficient way to retrofit by deriving new functions and features based on data analysis results.

The developed technical solution for retrofitting refers to the field of mechanical engineering of special purpose machines for the production of interfolded. Given the current presence of semi-automatic INTERFOLDERS in plants around the world and the lack of interest of manufacturers of original machines for their full automation, there was a need to retrofit them. Retrofitting was realized by introducing an automatic transfer of a log with a given number of wipes. It is a prerequisite for further automatic transport by conveyor to the packing and sawing section. The construction, manufacturing, assembly and verification of modules for automatic counting and separation of logs of self-folding wipes were realized through the cooperation of the company ENERTEH, the Consulting Agency 25. MAJ and the Department of Production Engineering, Faculty of Mechanical Engineering in Belgrade.

16.2 Problem Interpretation

There are several manufacturers of Interfolder machines on the market, such as: PERINI Italy, PCMC Italy, DCM France, Dechang Yu China, BaoSuo China,

HINNLI Taiwan, [8–13] and others. Interfolders, which process raw material from two roll widths up to 1500 mm, requires the constant presence of an operator who manually transfers a log of wipes up to 1500 mm wide to the conveyor for further processing (log sawing, packaging). This manual work is a bottleneck in the production process and an obstacle to automating transport.

The answer to the increased market demands and the reduction of unit product costs is the automation of the transfer of a log of interfolded wipes from the machine to the conveyor. The only offered construction solution for new INTERFOLDER type machines with a working width of 1500 mm counts the sheets just below the machine heads, lowers the stack and removes it to the side, perpendicular to the direction of the paper, through the side opening. Such machines are from the manufacturers Ocean Taiwan [14] and OMET Italy [15]. The described solution is not applicable to already existing semi-automatic machines for constructive reasons, primarily due to the lack of a side opening in the supporting structure of the machine. Currently for the improvement of the existing semi-automatic machines of the INTERFOLDER type for roll widths up to 1500 mm, there is no adequate engineering solution on offer. Therefore, we offer retrofit solution that is modular for automatic counting and separation of packages of inter-folded wipes, which provides automation of production lines by introducing automatic transfer of logs with a given number of wipes, as a precondition for further automatic transport by conveyor to the packing and cutting section.

16.3 INTERFOLDER Machine Retrofitting

The offers of INTERFOLDER machines indicate similar construction solutions, which include the following working sections of the machine:

- Unwinders,
- Lamination sections or joining two layers of paper over the entire surface by using glue,
- Embossing sections, i.e. edge joining layers of paper,
- Paper ironing sections or calendars,
- Towing rollers,
- Wrapping section,
- Automatic towel cutting and stacking heads, and
- Longitudinal paper cutting sections.

The tissue paper unwinds with pneumatic paper roll loading system, by use of pneumatic control for driving paper roll unwinding, and passes through vacuum based folding heads to the automatic counting and transfer system. Folding heads are coupled by gears powered by one AC motor equipped with independent frequency inverter control. Vacuum based folding heads consist of vacuum blade roller with three movable bottom blades with holes to hold the paper firmly while cutting.

Therefore the each sheet is same size and good cutting. Anti-paper twist device prevents a big scrap of paper to go into the folding system, and protect the blades.

Vacuum based folding heads form a log of wipes by coupling two sheets of paper that come from the parent rolls. Each folding head has three spiral cross-cutting blades and three stator blades placed on the perimeter of the head in alternating order. This enables alternating cut of each sheet of paper and fold of other sheet at the same time. It is the point where sheets of paper become inter-folded. Sensor is reading the number of turns/revolutions of the folding heads enabling the successive countdown of wipes with permanent lowering of the log.

Pulling belts and pushers ensure infallible transport of logs towards infeed conveyor of the wrapping unit. After wrapping, logs are transported into the log saw and output table. Sensors check the paper break, wrap up or jams on the machine, stopping it automatically when paper is broken. Emergency stop buttons are installed on each part of the machine.

At the end of the converting process, a continuous series of folded wipes up to 1500 mm wide was obtained, which slides through the slope straight and accumulates on the table. To achieve it, we produced a module that through complete automation provides a reduction in production costs, because the presence of a man who manually moves logs is unnecessary. This allows easy handling and automatic transport of the wipe stack to the packing and cutting sections. Automation is the necessary basis for the introduction of Industry 4.0 by setting up a sensor that provides the necessary data on the number of folded wipes. To achieve this retro-fitted machine performs three main functions schematically presented Fig. 16.1:

- forming a set of wipes just below the heads for cutting and stacking using front and rear claw (Fig. 16.1-left),
- successive countdown of wipes with permanent lowering of the log standing at bench and separation from following log using claws (Fig. 16.1-middle), and
- tracing the log in a shorter of paper movement through the machine using pusher (Fig. 16.1-right).

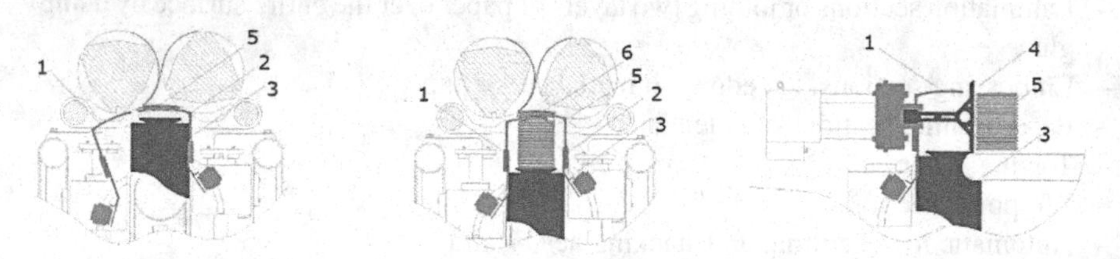

Fig. 16.1 Schematic presentation of forming a log of wipes just below the heads for cutting and stacking (left), successive countdown of wipes with permanent lowering of the log (middle) and removal of the log through the machine to the conveyor (right); Legend: 1—front claw, 2—rear claw, 3—bench, 4—pusher, 5—current log, 6—next log

16.3.1 Modular Design and Retrofitting

The last two working sections from the original machine form the base of the machine and they are retained, but the section for longitudinal paper cutting and the table for lowering the inter-folded wipes have been removed. The design solution presented here is **parametrically designed** in SolidWorks (see Fig. 16.2) and the **principle of modularity** is applied primarily for the upcoming rapid reactions to market demand, i.e. efficient and effective implementation of this module for retrofit semi-automatic machines from other manufacturers with different working widths. In this way, it is achieved that the design process does not go back to the beginning, but only to certain final stages that are automated with the application of contemporary CAD software.

The CAD model of the module is a digital prototype, which enabled the verification of the adequacy of overall dimensions, fitting into the existing structure and assembly with the rest of the structure with physical limitations arising from the folding heads, under which it is installed. Special attention during the design was required by the front and rear claws due to the complex movement of the claw tips (see Fig. 16.1).

On the existing construction of the semi-automatic machine type INTERFOLDER manufactured by Dechang Yu China, new working sections have been designed, manufactured and installed. The main parts of working sections are the front claw, the rear claw and the bench. Front and rear claw are engaged in automating separation of logs while the log permanently takes down standing at the bench. Physical representations of them are given in Fig. 16.3.

Process of automatic counting and separation of a set of self-folding wipes is consist of following functions:

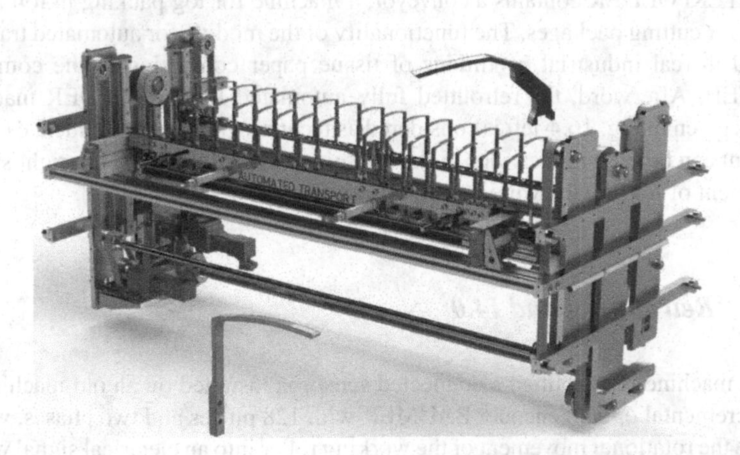

Fig. 16.2 CAD model of module for automatic counting and separation of a log of inter-folded wipes with two versions of claw tips

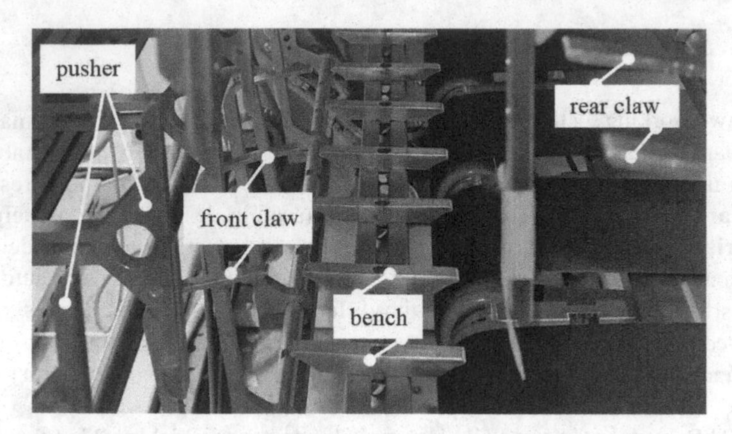

Fig. 16.3 Physical representation of the module for automatic counting and separation of a set of self-folding wipes

- Automatic counting of inter-folded wipes using mounted sensor which collects data directly from machine useful for enhancing productivity,
- Formation of wipe logs using the front claw, back claw and bench (see Fig. 16.1-left),
- Logs lowering on the bench (see Fig. 16.1-middle), and
- Log transferring onto the conveyor using the pusher (see Fig. 16.1-right).

In order to comply with the strict requirements of the CE mark, the improvement of the machine included the replacement of certain machine parts, pneumatic and electrical components. Checking the functioning of the module required the new program implementation for managing the complete production line, which along with INTERFOLDER contains a conveyor, a machine for log packing in foil and a log saw for cutting packages. The functionality of the module for automated transfer is tested in real industrial conditions of tissue paper converting in the company ENERTEH. Afterword, the retrofitted fully automated INTERFOLDER machine which is given in Fig. 16.4-left is considered as technological ready. Evidence of this statement can be seen in the video at the link given in [16]. Figure 16.4-right shows the moment of folding the wipe captured from the video.

16.3.2 Retrofitting and I4.0

When a machine is retrofitted a connected sensor is installed on an old machine. It is an incremental optical encoder BAUMER with 128 pulses and two phases, which converts the rotational movement of the working roller into an electrical signal which represents the feedback for the control system. As the working roller prepares the wipe sheets for V stacking, the exact position of the individual sheet is known at any time via the encoder, on the basis of which information on the number of sheets is

Fig. 16.4 Upgraded machine INTERFOLDER 1500, left: front view from where the paper enters, right: section for folding and folding wipes

obtained, but also the right moment when the front and rear claws should entry, in order to achieve proper separation of the two logs (see Fig. 16.1-middle).

Three sheets of two rolls of paper are cut through each of the two coupled work rollers, which have paper cutting knives on them. In this way, during one turn, a total of six sheets are formed, mutually inter-folded in a V within the log. The incremental encoder allows the movement to be divided into 512 segments during this rotation, which has been empirically proven to be quite sufficient for precisely determining the moment of claw entry. That moment depends on the speed of stacking, because the speed of rotation of the rollers changes, and the time required for the claws to come out of the extended position is constant. Specifically in this case, it depends on the speed of the rotating pneumatic cylinder and the working air pressure. Based on that, the speeds and moments of claw entry, which correspond to the counted number of wipes, were empirically determined. It is software-regulated that this happens in the first free moment after the last sheet has been selected.

The encoder whose pulses enable the counting of slips is the basis for the production process monitoring and the entire production line, with the possibility of making decisions based on data collected directly from the machine. Data on the number of folded wipes and the number of records collected during the hour, shift, day or year with their digital processing in the manner of Industry 4.0 are necessary for optimizing **OEE** (Overall Equipment Effectiveness) and provide insight into the condition of the machine without interrupting its operation. This enables preventive maintenance and appropriate reaction to the occurrence of irregular conditions.

The team responsible for retrofitting process was able to improve productivity (12 logs per minute, number of sheets per logs 50–200), efficiency and working speed up to 170 m/min, increase the life span of the machine and reduce maintenance costs. The performance of a retrofitted automatic machine of the INTERFOLDER type is given in [17].

16.4 Conclusion

Most production equipment is designed to last for decades. Although the appearance of advanced technologies cannot be ignored during that exploitation period, the procurement of completely new and modern equipment cannot always be realized. One of the possibilities to simultaneously realize the benefit that comes from the introduction of advanced technologies, while maintaining the existing equipment is the retrofit of old machines. This implies the replacement of existing and the addition of the necessary more advanced components, primarily sensors, with designed changes to individual sections of the machine to support the introduction of advanced technologies. The upgrade of the semi-automatic machine of the INTERFOLDER type enabled complete automation of the production line in the ENERTEH plant, which fulfilled the necessary precondition for improving productivity, safety at work and the final quality of the product. Retrofit paper processing lines can become more than a series of insulated machines between which paper is transferred. The developed technical solution refers to the field of industrial machine retrofitting. This is an example of the necessary design changes for the technological improvement of wipes INTERFOLDER machines with working space up to 1500 mm.

The physically representation of the module for automated transfer are tested in real industrial conditions of tissue paper converting in the company ENERTEH, assessment by **rate TR9** of the highest level of technological readiness. Retrofitting of the special purpose machines can be easily commercialized via fitting the module for automated transfer into semi-automatic machines, which are installed in plants around the world.

Acknowledgements The results presented here are the result of a survey supported by the RS MPTR under Contract 451-03-9/2021-14/200105.

References

1. Harmony Homepage: https://www.shpgroup.eu/tips/folded-toilet-paper-saves-the-space-and-money/. Last accessed 21 Mar 2021
2. Policartagico Homepage: https://policartagico.com/en/machines/tissue-and-paper-wipes-converting-machines-21. Last accessed 25 Mar 2021
3. ACelli e-book page: The Ultimate Industry 4.0 Guide for the Tissue and Nonwovens Market. https://www.acelli.it/en/industry-4-0. Last accessed 21 Feb 2021
4. Stock, T., Selinger, G.: Opportunities of sustainable manufacturing in industry 4.0. Proc. CIRP **40**, 536–541 (2016)
5. Al-Maeeni, S., Kuhnhen, C., Engel, B., Schiller, M.: Smart retrofitting of machine tools in the contex of industry 4.0. Proc. CIRP, **88**, 369–374 (2020)
6. Lins, T., Rabelo, R., Correia, L., Sá Silva, J.: Industry 4.0 retrofitting. In: Proceedings of the VIII Brazilian Symposium on Computing Systems Engineering, SBESC 2018, pp. 8–15. IEEE Press (2019)
7. Meyer, M., Frank, M., Massmann, M., Wendt, N., Dumitrescu, R.: Data-driven product generation and retrofit planning. Proc. CIRP **93**, 965–970 (2020)

8. PERINI, Körber Tissue Homepage: https://www.koerber-tissue.com. Last accessed 21 Mar 2021
9. PCMC Homepage: https://www.pcmc.com/our-products/product-groups/converting. Last accessed 21 Mar 2021
10. DCM Homepage: http://www.dcm.fr/en/machines/e-rdc-2/. Last accessed 21 Mar 2021
11. Dechang Yu Homepage: https://www.dechangyu.com/v-fold-facial-tissue-and-towel-machines.html. Last accessed 21 Mar 2021
12. BaoSuo Homepage,. https://www.baosuo.com/product/1500mm-2200mm-automatic-facial-tissue-production-line. Last accessed 21 Mar 2021
13. HINNLI: http://www.hinnli.com/product_en.php?id=71. Last accessed 21 Mar 2021
14. Ocean Homepage: http://www.ocn.com.tw/. Last accessed 21 Mar 2021
15. OMET: https://www.paperfirst.info/omet-introduces-the-new-asv-storm-line-for-interfolded-towel-and-facial-tissue/. Last accessed 21 Mar 2021
16. Video Record: https://www.youtube.com/watch?v=ioaBO5soPUQ&feature=emb_logo&ab_channel=TiCo. Last accessed 21 Mar 2021
17. ENERTEH Homepage: http://enerteh.rs/en/fully-automated-interfolder/. Last accessed 21 Mar 2021

Chapter 17
Equipment for Catalytic Water Purification with a Gas-Saturated Reactor

Aliaksandr Ph. Ilyushchanka, Aleksey R. Kusin, Iryna M. Charniak, Dzmitryi I. Zhehzdryn, Ruslan A. Kusin, Natalia V. Rutkovskaya, Evgeniy N. Eremin, and Tsanka D. Dikova

Abstract The design of the developed equipment for catalytic water purification with a gas-saturated reactor is described. Promising composite materials for the manufacture of the main structural elements of its units are presented. The results of water purification are presented, indicating the high efficiency of the equipment.

Keywords Equipment · Catalytic water purification · Gas-saturated reactor · Water deferrization · Tests

17.1 Introduction

Providing the population with high-quality drinking water is a priority social and environmental problem of any country, the solution of which is aimed at achieving the main goal, i.e. improving and maintaining human health [1]. At the same time, groundwater in Belarus, which is the main source of drinking and process water for most regions of the republic, is characterized by an iron content that exceeds

---

A. Ph. Ilyushchanka
State Research and Production Powder Metallurgy Association, Platonov Str. 41, 220005 Minsk, Republic of Belarus

A. Ph. Ilyushchanka · A. R. Kusin · I. M. Charniak (✉) · D. I. Zhehzdryn
State Scientific Institution "O.V. Roman Powder Metallurgy Institute", Platonov Str. 41, 220005 Minsk, Republic of Belarus
e-mail: irinacharniak@tut.by

R. A. Kusin · N. V. Rutkovskaya
Belarus State Agrarian and Technical University, Nezavisimosti Ave. 99, 220023 Minsk, Republic of Belarus

E. N. Eremin
Omsk State Technical University, Mir Ave. 11, 644050 Omsk, Russian Federation

T. D. Dikova
Medical University, Varna 55 "Marin Drinov" Str, 9000 Varna, Bulgaria

M. Rackov et al. (eds.), *Machine and Industrial Design in Mechanical Engineering*, Mechanisms and Machine Science 109,
https://doi.org/10.1007/978-3-030-88465-9_17

the current standards [2]. Therefore, the removal of iron from water is one of the important tasks in water purification. Accordingly, the development of high-quality and cheaper equipment for water deferrization in the foreseeable future will not lose its relevance.

The purpose of this work is to discuss the results of the development of the design of water deferrization equipment.

17.2 Experimental Results and Discussion

When developing the design of the equipment for catalytic water purification with a gas-saturated reactor, the most widespread method of catalytic purification was adopted as a basis [3, 4]. The developed design of the equipment is protected by the patent of the Republic of Belarus [5], the diagram, 3-D model and the appearance of the equipment are presented, respectively, in Figs. 17.1, 17.2 and 17.3.

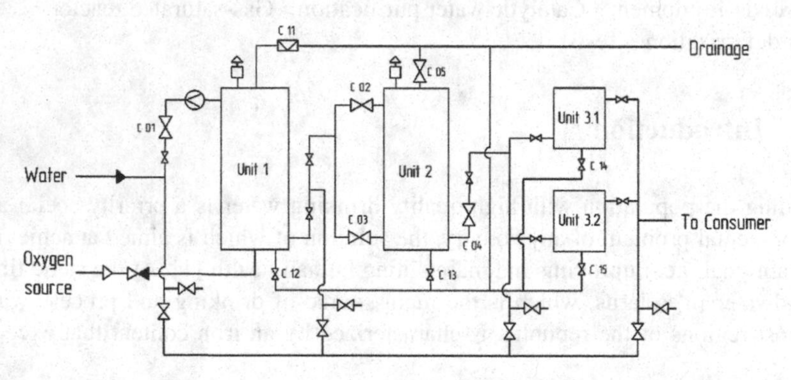

Fig. 17.1 Diagram of equipment for catalytic water purification with a gas-saturated reactor

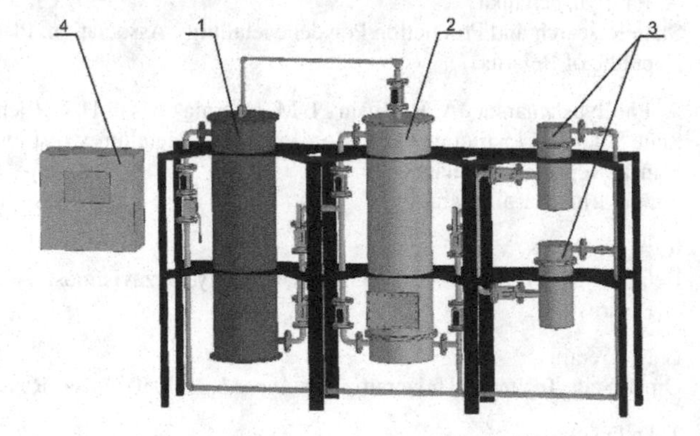

Fig. 17.2 3-D model of equipment for catalytic water purification with a gas-saturated reactor: 1—a unit with a gas-saturating reactor; 2—catalytic purification unit; 3—fine purification unit; 4—control unit

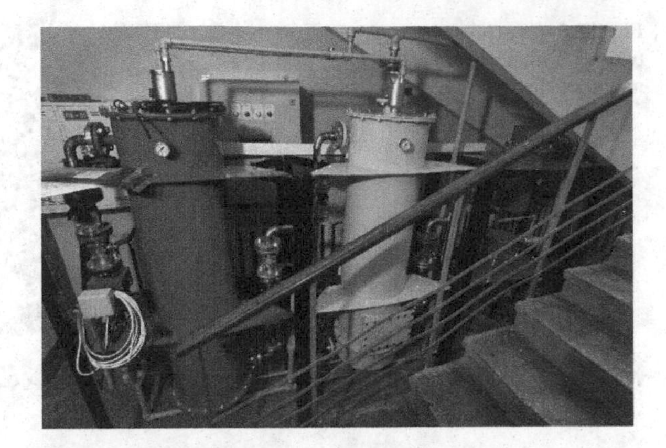

Fig. 17.3 Appearance of equipment for catalytic water purification with a gas-saturated reactor

The equipment for catalytic water purification with a gas-saturated reactor consists of three main units: a gas saturation unit (1), a catalytic purification unit (2) and a fine water purification unit (3) (Figs. 17.1 and 17.2). The equipment consumes no more than 2 kW of power, the capacity of the equipment is up to 10 m^3/h of water; the residual content of iron dissolved in water after treatment in the equipment should not exceed 0.2 mg/l. The maximum allowable water pressure is 10 atm.

The main task of the gas saturation unit (1) is to saturate water with oxygen in order to increase the efficiency of the oxidation process of iron dissolved in water, carried out in the catalytic purification unit (2) (Fig. 17.2). At the same time, in the gas saturation unit, directly in the process of saturating water with oxygen, a partial additional oxidation of ferrous iron into ferric iron occurs, that is, the transfer of dissolved iron into a solid precipitate. The main structural element of the gas saturation unit is a gas flow disperser that evenly distributes the gas flow in the liquid. The 3-D model of the disperser installed in the unit is shown in Fig. 17.4.

A promising material for the manufacture of a disperser is a two-layer filter material made of titanium powders with a particle size (minus 1000 + 400) μm and (minus 100 + 40) μm, the structure of which is shown in Fig. 17.5. Its peculiarity is the

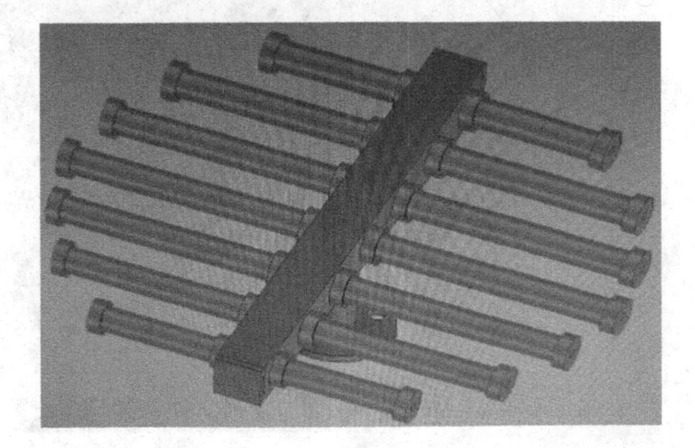

Fig. 17.4 3-D model of gas flow disperser

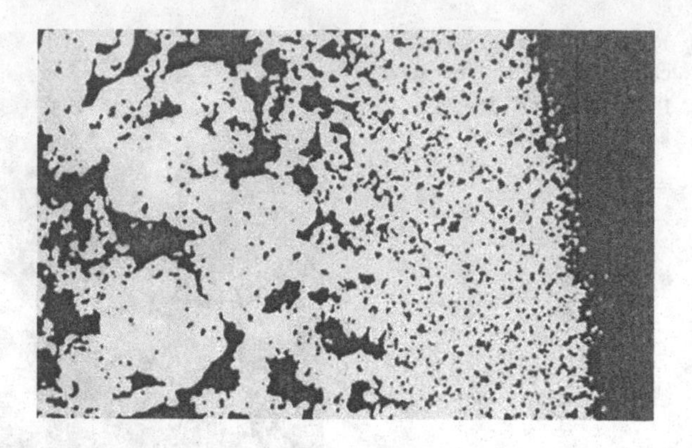

Fig. 17.5 Structure of a two-layer filter material for making a disperser

manufacturing technology, which includes only one sintering, despite the significant difference in the average particle size (up to about 10 times). The thickness of the finely dispersed layer (1.2–1.5 mm), a rather large thickness in comparison with the thickness of the finely dispersed layer (0.3–0.4 mm) for the filters below, provides a higher uniformity of filtering characteristics over the filtration surface, it is essential for a disperser.

The main role in the conversion of ferrous iron into ferric iron is played by the Birm brand fill, which is located in the body of the catalytic purification unit. To prevent the carry-over of fill particles, which have dimensions of 0.2–0.4 mm, a drainage element is installed in the lower part of the body (Fig. 17.6).

The main structural part of the drainage element is 7 filter elements with an orthotropic structure based on woven meshes made of corrosion-resistant steel with a fiber diameter of 0.5 mm and a cell size of 1 mm.

Compared to traditional filtration of the medium to be purified through the cells of woven meshes, the use of orthotropy of their structure allows to reduce the fineness

Fig. 17.6 A drainage element

Fig. 17.7 **a** Material structure of filter elements and **b** Water before (left) and after (right) purification using equipment for catalytic water purification with a gas-saturated reactor

of purification and ensure the regeneration process: when disassembling the filter element, the circuit of the elementary filter cell becomes open, and contamination on the filter material during disassembly of the filter element is retained only by adhesion forces. The developed material is protected by the patent of the Republic of Belarus [6].

The fine water purification unit (3), consisting of two parallel-connected filters (unit 3.1 and unit 3.2 in Fig. 17.1 and unit 3 in Fig. 17.2), provides water purification after the catalytic unit from solid particles of ferric iron using five tubular filter elements installed in its filters.

A promising material for the manufacture of filtering elements for the fine water purification is a two-layer filter material made of titanium powders with a particle size (minus 400 + 315) μm and (minus 80 + 40) μm, the structure of which is shown in Fig. 7a. Its production method is protected by the patent of the Republic of Belarus [7].

The developed technology makes it possible to produce two-layer powder filter materials with a filter layer thickness of 0.3–0.4 mm, with high throughput capacity and purification fineness. Implementation of the technology does not require expensive tooling and non-standard equipment and is quite simple to implement.

Tests of the equipment for catalytic water purification with a gas-saturated reactor were carried out while purifying water coming from a well with a depth of 95 m in the district of Ostroshitsky town (Minsk region). The sample volume was 2 L, the test conditions : temperature 21.9–22.4 °C, air humidity 25.6–26.7%, pressure 751 mm of mercury. The content of iron in water was determined by atomic spectrometry [8], the lower limit of measurements in accordance with the method was 0.05 mg/dm^3. As a result of the studies, the iron content in the non-purified water was 2.74 mg/dm^3; in the purified water, iron was not found (i.e. less than 0.05 mg/dm^3). Figure 7b shows images of water before and after purification.

17.3 Conclusion

The design of the developed equipment for catalytic water purification with a gas-saturated reactor is described. Promising composite materials for the manufacture of the main structural elements of its units are presented: two-layer powder filter materials with different layers of fine powders for gas saturation and fine water purification units and a filter material with an orthotropic structure based on woven meshes for a catalytic purification unit. The results of water purification from an underground source are presented, which testify to the high efficiency of the equipment.

References

1. Klimkov, V.: Quality water for the rural population. Belarus. Agr. **5**(73), 104–106 (2008)
2. The state of the natural environment in Belarus. National report—Ministry of natural resources and environmental protection of the Republic of Belarus. Beltamozhservice, Minsk (2009)
3. Ivanets, A., Kuznetsova T., Voronets E.: Oxidation of ferrous iron in water on manganese and copper oxide catalysts. In: Proceedings of the 6th International Conference on Chemistry and Chemical Education, Sviridov Readings 2012, pp. 30–36. Publishing Center of BSU, Minsk (2012)
4. Tumilovich, M., Pilinevich, L., Savich, V., Smorygo, O., Galkin, A.: Porous Powder Materials and Products Based on them for Protecting Human Health and Environmental Protection: Production, Properties, and Application. Belaruskaya Nauka, Minsk, (2010)
5. Ilyushchanka, A., Charniak, I., Kusin, A., Zhegzdrin, D., Kusin, R., Zakrevsky, I., Yakimovich, N.: Water deferrization equipment. Patent for Utility Model No. 11495 Republic of Belarus. Application No. 20160298, Oct 2017
6. Vityaz, P., Kusin, R., Valkovich, I., Kaptsevich, V., Krugley, V.: Slotted filter. Patent No. 4811 Republic of Belarus. Application No. 19980607, July 1998
7. Ilyushchanka, A., Kaptsevich, V., Kusin, R., Charniak, I., Zhegzdrin, D.: A method of obtaining two-layer porous powder filters. Patent No. 9898 Republic of Belarus. Application No. 20041036, Oct. 2017
8. GOST 31820-2021. Drinking Water. Determination of the Content of Elements by Atomic Spectrometry. Standardinform, Moscow (2019)

Chapter 18
Mechanical Design of the Bicycle Inner Tube Assembly Tool Based on the Reverse Engineering Methodology

Dušan Ćirić, Aleksandar Miltenović, Jelena Mihajlović, and Miroslav Mijajlović

Abstract Reverse engineering has an important role in product design and manufacturing. Reverse engineering represents a concept that generates required design data from existing components. It also describes the process in which product development goes in the reverse order comparing to the conventional product development process. That means that reverse engineering is using the existing product as the starting point rather than the conventional technical drawing. This paper gives a description of some postulates of Reverse Engineering for the mechanical design of the tool responsible for the bicycle inner tube assembly process. This tool, valve applicator, is obtaining proper valve positioning and application on the inner tube profile. The valve applicator must have the exact geometry of the valve—the valve must fit perfectly inside of its structure. How the valve is positioned and applied to the tube profile is essential for the proper and safe usage of the product. Since the technical documentation of the valves is not available the Reverse Engineering methodology must be applied.

Keywords Reverse engineering · Product development · Mechanical design · Inner tube · Tool-valve applicator

18.1 Introduction

Engineering represents the application of the scientific principles and knowledge for the design, analysis, retrofitting or construction of different kinds of technologies, which could be applied for practical purposes [1].

One of the oldest, broadest and most diverse and versatile discipline is mechanical engineering. Mechanical Engineering represents the discipline which scope of study

D. Ćirić (✉) · A. Miltenović · J. Mihajlović · M. Mijajlović
Faculty of Mechanical Engineering, University of Niš, Aleksandra Medvedeva 14, 18000 Niš, Serbia

M. Rackov et al. (eds.), *Machine and Industrial Design in Mechanical Engineering*, Mechanisms and Machine Science 109,
https://doi.org/10.1007/978-3-030-88465-9_18

are objects and their motion. This discipline uses the mathematical and physical principles for the design and manufacturing purposes of systems, as well as their maintenance [1]. Basically, there are two concepts of engineering: direct (conventional) engineering and reverse engineering.

Direct Engineering as a phase of product development process consists of the design, manufacturing, assembly and maintenance. The design process, probably the most important phase, is the process in which the information of lower-level transforms into higher-level information by calculations, analysis, simulations and graphical representations. Early design decisions could if not properly analyzed, significantly impact products functionality. As a result, the main goal is to find optimal solutions for every technical problem or specific production, development and recycling requests. The final steps of this phase are represented in a complete production specification (technical documentation) by which the specific product is going to be produced.

Reverse Engineering (RE) is used to describe the process in which product development follows a reverse order in comparison with direct engineering [2]. Consequently, reverse engineering represents a production process of individual parts, subassemblies, and assemblies-products when there is no required technical documentation, CAD model or drawing based on the existing physical model. That means that the starting point is the existing product rather than conventional technical drawing. RE can be identified as the analysis process in which the systems' components and their interrelationships are identified, representations of a system in a new modified form is created, and at the end physical model of the damaged part is obtained. This way the use of RE will largely decrease the manufacturing time and costs [3].

This paper provides a reverse engineering approach to obtain the original design specification of the mechanical component—specifically, the valve for the inner bicycle tube in order to create and manufacture the tool-applicator responsible for the proper valve application on the inner tube profile.

After the valve geometry is fully reconstructed the creation of the tool-applicator 3D CAD model could be done.

18.2 Literature Review

Reverse Engineering (RE) was defined as "the process of developing the set of specifications for a complex hardware system by an orderly examination of specimens of that system" [4, 5]. Some authors defined it as "the process of analyzing a subject system to identify the systems components and their relationships and to create representations of the system in another form or at a higher level of abstraction" [4, 6].

While developing the RE methodology researchers have been observing from multiple perspectives in order to create a model and its parameters from experimental data [7]. Thus the variety of mentioned applications is present today.

There are researchers which used reverse engineering methodology for recovery of broken, worn and damaged parts [3], such as: mechanical shafts [8], crankshafts [2], turbine blades [9], blade runner [10], etc. They have all used the same methodology but vary the principles on which they analyzed the existing real model to create the virtual reconstructed model on which proper tests could be furthermore obtained.

Also, some authors used hybrid and reverse engineering to design different kinds of technical systems [11], while others were focused on hardware reverse engineering [12]. Several authors specifically focused their research on scanning methods (they have talked about the pros and cons of various systems) [2, 3, 13, 14]. Then some of them discussed scanning path planning [15], while for others data-point preprocessing and reduction methods were crucial [16]. There were those researchers who deal with the integration with rapid prototyping and some other processes [17].

Because of the diversity in the RE applications, some researchers have applied its methodology for medical purposes, such as the creation of the geometrical model of the human fibula [18] and a 3D CAD model of the knee implant [19].

18.3 Case Study

This paper provides a RE approach to acquire the original design specification of the existing mechanical component (valve for the inner tubes) to help design the required tool-valve applicator.

The process of building inner tubes consists of a couple of phases: preparation of the rubber mixture and creating an inner tube profile, assembly of the inner tube profiles and valve application, after which the curing or vulcanization process is carried out. Each phase is crucial for the products' proper and safe usage. Every inner tube with flaws, imperfections, and defects is being declared as scrap and immediately discarded. The phase which regards this paper's research is the second phase, the assembly phase. This phase has two stages, butt weld joint, and valve placement. Both stages are obtained on the same machines, simultaneously, but with different tools. The so-called "upper operations" are responsible for the proper valve placement on the inner tube profile (Fig. 18.1).

The process of placing a valve has three steps. Each of these steps must be performed "perfectly" to avoid the possibility of flaws occurrences. The hole on the upper side of the profile is made in the first step so the compressed air can flow through the valve into the tube. After making a hole, the area around it must be cleaned. A clean surface enables the best conditions for the inner tube and valve cohesion. The final step is proper valve application which is obtained by the tool called applicator (Fig. 18.1).

The working surface of the applicator needs to have the exact geometrical characteristics as the foot of the valve. This way applied pressure that applicator does while placing the valve is continual on the inner tube. Thus, the application of the valve is performed. Any uneven contact between the applicator and the valve and

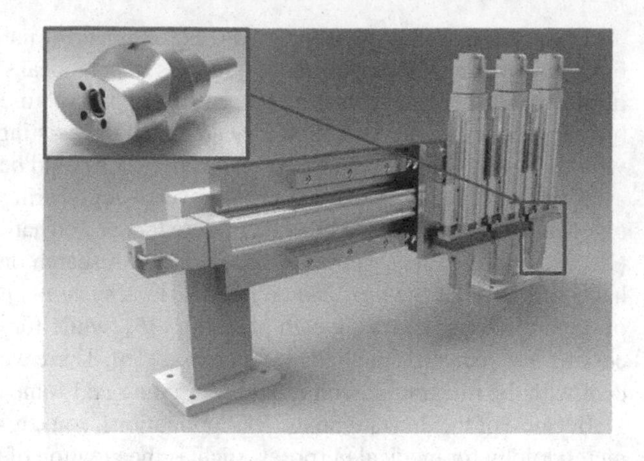

Fig. 18.1 The upper operations with the valve applicator

between the valve and the inner tube profile will cause the occurrence of flaws and furthermore, the scarp tubes.

The RE framework consists of a couple of phases. Those phases are not mandatory and they could vary in the available literature. Generally, those phases are data acquisition-digitization process (a process of acquiring point coordinates from the part surface), mesh processing (defects removing process), segmentation of the point clouds/meshes, feature classification (classification of the regions identified in the previous phase), modelling (finishing operations) and CAD model reconstruction [4, 20].

18.3.1 Data Acquisition—Digitization Process and Pre-Processing

Digitization is the process of acquiring point coordinates from the exiting part surface [13]. Numerous 3D acquisition technologies have been developed and used for product digitization (different types of measuring and scanning devices).

Every technology or device has its advantages and disadvantages regarding the principles on which they work. As the most frequent technique of digitization in use, there are optic systems (non-contact methods) and mechanical systems (contact methods). Non-contact methods use the sensors such as digital camera and have much faster data acquisition, but usually are less accurate and could be surface affected in comparison with contact methods (coordinate measurement machines (CMM)). Nevertheless, non-contact/optical systems are the most common devices (magnetic, acoustic, and optical scanners).

The 3D scanner used for geometrical data acquisition for this paper's purposes is HDI Advance R4X Optic Scanner. This scanner has 2 × 4.1-megapixel monochrome USB 3.0 cameras with 12 mm lenses with the field of view ranging from 212 to

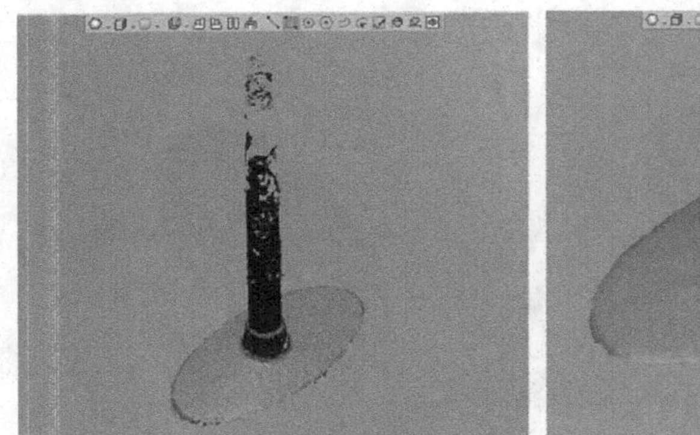
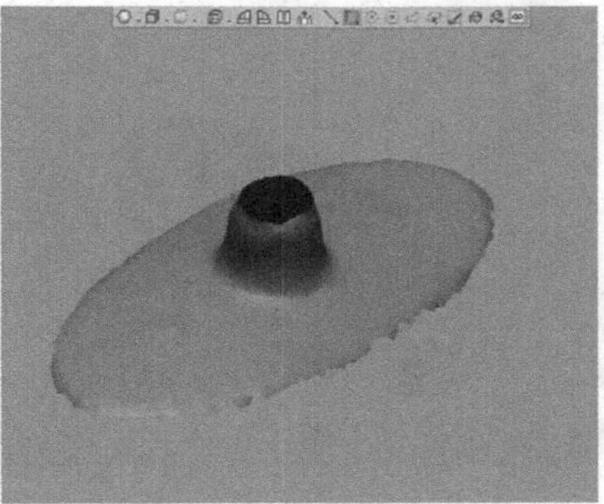

Fig. 18.2 The complete scan of the valve obtained in the FlexScan3D software (left) and the end stage of the pre-processing phase obtained in the Geomagic Design X software (right)

676 mm. The accuracy at 212 mm is 36 μm and 84 μm at 676 mm with 1.3 s per scan [21]. Scanning software is FlexScan3D.

The process of the digitization of the valve obtained by the 3D optical scanner is a non-contact technique. Frequently the object's surface needs to be prepared with proper markers or powders which decrease the light reflections. Specifically, there was no need for surface preparation because the focus of the scanning process was the foot of the valve (made from rubber) rather than the scape of the valve (made from steel). The characteristic of the optical method is to scan the object from different angles. This way a set of incomplete scans is made. The scan consists of a point cloud and represents the digital raster object. Those individual scans need to be joined in a one complete digital meshed object type which represents the valve (Fig. 18.2). Depending on the quality of the scanner, the scanning conditions, and the type of the object (its surfaces) the quality of the mesh is defined.

The next step, pre-processing, is to reduce the number of points and to obtain the noise reduction with surface management (geometrical surface corrections), so that mesh could be generated. In order to reconstruct product's topology the cloud of points must be reorganized by the mesh construction [4]. This process is mandatory in every RE process and it was supported by the Geomagic Design X software (Fig. 18.2).

18.3.2 Segmentation and Feature Classification

Segmentation is the process of subdividing the mesh generated object into separate regions in order to achieve a structure of regions that is similar as possible to the set of geometrical features and surfaces composing the model to be reconstructed [20].

Since only the foot of the valve is important for this research, the scape needs to be erased from the rest of the scanned object (Fig. 18.3). Also, only the upper surface

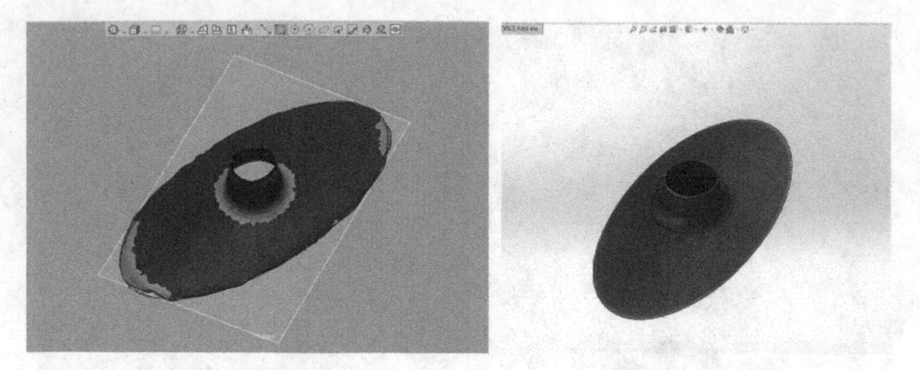

Fig. 18.3 Segmentation phase obtained in the Geomagic Design X software and CAD model generation in SolidWorks software

geometry of the foot of the valve is observed. Furthermore, segmentation is done, but not in a classical way with the feature classification which is necessary for CAD reconstruction.

The foot of the valve with fixed geometry and proper segmentation is exported to one of the required formats so that the 3D CAD model could be used for the determination of the applicators (tools) geometrical characteristics in order to satisfy production requirements.

18.3.3 3D CAD Model Generation

The final step of the RE framework is to generate the 3D CAD solid model of the foot of the valve. The SolidWorks application is selected for its wide range of possibilities to operate with mesh and surface models [19].

Once the imported model is generated into the software and surface reconstructed it could be used for the creation of the applicator's working surface. The overall construction of the applicator with all necessary features (mounting and positioning characteristics, compressed air and vacuum canals for valve manipulation, etc.) is carried out using the same software, SolidWorks. Now, when two models are properly aligned and intersecting each other, the imported model needs to be subtracted from the construction of the tool-applicator (Fig. 18.4). The rough shape of the working surface needs minor corrections to obtain the "perfect" geometrical characteristics of the valve (Fig. 18.4).

Thus, the best possible application of the valves on inner tube profiles is ensured. After the 3D CAD model is finished (Fig. 18.5), the technical documentation of the tool is created.

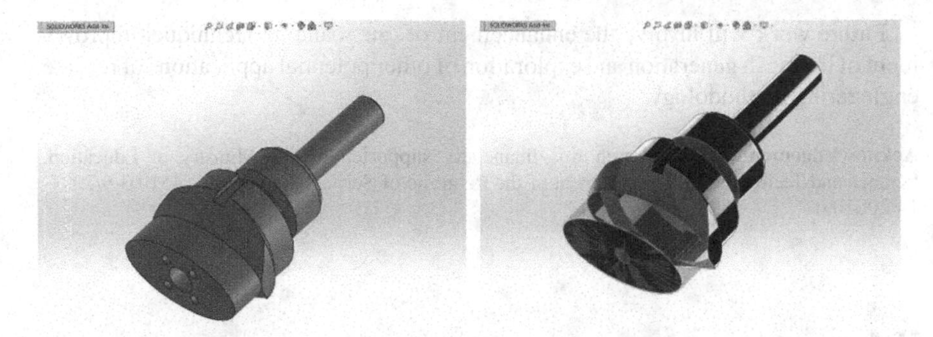

Fig. 18.4 Creation of the working surface of the tool-valve applicator

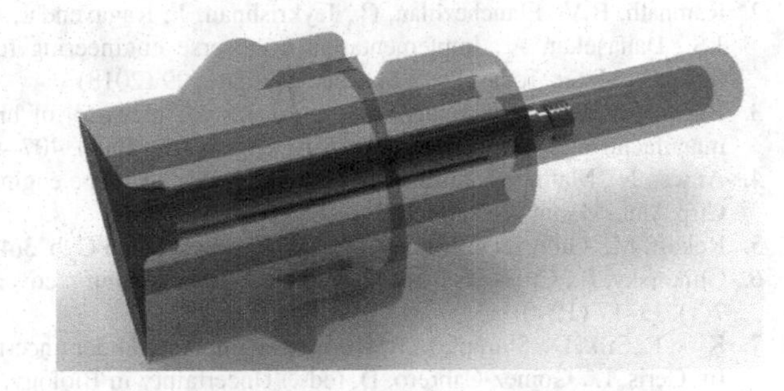

Fig. 18.5 Final 3D CAD model of the tool-valve applicator compare to the valve

18.4 Conclusion

The rapid development of the 3D scanners and their increased availability has led to the more frequent use of the RE methodology in various applications. CAD models are the most common mediums used by engineers in their everyday practice to convey dimensional and geometrical information on constructed parts, machines, etc. In the cases in which there is no CAD model available or that it does not exist, or it does not correspond to the real geometry of the manufactured object, the RE approach could be successfully applied.

This paper described the reverse engineering approach for extracting the mechanical components geometrical parameters (inner tube valve) to design the working surface of the tool (applicator) responsible for its proper application on the inner tube profile during the assembly production phase in the case that the original design specifications of the valve are not available. The outcome of this research is the technical documentation required for the manufacturing of the tool-applicator.

After the described solution is manufactured, tested and verified and because there are dozens of valve types in use (in the inner tube portfolio), the proper RE strategy needs to be developed to cover all the rest of the types of tools (applicators).

Future work will involve the enhancement of our scanning technique, improvement of the mesh generation and exploration of other potential applications of reverse engineering methodology.

Acknowledgements This research was financially supported by the Ministry of Education, Science and Technological Development of the Republic of Serbia (Contract No. 451-03-9/2021-14/200109).

References

1. https://www.me.columbia.edu/what-mechanical-engineering. Last accessed 15 Feb 2021
2. Ramnath, B.V., Elanchezhian, C., Jeykrishnan, J., Ragavendar, R., Rakesh, P.K., Dhamodar, J.S., Danasekar, A.: Implementation of reverse engineering for crankshaft manufacturing industry. Mater. Today-Proc. **5**(1, Part 1), 994–999 (2018)
3. Bagci, E.: Reverse engineering applications for recovery of broken or worn parts and re-manufacturing: three case studies. Adv. Eng. Softw. **40**(6), 407–418 (2009)
4. Anwer, N., Mathieu L.: From reverse engineering to shape engineering in mechanical design. Cirp Ann.-Manuf. Techn. **65**, 165–168 (2016)
5. Rekoff, M.: On reverse engineering. IEEE T. Syst. Man Cyb. **3**(4), 244–252 (1985)
6. Chikofsky, E., Cross, J.: Reverse engineering and design recovery: a taxonomy. IEEE Softw. **7**(1), 13–17 (1990)
7. Kirk, P., Silk, D., Stumpf, M.P.: Reverse engineering under uncertainty, uncertainty in biology. In: Geris, L., Gomez-Cabrero, D. (eds.) Uncertainty in Biology: Studies in Mechanobiology, Tissue Engineering and Biomaterials, vol. 17, pp. 15–32. Springer, Cham (2016)
8. Engel, B., Al-Maeeni, S.S.H.: An integrated reverse engineering and failure analysis approach for recovery of mechanical shafts. Proc. CIRP **81**, 1083–1088 (2019)
9. Chen, L.C., Grier, C.I.L.: Reverse engineering in the design of turbine blades a case studying applying the MAMDP. Robot. Cim.-Int. Manuf. **16**(2–3), 161–167 (2000)
10. Nedelcu, D., Bogdan, S.L., Pădurean I.: The reverse engineering of a blade runner geometry through photogrammetry. IOP Conf. Ser.: Mater. Sci. Eng. **393**, 012126-1–012126-10 (2018)
11. Ognjanović, M., Vasin, S., Ristić, M.: Hybrid and reverse engineering in actual design of technical systems. In: Proceedings of the 8th International Symposium Machine and Industrial Design in Mechanical Engineering, KOD 2014, pp. 33–40. Novi Sad (2014)
12. Fyrbiak, M., Strauβ, S., Kison, C., Wallat, S., Elson, M., Rummel, N., Paar, C.: Hardware reverse engineering: Overview and open challenges. In: Proceedings of the IEEE 2nd International Verification and Security Workshop, IVSW 2017, pp. 88–94. IEEE Press (2017).
13. Motavalli, S.: Review of reverse engineering approaches. Comput. Ind. Eng. **35**(1–2), 25–28 (1998)
14. Xinmin, L., Zhongqin, L., Tian, H., Ziping, Z.: A study of a reverse engineering systems based on vision sensors for free-form surfaces. Comput. Ind. Eng. **40**(3), 215–227 (2001)
15. Son, S., Kim, S., Lee, K.H.: Path planning of multi-patched freform surfaces for laser scanning. Int. J. Adv. Manuf. Tech. **22**, 424–435 (2003)
16. Huang, M.-C., Tai, C.-C.: The pre-processing of data points for curve fitting in reverse engineering. Int. J. Adv. Manuf. Tech. **16**, 635–342 (2000)
17. Lee, K.H., Woo, H.: Direct integration of reverse engineering and rapid prototyping. Comput. Ind. Eng. **38**(1), 21–38 (2000)
18. Tufegdžić, M., Trajanović, M., Vitković, N., Arsić, S.: Reverse engineering of the human fibula by the anatomical features method. FU Mech. Eng. **11**(2), 133–139 (2013)

19. Rajić, A., Desnica, E., Stojadinović, S., Nedelcu, D.: Development of method for reverse engineering in creation of 3D CAD model of knee implant. FU Mech. Eng. **11**(1), 45–54 (2013)
20. Buonamici, F., Carfagni, M., Furferi, R., Governi, L., Lapini, A., Yary, V.: Reverse engineering modelling methods and tools: a survey. Comput. Aided Des. Appl. **15**(3), 443–464 (2018)
21. http://lmi3d.com/products/hdi-advance. Last accessed 19 Feb 2017

Chapter 19
Product Redesign Using an Innovative Process: Application to a Case Study

Pedro Agustín Ojeda Escoto, Miguel Ángel Zamarripa Muñoz, and Gerardo Brianza Gordillo

Abstract Today, companies oriented to the manufacture of new products are continuously seeking to improve not only the manufacturing processes of their services, but also the transfer of design and engineering to production. In the process of developing products oriented to cover specific needs or requirements and speaking in terms of functionality is possible to achieve the interaction of two strategies: on the one hand, feedback from the end user and on the other the optimization of architecture by applying comparative analysis and reverse engineering. In the search for the consolidation of a new product and identifying opportunities for improvement, this paper presents the redesign of a forage-harvester (case study) in which the following premises were raised: (1) validation of the design proposal; (2) comparative and operational analysis; (3) geometry optimization for weight and production cost reduction. For the proposal of the new product, the theoretical framework is based on Concurrent Engineering and together with an innovative process the architectural configuration was defined and characterized. The procedure used to structure the design of the forage-harvester under the approaches of architectural analysis, quality improvement in production and overall cost reduction, and that allowed to define an architecture capable of being aligned to a mass production is also presented. Finally, the results obtained by finite element analysis for the geometry optimization of the new product are published.

Keywords Design · Product development · Optimization

19.1 Introduction

Design models are based on the execution and evaluation stages or on the optimization of an initial alternative. Following this context, the initial solution option is evaluated and improved taking into account different aspects such as: performance, cost, assembly, functionality, reliability, maintainability [1–3]. Products offered on

---

P. A. O. Escoto (✉) · M. Á. Z. Muñoz · G. B. Gordillo
Universidad Tecnológica de Aguascalientes, 20200 Aguascalientes, México
e-mail: pedro.ojeda@utags.edu.mx

M. Rackov et al. (eds.), *Machine and Industrial Design in Mechanical Engineering*, Mechanisms and Machine Science 109,
https://doi.org/10.1007/978-3-030-88465-9_19

the market are continuously updated to meet new requirements and satisfy new needs of the customers. The modification of the features of a product often spread in its structure and in the whole lifecycle, such as the production facilities or the end of life management. Current redesign techniques can limit product innovation. Besides, the impact of introducing new features should be limited and controllable, in order to speed the redesign phase and bound the uncertainty connected with the management of a new product. New product variants in product families are generally derived by adapting existing products to new requirements, scaling or changing their modules and components [4]. New innovative products are only introduced when major conflicts exist between customer needs and existing products. However, business success is also strongly related to product innovation. New ideas, new technologies, and new products capture customer interest and keep successful companies at the forefront of their industries. Some techniques to detect conflicts in a design, selection of functional dependencies, and the reuse of objective information to obtain new solutions are fundamental elements for any redesign process. However, all the methods analyzed focus on transforming an existing product, rather than developing a new product from designs already made [5]. Most of the new designs are refined versions of the existing designs developed through alterations and changes to the current design. Modifications in engineering design should be considered as a major focus for product improvement as users become more interested in functional products. Based on the above, all the information defined from the engineering update are very useful for new design proposals and product evolution [6].

This paper reports the procedure and results of the redesign of a forage-harvester (case study) structured with Concurrent Engineering (CE) criteria and using as approaches: architectural analysis, quality improvement in production and overall cost reduction. Likewise, the results obtained from finite element analysis of the mentioned harvester are reported in order to optimize its final architecture. Is also reported the steps and stages of the project that managed to consolidate the design of the new product.

19.2 Theoretical Framework

19.2.1 Concurrent Engineering

Once CE has been contextualized, which is currently closely related to the development of new products, it can be defined as the process of developing new products in which all areas must be involved, working accordingly in the creation of the product. This involvement ranges from the contribution of ideas to the matching and readjustment of information to achieve the agreed product. CE is a regular focus for both integrated and concurrent product design that also involves the related processes, as the manufacturing and support. This focus is intended to help designers consider all aspects of the product life cycle, from concept to decommissioning, including quality control, costs, time and user requirements, from the outset [7]. Finally, is a process in which the main functions that comprise the process of placing a product on the market are continuously involved in the development of the product, from its conception to final sale. The benefits of this technique are derived from the time a project is completed [8]. According to Borja [9], there are several definitions of CE or also called in some research as Simultaneous Engineering (Fig. 19.1). There are some other definitions, but all of them agree that SE belongs to the design process, and assigns parallel activities to decrease the product development time, improving its quality by integrating the product and its manufacturing process. Common problems that make it necessary to implement the CE in the design or redesign of a new product [10]: (1) Increased product variety and technical complexity that prolongs the product development process and makes it difficult to predict the impact of design decisions on the functionality and performance of the final product; (2) Increase the global competitive pressure resulting from the emerging concept of re-engineering. The need for a rapid response to changing consumer demand and a shorter product

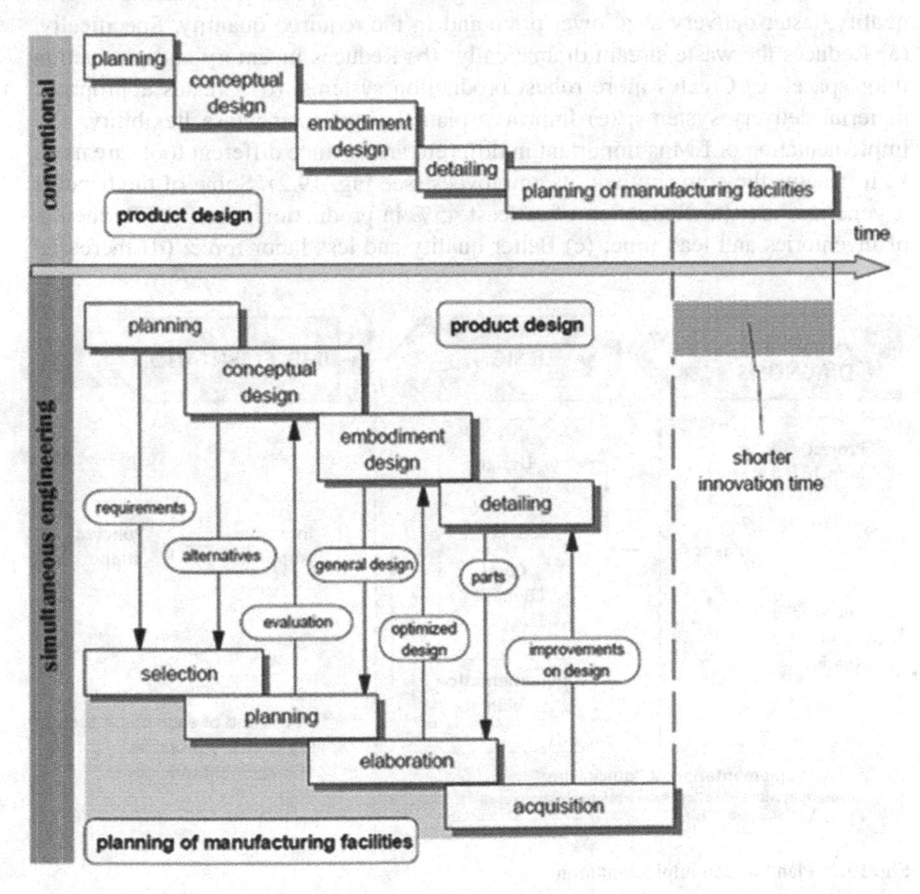

Fig. 19.1 Simultaneous engineering (Adapted by Borja [9])

life cycle; (3) Organizations with several departments working on the development of numerous products at the same time. New and innovative technologies that emerge at a very high rate, making the new product technologically obsolete in a short period of time.

Lean Manufacturing

Lean Manufacturing (LM) involves several tools that help eliminate all operations that do not add value to the product/service and/or process, increasing the value of each activity performed and eliminating what is not required, reducing waste and improving operations. The LM system has been defined as a philosophy of manufacturing excellence, based on: (a) The planned elimination of all types of waste; (b) Continuous improvement; (c) The permanent improvement of productivity and quality. The main objectives of LM are to implement a philosophy of continuous improvement that allows companies to reduce costs, improve processes and eliminate waste to increase customer satisfaction and maintain the profit margin. It provides companies with the tools to survive in a global marketplace that demands higher quality, faster delivery at a lower price and in the required quantity. Specifically: (a) Reduces the waste stream dramatically; (b) Reduces inventory and production floor space; (c) Creates more robust production systems; (d) Creates appropriate material delivery systems; (e) Improves plant layouts to increase flexibility. The implementation of LM is important in different areas, since different tools are used, so it benefits the company and its employees (see Fig. 19.2). Some of the benefits it generates are: (a) Reduction of at least 45% in production costs; (b) Reduction of inventories and lead time; (c) Better quality and less labor force; (d) Increased

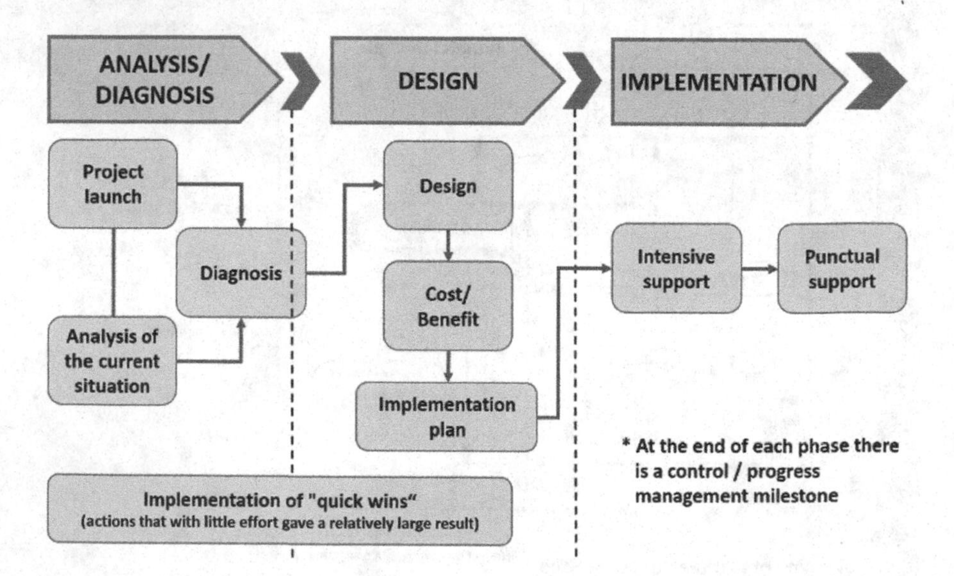

Fig. 19.2 Plan for lean implementation

equipment efficiency and reduced waste; (e) Overproduction and decrease in waiting time; (f) Transport, processes, inventories and movement of final product.

19.3 Research and Proposal Procedure

19.3.1 Research Methodology

Design is part science and part art. The scientific part of design can be learned through the different philosophies, methodologies and tools that exist and that try to systematize this task. However, the art part of the design process, nowadays, cannot be taught systematically. Therefore, many designers argue that the only way to learn the artistic part of design is by designing. The development of any investigation consists of extracting the information by following a methodology, the architecture of the product in question and the procedural details of the product in order to understand it. Figure 19.3 presents the methodology proposed for the development of the paper. Next, the main activity of each of the phases of the methodology used is presented in a very general way: NEW PRODUCT. Model systematization (development of specification and conceptualization): Conceptualization of the final product architecture; Performance analysis: Definition of FEA analysis to locate the best new design solution; Configuration testing: Test stage for assembly and disassembly of components; Comparative analysis: Analysis carried out to define the new concept of the forage-harvester. REDESIGN. Manufacturing: Implementation of manufacturing processes and continuous improvement; Costs: Cost analysis in design and manufacturing phases; Quality: Analysis and implementation of quality in manufacturing

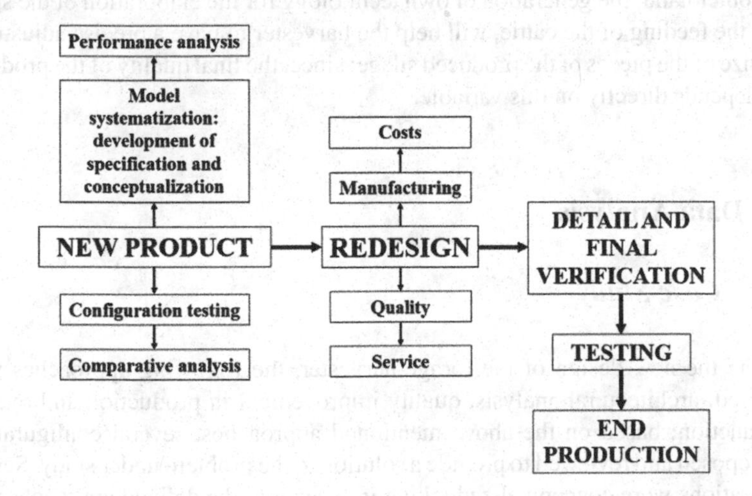

Fig. 19.3 Research methodology

processes; Service: Implementation of the necessary conditions for the application of manufacturing processes. DETAIL AND FINAL VERIFICATION: Numerical analysis of FEA results and first phase of physical tests; TESTING: Second phase of physical testing under real operating conditions; END PRODUCTION: Alignment of product manufacturing to final production taking into account some strategies.

Functional and innovation analysis

The application of this research focused on Mexican agricultural producers who were looking to increase the profitability of their crop, as well as agricultural producers who were looking to give added value to their crops by making their silage more efficient. The degree of innovation defined for the proposed prototype was of the incremental type since considerable improvements were added to the functional capacity of the harvester. Among these, the following can be mentioned:

- Special design of mechanical gearbox. Special design of gearbox that provides good ratio of speed and torque needed for optimum operation of gear train and thus improve the performance of cutting and crushing system.
- Special design of the geometry of the chopping and shredding rollers. Having an optimum geometry in the roller shredding elements will help to have a better ratio in the determined final sizes of the material and will also support a constant flow of the material on its way to the mechanical rotor.
- Special design for the ergonomic improvement of the hitch. This improvement is intended to make the hitch on the agricultural tractor more practical, as well as to make the levelling of the hitch faster and more efficient.

The forage-harvester will be developed with its own technology to add a plus in the part of the corrective and preventive maintenance and also to be easily adapted to the needs of the national field, since, it is sought that it has a practical operation. On the other hand, the generation of own technology for the elaboration of the silage used in the feeding of the cattle, will help the harvester to have a precise adjustment on the size of the pieces of the produced silage; since, the final quality of the produced silage depends directly on this variable.

19.4 Data Analysis

19.4.1 Case Study

To define the new design of the forage-harvester, the following approaches were considered: architectural analysis, quality improvement in production and over-all cost reduction; based on the above mentioned approaches, several configurations were proposed and reviewed to provide a solution to the problem under study. Several configurations were conceptualized taking into account the defined approaches and finally the final architecture of the harvester was generated (see Fig. 19.4a).

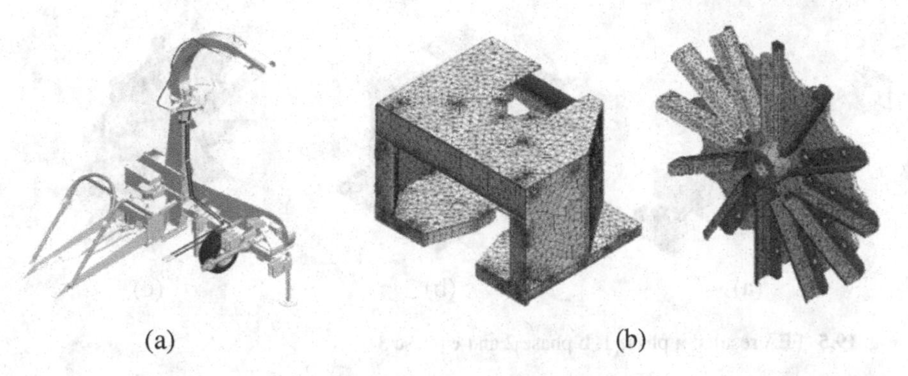

Fig. 19.4 Proposal for solution: **a** case study: forage-harvester and **b** application of border conditions

Proposal for solution

Once the final architectural geometry was selected, finite element analyses were performed to validate and optimize (for assembly purposes) the architecture. The mesh of the models was generated based on the existing dimensional relationships between the components of the assembly and the load conditions proposed for the analysis were determined based on the solicitations that are presented in the normal work of the harvester (see Fig. 19.4b). Based on the above mentioned approaches, several configurations were proposed and reviewed to provide a solution to the problem under study. Several configurations were conceptualized from the general characterization of this type of implement. Taking into account the defined premises: (a) validation of the design proposal, (b) comparative and operational analysis and (c) geometry optimization for weight and production cost reduction. Another consideration that was made for the product design was the analysis of the attachments for coupling the forage-harvester to the tractor.

Geometry optimization

The analysis by finite element was done in several stages, in each of them the different criteria were taken for such analysis and several punctual loads on the harvester were also defined based on the normal work it performs. Figure 19.5 shows the results obtained from the final architecture.

19.5 Results and Discussion

The stress study carried out on the harvester model allows us to demonstrate that each and every one of the elements that make up the model present stress magnitudes below the permissible elastic limit of the material. It was even possible to optimize some elements and comply with the design approaches that were taken into account to define the final architecture, these are:

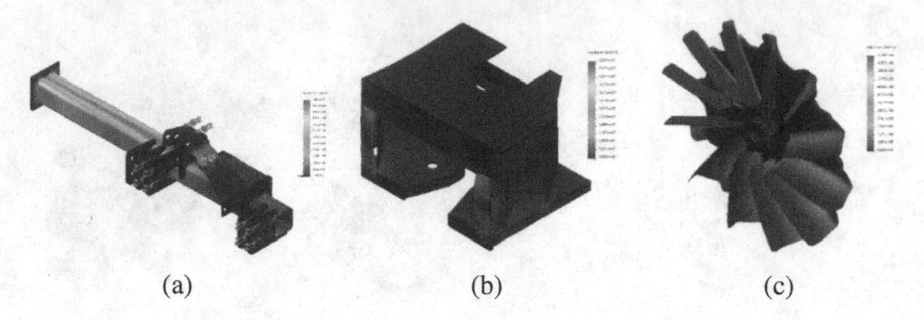

Fig. 19.5 FEA results: **a** phase 1, **b** phase 2 and **c** phase 3

- Architectural analysis—an improvement in manufacturing processes was achieved by having a simple architecture and free of assembly interferences.
- Quality improvement in production—when a final evaluation of the weight was made, it was possible to decrease by 33% derived from the optimization of geometry and the use of materials.
- Overall cost reduction—the percentage of cost reduction that was achieved ranged around 37%.

In the chassis zone of the implement is where the highest magnitudes of efforts are presented having concentration of efforts in some points, this is due to the geometry and functions under which the elements of this zone work. Is also important to mention that the magnitudes of efforts are within the normal ranges of deformation so, for the moment, they can be neglected to continue with the manufacture of the final prototype. It was also possible to define a successful process for the assembly of the harvester sub-systems, optimizing times and generating better routes for their alignment in mass production. Once the new product architecture was defined and the concept tests were per-formed, the strategies for aligning it to production were defined taking into account product image and marketing (factors defined by the experience of the manufacturer's sales area).

19.6 Conclusion

Advances in computer and technology analysis allow engineers and researchers to have effective diagnostic and simulation tools that facilitate, at a given time, the design, redesign or optimization of a mechanical system. The development and implementation of methodologies for design or redesign of products based on new technologies, increases the probability of alignment in the market and the increase of knowledge and technology transfer to the industrial sector. In order to define a new design of a forage-harvester, several configurations were proposed and reviewed to provide a solution to the problem under study. Several configurations were also conceptualized taking into account the defined approaches and finally the architecture of the product was generated. In this paper, the optimization of the geometry of a harvester and the results of the finite element analysis that helped to define the redesign of the final architecture were presented as a case study. It was also presented the strategies defined for the alignment of the new product developed to production, taking into account sales projections, product image and marketing. Finally, the impacts obtained with the development of the present investigation were the following:

- Generate a harvester design, with its own technology and easy to acquire by national producers in the field.
- To improve the profitability and competitiveness of the agricultural sector in Mexico.
- Contribute to the conservation and installation of new agricultural businesses, resulting in the growth of green areas such as pastures, fodder trees, fodder cereals, etc., improving climatic conditions.
- To improve the profitability and competitiveness of agricultural businesses in Mexico.

References

1. Ulrich, K.T., Eppinger, S.D.: Product Design and Development. McGraw-Hill (2004)
2. Pahl, G., Beitz, W., Feldhusen, J., Grote, K.H.: Engineering Design: A Systematic Approach. Springer (2007)
3. Ulrich, K.T., Eppinger, S.D.: Diseño y desarrollo de productos. McGraw-Hill (2013)
4. Smith, S., Smith, G., Shen, Y.: Redesign for product innovation. Des. Stud. **33**(2), 160–184 (2012)
5. Howard, T.J., Culley, S.J., Dekoninck, E.A.: Reuse of ideas and concepts for creative stimuli in engineering design. J. Eng. Design **22**(8), 565–581 (2011)
6. Ullah, I., Tang, D., Yin, L.: Engineering product and process design changes: A literature overview. Proc. CIRP **56**, 25–33 (2016)
7. Ullah, I., Tan, D., Wang, Q., Yin, L., Hussain, I.: Managing engineering change requirements during the product development process. Concurr. Eng. Res. A **26**(2), 171–186 (2018)
8. Chase, R.B., Aquilano, N.J., Jacobs, F.R.: Production and Operations Management. McGraw-Hill (2006)
9. Borja, R.V.: Redesign supported by data models with particular reference to reverse engineering. PhD Thesis. Loughborough University (1997)
10. Dongre, A., Jha, B., Aachat, P., Patil, V.: Concurrent engineering: A review. Int. Res. J. Eng. Tech. **4**, 2766–2770 (2017)

Chapter 20
Product Design Involving Culture, Sustainability and User Experience

Branislav Petrovic and Djordje Kozic

Abstract It is well known that design of a product is a key driver of the success or failure in this field. In industrial design, product design process is not entirely controlled by physical conditions as material properties, the structural strength or production constrains. Besides usability, ergonomics and functionality it is also influenced by many known and unknown factors. One of them is culture, because culture affects the way users respond to the product. And not only the culture of users, but also the designer's own culture, their aesthetic preferences, emotions and other non-physical aspects. For that reason, any activity of product design must include the knowledge of culture. For understanding the relation between culture, designer, users and the product itself—modern product design has to involve also the concept of sustainability and (especially) the connection between culture and sustainability. In this respect, there are nowadays a lot of talk about the so-called UX (user experience) design, which also be discussed in this article.

Keywords Product design · Culture · Culture-oriented design · Using experience · Sustainable design · Cultural dimensions

20.1 Introduction

In contemporary design the term product design is often used interchangeably with the term industrial design. Even though they are almost synonymous, there are certain differences between the two.

Product design, in the strict sense of the word, refers to product creation, from an idea and sketches to final design, which is generally completed with construction of a certain limited number of final products. On the other hand, the term industrial design usually refers to activities associated with mass production and concepts linked to pre-manufacturing, manufacturing, transportation/distribution, consumer

B. Petrovic (✉) · D. Kozic
Faculty of Mechanical Engineering, University of Belgrade, Belgrade, Serbia
e-mail: bpetrovic@mas.bg.ac.rs

M. Rackov et al. (eds.), *Machine and Industrial Design in Mechanical Engineering*, Mechanisms and Machine Science 109,
https://doi.org/10.1007/978-3-030-88465-9_20

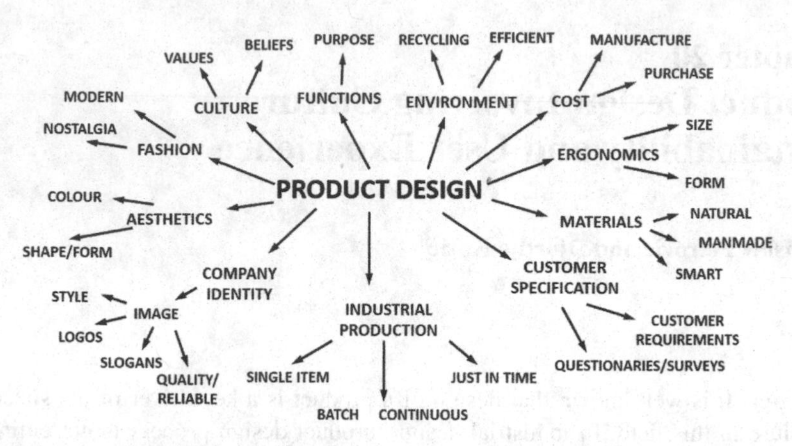

Fig. 20.1 Factors that influence product design

market (user) and disposal stages, all of which imply a different set of knowledge and skills. In that regard, industrial design is always part of the product design.

However, in Anglo-Saxon's literature, it is becoming increasingly common to use the term product design to refer to practically all activities associated with putting a new product on the market. In that sense, it is an illustrative practice to list nomenclatures and other designer attributes (like for example "Job titles"), such as: Product Designer, UX Designer; UI Designer, UX/UI Designer, Interaction Designer, Information Architect, UX Architect, UX Strategist, Digital Designer, etc.

What product design depends on is illustrated in Fig. 20.1 [1], in which, if one represents a product design as a diamond with many facets, these facets paint an extremely complex nature of dependence associated with the term.

20.2 Influence of Culture on Product Design

All aspects of human life are influenced by culture. In investigating the cultural impact on design, a clarification how culture is defined is far from being trivial. This also shows in the fact that there are numerous definitions of culture (there are as much as two hundred of them!). This is the reason why the relationship between culture and design ought to be examined in a higher academic level.

There are several theoretical attempts to define cultural dimensions that can be used to describe differences between culture groups. One of the notable researches in the field was Dutch scientist Geert Hofstede (1928–2020). In 2001, as a result of his research of many years, he published a concept of national cultural dimensions. At the time there were only four of them, but later after comparison with other research, this number increased to a total of six dimensions: **Power Distance Index (PDI)**—the extent of inequalities and hierarchy that a society tolerates, or level of acceptance of an unequal power distribution in a society; **Individualism versus Collectivism**

(IDV)—how strong individuals are integrated in and feel responsible for a social group, or extent of which members of a culture prioritize their individual goals over the goals of the group; **Masculinity versus Femininity (MAS)**—the degree of differentiation of gender roles, where a masculine culture is mainly driven by competition, while in a feminine culture cooperation and carrying for others are the more important values; **Uncertainty Avoidance (UAI)**—the level of stress that is caused by unclear and ambiguous situations, or desire to accept or avoid unknown or uncertain situations; **Long-term versus Short-term Orientation (LTO)**—the preferred focus of people's time orientation, in the future or the present, where planning and action are more based on long or short term goals; **Indulgence versus restraint (IVR)**—the degree to which people try to have power over their impulses and drives, or the extent to which natural human instincts related to having fun and enjoying life are permitted by the society.

Hofstede provides concrete scores for these dimensions for numerous countries. The data is available at the link [2], and a specific example in Fig. 20.2 is based on the data. Using this data it can be seen that for example for PDI—high score have China, Mexico, France, Brazil and India; low score have Australia, Canada, Germany, Denmark, Finland, Japan, U.K. and U.S. It is possible to find these data for other dimensions and countries.

One can pose a question: how the data can be used in solving challenges of product design? For example, in societies with high PDI values, status and age are very important, and people tend to be less innovative. This is why in such cultural environments customers are less open to new ideas and products. It means that high PDI has a negative effect on the acceptance rates of new products. On the other hand, it can be shown that the high value of IDV, due to the fact that then there are attitudes towards differentiations and uniqueness, and a member of individualistic cultures tend to see themselves as independent, unique persons separate from others—that gives the positive effects on penetration rates on new products, and so on [3].

The author of the second set of dimensions is Edward T. Hall (1914–2009). He introduced **perception of space** (the physical distance that is perceived as comfortable) and the **perception of time** (cultures that prefer to complete one task or more tasks simultaneously, designated as monochronic or polychronic cultures).

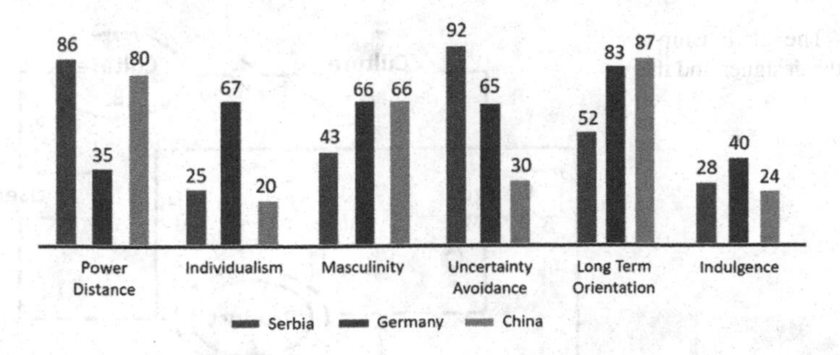

Fig. 20.2 Comparison between cultural dimensions

The dimensions of other investigators of this subject do not distinctly differ from Hofstede's and Hall's dimensions, besides their naming.

The influence of culture on culture-bound product design comes to the fore in many different ways, which can be categorised into two main groups: practical and theoretical:

- The practical group comprises aspects from the design process (methodology, procedures), design education (transfer of design knowledge among other cultures), strategy (business strategy and marketing products in other cultures), and designers (cultural influences on the designer himself);
- The theoretical group consists of aesthetic aspects (preference for design varying by country), semantics (understanding of function and design), and human-product interaction (how products are actually used in different cultures).

In the literature on the relationship between culture and design a special consideration is given to the influence of the designer's own culture on the design aspects of products [4]. It means that designer's own cultural values influence their design values.

Figure 20.3 indicates two possibilities: (1) the designer's culture is different from the user's culture, and (2) both user's and designer's culture, are similar or the same.

In the first case there is no link among theirs principle cultural dimensions, and a role of designer is more significant. As an example in which one can see how cultural environment can't be transferred from one culture to another, is the example of a Siemens washing machine exported to India. The users of the machine in India had major difficulties because of a different way of signing relevant functions. Another example is when products don't connect with the users, because they tend to reject or use them differently from the designers intention. A universal device for breaking or connecting an electric circuit is turned on and off by flipping a switch up or down but the direction varies between different countries. For instance, in the U.S., power is connected if switches are flipped up, whereas in the U.K. and Australia, the power is on if switches are flipped down. Flipping switches up or down can be a source of danger in emergency situations when people tend to act instinctively, intuitively or habitually rather than rationally or logically.

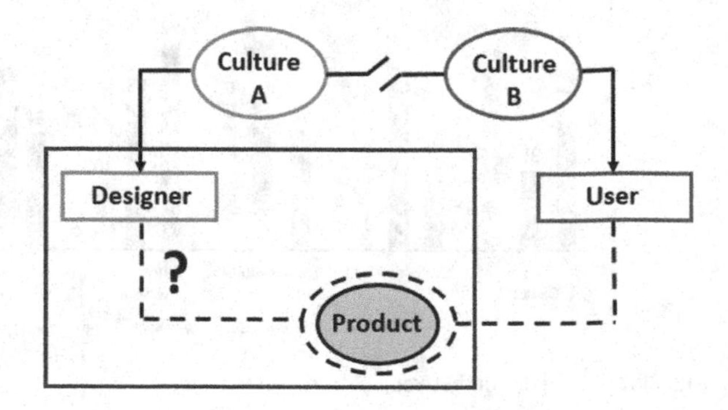

Fig. 20.3 The relationship between the designer and the user

In second case, he is in a possibility to include that their unique and rich cultural background which is a valuable resource of inspiration, and the question is, how to apply this resource into product design? For instance, prestigious design awards have been given to some products which included cultural characteristics; they are also an example of a possible trend in the market. Clearly, as this observation shows, designers have begun to apply national or exotic cultural elements by not simply copying, but transferring these elements and making a more sophisticated and inventive product design. Culture resources and information may be applied into product design and it indicates that political and religious ideologies, historic clothing, art movement and scientific innovation could all inspire designers.

20.3 Influence of Sustainability on Product Design

In the last decade, level of awareness of protection of environment and natural resources has risen. The idea of sustainable development has gained attention worldwide, and designers and engineers must include it in the process of designing products. In order to contribute to global ecological balance and achieve sustainable development—the relationship between product design and society must be change. Sustainability within product design is usually classified in two wider domains which are crucial and extremely useful for leading a designer through the process of design of sustainable products. These are eco-design and sustainable design [5]. Sustainable design aims to eradicate adverse environmental effects through adept and careful design [6].

Sustainable design is mostly a general reaction to global environmental crises, the rapid growth of economical activity and human population, depletion of natural resources, damage to ecosystems and loss of biodiversity. It does not require nonrenewable resources; it has a minimal impact on the environment, while connecting the people with the environment. Besides product design and industrial design, sustainable design can be applied in engineering, graphic design, interior design and fashion design, but also in architecture, landscape architecture, urban design and urban planning. The sustainable design we can also named the belief design, because we believe that the world will recognize the importance of themes on environment protection.

Sustainable design is widely defined as design of ecologically harmless products, in such a way that the environment can be sustained with minimal negative effects of the product in all of the stages of its life cycle: starting with design, production, distribution, use and expiry of product life.

Apart from achieving necessary technical product performance and defining product's costs, a designer must include other aspects associated with sustainable product design. Achieving sustainable design by improving process efficiency does not necessarily mean that it will positively impact on sustainability due to complexity while increasing efficiency. Majority of research related to sustainable development

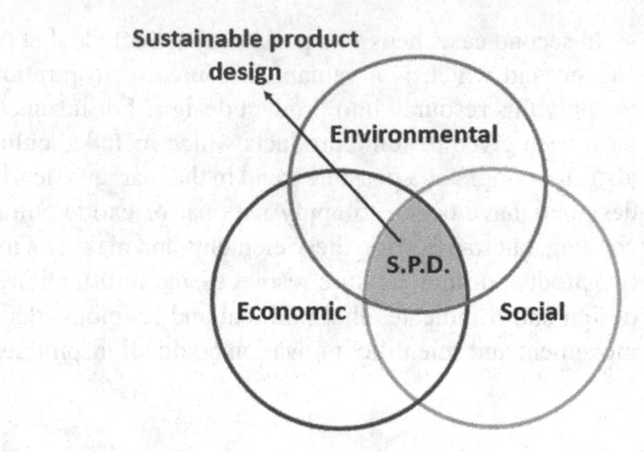

Fig. 20.4 Intersection of three sets

so far is focused on the aspect related to decreased use of materials, resources and energy which leads to less impact on the environment [5].

Sustainable product design can be accessed by classification within three aspects which are strongly interconnected: social, economical and ecological aspect (Fig. 20.4) [7]. Negative impact of individual aspects on product design can be reduced by their combination. Design requires a detailed analysis which takes into consideration risks and benefits of each of the aspects of sustainability. An ideal product is one that encompasses all of the three aspects by being ecologically acceptable, profitable for the company and beneficial for the society. Economical sustainability is easily quantified. Social sustainability, on the other hand, is somewhat more difficult to measure due to its non-material nature and subjective factors. Environmental sustainability, from the perspective of environmentally friendly products with lower negative impact on the environment, is also harder to quantify because it is necessary to take into account whole life cycle of the product which can be a complex and long process.

In order to quantify impact of a product on the environment, different methods have been developed: product analysis "from cradle to grave", product analysis "from cradle to cradle" and life cycle analysis. It is of paramount importance for a designer to choose raw materials which have the lowest impact on the environment, to influence production and distribution systems to minimize ecological and social influences, but also to define the use and disposal of unusable products.

Most products are cheaply made, they're not great quality, we don't use them very long, they're not easy to repair; we throw them away and buy new ones which are usually made with toxic and unhealthy materials which do not contribute to sustainable development. Consumerist "throwing away" of products (from single use lighters to cameras and expensive components in the energy system) is connected to planned obsolescence and has an extremely harmful effect on increasing level of environmental pollution [5]. All of these leads to lower ecological sustainability. The term of planned obsolescence and environmental sustainability are contradictory. It is a step during product design which makes the product obsolete on purpose or not functional after a specific period of use, in a way planned or created by the production company. Potential economical benefits for the production company are huge given that the consumer must buy the product again. Nowadays, unfortunately, too few products that are not designed with planned obsolescence in mind, but designers surely can influence mitigation of these problems within the process of designing new products that go beyond planned obsolescence.

In order to connect the aforementioned three major aspects of sustainability with sustainable product design, there are models which are focused mainly on ecological and economical aspects. Special attention should be given to the "quality design" as a factor that affects the longevity of a product. This also affects decrease of negative effect of the product on the environment and by that on the achievement of product sustainability and sustainable development. Designers may improve desirability of a product through the quality of its design, increase customer satisfaction and deepen product attachment in order to extend life of the product and improve its sustainability [8]. If one considers the life cycle of an ecologically sustainable hybrid car, one may notice that the quality of design of the car contributes its not being thrown away in the future, which may happen with a conventional car. By taking care of its maintenance—it could last for several generations. One of challenging roles of a designer is to change consumer's behavior for the benefit of all of us. This change can make a product more sustainable and can be very small—like a choice of one of the materials or big—decreased energy consumption and waste production.

20.4 Influence of UX on Product Design

One of the most prominent areas in contemporary design is the one—which Donald Norman in 1995 named UX (User Experience) design. The term denotes a wide range of various activities of designer which are associated with new products and services. In the beginning, the activities associated with this type of design were linked to the ancient Chinese tradition of Feng Shui (the importance of space), which among others was adopted by philosophers of Ancient Greeks (ergonomic principles), and then in the modern era, with the rise of industry, for example Frederick W. Taylor (the quest for workplace efficiency). Other notable names include Toyota (the value of human input), Henry Dreyfuss (the art of designing for people), Xeroxs, Apple and the PCera all the way up to the aforementioned D. Norman [9]. From that period there are many new activities with the perspective of further development (for example virtual riality).

In the current understanding, UX design is mainly associated with applications and web design, but it is so much more than just designing for a screen. Today, UX design has multiple interpretations, but it's really all about keeping users at the center of everything created, and also the process used to determine what the experience will be like when a user interacts with the product, making the decision about how the human and the product will interact. UX designers would take the principles that state how to make a product accessible, and actually embody those principles in the design process of a system so that a user that is interacting with it would find it as being

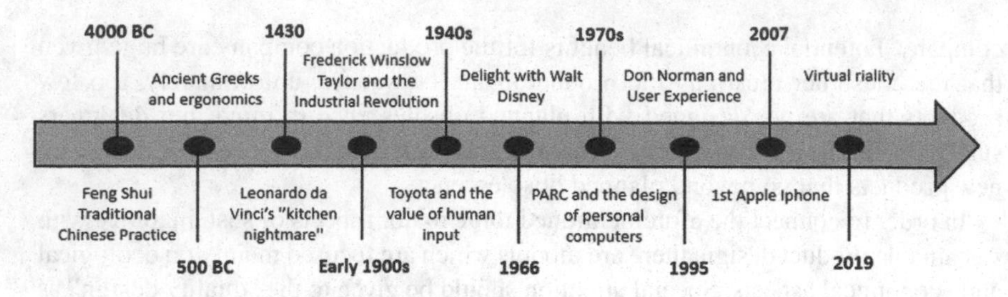

Fig. 20.5 History of user experience design

accessible. Relating to that, there is no commonly accepted definition for UX design. There are many aspects of user experience design and it has various disciplines—for example information architecture, usability, human–computer interaction, visual design, interaction design.

According to a study from the Oxford Journal Interacting With Computers: The goal of UX design in business is to "improve customer satisfaction and loyalty through the utility, ease of use, and pleasure provided in the interaction with a product". That is to say, UX design is defined as the process of designing (physical or digital) products that are useful and enjoyable and easy to use. This refers to enhancing people's experience in the interaction with products.

Figure 20.5 also indicates permanent never-ending process of seeing the world from the customer's perspective and working to improve the quality of their lives. Also, it is the process of maintaining the health of the business and finding new ways to help it grow sustainably. It is about the perfect balance between making money and making meaning. As one of the characteristics of this type of design is that it enables to identify what makes a good experience versus a bad one. And when done well, the designed elements of an experience become invisible and the users are delighted because the designers have anticipated their needs to give them something they don't think to ask for. Good UX design happens when the decision is made in a way that fulfills the needs of both product users and company business.

Very close to UX design is so-called UI (User Interface) design, as the dimensions that lies between humans and machines. Briefly, UX design consists of all the elements that make it possible for someone to interact with a service or product. Unlike UX design, UI design is the digital products only, and it is based on visual touch points.

20.5 Conclusion

The paper represents the basic issues of the multi-facets term product design, with an emphasis on culture, UX design, and sustainability design. It has been shown that the work of designers nowadays must take into account the increasingly demanding characteristics of products and services.

Acknowledgements The paper is supported by grants from the National Research Projects of Republic of Serbia Ministry of Education, Science and Technological Development and contract 451-03-9/2021-14/200105 (subproject TR33048).

References

1. https://technologystudent.com/PDF3/prod_dev1.pdf. Last accessed 2021/03/08
2. https://www.hofstede-insights.com/. Last accessed 2021/05/12
3. Yeniyurt, S., Townsend, J.D.: Does culture explain acceptance of new products in a country? An empirical investigation. Int. Market. Rev. **20**(4), 377–396 (2003)
4. Ramirez, M., Razzaghi, M.: The influence of the designers' own culture on the design aspects of products. In: Proceedings of the European Academy of Design International Conference, EAD06, pp. 1–15. Bremen (2005)
5. Diegel, O., Singamneni, S., Reay, S., Withell, A.: Tools for sustainable product design: Additive manufacturing. J. Sustain. Dev. **3**(3), 68–75 (2010)
6. McLennan, J.F.: The Philosophy of Sustainable Design. Ecotone Publishing Company LLC, Kansas City (2004)
7. Aoyama, H., Ghazali, I., Rashid, S.H.A., Dawal, S.Z., Tontowi, A.E.: Sustainable manufacturing: A framework of cultural aspects for sustainable product design. In: Proceedings of the 4th International Product Design and Development, pp. 1–7. Yogyakarta (2011)
8. Van Nes, N., Cramer, J.M.: Influencing poduct lifetime through product design. Bus. Strateg. Environ. **14**(5), 286–289 (2005)
9. https://www.yukti.io/what-is-ux-design-process-complete-guide-for-beginners/. Last accessed 2021/03/30

Chapter 21
Design Recommendations for FFF Parts

Jelena Djokikj, Tatjana Kandikjan, and Ile Mircheski

Abstract Additive Manufacturing (AM) technologies since their occurrence, they are under constantly development and improvement resulting in new applications. The unique working manner of these technologies ensures them wide range of possibilities but the lack of knowledge depicts them of wide spread use. The only way fully exploit AM is to know their advantages and restrictions in other words their characteristics. Design for AM (DfAM) is the field covering all the tools, rules and guidelines developed by various authors. With this paper we are making a contribution into this field by proposing design recommendations when designing for Fused Filament Fabrication (FFF). We choose the FFF process because is the AM process with most users, according to the sold machines. We are designing samples which are analyzed for deviations in the profile and shape, and based on the results we create the design recommendations. For the fabrication of the samples we are using open-source system.

Keywords Design for Additive Manufacturing (DfAM) · Fused Filament Fabrication (FFF) · Open-source systems · Fabrication quality · Three-dimensional scanning

21.1 Introduction

Additive manufacturing (AM) technologies consist of group of processes that work (build the model) by adding material in layers. The working manner of adding material provides these technologies with unique characteristics. These unique characteristics expressed mainly in the freedom of creation, made AM interesting for wide range of applications.

Among various AM processes, the process of Material Extrusion, more precisely FFF gained major popularization. The FFF works with thermoplastic, which is heated and extruded through nozzle onto the working platform (bed). The reason for the

J. Djokikj (✉) · T. Kandikjan · I. Mircheski
Faculty of Mechanical Engineering, Ss. Cyril and Methodius University, Skopje, Republic of North Macedonia
e-mail: jelena.djokikj@mf.ukim.edu.mk

M. Rackov et al. (eds.), *Machine and Industrial Design in Mechanical Engineering*, Mechanisms and Machine Science 109,
https://doi.org/10.1007/978-3-030-88465-9_21

FFF's status lies in its attributes: straightforward working process, no messy procedures, relatively good fabrication quality and low running costs. This process has been made especially affordable with the introduction of the open-source machines for FFF. Open-source machines make the FFF available to the public, but they are challenging when it comes to setting up for the initial startup, especially for novice users. Main reason are the numerous possibilities that has to be set up by the user, additionally most of these machines come as a kit that needs to be assembled.

With this paper fabrication quality and deviation in profile and shape of an open-source machine for FFF are being analyzed in order to draw recommendation for design for FFF. For the fabrication quality we are analyzing what are the minimum values of specific features that can be fabricated. For the dimensional accuracy we are applying reverse engineering as a technique for comparison and estimation of the overall geometry. With these analysis we want to see whether the fabrication possibilities and the quality of the open-source machines and give recommendations for application.

21.1.1 Background Research

Lack of knowledge of DfAM restricts the use these technologies and their application in the industry [1], slows down the use of AM for manufacturing of final parts [1, 2], prevents the designers of fully exploiting the possibilities of AM [3] and generally disenable the AM technologies reaching their full potential [4]. In order to create comprehensive methodology for DfAM extensive analysis of each one of the AM processes needs to be conducted.

Dimensional accuracy is one element that has huge impact on whether or not AM processes can be applicable in industry. Wang et al. [5] reported that the build orientation is the most significant process parameter affecting dimensional accuracy. Sood et al. [6] found that the dimensions of the FFF fabricated part are more in the Z-axis, while lesser in the X- and Y-axis when compared with the CAD model of the component. Chua et al. [7] state that higher deviations are noted with parts with small dimension. In research study conducted by Singh [8] he analyzes different parts and concludes that parts fabricated with FFF can be positioned in four IT classes according EN ISO 286–1. Although some studies positioned FFF fabricated parts from IT09 to IT14 or from IT11 to IT16 in other studies [9, 10]. Boschetto and Bottini [11] have developed a design for manufacturing methodology to improve the dimensional accuracy of parts fabricated with FFF. Zuowei et al. [12] introduced Skin Model Shapes paradigm as new promising method for modeling for in order to ensure better dimensional accuracy. Armillotta and Cavallaro [13] provided an experimental estimation of geometric errors on the edges of parts fabricated with FFF. Mohamed et al. [14] conducted analysis for stability dimensions of FFF parts and concluded that FFF process is capable of fabricating parts with high accuracy.

In the area of DfAM there are also numerous publication and as Thompson et al. state is one of the key challenges concerning AM [15]. Researcher work in order to create tools, rules and guidelines that will help users in order for better understanding of AM and designing for them. Filippi and Cristofolini [16] present guidelines for designing parts for fabrication with different AM process, where they work with specific AM processes. Teitelbaum et al. [17] develop list of rules based on experience. Their rules are tested through set of different parts and show positive results. Other researchers [18] make attempt in creation of rules for manufacturing parts with AM. They conduct extensive analysis for the mechanical characteristics of the materials used for AM and based on that make conclusions for the parts fabricated with AM. Adam and Zimmer [19] conduct extensive research in order to create design rules for three different processes: LM, SLS, and FFF. They design models with basic geometries in order to create ground rule knowledge. Urbanic and Hedrick [20] create design rules for FFF, which are infect set of guidelines for designing.

Knowledge gathered from the background review is used for planning and setting up the experimental analysis.

21.2 Design Framework

As mentioned above, the AM technologies require different approach in the design process that is not familiar to most of the users. This is why when designing for FFF of any other AM process, they encounter problems with particular features. For this research we compiled a list of the most common design feature problems advising various sources: research paper, AM community web sites, survey of FFF users, and author's years of experience in AM. Following is the list of the design feature problems explained in detail.

Wall thickness—wall thickness is critical element in the FFF process. Whether or not certain wall thickness will be fabricated needs to be examined. Two different cases of wall thickness are analyzed: integrated wall and stand-alone wall.

Gap—choosing the right gap thickness is critical for proper fabrication of the designed parts. If the gap is two small there is a possibility for it to not be fabricated properly i.e. the two adjusting features will be joined.

Feature thickness—choosing the feature thickness is important because it determines whether certain feature is fabricated. The feature thickness can influence to the strength of the part.

Cylindrical features—studying what is the right diameter of the cylindrical features in order for them to be fabricated properly (maintaining cylindricity).

Circular holes—circular holes same as cylindrical featured, needs to be studied not just whether or not are fabricated, but also is their circularity maintained.

Inclined features—inclined feature are important because with the appropriate designing they can be fabricated without supports. Appearance of the fabricated inclined features is also influenced by the staircase effect.

Overhangs and bridges—are features that need to be studied since their appropriate design will enable fabrication without supports. Elimination of the supports can result in stringing of the material.

All of the above features can be fabricated with any FFF machine, if they are adequately design. But our interest is to determine the minimal values for each feature or the maximum values without use of supports. Elimination of the supports is important since it can cause many problems in the post processing [21]. Ensuring repeatability of the results is another critical element of the FFF processes, this is why we fabricated the parts with repetition.

For ensuring good fabrication quality when working with open-source machines for FFF, setting the process parameters is of great influence. But for this research we are interested in the possibilities of the process as it is and not optimizing it, so we will keep them fixed thought out the whole fabrication process.

21.2.1 Design Samples

The samples used for the experimental analysis are embodiment of the design feature problems, explained in the paragraph above. The design of the samples is well thought, consisted of basic features, so that the results from the experimental analysis can be clear and unambiguous. We are particularly interested in the possibility to fabricate parts with small dimensions and fine details in order to determent the boundary conditions for the process.

The ten designed samples are designed using DS SolidWorks (Fig. 21.1). Every one of the samples has rectangular platforms with height of 1 mm. Dimensions of the rectangular platforms vary depending of the studied features. Text indicating the values of the analyzed features is placed on the right side with 3 mm in height and 2.5 mm in depth.

In Fig. 21.1 all the samples are graphically and textually explained in detail. Samples 1–1, 1–2, 1–3, 1–4 and 1–5 are intended to determine the minimal values that can be fabricated correctly. On the other hand, samples 1–6, 1–7 and 1–8 are intended to determine the maximal values that can be fabricated without the use of support structures.

21.3 Fabrication

Samples intended for the experimental analysis are fabricated using open-source machine for FFF (Prusa MK3) and PLA material. Detail process parameters used for the fabrications are presented in Table 21.1. It is important to be noted that the infill is used only for the platform, since the features are with small dimensions.

During the fabrication all of the parts are placed with the platform parallel to the bed which means that most of the features are perpendicular to the bed. Features placed perpendicular to the bed ensure best quality and accuracy especially for the cylindrical features and holes [22]. At the same time overhangs and bridges need to be parallel to the bed so that effect of the supports is analyzed.

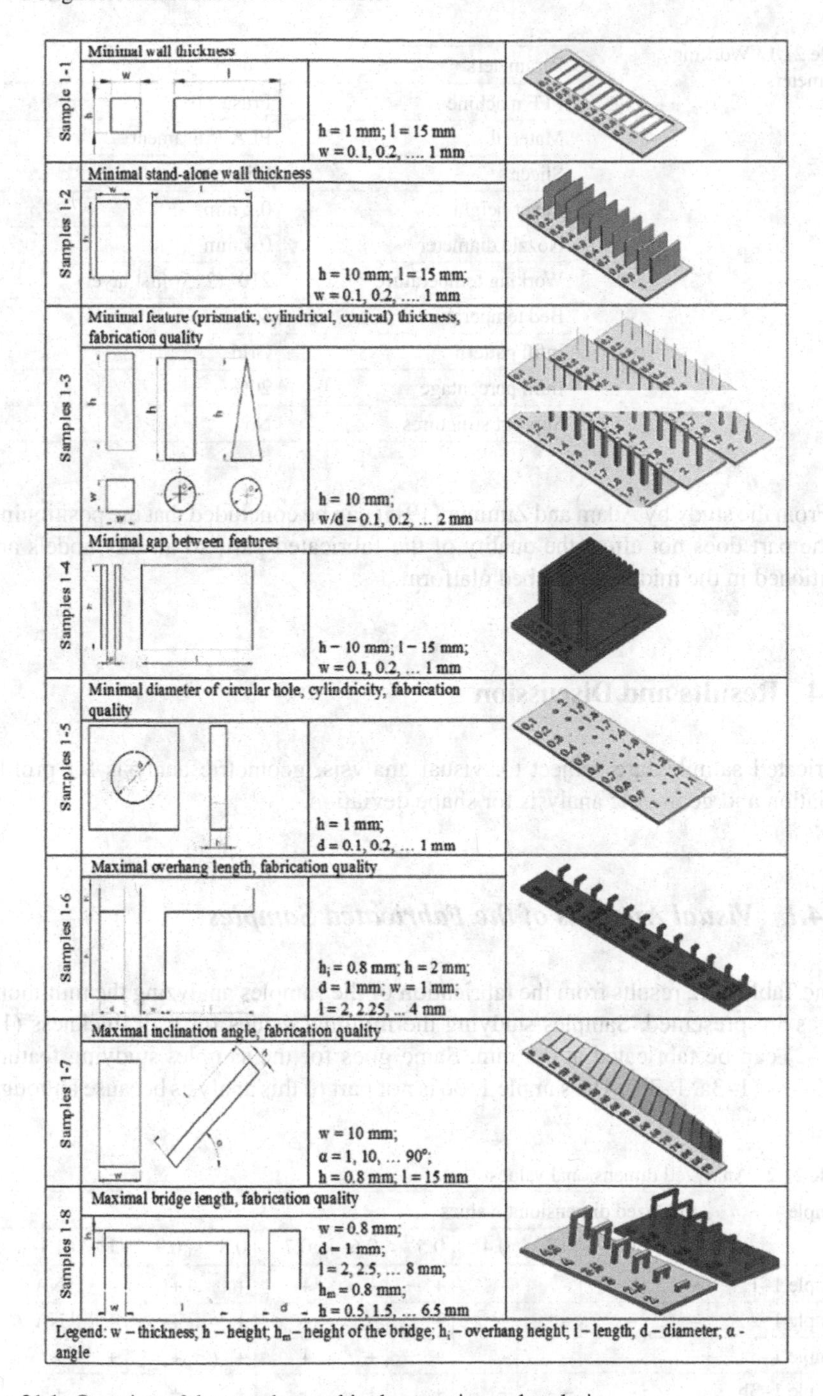

Fig. 21.1 Overview of the samples used in the experimental analysis

Table 21.1 Working parameters

Parameters	Values
FFF machine	Prusa Mk3
Material	PLA Prusament
Slicer	Slic3r
Layer height	0.2 mm
Nozzle diameter	0.4 mm
Working temperature	210° (215° first layer)
Bed temperature	60°
Infill pattern	Grid
Infill percentage	20%
Support structures	No

From the study by Adam and Zimmer [19] it can be concluded that the positioning of the part does not affect the quality of the fabricated part, so all the models are positioned in the middle of the bed platform.

21.4 Results and Discussion

Fabricated samples are subject to: visual analysis, geometric analysis for profile deviation and geometric analysis for shape deviation.

21.4.1 Visual Analysis of the Fabricated Samples

In the Table 21.2 results from the fabrication of the samples analyzing the minimum values are presented. Samples studying the minimal values for wall thickness (1–1; 1–2) can be fabricated at 0.5 mm. Same goes for the samples studying feature thickness (1–3a, 1–3b). The sample 1–3c is not part of this analysis because although

Table 21.2 Analyzed dimensional values

Sample	Analyzed dimensional values										
	0.1	0.2	0.3	0.4	0.5	0.6	0.7	0.8	0.9	1	1.1–2
Sample 1–1	–	–	–	–	+	+	+	+	+	+	NA
Sample 1–2	–	–	–	–	+	+	+	+	+	+	NA
Sample 1–3a	–	–	–	–	+	+	+	+	+	+	+
Sample 1–3b	–	–	–	–	+	+	+	+	+	+	+
Sample 1–5	–	–	–	–	–	–	+	+	+	+	NA

the features from 0.5 mm to 1 mm are fabricated, they are significantly shorter than the designed features. It is important to stress that although wall with thickness of 0.5 mm can be fabricated the users has to know that this wall is very fragile and cannot withstand any pressure or force. Nevertheless if the users need wall with thickness smaller than 1 mm with reservation. Under different working parameters, smaller nozzle or different part orientation even wall with thickness of 0.2 mm can be fabricated. Samples 1–3a and 1–3b cannot be fabricated with thickness/diameter under 0.5 mm, however everything under 1 mm, need to be avoided since they are fragile and can be damaged during the application and sometimes even during the fabrication process.

Samples studying the minimal diameter of circular holes (1–5) shows that the circular holes starting with 0.5 mm can be fabricated, although to maintain the circularity, diameter of 1 mm needs to be applied. Also, circular holes should always be positioned with the axis perpendicular to the bed, otherwise the circularity cannot be maintained.

All of the other samples (1–4, 1–6, 1–7 and 1–8) are fabricated complete, since their aim is to check the profile and shape and they are omitted from the Table 21.2. However they showed some imperfections during the fabrication. Sample 1–4 is fabricated with satisfying quality are 0.2 mm, the gap with 0.1 mm is not completely hollow all the way through. It is important to stress that features of the sample 1–7 where $\alpha \leq 30°$ have bad surface quality, which is result to the lack of the support material. We can note that samples 1–6 and 1–8 intended to detect maximum bridge and overhang lengths, with higher vales stringing occurs.

21.4.2 Geometric Analysis for Profile Deviation of the Fabricated Samples

Fabricated samples are analyzed for profile deviation using the software package Datinf Measurement. One of the expected occurrences is the roundness of the rectangular elements that are parallel to the bed, as shown on Fig. 2a and inner profile in Fig. 2b. Reason for this is the diameter of the nozzle which in our case is 0.4 mm.

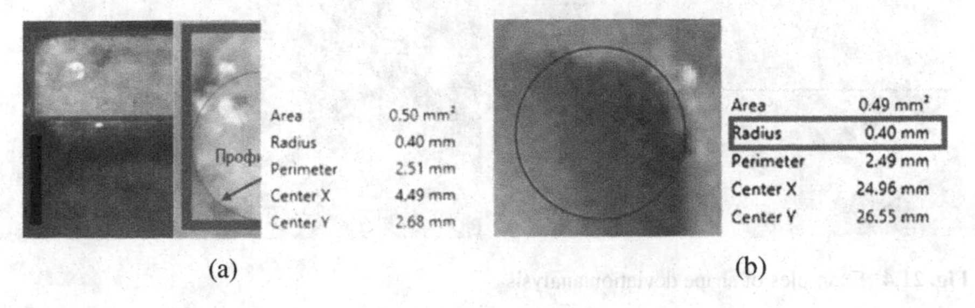

Fig. 21.2 Shape deviation in the profile: **a** outer and **b** inner

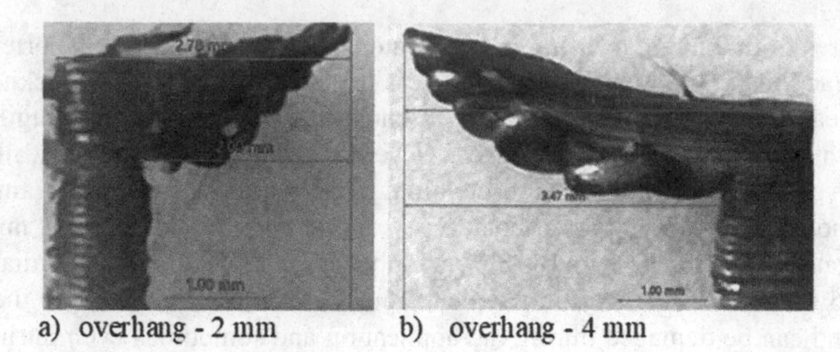

Fig. 21.3 Shape deviation of the overhangs

Consequently, this kind of deviation in the shape occurs only in the planes parallel to the bed. In cases where rounded angles are not permitted, different orientation of the part must be used.

Circular holes have best quality when their axis in perpendicular to the bed. But still, holes with small diameter can be fabricated badly. Although holes with diameter of 0.5 can be fabricated, circularity is maintained with diameter of 1 mm.

Another problem that can occur when working with FFF is warping, which happens due to uneven cooling of the part and is visible in large flat surfaces. But it is also visible in small and thin features, as in our sample 1–6 (Fig. 21.3). As it can be seen from the Fig. 21.3, for longer overhangs the deformation is higher. This can be avoided using support material, or designing thicker features.

21.4.3 *Geometric Analysis for Shape Deviation of the Fabricated Samples*

Fabricated samples are scanned using optical three-dimensional scanner by NewWay for the shape deviation analysis. The resulted point cloud is compared with the equivalent CAD model using the software package Geomagic Control X (Fig. 21.4).

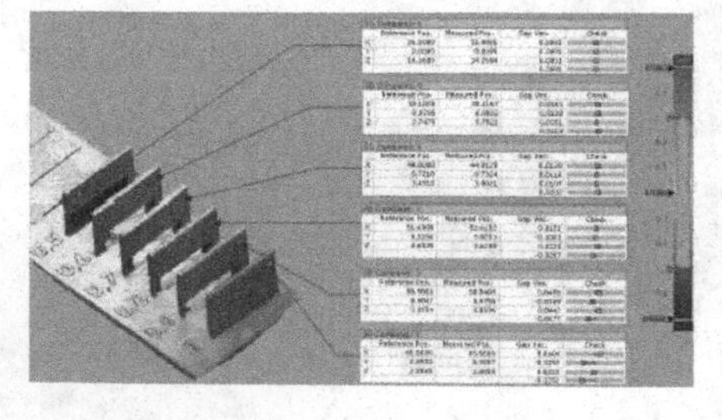

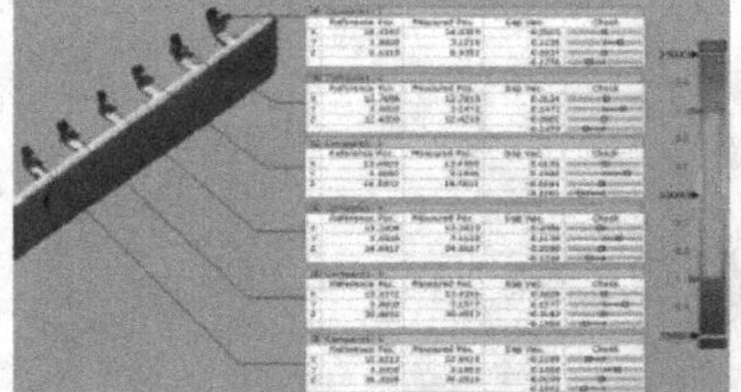

Fig. 21.4 Examples of shape deviation analysis

The used scale for the deviation is ±0.5 mm which is the precision of the open-source machines for FFF.

According the conducted analysis the shape deviation are in the range of ±0.1 mm, which is in the range of the professional FFF machines. Only exclusions are sample 1–3b with deviation of ±0.3 mm, and sample 1–6 with deviation of ±0.2 mm. The deviation at sample 1–3b occurs because the features are too thin and flexible so they oscillate during the fabrication, as explained in the text above. When it comes to the sample 1–6 which studies the maximal overhang without supports, the deviation occurs because of the thin elements that deformed (Fig. 21.3) during the fabrication.

If we compare the results for the shape deviation we can conclude that deviation occurs in the features with small dimension for the cross-section. This is why for the features dimension under 1 mm are not advised. For overhangs and bridges with higher lengths it is advised to use support material, also for the inclined elements where $\alpha \leq 30°$. For parts where the surface quality is of importance, it is advised to use soluble supports, which do not damage the surface of the part.

All in all the results are satisfying, apart from the two samples all of the others are in the range of ±0.1 mm which is the range for the professional machines for FFF.

21.5 Conclusion

With this research we wanted to explore the possibilities for open-source machines for FFF. We choose these machines since their popularity is increasing and the numbers of user is constantly rising. Also these machines are characterized as inferior to the professional one, concerning the quality and accuracy, so that was the additional motive. It should be emphasized that the experiments and conducted on one one-source machine Prusa Mk3, but we believe that is applicable to all open-source machines.

The lack of standardization for AM processed is another reason why we thought that this kind of research will be valuable for the research community. These processes have so many variables in the preparation and during the fabrication that really detailed analysis are needed.

The focus on features with small dimension and fine detail is because we believe is important to test the limits of the machines. Also during our work we did not come across such studies, so we decided to conduct such research.

For further research, more complex geometries needs to be analyzed. With this basic geometries we are setting ground rules which needs to be supplemented by new studies of this kind. Also deeper analysis in the fabrication of assemblies' has to be made.

References

1. Gausemeier, I.J.: Thinking ahead the Future of Additive Manufacturing—Analysis of Promising Industries. DMRC, Paderborn (2011)
2. Adam, G.A.O., Zimmer, D.: Design for additive manufacturing-element transitions and aggregated structures. CIRP J. Manuf. Sci. Technol. **7**, 20–28 (2014)
3. Geraedts, J., Doubrovski, Z., Verlinden, J.C., Stellingwerff, M.C.: Three views on additive manufacturing: Business, research, and education. In: Proceedings of the 9th International Symposium on Tools and Methods of Competitive Engineering, TMCE 2012, pp. 1–15. Karlsruhe (2012)
4. Azman, A.H., Vignat, F., Villeneuve, F.: Evaluating current CAD tools performances in the context of design for additive manufacturing. In: Proceedings of Joint Conference on Mechanical, Design Engineering and Advanced Manufacturing, pp. 44–1–44–7. Toulouse (2014)
5. Wang, T.M., Xi, J.T., Jin, Y.: A model research for prototype warp deformation in the FDM process. Int. J. Adv. Manuf. Tech. **33**(11–12), 1087–1096 (2007)
6. Sood, A.K., Ohdar, R.K., Mahapatra, S.S.: Parametric appraisal of mechanical property of fused deposition modelling processed parts. Mater. Des. **31**(1), 287–295 (2010)
7. Chua, C.K., Leong, K.F., Lim, C.S.: Rapid Prototyping: Principles and Applications. World Scientific, River Edge (2010)
8. Singh, R.: Some investigations for small-sized product fabrication with FDM for plastic components. Rapid Prototyping J. **19**(1), 58–63 (2013)
9. Lieneke, T., Adam, G.A., Lauders, S., Knoop, F., Josupeit, S., Delfs, P., Funke, N., Zimmer, D.: Systematical determination of tolerances for additive manufacturing by measuring linear dimensions. In: Proceedings of the Solid Freeform Fabrication Symposium, pp. 371–384. Austin (2015)
10. Lieneke, T., Denzer, V., Adam, G.A.O., Zimmer, D.: Dimensional tolerances for additive manufacturing: Experimental investigation for fused deposition modeling. Proc. CIRP **43**, 286–291 (2016)
11. Boschetto, A., Bottini, L.: Design for manufacturing of surfaces to improve accuracy in fused deposition modeling. Robot. Cim. Int. Manuf. **37**, 103–114 (2016)
12. Zuowei, Z.H.U., Keimasi, S., Anwer, N., Mathieu, L., Lihong, Q.I.A.O.: Review of shape deviation modeling for additive manufacturing. In: Eynard, B., Nigrelli, V., Oliveri, S., Peris-Fajarnes, G., Rizzuti, S. (eds.) Advances on Mechanics, Design Engineering and Manufacturing. LNME, pp. 241–250. Springer, Cham (2017)
13. Armillotta, A., Cavallaro, M.: Edge quality in fused deposition modeling: I Definition and analysis. Rapid Prototyping J. **23**(6), 1079–1087 (2017)
14. Mohamed, O.A., Masood, S.H., Bhowmik, J.L.: Investigation of dimensional variations in parts manufactured by fused deposition modeling using gauge repeatability and reproducibility. IOP Conf. Ser.: Mater. Sci. Eng. **310**, 012090–1–012090–6 (2018)
15. Thompson, M.K., Moroni, G., Vaneker, T., Fadel, G., Campbell, R.I., Gibson, I., Bernard, A., Schulz, J., Graf, P., Ahuja, B., Martina, F.: Design for additive manufacturing: Trends, opportunities, considerations, and constraints. CIRP Ann. **65**(2), 737–760 (2016)
16. Filippi, S., Cristofolini, I.: The Design Guidelines (DGLs), a knowledge-based system for industrial design developed accordingly to ISO-GPS (Geometrical Product Specifications) concepts. Res. Eng. Des. **18**(1), 1–19 (2002)
17. Teitelbaum, G.A., Schmidt, L.C., Goaer, Y.: Examining potential design guidelines for use in fused deposition modeling to reduce build time and material volume. In: Proceedings of the ASME International Design Engineering Technical Conferences and Computers and Information in Engineering Conference IDETC/CIE 2009, pp. 73–82. San Diego (2009)
18. Gibson, I., Goenka, G., Narasimhan, R., Bhat, N.: Design rules for additive manufacture. In: Proceedings of the Solid Freeform Fabrication Symposium, pp. 705–716. Austin (2010)
19. Adam, G.A.O., Zimmer, D.: On design for additive manufacturing: Evaluating geometrical limitations. Rapid Prototyping J. **21**, 662–670 (2015)

20. Urbanic, R.J., Hedrick, R.: Fused deposition modeling design rules for building large, complex components. Comput. Aided Des. Appl. **13**(3), 348–368 (2016)
21. Jiang, J., Xu, X., Stringer, J.: Support structures for additive manufacturing: A review. J. Manuf. Mater. Process **2**(4), 64–1–64–4 (2018)
22. Garg, A., Bhattacharya, A., Batish, A.: On surface finish and dimensional accuracy of FDM parts after cold vapor treatment. Mater. Manuf. Process. **31**(4), 522–529 (2016)

Chapter 22
Adaptive Neuro Fuzzy Estimation of Processing Parameters Influence on the Performances of Plasma Arc Cutting Process

Dalibor Petković and Milos Milovancevic

Abstract Plasma arc cutting process is one of the most advances non-conventional machining process which is mainly using for cutting purposes. Surface quality of the final product and dimensional accuracy are very important factors for particular applications in manufacturing industry. There are different input factors which need adjustment in order to find the optimal combinations for the best final product. Therefore the main aim of the study was to establish predictive models for the plasma arc cutting in order to determine the input parameters influence on cutting quality of the plasma arc cutting. As output factors for the cutting quality, material removal rate—MRR, heat affected zone—HAZ and kerf taper, are used. As the input processing parameters, cutting speed, gas pressure, arc current and stand-off distance, are used. The main aim is to evaluate the inputs' influence on the output factors based on the predictive performances. Predictive models are created based on adaptive neuro fuzzy inference system—ANFIS, which is suitable for nonlinear and redundant dataset. As cutting material Quard-400 was used. Results shown that the arc current has the highest relevance for the product quality by plasma arc cutting process.

Keywords Plasma arc cutting · Prediction · Cutting quality · ANFIS

22.1 Introduction

Plasma state is the fourth state of material after solid, liquid and gaseous states. Plasma state occurs after very high heating of the material. In other words the first state is solid state which converts in the liquid state after heating. The liquid state converts further in gaseous state after more heating. And finally the gaseous state converts into plasma state after more heating. Plasma represent in ionized gas which is electro conductive and operated on temperatures between 10,000 °C and 14,000 °C.

D. Petković
Pedagogical Faculty in Vranje, University of Niš, Partizanska 14, 17500 Vranje, Serbia

M. Milovancevic (✉)
Faculty of Mechanical Engineering, University of Niš, Vranje, Serbia
e-mail: milovancevic@masfak.ni.ac.rs

M. Rackov et al. (eds.), *Machine and Industrial Design in Mechanical Engineering*, Mechanisms and Machine Science 109,
https://doi.org/10.1007/978-3-030-88465-9_22

Plasma arc cutting is based on the ionized gas and it represents one type of thermal cutting process which uses a jet of the plasma gas in order to melt and cut material. The plasma arc cutting is very attractive process for material removal or cutting because of high quality of the final product. However the plasma arc cutting process is very expensive process in comparison to laser cutting or cutting by oxygen fuel.

Plasma arc cutting (PAC) is a well-known non-conventional machining technology for fabricating complicated part profiles for a variety of electrically conductive materials, such as superalloys and composites [1]. Stainless steel, alloy steel, aluminum, and other materials are often cut using the plasma arc cutting technique [2]. In article [3], the teaching learning-based optimization (TLBO) algorithm was used to investigate the effect of process parameters on surface roughness in plasma arc cutting of AISI D2 steel. The optimal selection of process parameters is critical for smoother and faster cutting, and experimental investigation of plasma arc cutting was carried out in research work [4], where Taguchi based desirability analysis (TDA) was used to find the optimal cutting conditions for improving the plasma arc cutting process' quality characteristics. The quality of the cut of the plasma arc cutting procces has been monitored by measuring the kerf taper angle (conicity), the edge roughness and the size of the heat-affected zone (HAZ) [5]. Results in paper [6] have been shown that this fuzzy control and PID neural network improves the precision, ripples, finish and other comprehensive indexes. Plasma arc cutting (PAC) is a thermal cutting process that makes use of a constricted jet of high-temperature plasma gas to melt and separate (cut) metal [7]. In work [8], the microstructural modifications are elaborated in detail. The paper [9] pointed out that high quality parts of the plasma arc cutting can be obtained as a result of an experimental investigation aimed at selecting the proper values of process parameters. There is need to develop and optimize novel plasma arc heat source such as cross arc and coupling arc [10]. In study [11] has been studied the influences of plasma arc remelting on the microstructure and properties of thermal sprayed Cr3C2-NiCr/NiCrAl composite coating. To reduce the kerf width and to improve the kerf quality, the hydro-magnetically confined plasma arc was used to cut engineering ceramic plates [12]. The quality of cuts performed on titanium sheets using high tolerance plasma arc cutting (HTPAC) process was investigated in [13].

In order to decrease price of the plasma arc cutting process there is need to establish predictive models of the process. The predictive models could help in estimation of the input parameters influence on the product quality of the plasma arc cutting process. In order to do the predictive models there is need for an advanced computational models like soft computing or computational intelligence. Therefore in this article is used adaptive neuro fuzzy inference system or ANFIS [14–18] in order to estimate the influence of the input parameters on the output quality factors of the plasma arc cutting process. As output factors for the cutting quality, material removal rate—MRR, heat affected zone—HAZ and kerf taper, are used. As the input processing parameters, cutting speed, gas pressure, arc current and stand-off distance, are used.

22.2 Methodology

22.2.1 Experimental Procedure

Figure 22.1 shows the total experimental procedure of the plasma arc cutting. As can be seen there are nine main steps of the procedure. The torch body contains cathode which is non-melting. Working material represent anode or positive electrode where high temperature plasma gas or primary gas will be impinged. Kerf represent width of material removal during cutting process where molten metal is removed. For the cooling purpose secondary gas is used.

As working material Quard-400 is used which is abrasion resistant steel. This material has optimal combination of hardness, ductility and strength and it is very suitable for cutting process. Chemical composition of the material has following elements: C, Mn, P, Si, Al, Cu, Nb, Ni, Cr, V, Ti, N_2, B and Fe.

For the experimental procedure CNC plasma arc cutting machine is used. Cutting specimens are dimension of 20 × 20 × 10mm. As cutting gas oxygen is used. Table

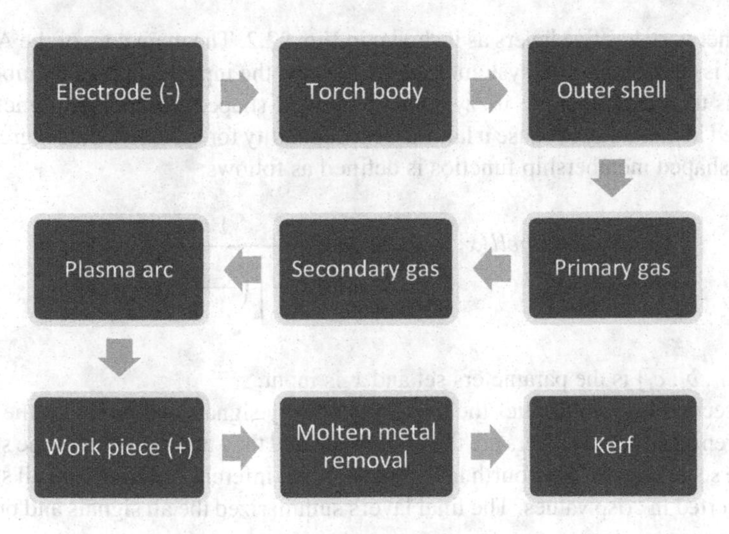

Fig. 22.1 Experimental procedure of the plasma arc cutting process

Table 22.1 Input and output factors of the plasma arc cutting process

Input factors				Output factors		
in1: Cutting speed (mm/min)	in2: Gas pressure (Bar)	in3: Arc current (A)	in4: Stand-off distance (mm)	Material removal rate (gm/sec)	Kerf taper (°)	Heat affected zone (mm)
2200–2600	3–4	45–55	2–3	24.32–46.37	2.52–8.56	2.48–6.21

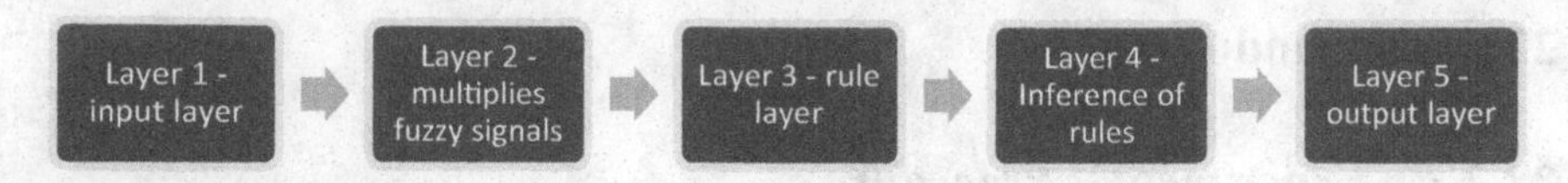

Fig. 22.2 ANFIS layers

22.1 shows the numerical values (minimum and maximum) of the input and output factors which are used and obtained during cutting process.

`Material removal rate or MRR is calculated based on the weight of the final product after cutting process. Kerf taper is calculated based on top and bottom kerf width at three different places on the length of cut. Heat affected zone—HAZ represents intensity of plasma around the cut surface region which is measured experimentally by microscope.

22.2.2 *ANFIS Methodology*

ANFIS network has five layers as it shown in Fig. 22.2. The main core of the ANFIS network is fuzzy inference system. Layer 1 accepts the inputs and uses membership functions to convert them to fuzzy values. The bell-shaped membership function is employed in this study because it has the best capability for nonlinear data regression.

Bell-shaped membership functios is defined as follows:

$$\mu(x) = bell(x; a_i, b_i, c_i) = \frac{1}{1 + \left[\left(\frac{x-c_i}{a_i}\right)^2\right]^{b_i}} \tag{22.1}$$

where $\{a_i, b_i, c_i\}$ is the parameters set and x is input.

The second layer multiplies the first layer's fuzzy signals and produces the rule's firing strength. The rule layers are the third layer, and they normalize all of the signals from the second layer. The fourth layer provides the inference of rules and all signals are converted in crisp values. The final layers summarized the all signals and provide the output crisp value.

22.3 Results

22.3.1 *Accuracy Indices*

Performances of the proposed models are presented as root means square error (RMSE) as follows:

$$RMSE = \sqrt{\frac{\sum_{i=1}^{n} (P_i - O_i)^2}{n}} \quad (22.2)$$

where P_i and O_i are known as the experimental and forecast values, respectively, and n is the total number of dataset.

22.3.2 ANFIS Results

ANFIS networks is trained with acquired experimental input/output dataset in order to determine RMSE values for each combinations of input and output pairs. The following inputs are used:

- in1: Cutting speed (mm/min)
- in2: Gas pressure (Bar)
- in3: Arc current (A)
- in4: Stand-off distance (mm)

The inputs influence on the material removal rate is calculated by ANFIS network and the following results are obtained:

- ANFIS model 1: in1 – > trn = 5.3746, chk = 4.6647
- ANFIS model 2: in2 – > trn = 4.9470, chk = 4.6068
- **ANFIS model 3: in3 – > trn = 4.7669, chk = 5.5596**
- ANFIS model 4: in4 – > trn = 4.7968, chk = 4.7504

As can be seen above the input 3 or arc current has the strongest influence on the MRR based on the training RMSE (trn). The checking RMSE (chk) is used for overfitting tracking between training and checking data. If two inputs are combined the following results are obtained for MRR prediction where one can see that the combination of arc current and stand-off distance is the most influential combination for MRR prediction:

- ANFIS model 1: in1 in2 – > trn = 4.2217, chk = 8.4118
- ANFIS model 2: in1 in3 – > trn = 4.1433, chk = 6.5070
- ANFIS model 3: in1 in4 – > trn = 3.7734, chk = 10.3866
- ANFIS model 4: in2 in3 – > trn = 3.8618, chk = 5.3528
- ANFIS model 5: in2 in4 – > trn = 3.8562, chk = 4.7520
- **ANFIS model 6: in3 in4 – > trn = 2.6716, chk = 5.1528**

The inputs influence on the kerf taper degree is calculated by ANFIS network and the following results are obtained:

- ANFIS model 1: in1 – > trn = 1.6369, chk = 1.8296
- ANFIS model 2: in2 – > trn = 1.6556, chk = 1.8746
- **ANFIS model 3: in3 – > trn = 1.3572, chk = 1.4332**

- ANFIS model 4: in4 – > trn = 1.4874, chk = 2.2487

As can be seen above the input 3 or arc current has the strongest influence on the kerf taper degree based on the training RMSE (trn). There is no overfitting between training and testing data. If two inputs are combined the following results are obtained for kerf taper degree prediction where one can see that the combination of arc current and stand-off distance is the most influential combination for kerf taper degree prediction:

- ANFIS model 1: in1 in2 – > trn = 1.5654, chk = 2.2506
- ANFIS model 2: in1 in3 – > trn = 1.1898, chk = 4.3314
- ANFIS model 3: in1 in4 – > trn = 1.2320, chk = 3.9265
- ANFIS model 4: in2 in3 – > trn = 1.1885, chk = 2.0542
- ANFIS model 5: in2 in4 – > trn = 1.3740, chk = 2.4450
- **ANFIS model 6: in3 in4 – > trn = 0.7243, chk = 1.6536**

The inputs influence on the HAZ is calculated by ANFIS network and the following results are obtained:

- ANFIS model 1: in1 – > trn = 0.8779, chk = 0.6894
- ANFIS model 2: in2 – > trn = 0.7672, chk = 0.5642
- **ANFIS model 3: in3 – > trn = 0.7570, chk = 0.8989**
- ANFIS model 4: in4 – > trn = 0.8720, chk = 0.6312

As can be seen above the input 3 or arc current has the strongest influence on the HAZ based on the training RMSE (trn). There is no overfitting between training and testing data. If two inputs are combined the following results are obtained for kerf taper degree prediction where one can see that the combination of arc current and stand-off distance is the most influential combination for kerf taper degree prediction:

- ANFIS model 1: in1 in2 – > trn = 0.6557, chk = 0.9558
- ANFIS model 2: in1 in3 – > trn = 0.6201, chk = 1.2745
- ANFIS model 3: in1 in4 – > trn = 0.6738, chk = 1.6608
- ANFIS model 4: in2 in3 – > trn = 0.5292, chk = 0.7530
- ANFIS model 5: in2 in4 – > trn = 0.6553, chk = 0.5633
- **ANFIS model 6: in3 in4 – > trn = 0.5264, chk = 0.8483**

22.4 Conclusion

In this paper was investigated predictive performance of adaptive neuro fuzzy inference system or ANFIS for prediction of output factors of plasma art cutting process. As output factors for the cutting quality, material removal rate—MRR, heat affected zone—HAZ and kerf taper, are used. As the input processing parameters, cutting speed, gas pressure, arc current and stand-off distance, are used. The main purpose of the ANFIS predictive models was to determine the inputs influence on the given outputs. ANFIS can eliminate the vagueness in the process in order to produce the best prediction conditions. In other words ANFIS network was used to convert the multiple performance characteristics into the single performance index. Results shown that the arc current has the highest relevance for the product quality by plasma arc cutting process.

References

1. Ananthakumar, K., Rajamani, D., Balasubramanian, E., Davim, J.P.: Measurement and optimization of multi-response characteristics in plasma arc cutting of Monel 400™ using RSM and TOPSIS. Measurement **135**, 725–737 (2019)
2. Bhowmick, S., Basu, J., Majumdar, G., Bandyopadhyay, A.: Experimental study of plasma arc cutting of AISI 304 stainless steel. Mater. Today-Proc. **5**(2), 4541–4550 (2018)
3. Patel, P., Nakum, B., Abhishek, K., Kumar, V.R., Kumar, A.: Optimization of surface roughness in plasma arc cutting of AISID2 steel using TLBO. Mater. Today-Proc. **5**(9), 18927–18932 (2018)
4. Kumar Naik, D., Maity, K.P.: An optimization and experimental analysis of plasma arc cutting of Hardox-400 using Taguchi based desirability analysis. Mater. Today-Proc. **5**(5, Part 2), 13157–13165 (2018)
5. Salonitis, K., Vatousianos, S.: Experimental investigation of the plasma arc cutting process. Proc. CIRP **3**, 287–292 (2012)
6. Deli, J., Bo, Y.: An intelligent control strategy for plasma arc cutting technology. J. Manuf. Process. **13**(1), 1–7 (2011)
7. Chamarthi, S., Reddy, N.S., Elipey, M.K., Reddy, D.R.: Investigation analysis of plasma arc cutting parameters on the unevenness surface of Hardox-400 material. Procedia Engineer. **64**, 854–861 (2013)
8. Rotundo, F., Martini, C., Chiavari, C., Ceschini, L., Concetti, A., Ghedini, E., Dallavalle, S.: Plasma arc cutting: Microstructural modifications of hafnium cathodes during first cycles. Mater. Chem. Phys. **134**(2–3), 858–866 (2012)
9. Bini, R., Colosimo, B.M., Kutlu, A.E., Monno, M.: Experimental study of the features of the kerf generated by a 200 A high tolerance plasma arc cutting system. J. Mater. Process. Tech. **196**(1–3), 345–355 (2008)
10. Chen, S., Zhang, R., Jiang, F., Dong, S.: Experimental study on electrical property of arc column in plasma arc welding. J. Manuf. Process. **31**, 823–832 (2018)
11. Ji-yu, D., Fang-yi, L., Yan-le, L., Li-ming, W., Hai-yang, L., Xue-ju, R., Xing-yi, Z.: Influences of plasma arc remelting on microstructure and service performance of Cr3C2-NiCr/NiCrAl composite coating. Surf. Coat. Tech. **369**, 16–30 (2019)
12. Xu, W.J., Fang, J.C., Lu, Y.S.: Study on ceramic cutting by plasma arc. J. Mater. Process. Tech. **129**(1–3), 152–156 (2002)
13. Gariboldi, E., Previtali, B.: High tolerance plasma arc cutting of commercially pure titanium. J. Mater. Process. Tech. **160**(1), 77–89 (2005)
14. Jang, J.-S.R.: ANFIS: Adaptive-network-based fuzzy inference systems. IEEE T. Syst. Man Cyb. **23**, 665–685 (1993)
15. Petković, D., Issa, M., Pavlović, N.D., Pavlović, N.T., Zentner, L.: Adaptive neuro-fuzzy estimation of conductive silicone rubber mechanical properties. Expert Syst. Appl. **39**(10), 9477–9482 (2012)
16. Petković, D., Ćojbašić, Ž: Adaptive neuro-fuzzy estimation of automatic nervous system parameters effect on heart rate variability. Neural Comput. Appl. **21**(8), 2065–2070 (2012)
17. Kurnaz, S., Cetin, O., Kaynak, O.: Adaptive neuro-fuzzy inference system based autonomous flight control of unmanned air vehicles. Expert Syst. Appl. **37**(2), 1229–1234 (2010)
18. Petković, D., Issa, M., Pavlović, N.D., Zentner, L., Ćojbašić, Ž: Adaptive neuro fuzzy controller for adaptive compliant robotic gripper. Expert Syst. Appl. **39**(18), 13295–13304 (2012)

[illegible] reduction [illegible] (ANFIS) [illegible] the multiple [illegible] characteristics [illegible] pattern [illegible]. Results [illegible] current [illegible] the highest relevance [illegible] quality [illegible] plasma [illegible].

References

1. [illegible], K., Kozuma[illegible] [illegible] On-line [illegible] multiple [illegible] measurement [illegible] 125, [illegible] (2011)
2. [illegible] Experimental [illegible] cutting [illegible] (2011)
3. [illegible] arc cutting [illegible] [illegible] (2018)
4. [illegible] [illegible] 152, [illegible] (2018)
5. [illegible] experimental investigation of the plasma arc cutting process. [illegible] (2011)
6. [illegible] intelligent control [illegible] cutting technology. [illegible] Process. [illegible] (2011)
7. [illegible] [illegible] surface of [illegible] steel [illegible]
8. [illegible] Garlin, C.L. [illegible] [illegible] alloy [illegible] [illegible] 1541, [illegible] (2018)
9. [illegible] [illegible] (1998)
10. Chen, [illegible], Pang, S.J. [illegible] [illegible] (2018)
11. [illegible] [illegible] (2019)
12. [illegible] [illegible] plasma [illegible] 122, [illegible]
13. [illegible] [illegible] (2008)
14. [illegible] A [illegible] network-based fuzzy inference system. IEEE Trans. Syst. Man Cybern. 23, [illegible] (1993)
15. [illegible] [illegible] (2013)
16. [illegible] [illegible] (2012)
17. [illegible] [illegible] (2019)
18. [illegible] [illegible] (2013)

Chapter 23
Identification of Important Parameters for Laser Photoresist Removal Process by ANFIS Methodology

Milos Milovancevic and Dalibor Petković

Abstract Laser photoresist removal in gaseous media is based on different parameters or factors like energy of laser pulse, frequency of pulse repetition and flow rate of gas. A statistical soft computing approach was applied in this article in order to determine which parameters have the most influence on the photoresist removal process by laser. As statistical approach adaptive neuro fuzzy inference system (ANFIS) was used since the methodology can handle strongly nonlinear data. By selected the most important factors one can adjust the photoresist removal process in order to produce the best final product. For the selecting process three input parameters are used: laser energy, rate of pulse repetition of laser and flow rate of hydrogen gas. These parameters are selected for the analyzing since these are independent variables. Pulse of laser repetition is selected as the most important factor for the photoresist removal process. Predictive models were created based on ANFIS network and corresponding results are compared with standard conventional approaches.

Keywords Laser · Photoresist removal · Prediction · Gas media · ANFIS · Running head · Laser photoresist removal process

23.1 Introduction

In order to fabricate microsystem there is need for perform several steps which include deposition of thin-film, patterning of polymer resist and etching. There are positive-tone and negative-tone resists process. Photoresist removal is performed by wet cleaning by chemicals. However the chemicals are very expensive and could harm environment. There are many attempts to reduce cost and damage of environment of the photoresist removal process.

M. Milovancevic (✉)
Faculty of Mechanical Engineering, University of Niš, Vranje, Serbia
e-mail: milovancevic@masfak.ni.ac.rs

D. Petković
Pedagogical Faculty in Vranje, University of Niš, Partizanska 14, 17500 Vranje, Serbia

M. Rackov et al. (eds.), *Machine and Industrial Design in Mechanical Engineering*, Mechanisms and Machine Science 109,
https://doi.org/10.1007/978-3-030-88465-9_23

A concept for the laser-assisted removal of material has been proposed in article [1] for achieving nanoscale correction in next-generation functional microstructures such as nanostructured photoresist surfaces and micro 3-D objects fabricated using microstereolithography. Positive photoresists are photosensitive materials widely used in lithographic processes in microelectronics and optics for component relief manufacturing and when exposed to ultraviolet radiation, chemical reactions are induced that modify their physical–chemical properties [2]. Details of fabrication route for photoresist template has been presented in article [3], which is useful in surface texturing of technologically important thin films such as solar absorbers, transparent conducting oxides (TCOs) and metals. The highly crosslinked epoxy remaining after development is difficult to remove reliably from high aspect ratio structures without damaging or altering the electroplated metal [4]. SU-8 photoresist maintained good resolution in thick film. A new route for pattern formation based on atmospheric pressure plasma directed assembly during photoresist removal has been presented in article [5] where the results shown the potential of atmospheric plasma directed assembly for uniform, large-area and open-air pattern definition for application in modern nanofabrication. In the [6–8] further consideration are explained in detail. Hydrogen radical process for photoresist removal by use of hot W catalyst has been investigated for a possible application to advanced Cu/low-k dielectric interconnects in LSI and results suggested that the hydrogen radical process for resist removal with W catalyst is promising for production of advanced interconnects [9]. Ozonated water cleaning systems have been used to remove photoresist and organic contaminants on silicon wafers and natural organic matter and bacteria in drinking water [10]. All-wet processes are gaining a renewed interest for the removal of post-etch photoresist (PR) in semiconductor manufacturing but changes in regulations call for a reduction in the environmental, safety & health (ESH) impact of solvents used [11].

In this study was analyzed a technique of photoresist removal without hazard chemical and without high temperature. This is a dry process by laser ablation and under presences of hydrogen gas medium. The main goal was to investigate how the three operating parameters, laser energy, rate of pulse repetition and rate of gas flow, have impact on the photoresist removal process by laser in medium of hydrogen gas. In order to do the analyzing there is need for an advanced computational models like soft computing or computational intelligence. Therefore in this article is used adaptive neuro fuzzy inference system or ANFIS [12–16]. The output factor is etch depth of the photoresist surface.

Fig. 23.1 Experimental setup of photoresist removal

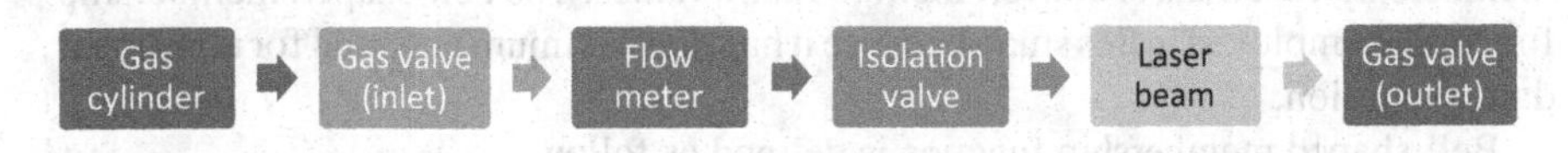

Fig. 23.2 Experimental procedure of gas handling for photoresist removal

Table 23.1 Input and output factors of the plasma arc cutting process

Input factors			Output factor
in1: Flow rate (lpm)	in2: Laser pulse repetition (Hz)	in3: Laser Energy (mJ)	Etch Depth (μm)
0.6–1.2	10–40	40–60	13.4–122.7

Fig. 23.3 ANFIS layers

23.2 Methodology

23.2.1 Experimental Procedure

Figure 23.1 shows the total experimental procedure of the photoresist removal by laser ablation system. The laser pulse has wavelength of 248 nm with pulse duration of 25 ns. Frequency of pulse repetition is in range 10–50 Hz. In this study is performed mask projection technique. Gas handling setup is shown in Fig. 23.2. For the photoresist removal process cylindrical chamber is used which has inlet and outlet connections with valves. Photoresist polymer E1020 is used in this study. Table 23.1 shows the numerical values (minimum and maximum) of the input and output factors which are used and obtained during photoresist removal process.

23.2.2 ANFIS Methodology

As shown in Fig. 23.3, the ANFIS network contains five levels. The fuzzy inference system lies at the heart of the ANFIS network. Layer 1 accepts the inputs and uses membership functions to convert them to fuzzy values. The bell-shaped membership function is employed in this study because it has the maximum potential for non-linear data regression.

Bell-shaped membership functios is defined as follows:

$$\mu(x) = bell(x;\, a_i,\, b_i,\, c_i) = \frac{1}{1 + \left[\left(\frac{x-c_i}{a_i}\right)^2\right]^{b_i}} \tag{23.1}$$

where $\{a_i, b_i, c_i\}$ is the parameters set and x is input.

The second layer multiplies the first layer's fuzzy signals and produces the rule's firing strength. The rule layers are the third layer, and they normalize all of the signals from the second layer. The fourth layer does rule inference and converts all signals to crisp values. The last layers summed all of the signals and provided a clean output value.

23.3 Results

23.3.1 Accuracy Indices

Performances of the proposed models are presented as root means square error (RMSE), Coefficient of determination (R^2) and Pearson coefficient (r) as follows:

- RMSE:

$$RMSE = \sqrt{\frac{\sum_{i=1}^{n}(P_i - O_i)^2}{n}} \tag{23.2}$$

- Pearson correlation coefficient (r):

$$r = \frac{n\left(\sum_{i=1}^{n} O_i \cdot P_i\right) - \left(\sum_{i=1}^{n} O_i\right)\cdot\left(\sum_{i=1}^{n} P_i\right)}{\sqrt{\left(n\sum_{i=1}^{n} O_i^2 - \left(\sum_{i=1}^{n} O_i\right)^2\right)\cdot\left(n\sum_{i=1}^{n} P_i^2 - \left(\sum_{i=1}^{n} P_i\right)^2\right)}} \tag{23.3}$$

- Coefficient of determination (R^2):

$$R^2 = \frac{\left[\sum_{i=1}^{n}\left(O_i - \overline{O_i}\right)\cdot\left(P_i - \overline{P_i}\right)\right]^2}{\sum_{i=1}^{n}\left(O_i - \overline{O_i}\right)\cdot\sum_{i=1}^{n}\left(P_i - \overline{P_i}\right)} \tag{23.4}$$

where P_i and O_i are known as the experimental and forecast values, respectively, and n is the total number of dataset.

23.3.2 ANFIS Results

Table 23.2 shows the prediction errors for the photoresist removal process for single input parameter and two parameters combinations as well. Training errors (trn) shows influence of the inputs for the photoresist removal process. Smaller training error more influence on the photoresist removal process. Checking errors (chk) is used for overfitting tracking between training and checking data.

Here one can see there is no overfitting since checking errors track training errors (Figs. 23.4 and 23.5). As can be seen input factor 2 or laser pulse repetition has the smallest training error or the most impact on the photoresist removal process. Furthermore combinations of 2 and 3 or laser pulse repetition and laser energy is the optimal combination for the photoresist removal process. This combination has the most influence on the photoresist removal process.

Table 23.2 Prediction errors for the photoresist removal process

ANFIS model 1: in1 – > trn = 34.1, chk = 30.2
ANFIS model 2: in2 – > trn = 12.3, chk = 6.5
ANFIS model 3: in3 – > trn = 29.6, chk = 38.2
ANFIS model 1: in1 in2 – > trn = 11.2, chk = 7.1
ANFIS model 2: in1 in3 – > trn = 29.6, chk = 38.1
ANFIS model 3: in2 in3 – > trn = 2.7, chk = 9.7

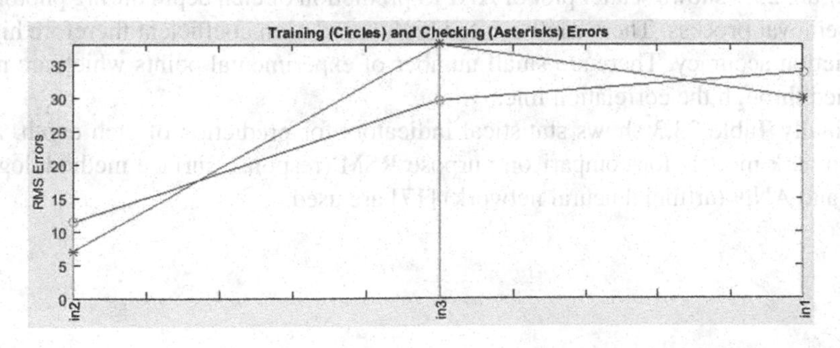

Fig. 23.4 Single input RMS errors

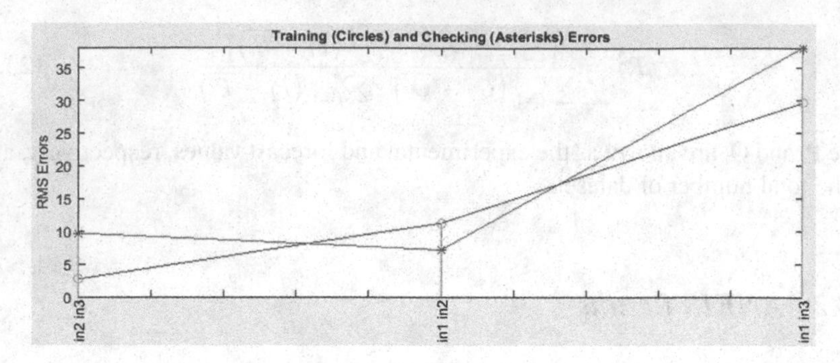

Fig. 23.5 Two inputs combinations RMS errors

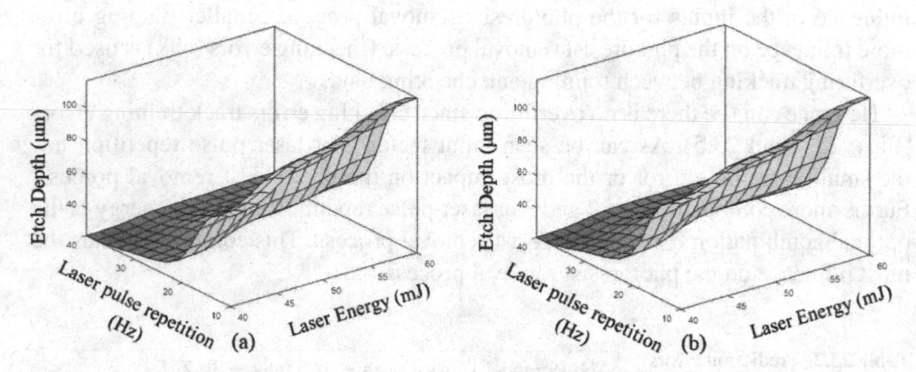

Fig. 23.6 ANFIS prediction of etch depth based on different parameter combination: **a** flow rate: 0.6 lpm and **b** flow rate: 0.8 lpm

Figures 23.6, 23.7 and 23.8 shows ANFIS prediction of etch depth based on combination of different input parameters. One can observe etch depth variation in depend on different input parameters.

Figure 23.9 shows scatter plot of ANFIS prediction of etch depth during photoresist removal process. There can be noted high correlation coefficient therefore high prediction accuracy. There are small number of experimental points which are not aligned through the correlation line.

Finally Table 23.3 shows statistical indicators for prediction of etch depth. As benchmark models for comparison purpose RSM (response surface methodology) [17] and ANN (artificial neural network) [17] are used.

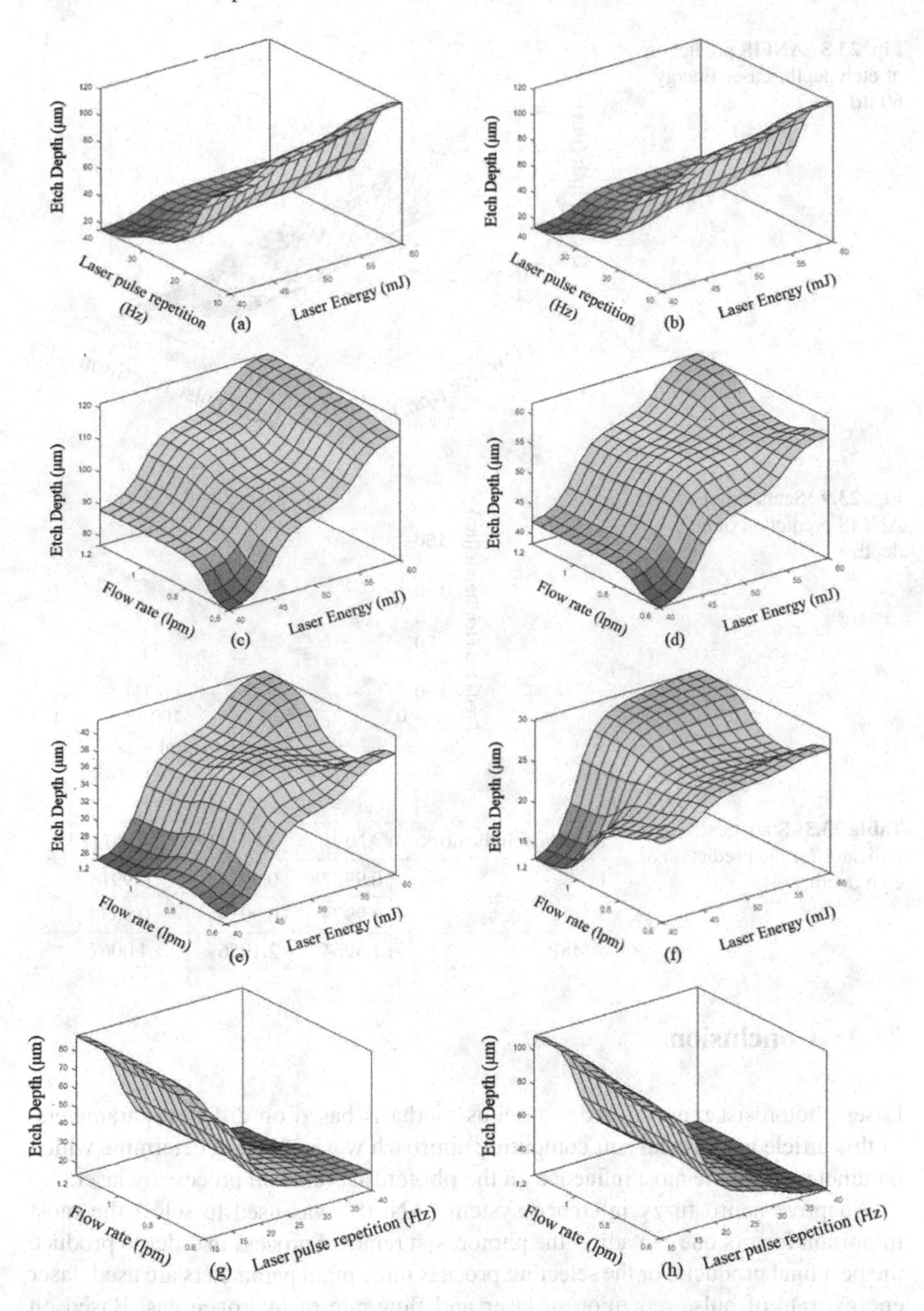

Fig. 23.7 ANFIS prediction of etch depth based on different parameter combination: **a** flow rate: 1 lpm, **b** flow rate: 1.2 lpm, **c** laser pulse repetition: 10 Hz, **d** laser pulse repetition: 20 Hz, **e** laser pulse repetition: 30 Hz, **f** laser pulse repetition: 40 Hz, **g** laser Energy: 40 mJ and **h** laser Energy: 50 mJ

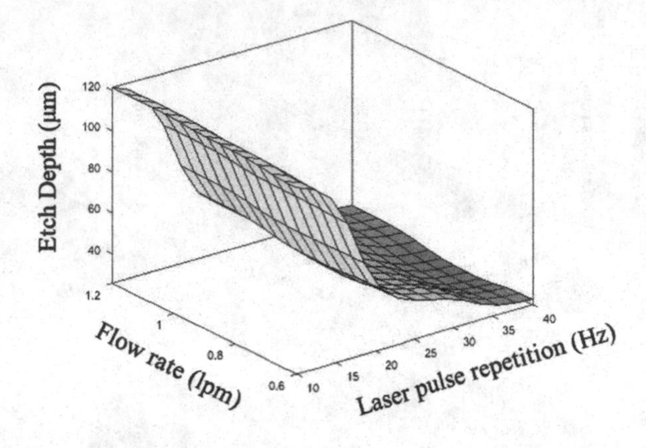

Fig. 23.8 ANFIS prediction of etch depth: Laser Energy 60 mJ

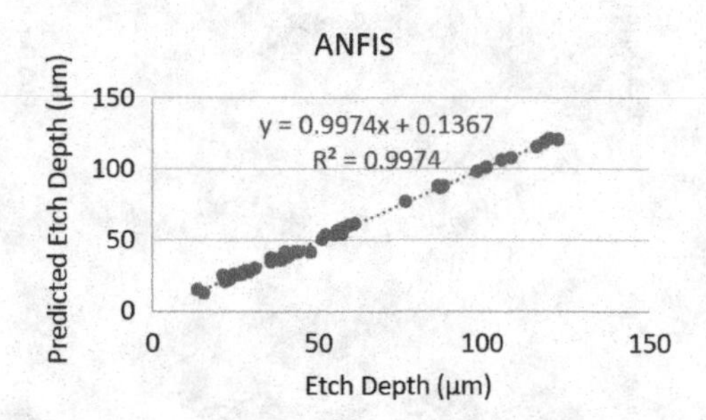

Fig. 23.9 Scatter plot of ANFIS prediction of etch depth

Table 23.3 Statistical indicator for the prediction of etch depth

Statistical indicators	ANFIS	ANN [17]	RSM [17]
r	0.9987	0.9977	0.9918
R^2	0.9974	0.9954	0.9838
RMSE	1.5964	2.1956	4.0067

23.4 Conclusion

Laser photoresist removal under gaseous media is based on different parameters. In this article a statistical soft computing approach was applied to determine which parameters have the most influence on the photoresist removal process by laser.

Adaptive neuro fuzzy inference system (ANFIS) was used to select the most important factors one can adjust the photoresist removal process in order to produce the best final product. For the selecting process three input parameters are used: laser energy, rate of pulse repetition of laser and flow rate of hydrogen gas. Based on obtained results pulse of laser repetition is selected as the most important factor for the photoresist removal process. ANFIS can eliminate the vagueness in the process in order to produce the best prediction conditions. In other words ANFIS network was used to convert the multiple performance characteristics into the single performance index.

References

1. Takahashi, S., Horita, Y., Kaji, F., Yamaguchi, Y., Michihata, M., Takamasu, K.: Concept for laser-assisted nano removal beyond the diffraction limit using photocatalyst nanoparticles. CIRP Ann. **64**(1), 201–204 (2015)
2. Martins, J.S., Borges, B.G.A.L., Machado, R.C., Carpanez, A.G., Grazul, R.M., Zappa, F., Lima, C.R.A.: Evaluation of chemical kinetics in positive photoresists using laser desorption ionization. Eur. Polym. J. **59**, 1–7 (2014)
3. Sathiamoorthy, S., Tiwari, K.J., Devi, G.R., Rao, M.R., Malar, P.: Photoresist template fabrication and template assisted growth for surface patterning of technologically important Cu2ZnSnSe4 thin films. Mater. Design **127**, 126–133 (2017)
4. Dentinger, P.M., Clift, W.M., Goods, S.H.: Removal of SU-8 photoresist for thick film applications. Microelectron. Eng. **61**, 993–1000 (2002)
5. Dimitrakellis, P., Smyrnakis, A., Constantoudis, V., Tsoutsou, D., Dimoulas, A., Gogolides, E.: Atmospheric pressure plasma directed assembly during photoresist removal: A new route to micro and nano pattern formation. Micro Nano Eng. **3**, 15–21 (2019)
6. Oh, E., Na, J., Lee, S., Lim, S.: Removal of ion-implanted photoresists on GaAs using two organic solvents in sequence. Appl. Surf. Sci. **376**, 34–42 (2016)
7. Yun, H., Lee, S., Jung, D., Lee, G., Park, J., Kwon, O.J., Park, C.Y.: Removal of photoresist residues and healing of defects on graphene using H_2 and CH_4 plasma. Appl. Surf. Sci. **463**, 802–808 (2019)
8. Hashimoto, K., Masuda, A., Matsumura, H., Ishibashi, T., Takao, K.: Systematic study on photoresist removal using hydrogen atoms generated on heated catalyzer. Thin Solid Films **501**(1–2), 326–328 (2006)
9. Takata, M., Ogushi, K., Yuba, Y., Akasaka, Y., Tomioka, K., Soda, E., Kobayashi, N.: Photoresist removal process by hydrogen radicals generated by W catalyst. Thin Solid Films **516**(5), 847–849 (2008)
10. Lee, J., Park, K., Lim, S.: Improvement of photoresist removal efficiency in ozonated water cleaning system. J. Ind. Eng. Chem. **14**(1), 100–104 (2008)
11. Kesters, E., Claes, M., Le, Q., Barthomeuf, K., Lux, M., Vereecke, G., Durkee, J.B.: Selection of ESH solvents for the wet removal of post-etch photoresists in low-*k* dielectrics integration. Microelectron. Eng. **86**(2), 160–164 (2009)
12. Jang, J.-S.R.: ANFIS: Adaptive-network-based fuzzy inference systems. IEEE T. Syst. Man Cyb. **23**, 665–685 (1993)
13. Petković, D., Issa, M., Pavlović, N.D., Pavlović, N.T., Zentner, L.: Adaptive neuro-fuzzy estimation of conductive silicone rubber mechanical properties. Expert Syst. Appl. **39**(10), 9477–9482 (2012)
14. Petković, D., Ćojbašić, Ž: Adaptive neuro-fuzzy estimation of automatic nervous system parameters effect on heart rate variability. Neural Comput. Appl. **21**(8), 2065–2070 (2012)
15. Kurnaz, S., Cetin, O., Kaynak, O.: Adaptive neuro-fuzzy inference system based autono-mous flight control of unmanned air vehicles. Expert Syst. Appl. **37**(2), 1229–1234 (2010)
16. Petković, D., Issa, M., Pavlović, N.D., Zentner, L., Ćojbašić, Ž: Adaptive neuro fuzzy con-troller for adaptive compliant robotic gripper. Expert Syst. Appl. **39**(18), 13295–13304 (2012)
17. Jacob, J., Shanmugavelu, P., Balasubramaniam, R.: Investigation of the performance of 248 nm excimer laser assisted photoresist removal process in gaseous media by response surface methodology and artificial neural network. J. Manuf. Process. **38**, 516–529 (2019)

[illegible] Figure [illegible]

Chapter 24
Identifying the Structure and Tribological Characteristics of the Material of Uncoated and Coated Model Single-Tooth Hob Milling Tools Using the Original Modular Fixture

Ivan Sovilj-Nikić, Bogdan Sovilj, Sandra Sovilj-Nikić, Bojan Podgornik, Matijaž Godec, and Radomir Đokić

Abstract The research of the gear cutting process of cylindrical gears by hob milling is very complex. The identification of tribological and other significant material characteristics is very important in order to be able to determine the optimal material for a model single-tooth hob milling tool for gear cutting of cylindrical gears. The modular fixture is necessary when testing the topographic, tribological and other significant material characteristics of model single-tooth hob milling tool. The paper presents a part of the research results related to the identification of microstructure, composition, friction coefficient and wear volume on the basis of which it is possible to determine the optimal material for model single-tooth hob milling tools. The research was carried out using original modular fixture specially designed and manufactured for this purpose. The research results show that the optimal material for a model single-tooth hob milling tool is the material HS-6-5-2-5 coated with TiAlN.

Keywords Modular fixture · Material structure · Tribological characteristics of the material.

24.1 Introduction

The problem of gear production is analyzed in science and application in various ways, identifying it once as an element of the machine, and the second time as a part

I. Sovilj-Nikić · B. Sovilj · R. Đokić
Faculty of Technical Sciences, University of Novi Sad, Novi Sad, Serbia

S. Sovilj-Nikić (✉)
Iritel a.d. Beograd, Batajnički put 23, 11080 Belgrade, Serbia
e-mail: sandrasn@eunet.rs

B. Podgornik · M. Godec
Institute of Metallic Materials and Technologies Ljubljana, University of Ljubljana, Ljubljana, Slovenia

M. Rackov et al. (eds.), *Machine and Industrial Design in Mechanical Engineering*, Mechanisms and Machine Science 109,
https://doi.org/10.1007/978-3-030-88465-9_24

Fig. 24.1 Basic forms of serration, basic gear cutting tools and gear cutting methods

of production, i.e. the finished product. Designed and properly manufactured gears are key components in most complex systems. Gears with involute serration are among the most important components of efficient modern transmission techniques, in accordance with the relevant legislation on pollution and energy conservation [1, 2]. Figure 24.1 shows the basic forms of serration, basic gear cutting tools and methods for gear cutting [2, 3].

Several different methods have been developed for gear cutting of cylindrical gears. Hob milling as one of the most complex cutting processes finds the widest application in the processing of cylindrical gears due to the high productivity of the process. Complicated kinematic and geometric relationships between the hob milling tool and the workpiece create a number of difficulties and problems that prevent the optimal use of the hob milling tool and the hob milling machine [2–4]. In the case of hob milling, in comparison with other types of cutting, high costs have to be taken into account, and this is conditioned by the complex production of the hob milling tool, high accuracy requirements and extremely expensive material of tool blank. Improving the gear cutting process by hob milling is significant and useful for both gear manufacturers and hob milling tool manufacturers. Quality materials for manufacturing of hob milling tools and gears are necessary in order to increase productivity and economy of their application [2–6].

It is important to know the influence of the composition and microstructure of materials and technological procedures for obtaining blanks and workpieces on the properties and behavior of materials, as well as their interaction [2, 5, 7]. In laboratory research, the tangential method of machining with a model single-tooth hob milling tool fixed in a special modular device was applied to optimize the process of gear cutting of cylindrical gears by hob milling.

24.2 Modular Fixture

It is very important in the interest of valid results of measuring tribological and topographic characteristics, as well as measuring significant characteristics of the base material and the material of the applied layers, to adequately place, position and clamp the model single-tooth hob milling tool in modular fixture. Since the body of the model single-tooth hob milling tool is semi-cylindrical shape, it is necessary to design and manufacture a modular fixture that will fulfill all the previously listed requirements. The modular fixture, which is presented in this paper, as a result of creation, represent the outcome of the creative process and the work of the creator. Designed and manufactured modular fixture is used in determining the topographic, tribological and other significant characteristics of the base material and applied layers on the cutting surfaces of model single-tooth hob milling tools [2, 4].

In Fig. 24.2 a three-dimensional representation of the model of modular fixture for the case of testing the structure, topography and wear of the face of the model single-tooth hob milling tool is shown. Figure 24.3a shows a position of the three-dimensional model for the case of determining the topographic and tribological characteristics of the input lateral flank of the model single-tooth hob milling tool at a certain angle of the tool profile α. The position of the three-dimensional model of the modular fixture for determining the output lateral flank of the model single-tooth hob milling tool at a certain angle of the tool profile α is given in Fig. 24.3b. Figure 24.3c shows a position of the three-dimensional model of the modular fixture

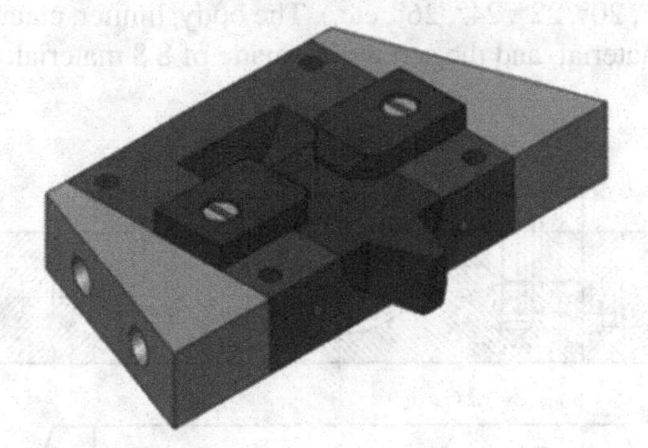

Fig. 24.2 Model of modular fixture with model single-tooth hob milling tool for the case of testing the structure, topography and wear of the tool face

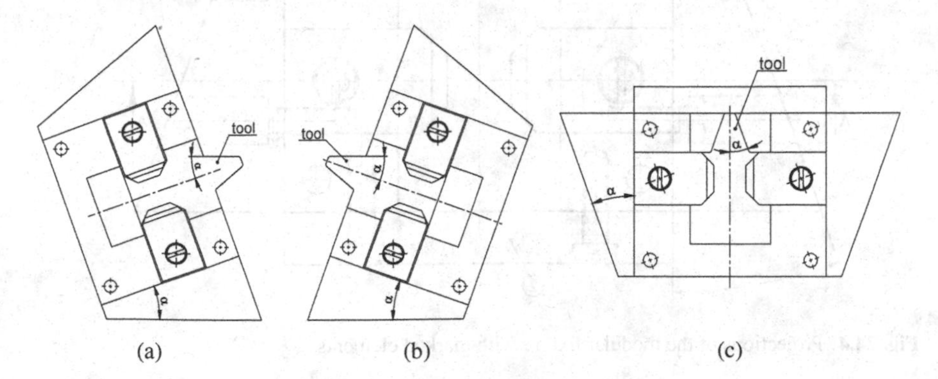

Fig. 24.3 Various position of three-dimensional model of the modular fixture

for the case of measuring the adhesion of the applied layer to the basic material and analysis of material structure of a model single-tooth hob milling tool.

In Fig. 24.4 two projections of modular fixture are given. The modular fixture consists of a body (1), a limiter (2), a clamp (3), an adjuster (4) and a screw (5). The body (1) of the modular fixture with dimensions 60f7 × 50f7 × 16f8 is box-shaped (Fig. 24.3). The body has a bearing in the shape of a semi-cylinder with dimensions R12.5 × 40H8. The bearing was finished by reaming in the quality $R_a = 6.3$ μm. On the upper surface of the body, two holes were made with a tapped thread M4 for fixing the clamps (3). Also, four openings were made on the upper surface $\phi =$ 4.25 mm through which the screws for fixing the modular fixture to the base plates of appropriate devices for measuring topographic, tribological and other significant characteristics of the base material and the applied layers on the model single-tooth hob milling tool should pass. On the front side of the body of the modular fixture, two holes were made with a tapped thread M4 for fixing the limiter (2) to the body (1). On the sides of the body of the modular fixture, two holes are made on each side with a tapped thread M4 for fixing the adjuster (4) to the body at different angles of the profile ($\alpha = 17°, 20°, 22°, 24°, 26°$, etc.). The body, limiter, clamps and adjusters are made of C60 material, and the screws are made of 8.8 material.

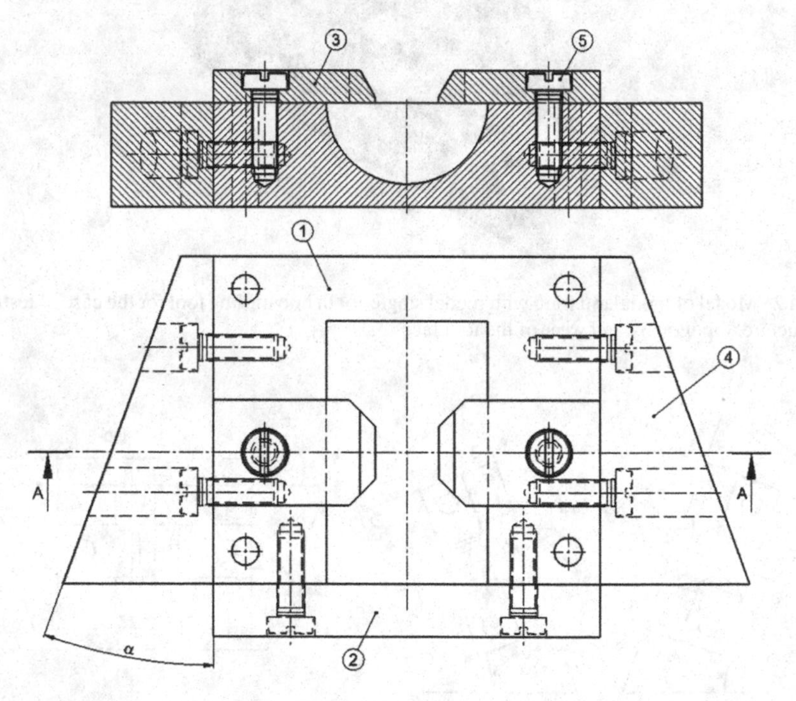

Fig. 24.4 Projections of the modular fixture with marked elements

24.3 Experimental Results and Discussion

In the laboratories of the Institute of Materials and Technologies in Ljubljana, modular fixture was used to test the structure and tribological characteristics of materials of uncoated and coated model single-tooth hob milling tool of modules $m = 3$ mm and $m = 5$ mm.

Figure 24.5 shows a model hob milling tool of module $m = 3$ mm made of HS-6-5-2-5 coated with TiN without modular fixture. The tool placed on the base plate in this way is unstable because its body is semi-cylindrical in shape, which is also one of the reasons for designing and manufacturing modular fixture. Modular fixture with a model single-tooth hob milling tool made of HS-6-5-2-5 coated with TiN module $m = 3$ mm is located on the basic movable plate and is ready to be placed in the device for analysis of material structure (Fig. 24.6a). Figure 24.6b shows the position of a modular fixture with a model single-tooth hob milling tool module $m = 5$ mm made of HS-6-5-2-5 coated with TiAlN immediately before the structure analysis.

Material microstructure images for model single-tooth hob milling tool were made using a JEOL JSM-6500F line electronic microscope with VDS, EDS and EBSD analyzer. Images taken at a magnification of $\times$10,000, $\times$15,000, and $\times$25,000 times were taken at a voltage acceleration of 15 kV and a working distance of 10 mm. Metallographic samples were prepared by grinding and polishing and final abrasion with nital.

Figure 24.7 shows the microstructure of high-speed uncoated tool steel HS-6-5-2-5 made by powder metallurgy. The microstructure consists of a martensitic base with a pre-austenitic grain size ~2 μm and homogeneously distributed round eutectic carbides (type MC—gray, type M6C—white). Also, images of PVD coatings applied to high-speed tool steel HS-6-5-2-5 were taken using a JEOL JSM-6500F line electron microscope with VDS, EDS and EBSD analyzer. The images were taken at a magnification of $\times$10,000, an acceleration voltage of 15 kV and a working distance

Fig. 24.5 Model single-tooth hob milling tool module $m = 3$ mm made of HS-6-5-2-5 coated with TiN without modular fixture on the basic movable plate of the SEM device

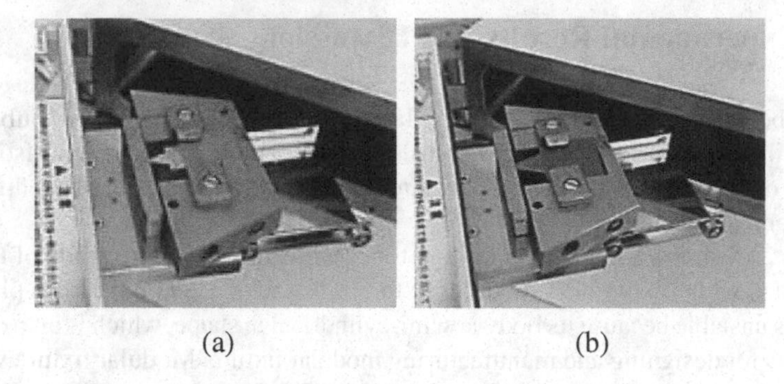

Fig. 24.6 **a** Model single-tooth hob milling module m = 3 mm made of HS-6-5-2-5 coated with TiN in modular fixture on the basic movable plate of the SEM device and **b** Model single-tooth hob milling tool module m = 5 mm made of HS-6-5-2-5 coated with TiAlN in modular fixture on the basic movable plate of the SEM device

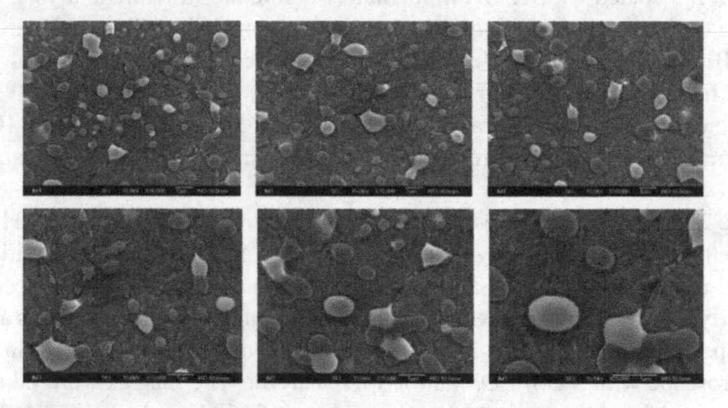

Fig. 24.7 Microstructure of high-speed tool steel HS-6-5-2-5

of 10 mm. Transverse fracture samples were prepared by cooling the sample in liquid nitrogen and brittle fracture. The first Fig. 24.8a shows the PVD fracture of

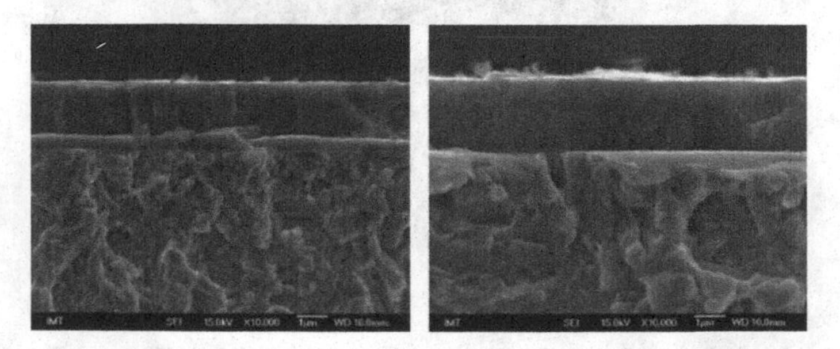

Fig. 24.8 **a** PVD HS-6-5-2-5 fracture coated with TiN and **b** PVD HS-6-5-2-5 fracture coated with TiAlN

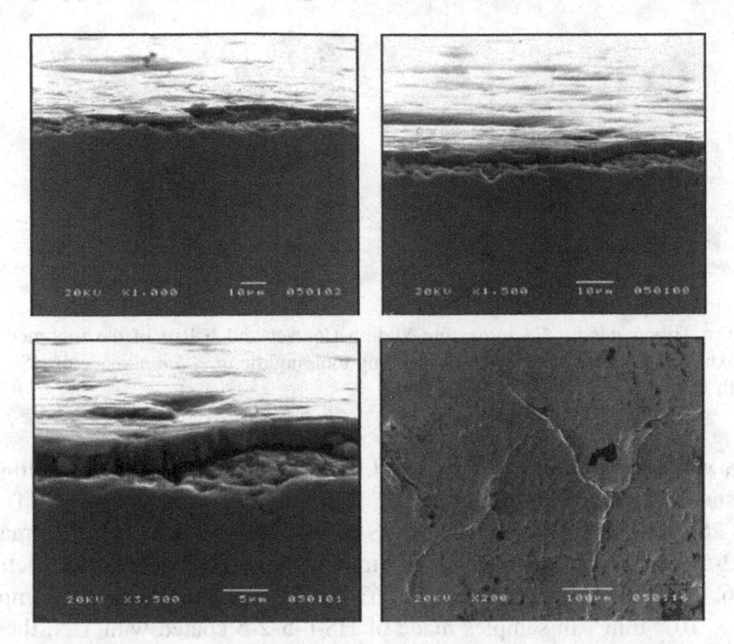

Fig. 24.9 Damage and peeling of the layer at magnification ×1000 (**a**), ×1500 (**b**) and ×3500 (**c**), formation of fatigue cracks on the surface of the layer at magnification ×200 (**d**)

the applied single-layer TiN layer on the base structure HS-6-5-2-5 with a thickness of 2 μm. In the second Fig. 24.8b, PVD is a multilayer TiAlN layer with a thickness of individual lamellae ~50 nm, the same base structure and a total thickness of 2 μm.

Figures 24.9a–c worn edges and surfaces of the TiN coated model single-tooth hob milling tool were recorded using a JEOL T330A line electronic microscope at a voltage acceleration of 20 kV and a working distance of 20 mm. The damage and peeling of the TiN coating on the cutting edge at a magnification of ×1000, ×1.500 and ×3.500 times and the formation of fatigue cracks on the surface of the coating at a magnification of ×200 times are shown in Fig. 24.9d.

Tribological tests of model single-tooth hob milling tool for gear cutting of cylindrical gears were also performed in the Laboratory of the Institute of Materials and Technologies in Ljubljana. The tribological characteristics of model single-tooth hob milling tool were determined on the device Tribometer for Reciprocating Sliding Contact 2014 (Fig. 24.10a). Figure 24.10b shows a part of a tribometer with a modular fixture in which is a model single-tooth hob milling tool module m = 3 mm made of HS-6-5-2-5 coated with TiAlN.

Five model single-tooth hob milling tools were tested, two uncoated made of HS-6-5-2-5 (no. 263 and 34), two made of HS-6-5-2-5 coated with TiN (no. 6 and 39) and one made of HS-6-5-2-5 coated with TiAlN (no. 22). The tests were performed in conditions of dry sliding contact, alternating sliding contact (contact pin-disk), normal force 44 N (contact pressure 1.8 GPa), sliding speed 0.01 m/s (feed 4 mm, frequency 1 Hz) and a total sliding track of 15 mm. A 10 mm diameter ball made

(a) (b)

Fig. 24.10 **a** Tribometer for Reciprocating Sliding Contact and **b** Part of the tribometer with modular fixture and model single-tooth hob milling tool module m = 3 mm made of HS-6-5-2-5 coated with TiAlN

of Al_2O_3 was used as a opposite material. The lowest coefficient of friction 0.38 and the smallest wear volume 0.44×10^{-3} mm^3 are shown by sample 22 (TiAlN). Samples 263 and 34 made of uncoated HS-6-5-2-5 and samples 6 and 39 made of HS-6-5-2-5 coated with TiN have approximately the same coefficient of friction of about 0.6. The wear volume for sample 263 is 2.1×10^{-3} mm^3, and for sample 34 it is 3.8×10^{-3} mm^3. In samples made of HS-6-5-2-5 coated with TiN, the wear volume of sample 6 is 5.9×10^{-3} mm^3 on the part of the face where the applied TiN layer was removed. The wear volume for sample 39 is 3.1×10^{-3} mm^3. Profile and groove images are given in Figs. 24.11, 24.12, and 24.13.

Figure 24.14a shows a diagram of the friction coefficient for five types of materials of model single-tooth hob milling tools. Figure 24.14b shows a diagram of the wear volume for the materials of the model single-tooth hob milling tools shown in Fig. 24.14b.

The numerous parameters and their alternating action significantly complicate the research of the wear process of the hob milling tool. The materials of the gear and the hob milling tool are, among other things, wear regulators of the elements of the tribological system. The composition and microstructure of the material limit the properties and behavior of the material in use. The microstructure of the material of uncoated and coated model single-tooth hob milling tool was determined using a

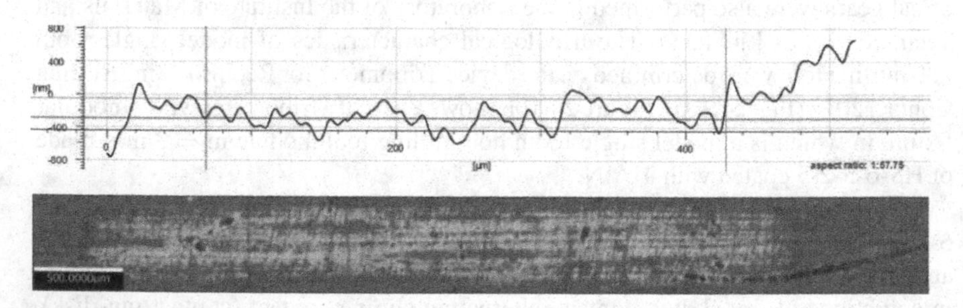

Fig. 24.11 Profile and groove for sample 22 (TiAlN)

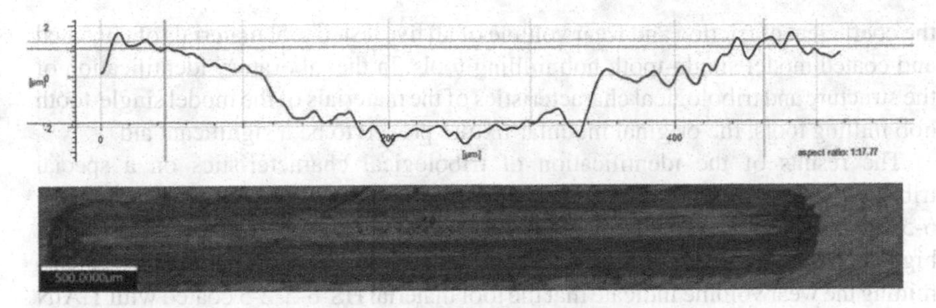

Fig. 24.12 Profile and groove for sample 263 (HS-6-5-2-5 uncoated)

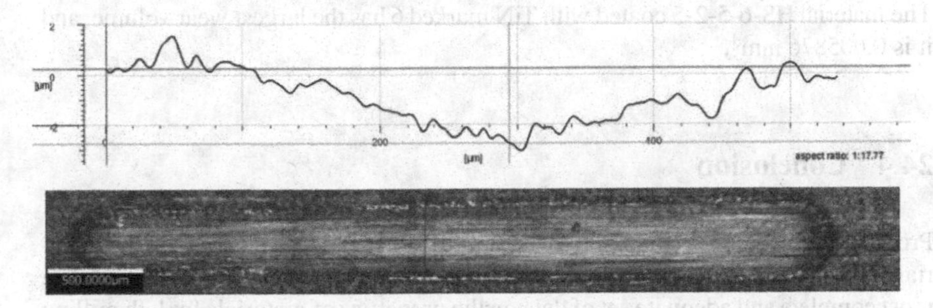

Fig. 24.13 Profile and groove for sample 39 (TiN)

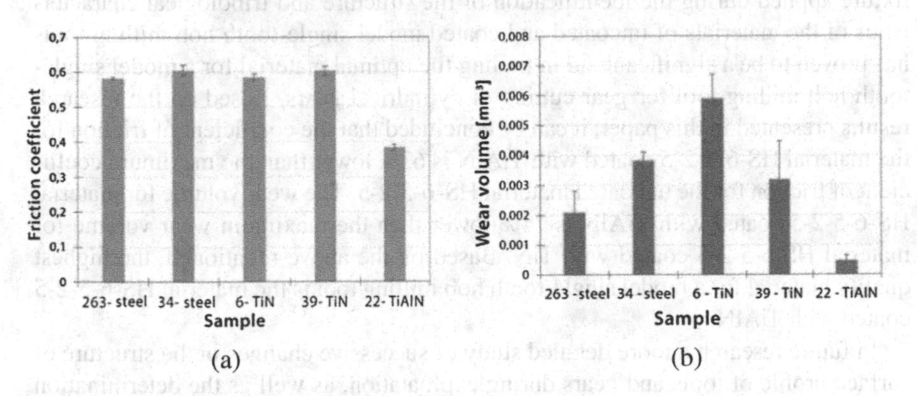

Fig. 24.14 a Friction coefficient diagram for five tools and b Wear volume diagram for five tools

line electron microscope, and the tribological characteristics of model single-toothed milling cutters were determined on a special tribometer.

The microstructure of uncoated tool steel HS-6-5-2-5 marked 263 and 34 made by powder metallurgy was identified, and images of PVD coatings TiN marked 6 and 39 and TiAlN marked 22 applied to high-speed tool steel HS-6-5-2-5 were made using a JEOL JSM-6500F line electron microscope with VSD, EDS and EBSD analyzer. The tool made of HS-6-5-2-5 coated with TiAlN had the best tribological characteristics the coefficient of friction and wear volume of all five tested tool materials of uncoated and coated model single-tooth hob milling tools. In the laboratory identification of the structure and tribological characteristics of the materials of the model single-tooth hob milling tools, the original modular fixture proved to be a significant aid.

The results of the identification of tribological characteristics on a special tribometer show that the coefficient of friction is the lowest for the material HS-6-5-2-5 coated with TiAlN marked 22 and it is 0.3829, and the highest for uncoated high-speed tool steel HS-6-5-2-5 marked 34 and it is 0.6048. The results of determining the wear volume indicate that the tool material HS-6-5-2-5 coated with TiAlN marked 22 had the smallest wear volume of 0.00044 mm^3. The uncoated material HS-6-5-2-5 marked 34 has a significantly higher wear volume, and it is 0.00379 mm^3. The material HS-6-5-2-5 coated with TiN marked 6 has the largest wear volume, and it is 0.005876 mm^3.

24.4 Conclusion

Productivity, economy and quality of gear cutting depend significantly on the material of the hob milling tool. The optimal tool material depends on various factors. The most complete and adequate set of data on the properties of materials for hob milling tools can be obtained by laboratory testing of the samples. The original modular fixture applied during the identification of the structure and tribological characteristics of the materials of uncoated and coated model single-tooth hob milling tools has proven to be a significant aid in finding the optimal material for a model single-tooth hob milling tool for gear cutting of cylindrical gears. Based on the research results presented in this paper, it can be concluded that the coefficient of friction for the material HS-6-5-2-5 coated with TiAlN is 63% lower than the maximum coefficient of friction for the uncoated material HS-6-5-2-5. The wear volume for material HS-6-5-2-5 coated with TiAlN is 74% lower than the maximum wear volume for material HS-6-5-2-5 coated with TiN. Based on the above mentioned, the highest quality material for a model single-tooth hob milling tool is the material HS-6-5-2-5 coated with TiAlN.

In future research a more detailed study of successive changes in the structure of surface profile of tools and gears during exploatation, as well as the determination of the relationship of these changes will certainly help in developing new methods of finishing of gear serration that will reduce wear as well as extend the tool life of hob milling tools and gear life.

References

1. Tanasijević, S.: Tribološki ispravno konstruisanje. Faculty of Engineering University of Kragujevac (2004)
2. Sovilj-Nikić, I.: Modelovanje i optimizacija procesa odvalnog glodanja. Faculty of Technical Sciences, Novi Sad, Serbia. PhD thesis (unpublished)
3. Bouzakis, K.D., Lili, E., Michalidis, N., Friderikos, O.: Manufacturing of cylindrical gears by generating cutting processes: A critical synthesis of analysis methods. CIRP Ann. Manuf. Techn. **57**, 676–696 (2008)
4. Sovilj-Nikić, I., Sovilj, B., Kandeva, M., Gajić, V., Sovilj-Nikić, S., Legutko, S., Kovač, P.: Tribological characteristics of hob milling tools from economic aspect. J. Balk. Tribol. Assoc. **18**(4), 577–585 (2012)
5. Sovilj, B., Sovilj-Nikić, I., Ješić, D.: Measurement methodology of characteristics and election of materials of elements of tribomechanical systems. Metalurgija **50**(2), 107–112 (2011)
6. Bouzakis, K.D., Skoridatis, G., Michalidis, M.: Innovative methods for characterizing coatings mechanical properties. In: Proceedings of the 7th International Conference "THE" Coatings in Manufacturing Engineering and EUREKA Partnering Event. International Academy for Production Engineering, Kallithea (2008)
7. Kandeva-Ivanova, M., Vencl, A., Karastoyanov, D.: Advanced Tribological Coatings for Heavy-Duty Applications: Case Studies. Prof. Marin Drinov Publishing House of BAS, Bulgaria (2016)

Chapter 25
Investigation of Machined Surface Roughness and Cutting Tool Wear from Prepared Tool Geometry

Róbert Straka and Jozef Peterka

Abstract The article deals with cutting edge preparation and its influence on the tool wear of cemented carbide cutting inserts and surface roughness of the workpiece. The effect of the cutting edge radius sizes on the tool life will be examined. The paper describes the influence of cutting edge preparation on the turning process, the surface roughness of the workpiece, and on the tool life. The cutting inserts were manufactured by Dormer Pramet and they also prepared them for the experiment. The preparation method used for cutting inserts was drag finishing. For the maintaining of the same conditions in each step, the cutting parameters did not change. These were determined based on numerous tests for a semi-finishing turning operation. The final cutting parameters, such as cutting speed, feed, and depth of cut were determined in consultation with the cutting inserts manufacturer to achieve a controlled development of tool wear with a tool life of 25 to 30 min. The material used for the workpiece was the Inconel 718 nickel alloy.

Keywords Cutting edge preparation · Turning · Tool wear

25.1 Introduction

Usage of cemented carbide as a material for cutting tools has become possible thanks to the development of new coating materials and technologies and also due to the application of proper micro-geometries [1]. The most important properties for the performance and life of the cutting tools are the geometry and condition of the cutting edge. Tool wear, the surface roughness of the machined material, tool life, residual stress of the machined surface, cost, productivity, and quality of the machining process is influenced by the geometry of the cutting edge [2]. Improvement of the cutting tool design is possible in its macro or micro-geometry area. Micro-geometry is influenced by the cutting material and the deposited thin layer, which also influences the surface roughness and the radius of the cutting edge [3]. There are several

R. Straka (✉) · J. Peterka
Slovak University of Technology in Bratislava, Vazova 5, 811 07 Bratislava, Slovakia
e-mail: robert.straka@stuba.sk

M. Rackov et al. (eds.), *Machine and Industrial Design in Mechanical Engineering*, Mechanisms and Machine Science 109,
https://doi.org/10.1007/978-3-030-88465-9_25

goals of the controlled application of cutting edge preparation such as increasing the strength of the edge, elimination of the defects created during the manufacturing of the tool, enhancing tool life, preparing it for coating, and to minimize the chipping of the cutting edge. It also influences the mechanical and thermal loads generated during the machining process [4]. Micro-shaping of cutting tool edges is called edge preparation and it removes the initial edge defects from the sharp edge and replaces them with a smooth profile, which increases the tool quality. It creates a cutting edge with better stability and improved tool surface integrity, which leads to enhanced chip flow and cutting process [5]. Preparation is used for different types of cutting tools and different machining technologies such as turning [6], drilling [7], and milling [8, 9]. Rounding of the cutting edges results in a reduction of the local loads in the cutting tools and also an incidention of chipping. Cemented carbide tools have the highest potential for this when machining steel [10–12]. Zhang and Zhuang researched the effect of cutting edge microgeometry on surface roughness in turning AISI 52,100 steel. They found out that increasing chamfer width can reduce surface roughness but reasonable chamfer geometry must be selected because the increase of chamfer angel leads to the extreme change in the surface roughness value. They also found out that the enhancement of chamfer length and angle can contribute to white layer formation [13]. Ventura et al. found out that asymmetric geometry increases tool life. They demonstrated the application of the prepared cutting inserts and proved that the use of those micro geometries is adequate in turning [14].

There are many preparation methods used to modify cutting edge geometry to manufacture the cutting tools with the required geometry. The most common ones are brushing, drag finishing, grinding, and abrasive machining. There is also the possibility to use electric discharge machining or plasma discharges in an electrolyte for cutting edge preparation, but they are not widely used, even though the material removal mechanism is based on melting and evaporation instead of grinding which leads to material removal is not affected by the hardness of the workpiece, but needs to have the required electrical conductivity [15, 16]. Uhlmann et. al. compared brush polishing, polish blasting, magnet finishing, and immersed tumbling as preparation methods for micro-milling tools. They found out that magnet finishing and immersed tumbling improved the tool performance in a matter of resultant forces F_z by 7.5%, and 14% respectively, compared to brush polishing and polish blasting. Tool wear was reduced by 9% compared to brush polishing and by 13.2% compared to polish blasting [17]. The purpose of this article is to investigate the influence of cutting edge preparation on tool life during the turning of difficult-to-cut material Inconel 718 nickel alloy.

25.2 Materials and Methods

Tools used for this research were double-sided negative cutting inserts by the producer Dormer Pramet. They were made from fine-grained WC–Co carbide within ISO ranges from P25 to P40 and M20 to M35 and were PVD coated with TiAlN coating.

Table 25.1 Cutting insert parameters

Tool parameter	Value
Length [mm]	12.90
Thickness [mm]	4.76
Nose radius [mm]	0.8
Feed (min–max) [mm]	0.20–0.45
Depth of cut (min–max) [mm]	0.80–4.00
Cutting edge angle [°]	80.00

Their product name is CNMG 120408E-SM: T8330 and is suitable for finishing and semi-roughing operations in machining steel, superalloys, and hard materials. Four cutting edge radii were used, ranging in size from 5 to 50 μm. The parameters of the tool are in Table 25.1.

The machined material used in the experiment was Inconel 718. It is a precipitation-hardenable superalloy based on nickel and chromium with high strength. Used at temperatures up to 648 °C.

25.2.1 Experiment Design

The experiment was carried out on CTX Alpha 500 5-axis turning center. The length of the cylindrical workpiece was 60 mm and its diameter was 101.6 mm. The machining process used was longitudinal turning. To maintain the same conditions in each turning step, the cutting parameters were unchanged and were determined based on many tests for the semi-finishing turning operation. The final cutting parameters, such as cutting speed $v_c = 40 \text{ m} \cdot \text{min}^{-1}$, feed f = 0,15 mm, and depth of cut $a_p = 1$ mm, were used for the controlled development of wear with a tool life of 25 to 30 min. Cutting conditions were determined from multiple tests for a semi-finishing turning operation. Even though not all final cutting parameters used for the research are similar to the parameters recommended by the supplier listed in Table 25.1, they were determined in consultation with Dormer Pramet to achieve a controlled development of tool wear.

A Dino-Lite Edge microscope with a resolution of 1280 × 960 pixels and a magnification range of 20–220 times was used to measure the wear area VB_N and VB_B on the flank surface area of the cutting insert. DinoCapture 2.0 was used to capture the development of wear. This program is used to capture photos and record videos with Dino-Lite microscopes. Surface roughness was measured by portable surface roughness gauger Mitutoyo SJ-210.

The tools were first inspected and cleaned of impurities and photographed using DinoCapture 2.0 before turning to see where the wear is formed during the experiment. The workpiece clamped in the three-jaw chuck of the main spindle was turned with 3 cutting inserts with the same cutting edge radius. Each insert traversed a path of 20 mm, which stands for the cutting length of each insert so that the first cutting

Table 25.2 Measured values of the wear on the cutting inserts with cutting edge radius $r_n = 5\ \mu m$

Time of machining [min]	$r_n = 5\ \mu m$ sample 1		$r_n = 5\ \mu m$ sample 2		$r_n = 5\ \mu m$ sample 3	
	VB_N [mm]	VB_B [mm]	VB_N [mm]	VB_B [mm]	VB_N [mm]	VB_B [mm]
0.00	0.000	0.000	0.000	0.000	0.000	0.000
1.04	0.082	0.000	0.078	0.000	0.097	0.000
3.05	0.167	0.082	0.174	0.130	0.163	0.052
4.97	0.185	0.111	0.185	0.152	0.182	0.067
6.82	0.211	0.134	0.219	0.152	0.185	0.093
9.42	0.219	0.134	0.204	0.159	0.200	0.093
11.84	0.234	0.134	0.215	0.159	0.208	0.100
14.07	0.256	0.148	0.223	0.163	0.230	0.108
16.12	0.260	0.152	0.234	0.163	0.260	0.108
18.55	0.267	0.152	0.241	0.167	0.263	0.108
20.64	**0.286**	0.152	0.256	0.171	**0.757**	0.130
22.79	0.627	0.672	**0.297**	0.174	–	–

insert went from 0 to 20 mm on the workpiece, the second went from 20 to 40 mm, and the third one from 40 to 60 mm, so that they were not affected by each other and to measure the surface roughness that was created by each tool separately. After the first minute, the tools were removed from the machine, cleaned of process fluid and impurities, and placed in the field of view of the microscope. After focusing the microscope, the cutting edge was photographed to prevent the tool from being shifted by external influences, and the length of the wear area VB_N and VB_B was measured in the photo. Then the surface roughness parameters Ra, Rq, and Rz were measured with a surface roughness gauger, that was connected to the turret tool head in the machine and brought over the machined surfaces that were created on the workpiece after turning. After the measurement, the tools were returned to their positions in the machine and machining continued. This measurement was repeated at the earliest after 1 min, later after 2 min, and then after 3 min, until the wear on the cutting insert reached the criterion of dullness on the flank surface VB_k, which had a value of 0,27 mm. This value was chosen because the wear started to increase significantly after reaching it. The cutting process was stopped after each time sequence to measure the wear properly, as seen in Table 25.2.

25.3 Results and Discussion

As mentioned before, measured values were the length of wear surfaces on the flank face area, and surface roughness parameters of the machined part Ra, Rq, and Rz. The result of these measurements is a graphical comparison of the effect of cutting edge microgeometry on the tool life of the cutting inserts when machining difficult-to-cut superalloy Inconel 718 and changing surface roughness through time. For each specific size of the cutting edge, radius r_n were performed 3 durability tests, so the durability was measured three times for each tool. Table 25.2 and Fig. 25.1 show how was the wear developed during turning with 3 cutting inserts with cutting edge radius $r_n = 5\ \mu m$. Each insert is labeled as a sample. Such a graphical development was subsequently made for all the cutting edge radii used during the experiment.

Values of the measured surface roughness parameters are in Fig. 25.2. The peaks seen in the figure were created by winding the continuous chip on the workpiece during the turning process. This effect is not desired and has affected the values of roughness parameters.

Statistical equations for the dependence of the mean deviation of the evaluated profile Ra on the time were calculated for each cutting edge radius r_n. The equations were calculated using the least-squares method in Excel. The form of the equation for each radius is:

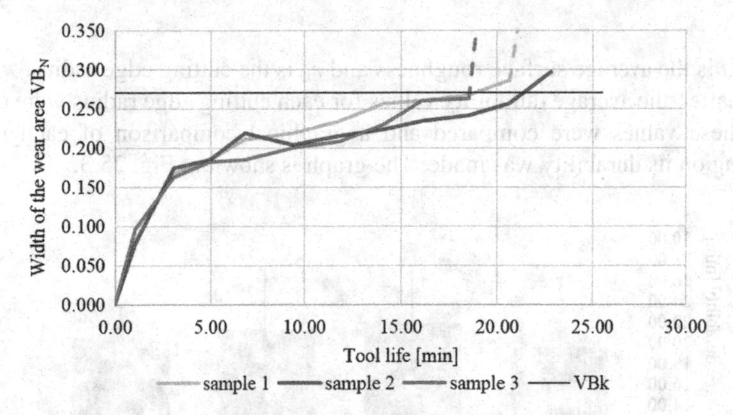

Fig. 25.1 Development of the width of the wear area VB_N in time for cutting insert with cutting edge radius $r_n = 5\ \mu m$

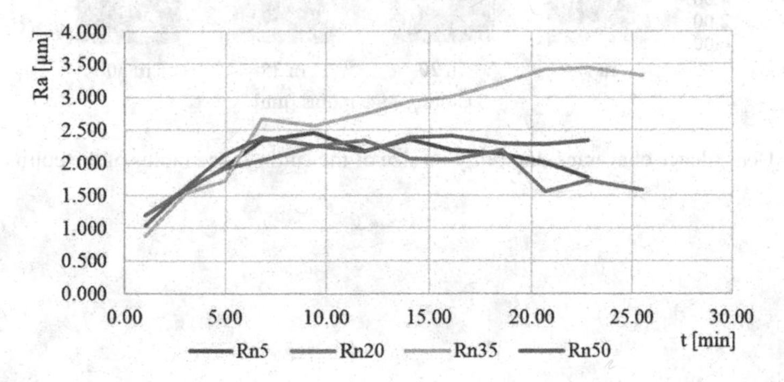

Fig. 25.2 Graphical dependence of the arithmetical mean deviation of the evaluated profile Ra on time

$$r_n = 5\mu m : Ra = 0.019 \cdot t + 1.7868[\mu m; min] \quad (25.1)$$

$$r_n = 20\mu m : Ra = 0.0036 \cdot t + 1.8294[\mu m; min] \quad (25.2)$$

$$r_n = 35\mu m : Ra = 0.0934 \cdot t + 1.4145[\mu m; min] \quad (25.3)$$

$$r_n = 50\mu m : Ra = 0.0394 \cdot t + 16552[\mu m; min] \quad (25.4)$$

where Ra is the average surface roughness and t is the time of machining.

Next, the statistical equation for the dependence of the surface roughness parameter Ra on the cutting edge radius was calculated. This was done the same way as the equations above. The form of this equation is:

$$Ra = 0,0071 \cdot r_n + 1,9602[\mu m; \mu m] \quad (25.5)$$

where Ra is the average surface roughness and r_n is the cutting edge radius.

Thereafter, the average durability values for each cutting edge radius were calculated. These values were compared and a graphical comparison of each radius depending on its durability was made. The graph is shown in Fig. 25.3.

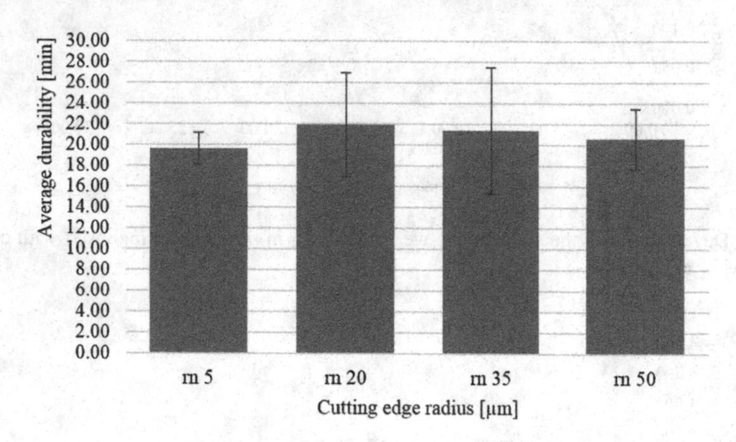

Fig. 25.3 Dependence of average durability on size of the cutting edge radius of the cutting insert

25.4 Conclusion

The paper researched how the microgeometry of the cutting edge affects the durability of cutting tools when turning the difficult-to-cut material Inconel 718. The durability was measured in terms of tool wear and terms of surface roughness.

The dulling criterion was chosen to be $VB_k = 0.27$ mm. As seen on the graph in Fig. 25.3 tool with the cutting edge radius $r_n = 20$ μm seems to be the one with the highest durability, which was 21.9 min. Compared to the tool with the radius $r_n = 35$ μm, it was about 2.30% higher durability. Compared to radii of 50 μm or 5 μm, the durability was higher about 6.47% and 11.60% respectively. The author thinks that a cutting tool with a radius between 20 μm and 35 μm should be used to achieve greater durability, because in the range of these radii the cutting forces and mainly residual stress are kept under control, but more research should be done in this area in future to confirm or refute this claim.

It should be noted that none of the tools reached the required durability between 25 and 30 min. Therefore, it would be appropriate to repeat the experiment with a differently chosen dulling criterion. More research needs to be done for the machined material as this can also affect the determination of the tool's life.

Three surface roughness parameters were measured on the machined part. The surface roughness was classified as unsatisfactory if the values of a parameter Ra and Rq exceeded the value of 3 μm and if the values of the parameter Rz exceeded the value of 15 μm. Deteriorated surface roughness was manifested on the workpiece during turning with the cutting insert that has a cutting edge radius of 35 μm from the point of view of the parameter Ra because in this case the values exceeded the criterion and stayed above it through the process.

The values of the surface roughness parameters measured during turning with the remaining 3 cutting edge radii did not exceed the selected criterion and therefore were appropriate. There can be seen peaks in the graphs and these were caused by winding the continuous chip around the workpiece. This led to a jump in surface roughness at some times.

Another effect can be seen on the graph of surface roughness parameter Ra. After the initial increase, the surface roughness begins to decrease or has remained the same. The author thinks that because of this it is possible to assume that there is no relationship between the main flank face area and the surface roughness. Therefore, wear on the secondary flank face area should be measured in the future, as this should also affect the roughness of the machined surface. There is also the assumption that the surface roughness should be affected by the built-up that occurred at the cutting edge.

Twelve different cutting inserts were used for the experiment, but all were prepared by one preparation method. In the future, it would be appropriate to repeat the experiment for cutting inserts prepared by other methods and thus compare the durability from the point of view of different methods of preparation of cutting edges.

Acknowledgements This research was funded by research project VEGA 1/0019/20 "Accurate calculations, modeling and simulation of new surfaces based on physical causes of machined surfaces and additive technology surfaces in machinery and robotic machining conditions".

References

1. Denkena, B., Köhler, J., Breidenstein, B., Abrão, A.M., Ventura, C.E.H.: Influence of the cutting edge preparation method on characteristics and performance of PVD coated carbide inserts in hard turning. Surf. Coat. Technol. **254**, 447–454 (2014)
2. Bergs, T., Schneider, S.A.M., Amara, M., Ganser, P.: Preparation of symmetrical and asymmetrical cutting edges on solid cutting tools using brushing tools with filament-integrated diamond grits. Proc. CIRP **93**, 873–878 (2020)
3. Fulemova, J., Janda, Z.: Influence of the cutting edge radius and the cutting edge preparation on tool life and cutting forces at inserts with wiper geometry. Procedia Engineer. **69**, 565–573 (2014)
4. Pokorný, P., Pätoprstý, B., Vopát, T., Peterka, J., Vozár, M., Šimna, V.: Cutting edge radius preparation. Mater. Today-Proc. **22**, 212–218 (2020)
5. Wang, W., Biermann, D., Aßmuth, R., Arif, A.F.M., Veldhuis, S.C.: Effects on tool performance of cutting edge prepared by pressurized air wet abrasive jet machining (PAWAJM). J. Mater. Process. Tech. **277**, 1164561-1–1164561-12 (2020)
6. Vopát, T., Kuruc, M., Šimna, V., Zaujec, R., Peterka, J.: Cutting edge microgeometry and preparation methods. In. Proceedings of the Annals of DAAAM and Proceedings of the International DAAAM Symposium, pp. 384–391. DAAAM International, Vienna (2017)
7. Vopát, T., Kuruc, M., Šimna, V., Necpal, M., Buranský, I., Zaujec, R., Peterka, J.: The influence of cutting edge radius size on the tool life of cemented carbide drills. In: Proceedings of the Annals of DAAAM and Proceedings of the International DAAAM Symposium, pp. 421–425. DAAAM International, Vienna (2018)
8. Vozar, M., Pätoprstý, B., Vopát, T., Peterka, J.: Overview of methods of cutting edge preparation. In: Katalinic, B. (ed.) DAAAM International Scientific Book 2019, pp. 251–264. DAAAM International, Vienna (2019)
9. Peterka, J., Pokorny, P., Vaclav, S., Patoprsty, B., Vozar, M.: Modification of cutting tools by drag finishing. MM Sci. J. **2020**(3), 3822–3825 (2020)
10. Bassett, E., Köhler, J., Denkena, B.: 'On the honed cutting edge and its side effects during orthogonal turning operations of AISI1045 with coated WC-Co inserts. CIRP J. Manuf. Sci. Technol. **5**(2), 108–126 (2012)
11. Denkena, B., Michaelis, A., Herrmann, M., Pötschke, J., Krödel, A., Vornberger, A., Picker, T.: Influence of tool material properties on the wear behavior of cemented carbide tools with rounded cutting edges. Wear, **456–457**, 203395–1– 203395–8 (2020)
12. Bergmann, B., Grove, T.: Basic principles for the design of cutting edge roundings. CIRP Ann. **67**(1), 73–78 (2018)
13. Zhang, W., Zhuang, K.: Effect of cutting edge microgeometry on surface roughness and white layer in turning AISI 52100 steel. Proc. CIRP **87**, 53–58 (2020)
14. Ventura, C.E.H., Köhler, J., Denkena, B.: Cutting edge preparation of PCBN inserts by means of grinding and its application in hard turning. CIRP J. Manuf. Sci. Technol. **6**, 246–253 (2013)
15. Zimmermann, M., Kirsch, B., Kang, Y., Herrmann, T., Aurich, J.C.: Influence of the laser parameters on the cutting edge preparation and the performance of cemented carbide indexable inserts. J. Manuf. Process. **58**, 845–856 (2020)

16. Vopát, T., Podhorský, Š, Sahul, M., Haršáni, M.: Cutting edge preparation of cutting tools using plasma discharges in electrolyte. J. Manuf. Process. **46**, 234–240 (2019)
17. Uhlmann, E., Oberschmidt, D., Kuche, Y., Löwenstein, A., Winker, I.: Effects of different cutting edge preparation methods on micro milling performance. Proc. CIRP **46**, 352–355 (2016)

Chapter 26
Residual Stress in Pipe Joints Welded with Base and Cellulose Coated Electrode

Amna Bajtarević, Nedeljko Vukojević, Fuad Hadžikadunić, Josip Kačmarčik, and Amra Talić-Čikmiš

Abstract The residual stress, that often occurs as consequence of the uneven heat distribution is common phenomenon for welded constructions. Possible consequences of residual stress are construction deformation, stress corrosion and fractures. Appearance of residual stress and deformation cannot be eliminated in total, but its intensity could be controlled by means of design and technological solutions. Thus, it is important to optimize the parameters of construction during designing phase. In this paper, a study conducted in order to find optimized technological solution for a welded pipe joint is presented. In the study, two technological solutions for the joint were compared by residual stress measuring with the "hole-drilling" method. The major difference between two technological solutions was the electrode type used during electric arc welding. Measurements were conducted on samples welded with base coated electrode and with cellulose coated electrode. According to the experts in this field, cellulose coated electrodes are more appropriate from economic and technological point of view. The purpose of this paper is the validation of cellulose coated electrode welded joints quality based on residual stress. Intensity and direction of residual stress at three measuring points of each sample were measured and compared.

Keywords Pipe joint · Residual stress · Weld · Coated electrodes

26.1 Introduction

One of the most important disadvantage of welding is appearance of residual stress in the welded constructions. The residual stress as a condition in a construction is caused by elastic deformations that are balanced inside of construction. Residual stresses can be superimposed with external stresses, and cause a change in shape or fracture of the structure, increase the tendency of the structure to brittle fracture, promote material fatigue or some corrosion processes in the material (stress corrosion, corrosion

---

A. Bajtarević (✉) · N. Vukojević · F. Hadžikadunić · J. Kačmarčik · A. Talić-Čikmiš
University of Zenica, Fakultetska 1, 72000 Zenica, Bosnia and Herzegovina
e-mail: amna.bajtarevic@unze.ba

M. Rackov et al. (eds.), *Machine and Industrial Design in Mechanical Engineering*, Mechanisms and Machine Science 109,
https://doi.org/10.1007/978-3-030-88465-9_26

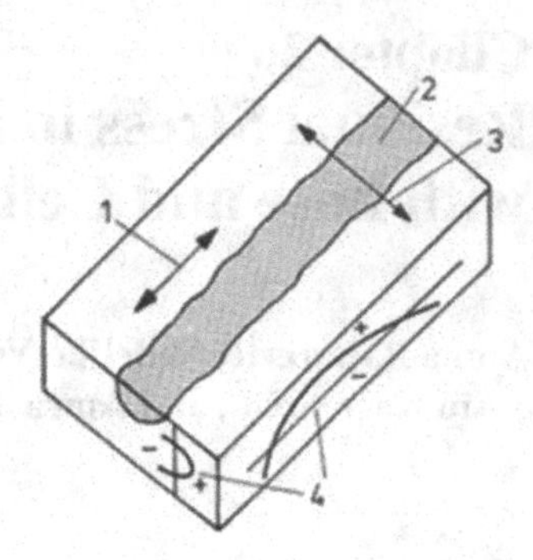

Fig. 26.1 The residual stresses caused by welding process: 1—longitudinal contraction, 2—weld metal, 3—transverse contraction, 4—stress [2]

due to fatigue), and adversely affect plastic properties [1]. Accordingly, the welded joints disadvantages, i.e. its impact on the welded joints durability and safety is the subject of the numerous scientific researches. Phenomenon of residual stresses in the structure is very complex due to thermal-metallurgical-mechanical processes. Also, residual stress occurs due to geometrical mismatch and thermal or plastic deformations. Within structures, large variations of residual stresses are usually observed depending on geometric position. The occurrence of residual stresses during the welding process is shown in Fig. 26.1. The colder surrounding metal limits the contraction of the molten weld metal during the cooling process that causes stresses.

The piping systems usually contain a lot of welded joints. Basically, steel welded pipe joints are an indispensable part of a large number of pipelines and due to a large number of impact factors, the analysis of their structural integrity can be observed from different aspects. A large number of papers is based on the issue of residual stresses and strains in these joints, where the authors emphasize the impact of the type of welding process, but also other parameters such as dimensional characteristics or object materials to be welded [3].

In order to analyze the impact of the electrode coat material type on the direction and intensity of occurred residual stresses, experimental measurement on two welded samples was performed. So, according to that results, optimal technological solution of welded pipe joint based on residual stress is determined.

26.2 Technical Description

In the preparation phase, two pipe joint samples with identical construction were welded. Geometrical characteristics are shown in Fig. 26.2. The fillet shape of weld was achieved by stabbing a smaller pipe with nominal diameter DN 80 to a DN 200 pipe. Both pipes are made of steel P235GH according to the standard EN 10,028 [4]. The welding process was performed in two passes in order to achieve optimal quality of weld. Table 26.1 shows welding parameters. A sample named "B" was welded by using base coated electrode marked as EVB 50, which producer is "Elektrode Jesenice", Slovenia and weld of sample "C" was achieved by using cellulose coated

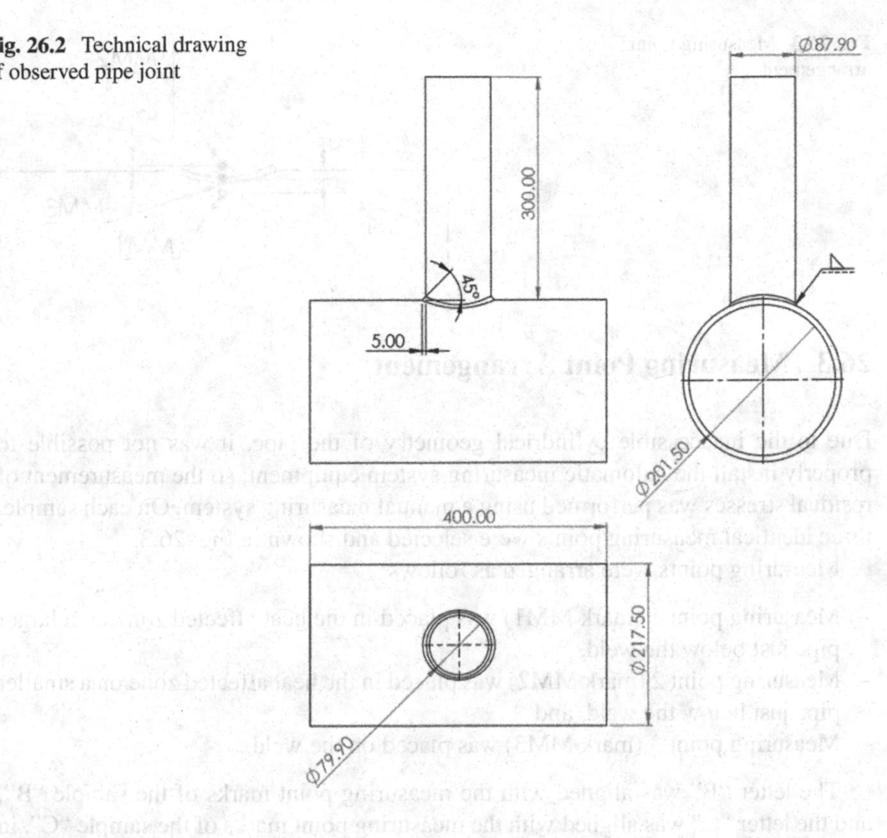

Fig. 26.2 Technical drawing of observed pipe joint

Table 26.1 Measured and calculated welding parameters

Electrode coat	Electrode diameter [mm]	Pass no	Current [A]	Voltage [V]	Electrical arc duration [s]	Weld length [mm]	Welding Speed [cm/min]
Base	2.5	1	91	23.64	171	276.15	9.69
Base	2.5	2	94	23.76	205	276.15	8.08
Cellulose	2.5	1	70	22.8	135	276.15	12.27
Cellulose	3.2	2	80	23.2	115	276.15	14.41

electrode marked as Bohler Fox Cell, AWS A5.1:E6010, which producer is “Bohler Welding” from Germany. In order to eliminate human impact, both samples were welded by the same operator. Welding parameters (electrode diameter and current) were defined empirically.

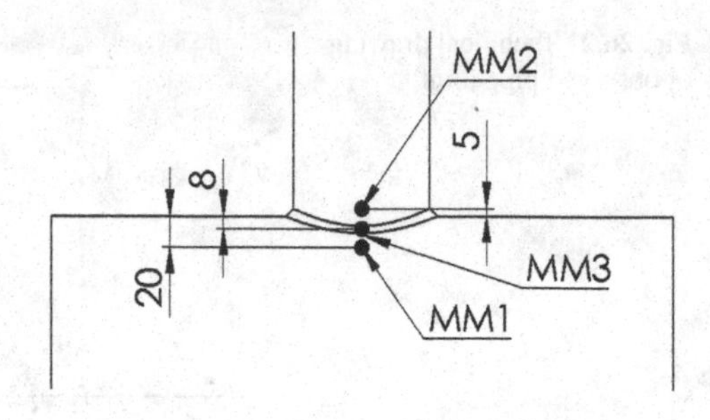

Fig. 26.3 Measuring point arrangement

26.3 Measuring Point Arrangement

Due to the inaccessible cylindrical geometry of the pipe, it was not possible to properly install the automatic measuring system equipment, so the measurement of residual stresses was performed using a manual measuring system. On each sample, three identical measuring points were selected and shown in Fig. 26.3.

Measuring points were arranged as follows:

- Measuring point 1 (mark MM1) was placed in the heat affected zone on a larger pipe just below the weld,
- Measuring point 2 (mark MM2) was placed in the heat affected zone on a smaller pipe just below the weld, and
- Measuring point 3 (mark MM3) was placed on the weld.

The letter "B" was aligned with the measuring point marks of the sample "B", and the letter "C" was aligned with the measuring point marks of the sample "C", in order to distinguish measuring point of different samples.

26.4 Preparation of an Experiment

After the surface preparation was completed, the strain gauges were positioned to the defined measuring points. The variant of used strain gauges was 0°/45°/90°, marked as RY 61 whose producer is HBM, Germany (renamed in 2020 to HBK Company, Germany). The strain gauges were connected to amplifier making 1/4 Wheatstone bridge.

Prepared measuring points are shown in Fig. 26.4. Deformation measurement directions were defined as shown on Fig. 26.5.

Fig. 26.4 Prepared measuring points

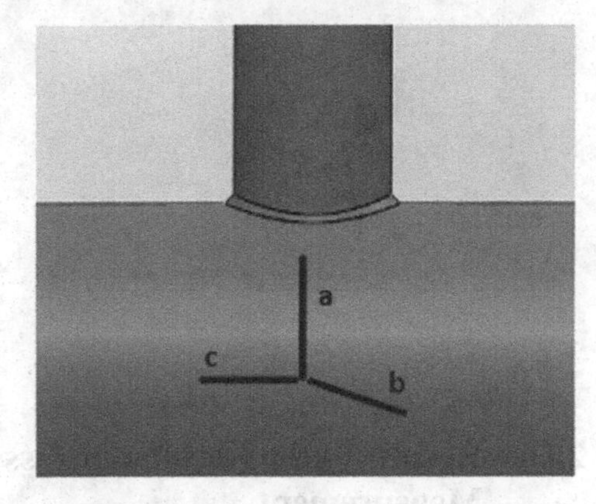

Fig. 26.5 Defined deformation measurement directions

26.5 Conducting an Experimental Measurement

After the preparation, the measurement of residual stresses was performed with the hole-drilling method by a manual measuring system, at six measuring points, as follows:

- three measuring points at pipe joint sample that was welded using base coated electrode (MM1B, MM2B and MM3B), and
- three measuring points at pipe joint sample that was welded using cellulose coated electrode (MM1C, MM2C and MM3C).

The values of deformations in a, b and c directions were calculated by Catman software. Drilling was performed with a hand drill. The diameter of drill tool was 1.5 mm, so total depth of drilling was also 1.5 mm. Figure 26.6 shows measurement process.

Fig. 26.6 Conducting an measurement

26.6 Results of the Residual Stress Experimental Measurement

When the experiment was completed, the values of deformations in three directions (a, b and c) were measured for a large number of drill positions. The deformation values for 40 drilling depths (40 material layers at different depths) were registered for data processing.

Using the measured values, diagrams of the deformation value change for each measuring direction depending on the position of the drill were created. Above mentioned diagrams for six measuring points are shown in Fig. 26.7. The deformation changes in three directions are similar for the two different pipes, as it is shown in Fig. 26.7, thus confirming the measurement results validity.

The principal stresses values and orientation were calculated from the deformation values. Both stress and angle calculations were performed according to the [5].

As an example, principal stresses values and orientation angle change with the drilling depth for measuring point MM2B is plotted in Fig. 26.8. Figures 26.9 show the maximum principal stress directions observed at both pipe joints.

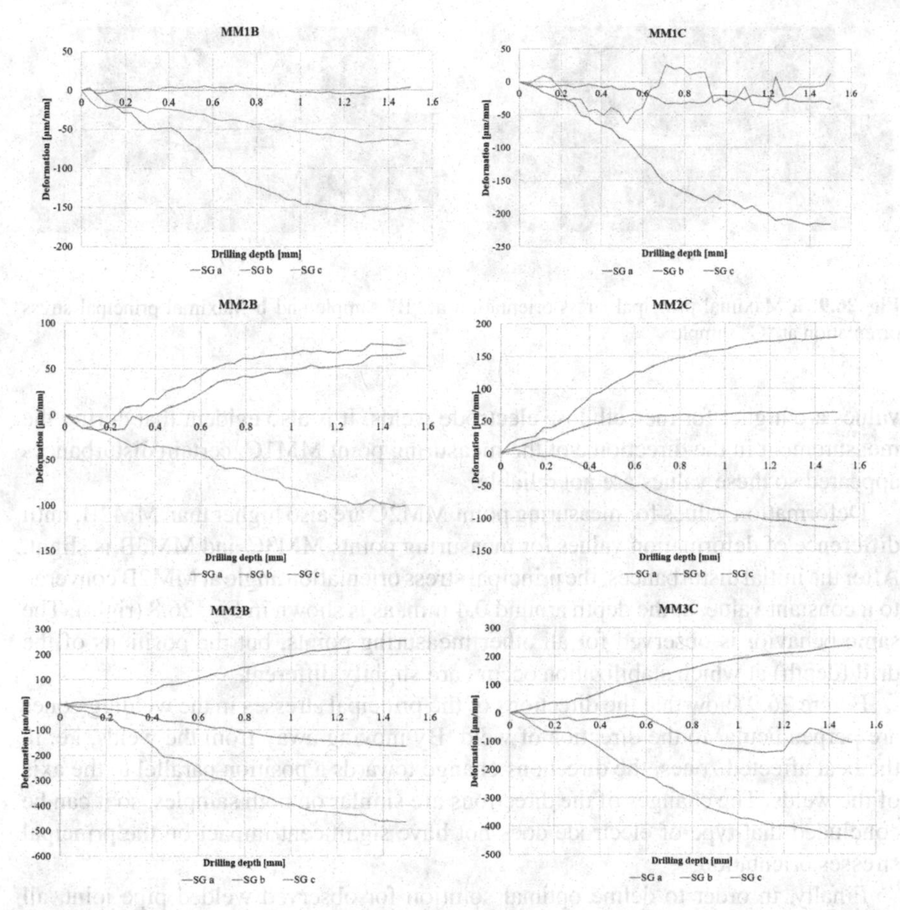

Fig. 26.7 Measured values of deformations in three directions at six measuring points

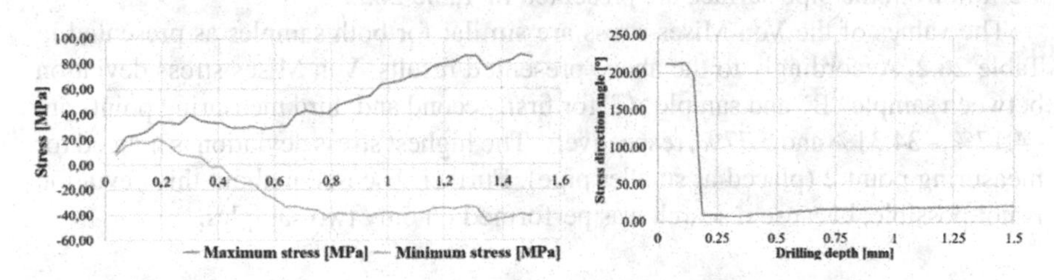

Fig. 26.8 The principal stresses values and orientation at the measuring point MM2B

26.7 The Experimental Research Results Analysis

By analyzing the deformation diagrams for measuring points MM1B and MM1C in Fig. 26.7, it can be concluded that for direction a and direction b, the deformation

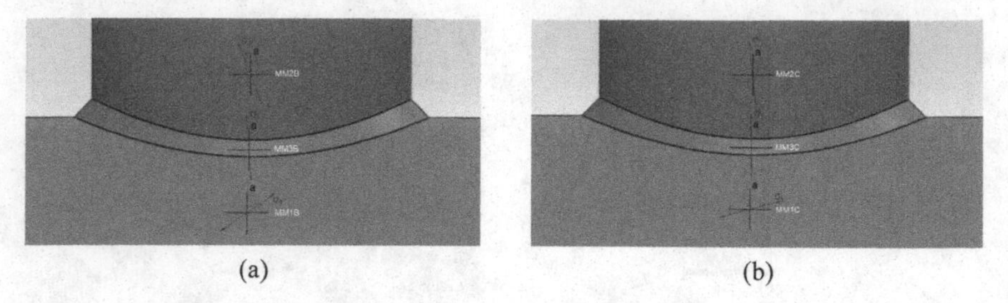

(a) (b)

Fig. 26.9 **a** Maximal principal stress orientation at "B" sample and **b** Maximal principal stress orientation at "C" sample

values are higher for the cellulose electrode welds. It is also evident that during the measurement in the direction c of the measuring point MM1C, certain disturbances appeared so these values are not reliable.

Deformation values for measuring point MM2C are also higher than MM2B, until difference of deformation values for measuring points MM3C and MM3B is slight. After the initial disturbances, the principal stress orientation angle at MM2B converge to a constant value, at the depth around 0.4 mm, as is shown in Fig. 26.8 (right). The same behavior is observed for all other measuring points, but the positions of the drill (depth) at which stabilization occurs are slightly different.

Figure 26.9 show that the directions of the principal stresses in the welding zones are perpendicular to the direction of weld. By moving away from the welds, i.e. in the heat affected zones, the directions change towards a position parallel to the axis of the welds. The changes of the directions are similar on both samples, so it can be concluded that type of electrode does not have significant impact on the principal stresses orientation.

Finally, in order to define optimal solution for observed welded pipe joint, all parameter values at the final moment of experiment, i.e. when depth of drill was 1.5 mm from the pipe surface are presented in Table 26.2.

The values of the Von Mises stress are similar for both samples as presented in Table 26.2. Accordingly to the above presented results, Von Mises stress deviation between sample "B" and sample "C" for first, second and third measuring points are –4.17%, –34.31% and 5.77%, respectively. The highest stress deviation is noticed for measuring point 2 (placed at smaller pipe). Further discussion about this deviation is not possible, because research was performed on only two samples.

26.8 Conclusion

It can be concluded that the type of electrode coat does not cause important impact to welded pipe joint residual stresses. It is known, considering to the experts for production of welded structures opinion, cellulose coated electrodes are more appropriate from economic and technological point of view. So, according to this research,

Table 26.2 Measured and calculated parameters on the end of the experiment

Measuring point	ε_a [μm/mm]	ε_b [μm/mm]	ε_c [μm/mm]	σ_1 [N/mm^2]	σ_2 [N/mm^2]	Von Mises stress [N/mm^2]	Principal stress angle [°]
MM1B	−150.96	3.36	−65.04	263.23	114.00	228.64	124.45
MM2B	75.36	66.48	−100.80	85.15	−40.72	111.25	20.98
MM3B	253.92	−165.12	−465.36	412.92	−43.66	436.39	184.69
MM1C	−217.68	−34.08	12.24	270.21	88.57	238.59	105.42
MM2C	177.84	62.40	−113.28	35.83	−148.58	169.36	5.85
MM3C	237.84	−107.12	−437.52	390.26	−41.54	412.60	180.62

residual stress is not restriction factor for using cellulose coated electrode during welding of pipe joints.

References

1. Pašić, F.: Zavarivanje. IP "Svjetlost" D.D. Sarajevo (1998)
2. Bajtarević, A.: Structural integrity analysis of the pipe joints in the complex stress states. MSc Thesis. University of Zenica, Zenica (2019)
3. Noyan, I., Cohen, J.: Residual Stress: Measurement by Diffraction and Interpretation. Springer-Verlag, New York (1987)
4. Standard EN 10028: Part 2: Non-alloy and alloy steels with specified elevated temperature properties
5. HBM Homepage. https://www.hbm.com/en/2073/strain-gauge-pdf-catalog/. last accessed 2021/03/25

Chapter 27
Comparison Analysis Between Different Technologies for Manufacturing Patient-Specific Implants

Georgi Todorov, Yavor Sofronov, and Krasimira Dimova

Abstract The demand for improved, fast and effective way to produce patients-specific implants has led to the developing new technologies in Implantology. The increasingly advanced development of methods for materialization of products used in Medicine allows the introduction of new materials with better characteristics and more efficient usage of the resources for manufacturing. The Virtual Prototyping and Rapid Prototyping technologies are used in the manufacturing process of the products such as personalized implants and prostheses, dental implants and joint implants, intervertebral discs, medical instruments and devices for fixation and guidance. Applying those technologies improves the process of designing and producing patient-specific implants. Actually, the goal of the process is to minimize the extra removal of tissues and to improve the precision of the process suggesting exact Prototyping technology for personal case. The research compares in few aspects the manufacturing technologies of cage type spinal implant. A patient-specific cage type implant is in the area of interest of the paper especially a comparison of the manufacturing technologies for producing it.

Keywords Technologies · Manufacturing · Patient-specific implants

27.1 Introduction

In the field of Implantology not only surgical implementation it is important part of the process but the implant's manufacturing. With the developing the Medicine more and more technologies are involved in the innovations. Nowadays, Virtual and Rapid prototyping are developing as research areas that use basic engineering principles for solving complex problems in the human body in a medical environment. The physical prototyping could be accomplished with additive technology—in the current research were chosen SLM and FFF technology or subtractive technology. In

G. Todorov · Y. Sofronov · K. Dimova (✉)
FIT-Faculty of Industrial Technology, Technical University of Sofia, Sofia, Bulgaria
e-mail: krdimova@tu-sofia.bg

Laboratory "CAD/CAM/CAE in Industry", Sofia, Bulgaria

M. Rackov et al. (eds.), *Machine and Industrial Design in Mechanical Engineering*, Mechanisms and Machine Science 109,
https://doi.org/10.1007/978-3-030-88465-9_27

certain clinical cases, the additive technologies are very suitable for creating patient-specific implants, and it is possible to select a suitable material for producing, which must be biocompatible [1]. At the other hand Subtractive technologies might be better option in some clinical cases. Additive technologies, both metal and plastic printing, continue to expand their design capabilities, improving the functions and complexity of implants. Traditional methods, such as machining, are also being improved especially the control and precision, but for some cases and parts, which have a complex geometry, they still require a wide range of machines and highly qualified personnel. The aim of the research is make comparison analysis between different technologies for manufacturing of cage type spinal implant considering the implants material, complexity, dimensions and other factors and using Virtual prototyping [2].

27.2 Methods and Materials

For the aims of the research was used a 3D model of cage type spinal implant. With some of the existing Additive and Subtractive technologies and the help of CAM (Computer Aided Manufacturing) software and a 3D slicing software were evaluated the implant's precision and cost. For the current case was chosen the optimal material and technology for producing. The materials in Implantology should have specific characteristics like bio-compatibility, strength, corrosion resistance, fatigue strength. The most common materials in Implantology are titanium alloys and plastics, rarely used material is ceramic. For the research are compared Ti6Al4V, PEEK (Polyether ether ketone) and ceramic [3]. With increasing the complexity of the part its cost is increasing too, both in Subtractive manufacturing and in Additive manufacturing. But in Additive manufacturing there is a point beyond which the levels of complexity and customization is free, which make the technology more suitable for complex shapes as its shown on Fig. 27.1 [4]. The total number of required parts is a key design consideration when selecting a manufacturing technology [5]. The patient-specific implants require unique design for that reason the Additive technology is competitive to Subtractive technology in that case. In cases where are required large number of parts the Subtractive manufacturing might be more suitable as it shown on Fig. 27.2.

These advantages of the Additive manufacturing include less material consumption, increasing design complexity, opportunity to produce direct assembly, variety of materials [8]. Also applying Subtractive manufacturing technology results in a lot of raw material waste, which is often expensive titanium alloy [9]. One of the flaws of the Additive process is quite rough surfaces, which could require additional finishing operation like sanding or blowing [10]. The right chose of technology depends also on number of required series, the complexity of the shape, the material and technical requirements.

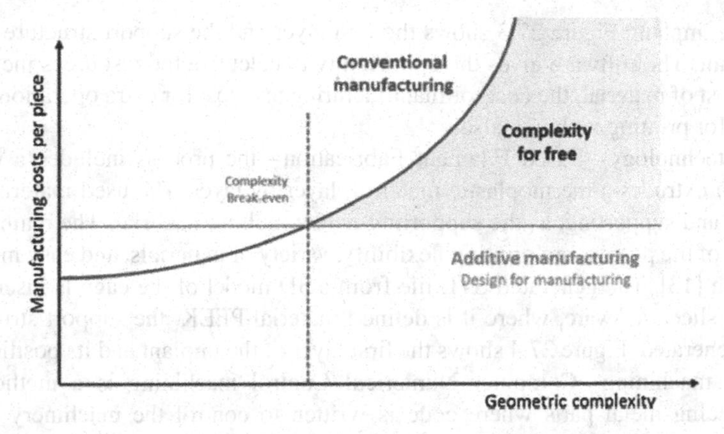

Fig. 27.1 Cost of the part considering its complexity in traditional and in additive manufacturing [6]

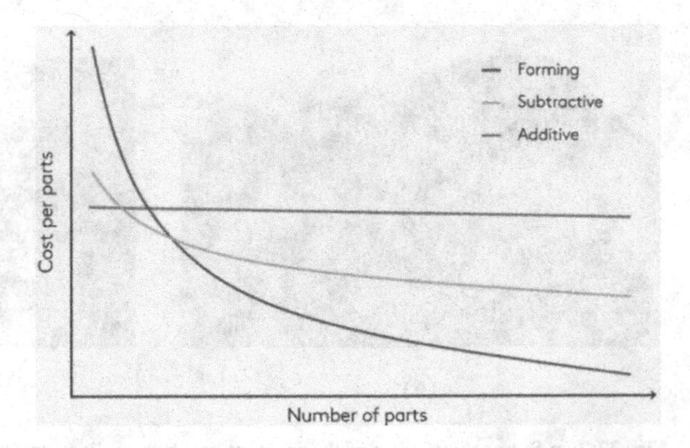

Fig. 27.2 Cost per parts considering the number of parts [7]

The involved additive technologies in the research are: SLM (Selective Laser Melting), FFF (Fused Filament Fabrication) and are compared to the CNC technologies. The study compares the manufacturing technologies by which the cage could be accomplished using only different Virtual systems for CAM and software for slicing:

- SLM technology—Selective Laser Melting is method which melts metallic powder with laser to produce metal with unlimited complex shape. Its main advantage is the independence with respect to material selection [11]. It is a complex thermo-physical process which requires delicate process determination to achieving high precision [12]. The first step is importing the generated STL file of the implant in a slicer software where is chosen the technology and the material, in the current case- the used material is Ti6Al4V. The material is often used because of its corrosion-resistance, wear resistance and a good osseointegration. The main parameters of the process as power of beam, scan strategies, layer thickness and other. The metal shrinking should be considered and the positioning of the implant. Figure 27.3 shows the first layer and the support structure of the implant. The software gives the opportunity to calculate the resources including the cost of material, the cost for manufacturing and cost for extra operations. The time for printing is shown also.

- FFF technology—Fused Filament Fabrication—the process includes a nozzle which extrudes a thermoplastic material, layer by layer. The used materials are main and supporting, as the supporting is low-melting material. The main privileges of the process are printing flexibility, variety of materials, and easy material switch [13]. The generated STL file from a 3D model of the cage is inserted in other slicer software, where it is defined material PEEK, the support structures are generated. Figure 27.4 shows the first layer of the implant and its positioning.
- CNC machining—Computer Numerical Control machining is a method for producing metal parts where code is written to control the machinery in the process of subtraction. The code determines the whole machining process. As main factors that could limit the CNC manufacturing process are tool access and clearances, and also the level of geometry complexity [7].

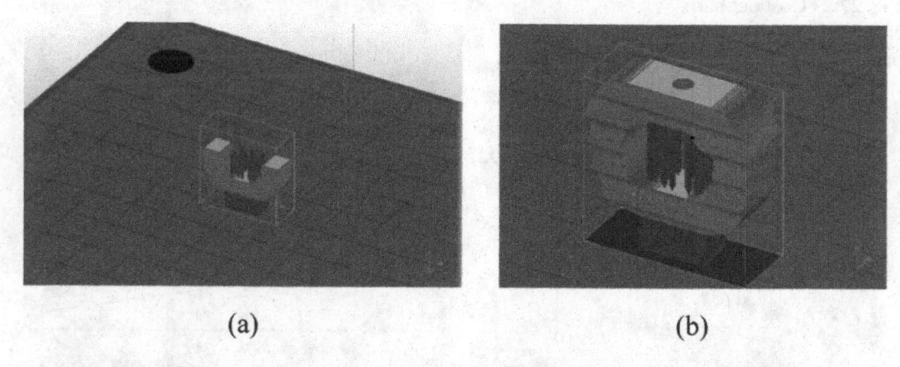

(a) (b)

Fig. 27.3 **a** Support structure of the implant and **b** visualization of 3D printed implant

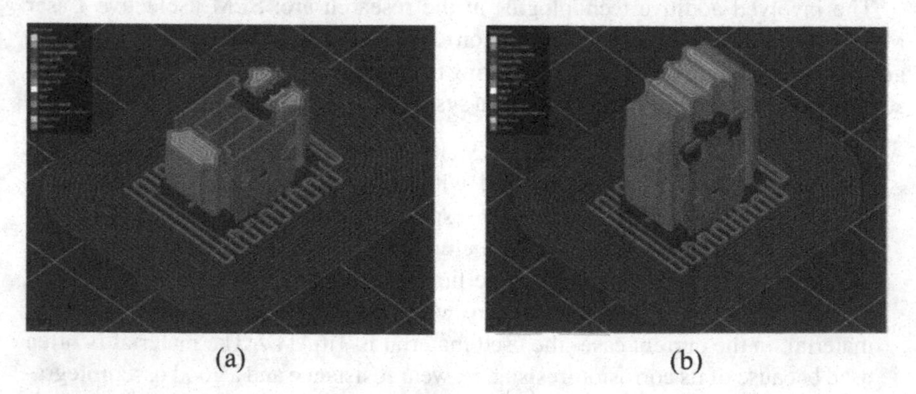

(a) (b)

Fig. 27.4 **a** Supporting structure of the implant and **b** visualization of 3D printed implant

In current case is used a 3D model of the implant and CAM module of software. The appropriate milling machine is chosen, with 3 or 5 axis. The model's coordinate system is adjusted to the machine's coordinate system. Then it is generated a workpiece. Titanium alloy is defined as material. The technology includes few operations, as the main operations are roughing, finishing and the last CNC operation is cutting. The cutting speed for current material was 100 m/min for Ti6Al4V. The calculated time for the whole process included all operations was 134 min. Figure 27.5 shows the process of planning the CNC manufacturing using CAM software.

The CNC machining of a workpiece form PEEK is also part of the research. The PEEK workpiece has the same dimensions as the one from titanium alloy. The mode of the technological operations are different also the material cost. When applying FFF technology the PEEK material is powdered while applying CNC manufacturing the material is rod workpiece.

In Table 27.1 are compared the three cases and it was chosen FFF technology with PEEK material, which is an expensive material, but the technology allows the inside structure to be half-full, the accuracy of FFF technology is low but enough for manufacturing the implant. This additive technology is working with main and supporting material. Considering the results could be summarized that the implant's

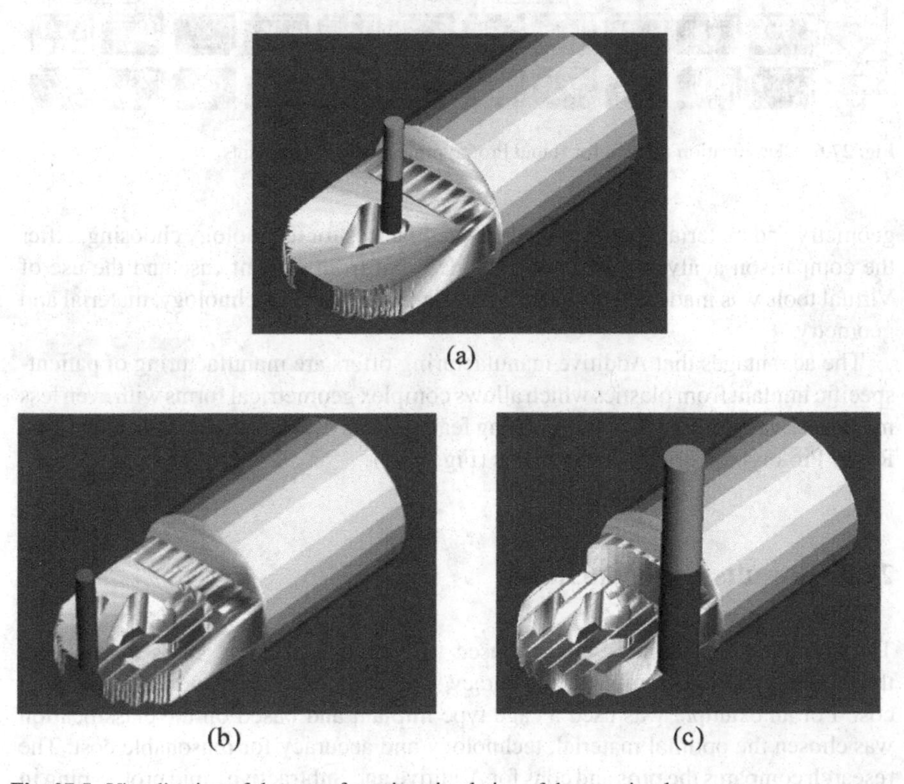

Fig. 27.5 Visualization of the manufacturing simulation of the cage implant

Table 27.1 Comparison analysis

	Material type	Mass [g]	Material cost [€/kg]	Manufacturing cost [€]	Time for manufacturing [min]
SLM	Ti6Al4V	2	300	100	120
FFF	PEEK	3	500	40	30
CNC manufacturing	Ti6Al4V	40	150	90	134
CNC manufacturing	PEEK	12	350	70	96

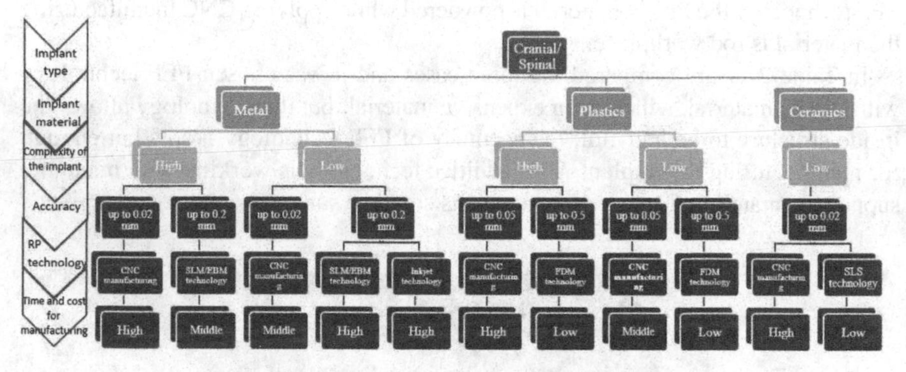

Fig. 27.6 Classification method for Rapid Prototyping choice for implants

geometry and material are the main depended factors for technology choosing. After the comparison analysis could be concluded that from current case and the use of Virtual tools was made an optimal chose for manufacturing technology, material and geometry.

The advantages that Additive manufacturing offers are manufacturing of patient-specific implant from plastics which allows complex geometrical forms with even less material. Evaluating each manufacturing features was made classification method for Rapid Prototyping choice for implants (Fig. 27.6).

27.3 Results and Conclusion

The classification method could be used in cranial and spinal cases considering the implant material, complexity, accuracy, RP technology, time and manufacturing cost. For an example was used a cage type implant and based on the classification was chosen the optimal material, technology and accuracy for reasonable cost. The research compares the pros and cons for Additive and Subtractive rapid prototyping in the process of manufacturing the implant. The major role of the comparison analysis played the applied Virtual tools which reduces the time for evaluating the manufacturing parameters. The results for current case show that the most suitable technology is SLM technology with Ti4Al6V considering the main factors. The method could be applied in cranial and spinal reconstructions in order to define the best technology for current clinical case. This classification could be used for operation planning. To summarize by the initial cage type implant it was made a classification method which could be applicable in cranial and spinal clinical cases.

Acknowledgements This study is performed by the support of project by the European Regional Development Fund within the Operational Programme "Science and Education for Smart Growth 2014–2020" under the Project CoE "National center of mechatronics and clean technologies" BG05M2OP001-1.001-0008-C01.

References

1. Todorov, G., Nikolov, N., Sofronov, Y., Gabrovski, N., Laleva, M., Gavrilov, T.: Additive/subtractive computer aided manufacturing of customized implants based on virtual prototypes. In: Poulkov, V. (ed.) Future Access Enablers for Ubiquitous and Intelligent Infrastructures: FABULOUS 2019. LNICS-SITE, vol. 283, pp. 347–360. Springer, Cham (2019)
2. Todorov, G., Kamberov, K.: Virtual Engineering (2017)
3. Gavrilov, T., Sofronov, Y.: Selection and comparative study of RP technologies for personalized implant manufacturing. In: Proceedings of the International Conference on Creative Business for Smart and Sustainable Growth, CREBUS 2019, pp. 1–8, Sandanski (2019)
4. Song, X., Zhai, W., Huang, R., Fu, J., Fu, M., Li, F.: Metal-based 3D-printed micro parts & structures. In: Reference Module in Materials Science and Materials Engineering. Elsevier (2020)
5. Conner, B.P., Manogharan, G.P., Martof, A.N., Rodomsky, L.M., Rodomsky, C.M., Jordan, D.C., Limperos, J.W.: Making sense of 3-D printing: Creating a map of additive manufacturing products and services. Addit. Manuf. **1–4**, 64–76 (2014)
6. Kirchheim, A., Zumofen, L., Hans-Jörg, D.: Why education and training in the field of additive manufacturing is a necessity: The right way to teach students and professionals. In: Proceedings of the Industrializing Additive Manufacturing: Additive Manufacturing in Products and Applications, AMPA 2017, pp. 329–336. Zurich (2018)
7. Varotsis, A.B.: 3D printing vs CNC machining. https://www.3dhubs.com/knowledge-base/3d-printing-vs-cnc-machining/. Last accessed 2021/01/01
8. Gaget, L.: Comparison between 3D printing and traditional manufacturing processes for plastics (2019). https://www.sculpteo.com/blog/2019/07/16/comparison-between-3d-printing-and-traditional-manufacturing-processes-for-plastics-3
9. Uckelmann I.: Metal 3D printing vs CNC machining? https://www.materialise.com/en/blog/metal-3d-printing-vs-cnc-machining-theres-no-competition. Last accessed 2020/12/01
10. 3D e-shop, Additive vs subtractive manufacturing: difference, pros & cons. https://www.3de-shop.com/additive-vs-subtractive-manufacturing-difference-pros-cons/. Last accessed 2021/01/01
11. Kruth, J.P.: Material incress manufacturing by rapid prototyping technologies. CIRP Ann. **40**(2), 603–614 (1991)

12. De Viteri, V.S., Fuentes, E.: Titanium and titanium alloys as biomaterials. In: Gegner, J. (ed.) Tribology: Fundamentals and Advancements, pp. 155–181. IntechOpen (2013)
13. Singh, S., Singh, G., Prakash, C., Ramakrishna, S.: Current status and future directions of fused filament fabrication. J. Manuf. Process. **55**, 388–306 (2020)

Chapter 28
Effects of Additives on the Mechanical Properties of Aluminum Foams

Szilvia Gyöngyösi, András Gábora, Gábor Balogh, Gábor Kalácska, Tamás Bubonyi, and Tamás Mankovits

Abstract The application area of aluminum foams is continuously widening. Large parts of these application fields rely on the functional properties of metallic foams. The aluminum material and the special cell structure of the foams, however, limit the functional properties. One possible solution for improvement is the use of different additives in the material, which modify the mechanical properties and the cell structure of the foam. Different additives are selected and added to the material, and the structure and mechanical properties of the foam are tested and evaluated to study the effect.

Keywords Aluminum foam · Mechanical properties · Additives · Zirconia

28.1 Introduction

Metallic foams, especially aluminum foams, are advanced materials with continuously developing application areas [1, 2]. Aluminum foams are used as structural materials, but in most cases the functional properties also important [3]. The scope of the development is to improve the functional properties where these play a significant role in the use of the material.

Currently the utilization of the aluminum, aluminum alloy and aluminum structures is continuously widened due to the continuous development of the alloys and processing technologies [4, 5]. As the aluminum alloys the aluminum based composite structures are intensively studied [6, 7]. The material testing of special

S. Gyöngyösi (✉) · A. Gábora · G. Balogh · T. Mankovits
Faculty of Engineering, University of Debrecen, Ótemető 2-4, 4028 Debrecen, Hungary
e-mail: gyongyosi.szilvia@eng.unideb.hu

G. Kalácska
Faculty of Mechanical Engineering, Hungarian University of Agriculture and Life Sciences, Gödöllő, Hungary

T. Bubonyi
Institute of Physical Metallurgy, Metalforming and Nanotechnology, University of Miskolc, Egyetemváros, 3515 Miskolc, Hungary

M. Rackov et al. (eds.), *Machine and Industrial Design in Mechanical Engineering*, Mechanisms and Machine Science 109,
https://doi.org/10.1007/978-3-030-88465-9_28

composite structures needs new testing methods and equipment as well [8]. The detailed knowledge and test results opens the mentioned new application areas. A typical example is the development of aluminum foams.

The foams have a special structure which basically determines the mechanical and functional properties. The average cell size and the size distribution are basic parameters besides the material properties of the cell walls [9]. The most effective testing method of the cell structure is computer tomography (CT), which reveals the whole 3D structure. On the CT images the evaluation of the cell structure is possible in a detailed way [10].

However, if the functional properties are important for a given application, the strength of the foams would also be significant. It can be evaluated simply with a compression test, but based on the CT imaged structure, a FEM study is possible [11, 12].

The base aluminum alloy determines the essential properties of the foam. If some functional properties must be improved, additives are necessary. The base alloy is generally a composite material, so the effect of the potential additives can basically modify its structure and properties. The main question is how these additives work in relation to the material and size. This study introduces some special additives. Different foams are produced with the same technique, and the cell structure and the mechanical behavior are compared to a basic foam. The test results show how these additives change the basic foam.

28.2 Material and Methods

Aluminum foams are produced by direct method. The raw material is Duralcan® (A359 + 20%SiC). The raw material was melted at 800 °C and 1.5 V/V% TiH_2 powder was added to the melt. The melt and the TiH_2 was mixed while the foam was forming. After foaming the material was cooled down in water.

Two different additives were mixed to the melt before the foaming: ZrO_2 particles (24 V/V%) and WO_3 particles (1.5 V/V%). Both additives are heavier than the aluminum. The size of the particles was significantly different. The average diameter of the ZrO_2 particles was 1 mm, while the WO_3 was a powder with 10-30 μm size. Both additives as solid particles were easily mixed with the molten Duralcan® material. In the following, Al denotes the original foam, Al(ZrO_2) denotes the foam with ZrO_2 addition. According to this scheme Al(WO_3) represents the foam which contained WO_3 particles.

The structure of all three foams was checked with computer tomography (CT). The main scope of the CT observation to study the effects of the additives on the cell structure, as the structure has an essential effect to the mechanical properties of the foams.

Small cubes (30 × 30 × 30mm) were prepared from the produced foams and compression tests (5 mm/min) were made to determine the strength of the foams. An Instron 8801 universal material tester was used for the mentioned compression tests.

28.3 Results and Discussion

The cell structure of the base aluminum foam is inhomogeneous (Fig. 28.1). It contains few large pores covered by a lot of small-sized cell. The aspect ratio of the small pores is nearly one, but in the case of large pores elongation is visible. These probably formed with the damage of cell walls during the foaming process.

The average size of the ZrO2 particles is 1 mm in a narrow distribution (Fig. 28.2.). These are Ce-doped spherical particles which are originally used in ball mills. The particles were mixed to the melt before the foaming. The spheres were heavy compared to the aluminum, therefore the sedimentation in the aluminum melt was a fast process.

The foaming process made an extra convection in the melt which helped to develop a relatively homogenous distribution of solid particles. To achieve a good quality these two effects give only a short time between the end of mixing of the solid particles into the melt and the foaming.

The large density of the ZrO_2 spheres suppress the visibility of the cell walls on the CT image. On areas where the particles are relatively far from each other, the cell structure can be observed. The ZrO_2 addition did not modify the average size of the cells, and the mentioned inhomogeneity also can be seen.

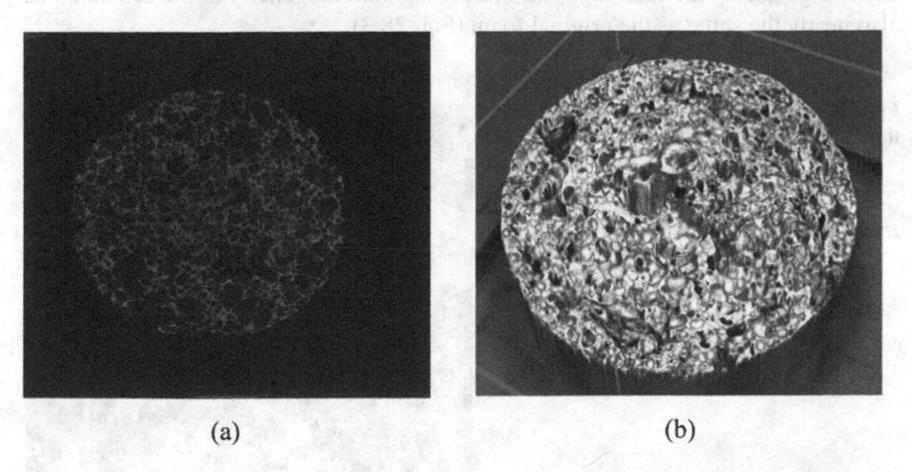

Fig. 28.1 Reconstructed CT cross section of the basic foam (a) 2.5D image of the foam made with a Zeiss Smartzoom5 microscope (b)

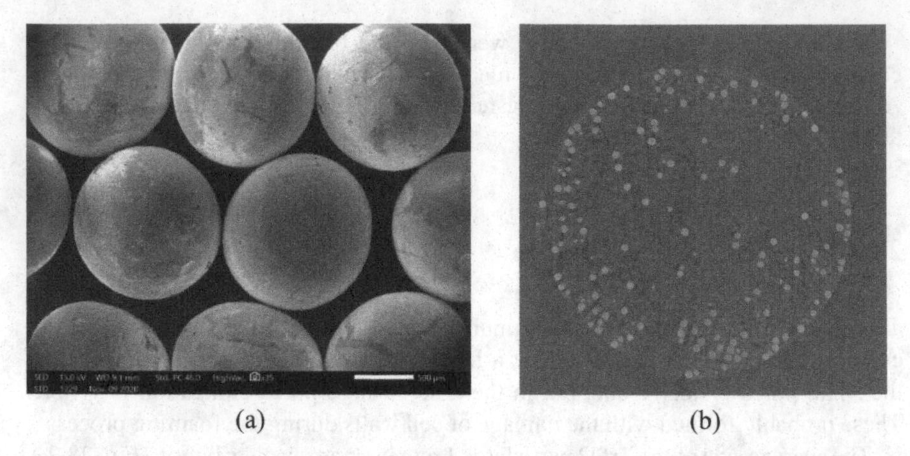

(a) (b)

Fig. 28.2 SEM image taken from the ZrO_2 spheres (a), and a CT section of the foam with ZrO_2 addition (b)

Clustering was observed in the spatial distribution of the ZrO_2 spheres. This phenomenon also shows the inhomogeneity of the pore size. The volumes filled with small-sized cell contain more spheres. Of course, the sedimentation effect also can be observed along the height of the foam in a smaller extent. Deeper observation revealed that the spheres in the structure are covered by aluminum, so the mechanical connection between the cell walls and the spheres are perfect.

The size of the WO_3 particles were the same as the SiC particles in the Duralcan material. The same amount was added to the melt as the foaming agent. The WO_3 particles mixed homogeneously in the molten aluminum, and this homogeneity did not change during the foaming. The size and the inhomogeneity of the cell size was also nearly the same as the original foam (Fig. 28.3).

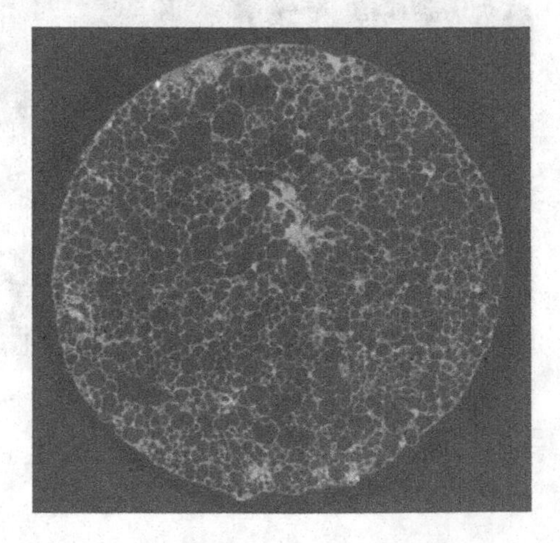

Fig. 28.3 The CT section of the Al(WO_3) foam

Fig. 28.4 A sample and a compressed foam in the testing apparat

The CT investigation shows that the structure of the foams has not changed significantly with the addition, so the results of mechanical tests are comparable. Small cubes were manufactured from the foams, and compression tests were made (Fig. 28.4). The compression tests were made till the samples fully collapsed.

The results of the compression tests are showed on Fig. 28.5. All curves have an elastic section at the beginning. After the elastic part the damage of the cell walls starts, so a plateau forms on the curve. Eventually, the foam becomes dense enough due to the damages the compresson that the force increases intensively. The different additives modify the compression curve in different ways and extents. The elastic limit of the original foam is 2.2 N/mm2, the length of plateau is 56.7% of the height of the samples (30 mm).

The ZrO_2 particle addition reduced the elastic limit of the foam (1.8 N/mm^2), and additionally "straightened" the plateau. This shows that the large particles make the

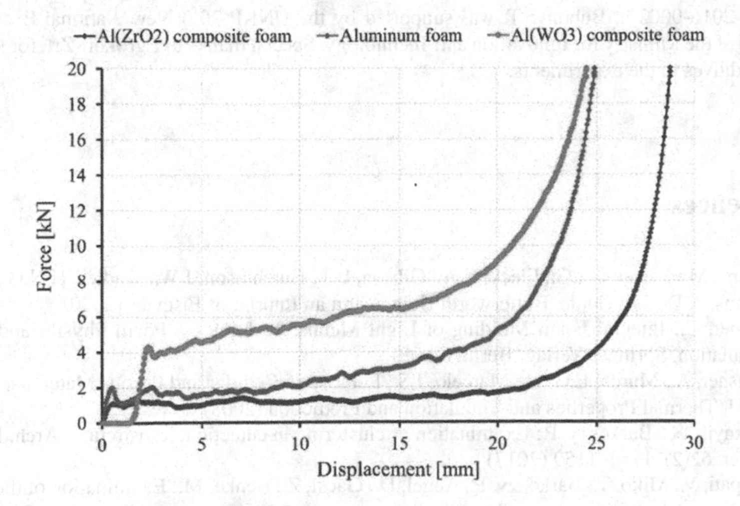

Fig. 28.5 Typical compression curves of the examined foams

cell walls weaker. The plateau ends at larger displacement (80%). This also points out that the cell walls damaged more easily than the original foam.

WO3 small particle addition resulted in a stronger foam. The elastic limit is doubled compared to the original foam (4.5 N/mm^2) however the displacement is significantly smaller in this case (43.3%). From a mechanical point of view, with the addition of WO_3 a stronger foam can be created.

28.4 Conclusion

Aluminum foams were produced from Duralcan material by direct foaming technique. The basic foaming process was made with two different additives: ZrO_2 large spheres, and WO_3 small particles. The main scope of the foam production was to enhance different functional properties of the material, but the mechanical behavior was also examined through compression tests. Besides the compression tests the structure of the foams was also studied via CT imaging. The results of the observations and tests show that the cell structures of the foams are inhomogeneous, large pores can be found in the structure next to smalls ones. The additives did not modify this feature and the pore sizes significantly. The mechanical properties on the other hand, changed. The ZrO_2 addition decreased the elastic limit in a small extent while WO_3 particle addition doubled it. The effect of the additives on the displacement was the opposite compared to each other. The ZrO_2 sphere addition increased the displacement in a large extent. So, the ZrO_2 addition in this way makes a weaker foam material which must be taken into in account in its application. The WO_3 addition aside from the functional advances improved the strength of the foam.

Acknowledgements The research was financed by the Debrecen Venture Catapult program (EFOP-3.6.1-16-2016-00022). Bubonyi T. was supported by the ÚNKP-20-3 New National Excellence Program of the Ministry for Innovation and Technology. Special thanks to Egrokorr Zrt. for support some additives to the experiments.

References

1. Ashby, M.F., Evans, A.G., Fleck, N.A., Gibson, L.J., Hutchinson, J.W., Wadley, H.N.G.: Metal Foams: A Design Guide. Butterworth Heinemann an Imprint of Elsevier (2000)
2. Körner, C.: Integral Foam Molding of Light Metals, Technology. Foam Physics and Foam Simulation. Springer-Verlag, Berlin (2008)
3. Öchsner, A., Murch, E.G., de Marcelo, J.S., Lemos, S.: Cellular and Porous Materials. Wiley-VCH, Thermal Properties and Simulation and Prediction (2008)
4. Bortnyik, K., Barkóczy, P.: Examination of clustering in eutectic microstrcture. Arch. Metall. Mater. **62**(2), 1155–1159 (2017)
5. Karpati, V., Miko, T., Barkoczy, P., Angel, D., Gácsi, Z., Benke, M.: Examination of the effect of homogenization processes through compression tests in aluminum alloys. IOP Conf. Ser.: Mater. Sci. Eng. **426**, 012023–1–012023–8 (2018)

6. Corrochano, M L., Ibanez, J., Gácsi, Z., Tomolya, K., Svéda, M., Barkóczy, P.: Reiforcement distribution in AA6061/MoSi2/5p, 15p, 25p composites obtained by extrusion of the powders. In: Proceedings of the 5th International Powder Metallurgy Conference, pp. 1–5. TPMA, Ankara (2009)
7. Gacsi, Z., Kovács, J., Barkóczy, P., Piezonka, T.: Arrangement of Ceramic Particles in PM Composites In: Proceedings of the Powder Metallurgy World Congress & Exhbition, PM 2004, pp. 257–262. EPMA (2004)
8. Bubonyi T., Barkóczy P., Gácsi Z.: Comparison of CT and metallographic method for evaluation of microporosities of dye cast aluminum parts. IOP Conf. Ser.: Mater. Sci. Eng. **903**, 012038–1–012038–6 (2020)
9. Májlinger, K., Bozóki, B., Kálácská, G., Keresztes, R., Zsidái, L.: Tribological properties of hybrid aluminum matrix syntactic foams. Tribol. Int. **99**, 211–223 (2016)
10. Characterisations of ALUHAB aluminium foams with micro-CT 2014 Bábcsán, N., Beke, S., Számel, Gy., Borzsonyi, T., Szábo, B., Mokso, R., Kádár, Cs., Kiss, B.J Proc. Mat. Sci. **4**, 67–72 (2014)
11. Mankovits, T., Varga, T., A., Manó, S., Kocsis, I.: Compressive response determination of closed-cell aluminum foam and linear elastic finite element simulation of μCT based directly reconstructed geometrical models. Stroj. Vestn.-J. Mech. E. **64**(2), 105–113 (2018)
12. Mankovits, T., Budai, I., Balogh, G., Gábora, A., Kozma, I., Varga, T.A., Manó, S., Kocsis, I.: Structural analysis and its statistical evaluation of closed-cell metal foam. Int. Rev. Appl. Sci. Eng. **5**(2), 135–143 (2014)

Chapter 29
The Influence of the Hardening and Tempering Heat Treatment on the Strength of the CuZn39Pb3 Brass to Cavitation Erosion

Iosif Lazar, Ilare Bordeasu, Liviu Daniel Pirvulescu, Mihai Hluscu, Dumitru Viorel Bazavan, Lavinia Madalina Micu, and Cristian Ghera

Abstract The most common procedures for changing the metal structures exposed to the destructive demands of cavitation erosion are the volumic heat treatments. Therefore, the paper presents the behavior and resistance results of the CuZn39Pb3 brass structure, obtained by volumic heat treatment at 800 °C, with water quenching, followed by tempering at 600 °C, with air quenching, to microjets generated by the hydrodynamic mechanism of vibrating cavitation. The comparison with the delivered state and with the standard laboratory samples, used in manufacturing ships propellers (naval brass and CuNiAl I-RNR bronze), shows that the new structure and the increased hardness, resulting from the heat treatment, lead to a significant increase in the resistance to vibratory cavitation demands.

Keywords Cavitation erosion · Microstructure · Hardness · Heat treatment · Erosion depth · Resistance to cavitation · Brass · Volumic quenching heat treatment · Tempering

29.1 Introduction

Brass CuZn49Pb3 is widely used for components operating in low intensity cavitation conditions, such as: pipe fittings (couplings, elbows, tees), and the gates of the valves used for plumbing.

Due to its excellent thermal conductivity, it is also used for the pipes of the heat exchangers and coolers. In order to be used for the components operating in higher intensity cavitation regimes, such as sailboat propellers or the bushings securing them on the motor shaft, it is necessary to increase the mechanical properties of the

Present Address:
I. Lazar · I. Bordeasu (✉) · L. D. Pirvulescu · M. Hluscu · D. V. Bazavan · C. Ghera
Politehnica University of Timisoara, PiaţanVictoriei Nr. 2, 300006 Timisoara, Romania

L. M. Micu
Banat University of Agricultural Sciences and Veterinary Medicine of Timişoara, Calea Aradului nr. 119, Timisoara, Romania

M. Rackov et al. (eds.), *Machine and Industrial Design in Mechanical Engineering*, Mechanisms and Machine Science 109,
https://doi.org/10.1007/978-3-030-88465-9_29

surfaces exposed to the cavitation microjets, especially the hardness, simultaneous with an adequate structure, different from the regular one, made up of a solid solution α and a β‘ electronic compound [1–3]. The fastest technological process to achieve these requirements is the volumic heat treatment. Such a heat treatment is quenching followed by tempering, the same process used in the research, of whose results are described in this paper.

29.2 Material Used and Heat Treatments Applied

The researched material is brass (CuZn39Pb3) used in pressure and flow control regulation found in plumbing installations components, where the intensity of the cavitation-erosion process is not so high. In order for this material to be used for bushings that secures the ship impellers on the motor shaft, as well as the propellers for recreational craft, such as sailboats [1, 4], volumic heat treatments are needed in order to change the initial structure, as well as the hardness of the surface exposed to the vortex cavitation.

The volumic heat treatment is imposed by the necessity for increasing the resistance to the erosion action caused by the vortex cavitation, limited in time and of moderate intensity, resulted from a low engine speed, when compared to the ships propelled at high speeds, where the intensity of the created vortex is extremely high.

The chemical composition, the mechanical properties and the structure of the semi-fabricated samples, as determined in the Materials Engineering Laboratory at the Polytechnic University of Timisoara, are as follows [1, 2]:

- chemical composition: 57.7% Cu, 38.49% Zn, 3.3% Pb, 0.2% Fe, 0.1% Ni, 0.2% Sn, 0.01% Al;
- the two-phase structure, made up of the solid solution α, with a c.f.c. grid; and the β′ electronic compound, with a c.v.c. grid;
- R_m = 502 MPa, fluid flow limit $Rp_{0.2}$ = 365 MPa, elongation yield point A5 = 18%, longitudinal young's modulus of elasticity E = 97 GPa, density ρ = 8.47 g/cm^3, hardness (average of 8 measurements) 124 HV0.5.

The applied heat treatment consists in heat treatment at 800 °C (for 40 min) with water quenching, followed by tempering at 600 °C (for 60 min) with air quenching.

In order to simplify the analysis and distinguish from the delivered state samples, the annotation used in this paper for the brass samples subjected to heat treatment is C800/R600.

Out of the mechanical characteristic resulted from the heat treatment, only the hardness was assessed, as this is the mechanical property that has a significant impact on the resistance to the destructive demands of the cavitation [5, 6]. The average arithmetic value of the eight measurements performed on the surface of one of the three samples subject to testing is equal to 133.8 HV05, representing a 7.9% increase compared to delivered state.

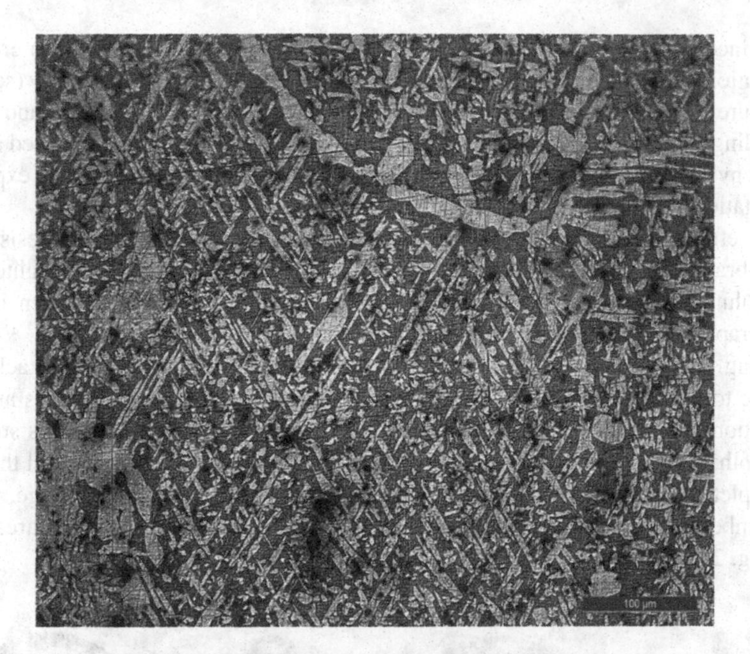

Fig. 29.1 Image taken with the microscope before the exposure to cavitation erosion

The image (see Fig. 29.1), obtained with the OLIMPUS SYX7 optical microscope, shows that tempering at 600 °C, after hardening at 800 °C, has led at two processes occurred in parallel: one that increased the α-phase solubility into β-phase, causing a slightly decrease in the amount of α-phase grains, and another one of a more intense coalescence of the small acicular α-phase grains, forming larger, polygonal grains.

29.3 Experimental Results and Analysis

The research to cavitation erosion was performed on the standard vibrating device with piezoceramic crystals, of the Cavitation Research Laboratory from Polytechnic University of Timisoara. The used procedure, specific to the laboratory protocols, is described in [7, 8] according to ASTM G32-2010 international standards and norms [9].

For consistency, three samples were exposed to cavitation erosion for a total duration of 165 min, divided in 12 intermediary durations, as follows: one of 5 and one of 10 min, and 10 durations of 15 min each. At the beginning of the cavitation test and after the end of each testing period, the probes were weighted using an analytical balance, with a precision of 0.01 mg. in order to obtain the mass losses. Also, for accuracy, at the end of the cavitation exposure, before weighing, the samples were washed in a drinking water jet, double distillated water and finally with acetone and there were also dried in a stream of warm water. From the mass losses was determined the mean depth of the cumulative erosion MDE and the mean erosion penetration velocity MDER, using the relations established in our laboratory (see the procedure in [3, 4]) values employed for the diagrams of Figs. 29.5, 29.6 and 29.7. According to ASTM G32-2010 requirements, double-distilled water was used as the liquid environment, at a constant temperature of 22 ± 1 °C, and the surface exposed to cavitation was polished to a roughness of Ra = 0.2 μm.

The effect of the heat treatment (C800/R600) on the behavior and resistance of the brass (CuZn39Pb3) when exposed to cavitation erosion, is highlighted by the evolution of wear in the exposed areas of the three samples, as seen in the photographic images (see Fig. 29.2), and in the SEM image (see Fig. 29.3).

In Fig. 29.2 it can be seen that, after the first 15 min of cavitation attack, the damage to the surface area is mostly characterized by plastic deformations and by generation of pitting/cracks and, after 90 min, the eroded area expands its surface and depth. As the duration of the exposure to cavitation increases, and until the test is complete, the erosion continues mainly with the depth increasing damage, while the number of crevices increases mainly towards the edges of the eroded area (see also Fig. 29.3).

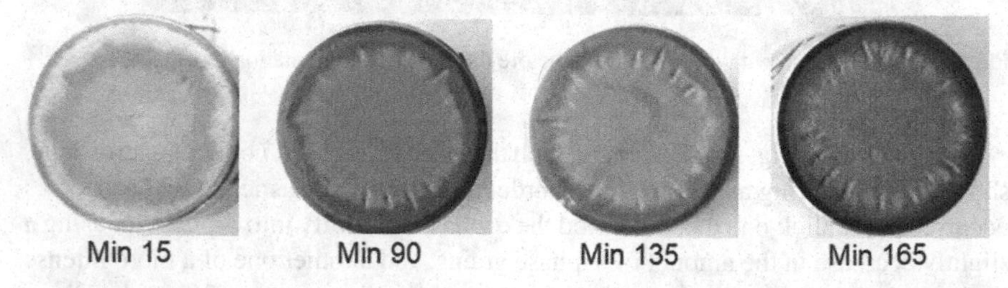

Fig. 29.2 The evolution of the eroded area according to the cavitation exposure duration (recorded with a Canon A480 camera)

Fig. 29.3 SEM and macroscopic images of the eroded microstructure, after 165 min of exposure to cavitation (images taken from the central area of the surface exposed to cavitation)

The SEM image (see Fig. 29.3) shows the crack and tearing propagation, by the formation of crevices, as a response of the structure to strain caused by the impact of the microjets, generated by the implosion of the cavitation bubble.

In Figs. 29.4 and 29.5 are presented the results of the cavitation experiments. Those are expressed by the experimental values of the three tested samples and by mediation curves MDE(t) and MDER(t), developed analytically with the equations of the laboratory [10, 11]. These diagrams also show the values of the parameters MDE_{max}, $MDER_{max}$ and $MDER_s$, as recommended by the ASTM G32-2010 international norms, and used in the laboratory protocol [12].

By analyzing the dispersion of the experimental values, for an average cumulated depth MDE (see Fig. 29.4) and for the average erosion rate MDER (see Fig. 29.5), it is shown that the hardness of 133.8 HV 0.5 and the modified structure, obtained after the heat treatment, determine this behavior and resistance. The 0.16 μm/min difference (see Fig. 29.5) between the final $MDER_s$ value and the maximum $MDER_{max}$ value it is

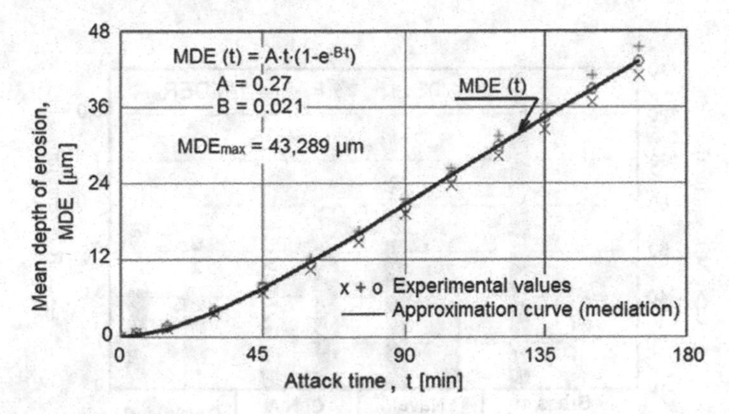

Fig. 29.4 The evolution of the mean depth of erosion with the duration to cavitation exposure

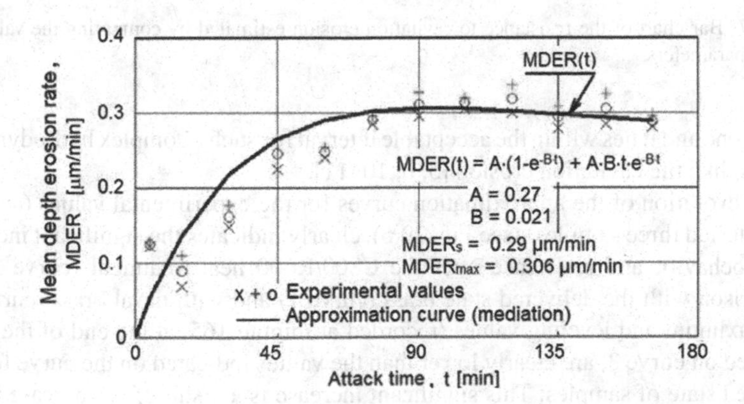

Fig. 29.5 The evolution of the mean depth of penetration rate, with the duration to cavitation exposure

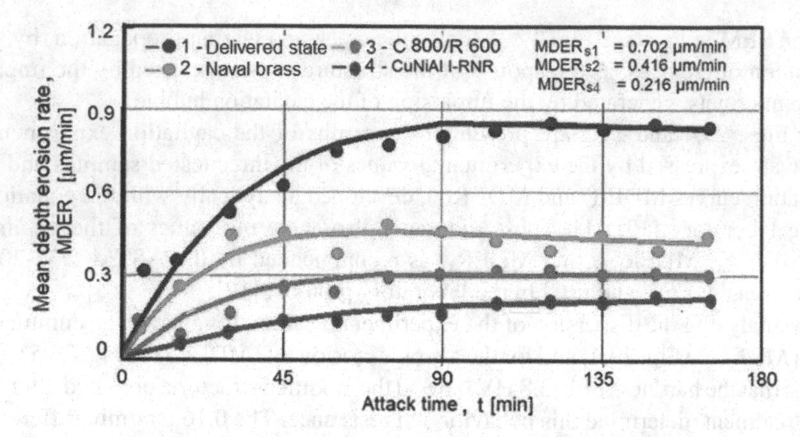

Fig. 29.6 Comparison of the mean depth erosion rates variations with the cavitation exposure duration, only for the naval brass

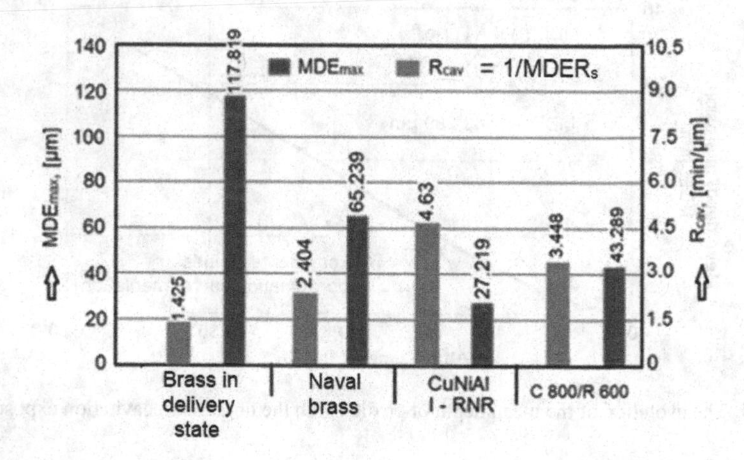

Fig. 29.7 Bar chart of the resistance to cavitation erosion estimated by comparing the values of specific parameters

normal one and it lies within the acceptable interval for such a complex hydrodynamic process, like the cavitation erosion [5, 6, 10, 11].

The evolution of the approximation curves for the experimental values (average for the tested three samples) (see Fig. 29.6), clearly indicates the significant increase in the behavior and resistance after the C800/R600 heat treatment (curve 3), in comparison with the delivered state brass (curve 1) and with naval brass (curve 2). The maximum and leveling values (recorded at minute 165, at the end of the test), indicated on curve 3, are clearly lower than the values indicated on the curve for the delivered state of samples. This significant increase is a result of an increase in the HV0.5 hardness and the structure resulted from the heat treatment.

The bar chart (see Fig. 29.7) shows the values of the parameters used in the Cavitation Research Laboratory at the Polytechnic University from Timisoara [12], and recommended by the ASTM G32 [9] international norms, in order to evaluated by comparison the resistance of a surface to cavitation erosion.

The comparisons (see Fig. 29.6) and the bar chart (see Fig. 29.7) shows that, compared to the characteristics of the delivered state, the average erosion depth MDE_{max} is around 2.72 lower, and the speed of cavitation erosion $MDER_s$, respectively the resistance to cavitation erosion $R_{cav} = 1/MDER_s$ are around 2.41 lower.

When compared with the two reference materials, naval brass and bronze CuNiAl I-RNR, (used for the propellers of the river and sea ships), the heat treated brass (C800/600) has a distinct behavior. Therefore:

- the comparison with the naval brass shows that the average erosion depth MDE_{max} decreases by about 51%, while the resistance to cavitation erosion-corrosion R_{cav} increases by about 43%.
- the comparison with the bronze CuNiAl I-RNR shows that the average erosion depth MDE_{max} increases by about 59%, while the resistance to cavitation erosion R_{cav} decreases by about 34%.

This observation indicates the need to pursue further studies, with other types of treatments, using different parameters of the technological process that will lead to structures and mechanical properties that ensures an increase of the CuZn39Pb3 resistance to cavitation erosion generated by the high intense regimes of the hydrodynamic cavitation [6], superior to the naval bronze CuNiAl I-RNR.

29.4 Conclusions

The evolution of the eroded area in relation to the duration of the cavitation test, with the formation of the outer ring of erosion and of the star-shaped crevices in the exposed sample surface, respects the specific mechanism of the vibratory cavitation.

The SEM images show that the crack grids start at the boundary between the grains. The degree of damage in the surface exposed to cavitation depends on the structure type, as well as on the hardness level resulted from the applied heat treatment.

By comparing the state obtained by volumic heat treatment, with the delivered state (semi-finished) brass CuZn39Pb3 and with naval brass, by the evolution of the characteristic curves and of the reference parameters values, the applied heat treatment leads to an increased resistance to cavitation demands.

The comparison of the state resulted from the volumic heat treatment, with the naval bronze CuNiAl I-RNR indicates the need to pursue further analysis using different parameters for the volumic heat treatment, in such a way that the increased resistance allows the use of the heat-treated brass in components operating in intense cavitation conditions, such as those occurring at ship propeller operation.

The research results and the assessments performed based on the specific curves and on the values of the reference parameters, MDE_{max} and R_{cav}, confirms that, after applying this heat treatment, the brass CuZn39Pb3 can be used for components operating in moderate-cavitation hydrodynamic conditions.

References

1. Lazar, I.: Techniques for optimizing the resistance to cavitation erosion of Cu-Zn and Cu-Sn alloys. PhD thesis. Timisoara (2020)
2. Lazar, I., Bordeasu, I., Popoviciu, M.O., Mitelea, I., Craciunescu, C.M., Pirvulescu, L.D., Sava, M., Micu, L.M.: Evaluation of the brass CuZn39Pb3 resistance at vibratory cavitation erosion. IOP Conf. Ser.: Mater. Sci. Eng. **477**, 012002-1–012002-9 (2019)
3. Bordeaşu, I., Popoviciu, M.O., Patrascoiu, C., Bălăsoiu, V.: An analytical model for the cavitation erosion characteristic curves. Trans. Mech. **49**(63), 253–258 (2004)
4. Micu, L.M., Bordeasu, I., Popoviciu, M.O.: A new model for the equation describing the cavitation mean depth erosion rate curve. Rev. Chim. **68**(4), 894–898 (2017)
5. Mitelea, I., Ghera, C., Bordeasu, I., Craciunescu, C.M.: Ultrasonic cavitation erosion of a duplex treated 16MnCr5 steel. IJMR **106**(4), 391–397 (2015)
6. Bordeaşu, I.: Cavitation erosion on materials used in the construction of hydraulic machines and naval propellers. Scale effects. PhD thesis. Timişoara (1997)
7. Standard method of vibratory cavitation erosion test. ASTM, Standard G32 (2010)
8. Mitelea, I., Bordeasu, I., Hadar, A.: The effect of nikel from stainless steels with 13% chromium and 0.10% carbon on the resistance of erosion by cavitation. Rev. Chim. **56**(11), 1169–1174 (2005)
9. Franc, J.P., Kueny, J.L., Karimi A., Fruman, D.H., Fréchou, D., Briançon-Marjollet, L., Yves Billard, J.Y., Belahadji, B., Avellan, F., Michel, J.M.: La cavitation. Mécanismes physiques et aspects industriels. Press Universitaires de Grenoble, Grenoble (1995)
10. Li, X.-Y., Yan, Y.-G., Xu, Z.-M., Li, H.-G.: Cavitation erosion behavior of nickel-aluminum bronze weldment. Trans. Nonferrous Met. Soc. China **13**, 1317–1324 (2003)
11. Hobbs, J.M.: Experience with a 20 – KC cavitations erosion test, erosion by cavitations or impingement. ASTM STP 408, Atlantic City (1960)
12. Bordeasu, I.: Monograph of the Cavitation Erosion Research Laboratory of the Polytechnic University of Timisoara (1960–2020). Editura Politehnica, Timisoara (2020)

Chapter 30
MCDM Approach in Choosing the Optimal Composite Shaft Material—Application of SAW Method

Zorica Djordjevic, Sasa Jovanovic, Sonja Kostic, Amra Talic-Cikmis, and Danijela Nikolic

Abstract In this paper, the selection of materials for production of shafts is considered from the aspect of the influence of a large number of parameters (mechanical characteristics of the material, density or mass of the shaft, fiber orientation angle in composite materials, shaft bending displacement values, free frequencies, etc.). In order to include all the above criteria in the selection of materials, the approach of Multi-Criteria Decision-Making (MCDM) using the Simple Additive Weighting Method (SAW) was applied. The analysis was performed for four different shaft materials (steel, aluminum, Carbon Fiber Reinforced Polymers CFRP and Glass Fiber Reinforced Polymers GFRP). The results of the analysis showed that the best characteristics, in addition to steel shafts, have composite shafts made of CFRP and GFRP with fiber orientation 0° and that they can be their adequate replacement, especially in systems where the weight of the structure is to be as small as possible.

Keywords Composite materials · Shafts · SAW method

30.1 Introduction

Composite materials are a combination of two or more materials with different mechanical and physico-chemical properties in order to obtain materials with improved characteristics in relation to the constituent components. Some of the advantages of composite concerning to traditional metallic materials are reflected in increased tensile strength, impact resistance, vibration, fatigue, temperature changes,

---

Z. Djordjevic · S. Jovanovic · D. Nikolic
Faculty of Engineering, University of Kragujevac, Sestre Janjic 6, 34000 Kragujevac, Serbia

S. Kostic (✉)
Department in Kragujevac, Academy of Professional Studies Sumadija, Kosovska 8, 34000 Kragujevac, Serbia
e-mail: skostic@asss.edu.rs

A. Talic-Cikmis
Faculty of Mechanical Engineering, University of Zenica, Fakultetska 1, 72000 Zenica, Bosnia and Herzegovina

M. Rackov et al. (eds.), *Machine and Industrial Design in Mechanical Engineering*, Mechanisms and Machine Science 109,
https://doi.org/10.1007/978-3-030-88465-9_30

reduced weight and the like. The specificity of composite materials is that these characteristics can vary depending on the amount and type of material used, and the angle of orientation of the fibers in the composite.

Composites consist of fibers and a matrix. Fibers (carbon, glass, aramid, etc.) form the supporting part of the composite, while the matrix (metal, polymer, ceramic, etc.) serves to bond and transfer loads between the fibers, gives shape to the structure, prevents damage to the fibers, etc.

Composite shafts, in comparison with steel, are characterized by increased load capacity, reduced weight, higher fatigue endurance, extremely harmonious vibration damping, the increased value of critical speed, etc. Carbon or glass fibers in combination with epoxy or polyester resin are most often used to make shafts. Also, hybrid shafts obtained by a combination of metal (steel or aluminum) and a composite material give good results.

When choosing an adequate material for the production of shafts, a large number of material characteristics, ie criteria must be taken into account (physical and mechanical characteristics should be maximum, while the weight, price and weightiness of the production organization should be minimal). Since the problem of material selection is very complex, it is recommended to apply some of the methods of MCDM.

Not so many researchers have proposed a specific analysis model to select the right material for a particular application. In study [1], the selection of suitable natural fibers for resin reinforcement for the production of roto molded product was analyzed, using the technique of MCDM. In particular, the MOORA and TOPSIS methods were used. It was concluded that, as far as natural fibers are concerned, the best properties for this purpose were shown by coconut fibers.

Researchers in [2] dealt with the analysis of mechanical properties of polyamide-based composites in combination with carbon fibers from the aspect of engineering application. An integrated multicriteria approach in material selection was applied, using the fuzzy best–worst method and the fuzzy G-VIKOR method.

The analysis of polymer composite materials reinforced with sugar palm fibers using three different multicriteria methods (AHP, TOPSIS and ELECTRE) was performed by researchers in [3].

Researchers have shown that the application of the method of MCDM can significantly facilitate the choice of materials in the automotive industry [4]. For this purpose, they used the methods of MOORA, TOPSIS and VIKOR.

30.2 Selection of the Composite Shaft Material

The analyzed shaft has an annular cross section 1 m long, 0.05 m middle radius, and 0.005 m wall thickness. The influence of the shaft material and the angle of orientation of the composite fibers on the values of maximum displacements due to bending and free frequencies of the shaft is considered.

Table 30.1 Material property values

Material	Elastic modulus, E_1 (MPa)	Elastic modulus, E_2 (MPa)	Shear modulus, G_{12} (MPa)	Density, ρ (kg/m^3)	Max. deflection, f (mm)	Free frequencies, f_s (Hz)
Steel	210,000	210,000	83,000	7830	0.087	277.6
Aluminum	70,000	70,000	2800	2600	0.262	278.2
$CFRP(0)_4$	130,000	10,000	7000	1500	0.440	426.7
$GFRP(0)_4$	40,300	6200	3000	1900	0.984	232.4
$CFRP(90)_4$	130,000	10,000	7000	1500	1.495	139.8
$GFRP(90)_4$	40,300	6200	3000	1900	2.521	111.7
Type of criteria	Max	Max	Max	Min	Min	Max

The results were obtained numerically using the software FEMAP and NeNastran. The shaft is modeled by isoparametric quadrangular finite elements in the form of multilayer shells. Composite shafts are made as laminates consisting of 4 layers—lamina.

The analysis was performed for the following materials: steel, aluminum, CFRP and GFRP. Composite shafts were considered with a fiber orientation angle of 0° ($CFRP(0)_4$ and $GFRP(0)_4$) and 90° ($CFRP(90)_4$ and $GFRP(90)_4$).

The mechanical characteristics of the analyzed materials as well as the obtained values of maximum displacements and free frequencies of the shaft, are given in Table 30.1.

30.2.1 Application of MCDM Methods in the Selection of Optimal Material

The decision-making process is used in management theory as part of the process of solving a certain problem. Many authors of management theory consider decision-making to be one of the most important tasks, both at the strategic and operational levels. Multi-criteria decision-making (MCDM) is a relatively new discipline, which through its development, should provide support to decision makers who face numerous and very often conflicting influencing factors. This method aims to maximize the quality of the decision following the selected criteria.

Methods of multi-criteria decision-making in the conceptual sense are not particularly complex and are easier to understand than classical single-criteria optimization. The most common division of MCDM procedures presupposes the existence of two basic groups of these methods [5–8], namely the Multiple Attribute Decision Making method (MADM) and the Multiple Objective Decision Making method (MODM). As the choice of one (best) of alternatives (materials) from a limited number of alternatives (materials) is sought within the set problem analyzed in this paper, we opted for the SAW method, which belongs to the group of MADM methods. The SAW (Simple Additive Weighting Method) method is a relatively simple method of multi-criteria decision-making, which takes into account the weight of the selected criteria. For each possible alternative solution, the so-called summary characteristic (A_i)—the value obtained by summing the products of relative weighting factors (W_j') and normalized performance values according to all selected criteria (r_{ij}). The alternative that corresponds to the highest calculated value of the summary characteristic (A^*) is the optimal solution:

$$A^* = \left\{ A_i \middle| \max_i \sum_{j=1}^{n} W'_j r_{ij} \right\} \tag{30.1}$$

30.2.2 Multicriteria Analysis of Material Characteristics

The paper analyzes a total of six different materials (alternatives, solutions) for six different characteristics (criteria), in fact for four materials, two of which have different fiber orientations. In the Table 30.2, the values of the selected characteristics for the considered materials are presented. At the same time, the types of criteria-characteristics are indicated and classified into those of maximization (max, the higher the value, the better the material characteristic) and minimization (min, the smaller the value, the better the material characteristic). Table 30.2 shows the values of normalized factors r_{ij}, obtained by one of the pre-ordered methods of data normalization.

The values of weight coefficients were determined using the Seaty procedure (Tables 30.3, 30.4 and 30.5). In order to examine the stability of the solution—the

Table 30.2 of normalized values of material characteristics (r_{ij})

material	Elastic modulus, E_1 (MPa)	Elastic modulus, E_2 (MPa)	Shear modulus, G_{12} (MPa)	Density, ρ (kg/m^3)	Max. deflection, f (mm)	Free frequencies, f_s (Hz)
Steel	1	1	1	0	1	0.53
Aluminum	0.18	0.31	0	0.83	0.93	0.53
CFRP(0)$_4$	0.53	0.02	0.05	1	0.85	1
GFRP(0)$_4$	0	0	0	0.94	0.63	0.38
CFRP(90)$_4$	0.53	0.02	0.05	1	0.42	0.09
GFRP(90)$_4$	0	0	0	0.94	0	0
Type of criteria	1	1	1	0	1	0.53

Table 30.3 Values of weighting coefficients (Seaty scale-procedure), variant 1

	Variant 1 (W_{i1}')							
	k_1	k_2	k_3	k_4	k_5	k_6	$\sum$	W_{i1}'
k_1	1	3	3	1	0	1	9	0.15
k_2	0	1	1	0	0	0	2	0.033333
k_3	0	1	1	0	0	0	2	0.033333
k_4	1	3	3	1	0	1	9	0.15
k_5	5	7	7	5	1	5	30	0.5
k_6	1	3	3	1	0	1	8	0.133333
$\sum$							60	1

Table 30.4 Values of weighting coefficients (Seaty scale-procedure), variant 2

	Variant 2 (W_{i2}')							
	k_1	k_2	k_3	k_4	k_5	k_6	$\sum$	W_{i2}'
k_1	1	1	1	1	1	1	6	0.166667
k_2	1	1	1	1	1	1	6	0.166667
k_3	1	1	1	1	1	1	6	0.166667
k_4	1	1	1	1	1	1	6	0.166667
k_5	1	1	1	1	1	1	6	0.166667
k_6	1	1	1	1	1	1	6	0.166667
$\sum$							36	1

Table 30.5 Values of weighting coefficients (Seaty scale-procedure), variant 3

	Variant 3 (W_{i3}')							
	k_1	k_2	k_3	k_4	k_5	k_6	$\sum$	W_{i3}'
k_1	1	3	3	0	0	0	7	0.090909
k_2	0	1	1	0	0	0	2	0.025974
k_3	0	1	1	0	0	0	2	0.025974
k_4	5	7	7	1	1	1	22	0.285714
k_5	5	7	7	1	1	1	22	0.285714
k_6	5	7	7	1	1	1	22	0.285714
$\sum$							77	1

choice of the optimal material, an analysis was performed for three different variants of weight coefficients, depending on the preference of certain characteristics.

On Figs. 30.1 and 30.2 the summary characteristics (A_i) for all three variants of weight coefficients for all six potential shaft materials are given. By analyzing both diagrams, it can be seen that steel has the best cumulative characteristic, in two

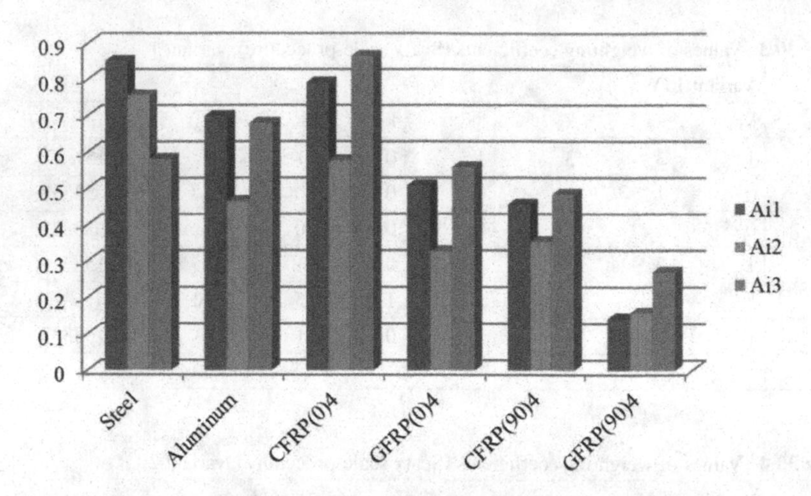

Fig. 30.1 Diagram representation of summary characteristics for three variants of performance weights—criteria for six shaft materials

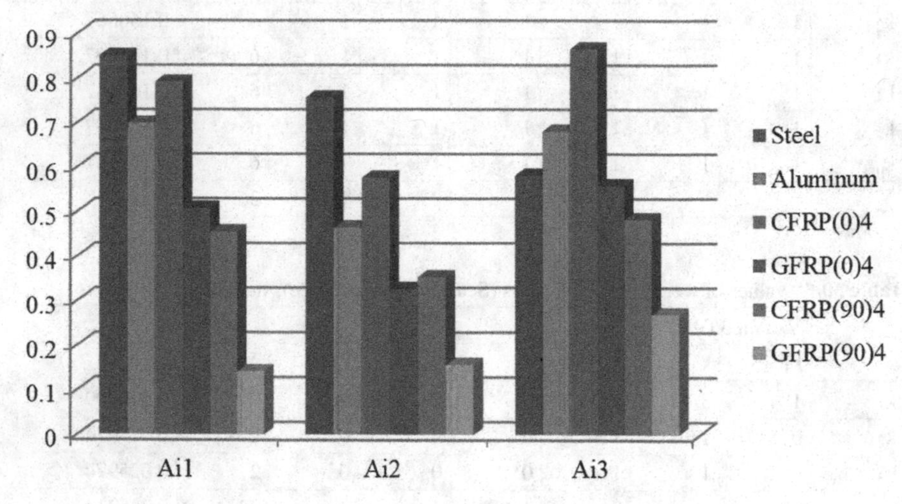

Fig. 30.2 Diagram representation of summary characteristics for three variants of performance weights—criteria

of the three cases, but that the average value is very close to the material made of polymer-reinforced carbon fibers ($CFRP(0)_4$).

30.3 Conclusion

The process of selecting the optimal material represents a significant and sensitive phase in the product design process. Within this phase, the constructor must, in accordance with the available material characteristics and the requirements of the potential construction, make an appropriate decision using an adequate method. Solving this type of design problem is most often done through the application of some of the methods of multi-criteria decision-making (MCDM). The paper specifically applies the method of additive weight factors known as the SAW method. Six selected materials (two "classic", steel and aluminum and four composite) presented with six different characteristics were analyzed. The characteristics of the material were treated as criteria, which, depending on the assessment and affinity of the constructor, were assigned different weight coefficients in the three presented variants. Based on a comparative analysis of the distribution of weight coefficients, it was concluded that of the six materials considered, the best characteristics (observing all three variants with different weight coefficients), during bending, showed shafts made of steel and composite of carbon fibers reinforced with polymers (CFRP$(0)_4$). In each of the three variants of weight coefficients these two materials showed better characteristics than the others.

Acknowledgements This is result of the TR33015 project, which is investigation of the Technological Development of the Republic of Serbia. The project is titled "Research and development of a Serbian net-zero energy house". We would like to thank to the Ministry of Education, Science and Technological Development of the Republic of Serbia for their financial support during this investigation.

References

1. Gupta, N., Ramkumar, P.L., Abhishek, K.: Material selection for rotational molding process utilizing distinguished multi criteria decision making techniques. Mater. Today-Proc. **44**(Part 1), 1770–1775 (2021)
2. Zhang, H., Wu, Y., Wang, K., Peng, Y., Wang, D., Yao, S., Wang, J.: Materials selection of 3D-printed continuous carbon fiber reinforced composites considering multiple criteria. Mater. Design 196, 109140-1–109140-13 (2020)
3. Alaaeddin, M.H., Sapuan, S.M., Zuhri, M.Y.M., Zainudin, E.S., AL-Oqla, F.M.: Polymer matrix materials selection for short sugar palm composites using integrated multi criteria evaluation method. Compos. Part B-Eng. **176**, 107342-1–107342-11 (2019)
4. Moradian, M., Modanloo, V., Aghaiee, S.: Comparative analysis of multi criteria decision making techniques for material selection of brake booster valve body. J. Traffic Transp. Eng. **6**(5), 526–534 (2019)
5. Chai, J., Liu, J., Ngai, E.: Application of decision making techniques in supplier selection: a systematic review of literature. Expert Syst. Appl. **40**(10), 3872–3885 (2013)
6. Kaplan, P.O.: A New Multiple Criteria Decision Making Methodology for Environmental Decision Support. PhD Thesis. Faculty of North Carolina State University, USA. (2006)

7. Polatidis, H., Haralambopoulos, D.A., Munda, G., Vreeker, R.: Selecting an appropriate multi-criteria decision analysis technique for renewable energy planning. Energy Source Part B **1**(2), 181–193 (2006)
8. Yoon, K.: Systems Selection by Multiple Attribute Decision Making. PhD Thesis. Kansas State University (1980)

Chapter 31
Experimental Analysis for Defining Mechanical Properties of Steel Sheet Metal on Different Material Thickness

Florinda Sejfullai, Tasuli Taleski, Bojan Mitev, and Atanas Kochov

Abstract The present study investigates the tensile properties of DX51D on different specimen thicknesses. The experimental study was conducted using rectangular specimens with different thicknesses ranging from 0.6 to 1.2 mm, a gauge length of 100 mm and a width of 12 mm. The results show that the ultimate tensile strength and yield strength or the material are not significantly affected by the difference in specimen thickness. For greater specimen thickness, the elongation of the material appears to increase.

Keywords Tensile test · Tensile properties · Steel

31.1 Introduction

Tensile properties such as strength parameters, like ultimate tensile strength and yield strength and ductility parameters, like elongation and area reduction are often necessary to be measured in order to determine the tensile properties of materials, and for this, specimens with a rectangular cross-section are commonly used in standard methods of testing. The shape and geometric size of the specimens can be chosen according to the ASTM standards, however, in many cases, the specimens don't always correspond to the ASTM standard, and non-standard tensile specimens which vary in thickness and geometry (shape) are often used to examine the tensile behavior of the material. The ASTM standard does not give an exact requirement for the thickness of the specimen, and the thickness can vary over a wide range [1].

The tensile strength is determined with a tensile test (a standard method of testing in accordance with the ISO 6892 series of standards for metals) [2].

The tensile properties and the effects of the thickness of a rectangular specimen have been investigated. The data results are plotted in a stress–strain curve which shows the material's reaction to the forces being applied. Rectangular specimens

F. Sejfullai (✉) · T. Taleski · B. Mitev · A. Kochov
Faculty of Mechanical Engineering Skopje, Ss. "Cyril and Methodius" University Skopje, St. Rugjer Boskovic 18, Skopje, Republic of North Macedonia

M. Rackov et al. (eds.), *Machine and Industrial Design in Mechanical Engineering*, Mechanisms and Machine Science 109,
https://doi.org/10.1007/978-3-030-88465-9_31

with different thicknesses have been used to examine the tensile properties and the effects of the thickness of a rectangular specimen on the results.

31.2 Materials and Experimental Method

In this study, the tensile behavior of rectangular (dog-bone) tensile specimens with different thicknesses of 0.6, 0.8, 1.0 and 1.2 mm were investigated. Because of its technical application, the material used in this study is DX51D steel. The parameters of the specimens investigated in the study are given in Table 31.1.

The quasi-static tensile tests were performed in the accredited laboratory for testing of mechanical properties LT-04 at the Faculty of Mechanical Engineering in Skopje on a SHIMADZU AGX-V series material testing machine with a load cell capacity of 10 kN, measurement accuracy of ±1% of indicated test force, displacement speed of 5 mm/min, at room temperature. The specimens were clamped using pneumatic grips equipped with flat serrated jaw faces, as shown in Fig. 31.1.

Table 31.1 Specimen parameters

Group	Steel grade	Thickness (mm)	Gauge length (mm)	Width (mm)
I	DX51D	0.6	100	12
II	DX51D	0.8	100	12
III	DX51D	1.0	100	12
IV	DX51D	1.2	100	12

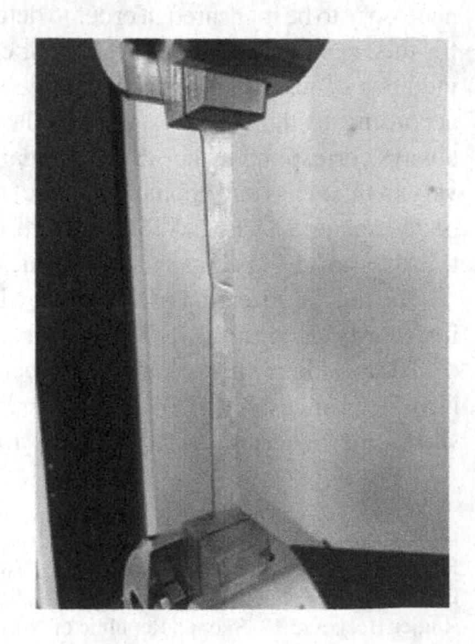

Fig. 31.1 Specimen alignment using pneumatic grips

During testing the value of the load and elongation were recorded continuously. The recorded data for each specimen were obtained from the data gathering software TRAPEZIUMX. The linear relationship of load and extension was then used for the calculation of the engineering stress and strain and plotting the Stress–Strain diagrams. The main parameters that were calculated are: tensile strength (maximum engineering stress), yield strength (defined as the 0.2% offset yield stress) and elongation.

31.3 Experimental Results and Discussion

The tensile tests were performed on the specimens shown in Fig. 31.2. Figure 31.3 shows the experimental results recorded by the software TRAPEZIUMX of the specimens with a thickness of 0.6 mm. The average displacement (elongation) for the tested specimens is 22.7 mm.

The obtained results for all the specimens in the first group are plotted in the Stress–Strain curves shown in Fig. 31.4 and Table 31.2.

The Yield Strength of the tested specimens is in the range from 346 to 356 N/mm^2. The average values for Ultimate Tensile Strength and Yield Strength are 397 MPa and 350 MPa, respectively. It is evident that all four specimens with a thickness of 0.6 mm showed almost identical mechanical characteristics with minimal differences.

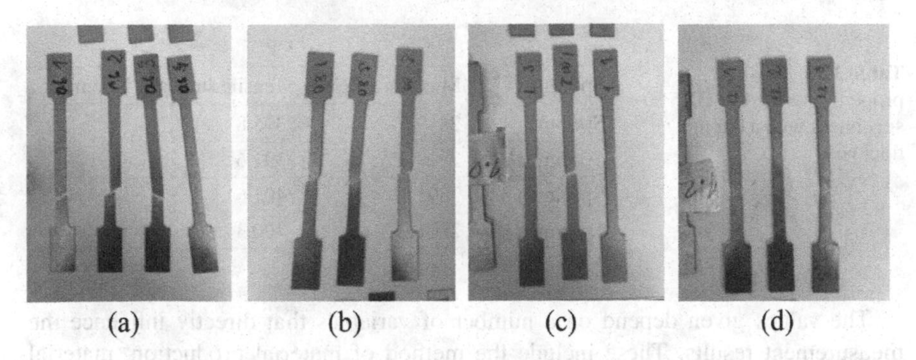

Fig. 31.2 Snapshots after the tensile stress of specimens with different thickness: **a** 0.6 mm; **b** 0.8 mm; **c** 1.0 mm; **d** 1.2 mm

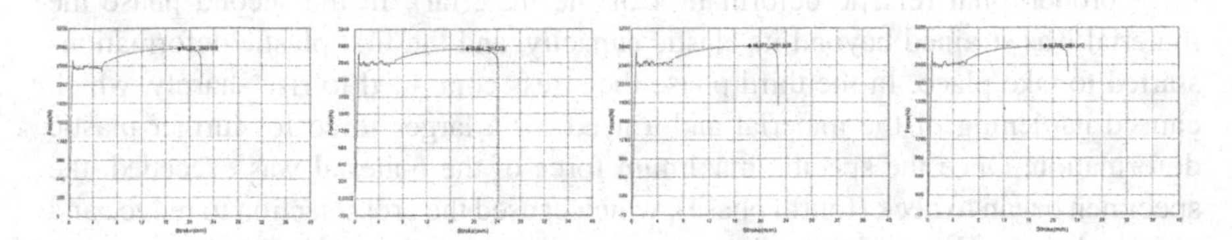

Fig. 31.3 Group I: Force and stroke history plots for DX51D specimens with a thickness of 0.6 mm

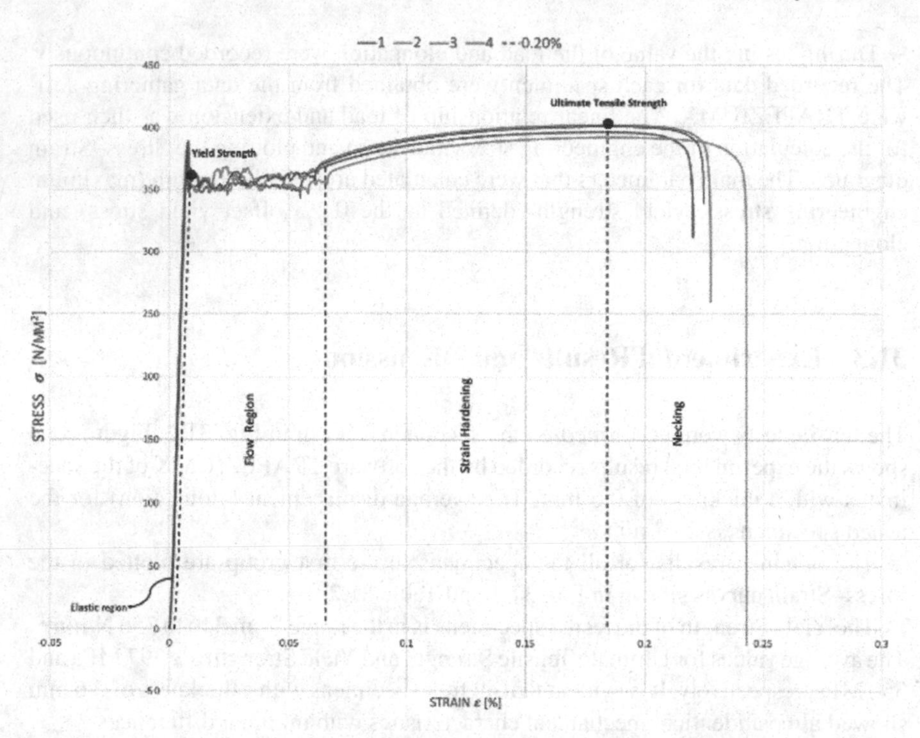

Fig. 31.4 Stress versus strain curves for DX51D specimens with a 0.6 mm thickness

Table 31.2 Tensile properties results for DX51D specimens with a 0.6 mm thickness

Specimen	Max. force (N)	Tensile strength (N/mm^2)
Specimen 1	2849.5	395.8
Specimen 2	2819.3	391.6
Specimen 3	2891.8	401.6
Specimen 4	2881.9	400.3

The values given depend on a number of variables that directly influence the measurement results. These include the method of material production, material composition, microscopic imperfections and temperature. For this reason the diagram of each specimen has slight differences. In the first phase of the test, stress and strain were proportional (elastic deformation of the material). In the second phase the material was strained beyond its elastic capacity, and the first plastic deformations started to take place. In the third phase the stress continued to rise sharply, which caused hardening of the material and a need for a larger force for further plastic deformation. Once the specific maximum force of the material was exceeded, the specimen began to neck (fourth phase), which caused the cross-section to reduce and ultimately tear. The four main phases were present in each group of specimens.

Figure 31.5 shows the experimental results recorded by the software TRAPEZIUMX of the specimens with a thickness of 0.8 mm.

The average displacement (elongation) for the tested specimens is 26.2 mm. The obtained results for all the specimens in the second group are plotted in the following Stress–Strain curves, shown in Fig. 31.6 and Table 31.3.

The Yield Strength of the tested specimens is in the range from 332 to 345 N/mm^2. The average values for Ultimate Tensile Strength and Yield Strength are 396 MPa and 340 MPa, respectively.

Figure 31.7 shows the experimental results recorded by the software TRAPEZIUMX of the specimens with a thickness of 1.0 mm. The average displacement (elongation) for the tested specimens is 29.7 mm.

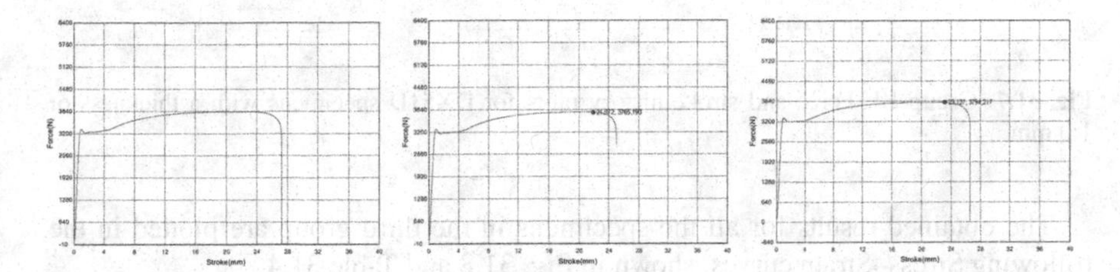

Fig. 31.5 Group II: Force and stroke history plots for DX51D specimens with a thickness of 0.8 mm

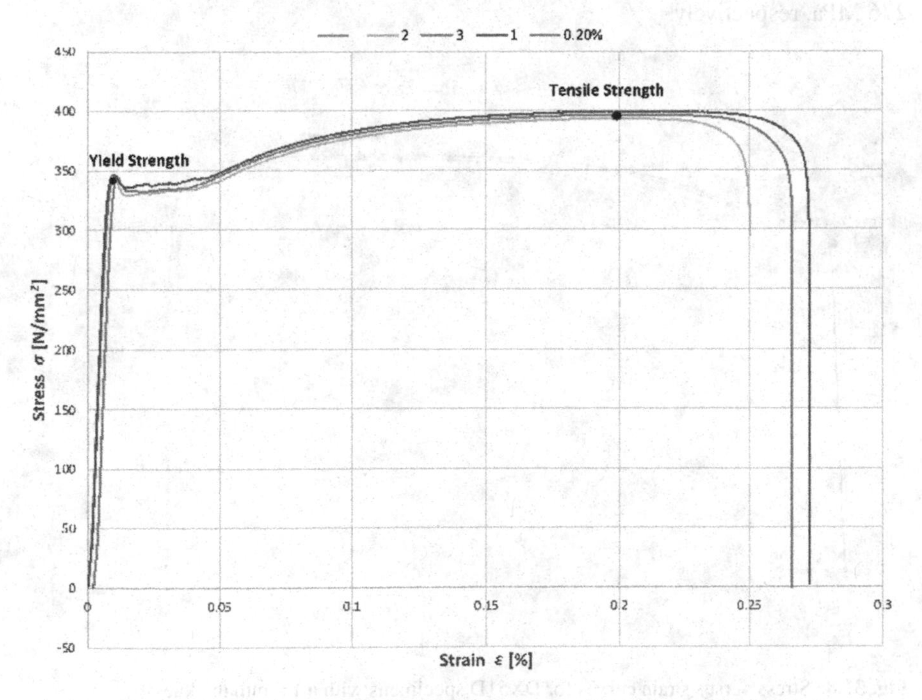

Fig. 31.6 Stress versus strain curves for DX51D specimens with a 0.8 mm thickness

Table 31.3 Tensile properties results for DX51D specimens with a 0.8 mm thickness

Specimen	Max. force (N)	Tensile strength (N/mm²)
Specimen 1	3832.4	399.2
Specimen 2	3774.8	393.2
Specimen 3	3805.1	396.4

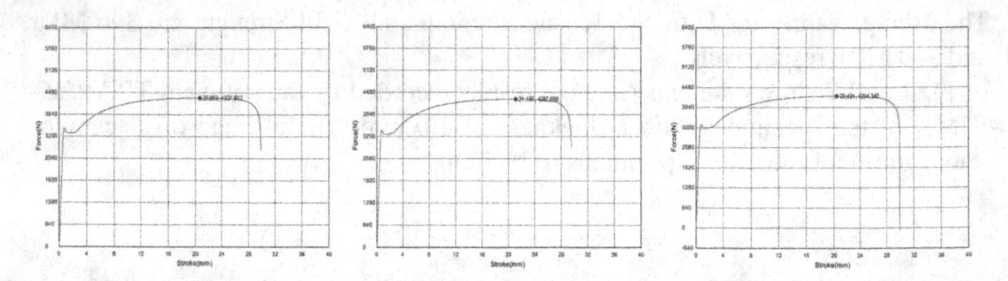

Fig. 31.7 Group III: Force and stroke history plots for DX51D specimens with a thickness of 1.0 mm

The obtained results for all the specimens in the third group are plotted in the following Stress–Strain curves, shown in Fig. 31.8 and Table 31.4.

The Yield Strength of the tested specimens is in the range from 272 to 280 N/mm^2. The average value for Ultimate Tensile Strength and Yield Strength is 356 MPa and 276 MPa, respectively.

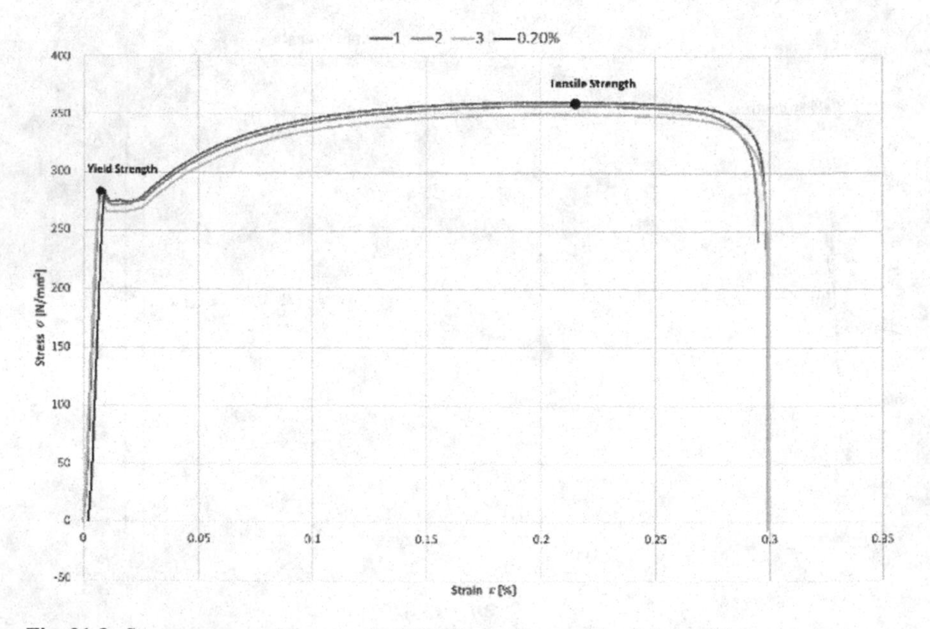

Fig. 31.8 Stress versus strain curves for DX51D specimens with a 1.0 mm thickness

Table 31.4 Tensile properties results for DX51D specimens with a 1.0 mm thickness

Specimen	Max. force (N)	Tensile strength (N/mm^2)
Specimen 1	4327.9	360.6
Specimen 2	4287.6	357.3
Specimen 3	4204.3	350.4

Figure 31.9 shows the experimental results recorded by the software TRAPEZIUMX of the specimens with a thickness of 1.2 mm. The average displacement (elongation) for the tested specimens is 32.7 mm.

The obtained results for all the specimens in the fourth group are plotted in the Stress–Strain curves shown in Fig. 31.10 and Table 31.5.

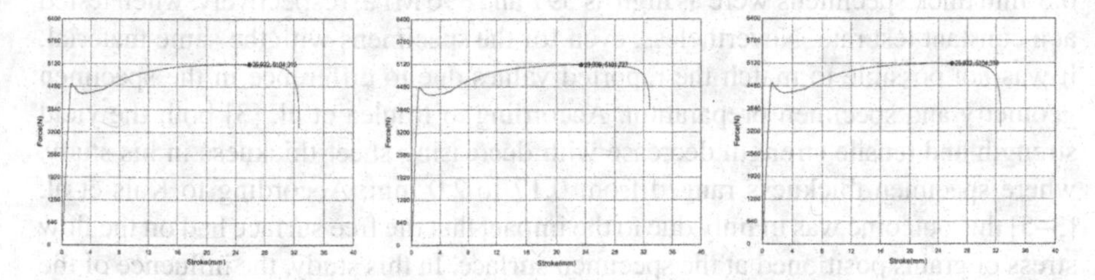

Fig. 31.9 Group IV: Force and stroke history plots for DX51D specimens with a thickness of 1.2 mm

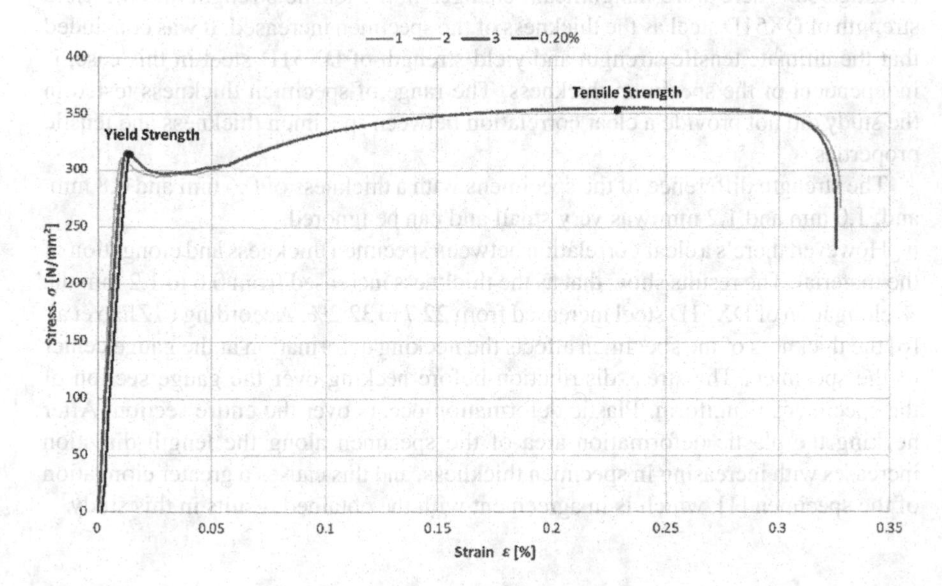

Fig. 31.10 Stress versus strain curves for DX51D specimens with a 1.2 mm thickness

Table 31.5 Tensile properties results for DX51D specimens with a 1.2 mm thickness

Specimen	Max. force (N)	Tensile strength (N/mm^2)
Specimen 1	5111.2	355.9
Specimen 2	5101.7	354.3

The Yield Strength of the tested specimens is in the range from 308 to 312 N/mm^2. The average value for Ultimate Tensile Strength and Yield Strength is 355 MPa and 310 MPa, respectively.

The experimental results indicate that for a constant test rate, the tensile strength decreases with increasing the specimen thickness. However, evaluating the results' plots, it is possible to note that the maximum tensile strength for the 1.0- and 1.2-mm specimen was 356 and 355 MPa, respectively. The tensile strengths for 0.6- and 0.8-mm thick specimens were as high as 397 and 396 MPa, respectively, when tested at a constant test rate. Nevertheless, even for the specimens with the same material, it was not possible to match the reported values due to difference in the specimen geometry and specimen preparation. According to Raulea et al. [3] both the yield strength and tensile strength decrease with decreasing sheet thickness in his study, where specimen thickness ranged from 0.17 to 2.0 mm. According to Kals et al. [3–5] this outcome was mainly due to the impact that the free surface had on the flow stress of grains positioned at the specimen surface. In this study, the influence of the free surface on the tensile properties was not significant for the range of thicknesses tested, therefore the effect of surface stress on the tensile properties is not further investigated. There were insignificant changes in the tensile strength and the yield strength of DX51D steel as the thickness of the specimen increased. It was concluded that the ultimate tensile strength and yield strength of DX51D steel in this case, is independent of the specimen thickness. The range of specimen thickness tested in the study did not provide a clear correlation between specimen thickness and tensile properties.

The strength difference of the specimens with a thickness of 0.6 mm and 0.8 mm, and, 1.0 mm and 1.2 mm was very small and can be ignored.

However, there's a clear correlation between specimen thickness and elongation of the material. The results show that as the thickness increased from 0.6 to 1.2 mm the % elongation of DX51D steel increased from 22.7 to 32.2%. According to Zhao et al. [6] the thickness of the specimen affects the necking deformation at the gauge center of the specimen. The stress distribution before necking over the gauge section of the specimens is uniform. Plastic deformation occurs over the entire section. After necking the plastic deformation area of the specimen along the length direction increases with increasing in specimen thickness, and this causes a greater elongation of the specimen [1], which is in agreement with the obtained results in this study.

Table 31.6 Summary table—average values of the tensile properties

Group	Tensile strength (N/mm^2)	Yield strength (N/mm^2)	Displacement (mm)
I	397	350	22.7
II	396	340	26.2
III	356	276	29.7
IV	355	310	32.7

31.4 Conclusion

In this study, uniaxial tests were performed on DX51D steel in order to measure the tensile properties and address different specimen thicknesses to investigate the effect of material thickness on the tensile properties.

The different specimen thicknesses in the tested range did not cause a significant difference in the obtained results for the ultimate tensile strength and yield strength of the material, therefore we can conclude that the specimen thickness did not influence the ultimate tensile strength and yield strength of DX51D steel. The average values are summarized in Table 31.6.

References

1. Yuan, W.J., Zhang, Z.L., Su, Y.J., Qiao, L.J., Chu, W.Y.: Influence of specimen thickness with rectangular cross-section on the tensile properties of structural steels. Mat. Sci. Eng. A Struct. **532**, 601–605 (2012)
2. ISO 6892-1:2019. https://www.iso.org/standard/78322.html, last accessed 2019/11/01
3. Raulea, L.V.: Size effects in the processing of thin metal sheets. J. Mater. Proc. Technol. **115**, 44–48 (2001)
4. Kals, R., Vollertsen, F., Geiger, M.: Scaling effects in sheet metal forming. In: Proceedings of the 4th International Conference on Sheet Metal, SheMet 1996, pp. 65–75. Enschede (1996)
5. Kals, R.T.A., Eckstein, R., Geiger, M.: Miniaturization in metal working. In: Proceedings of the 6th International Conference on Sheet Metal, SheMet 1998, pp. 15–24. Enschede (1998)
6. Zhao, Y.H., Guo, Y.Z., Wei, Q.: Influence of specimen dimensions and strain measurement methods on tensile stress–strain curves. Mater. Sci. Eng. A-Struct. **525**, 68–77 (2009)

Chapter 32
Determination of LCF Plastic and Elastic Strain Components of Steel

Vujadin Aleksić, Ljubica Milović, Srđan Bulatović, Bojana Zečević, and Ana Maksimović

Abstract The behavior of steel in low-cycle fatigue (LCF) is tested experimentally, in accordance with ISO 12106:2017 (E) and/or ASTM E 606-04. For this purpose, smooth specimens which are exposed to low-cycle fatigue at several levels of regulated strains and/or loads at room, elevated or reduced temperatures are used. Stress–strain response at LCF has the shape of an ideal hysteresis loop. The strain range $\Delta\varepsilon$ corresponds to overall loop width, while the stress range $\Delta\sigma$ corresponds to its overall height. The paper presents a method for determining the intersection of the idealized hysteresis loop and the positive part of the strain axis in order to determine the values of elastic, $\Delta\varepsilon_e/2$, and plastic, $\Delta\varepsilon_p/2$, components of the strain amplitude to characterize the behavior of steel under low cyclic fatigue. The values of elastic and plastic components of the strain amplitude are needed to determine the characteristic curves of low-cycle fatigue, which describe the behavior of steel under the loading of low-cycle fatigue.

Keywords LCF—Low cycle fatigue · Strain

32.1 Introduction

Fatigue has the most important role in the fracture of machine parts and steel structures [1]. A large number of structural damages, caused by steel fatigue, leads to catastrophic fractures. It is estimated that such damages represent 50–90% of all damages in operation [2, 3]. Therefore, extensive research has been dedicated to the study of fractures caused by fatigue load and conditions in which cracks appear and grow for more than 160 years [4–8].

V. Aleksić (✉) · S. Bulatović
Institute for Testing of Materials, Bulevar Vojvode Mišića 43, Belgrade, Serbia
e-mail: vujadin.aleksic@institutims.rs

L. Milović
Faculty of Technology and Metallurgy, University of Belgrade, Karnegijeva 4, Belgrade, Serbia

B. Zečević · A. Maksimović
Faculty of Technology and Metallurgy, Innovation Centre, Karnegijeva 4, Belgrade, Serbia

M. Rackov et al. (eds.), *Machine and Industrial Design in Mechanical Engineering*, Mechanisms and Machine Science 109,
https://doi.org/10.1007/978-3-030-88465-9_32

Although more than 10 publications related to material fatigue are published in the world every day [3, 4], the economic impact of fractures of machine parts and structures in the world is significant and amounts to approximately 4% of the gross national product [1, 3]. Numerous examples of catastrophic accidents are known, especially oilfield platform failures [9].

Rapid progress in the field of material fatigue can be tracked by various databases that can contain millions of records [3]. In Serbia, this is made possible through the database of the Consortium of Libraries of Serbia for Unified Procurement (KoBSON).

32.2 Low Cycle Fatigue of Material

Fatigue of material is a phenomenon of gradual destruction, which occurs due to the long-term action of periodic or cyclic variable stress [10], and results in the crack initiation. The area of fatigue with high stress values, corresponding macro-strains and with expressed small number of cycles until the initiation of a fatigue crack is called low-cycle fatigue (LCF). It ranges from just a few cycles to (1–5) $\times 10^4$ cycles (when macro plastic strain become dominant in the fatigue process).

The boundary between low-cycle and high-cycle fatigue is not defined by a precisely defined number of cycles [11]. The biggest difference is that low-cycle fatigue is associated with macro plastic strain in each cycle, while high-cycle fatigue is more related to the elastic behavior of the material on the macro scale.

32.2.1 Low Cycle Fatigue Test of Steel

Low-cycle fatigue behavior of steel is tested experimentally [12–19], in accordance with ASTM E 606-04 (USA) [20] and/or ISO 12106: 2017 (EU) [21]. For this purpose, smooth test specimens which are exposed to low-cycle fatigue at several levels of regulated strains, with an asymmetry factor $R_\varepsilon = \varepsilon_{min}/\varepsilon_{max} = -1$, at room, elevated or reduced temperatures are used.

Stress–strain response at LCF has the shape of an ideal hysteresis loop [11, 22] shown in Fig. 32.1. The strain range $\Delta\varepsilon$ corresponds to overall loop width, while the stress range $\Delta\sigma$ corresponds to its overall height. The stress amplitude equals the stress half-range, $\Delta\sigma/2$.

The strain range $\Delta\varepsilon$ is equal of the total elastic component, $\Delta\varepsilon_e$ and plastic component, $\Delta\varepsilon_p$. By introducing the strain amplitudes expressed over the corresponding half-ranges, the following equation is reached:

$$\frac{\Delta\epsilon}{2} = \frac{\Delta\epsilon_e}{2} + \frac{\Delta\epsilon_p}{2} \tag{32.1}$$

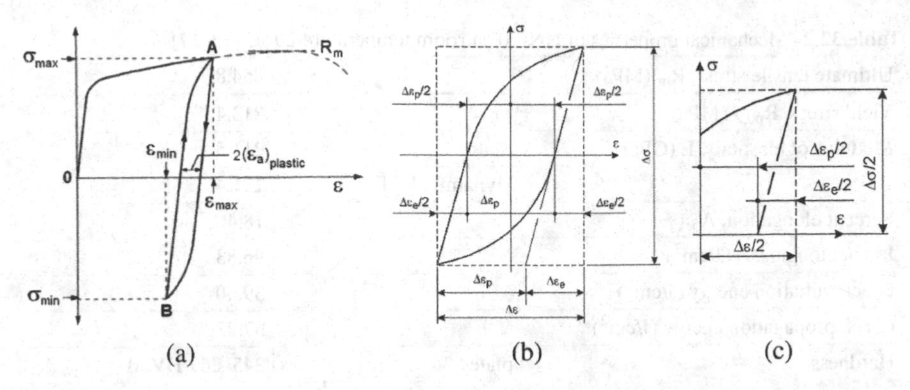

Fig. 32.1 Ideal hysteresis loop [18]: **a** σ_{max} and N_1 [11], **b** σ-ε response and **c** Positive part of loop

32.2.2 Characteristics of Steel NN-70

High strength low alloy steel NIONIKRAL 70 (NN-70), Yugoslav version of American steel HY-100, was used in scientific research for the publication of professional and scientific papers [14–20], which results were also used in this paper in order to present the methodology for determining the plastic and elastic part of LCF strain of steel exposed to low-cycle fatigue. The results used are related to the characteristics of NN-70 steel, and refer to the chemical and mechanical properties, as well as the weldability of this type of steel and are given in Tables 32.1 and 32.2.

For this experiment of low cycle fatigue 10 specimens (dimensions 11 × 11 × 95 mm) were investigated, shown in Fig. 32.2a, b. Also, experiment on total strain amplitudes ranging from 0.35% to 0.80% was performed.

Low cycle fatigue test, in accordance with ISO 12106:2017 (E) [21], was performed on a universal servo-hydraulic MTS machine (rating 500 kN), in the Military Technical Institute in Žarkovo.

The test results of 4 specimens with controlled strain regimes shown in Table 32.3 were considered.

Table 32.1 Chemical composition of NN-70 (%wt.) [12–17]

C	Si	Mn	P	S	Cr
0.106	0.209	0.220	0.005	0.0172	1.2575
Ni	Mo	V	Al	As	Sn
2.361	0.305	0.052	0.007	0.017	0.014
Cu	Ti	Nb	Ca	B	Pb
0.246	0.002	0.007	0.0003	0	0.0009
W	Sb	Ta	Co	N	C_{eq}
0.0109	0.007	0.0009	0.0189	0.0096	0.542

$C_{eq} = C + Mn/6 + Si/24 + Ni/40 + Cr/5 + Mo/4 + V/14$ s.

Table 32.2 Mechanical properties of NN-70, at room temperature 20 °C [12–17]

Property		Value
Ultimate tensile stress, R_m (MPa)		854.8
Yield stress, $R_{p0.2}$ (MPa)		813.4
Modulus of elasticity, E (GPa)	Static	211.5
	Dynamic, LCF	**221.4**
Percent elongation, A_5 (%)		18.4
Impact toughness (J/cm^2)		96.83
Crack initiation energy (J/cm^2)		39.60
Crack propagation energy (J/cm^2)		57.23
Hardness	plate	245–269 HV30
	LCF specimen	252–262 HV10

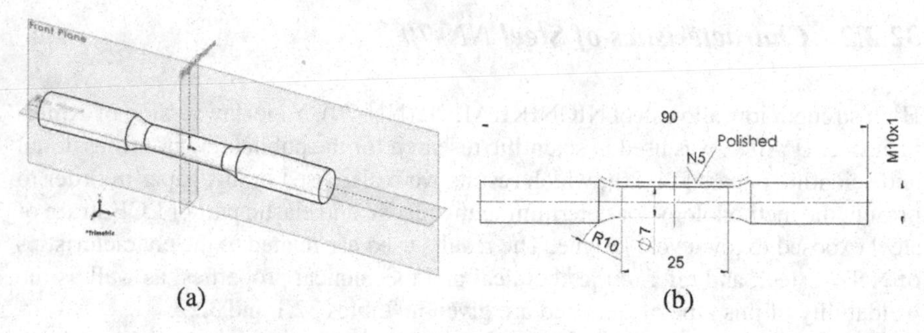

Fig. 32.2 Specimen for LCF test of steel NN-70 [18]: **a** Symmetry in three planes. **b** Type of specimen NN-70 employed in test

Table 32.3 Basic data on controlled strain regimes of LCF test NN-70

Specimen	1	2	3	4	5	6	7
	Δε/2 (%)	Δε/2 (V)	Δε/2 (mm/mm)	Δl (mm)	Δε (%)	T (s)	F (Hz)
	experiment	$\varepsilon[\%] = \varepsilon[V] \cdot 0.2$	1/100	3*25	1*2	experiment	1/6
09	0.35	1.75	0.0035	0.0875	0.70	4.30	0.2326
03	0.50	2.50	0.0050	0.1250	1.00	4.30	0.2326
06	0.60	3.00	0.0060	0.1500	1.20	4.30	0.2326
08	0.80	4.00	0.0080	0.2000	1.60	4.30	0.2326

32.3 Analysis of Low-Cycle Fatigue Test Results

As a result of low-cycle fatigue test on one specimen (one amplitude level of strain), a record in EXCEL, can be further processed according to requirements by using the tools available in EXCEL. Tables 32.4 and 32.5 and Figs. 32.3 and 32.4 were derived from this process.

Figure 32.3a, b show the maximum and minimum values of the specimen load during the N cycle of exposure to low-cycle fatigue. In Fig. 32.3c–f, a linear dependence of the maximum values of the load (stabilization area) was established and the characteristic hysteresis cycles of the beginning of stabilization, N_{bs}, end of stabilization, N_{es} and N_s stabilization cycle (defined by standard [21]) from which all other characteristics of LCF steel are determined. The cycle of significant specimen failure, N_f (defined by the standard [21]) was also determined. Table 32.4 is made from graphic part of Fig. 32.3.

Figure 32.4a shows the characteristic cycles from cycle $N_{1/4}$ to cycle N_f for the total strain amplitude $\Delta\varepsilon/2 = 0.5\%$. In Fig. 32.4b are stabilized hysteresis, N_s, for the total strain amplitudes $\Delta\varepsilon/2 = 0.35$, 0.50, 0.60 and 0.80%. Figure 32.4c shows the relationship between the maximum load and the stabilization cycle, N_s, for each considered amplitude level of strain. From each LCF cycle it's possible to determine the value of plastic and elastic strain (Fig. 32.4d). Figure 32.4d shows linearized sections of the positive part of the x-axis and the part of the hysteresis between the two measuring points. Using the dependence from Eq. 32.1 and the equation from Fig. 32.4d, and finally using the following equations:

Table 32.4 Characteristic processed test data of LCF steel NN-70

LCF NN-70		Stabilization regions		Characteristic cycles of stabilization			
Specimen	$\Delta\varepsilon/2$ (%)	y = F, kN; x = N	R^2	N_{bs}	N_{es}	N_f	$N_s = N_f/2$
09	0.35	$F = -0.0002\,N + 24.30$	0.95	812	6740	8329	4165
03	0.50	$F = -0.0022\,N + 28.57$	0.97	256	1271	1402	701
06	0.60	$F = -0.0057\,N + 29.66$	0.94	127	415	501	251
08	0.80	$F = -0.0162\,N + 30.83$	0.94	50	165	207	104

Table 32.5 Calculated LCF elastic and plastic strain amplitude components of NN-70 steel at characteristic N_s

Sp	y = mx – b; y = 0; x = b/m	N_s	$\Delta\varepsilon/2$ (%)	$\Delta\varepsilon_p/2$ (%)	$\Delta\varepsilon_e/2$ (%)	σ_{max} (MPa)	σ_{min} (MPa)	$\Delta\sigma/2$ (MPa)
09	$\Delta\varepsilon_p = 3.04/61.38$	4165	0.35	0.0495	0.3005	608.14	–689.48	648.81
03	$\Delta\varepsilon_p = 18.74/109.15$	701	0.50	0.1717	0.3283	702.84	–707.19	705.01
06	$\Delta\varepsilon_p = 16.93/76.92$	251	0.60	0.2201	0.3799	736.15	–698.00	717.07
08	$\Delta\varepsilon_p = 27.97/65.04$	104	0.80	0.4301	0.3699	761.87	–709.04	735.46

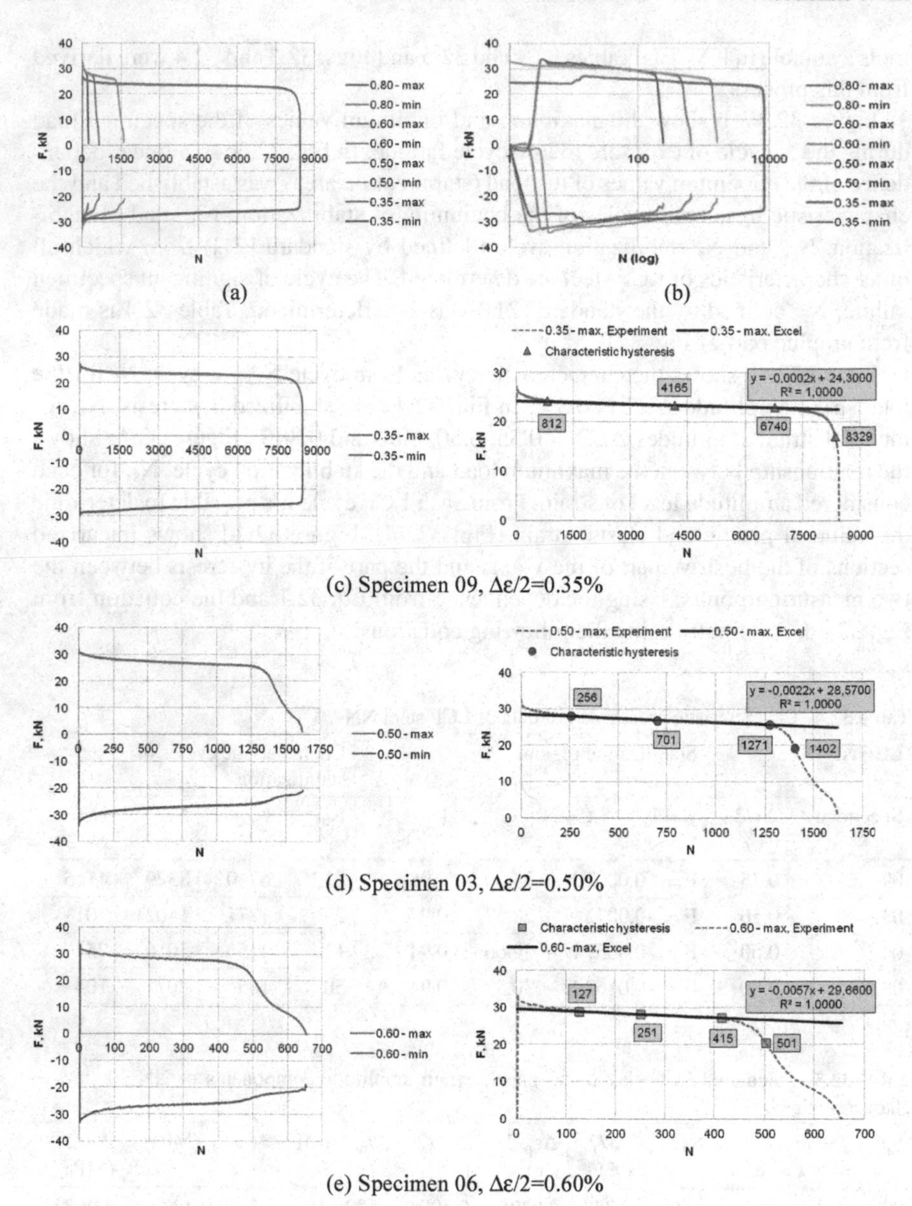

Fig. 32.3 Graphical processing of LCF test results of NN-70 steel

$$y = mx - b \tag{32.2}$$

$$y = F,\ kN;\ x = \Delta\varepsilon_p/2 \tag{32.3}$$

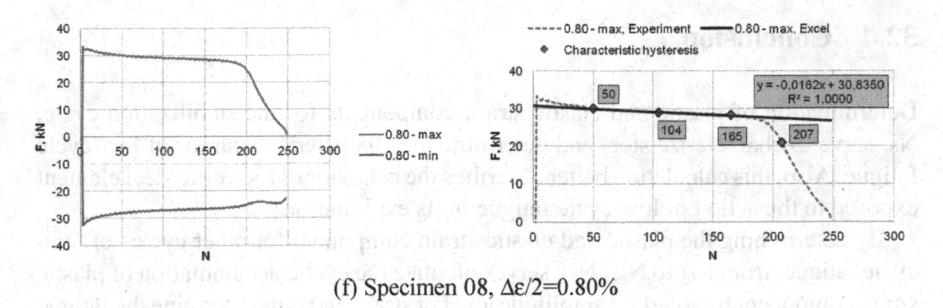

(f) Specimen 08, Δε/2=0.80%

Fig. 32.3 (continued)

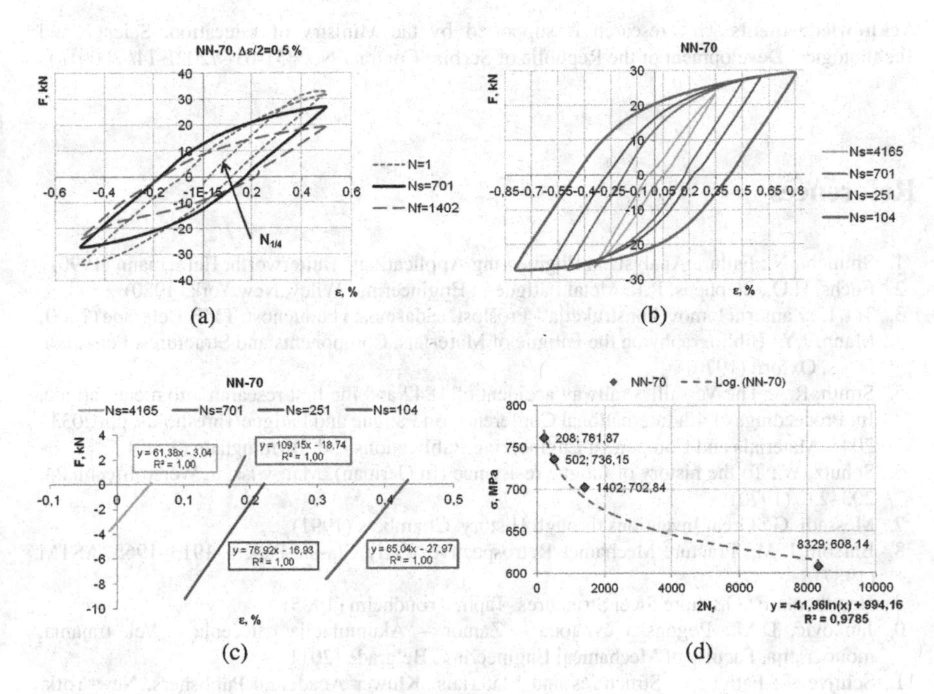

(a) (b) (c) (d)

Fig. 32.4 Graphical results of LCF test of NN-70 steel specimens

$$F = 0;\ \Delta\varepsilon_p/2 = b/m \tag{32.4}$$

$$\Delta\varepsilon_e/2 = \Delta\varepsilon/2 - \Delta\varepsilon_p/2 \tag{32.5}$$

Based on the above relations, plastic strain amplitude and elastic strain amplitude are obtained for each specimen and its characteristic stabilization cycle, N_s, shown in Table 32.5. The presented methodology can be applied to each LCF cycle.

32.4 Conclusion

Determination of plastic and elastic strain components for the stabilization cycle, Ns, serve to characterize steel and determine the characteristic curves of low-cycle fatigue. Also, this calculation better describes the behavior of a specific steel element exposed to the action of low-cycle fatigue in its exploitation.

By determining the plastic and elastic strain component for other cycles of low-cycle fatigue, from N_{bs} to N_{es} (N_f), serve calculate the cyclic accumulation of plastic strain component for a certain amplitude level of strain and thus determine the fatigue life of the steel part depending on its purpose and loading.

Acknowledgements This research is supported by the Ministry of Education, Science and Technological Development of the Republic of Serbia (Contract No. 451-03-9/2021-14/ 200012).

References

1. Shinichi, N.: Failure Analysis in Engineering Applications. Butterworth, Heinemann (1990)
2. Fuchs, H.O., Stephens, R.I.: Metal Fatigue in Engineering. Wiley, New York (1980)
3. Tot, L.: Zamorni lomovi konstrukcija – Prošlost, sadašnjost i budućnost. TMF, Belgrade (2000)
4. Mann, J.Y.: Bibliography on the Fatigue of Materials. Components and Structures. Pergamon Press, Oxford (1970)
5. Smith, R.A: The Versailles railway accident of 1842 and the first research into metal fatigue. In: Proceedings of 4th International Conference on Fatigue and Fatigue Thresholds, pp. 2033–2041. Materials and Component Engineering Publications Ltd, Birmingham (1990)
6. Schutz, W.: To the history of fatigue resistance (in German), Mat.-wiss. u. Werstofftechn 24, 203–232 (1970)
7. Messadi, G.: Great Inventions through History. Chambers (1991)
8. Barsom J. M.: Fracture Mechanics Retrospective: Early Classic Papers, 1913–1965. ASTM (1987)
9. Handbook, F.: Offshore Steel Structures. Tapir, Trondheim (1985)
10. Janković, D.M.: Pogonska čvrstoća – Zamor – Akumulacija oštećenja – Vek trajanja, monografija. Faculty of Mechanical Engineering, Belgrade (2011)
11. Schijve, J.: Fatigue of Structures and Materials. Kluwer Academic Publishers, New York, Boston, Dordrecht, London, Moscow (2004)
12. Aleksić, V., Aleksić, B., Milović, L.: Methodology for determining the region of stabilization of low-cycle fatigue. In: Proceedings of the 16th International Conference on New Trends in Fatigue and Fracture, NT2F 2016, pp. 189–190. Dubrovnik (2016)
13. Aleksić, V., Milović, L., Aleksić, B., Hemer, A.M.: Indicators of HSLA steel behavior under low cycle fatigue loading. Procedia Struct. Integr. **2**, 3313–3321 (2016)
14. Aleksić, V., Aleksić, B., Milović, L.: Metodologija određivanja pokazatelja ponašanja HSLA čelika pri delovanju niskocikličnog zamora. In: Proceedings of the V Međunarodni kongres "Inženjerstvo, ekologija i materijali u procesnoj industriji", pp. 1123–1135. Jahorina (2017)
15. Aleksić, V., Milović, L., Aleksić, B., Bulatović, S., Burzić, Z., Hemer, A.: Behaviour of nionikral-70 in low-cycle fatigue. Struct. Integr. Life **17**(1), 61–73 (2017)
16. Aleksić, B., Aleksić, V., Hemer, A., Milović, L., Grbović, A.: Determination of the region of stabilization of low-cycle fatigue HSLE steel from test data. In: Ambriz, R., Jaramillo, D., Plascencia, G., Nait Abdelaziz, M. (eds.) Proceedings of the 17th International Conference on New Trends in Fatigue and Fracture, NT2F 2017, pp. 101–113. Springer, Cham (2018)

17. Aleksić, V., Milović, Lj., Blačić, I., Vuhever, T., Bulatović, S.: Effect of LCF on behavior and microstructure of microalloyed HSLA steel and its simulated CGHAZ. Eng. Fail. Anal. **104**, 1094–1106 (2019)
18. Aleksić, V.: Low cycle fatigue of high strength low alloy steels. PhD thesis (in Serbian). University of Belgrade (2019)
19. Aleksić, B., Grbović, A., Milović, L., Hemer, A., Aleksić, V.: Numerical simulation of fatigue crack propagation: A case study of defected steam pipeline. Eng. Fail. Anal. **106**, 104165 (2019)
20. ASTM E606-80: Standard Recommended Practice for Constant Amplitude Low Cycle Fatigue Testing, ASTM Designation E606-80. Annual Book of ASTM Standards, vol. 3, p. 681 (1985)
21. ISO 12106:2017(E): Metallic materials-fatigue testing-axial-strain-controlled method. ISO, Geneva (2003)
22. Bannantine, J.A., Comer, J., Handrock, J.: Fundamentals of Material Fatigue Analysis. Prentice-Hall, Englewood Clifs, New Jersey (1990)

Part III
Management and Monitoring of Manufacturing Processes

Chapter 33
Energy and Utility Management Maturity Model for Sustainable Industry

Milena Rajic, Milan Banic, Rado Maksimovic, Marko Mancic, and Pedja Milosavljevic

Abstract Energy management has becoming a priority for organizations to reduce energy costs, to maximize their profits with minimal resources used, to comply legal requirements and to strive to sustainable business. Energy management standards, such as ISO 50001, represent the good practice, but have certain limitations. Therefore, the energy management maturity models could be used in different engineering sectors as a tool for continuous improvement. Energy maturity models have been used as an instrument to assess the current state of an organization, to rank the improvement measures and to manage and identify the progress. This paper proposes the energy management maturity model that is ISO 50001 processes based and Plan-Do-Check-Act cycle organized. The model was verified in production organizations in Serbia, which shows that all maturity levels exist in practice. The conducted study was performed in 24 organizations in two years: 2019 and 2021, where overall organizations' maturity levels had better ranks. The average maturity level in research sample was 3.44 (proposed model with 5 maturity levels). The Act process phase had average maturity level 3 and Check phase 4.31. In two years there is evident overall improvement of the maturity level scores for most of the analyzed organizations.

Keywords Energy management · ISO 50001 · Maturity model

33.1 Introduction

Industrial organizations represent significant energy consumers, especially electricity consumers. Energy costs become a major factor for managing the organization and business itself. Taking into account the necessity of clean manufacturing and proper energy management, sustainable industry becomes a priority. An increasing number of studies priorities the need to develop models for monitoring and proper managing

M. Rajic (✉) · M. Banic · M. Mancic · P. Milosavljevic
Faculty of Mechanical Engineering, University of Nis, 18000 Nis, Serbia
e-mail: milena.rajic@masfak.ni.ac.rs

R. Maksimovic
Faculty of Technical Science, University of Novi Sad, 21000 Novi Sad, Serbia

M. Rackov et al. (eds.), *Machine and Industrial Design in Mechanical Engineering*, Mechanisms and Machine Science 109,
https://doi.org/10.1007/978-3-030-88465-9_33

of energy flows, especially in manufacturing organizations [1–4]. To use energy rationally, but preserving the environment, represents an increasing challenge for organization. Studies based on energy and material flows in manufacturing sector, how these flows are flexible and applicable in different processes and/or in different production sectors would provide enough information to design a process with as much resource savings as possible and with a less negative impact on the environment. Industrial organizations, primarily production organizations, represent the largest energy consumers [5].

Manufacturing organizations are the most analyzed for the development of energy management models [6–11]. Energy prices and costs of energy flows are among the basic factors that determine production processes. Over 80% of energy needs in industry are compensated by using fossil fuels [12]. The use of energy in the industrial sector varies from 30 to 70% of the total energy in different countries [3] based on the energy efficiency analysis in industry. Energy management system offers many benefits, such as optimizing energy consumption, reducing costs, improving the corporate image of the organization, reducing the negative impact on the environment [13]. Therefore, organizations have developed more suitable approaches to energy management, aiming to reduce energy losses [14]. Energy management has becoming a critical approach, particularly considering the energy costs as the largest in relation to other costs in the manufacturing cycle. Analysis done in [15, 16] stated that certified and applied energy management ensures that end-users receive the necessary amount of energy, supplied with minimal costs, while preserving safety and reliability during operations and in the same manner the environmental protection. Energy management requires having a systematic and continuous approach and therefore should not be time-limited programs and/or projects. Energy Management System (EnMS) begins with an energy policy, defines energy goals and ways to achieve these goals, forms a system for monitoring energy performance and implements procedures for a continuous improvement of energy performance [17]. The ISO 50001:2011 standard does not define specific performance criteria related to energy consumption and efficiency. Instead, it proposes a management model that contributes to the development and implementation of the energy policy in order to achieve the objectives and action plans, taking into account legal requirements and information resulting from the analysis and management energy consumption data. The standard describes the ultimate goal, but not the way how to achieve it, and it does not allow the organization to fully understand the position on the way to the ultimate goal. Energy management system application through guidelines given within ISO 50001:2011 standard may be used with reliability for designing, implementing, maintaining and improving of an existing energy management system [18].

On the other side, maturity models were used in different engineering sectors and fields as a tool for continuous improvement. Maturity models may be used as an instrument to assess the current state of an organization, to rank the improvement measures and to manage the progress [19, 20]. There are previously defined maturity levels that represent an organizational plan for improvement, where the initial maturity level defines a state of an organization which has poor capabilities in certain domain and the highest maturity level defines a state of total maturity [21, 22]. In such case, the maturity is defined as a metric to evaluate organizations' properties and capabilities and their improving progress.

Maturity models have been used so far in different studies and industries. Literature data showed the application of maturity models in widespread sectors as: information technology [21, 22], healthcare [23], finance [24], production [25, 26], etc. Therefore, specific energy maturity models were developed for different purposes. ISO 50001 based energy maturity model was proposed by [20, 27, 28]. Different approach to deliver the qualitative metrics in the form of an energy management maturity model was proposed by [29].

In this paper the proposed Energy Management Maturity Model is presented, together with the evaluated study of presented model done by production organizations in 2019 and 2021. The general approach of presented model enables organization to have full insight into energy management practices, to follow the path for continuous improvement and to fulfill the needed requirements for energy management excellence.

33.2 Energy Management Maturity Models

As it was stated, Energy Management System (EnMS) may provide an approach for organization's energy efficiency and sustainability improvement, but there are not sufficient data how to achieve it. Due to [20], an energy management maturity model developed for organization would provide: structuring and improving the understanding of EnMS practices, understanding the roadmap for continuous improvement and successful energy management requirements implementation, benchmarking the current organization's state with other organizations and guiding through action measures. According to [30], energy management maturity models enable achieving the pre-defined maturity level of the analyzed organization, through of defining the current state, benchmarking and continuous improvement.

There are proposal of energy management maturity models through industrial guidelines, such as Carbon Trust Energy Management Matrix [31] with 5 maturity levels. Also as industrial guide, organized as 4 domain of the plan-do-check-act cycle is given in [32], as well as the proper survey conducted by [33].

In literature as scientific baseline was given 5-leveled model, which defines progress between maturity levels achieved by organizations [34]. For multi-site industrial organizations detailed energy maturity model was presented by [29]. PDCA cycle based maturity model with clearly defined activities and requirements was given in [14, 20, 28].

This paper proposed energy management maturity model for production organization which includes utility audits of production plants, facilities and units. The model was inspired by [29] but includes certain modification for production companies in Serbia.

33.3 Proposed Model

The proposed model is ISO 50001 based following the PDCA cycle and it was summarized in Table 33.1. The model levels was inspired by [28, 34] and described as:

Maturity level 1—Minimal: The identified processes are not EnMS comply; EnMS procedures and policies are not implemented; Energy profile and energy performance indicators are not established.

Maturity level 2—Planning: EnMS requirements are established and known; Energy demanding processes and equipment are monitored as well as significant energy users; Procedures for monitoring and analyses are established and followed; Some corrective actions within EnMS are applied.

Maturity level 3—Implementation: Requirements and practices within EnMS are standardized; Process indicators and parameters are monitored, documented and controlled; Employees are constant involved in communication, documentation within EnMs.

Maturity level 4—Monitoring: Energy demanding processes and equipment are monitored, measured and performed results are analyzed; Organization conducts internal audits within EnMS; Causes of energy efficiency measures failures are identify.

Table 33.1 ISO 50001 PDCA cycle model based

Plan	Do	Check	Act
EnMS identification	Involving employees in EnMS	Energy indicators monitoring and controlling	EnMS review
Commitment of management for EnMS	Communication regarding EnMS	Evaluating of applicable energy legal requirements	
Energy policy establishment	Documentation and records regarding EnMS	Energy audit within EnMS	
Energy planning	Process management within EnMS	Action plan for continuous improvement	
Legal requirement for EnMS	Energy efficient monitoring of plants, facilities, processes		
Energy management review			
Energy profile establishment			
Energy performance indicator defining			
Energy objective and target value defining			

Maturity level 5—Improvement: Management reviewing constantly the established procedures and action plans for continuous improvement within EnMS; Processes are optimized, energy efficient and organization operates in accordance with the sustainable principles; Action plans are constantly reviewed.

In the survey, the analyzed organizations would get one to five points. The questions are related to maturity level not only organization itself but also the maturity level of processes, equipment, plant, building, facilities.

33.4 Validation in industry: Case Study

Implementing the proposed model was the next step in this study, which covers two analyzed years 2019 and 2021. The surveyed organizations were interviewed according to the questionnaire given and proposed in [35, 36]. The survey was filled by each individual organization and for each production plant. The given results were collected and documented for future analysis. The surveyed data for each production plant, facility and utility was analyzed in according to have average implementation, maximal and minimal values, in accordance to PDCA model and the proposed maturity levels. The final results are compared within the research sample and with literature data. The proposed model was tested in 24 organizations selected according the requirements defined in [35]. All organizations were ISO 9001 certified.

The results included average values, standard deviation, minimum and maximum values and the average maturity level is 3.44. The processes where attention should be made are "Energy management system review" (average level 2.69), "Energy management system identification", "Involving employees in Energy management system" and "Energy audit within Energy management system" (less than 3.00). It should be emphasized that "Energy efficient monitoring of plants, facilities and processes" was very close to the maturity level 5 (average level 4.75). The limitation of the study is certainly small research sample. The future work would include SME in Serbia and wide research sample within production organizations. Figure 33.1 shows average, minimum and maximum values, for each stage of the presented model in analyzed organizations.

In Fig. 33.2a, the average maturity level of analyzed organizations within PDCA cycle based model was presented for two analyzed years (2019 and 2021), while Fig. 33.2b summarized the average results of analyzed organization for 2019 and 2021. Figure 33.2a, b show an overall improvement of the maturity level scores, both presented in PDCA cycle model and for each organization. It is significant that improvement is more present for 'Plan' and 'Check' phase.

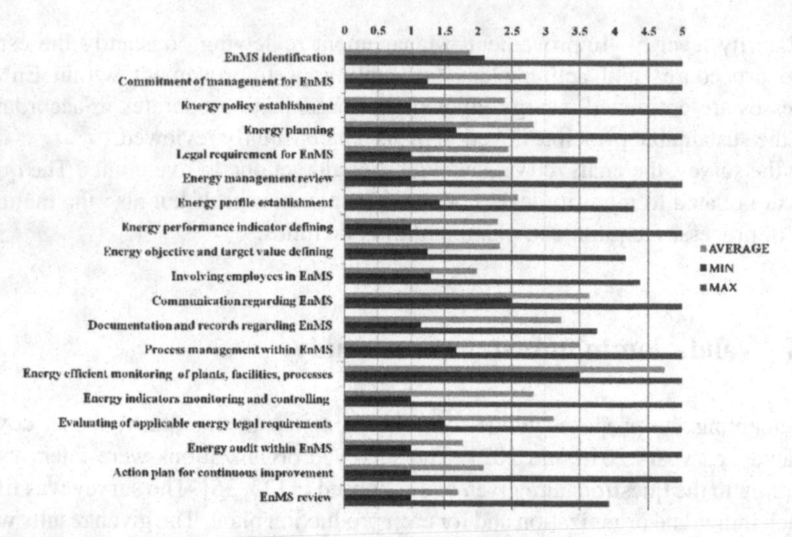

Fig. 33.1 Average, minimal and maximal maturity levels of analyzed organizations

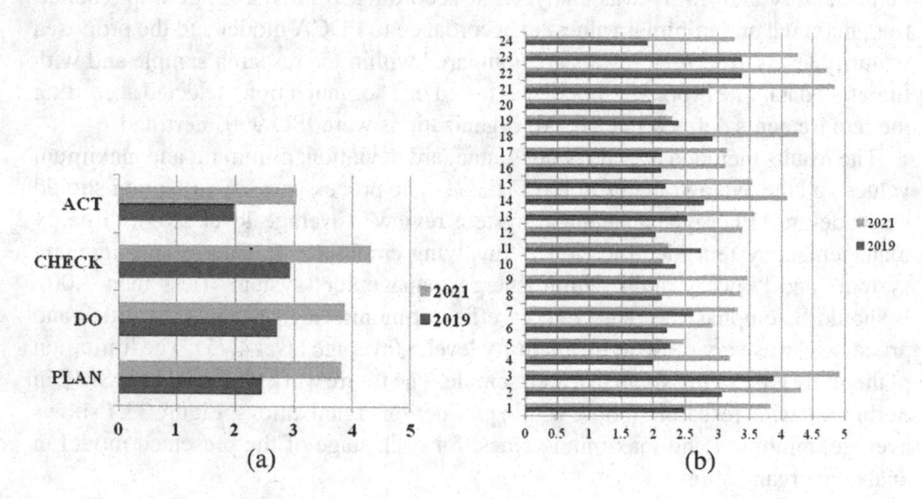

Fig. 33.2 **a** Averaged maturity levels in PDCA cycle of analyzed organizations for two analyzed years. **b** Maturity levels of 24 analyzed organizations for two analyzed years

33.5 Conclusions

In this paper the energy and utility management maturity model is proposed. The model is ISO 50001 process based model, PDCA cycle organized, and adopted and adjusted for production plant, facility and processes reviews. The model was validated using the questionnaire within the production organizations in Serbia (24 analyzed organizations) in two analyzed years: 2019 and 2021. The results indicate that all maturity levels exist in practice. Minimum and maximum maturity level scores shows that all maturity levels that are identified are achievable. The obtained results can also be used for benchmarking in different countries, sectors, industries. The presented model is considered to be universal and could be applicable both, in production and service organizations. The contribution of presented model is ISO 50001 process model organized with PDCA cycle. The proposed model could be used by industry or regulatory bodies to monitor energy management maturity level and their progress, for energy regulations revisions and basis for EnMS requirements certification.

Future work would include wide range of organizations in different sectors in Serbia, where SME would be involved. In that manner, the larger research sample would be analyzed, so the more accurate results would be achieved. In this paper the organizations submitted surveys where they should have to estimate their maturity level. But in the future, the independent auditors should be involved for model validation. Similar work could be done in neighboring countries so the proper results benchmarking could be performed.

Acknowledgements This research was financially supported by the Ministry of Education, Science and Technological Development of the Republic of Serbia (Contract No. 451-03-9/2021-14/200109).

References

1. Thiede, S., Posselt, G., Herrmann, C.: SME appropriate concept for continuously improving the energy and resource efficiency in manufacturing companies. CIRP J. Manuf. Sci. Technol. **6**(3), 204–211 (2013)
2. Ghadimi, P., Li, W., Kara, S., Herrmann, C.: Integrated material and energy flow analysis towards energy efficient manufacturing. Proc. CIRP **15**, 117–122 (2014)
3. Madlool, N.A., Saidur, R., Rahim, N.A., Kamalisarvestani, M.: An overview of energy savings measures for cement industries. Renew. Sust. Energy Rev. **19**, 18–29 (2013)
4. Rajić, M.N., Milovanović, M.B., Antić, D.S., Maksimović, R.M., Milosavljević, P.M., Pavlović, D.L.: Analyzing energy poverty using intelligent approach. Energy Environ. **31**(8), 1448–1472 (2020)
5. Laitner, J.A.: An overview of the energy efficiency potential. Environ. Innov. Soc. Trans. **9**, 38–42 (2013)
6. Yin, R.Y.: Metallurgical Process Engineering. Springer, Heidelberg (2011)
7. Mancic, M.V., Zivkovic, D.S., Djordjevic, M.L., Rajic, M.N.: Optimization of a polygeneration system for energy demands of a livestock farm. Therm. Sci. **20**(Suppl. 5), 1285–1300 (2016)
8. Yin, R.Y.: Analysis and integration of steel manufacturing process. Acta Metall. Sin. **36**(10), 1077–1084 (2000)
9. Milovanović, M.B., Antić, D.S., Rajić, M.N., Milosavljević, P.M., Pavlović, A., Fragassa, C.: Wood resource management using an endocrine NARX neural network. Eur. J. Wood Wood Prod. **76**(2), 687–697 (2018)
10. Yin, R.Y.: The Essence, functions, and future development mode of steel manufacturing process. Sci. Sin. Technol. **38**(9), 1365–1377 (2008)
11. Todorović, M.N., Živković, D.S., Mančić, M.V., Ilić, G.S.: Application of energy and exergy analysis to increase efficiency of a hot water gas fired boiler. Chem. Ind. Chem. Eng. Q. **20**(4), 511–521 (2014)

12. Saidur, R., Atabani, A.E., Mekhilef, S.: A review on electrical and thermal energy for industries. Renew. Sust. Energy Rev. **15**(4), 2073–2086 (2011)
13. Capehart, B.L., Turner, W.C., Kennedy, W.J.: Guide to Energy Management. The Fairmont Press, Inc (2006)
14. Introna, V., Cesarotti, V., Benedetti, M., Biagiotti, S., Rotunno, R.: Energy management maturity model: An organizational tool to foster the continuous reduction of energy consumption in companies. J. Clean. Prod. **83**, 108–117 (2014)
15. Petrecca, G.: Industrial Energy Management: Principles and Applications. Springer Science & Business Media (2012).
16. Piper, J.E.: Operations and Maintenance Manual for Energy Management. Routledge, Taylor & Francis Group (2016)
17. International Organization for Standardization (ISO): The 50001:2011 Standard. ISO. Geneva (2011)
18. Silva, V.R.G.R.D., Loures, E.D.F.R., Lima, E.P.D., Costa, S.E.G.D.: Energy management in energy-intensive industries: Developing a conceptual map. Braz. Arch. Biol. Tech. **62**, e19190017-1–e19190017-17 (2019)
19. De Bruin, T., Rosemann, M., Freeze, R., Kaulkarni, U.: Understanding the main phases of developing a maturity assessment model. In: Proceedings of the 16th Australasian Conference on Information Systems, ACIS 2005, pp. 1–10. AISeL, Sydney (2005)
20. Antunes, P., Carreira, P., da Silva, M.M.: Towards an energy management maturity model. Energ. Policy **73**, 803–814 (2014)
21. Becker, J., Knackstedt, R., Pöppelbuß, D.-W.I.J.: Developing maturity models for IT management. Bus. Inf. Syst. Eng. **1**(3), 213–222 (2009)
22. Becker, J., Niehaves, B., Poeppelbuss, J., Simons, A.: Maturity models in IS research. In: Proceedings of the 18th European Conference on Information Systems, ECIS 2010, pp. 1–12. AISeL, Sydney (2010)
23. Brooks, P., El-Gayar, O., Sarnikar, S.: A framework for developing a domain specific business intelligence maturity model: Application to healthcare. Int. J. Inf. Manag. **35**(3), 337–345 (2015)
24. Lederman, P.F.: Getting buy-in for your information governance program. Inf. Manag. J. **46**(4), 34–37 (2012)
25. Neff, A.A., Hamel, F., Herz, T.P., Uebernickel, F., Brenner, W., Vom Brocke, J.: Developing a maturity model for service systems in heavy equipment manufacturing enterprises. Inf. Manag. **51**(7), 895–911 (2014)
26. Backlund, F., Chroneer, D., Sundqvist, E.: Project management maturity models—a critical review: a case study within Swedish engineering and construction organizations. Procd. Soc. Behv. **119**, 837–846 (2014)
27. Jin, Y., Long, Y., Jin, S., Yang, Q., Chen, B., Li, Y., Xu, L.: An energy management maturity model for China: Linking ISO 50001: 2018 and domestic practices. J. Clean. Prod. **290**, 125168 (2021)
28. Jovanović, B., Filipović, J.: ISO 50001 standard-based energy management maturity model–proposal and validation in industry. J. Clean. Prod. **112**, 2744–2755 (2016)
29. Finnerty, N., Sterling, R., Coakley, D., Keane, M.M.: An energy management maturity model for multi-site industrial organizations with a global presence. J. Clean. Prod. **167**, 1232–1250 (2017)
30. Neuhauser, C.: A maturity model: Does it provide a path for online course design. J. Interact. Online Learn. **3**(1), 1–17 (2004)
31. Carbon Trust.: Energy Management Self-assessment Tool (2015)
32. O'Sullivan, J.: Energy Management Maturity Model (EM3). SEAI (2011)
33. EDF Climate Corps.: EDF Smart Energy Diagnostic Survey. EDF Climate Corps (2015)
34. Ngai, E.W.T., Chau, D.C.K., Poon, J.K.L., To, C.K.M.: Energy and utility management maturity model for sustainable manufacturing process. Int. J. Prod. Econ. **146**, 453–464 (2013)

35. Rajic, M.N., Maksimovic, R.M., Milosavljevic, P., Pavlovic, D.: Energy management system application for sustainable development in wood industry enterprises. Sustainability 12(1), 76-1–76-16 (2020)
36. Rajic, M.: The model of the energy flow management in industrial systems. (In Serbian: Model upravljanja tokovima energije u industrijskim sistemima). University in Novi Sad, Novi Sad (2020)

Chapter 34
Fundamental Ideas About Intelligent Manufacturing Systems

Vanessa Prajova, Peter Košťál, Sergiu-Dan Stan, and Štefan Václav

Abstract The current product quality and productivity-increasing trend are affected by time analysis of the entire manufacturing process. The primary requirement of manufacturing is to produce as many products as soon as possible. The production realizes at the lowest cost, but of course with the highest quality. Such requirements may be satisfied if all elements entering and affecting the production cycle are fully functional. The "Intelligent manufacturing system" it is a system that can respond to changes of internal and external processes status.

Keywords Intelligent manufacturing systems · Manufacturing process · Control system

34.1 Introduction

When we talk about flexible manufacturing systems, it is adequate to speak about the possible use of new generation manufacturing systems. The new generation of manufacturing systems is also called "intelligent" manufacturing systems (IMS) [1].

These systems are equipped with artificial intelligence and intelligent control. We are currently seeking new solutions for the implementation of artificial intelligence in machine production [2].

The main requirements in installing intelligent production systems are shortening production times, higher productivity, economic expediency, and worker hand intervention elimination.

V. Prajova · P. Košťál (✉) · Š. Václav
Faculty of Material Science and Technology, Institute of Manufacturing Technology, Slovak University of Technology, J. Bottu 25, 91724 Trnava, Slovak Republic
e-mail: peter.kostal@stuba.sk

V. Prajova
e-mail: vanessa.prajova@stuba.sk

S.-D. Stan
Faculty of Mechanical Engineering, Technical University of Cluj-Napoca, Cluj-Napoca, Romania

M. Rackov et al. (eds.), *Machine and Industrial Design in Mechanical Engineering*, Mechanisms and Machine Science 109,
https://doi.org/10.1007/978-3-030-88465-9_34

The basics needed for the realization of intelligence in manufacturing systems are monitoring, which can monitor the system's internal status and watch the environment's changing conditions. Monitoring systems are use sensors located at the whole system. The highly exposed parts of the manufacturing system, such as a rack for tools, machinery, or handling devices, usually place various sensors. Sensors were identifying parameters used as input data of the control system. Following these data is realized some, technological, manipulating or other supporting processes.

The similarity recognition of manufactured parts allows to assort them to the groups by machines required for its manufacturing. By manufacturing a single group of components, the production can achieve economic benefit near to mass production [3].

34.2 The Requirements for an Intelligent Manufacturing System

The manufacturing profile, material, and information flows must be taken into account when automating the intelligent production process. These fundamental elements of the manufacturing process are usually automated together in praxis. It is necessary to control and direct the main manufacturing processes that have the required functions. It could be to run more hard manufacturing processes together and individually in the manufacturing process, and all these processes controlled by a specific method. The name of this method is manufacturing process control [4].

Process control is characterized as the process organization that provides the required final state's achievement. But it is necessary to know its output for the manufacturing process control in the relatively closed system that it could be to influence the inputs back, eliminate the ineligible influences of the environment, and reorganize the internal system structure to achieve the required final state. The necessary element of the control system is feedback.

The utilization of this information for the manufacturing process control could be if known, the noted model of system behavior. It is possible to create the control algorithm and technical system for its realization by defined aim. The required final state must be honest, achievable, the algorithm realizable, and the information provision comfortable.

34.2.1 The Manufacturing Process Control

It is needed to know the answers to three fundamental questions for the proper control system realization. Those are the questions:

1. What has to be controlled?

In this case, it is necessary to define the controlled system's mathematical models by physical basis and identification. Simultaneously, it is required to analyze the manufacturing process in terms of the operative motion sequence. Also, it is necessary to know the manufacturing object's character before the definition of automated control of the manufacturing process. At once also to analyze the manufacturing process in term of several in sequence connected operative motions.

The analysis of the manufacturing process and dividing it into partial operations are the critical steps. The essential element of the manufacturing process is the operative motion. The operative motion may be defined as the individual part of the operative action by the existing regulation.

2. What shall we use for control?

In this case, it is necessary to select appropriate automatic control and energy devices in an automated production process. Automatic steering devices are a group of tools, equipment, tools. This knowledge about automated devices will allow us to realize practical control of the type of production process. The devices of the automatic control are, in general, all technical devices. These could enable to achieve, transfer, save, process and utilize the information and all supporting devices, and these could allow the activity of those named technical devices. Technological equipment is divided into several groups by individual aspects that are defined them.

3. What way shall we choose for control?

The manufacturing process control is often the integral member of the manufacturing process automation. The control is understood as the technical advance. The machines and devices with the limited system are influenced in their physical and technical values in terms of required method by reasonable regularity. The theory of control or the automatic control theory is engaged in the technical meaning of control. The theory of automatic control is the theory that checks the principles of control system functioning and designing.

The manufacturing process contains a more extensive number of elements, operations, and motions. It has existed many interacted connections. The control system used in the manufacturing process has a hierarchical structure. The hierarchical structure is one of the manufacturing process base features. It is necessary to give one's time to control individual motions and operations and interaction cooperation and communication between them and with the environment by control realization. The control system used more methods, systemic technologies, and programming instruments [5].

For a better understanding of the term "intelligent" manufacturing systems, it is the most suitable to compare its behavior with the classical ("non-intelligent") automated, flexible manufacturing system.

The automated manufacturing system is a manufacturing device with huge range of automation of its activities and with several subsystems integration (supervisory, manipulating, technological, transportation, controlling):

- it is specialized (they have a set of technological workstations),
- the transport and manipulating is realized by industrial robots or manipulators, and different type of conveyors [6, 7],
- the supervisory is included directly, and
- controlling it have own control systems of all devices [3].

Using intelligent production systems is conditioned by the efficiency of all subsystems in a given manufacturing system. Subsystems are developed together with automatic manufacturing systems to save the whole system parameters.

Automatic manufacturing systems designed for repetitive production demand a significant rate of flexibility called flexible manufacturing systems.

The category of FMS included one or more technological workstations, at which are all technological and material movement is automatic. The primary classification of automated manufacturing systems considers the number of machines in the manufacturing system and the flexibility of the production.

According to this classification, we distinguish three basic types of automated manufacturing systems:

- Flexible manufacturing cell—up to a maximum of 3 of the machine tools; the highest level of flexibility characterizes it,
- The lowest flexibility level characterizes a flexible manufacturing line; the range of goods is not huge and is produced in larger batches, and
- A flexible manufacturing system contains a minimum of 3 machines (and more) characterized by a lower flexibility level.

The intelligent manufacturing system presents a system that contained adaptation to unexpected changes, i.e., assortment changes, market requirements, technology changes, social needs, etc.

However, these systems' intelligence is frequently understood as control of the software product and not as an implementation of modern technologies of artificial intelligence.

The intelligent manufacturing systems consist a several subsystems like the automatic production systems (technological, supervisory, transportation, manipulating). These subsystems are equipped with aids, which give them a specific level of intelligence. It means that we can to consider it as a higher phase of flexible production systems. The intelligent manufacturing systems consist of (see Fig. 34.1).

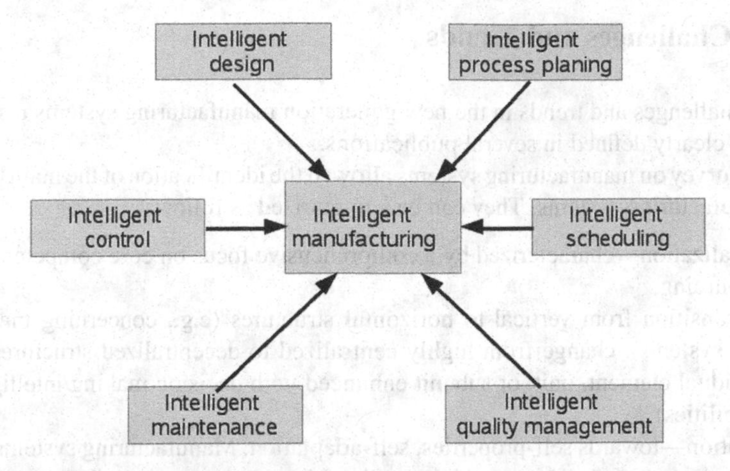

Fig. 34.1 Intelligent manufacturing content

34.3 The Intelligence Enhancement for the Mechanical Periphery

The monitoring of all of the actions inside the production process and its environment aimed at increasing this control system reliability and failure prevention of the manufacturing system itself or avoidance of defective products. We have several possibilities that can enhance the intelligence of the production system [8].

One of the application areas of monitoring systems is the area of robotized assembly. Equipping assembly systems with sensors is the basic level of increasing automation and machine intelligence. Sensoric systems provide scanning and monitoring of various functions of the assembly process, assembly technology, properties of assembly objects, mounted parts, and environment properties. The realization of monitoring functions provides suitable sensor sorts, at which point the supervisory systems provide the control interventions. Sensors are the primary devices for capturing information and its transformation. The present monitoring systems have reserved structures realized according to application or purpose.

Tactile sensors are used if technical realization instruments, mostly assembly robots, are in direct contact with the assembly's object. Important are especially sensors that enable control of the object presence, identification of grasp force, monitoring of the starting position of assembly tools, and other functions. With tactile sensors are equipped tentacles, position table, transporting units, different units, and devices. For recognizing the orientation, kind of objects, detection of edges is used various visual types of sensors [9].

34.4 Challenges and Trends

Today challenges and trends in the new generation manufacturing systems research are very clearly defined in several publications.

The survey on manufacturing systems allowed the identification of the main trends for manufacturing systems. They can be summarized as follows:

- specialization—characterized by a comprehensive focus on core competencies,
- outsourcing,
- the transition from vertical to horizontal structures (e.g., concerning management systems), change from highly centralized to decentralized structures (the individual element, unit, or subunit enhanced with decision making/intelligence capabilities),
- evolution—towards self-properties, self-adaptation. Manufacturing systems with these characteristics have a high level of integration, are easily upgradable, evolvable, and adaptable e.g., to new market conditions,
- the development of technologies and applications to support all the requirements of current distributed manufacturing systems,
- competitiveness—the manufacturer should remain competitive. It means adequate types of equipment and machines (e.g., sensors) acceptable to new manufacturing paradigms; sustainability (e.g., to consider environmental concerns into the design),
- technology, equipment and manufacturing systems' selection—Evaluate various systems configurations based on lifecycle economics, manufacturing quality, system reliability,
- integration of operators with software and machines; this means non-functional properties, e.g., fault tolerance,
- openness, self-adaptability; each unit/sub-unit/element of the manufacturing system should independently take optimal wise decisions (e.g., concerning resource utilization, incorporating scheduling algorithms, planning, and control execution techniques), having a goal-driven and cooperative behave, and
- performance assessment.

Concerning future trends: it is rather challenging to forecast long-term trends for manufacturing engineering systems [10].

34.5 Conclusion

IMS meets the challenge of integrating intelligence and flexibility at the highest level of the manufacturing control system. We can use the unique simulation tool "Digital tween" for a production configuration. This tool avoids reliably simulated the entire production line, the production process as a digital factory.

Intelligent manufacturing shifts the whole manufacturing process from a resource-intensive industry towards to knowledge-based and customer-driven approach.

IMS allows the implementation of a multi-variant system to have an adequate number of production lines to manufacture sufficient goods. These production lines using a decent number of machine tools and devices to meet the requirements of increasing product variants and producing at ever smaller lot sizes. Due to the IMS system's knowledge and responsibility segregation, the various production units are easily extendable and exchangeable and offer complete production functionality. We can produce different product variants with the same manufacturing units on the same production line. The new concept achieves high manufacturing equipment reusability and is fast, flexible, reconfigurable, and modular.

But it is also important to say that problematic of intelligent manufacturing systems is still in the stay research and development.

The concept of all production devices is controlled according to the character of production. New generation production systems vary from flexible production systems in construction, especially in properties. Undoubtedly, its implementation to the production process brings huge possibilities in increasing productivity and decreasing production costs.

With that, it also decreases additional and running times, mainly eliminating worker hand intervention in the production process.

Acknowledgements This publication has been written thanks to support of the research project ERASMUS+ Programme—Strategic Partnership: "Development of mechatronics skills and innovative learning methods for Industry 4.0", project Nr: 2019-1-RO01-KA203-063153.

References

1. Holubek, R., Delgado Sobrino, D.R., Košťál, P., Oravcova, J.: Incorporation, programming and use of an ABB robot for the operations of palletizing and despalletizing at an academic-research oriented intelligent manufacturing cell. Appl. Mech. Mater. **309**, 62–68 (2013)
2. Danišová, N., Velíšek, K.: Intelligent manufacturing and assembly system. In: Monograph machine design 2007: On the occasion of 47th anniversary of the Faculty of Technical Sciences: 1960–2007, pp. 413–416 (2007)
3. Košťál, P., Velíšek, K.: Flexible manufacturing system. World Acad. Sci. Eng. Technol. **77**, 825–829 (2011)
4. Horváth, Š, Hrušková, E., Mudriková, A.: Areas in flexible manufacturing-assembly cell. Ann. Fac. Eng. Hunedoara **6**(3), 123–127 (2008)
5. Meziane, F., Vadera, S., Kobbacy, K., & Proudlove, N. (2000). Intelligent systems in manufacturing: current developments and future prospects. Integr. Manuf. Syst.
6. Cselenyi, J., Shmidt, L., Kovács, G.: Evaluation methods of storage capacity between manufacturing levels of eees at changing product structure. In. Proceedings of the International Scientific Conference microCAD 2002, pp. 63–71. Miskolc (2002)
7. Kovács, G.: Methods for efficiency improvement of production and logistic processes. Res. Pap. Fac. Mater. Sci. Technol. Slovak Univ. Technol. **26**(42), 55–61 (2018)
8. Varga, G., Dudas, I.: Самоорганизующиеся производственные системы с точки зрения производства сложных поверхностных пар. Проблемы Машиностроения и Автоматизации **3**, 39–44 (2009)

9. Košťál, P., Mudriková, A., Kerak, P.: Clamping fixture for new paradigms of manufacturing. In: Proceedings of the 21st International DAAAM Symposium: Intelligent Manufacturing & Automation: Focus on Interdisciplinary Solutions, pp. 361–362. DAAAM International, Vienna (2010)
10. Chituc, C., José Restivo, F.: Challenges and Trends in Distributed Manufacturing Systems: Are Wise Engineering Systems the Ultimate Answer? Massachusetts, Cambridge (2009)

Chapter 35
Material Flow in the Flexible Manufacturing

Peter Košťál, Vanessa Prajova, Erika Hrušková, and Miriam Matúšová

Abstract Efficient production in flexible production systems depends on efficient information and material flow in these systems. It is in a central position in logistics. Before, all kinds of material movement could be and follow information. All these steps are necessary to information flow attend and bring that information possible to adopt decisions. The majority of problem resolution is in optimizing material and information flow and their application into the highest level of automation and mechanization into manipulation and transport processes. In the material flow projecting, it is necessary to know that the project aims not to transport and store material because the price of these operations is high, and the material rate is not higher.

Keywords Flexible manufacturing · Robotized manufacturing · Paperless manufacturing

35.1 Introduction

The strategy of consumer's individualization characterizes today's market. This is the individual consumer's requests-oriented strategy. Consumers want new products, and the time to request the consumer's satisfaction has a fundamental role. The production was broadening, the innovation cycle is shortening, and the products have a new shape, material, and functions. At this strategy is the most critical parameter the time and improvement is its shortening. The production strategy focused on time needs changes from traditional functional production structure to production by flexible manufacturing cells and lines. The most critical manufacturing philosophy in the last years appears the production of Flexible Manufacturing System (FMS) [1].

Achieving these classes requires an integrated approach to the design and production of continuous communication at each development process stage. Although there are many engineering tools, especially in CAD Computer Aided Design), there are often separate packages without interactivity [2].

P. Košťál (✉) · V. Prajova · E. Hrušková · M. Matúšová
Faculty of Material Science and Technology, Institute of Manufacturing Technology, Slovak University of Technology, J. Bottu 25, 91724 Trnava, Slovak Republic
e-mail: peter.kostal@stuba.sk

M. Rackov et al. (eds.), *Machine and Industrial Design in Mechanical Engineering*, Mechanisms and Machine Science 109,
https://doi.org/10.1007/978-3-030-88465-9_35

This philosophy of flexible manufacturing is based on similarity of:

- manufactured parts,
- process plans [3].

Recognize the similarity of manufactured parts allows grouping them into groups by machines required for its manufacturing. By manufacturing this group of parts, we achieve an economic effect near to mass production.

By some study is existing manufacturing capability of machines used only 30–40%. The other resources say that technological processes spend only 5% of the time needed for manufacturing. The rest of the time is spending on manipulation, transport, and storage [4].

The flexible manufacturing system consists of a NC machine, a material handling system based on industrial robots, and a control system for integration the NC machine and the Material Handling System (MHS). Integration of these devices generally involves using a controller, software, and an overall some kind of computer network that coordinates the machine tools actions, and the material handling actions [5]. This system design allows the manufacturing of a group of similar workpieces. Its internal material and information flow characterize the system. The manufacturing process represents a complex dynamical process that included technological, manipulation, and control operations.

The production process is planned and simulated at the CAD laboratory, and after a successful simulation, we are sending the production data to the production system by internet connection.

35.2 Material Flow Planning

In general, it is essential for the production process's full functioning to ensure a continuous flow of material, raw materials, semi-finished products, etc. The term flow of material means the organized movement of material from the production process's entrance through all periods of storage, transport, and production processes to produce final products. Material flow analysis is one of the main points of the final analysis of material handling. Another task of a complete material flow analysis in the production process is to specify the material flow. It means determining the type, quantity, weight, shapes, and dimensions of the handled materials. These data affect handling and determining the requirements for handling, transport, and storage of the handled material and product.

The material handling system is the most crucial part of today's manufacturing systems and increasingly plays an essential role in the plant's productivity. Selecting a suitable MHS is a complex and tedious task for manufacturing companies because of the considerable capital investment required. Furthermore, to minimize production costs and increase profits, the appropriate MHS has to be selected. Handling activities generally account for 30–40% of production costs [6].

Principles of rational manipulation with material:

- To create straight, as short as possible transport ways,
- To design manufacturing process and solution of disposition about optimal material flow,
- If it is possible to use gravitation for material moving,
- To make optimal quantity, sizes, and weights of transported units,
- To design all manipulation, transport, and storage operations as effectively as possible,
- To resolve manipulation completely, it means from the view of the final factory and in relation without plant material flows,
- To use the possibility of type manipulation device,
- To speculate about the optimal use of transport and manipulation devices (minimum 60%), and
- To respect ergonomic and safety requirements.

The description of material and information flow in the production of this flexible manufacturing cell must be defined to choose the type range of the workpiece [6].

The principle of solving complex problems of material flow and elimination of losses and waste during the production process is focused mainly on four primary areas:

- Minimizing the size of stocks and supply and freeing up financial resources, and reducing logistics costs for storage, handling, etc.,
- Continuous improvement of all the logical chain activities: suppliers—product—customer (reduction of production costs, shortening of production time, improvement of the working environment, etc.),
- Focus on essential places for quality, competitiveness, perspective, productivity, costs, etc.,
- Optimization of the system of material and information flows—eliminating losses that can cause irregularities or overload of production, the complexity of material flows, outages caused by the organization [7].

The solution of optimizing the production process requires detailed information about the whole process. The given technological process plan of production (it does not change); is necessary to adjust the material and data flow.

Waste is usually present in every workplace, but it can identify by simple tools and resolved using the 5S method, which is suitable for production and service-oriented organizations. The application of the 5S process makes it possible to improve and simplify the flow of material, the arrangement of machines and stocks. Other benefits are:

- Improving quality, productivity, and work safety,
- Better corporate culture, more positive mental approach of employees to fulfill their duties, and
- It improved work environment.

The 5S method is in development, and today, the sixth S, known as safety, is also defined. The material flow design aims to solve:

- Minimization of transport, handling, and storage,
- Simplification of the system to a minimum—minimum consumption of costs and time - > solution of essential links,
- Workplaces and capacities—incorrectly designed capacity causes unevenly distributed material flow, accumulation of stocks, the need for intermediate warehouses and buffer warehouses and other handling activities,
- Information flow and control system—correct control of inputs of production tasks, synchronization of purchasing, production, and dispatch, coordination of production control system with the transport system,
- All components of the production system must be designed concerning each other, and it is ideal if all of them are verified using a simulation model before installation.

The solution of a given material flow requires the specification of a given material flow's innovation goal.

The material flow is one of the most expensive production operations because it employs the most workers and takes the most significant time. For high efficiency of production, it is necessary to consider the increased usage of transport methods based on transported material and the exploitation of single production system devices at the project proposal.

The transport and handling subsystem is generally one of the most critical subsystems of the FMS. It is responsible for the transport and manipulation with primary and supply material, manufacturing tools (tools and tool holders, fixture, equipment, scales, carriers, and others), and waste. Used inter-operational transport and handling features are dependent on geometrical shape and weight of transported material, parts, and tools. Automation of handling in a manufacturing system can be realized by technological pallets (holders or another element with this function, e.g., fixtures). The part is clamped outside a machine workspace onto a specialized pallet and transported together. The piece is positioned and clamped into position for machining. Technological pallet serves at the same time as a transport pallet [8].

The other objects for transport and handling are production devices representing the needed technological equipment of the workstation. They are required to realize specific operations using a correspondent machine. These objects are transported to workstations in sets or parts when they create non-detachable pieces of standard workstation equipment.

In material flow planning, it is necessary to consider that the plan's aim is not the transport and storage of material because these activities are expensive and do not improve the material value. The selection of a suitable storage system is dealt with e.g. [9].

Current systems for handling, transportation, and storage provide many possibilities for applying costly and complex systems. The optimal design should contain minimum storage, transport, and handling. Hence, the suitable way before elaborating a detailed system solution is to reduce mentioned activities to a minimum.

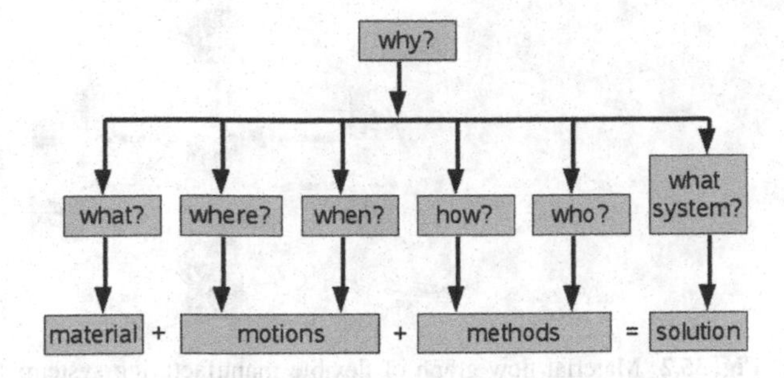

Fig. 35.1 Material flow equation

It is necessary to take into account the great importance of the dependence of material flow and the following elements of the manufacturing system:

- Workstation and its capacity—wrong capacity design induce unbalanced material flow, resources accumulating, and the need for additional buffer stocks, containers, and additional handling operations, and
- Material flow and the system controls, clean regulation of manufacturing tasks entering into system, purchase, manufacturing and expedition synchronization, coordination of manufacturing system and the transport system. All of them have a significant effect on the material flow plan.

The general sequence of material flow planning is the following:

- the volume of transport operations determination—material flow analysis,
- the variant layout design,
- analyze of existing devices—what device we have—what device we need,
- transport systems variants, and
- computer simulations of material transport and handling devices.

In this time, we solve the equations of material flow (Fig. 35.1).

35.3 The Material Flow in Flexible Manufacturing System

We want to produce (simulate production) various components of the shaft, flange, bracket, and box shape in this system. Each part made will represent piece production that means only one piece of this part will be made. Variability (dimensions and shape versions for each component) will be relatively broad. To that fact must be adapted planning and management of the production process in FMS.

The whole process must run automatically without human intervention from the design up to the final component storage. According to the program, the storage system's material will be automatically taken out, transported to individual machines, and put in the operating area by a handling device (industrial robot). The machine will execute separate technological operations to reach the component's final properties (shape and dimension). In case of simple part can be processed only by one machine.

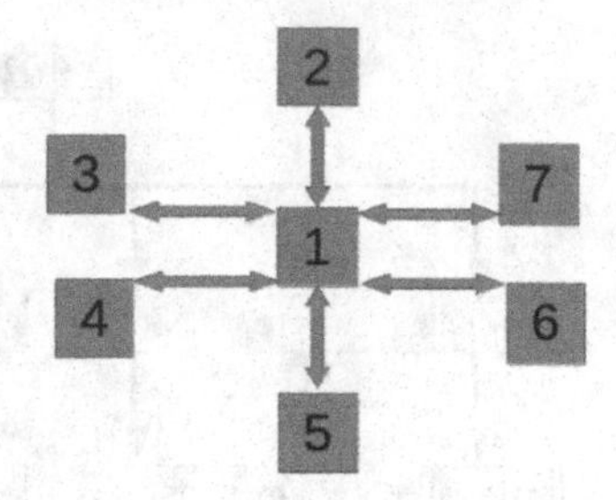

Fig. 35.2 Material flow graph of flexible manufacturing system: 1—conveyor, 2—storage, 3—pallet handling and quality station, 4—robot vision and assembly station, 5—robot feeder of machine tools, 6—CNC lathe, 7—CNC milling machine

The more complicated parts will have to be processed in one machine (e.g., turned to another position) or relocated to another machine to realize other necessary technological operations (sometimes, this relocation between individual devices needs to repeat several times).

After completing all necessary technological operations, the component made will relocate to the checking station for quality control. If quality control is successful, the finished and checked piece be automatically transferred to the FMS storage system. If the quality control is not successful, the component is also transferred to the storage system, where faulty products are stored.

The function graph of the whole flexible manufacturing system is shown in Fig. 35.2. Material flow in flexible manufacturing systems is described in [10] and in [11].

35.4 Conclusion

The project's target is to build up a flexible manufacturing system with robotized operation enabling drawing-free production. That means the by PC in a suitable 3D CAD program product design realized. In a base of 3D model, will generate the NC program for manufacturing. The flexible manufacturing system physically produces that component on the base of this program. In this way, it would make it possible to create all necessary parts for a specific product. In the final phase of production, the product was assembled.

In the last years, cell manufacturing becomes one of the most critical manufacturing types. This conception is based on the relationship between manufacturing cell and workpiece. Flexible manufacturing cells allow the manufacture of a small number of parts from a huge range of types and achieve appropriate economic effects near large batch or mass production. The manufacturing cell structure enables a connection between several machines and saves production time, space, and production costs. The machine's functions are coordinated, and the material flow can be fast.

Acknowledgements This publication has been written thanks to support of the research project ERASMUS+ Programme—Strategic Partnership: "Development of mechatronics skills and innovative learning methods for Industry 4.0", project Nr: 2019-1-RO01-KA203-063153.

References

1. Košťál, P., Krajčová, K., Ružarovský, R.: Material flow description in flexible manufacturing. ALS **4**, 104–108 (2010)
2. Vaclav, Š., Kostal, P., Michal, D., Lecky, Š.: Assembly systems planning with use of databases and simulation. IOP Conf. Ser.: Mater. Sci. Eng. **659**, 012023-1–012023-5 (2019)
3. Košťál, P., Mudriková, A., Sobrino, D.D.: Material flow in flexible production systems. Proc. Manuf. Syst. **5**(4), 213–216 (2010)
4. Mudriková, A., Delgado Sobrino, D.R., Košťál, P.: Planning of material flow in flexible production systems. In: Proceedings of the Annals of DAAAM 2010 & 21th International DAAAM Symposium, pp. 247–248. DAAAM International, Vienna (2010)
5. Um, I., Cheon, H., Lee, H.: The simulation design and analysis of a flexible manufacturing system with automated guided vehicle system. J. Manuf. Syst. **28**(4), 115–122 (2009)
6. Onut, S., Kara, S., Mert, S.: Selecting the suitable material handling equipment in the presence of vagueness. Int. J. Adv. Manuf. Tech. **44**(7), 818–828 (2009)
7. Kusá, M., Matúšová, M., Charbulová, M.: Optimalisation method of material flow at manufacturing process. In: Proceedings of the International symposium on Advanced Engineering & Applied Management-40th Anniversary in Higher Education, pp. 1–4. Hunedoara (2010)
8. Kostal, P., Mudrikova, A.: Material flow planing at flexible manufacturing in general. Sci. Bull. Series C: Fascicle Mech. Tribol. Mach. Manuf. Technol. **23**, 85–92 (2009)
9. Cselenyi, J., Shmidt, L., Kovács, G.: Evaluation methods of storage capacity between manufacturing levels of eees at changing product structure, pp. 63–71. Miskolc (2002)
10. Mudrikova, A., Hruskova, E., Velisek, K.: Logistics of material flow in flexible manufacturing and assembly cell. In: Proceedings of the Annals of DAAAM 2008 & 19th International DAAAM Symposium, pp. 919–920. DAAAM International, Vienna (2008)
11. Košťál, P., Mudriková, A.: Material flow in flexible manufacturing and assembly. Acad. J. Manuf. Eng. **1**, 185–191 (2008)

Chapter 36
Example of Good Maintenance Practice for Maintaining the Health of a Hydraulic System

Mitar Jocanović, Sladan Andrić, Milovan Lazarević, and Dejan Lukić

Abstract The maintenance of modern hydraulic systems is a very complex process. In addition to multidisciplinary knowledge and training requirements, additional procedures must be implemented and analysed to keep the hydraulic system functional, efficient, and in a failure-free state. Real-time maintenance of the hydraulic system (HS) includes managing a hydraulic system at peak operational efficiency. The article presents an example from practice that depicts the impact of preventive maintenance on the hydraulic system's functionality and discusses the contribution to improving real-time monitoring and benefits for advancing predictive maintenance practice.

Keywords Maintenance practice · Oil contamination · Filter management · Contamination control · Preventive maintenance

36.1 Introduction

The maintenance of hydraulic systems (HS) is a complex task for technicians since it requires multidisciplinary knowledge and training in fluid mechanics, tribology, reliability, and statistics. The most common failures of HS, around 70–85%, occur due to fluid contamination [1–3], consequently reducing system and components' reliability [4]. Therefore, an appropriate maintenance strategy, program and technology should be utilised to reduce both random and non-random failures by investigating underlying mechanisms that result in a stoppage. Numerous research papers are published on the topic of failures in hydraulic systems. Most of the research dedicates to the hydraulic system's oil analysis aspect and contamination control as presupposed.

M. Jocanović (✉) · M. Lazarević · D. Lukić
Faculty of Technical Sciences, University of Novi Sad, Trg Dositeja Obradovića 6, 21000 Novi Sad, Serbia
e-mail: mitarj@uns.ac.rs

S. Andrić
HYDROCAD, Zoltan Čuka 32, 21220 Bečej, Serbia

M. Rackov et al. (eds.), *Machine and Industrial Design in Mechanical Engineering*, Mechanisms and Machine Science 109,
https://doi.org/10.1007/978-3-030-88465-9_36

Different effects of contamination on HS wear have been described in scientific papers. For instance, Jocanović et al. [5] investigated the influence of inadequate maintenance on HS excavator components' wear rate. Karanović et al. [6] showed the influence of contaminants on the rate of degradation and wear of HS distribution valves. Nikolić et al. [7], in their work, derived equations for the temperature distribution in the bearing oil film with bearings to predict the bearing's thermal load. Orosnjak et al. [8] investigate silicon particles' presence as a trigger point for further wear and increase of metal (e.g., Fe, Cu) particles in the hydraulic, agricultural system tractor for improving prognostics and health management (PHM). Wakiru et al. [9] proposed fuzzy cluster analysis in determining the condition of a system through lubricant condition management (LCM). Singh et al. [10] explain preventive maintenance strategies in reducing failures by synthesising data from oil analysis and conducting maintenance activities accordingly. The research efforts dedicated to contamination control resonate compelling interest with constant efforts dedicated to reducing stoppages due to contamination. In this paper, the authors give a brief overview of maintenance practices for improving contamination control practice and associated filter management techniques of a hydraulic system.

The paper is formed as follows. Section 36.2 gives a brief explanation of ongoing maintenance methodologies and typologies of hydraulic system health preservation. The following section highlights the importance of real-time condition-monitoring of a hydraulic system and associated benefits for improving a hydraulic system's health. After depicting the importance of addressing the degradation or causality between a system's condition and a failure, the following section shows a practical example of addressing contamination in the hydraulic system and the discussion. The last section provides closing remarks and contributions to the literature.

36.2 Maintenance Methodology of Hydraulic Systems

For HS operation, several types of maintenance practice are used: Reactive maintenance (RM); Preventive maintenance (PM); Predictive maintenance (PdM) [11]; Proactive maintenance (PrM), maintenance within the concepts of flexible manufacturing system (FMS), or maintenance 4.0 (M4.0) as for the use of a remote monitoring techniques such as IoT and digital-based architectures, such as CPS (Cyber-Physical System models) [12] or Digital Twin (DT) [13] concepts for estimating the degradation process of impeding, overloading or exposing the system through the harsh working conditions (e.g., simulation through Accelerated Life Testing), where the Physics-of-Failure (PoF) can be estimated by the use of Deep Learning (DP) techniques [14].

Preventive maintenance (PM) encompasses scheduled or machine-based actions that detect, prevent, or mitigate component or HS failure to maintain or extend useful life through degradation control to an acceptable level. In this way, the HS is maintained in good condition and increases the reliability of the system. Predictive Maintenance (PdM), or commonly referred to as Condition-Based Maintenance

(CBM), uses sensor devices to gather information about HS and components. In this way, a signal is given to the maintainers to perform certain maintenance actions at the moment when it is needed. Predictive maintenance can best be described as a process that requires human technology and skills, combined with all available diagnostic and performance data, maintenance history, operator logs and design data to make timely decisions about main/critical maintenance requirements components or systems.

In the future, more massive applications of CBM can be expected, especially in systems that are more complex, expensive, and maintenance costs justify the use of expensive sensors. With Industry 4.0, CBM will be far more justified. The task will be for the machine to record operating parameters, control them and correct if necessary, leading to a self-sustainable way of maintaining the HS. However, maintenance still requires human resources' physical activity to replace the damaged or part in the failure. Proactive maintenance focuses on analysing the underlying cause, not just the symptoms. It seeks to prevent or eliminate failure from the source after identifying the root cause. Requires excessive knowledge of the maintainers, who, in addition to troubleshooting and successful maintenance, participate in collecting information sent to the design bureau to eliminate the system's design and operational deficiencies. Self-maintenance is an essentially new methodology in maintenance. It implies the functional ability of the system to monitor and diagnose itself. If a failure or malfunction in the system occurs, it should maintain its functionality for some time. Self-sustainability means that the system (machine) independently deals with monitoring and self-diagnosis.

36.3 HS and Their Maintenance in Real-Time

The consequences of reactive maintenance (RM) are reduced to large downtimes in the production, operation or service of HS and high production costs. These consequences are the result of an RM policy that involves reacting only after HS failure.

36.3.1 Real-Time Preventive Maintenance (PM)

In preventive maintenance, regardless of the actual state of the HS or the process in which the HS participates, the goal of the PM is to reduce the time of HS inactivity in the overhaul time interval and eliminate the HS, machine or process in failure. The PM program must focus on HS performance, not activity. Many organisations have good PM procedures but do not require maintenance staff to follow them.

For quality PM implementation of any HS, it is necessary to collect more data on the observed HS. These data refer to:

1. Identification of the system's state, i.e., (a) does the system work 24 h a day, 7 days a week? (b) does the system work with flow and pressure losses (consult technical support or documentation for components such as pumps and actuators)? (c) in what working environment is the HS (ambient temperature and environmental pollution)?
2. What requirements does the equipment manufacturer specifies for the preventive maintenance of HS?
3. Are the hydraulic fluid operating temperatures within the limit values (40–60 °C)?
4. The requirement of equipment manufacturer in the quality level of working fluid—oil in HS?
5. What requirements and operating parameters does the component manufacturer specify regarding the working fluid's purity—oil?
6. Identification of quality filter level in HS and, if necessary, correction with better filter cartridges?
7. HS time cycle monitoring and recording of delays and possible failures. What are the causes of such anomalies in the system?
8. Availability of technical documentation related to the HS's installed equipment to verify the observed HS's above procedures?
9. Depending on the HS's most sensitive component (pump, proportional or servo valve), it is required to use a secondary filtration system to achieve the required fluid purity.

These are just some of the steps used to implement PM in HS and depend on the complexity of the system, process and environment in which HS is used. The PM program involves creating a (written or software) procedure for each PM task in the HS. For each task, maintenance steps or procedures must be established with well-defined tasks for the maintenance staff. Note that the steps and procedures must be understandable and accurate regardless of the staff's training level (from beginners to masters).

36.3.2 *Real-Time Predictive Maintenance (PdM)*

Maintenance, according to the PdM model, is present nowadays and has been active since 2017. According to Mulders et al. [10], PdM is actively applied in the most developed EU countries in some percentages of 20–22% of the industry. In comparison, self-sustaining systems so-called PdM 4.0 are applied in 11% of companies. However, these data refer to different industries and applications of different systems compared to HS.

Predictive maintenance of HS means obtaining data from sensor devices in the system that measures some of the basic parameters such as temperature, pressure, vibration, rotation speed or torque, physical and chemical properties of the oil, the number of solid particles in the oil, etc. Also, by applying equipment such as the

Internet of Things (IoT) and the Industrial Internet of Things (IIOT), it is possible to monitor the condition of components and HS in real-time. However, in HS, this requires, in addition to standard equipment, additional equipment costs. On the other hand, it provides wide possibilities for monitoring the equipment and its condition in real-time. Predictive maintenance software allows users to store and analyse the critical results of their HS. One of the key things is the use of software and data obtained for improved HS's input parameters and maintenance. Using PdM, monitoring some of the HS parameters can prevent component failures or severe accidents. For example, a temperature sensor may indicate an increased temperature that could cause the oil to burn, create asphalt deposits, and cause components to malfunction or damage the pump. Vibration analysis can provide insight into possible failures. Increased vibration may indicate signs of damage and future component and HS failures.

Oil analysis includes analysis of lubricant properties to evaluate how the machine is depreciated in operation. The depreciation rate is estimated by measuring the number of suspended contaminants, wear residues in the lubricant, etc. While 50 years ago, this was the job of a tribologist, a machine wear experts, now the IoT provides the ability to sample oil and analyse it on-site. Table 36.1 shows how to maintain HS in the industry, using visual and instrumental inspection, Real-Time Condition Monitoring and PdM 4.0.

36.3.3 Advantages of Predicting the "Unpredictable"—Self-sustaining HS

The real potential offered by PdM 4.0 is already being proven where it started to be applied. Real-time monitoring only leads to a certain level of reliability, the level at which unpredictable and unexplained failures and delays occur. However, some of these failures are solved by analysing a large amount of data. Self-sustaining PdM 4.0 maintenance mode involves harnessing artificial intelligence's power to perform inspection and detection and detect patterns and anomalies, the detection of which is missed by even the most gifted people in the field maintenance.

The advantage of the self-sustaining PdM 4.0 maintenance method is that it maintains the maintenance problem from human to artificial intelligence. Artificial intelligence should recognise all possible ways and situations in which HS can be found based on HS's current and historical state. Also, the processing of large amounts of data from sensors and devices for analysis and detection of certain parameters, possible failures and system downtime can be analysed, assumed and concluded. In this way, when necessary to stop the system and perform the service without downtime and impact, the overall operation and production process are precisely defined. Also, by analysing the production data and the influential parameters for the HS operation, artificial intelligence plans the period in which the service should be performed.

Table 36.1 Maintenance HS by applying real-time condition monitoring and PdM 4.0 [15]

Capability	1. Inspections	2. Instrument	3. RTCM	4. PdM4.0
Processes	Periodic inspection (physical) Checklist Paper recording	Periodic inspection (physical) Instruments Digital recording	Continuous inspection (remote) Sensors Digital recording	Continuous inspection (remote) Sensors and other data Digital recording
Content	Paper-based condition data Multiple inspection points	Digital condition data Single inspection points	Digital condition data Multiple inspection points	Digital condition data Multiple inspection points Digital environment data Digital maintenance history
Performance Measurement	Visual norm verification Paper-based trend analyses Prediction by expert opinion	Automatic norm verification Digital trend analyses Prediction by expert opinion	Automatic norm verification Digital trend analyses Monitoring by CM software	Automatic norm verification Digital trend analyses Prediction by statistical software Advanced decision support
IT	MS Excel/MS Access	Embedded instrument software	Condition monitoring software Condition database	Condition monitoring software Big data platform Wifi network Statistical software
Organisation	Experienced craftsmen	Trained inspectors	Reliability engineers	Reliability engineers Data scientists

36.4 A Practical Example of Preventive Maintenance

Preventive maintenance applied to a complex mobile machine shows the possibilities and importance of preventive maintenance with these machines and HS. The example of a new excavator that worked on the construction site for only 900 working hours showed problems related to the realisation of working pressures during operation, increased hydraulic fluid's working temperature, and variable working speeds during operation. The authors placed diagnostic devices on the excavator's hydraulic installation (Hydrotechnik Multi Handy 3010 type monitoring equipment was used for measure pressure, temperature and flow and HIAC™ ROC Remote Online Counter for measurement solids particle). Based on the recorded data, the values of operating parameters related to pressure, temperature and the number of solids in the oil mass were obtained. It is noted that the interval of recording parameters lasted 301 working hours of the excavator in operation after replacing the filter and new oil filling by an authorised service. The excavator is driven by a piston-axial pump with LS regulation, proportional and servo valves, and this group represents the most sensitive components in HS. Filters for oil filtration in these excavators should have a degree of filtration fineness between 4 and 6 μm (c) and a beta factor $\beta x = 200$ according to the ISO 4406/99 standard. In the existing condition on the excavator, the built-in filters were of adequate beta factor. The filtration degree was too rough and amounted to 30 μm, which was not adequate to the built-in components in HS. Another problem was the occurrence of excessive pressure in the oil return line to the filter. Due to large pressure dilatations of $\Delta p = 3.5$ bar (according to the specifications of the excavator manufacturer $\Delta p_{\max} = 1.5$ bar), it led to deformation and tearing of the filter cartridge—see Fig. 36.1. This phenomenon has affected the filter's non-functionality and the quality of the return oil filtration.

Table 36.2 shows the solids measurements in HS excavators after 301 operating hours, after oil and filter change. From the attached Table 36.2, there is a noticeable

Fig. 36.1 Damage to the steel filter cartridge mesh

Table 36.2 Results of measuring solid particles in the excavator fluid mass after an oil change

Description of the sample	Number of operating hours (h)		Number of measured solid particles by size in the 1 ml sample			Purity class	
	Excavator	New oil	4 μm (c)	6 μm (c)	14 μm (c)	NAS 1638	ISO 4408/99
After filtering the new oil (2)	900	0	1.576	158	2	7	18/16/10
During the operation (3)	943	43	26.129	1.799	22	10	21/18/13
During the operation (4)	1201	301	43.112	11.898	338	13	23/21/17

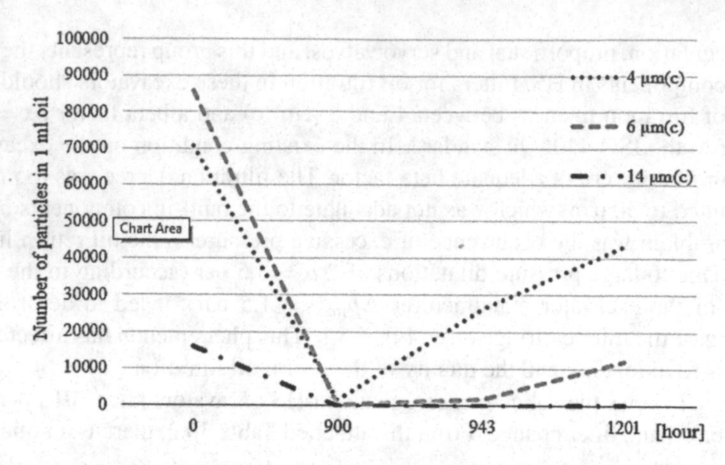

Fig. 36.2 Oil contamination level after 43 and 301 operating hours

increase in impurities from the NAS 7 class to the NAS 13 class from the enclosed. The consequence of this enormous input of particles was a faulty hammer on the excavator's arm, whose sealing system was poor and enabled contaminants' input through the hammer's working surfaces. The excavator's environment also contributed to particles' input because the excavator worked in the cement plant's environment on demolition work, with a large amount of surrounding dust. Figure 36.2 shows the ratio of particles at the beginning, after 43 and after 301 working hours of the excavator.

36.5 Conclusion

In conclusion, for maintaining a hydraulic system, it is necessary to apply specific diagnostic measures during the operation of a modern mobile hydraulic system with appropriate systems for keeping the pureness of the fluid at the required level. Applying different types of maintenance typologies strongly depends on availability and data acquisition techniques applied. The conditions for monitoring and defining the critical parameters of HS's in the operation of mobile machines are much more complex compared to stationary industrial machines. In the future, applying IoT concepts could be beneficial and, in fact, unavoidable in the process of providing autonomous maintenance techniques and DP approaches. The self-monitoring system still requires the participation of good experts in the field of HS's maintenance in order to make appropriate decisions in the maintenance process.

References

1. Tič, V., Edler, J., Lovrec, D.: Operation and accuracy of particle counters for on-line condition monitoring of hydraulic oils. Ann. Fac. Eng. Hunedoara **10**(3), 425–428 (2012)
2. Orošnjak, M., Peković, T., Jocanović, M., Karanović, V., Horvatić Novak, A.: Using contamination control in condition based maintenance of a hydraulic system. In: Proceedings of the Maintenance Forum on Maintenance and Asset Management Scientific conference maintenance of machinery and equipment, pp. 102–111. Budva (2017)
3. Norvelle, F.D.: Contamination control—the key to reliability in fluid power systems. Wear **94**, 47–70 (1984)
4. Orošnjak, M., Jocanović, M., Karanović, V.: Quality analysis of hydraulic systems in function of reliability theory. In: Proceedings of the 27th DAAAM International Symposium on Intelligent Manufacturing and Automation, pp. 569–577. DAAAM International, Vienna (2016)
5. Jocanović, M.T., Karanović, V.V., Ivanišević, A.V., Knežević, D.M.: Hydraulic hammer excavator failure due to solid particle contamination. Vojnoteh. Glas. **62**, 112–129 (2014)
6. Karanović, V., Jocanović, M., Baloš, S., Knežević, D., Mačužić, I.: Impact of contaminated fluid on the working performances of hydraulic directional control valves. Stroj. Vestnik/J. Mech. Eng. **65**, 139–147 (2019)
7. Nikolic, N., Antonic, Z., Doric, J., Ruzic, D., Galambos, S., Jocanovic, M., Karanovic, V.: An analytical method for the determination of temperature distribution in short journal bearing oil film. Symmetry (Basel) 12(4), 539-1–539-19 (2020)
8. Orošnjak, M., Jocanović, M., Karanović, V.: Applying contamination control for improved prognostics and health management of hydraulic systems. In: Ball, A., Gelman, L., Rao, B. (eds.) Advances in Asset Management and Condition Monitoring: Smart Innovation, Systems and Technologies, vol. 166, pp. 583–596. Springer, Cham (2020)
9. Wakiru, J.M., Pintelon, L., Karanović, V.V., Jocanović, M.T., Orošnjak, M.D.: Analysis of lubrication oil towards maintenance grouping for multiple equipment using fuzzy cluster analysis. IOP Conf. Ser. Mater. Sci. Eng. **393**, 012011-1–012011-9 (2018)
10. Singh, M., Lathkar, G.S., Basu, S.K.: Failure prevention of hydraulic system based on oil contamination. J. Inst. Eng. Ser. C. **93**, 269–274 (2012)
11. Shanbhag, V.V., Meyer, T.J.J., Caspers, L.W., Schlanbusch, R.: Failure monitoring and predictive maintenance of hydraulic cylinder: State-of-the-art review. IEEE/ASME Trans. Mechatronics. https://doi.org/10.1109/TMECH.2021.3053173
12. Zhang, C., Jiang, P., Cheng, K., Xu, X.W., Ma, Y.: Configuration design of the add-on cyber-physical system with cnc machine tools and its application perspectives. Proc. CIRP **56**, 360–365 (2016)
13. Luo, W., Hu, T., Zhu, W., Tao, F.: Digital twin modeling method for CNC machine tool. In: Proceedings of the IEEE 15th International Conference on Networking, Sensing and Control, ICNSC 2018, pp. 1–4. IEEE Press (2018)
14. Nurmi, J.: On Increasing the Automation Level of Heavy-Duty Hydraulic Manipulators with Condition Monitoring of the Hydraulic System and Energy-Optimised Redundancy Resolution. Tampere University of Technology, Tampere (2017)
15. Mulders, M., Haarman, M.: Predictive Maintenance 4.0—Beyond the Hype Report: PdM 4.0 Delivers Results. PWC (2018)

Chapter 37
Development of Portable IoT Diagnostic Device for Particle Contamination Monitoring

Goran Rodić, Velibor Karanović, and Dragana Oros

Abstract Every modern maintenance concept involves technical diagnostics. Technical diagnostics allow maintenance managers to be well informed about a specific problem or phenomenon, before making decisions. In this sense, for the maintenance of modern hydraulic systems, the counting of solid contaminants in hydraulic oil according to ISO 4406 standard is of great importance. The solid particles as contaminants in various ways adversely affect the efficient operation of the hydraulic system and therefore their concentration must be monitored. Different techniques are used for this purpose, and the most common being laser particle counting. This paper presents a developed portable IoT diagnostic device for particle contamination monitoring. The device can be used in online or offline mode and can transfer data in real-time via the internet, so it can immediately be analyzed. This significantly speeds up the process of decision-making by maintenance managers and eliminates human factor errors during measuring or transferring data.

Keywords Maintenance · Condition monitoring · Technical diagnostics · Automatic particle counter · Hydraulic power system · Development · Internet of Things (IoT)

37.1 Introduction

Modern maintenance concepts such as Condition-Based Maintenance (CBM), Predictive Maintenance (PdM), Reliability-Based Maintenance (RBM), or others, cannot be implemented without the use of technical diagnostics. Technical diagnostics uses methods and techniques such as vibrodiagnostic analysis, oil analysis, thermography, and others, to help maintenance managers to monitor degradational processes and make timely and knowledge-based decisions [1, 2].

The automatic particle counter (APC) is a diagnostic tool whose application is of great importance for modern, high precision, and high accuracy hydraulic systems.

G. Rodić · V. Karanović (✉) · D. Oros
University of Novi Sad, Faculty of Technical Sciences, Trg Dositeja Obradovića 6, 21000 Novi Sad, Republic of Serbia
e-mail: velja_82@uns.ac.rs

M. Rackov et al. (eds.), *Machine and Industrial Design in Mechanical Engineering*, Mechanisms and Machine Science 109,
https://doi.org/10.1007/978-3-030-88465-9_37

To achieve the maximum efficiency of such systems, the fluid condition is one of the basic requirements [3]. Fluid cleanliness is considered to be one of the main causes of hydraulic system failure [4]. The APC as a diagnostic device gives information about oil cleanliness and installed filtration system efficiency [5]. The usual way to apply oil contamination monitoring is to conduct laboratory analysis. This approach introduces some uncertainties into the final results, i.e. if sampling equipment is not suitable, or untrained personnel extracting a sample, or inexperienced laboratory technician preparing the sample and conducting the laboratory test. Also, the response time is not instantaneous, which delays management decisions. To avoid all these uncertainties, some solutions, include the installation of an APC unit on a hydraulic system, to have real-time monitoring. For many companies, this option is expensive if a large number of hydraulic systems have to be monitored. Another solution is to have a portable diagnostic unit. By regularly checking the cleanliness of the fluid in the machines, the user can plan and take certain maintenance actions on time, in case the particle contamination above the tolerated limits occurs. To improve monitoring, make better predictions, and act faster, Internet of Things (IoT) technology must be integrated. IoT can connect to sensors, machines, or smartphones, and exchange information over the internet [6]. There are a large number of studies, in different areas, that have proved that the application of these technologies significantly improves efficiency, accuracy, real-time remote monitoring, remote control [7–9].

Further, to perform online and offline testing, the concept of a portable APC unit has been developed. This device can be used as a maintenance tool within Industry 4.0. By transmitting the measurement results to the remote location in real-time, received data can be instantly analyzed.

37.2 Portable Monitoring Device—Conceptual Design

The portable diagnostic device consists of three integral parts: an electronic system, hydraulic system, and software. The system design is shown in Fig. 37.1.

37.2.1 Electronic System Components

The idea was to realize a portable monitoring device with software control on the PCB board. This board communicates with all the peripherals shown in Fig. 37.1. Besides the PCB board, other electronic hardware components of this portable monitoring device are the microcontroller, communication driver module, LCD, DC motor driver, voltage converter, and Li-ion battery.

The main task of the microcontroller is to communicate with the sensor via RS232 protocol, process the received data, and display them on the LCD. It also controls the el. motor speed, and creates a web server and displays the same data on the web page as on the LCD. The communication driver module converts RS232 communication

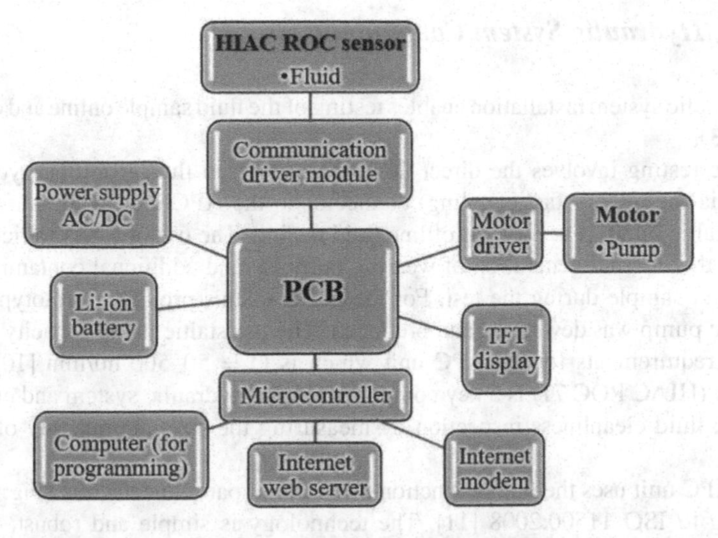

Fig. 37.1 A Portable monitoring device electronic hardware components

protocol to UART using a microcontroller, so the microcontroller can communicate with the APC sensor. The DC el. motor driver increases the PWM signal, of a certain amplitude, on the el. motor and to adjust the voltage level. The voltage converter lowers the mains voltage to 24VDC so the sensor can operate. Li-ion battery is used as a backup power supply to the system.

Figure 37.2 shows the conceptual design of a PCB board with power supply groups for each of the components of the proposed system.

All necessary paths with appropriate connections have been constructed on the board, and all voltage dividers, as well as decreasing voltage converters, have been implemented.

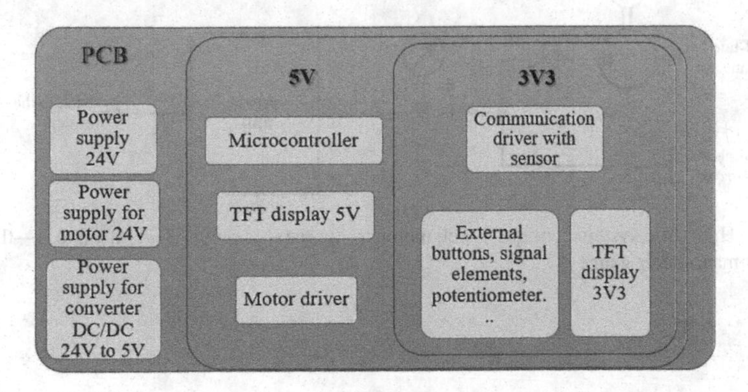

Fig. 37.2 PCB board concept

37.2.2 Hydraulic System Components

The hydraulic system installation enables testing of the fluid sample online and offline (Fig. 37.3).

Online testing involves the direct flow of pressurized fluid from the hydraulic system via the online (fast coupling) connector to the APC unit, while the use of the peristaltic pump is needed for offline fluid testing. The use of a peristaltic pump excludes the internal generation of wearing particles and additional contamination of the fluid sample during the test. For the needs of this project, a prototype of a peristaltic pump was developed and produced. The peristaltic pump capacity meets the flow requirements for the APC unit which is Q = 50–500 ml/min [10]. The APC unit (HIAC ROC 71) is a key component of the hydraulic system and enables hydraulic fluid cleanliness inspection by measuring the size and quantity of solid particles.

The APC unit uses the light extinction principle of particle detection (Fig. 37.4), according to ISO 11500:2008 [11]. The technology is simple and robust, which makes this detection technique suitable for use in the field. The APC has a sample cell through which the fluid sample passes. The sample cell includes a sapphire window, which allows a laser beam to pass through it and the fluid onto the photodiode (detector). The fluid flow through the sensor is predetermined, controlled, and calibrated. When the fluid is very clean, almost all the power of the laser reaches the photodiode, because nothing blocks the laser, and the output signal from the photodetector remains stable. However, when particles are present, they occasionally block some of the light from reaching the photodetector. The amount of blocked light is directly proportional to the particle size. The larger the particle, the more light is blocked, and consequently the greater the amplitude of the pulse coming out of the photodetector [12].

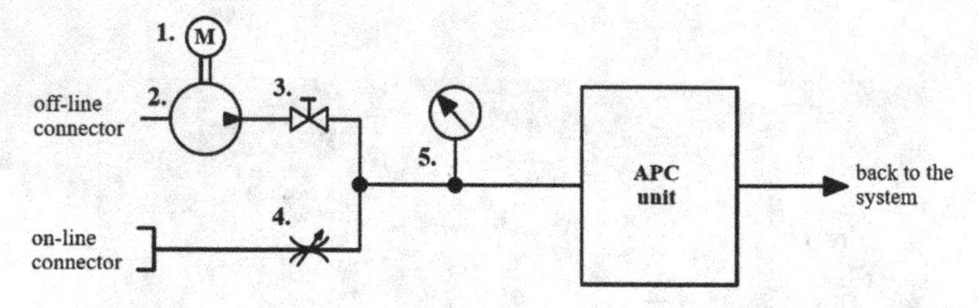

Fig. 37.3 Hydraulic system concept: 1—el. motor, 2—peristaltic pump, 3—ball valve, 4—throttle valve, 5 – manometer gauge

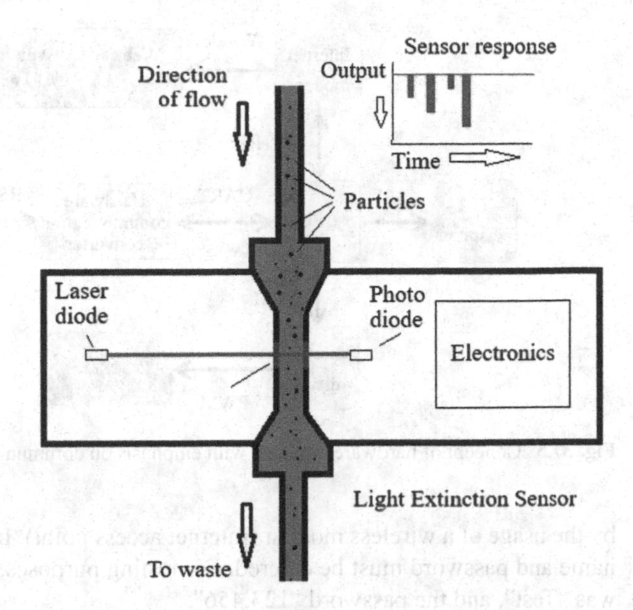

Fig. 37.4 APC operating principle [12]

37.2.3 Software Concept

For the system to operate in the way shown in Fig. 37.2, it is necessary to define all the data. Since the basis of the project is a particle sensor with which the microcontroller communicates in both directions, that communication can be a problem. First of all, the CPU should send request data, after which the sensor should respond to that request (full-duplex communication). During the period between sending this request data, the microcontroller is sending the response of the sensor to the LCD screen and to the website which is created on the webserver. In the meantime, the microcontroller also needs to generate a PWM for the motor on a defined connection pin. Figure 37.5 presents the concept of hardware modules of the diagnostic device.

In order to generate the appropriate signal to the sensor, it is necessary to know it's the communication protocol. In addition to the RS232 communication protocol defined by 8-bit, no parity, and one stop bit, the sensor uses the MODBUS protocol which can be seen in the technical documentation of the sensor. Since the sensor is intended for communication with ROC software [6], of which only the demo version was available to us, the communication protocol must first be identified.

This software will be the starting point in the firmware development for the IoT communication module. To do that, it is necessary to perform testing and recording due to the complexity and ignorance of the device being operated.

It is necessary to keep in mind that there is no appropriate technical documentation that would help solve this problem, i.e. the logic circuit, registers, and all other electrical components should be chosen.

In order to display the data on a web server, the user is obliged to provide access to the Internet. This can be done in two ways: by the mobile access point as access or

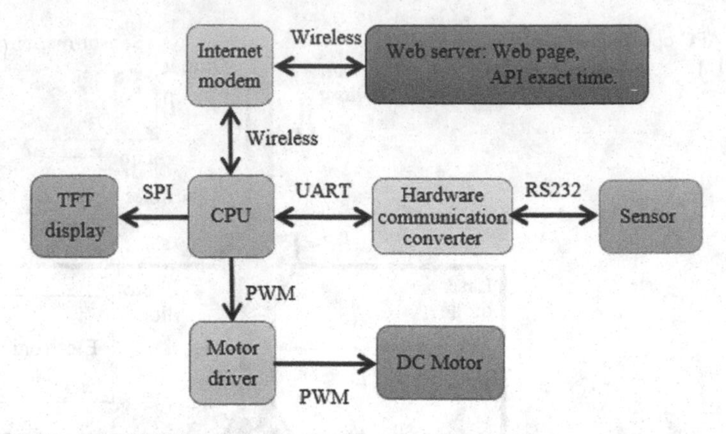

Fig. 37.5 Concept of hardware modules with emphasis on communication between them

by the usage of a wireless modem (Internet access point). In either way, the device's name and password must be entered. For testing purposes, the default device name was "Test", and the password "123,456".

After setting up the network on the mobile phone or modem, it is necessary to turn on the communication modem, after which the MCU (Microcontroller Unit) performs all the necessary initializations and connects to the Internet. While this is being done, the status of the microcontroller is displayed on the LCD. As soon as the microcontroller connects to the device and the Internet, a message appears on the screen showing the IP address of the device. This address should be copied to any internet browser. After loading the address, the microcontroller receives the information about the established connection via the Internet and automatically starts displaying the codes read from the sensor Fig. 37.6b. Also, the current time and date of the data readings are displayed on the screen Fig. 37.6a.

The user is not obliged to connect to the website, the connection time expires in 10 s and it is possible to continue recording data and displaying it only on the LCD.

37.3 Implementation and Testing of the Developed Device

The testing of the hardware system was performed in the following way. Since there is no appropriate technical documentation for the operation of the sensor, in terms of the operation of the hardware electronics contained in it, the decision to perform a test was made. Namely, it is known that the sensor works adequately with ROC software, using a USB TO SERIAL modem that connects to a computer. This knowledge was used to record communication between the two devices. The recording was performed using a logic analyzer. For the logic analyzer to be used, it was necessary to lower the logic levels of RS232 communication, from 24 V, below 5 V, which can

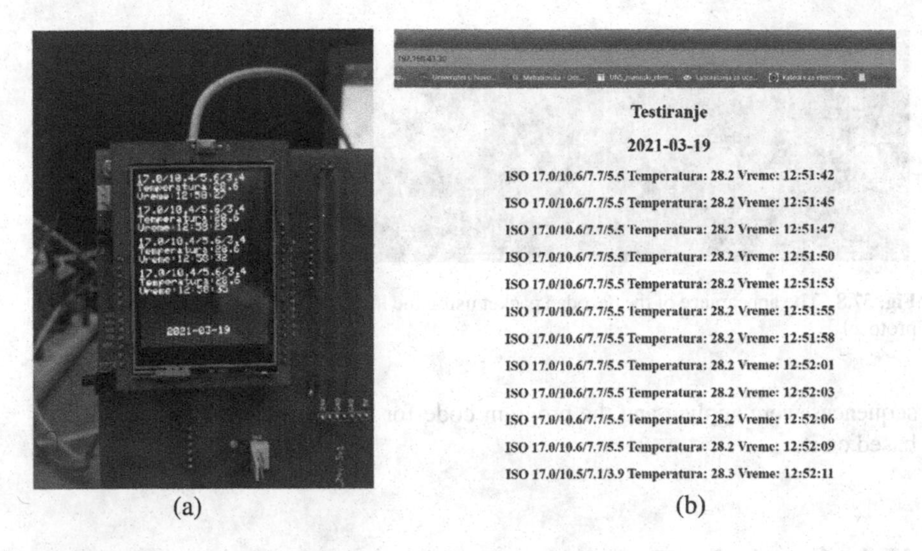

Fig. 37.6 Display of measured hydraulic oil cleanliness classes on LCD and computer screen via internet page

be read, without any faults. Lowering the logic level dealt with the application of additional PCBs.

After successful implementation, a test of recording communication between the sensor and the computer was performed, and the obtained result is shown in the Fig. 37.7.

After recording the communication protocol, the next step was the attempt to generate an identical signal using a microcontroller. However, the problem was in the amount of processor time that required precisely defined intervals, which is why it limited the operation of the microcontroller. Therefore, it was accessed using ready-made open source libraries for the MODBUS protocol.

After successful decoding of the signal Fig. 37.8, a map database of all signals was created, which took place during the recording. After creating the map, repetitive

Fig. 37.7 The result obtained by recording the communication 60 s: channel 1. is TX signal, channel 2. is RX signal

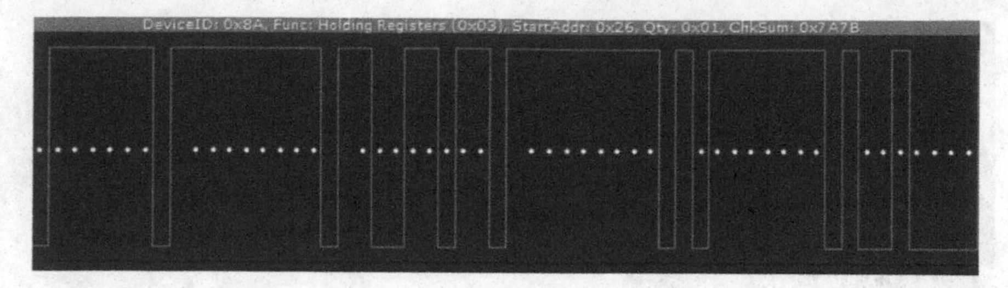

Fig. 37.8 The appearance of the decoded packet using the logic analyzer software for the Modbus protocol

sequences were applied and the program code for the microcontroller was created based on it.

37.4 Conclusion

By introducing regular control of the cleanliness of the hydraulic fluid, machine users will be more successful in carrying out maintenance actions, which directly affect greater efficiency and productivity, reduces the number of failures and maintenance costs. The proposed diagnostic device solution provides the following:

- Online and offline measuring,
- Real-time data transfer and representation via the internet,
- Portability and suitability for fleet monitoring, and
- Elimination of errors due to the human factor.

Future directions of action include the integration of other sensors into the assembly of this device, such as sensors for measuring humidity, viscosity, and others. This will provide more precise data about the current state of the system so managers can deeply understand system behavior. Also, software should include the development of a user-friendly interface and provide quality data representation.

Acknowledgements This paper is a result of the research project conducted in the Laboratory for Quality, Logistics, and Maintenance on the Faculty of Technical Sciences, during the realization of the Master thesis by Goran Rodić. The project is financed by the Provincial Secretariat for Higher Education and Scientific Research (project no. 142-451-1003/2020-02) and supported by donations from companies: Hansa Flex, MikroElektronika, and Tehnomark.

References

1. Karanovic, V., Jocanovic, M., Wakiru, J., Orošnjak, M.: Benefits of lubricant oil analysis for

maintenance decision support: a case study. IOP Conf. Ser.: Mater. Sci. Eng. **393**, 012013–1–012013–7 (2018)
2. Czichos, H.: Handbook of Technical Diagnostics: Fundamentals and Application to Structures and Systems. Springer, Berlin Heidelberg (2013)
3. Orošnjak, M., Jocanović, M., Karanović, V.: Applying contamination control for improved prognostics and health management of hydraulic systems. In: Ball, A., Gelman, L., Rao, B. (eds.) Advances in Asset Management and Condition Monitoring: Smart Innovation, Systems and Technologies, vol. 166, pp. 583–596. Springer, Cham (2020)
4. Tic, V., Lovrec, D., Edler, J.: Operation and accuracy of particle counters for online condition monitoring of hydraulic oils. Ann. Fac. Eng. Hunedoara **10**(3), 425–428 (2012)
5. Patel, K.K., Patel, S.M.: Internet of things-IOT: definition, characteristics, architecture, enabling technologies, application & future challenges. Int. J. Eng. Sci. Comput. **6**(5), 6122–6131 (2016)
6. Oros, D., Penčić, M., Šulc, J., Čavić, M., Stankovski, S., Ostojić, G., Ivanov, O.: Smart intravenous infusion dosing system. Appl. Sci. **11**(2), 513-1–513-26 (2021)
7. Tegeltija, S., Tejić, B., Šenk, I., Tarjan, L. and Ostojić, G.: Universal IoT vending machine management platform. In: Proceedings of the 19th International Symposium INFOTEH-JAHORINA 2020, pp. 1–5. IEEE Press (2020)
8. Nikoličić, S., Kilibarda, M., Atanasković, P., Duđak, L., Ivanišević, A.: Impact of RFID technology on logistic process efficiency in retail supply chains. Traffic **27**(2), 137–146 (2015)
9. Beckman Coulter, HIAC ROC Liquid Particle Counter. https://www.beckman.com/techdocs/DOC026.97.803045/wsr-213772, last accessed 2021/02/25
10. Stecki, J.: Contamination control and failure analysis. In: Gresham, R.M., Totten, G.E. (eds.) Lubrication and Maintenance of Industrial Machinery: Best Practices and Reliability, pp. 1–57. CRC Press, Taylor & Francis Group (2006)
11. ISO 11500:2008, Hydraulic fluid power—Determination of the particulate contamination level of a liquid sample by automatic particle counting using the light-extinction principle (2008)
12. https://www.beckman.com/resources/reading-material/application-notes/particle-counting-in-mining-applications, last accessed 2021/02/27

Chapter 38
New Method for Evaluation of Corrosive Resistivity of Impregnants

Miroslav Ďuračka, Tomáš Dérer, Zuzana Filová, Katarína Kocúrová, Alena Kozáková, and Michal Šištík

Abstract Nowadays for evaluation of resistivity against corrosion of electric rotating machines windings and transformers there are several methods used. These methods are assigned for samples of complete motors and stators. Problems of evaluating process (material and financial difficultness) are eliminated using a new method for evaluating of corrosive resistivity of impregnants. In this method there are used small testing objects with layer of impregnant, which is cured at conditions required for used impregnant. Impregnated object is exposed in a particular corrosive environment and interval of sample observing is set up. During this interval the level of corrosion of impregnated tested sample is observed until it reaches the specified degree of corrosion. An advantage of this method is taking photographic records of samples and evaluation of level of corrosion using graphic software. This softwarc improves visual evaluation of rusted part, because it analyses its photographic record and is usable for optical examination of wide scale of materials. For evaluation there was used the contrast of white and black colour, black coloured areas are without corrosion and white coloured are rusted areas. Using the evaluating concept

M. Ďuračka · T. Dérer · Z. Filová · K. Kocúrová · A. Kozáková (✉) · M. Šištík
VUKI a. s., Rybničná 38, 831 07 Bratislava, Slovakia
e-mail: kozakova@vuki.sk

M. Ďuračka
e-mail: duracka@vuki.sk

T. Dérer
e-mail: derer@vuki.sk

Z. Filová
e-mail: filova@vuki.sk

K. Kocúrová
e-mail: kocurova@vuki.sk

M. Šištík
e-mail: sistik@vuki.sk

M. Rackov et al. (eds.), *Machine and Industrial Design in Mechanical Engineering*, Mechanisms and Machine Science 109,
https://doi.org/10.1007/978-3-030-88465-9_38

of corrosive resistivity of impregnants according to this method is very practical during designation of anticorrosive system. In this process there is at first the selection of adequate components and their combinations are tested. It presents a huge amount of photographic records, which can be processed into graphic dependences after using the graphic software and there after analysed and evaluated.

Keywords Graphic program · Quality control · Resistivity against corrosion · Humidity · Optical examination

38.1 Introduction

Nowadays, the requirements for the properties of impregnating resins used in the construction of electrical machines [1] are being expanded in technical practice. The demand for increased corrosion resistance of the impregnated electrical rotating machines (ERM) components and transformers is increasingly emerging from their manufacturers. The simulation of a corrosive environment is relatively simple and can be performed according to standards (e.g. humidity tests according to EN 60,455-2 [2], EN 60,034-1: 2005 [3]) or under conditions determined individually, e.g. by immersion in water or NaCl solution. Exposure of impregnated machine parts or complete machines in a corrosive environment, in particular salt mist and NaCl solution shall demonstrate the resistance of the impregnants tested in a relatively short time. After exposure, electrical parameters are evaluated alongside with visual assessment of the corrosion extent. The results of comparing the impregnated stators with the different impregnants before and after exposure in the chamber are shown in Fig. 38.1.

Fig. 38.1 Practical application of anticorrosive additives on the stator on the left, after exposure according to EN 60,455-2 [2]

The difficulty of the evaluation process lies in the material demands and the price of the used parts and whole machines. Optional number of pieces cannot be used and an affordable range of tests should be considered.

We have designed a relatively simple method without demanding technical equipment for evaluating results in our laboratory. The purpose of this method is not a qualitative assessment of corrosion but quantitative. Optical method, this suggested is the optimal variant for us. According to our knowledge, similar methods are used to evaluate, for example, biological and mineralogical samples, but in the electrotechnical industry and the paint industry is not used. It was not yet published yet, therefore no professional references.

38.2 Monitoring of the Stage of Corrosion by a Graphic Program

Aforementioned evaluation method would be problematic when researching, developing and designing an anticorrosive system. The anticorrosive system is generally multi-component. When evaluating the efficacy of individual additives and their combinations with different component ratios, the number of test bodies on which the extent of corrosion is assessed is considerably increasing.

Visual evaluation (included in standardized tests [4–7]), an initial evaluation proves to be a practical and simple method that can also be done by directly comparing the photographic record, however the clarity of the results becomes worse due to a large number of records.

In this case, it is preferable to quantify the areas affected by corrosion what allows a suitably selected graphic program by which it is possible to express the extent of corrosion, for example, in percentage. For the evaluation of corrosion resistance of selected types of impregnants, NaCl solution was chosen as a corrosive medium, which is easy and safe to work under laboratory conditions. The test specimen is a 100 × 40 × 1.5 mm carbon steel uncoated sheet, degreased with acetone. After impregnation with the appropriate impregnant, it is placed in a 4% NaCl solution at laboratory temperature and photographic records are made at 24-h intervals. These are then visually compared and the extent of corrosion damage is not quantified (see Fig. 38.2).

The second option is an evaluation, using a graphic program that serves for image analysis and is useful for optical examination of a wide range of materials. Black-white contrast has been used in the evaluation. The results are shown in Fig. 38.3 and Table 38.1.

Figure 38.3 clearly shows the difference in the size of the corroded area in a sample without additives and the one using anti-corrosive additives. Applied graphic program [8] is able to analyse a black and white image and express how much of the total area of the sheet is corroded in percentage. On the basis of this analysis it

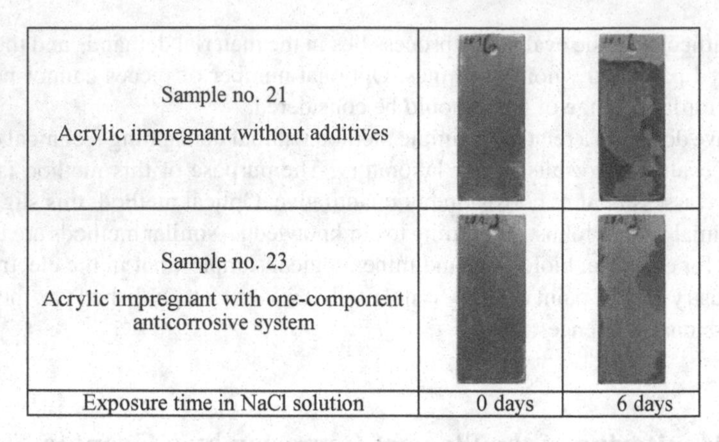

Fig. 38.2 Direct comparison of test samples before and after exposure in NaCl solution

Sample no. 21 Acrylic impregnant without additives		
Sample no. 23 Acrylic impregnant with one-component anticorrosive system		
Exposure time in NaCl solution	0 days	6 days

Fig. 38.3 Display of evaluated samples before and after exposure in salt solution for graphical evaluation

Table 38.1 Graphic evaluation of samples before and after exposure in salt solution

Sample	Total area (mm^2)	Average size of partial area (mm^2)	Colour homogenous area (%)	Correlation
0 days versus 21	974,186	12,989.147	99,248 (no corrosion)	0.918
6 days versus 21	627,046	784,788	77,525 (corroded)	0.926
0 days versus 23	870,203	72,516.917	99,367 (no corrosion)	0.916
6 days versus 23	226,301	37,523	28,010 (corroded)	0.919

is possible to monitor the progress of corrosion over time, to display the values in a chart and thus clearly evaluate the effect of anti-corrosive additives.

Two groups of anticorrosive additives were tested, organic labelled "H" and inorganic labelled "R". Each of two types of impregnants, labelled NAH and NAB. Samples were monitored for 25–28 days, whereas the degree of corrosion that the samples will reach over this time period was compared. The results are shown in Figs. 38.4, 38.5, 38.6, 38.7, 38.8 and 38.9.

Figures 38.4 and 38.5 show that the presence of organic "H" inhibitors in NAH impregnant has cut on the corrosion process by 4 days. The presence of silica "A" in the mixture set no. 17–20, which was added to adjust the flow properties, has not affected the corrosion progress of these samples.

Figure 38.6 displays the corrosion behaviour of samples with the impregnating agent NAH, but with the addition of inorganic inhibitors "R". Obviously, the additive samples exhibit a slower corrosion process and at the time when the non-additive sample reached the degree of complete corrosion, these samples were corroded to 80 or 90%.

The following figures show the behaviour of the samples with NAB impregnant. Figure 38.7 shows the effect of the organic inhibitors "H" in mixtures 10–2.

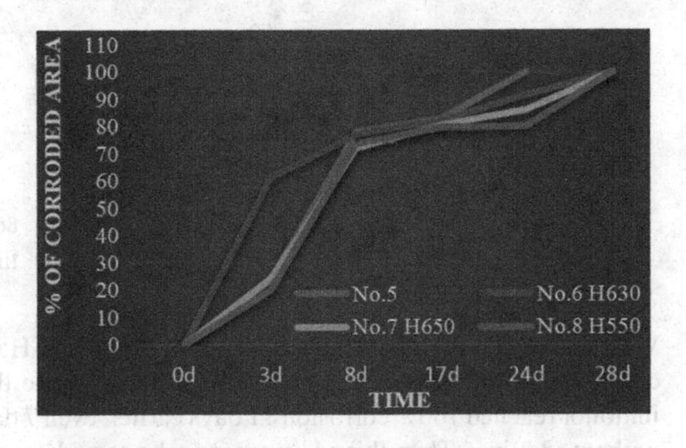

Fig. 38.4 NAH, corrosion development in mixtures with organic inhibitors without silica "A"

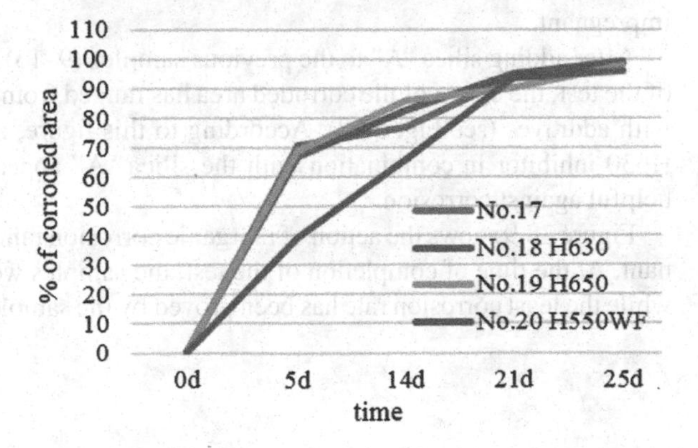

Fig. 38.5 NAH, corrosion development in mixtures with organic inhibitors with silica "A"

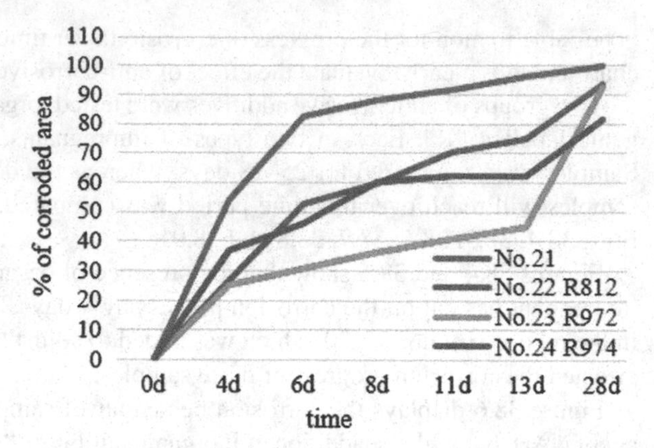

Fig. 38.6 NAH, corrosion development in mixtures with inorganic inhibitors (hydrophobic silica)

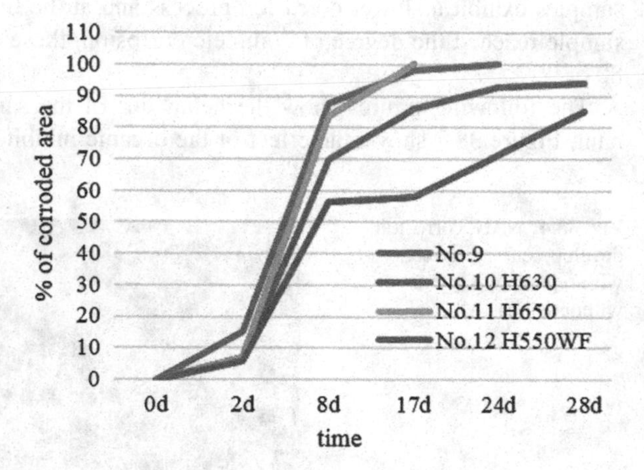

Fig. 38.7 NAB, corrosion development in mixtures with organic inhibitors

We observe that the samples with additives H630 and H550 have reached 80–90% corrosion, at the time of completion of the test, while the sample with the H650 inhibitor reached 100% corrosion 11 days earlier, even 7 days earlier than the sample without additives. Thus, this additive cannot be considered suitable for use in the NAB impregnant.

After adding silica "A" to the previous samples (9–13), at the time of completion of the test, the extent of the corroded area has ranged from 85 to 95% by the samples with additives (see Fig. 38.8). According to this figure, it can be assumed that the H630 inhibitor in combination with the silica "A" appears to be at least partially helpful against corrosion.

Figure 38.9 shows the action of inorganic corrosion inhibitors in the NAB impregnant. At the time of completion of the test, the samples were corroded to 80–100%, while the least corrosion rate has been proved by the sample inhibited by using R972.

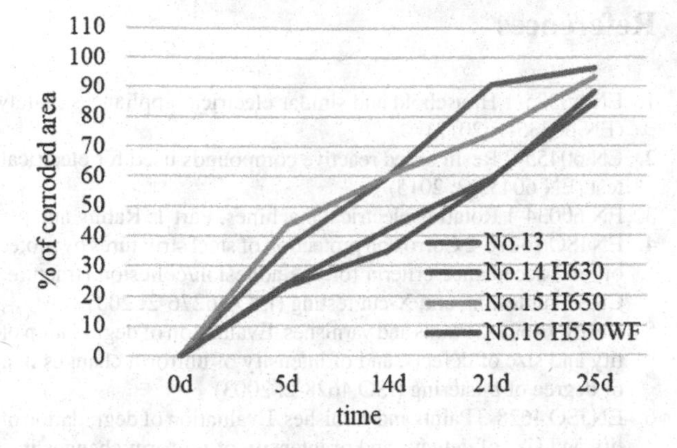

Fig. 38.8 NAB, corrosion development in mixtures with organic inhibitors with silica "A"

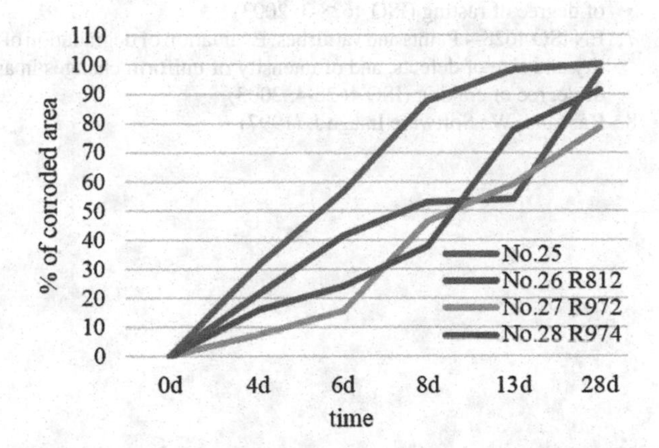

Fig. 38.9 NAB, corrosion development in mixtures with inorganic inhibitors (hydrophobic silica)

38.3 Conclusion

The results presented illustrate the evaluation of the effectiveness of corrosion inhibitors in two types of impregnants, where each inhibitor has been assessed separately in the same corrosive environment. This procedure may be used to assess the impact of any corrosive environment. Quantifying the size of the corroded area enables selecting the most effective substances and design multi-component anti-corrosion systems effectively. Searching and developing of a new method of corrosion assessment has resulted from the practical need to have a relatively simple tool for evaluating the effectiveness of anti-corrosion systems. Therefore, it is vital, especially for the first phase of research and formulae creation, when the largest amount of samples are processed and selected.

Acknowledgements This work was supported by the Slovak Research and Development Agency under the Contract no. APVV-17-0304, APVV-18-0028, APVV-18-0029 and APVV-19-0045.

References

1. EN 60335-1 Household and similar electrical appliances. Safety. Part 1: General requirements (EN 60335-1: 2012)
2. EN 60455-2 Resin.based reactive compounds used for electrical insulation. Part 2: Methods of test (EN 60455-2: 2015)
3. EN 60034-1 Rotating electrical machines. Part 1: Rating and performance (EN 60034-1: 2010)
4. EN ISO 16276-2 Corrosion protection of steel structures by protective paint systems. Assessment of, and acceptance criteria for, the adhesion/cohesion (fracture st5ength) of a coating. Part 2: Cross-cut testing and X-cut testing (ISO 16276-2: 2007)
5. EN ISO 4628-2 Paints and varnishes. Evaluation of degradation of coatings. Designation of quantity and size of defects, and of intensity of uniform changes in appearance. Part 2: Assessment of degree of blistering (ISO 4628-2: 2003)
6. EN ISO 4628-3 Paints and varnishes. Evaluation of degradation of coatings. Designation of quantity and size of defects, and of intensity of uniform changes in appearance. Part 3: Assessment of degree of rusting (ISO 4628-3: 2003)
7. EN ISO 4628-4 Paints and varnishes. Evaluation of degradation of coatings. Designation of quantity and size of defects, and of intensity of uniform changes in appearance. Part 4: Assessment of degree of craking (ISO 4628-4: 2003)
8. Rasband, W.: Software Image J. (1997)

Chapter 39
Analysis of Pressure Force in Robot Supported Sheet Metal Forming

Malik Čabaravdić, Dennis Möllensiep, Bernd Kuhlenkötter, and Alfred Hypki

Abstract In the area of sheet metal forming, as a very flexible manufacturing process, there is great potential for improvement, regarding to the economical and fast production of components in small quantities, such as small series or prototypes. For this reason, a new process was developed at the Chair of Production Systems (LPS) at the Ruhr University Bochum for research in the field of robot supported incremental sheet metal forming, so-called roboforming. Due to the heavy calculations and high speed of communication between the robots and external computer by the online stiffness compensation, which is necessary for the improvement of process accuracy, control units of the robots were sometimes not able to fulfill the given task, causing, especially by the smaller radii of work pieces, unsmooth movement of manipulator. This resulted in the further deviations of the final geometry. Therefore, there is a need to improve the existing method of processing or to test some other strategies by the robot supported incremental sheet forming. One of these strategies is force controlled incremental sheet forming. In this approach, the pressure force between tool and work piece is used as additional referral variable in the control loop (besides the position of the robot end-effector), enabling the system to reduce error in the final geometry of work piece. An analysis of the contact force in the robot supported metal sheet forming, which is necessary for the implementation of force control, is given in this paper.

Keywords Sheet metal forming · Robot force control · Pressure force

39.1 Introduction

The market behavior of the last few years shows a trend towards an increasing customization of the products to be manufactured. The associated increase in the

---

M. Čabaravdić (✉)
University of Zenica, Zenica, Bosnia and Herzegovina
e-mail: malik.cabaravdic@unze.ba

D. Möllensiep · B. Kuhlenkötter · A. Hypki
Lehrstuhl Für Produktionssysteme, Ruhr-Universität-Bochum, Bochum, Germany

M. Rackov et al. (eds.), *Machine and Industrial Design in Mechanical Engineering*, Mechanisms and Machine Science 109,
https://doi.org/10.1007/978-3-030-88465-9_39

number of variants and the shorter product life cycles require higher speed and flexibility in product development [1, 2]. While the transition from mass production to the manufacture of product variants took place in the last century, the economic production of individual products (lot size 1) is the task to be solved in the future [3]. As a result, continuous development of innovative production systems is required to meet these requirements [1].

Particularly in the area of sheet metal forming, there is great potential for improvement, regarding to the economical and fast production of components in small quantities, such as small series or prototypes. Classic forming processes (i.e. stretching and deep drawing) result in a low level of flexibility due to the use of massive and work piece-specific tools for geometrically different work pieces and prototypes, reflecting in the higher costs, increased effort and additional time required. Therefore, the research of incremental sheet metal forming processes, which have the potential for more flexible production, has been intensified in the last years. Incremental sheet metal forming is characterized by spatially limited, repeatedly occurring deformations for the step-by-step creation of a shape, what enables the use of tools that are independent of product geometry. In addition, these locally limited forming zones offer, in comparison to classic processes, the advantage that significantly lower process forces are required. As a result of the very small contact surface of the forming tool with the sheet metal to be processed, CNC machines or industrial robots are often used for tool movement, which are characterized by flexibly programmable movement paths [1, 4].

For this reason, a new process was developed at the Chair of Production Systems (LPS) at the Ruhr University Bochum for research in the field of incremental sheet metal forming, so-called ROBOFORMING.

ROBOFORMING is a new developed incremental sheet metal forming process supported by robots. It is used for the production of sheet metal components in small lot sizes and assumes the application of a robot cooperation system, consisting of two industrial robots for the implementation of the kinematical shape-generation. One robot drives the forming, the other the supporting tool. The forming of the final shape occurs by the incremental infeed of the tool in depth direction synchronized with the lateral movement along the contour while the sheet is sustained by the supporting tool at the opposite side. Compared to other incremental sheet metal forming processes this system is characterized by high flexibility of the workpiece geometry excluding the need for special tools, which are dependent on the final product shape. The main focus of the ROBOFORMING approach lies in the development and the implementation of an automated system design for the production of individual sheet metal parts following the approach "direct forming of the individual part on the basis of a CAD file" [5].

39.2 Stiffness Compensation

However, the potentials mentioned and other advantages, such as the production of complex components with undercuts, are also accompanied by process-related disadvantages. Due to their structure with rotational axes, industrial robots have a low mechanical rigidity, which means that when the load is greater, in this case large deformation forces, deflections occur in the joints. These lead to a displacement of the tool tip, mounted at the end of the respective robot arm, from the target position, causing significant inaccuracies in terms of shape and dimensional accuracy during processing [1]. In order to minimize the processing error, several investigations and developments have already been carried out at the Chair of Production Systems (LPS) at the Ruhr University Bochum, which result in various implementation variants and concepts of a so-called stiffness compensation. The function of the stiffness compensation is to calculate the process force-related displacements of the tool on the basis of information about the deformation forces in order to derive the corresponding correction values, so that the robot moves back to the original target points. In the implementations of Zhu [6] and Kang [7], these calculations or work steps take place before the start of the forming process, so-called offline method. Laurischkat [8] designed additional concepts in which the above-mentioned calculations for the correction values (partly) take place during the forming process (online method). Due to the limited computing power, however, only a partially online version has been implemented on the real robots. In the work of Gorlas, [9] a functional online stiffness compensation, which did not require separate calculations, was implemented for the first time. In detail, this means a determination of the deviations and calculation of the correction values taking place on a computer located next to the system during the forming process, as well as the subsequent implementation by the robot in real time.

Although this method improved the accuracy of the products obtained by the incremental sheet forming, the final result was still not on the required level for some complex geometries. Besides, due to the heavy calculations and high speed of communication between the robots and external computer by the online stiffness compensation, control units of the robots were sometimes not able to fulfill the given task, causing, especially by the smaller radii of work pieces, unsmooth movement of manipulator. This resulted in the further deviations of the final geometry.

Therefore, there is a need to improve the existing method of processing or to test some other strategies by the robot supported incremental sheet forming. One of these strategies is force controlled incremental sheet forming. In this approach, the pressure force between tool and work piece is used as additional referral variable in the control loop (besides the position of the robot end-effector), enabling the system to reduce error in the final geometry of work piece.

39.3 Pressure Force Modelling

In the first step of implementation of force control in sheet metal forming, the pressure forces from previous experiments, done at the Chair of Production Systems, were analyzed. The basis were the experiments for automatic parameter estimation of local support at all path points of the tool in incremental sheet forming supported by robots [10]. Main input parameters were: infeed depth, wall angle, support angle and support force.

A dedicated base geometry with different curvatures was designed and it was formed with various draft angles whereby work pieces with wall angles between 30° and 60° were created (see Fig. 39.1). The forming of the base geometry is not possible at higher draft angles since the surfaces intersect. Smaller draft angles are not applied since the process limits [11–13] would be exceeded. The geometry consists of concave, convex and straight geometric areas. All concave segments have unique arc length of 14 mm with the variation of the radius between 20 and 70 mm. All convex segments have length of 20 mm and the radius varies from 20 to 195 mm.

On the basis of existing research, the limits for all input parameters were determined. Support angles between 0 and 100% as well as support forces between 100 and 500 N were considered. At lower forces the effect of support can be neglected. At forces which are higher than these the tool actively participates in the process and the support-effect is then lost. Furthermore, the infeed depth ranges from 0.1 to 0.6 mm. An lower infeed depth will cause very long paths of the tool and is very hard to be executed by the industrial robots because of their positional accuracy (about 0.1 mm). Infeed depths higher than 0.6 mm will cause very high deviations in the workpiece geometry and therefore cannot be considered here.

Main output parameter for the analysis in this work is the pressure force. Through the RSI connection with the robot, the online data from the 6-axes force-torque sensor, mounted on the robot wrist, during the processing were collected. After the execution of every single experiment, the active pressure force was examined and its mean value was calculated.

Overall, 33 different experiments were considered. The values of input parameters for each experiment, as well as the mean values of the pressure force are given in the Tables 39.1a, b.

Using the obtained data a regression model was established, which predicts mean pressure force for a given set of input parameters and which can be applied in the

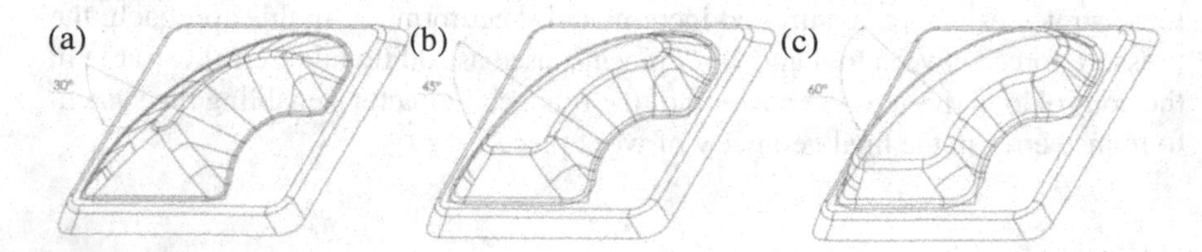

Fig. 39.1 Exemplary work piece geometries with wall angles of **a** 30, **b** 45 and **c** 60° [10]

Table 39.1 Input parameter combinations for 33 experiments and mean values of pressure forces

Exp. No	X_1 Support force (N)	X_2 Support angle (%)	X_3 Infeed depth (mm)	X_4 Wall angle (°)	Y Pressure force (N)
1	223	0.23	0.38	55	833.72
2	376	0.69	0.19	45	758.37
3	157	0.73	0.11	60	656.18
4	449	0.99	0.55	55	895.91
5	305	0.15	0.52	50	863.59
6	138	0.57	0.34	40	736.89
7	440	0.8	0.59	45	876.73
8	418	0.91	0.23	40	778.30
9	430	0.48	0.2	40	732.66
10	100	0.43	0.15	45	673.94
11	370	0.18	0.14	55	808.03
12	324	0.11	0.54	35	755.35
13	334	0.96	0.49	35	806.06
14	480	0.01	0.24	35	624.16
15	459	0.03	0.28	50	786.58
16	114	0.26	0.27	30	707.32
17	164	0.78	0.47	35	730.30
18	391	0.93	0.43	50	884.79
19	306	0.88	0.56	60	856.73
20	259	0.54	0.11	60	665.20
21	234	0.09	0.42	35	773.24
22	281	0.5	0.18	40	729.92
23	496	0.76	0.22	45	835.15
24	203	0.21	0.58	50	900.37
25	360	0.35	0.51	60	861.11
26	402	0.6	0.39	30	741.25
27	349	0.12	0.36	55	768.17
28	186	0.69	0.45	55	817.71
29	171	0.44	0.46	30	655.82
30	240	0.65	0.29	30	687.59
31	124	0.38	0.41	50	800.13
32	284	0.32	0.31	30	614.56
33	469	0.63	0.34	45	876.38

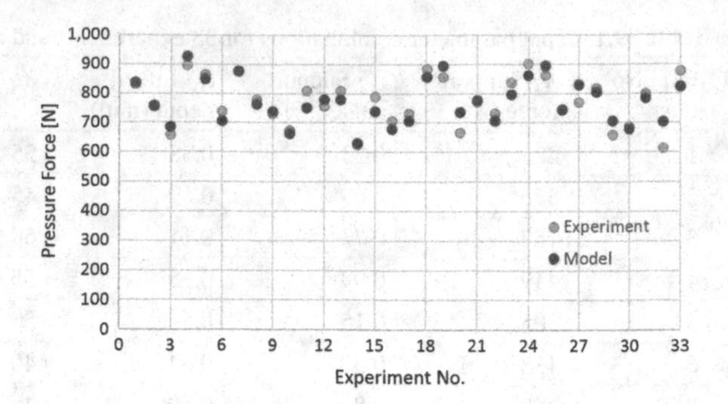

Fig. 39.2 Comparison experimental data—model data

force control of an industrial robot by incremental sheet forming. The best fit model (R-squared = 0.806; Adjusted R-squared = 0.77) was obtained through logarithmic linear regression. The equation which describes the influence of input parameters on pressure force in this case is:

$$Y = 141.36 \cdot X_1^{0.155} \cdot X_2^{-0.564} \cdot X_3^{0.112} \cdot X_4^{0.249} \cdot e^{0.098 \cdot \ln X_1 \cdot \ln X_2}. \quad (39.1)$$

Comparison of the data obtained in the experiment and model data is given in Fig. 39.2.

Obtained model of pressure forces can be considered as a starting point for implementation of the robot force control for sheet metal forming, giving the input values for control schemas for different process parameters. In the next step of the research, a hybrid force-position control of the robot will be applied to the incremental sheet metal forming process.

39.4 Conclusion

In order to improve the quality of products obtained by robot-based incremental sheet metal forming it is necessary to implement a new force control schema. To achieve this goal previous experiments, done at the Chair of Production Systems at the Ruhr University in Bochum, had to be analyzed in the first step. As the result of this analysis, a new mathematical model, describing dependence of the mean pressure force on the process parameters, was developed.

This model of pressure forces can be considered as a starting point for implementation of the robot force control for sheet metal forming, giving the input values for control schemas for different process parameters. The next step of research in this field should be introduction of a hybrid force-position control of the robot in the incremental sheet metal forming process.

References

1. Meier, H., Smukala, V., Buff, B.: Roboterbasierte inkrementelle Blechumformung: **Bauteilgenauigkeitssteigerung** in einem automatisierten industriellen Gesamtkonzept. wt Werkstatttechnik online 98(10), 831–836 (2008)
2. Groche, P., Kraft, M., Schmitt, S.O., Lorenz, U., Pokutta, S., Martin, A., Ziems, N.: Rechnet sich Flexibilität in der Umformtechnik? Ein wirtschaftlicher Vergleich konventioneller und flexibler Umformanlagen. wt Werkstatttechnik online **100**(10), 734–742 (2010)
3. Uhlmann, E.: Wandel der Fabrik durch Produktindividualisierung. Marktchance Individualisierung. Springer, Berlin (2013)
4. Kreimeier, D., Smukala, V., Magnus, C., Buff, B., Zhu, J.: Kraftgeregelte Umformung im Roboforming: Einfluss auf Bauteilgenauigkeit, Oberflächenqualität und multimodale Geometrien. wt Werkstatttechnik online 100(10), 772–778 (2010)
5. Ruhr-Universität Bochum Lehrstuhl für Produktionssysteme: "Roboforming"—Roboterbasierte Blechumformung komplexer Bauteile in kleinen Stückzahlen: Abschlussbericht des Verbundprojektes (2010)
6. Zhu, J.: Simulation of the load-dependent tool path deviations in robot based incremental sheet metal forming. Ruhr-Universität Bochum, Masterarbeit (2009)
7. Kang, J.: Weiterentwicklung und experimentelle Verifikation eines modellbasierten Ansatzes zur Genauigkeitssteigerung in der roboterbasierten inkrementellen Blechumformung. Ruhr-Universität Bochum, Masterarbeit (2010)
8. Laurischkat, R.: Kompensation prozesskraftbedingter Bahnfehler bei der roboterbasierten inkrementellen Blechumformung. Schriftenreihe des Lehrstuhls für Produktionssysteme. Shaker, Aachen (2012).
9. Gorlas, T.: Entwicklung einer Steifigkeitskompensation für die roboterbasierte inkrementelle Blechumformung. Ruhr-Universität Bochum, Masterarbeit (2020)
10. Störkle, D., Altmann, P., Möllensiep, D., Thyssen, L., Kuhlenkötter, B.: Automated parameterization of local support at every toolpath point in robot-based incremental sheet forming. Procedia Manuf. **29**, 67–73 (2019)
11. Duflou, J.R., Verbert, J., Belkassem, B., Gu, J., Sol, H., Henrard, C., Habraken, A.M.: Process window enhancement for single point incremental forming through multi-step toolpaths. CIRP Ann. **57**(1), 253–256 (2008)
12. Hussain, G., Gao, L.: A novel method to test the thinning limits of sheet metals in negative incremental forming. Int. J. Mach. Tool. Manu. **47**, 419–435 (2007)
13. Jeswiet, J., Micari, F., Hirt, G., Bramley, A., Duflou, J., Allwood, J.: Asymmetric single point incremental forming of sheet metal. CIRP Ann. **54**(2), 88–114 (2005)

Chapter 40
Optimization of Gears Production Processes by Application of Material Flow Simulation

Dragana Radakovic, Dejan Lukic, Sanja Bojic, and Mijodrag Milosevic

Abstract The modern production development is conditioned by the constant expansion of market demands, in terms of product range and quality on one side, and reduction of prices and delivery times, on other side. Such requirements impose solutions of production systems that are based on the effects of large-scale and mass production in terms of productivity and economy, and the effects of customized and small-scale production in terms of flexibility and mobility. For the successful development and operation of these production systems, it is necessary to perform optimal design and management of material flow through the production system. These are very complex activities that require application of up-to-date simulation solutions. In this paper, the production process in an experimental production plant of gears is optimized by application of the Tecnomatix Plant Simulation software.

Keywords Production process · Optimization · Group of gears · Modeling · Simulation · Tecnomatix plant simulation

40.1 Introduction

Intensive technologies development, as well as increasing competition on the market, bring challenges to companies in terms of the shortest possible period of product development, high quality, lower prices, customized production adapted to market demands, etc. As the flow of information and the development of innovations nowadays is taking place at an incredible rapidity, companies' existence in the market requires timeliness, up-to-date, and a quick response to consumer demands to ensure a safe place in the consumer society. In addition, the flexibility of production systems, high quality, continuous improvement, computer-integrated activities, etc. [1].

To accomplish these requirements, it is necessary to optimize technological and production processes, which can be achieved by modeling and simulation. Simulation is one of the recognized optimization methods and is an efficient tool for planning,

D. Radakovic (✉) · D. Lukic · S. Bojic · M. Milosevic
Faculty of Technical Sciences, University of Novi Sad, Trg Dositeja Obradovica 6, 21000 Novi Sad, Serbia

M. Rackov et al. (eds.), *Machine and Industrial Design in Mechanical Engineering*, Mechanisms and Machine Science 109,
https://doi.org/10.1007/978-3-030-88465-9_40

organizing, and optimizing production, warehousing, and logistics processes. Using a simulation model, it is possible to systematically observe and notice possible irregularities, bottlenecks, problems that occur in production systems, and the possibility of their elimination, not only in the case of existing production systems but also in the process of their planning and development [2].

The research goal in this paper is to investigate the possibility of applying the software package Tecnomatix Plant Simulation for modeling and simulation of manufacturing processes. The practical goal of the paper refers to the rationalization and optimization of manufacturing processes of developing the production program of a group of gears, as a basis for designing a production plant with a high level of productivity.

40.2 Modeling and Simulation of Production Processes

Modeling and simulation of production processes is a key factor in improving their design and optimal production management. In a very short time, modeling and simulation of these processes lead to very reliable results in creating rational layouts, based on which the production process is easier and faster to organize [3]. Advantages of using the simulation techniques can be [4]:

- the ability to respond to new "what if" circumstances,
- new policies, management and organizational procedures can be explored and virtually implemented without interrupting and disrupting the current system,
- utilization degree of new machines, factory layout, transport systems, etc. is possible to determine very easily without the need to purchase them,
- it is possible to determine under what conditions a certain phenomenon or malfunction occurred,
- it is possible to virtually speed up or slow down the process (compress the time interval), depending on what and which phenomenon is monitored during the simulation,
- gain insight into the importance of individual variables on the performance of the entire system,
- the so-called "Bottleneck Analysis" provides insight into where what and how much WIP (Work In Progress), materials, information, and other things are being disposed of—creating stocks, and
- simulations make it possible to obtain a proper, realistic insight into the work of the entire production system.

The world has developed a large number of simulation software such as Tarakos, NX, Solidworks, FlexSim®, and others, by which it is possible to create a virtual model of a plant and manage it (digital factory), while in this paper Tecnomatix Plant Simulation (TPS) is used. TPS represents the computer application developed by Siemens for modeling, simulation, analysis, visualization, and optimization of production systems and processes, as well as material flow, and logistics operations.

It also provides the ability to simulate discrete events and monitoring of statistical analysis to optimize material handling, use of machinery and manpower required [5, 6].

The application of software enables the creation of a digital model of logistics systems (production, logistics) like a real model, simulation of different scenario models: what happens, when, where, etc., evaluation of results through analytical, static, and graphical tools, visualization and animation of the complete project and alternative solutions, making easier, faster and more efficient decisions in the phases of considering new projects, production, etc. [5].

In addition to modeling and simulating production systems and their processes, Tecnomatix Plant Simulation provides the ability to optimize material costs, resource usage, and logistics for all levels of planning (from individual machines, production lines, through local factories with global production facilities) [7].

40.3 Modeling and Simulation of the Production Process/Case Study—Group of Gears

40.3.1 Input Data—Production Program Structure

The structure of the production program of the observed production plant consists of gears, which are shown with the corresponding data in the tables, where Q represents (quantity), m (mass), and V (value) of the product. These data represent the basis for the analysis of the production program and the selection of representative product.

40.3.2 Product Representative's Selection and Quantity Reduction

To rationalize the manufacturing processes planning, the methodology of reducing the production program to a representative product was chosen, which is selected based on the ABC analysis [8]. Within the ABC analysis, the calculation of quantitative, mass, and value involvement of products was performed, and based on the given ratios, the areas of the largest (A), significant (B), and insignificant (C) increments of the given quantities were determined. Since the production program is not structurally broad, area C is not defined in the observed case. Based on the last three columns from Table 40.1, ABC analysis diagrams are drawn, Fig. 40.1.

By analyzing the ABC analysis diagram, product 2 was singled out, which has the most significant quantitative and value share, and since it belongs to area A for mass participation, a representative of the observed group of gears was chosen as the product, Fig. 40.2. The reduction of the production program to the product representative was performed by applying the mass reduction coefficient r_{mj} and

Table 40.1 Production program data

No.	Product number	Q (pcs/y)	m (kg/pcs)	m (kg/y)	V (€/pcs)	V (€/y)	Q (%)	m (%)	V (%)
1	146.02.061.05	2000	0.85	1700	120	240,000	10.75	9.15	12.45
2	146.02.122.04	2500	1	2500	140	350,000	13.44	13.45	18.15
3	146.02.161.01	1700	0.75	1275	90	153,000	9.13	6.86	7.93
4	146.02.08.07	1500	2	3000	135	202,500	8.06	16.15	10.50
5	146.02.14.05	2100	1.05	2205	100	210,000	11.29	11.87	10.89
6	146.02.121.01	1400	0.9	1260	80	112,000	7.52	6.78	5.81
7	146.02.200.06	1800	0.7	1260	70	126,000	9.67	6.78	6.53
8	146.02.08.06	1900	1.5	2850	100	190,000	10.21	15.34	9.86
9	146.02.062.04	1500	0.95	1425	90	135,000	8.06	7.67	7.00
10	146.02.121.04	2200	0.5	1100	95	209,000	11.82	5.92	10.84
	Total	18,600		18,575	1020		100	100	100

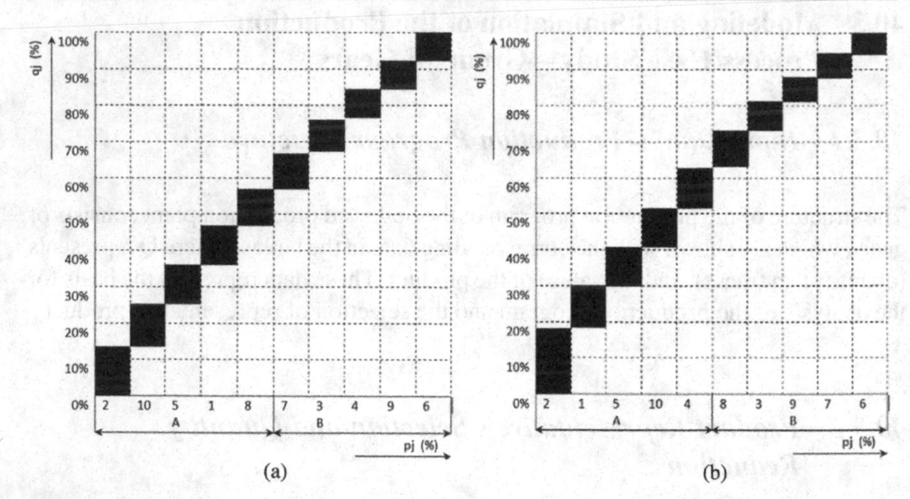

Fig. 40.1 ABC analysis of gear production program: **a** quantitative ABC analysis; **b** valuable ABC analysis

Fig. 40.2 3D model of a group of gears representative

Table 40.2 Reduction of the production program

Product	Q_j (pcs/y)	m_j (kg/pcs)	S_j	r_{mj}	r_{sj}	$Q_{red} = Q_j \cdot r_u$ (pcs/y)
146.02.061.05	2000	0.85	0.75	0.85	0.75	1275
164.02.122.04	2500	1	1	1	1	2500
164.02.161.01	1700	0.75	0.8	0.75	0.8	1020
146.02.08.07	1500	2	0.75	2	0.75	2250
146.02.14.05	2100	1.05	0.75	1.05	0.75	1654
146.02.121.01	1400	0.9	0.75	0.9	0.75	945
146.02.200.06	1800	0.7	0.65	0.7	0.65	819
146.02.08.06	1900	1.5	0.65	1.5	0.65	1853
146.02.062.04	1500	0.95	0.8	0.95	0.8	1140
146.02.121.04	2200	0.5	1	0.5	1	1100
Total quantity reduction Q_{red} =						14,555

the coefficient that takes into account the geometric and technological similarity of the product with the product representative r_{sj}. The procedure for determining the reduced quantity (Q_{red}) is shown in Table 40.2. Based on the performed reduction of the program, the reduced quantity Q_{red} = 14,555 (pcs/year) was determined and associated with the product representative.

40.3.3 Manufacturing Process Planning and Variant Selection of Production Flow

The manufacturing process for the representative product, designed based on input data (drawing, production volume, and available resources), is shown in Table 40.3. These data represent the basis for determining the type of flow of the production system, the calculation of the required production resources and are also important for the development of the production plant.

Based on the product quantities Q_{red} = 14.555 (pcs/year), processing time from the manufacturing process, and the effective time capacity for the projected number of working days and shifts, a flow variant was selected whose basic characteristics are continuity of flow, manufacturing systems of production character and workplace deployment in the order of operations from the manufacturing process. Based on the manufacturing process, workplaces were defined, of which there were 16 (1001–1016) in the observed case. For 2 shifts/day, calculations showed that the number of workplace/machines N = 2 for operations 2, 4, 5, 13, 15 and 16 (workplace 1002, 1004, 1005, 1013, 1015 and 1016), while for other operations number of workplace N = 1.

Table 40.3 The content of the manufacturing process of the representative product

No.	Operation name	Machine/workplace	t_k (min/oper)	T_{pz} (min/bat)
10	Cutting off	Metal saw machine/1001	5	20
20	Turning	CNC lathe/1002	15	60
30	Groove milling	Vertical milling machine/1003	7	30
40	Gear milling	Gear hobbing machine-Pfauter/1004	20	80
50	Rounding teeth	Tooth rounding machine/1005	15	30
60	Internal splines cutting	Broaching machine/1006	13	60
70	Adjustment and protection	Working table/1007	10	15
80	Cementation	Cementation furnace/1008	13	30
90	Case hardening	Hardening furnace/1009	13	30
100	Dismission	Furnace heat treatment/1010	13	30
110	Hardness control	Hardness control/1011	10	10
120	Cementation depth control	Measuring cementation depth/1012	10	10
130	Circular grinding	Circular grinding machine/1013	15	30
140	Forehead grinding	Flat grinding machine/1014	13	20
150	Gear grinding	Tooth grinding machine/1015	20	40
160	Final control	Coordinate measuring machine/1016	15	20

40.3.4 Simulation Study and Obtained Results

To achieve a production of 14.555 gears/year, several variants of the simulation model have been developed, in terms of work shifts, the number of batches, the number of buffers as well its appropriate capacities, and maximum production volume with increased product demand. Variants of the simulation model that have been developed:

- Model 1—without buffer, 2 shifts, 8 h/day, 250 days/year = 5 days per week, 10 series,
- Model 2—including buffer, 2 shifts, 8 h/day, 5 days per week, 10 series,
- Model 3—including buffer, 3 shifts, 5 days per week, 10 series,

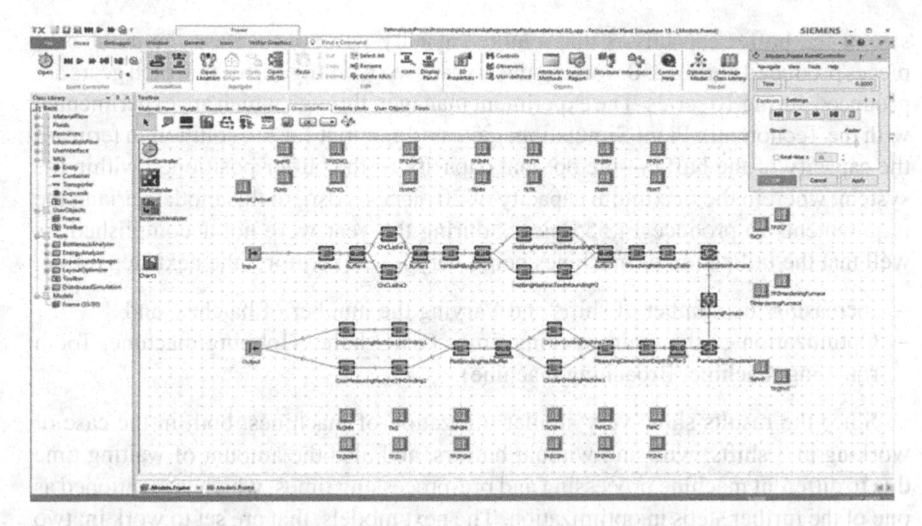

Fig. 40.3 Basic simulation model of gear production system (Model 1)

- Model 4—3 shifts in a system without buffer, 5 days per week, 10 series,
- Model 5—including buffer, 2 shifts per 8 h/day, 7 working days per week,
- Model 6—including buffer, 3 shifts, 7 working days per week, and
- Model 7 and 8—max. production volume during 2 and 3 shifts/day, 7 days/week.

The representation of the simulation model, with the necessary objects of the manufacturing system, appropriate tables for setting the operating time of processes, tools for checking and displaying the functioning of the system and machines utilization, as well as setting work shifts during production, is shown in Fig. 40.3.

40.4 Analysis of Results and Conclusion

In the first simulation model, the number of gears throughout a day is 11887, and the system works without buffers between machines that have significantly different main processing times (see Table 40.3 and Fig. 40.2). The next step was to insert an optimal number of buffers that will balance production flow, placed between machines where is main operating time shorter than operation time of the next machine. Since the machines are blocked for a long time during the simulation, primarily caused by the different preparatory end and main operating times of the machines, the first step towards optimization was to determine the buffer number and capacity to reduce the blockage in the system. However, after inserting buffers and starting with capacity optimization, it was determined by experiments that the solution of a larger buffer capacity to prevent blockage indicates that we need the capacity of some buffers even over 2500 pieces, which nowadays is not a realistic and optimal solution due to the accumulation of stocks in production. Another conclusion reached when varying the size of the buffer capacity is that a large capacity will not increase the number of pieces produced during 250 working days/year, two working shifts, in which gears are produced during 10 series. The experiment manager, through which are experimented with the Tecnomatix Plant Simulation, gave us an output for the solution in terms of the capacity of the buffer—the optimal capacity of the buffer positioned within the system, wherein the maximum capacity of 20 pieces. Also, in this model variant, the requirements to produce 14,555 pieces during the year were not accomplished, as well that the utilization of machines being very low. Therefore, the next steps were:

- Increasing the number of shifts and Varying the number of batches, and
- Optimization of the main working times (CNC lathe, Hobbing machine, Tooth rounding machine, Broaching machine).

Since the results show very similar utilization of machines, both in the case of working in 3 shifts; with and without buffers, and also the amount of waiting time due to different machine processing and pre-processing times, which is mentioned as one of the further steps in optimization. The next models, that are set to work in: two and three shifts; 354 days a year, i.e. 7 days a week, except holidays, non-working days, leaving 10 series in which gears are produced.

In this variant of the model, working in 2 or 3 shifts 7 days a week, the required number of pieces will be produced in that time. The utilization of the machines is approximately the same, so the optimal solution would be to work in 2 shifts 7 days a week and the deadline for the production of gears of one year will be met. In the following model, the goal was to check what the utilization and condition of the system would be if they had an increased demand for gear production. These results show how much the maximum number of gears can be produced in a year, working in 2 and 3 shifts. The first thing to notice is that the utilization of machines is much better when working in 3 shifts 7 days a week (analogous to that when working 5 days a week), and as for waiting—it is inevitable if the process times of these machines are not in some way optimized so that we would have a larger number of manufactured gears and the reliability of the production system to meet market demands, if there is, for example, variation in quantities/increased demand for products.

Variation in the number of batches during production has shown that increasing the number of batches reduces the productivity of the system during the year. This happens because we have to change tools on the machines more often, adjust the machines we need additional setup times of the machines, which at some machines takes up to an hour and a half. Reducing the number of batches slightly increases productivity. The results of simulation models in terms of the number of gears produced depending on the model variants are shown in Table 40.4. Based on the table, we can conclude that:

- In model 2, with the same number of operating hours and the same number of shifts as in variant 1, the number of gears produced increased by 0.2% compared to the previous variant, waiting parts in front of furnaces and system blockage decrease, which was achieved by inserting a buffer (intermediate storage) into the system,

Table 40.4 Obtained quantity of gears through simulation model variants

No.	Quantity of produced gears (pcs)	Shifts no.	No. days/w
Model 1	11,887	2	5
Model 2	11,910	2	5
Model 3	14,555	3	5
Model 4	14,555	3	5
Model 5	14,555	2	7
Model 6	14,555	3	7
Model 7	23,480	2	7
Model 8	35,250	3	7

- In model 3, with 3 shifts during the day, the desired number of gears is successfully produced, the utilization of machines is higher compared to the previous two models, waiting and system blocking are reduced,
- Is the model 7, with the operation of the system in 2 shifts of 8 h/day, 7 days a week it is possible to achieve productivity by ~38% higher than model with limited amount of production, and
- In model 8, when the system operating in 3 shifts, 7 days a week it is possible to achieve productivity by ~58% higher than required production.

So that, regarding of the utilization of machines; as model variants are developed, so the utilization is better and the production of gears is increasing. However, as the system is organized in that sense, it is necessary to adjust the amount of the buffer capacities. In this regard, the last developed model whose development aimed to show how much would be the maximum production volume, during the work of 3 shifts/day and 250 days/year, for the same buffer capacity as in previous variants, proves that this number was higher than the required production and amounts to 17.966 pcs/year.

Acknowledgements Research presented in this paper was supported by Ministry of Education, Science and Technological Development of Republic of Serbia, Title: "Innovative scientific and artistic research from the FTS (activity) domain".

References

1. Unger, D., Eppinger, S.: Improving product development process design: a method for managing information flows, risks, and iterations. J. Eng. Design. **22**(10), 689–699 (2011)
2. Chryssolouris, G.: Manufacturing Systems—Theory and Practice. Springer Verlag, New York (2006)
3. Law, A.M., Kelton, W.D.: Simulation Modeling and Analysis. McGraw-Hill, New York (2000)
4. Banks, J., Carson, J.S., Nelson, B.L., Nicol, D.M.: Discrete-Event System Simulation. Pearson (2005)

5. Todić, V., Lukić, D., Milošević, M., Borojević, S., Vukman, J.: Application of simulation techniques in the development and implementation of flexible manufacturing systems. In: Proceedings of the International Scientific Conference on Industrial Systems, IS 2011, pp. 23–28. Novi Sad (2011)
6. Bangsow, S.: Manufacturing Simulation with Plant Simulation and Simtalk: Usage and Programming with Examples and Solutions. Springer-Verlag, Berlin (2010)
7. www.axiomtech.rs. Last Accessed 01 Feb 2021
8. Lukić, D., Antić, A., Borojević, S., Jocanović, M., Kuric, I.: Evaluation of the technological effects of application of the FMS elements. IOP Conf. Ser. Mater. Sci. Eng. **749**, 012018-1–012018-8 (2020)

Part IV
Applied Modelling, Industrial Design and Virtual Engineering

Chapter 41
Influence of Active Vehicle Suspension to Maintain Transverse Stability in Bends

Hristo Uzunov, Stanimir Karapetkov, Lubomir Dimitrov, Silvia Dechkova, and Vasil Uzunov

Abstract Active control systems of today's cars have been used over a vast domain of applications, as they seem to represent a complex compromise between car handling, stability and ride comfort. Finding a balance between these three components is of paramount importance to stability, a well-known prerequisite, directly proportional, for road safety. Active suspension means active control of certain parameters of vehicle suspension and their changes over time in their equilibrium state. The aim is to maintain vehicle stability going round bends. Setting a tolerance for these parameters results in a compromise in ride quality of vehicle carbody. This is usually accomplished by changing the elasticity of the springs in suspension and increasing their elastic constant to hardening. Thus, a minimum tolerance in the rotation of the carbody around the transverse and longitudinal axes can be guaranteed, respectively a reduction of the centrifugal inertial forces as a function of the rotation defined. To solve this problem a dynamic study of a car model is needed, taking into account elasticity of spring suspension and wheel suspensions, dampers and tyre damping as well as tire-road friction forces. An indicator of this is the variable friction coefficient as a function of the velocity of the contact point.

Keywords Automobile · Active suspension · Dynamic model · Failure · Matlab

41.1 Introduction

Vehicle motion planning, generally, is spatial, defined by six generalized coordinates. For this purpose, coordinates of vehicle centers of mass x_C, y_C, z_C are used in a fixed coordinate system, as well as three Euler angles φ, ψ, θ. A fixed coordinate system

---

H. Uzunov · S. Karapetkov · S. Dechkova (✉) · V. Uzunov
Faculty and College, Technical University of Sofia, Sliven, Bulgaria
e-mail: si_yana@abv.bg

L. Dimitrov
Technical University of Sofia, Kliment Ohridski Blvd, Sofia 1000, Bulgaria

M. Rackov et al. (eds.), *Machine and Industrial Design in Mechanical Engineering*, Mechanisms and Machine Science 109,
https://doi.org/10.1007/978-3-030-88465-9_41

$Oxyz$ and a mobile one $Cx'y'z'$, constantly fixed to the vehicle, have been chosen. Euler's dependences are also used to transform the mobile coordinate system to the fixed one [1–4].

41.2 Dynamic Model of Active Car Suspension

Macro simulation of vehicle motion in case of loss of lateral stability is observed in an arbitrarily accepted absolute coordinate system $OXYZ$ [5–10]. To study the car motion, it has been assumed that its own coordinate system $Cx'y'z'$ is movable and permenantly connected to the vehicle center of mass C (Fig. 41.1a). In addition, a permanently connected $Cxyz$ coordinate system is attached to it, parallel to the absolute and translationally movable one.

Coordinates of the vehicle center of mass C x_c, y_c, z_c in the fixed coordinate system are selected for generalized coordinates of the car motion.

Rotational motion of the car is expressed by the Euler transformations and corresponding angles, namely ψ, θ and φ. The precession angle of ψ, taking into account the rotation around the axis Cz; respectively, the angular velocity of ψ is obtained; the angle θ of nutation, taking into account the rotation with respect to the axis $C\rho$, the intersection of the planes Oxy and $Cx'y$.

Therefore, the force of gravity $\vec{G}$ will lie on the axis Oz. The spatial arrangement model of the car is a plane located on four elastic supports, which are marked by $K_i(i = 1 \div 4)$ (Fig. 41.1b).

$\vec{F}_i(i = 1 \div 4)$ is elastic force generated by the elasticity of tires and springs; $\vec{N}_i(i = 1 \div 4)$ is normal reaction at the contact point of automobile tires, corresponding to elastic force; $\vec{V}_i(i = 1 \div 4)$ is velocity of the contact point P_i in the plane of the road Oxy; $\vec{T}_i(i = 1 \div 4)$ is friction force at the contact points that lies in the plane of the road Oxy; $\vec{R}_i(i = 1 \div 4)$ is resistance force generated by damping

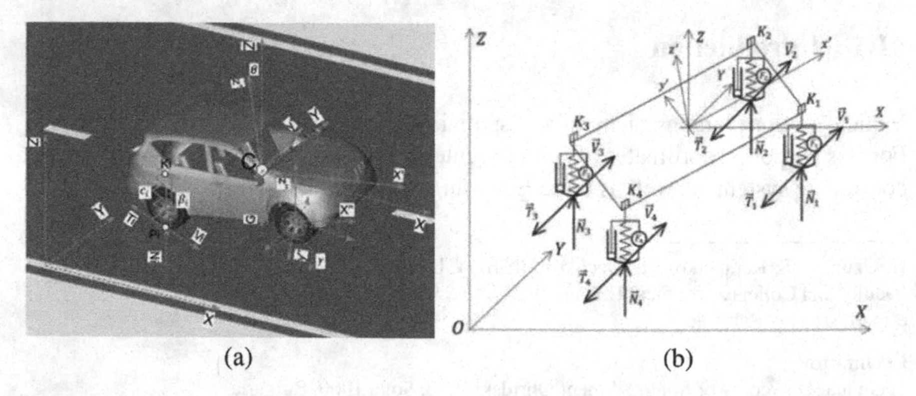

Fig. 41.1 **a** Spatial dynamic model of an automobile with elastic suspension and **b** model of the forces acting on a car in its spatial motion, taking into account the elasticity of tires (suspension)

elements in suspension; c_i, $\frac{\mathrm{N}}{\mathrm{m}}(i = 1 \div 4)$ elasticity of suspension, taking into account both coefficient of elasticity of tires and suspension; b_i, $\frac{\mathrm{N\,s}}{\mathrm{m}}(i = 1 \div 4)$ coefficient of linear resistance.

The car motion according to the studies of kinetic energy and generalized forces is defined by six differential equations with six generalized coordinates. These equations are valid if the friction force is in accordance with Coulomb's law and the wheels slide on the ground without rolling. According to (41.5), the wheels keep a continuous contact with the road.

Generalized forces and moments in the right-hand sides of the differential equation (41.1) are determined by assuming that the absolute coordinate system has a vertical axis of Oz [11–14]:

$$\begin{aligned} m \cdot \ddot{x} &= \left[\sum_{i=1}^{4} F_{xi}\right] \\ m \cdot \ddot{y} &= \left[\sum_{i=1}^{4} F_{yi}\right] \\ m \cdot \ddot{z} &= \left[-G + \sum_{i=1}^{4} N_i - \sum_{i=1}^{4} R_i\right] \end{aligned} \tag{41.1}$$

$$a_{11} \cdot \ddot{\varphi} + a_{12} \cdot \ddot{\psi} + a_{13} \cdot \ddot{\theta} = \left\{ \begin{array}{l} \sum_{i=1}^{4} N_i \cdot \delta_{\varphi i} + \sum_{i=1}^{4} \left(F_{xi} \cdot f_{\varphi_{xi}} + F_{yi} \cdot f_{\varphi_{yi}}\right) - \sum_{i=1}^{4} R_i \cdot \delta_{\varphi i} - \\ - b_{11} \cdot \dot{\varphi}^2 - b_{12} \cdot \dot{\psi}^2 - b_{13} \cdot \dot{\theta}^2 - c_{11} \cdot \dot{\varphi} \cdot \dot{\psi} - c_{12} \cdot \dot{\varphi} \cdot \dot{\theta} - c_{13} \cdot \dot{\psi} \cdot \dot{\theta} \end{array} \right\} \tag{41.2}$$

$$a_{21} \cdot \ddot{\varphi} + a_{22} \cdot \ddot{\psi} + a_{23} \cdot \ddot{\theta} = \left\{ \begin{array}{l} \sum_{i=1}^{4} \left(F_{xi} \cdot f_{\psi_{xi}} + F_{yi} \cdot f_{\psi_{yi}}\right) - b_{21} \cdot \dot{\varphi}^2 - b_{22} \cdot \dot{\psi}^2 - b_{23} \cdot \dot{\theta}^2 - \\ - c_{21} \cdot \dot{\varphi} \cdot \dot{\psi} - c_{22} \cdot \dot{\varphi} \cdot \dot{\theta} - c_{23} \cdot \dot{\psi} \cdot \dot{\theta} \end{array} \right\} \tag{41.3}$$

$$a_{31} \cdot \ddot{\varphi} + a_{32} \cdot \ddot{\psi} + a_{33} \cdot \ddot{\theta} = \left\{ \begin{array}{l} \sum_{i=1}^{4} N_i \cdot \delta_{\theta i} + \sum_{i=i}^{4} \left(F_{xi} \cdot f_{\theta_{xi}} + F_{yi} \cdot f_{\theta_{yi}}\right) - \sum_{i=1}^{4} R_i \cdot \delta_{\theta i} - \\ - b_{31} \cdot \dot{\varphi}^2 - b_{32} \cdot \dot{\psi}^2 - b_{33} \cdot \dot{\theta}^2 - c_{31} \cdot \dot{\varphi} \cdot \dot{\psi} - c_{32} \cdot \dot{\varphi} \cdot \dot{\theta} - c_{33} \cdot \dot{\psi} \cdot \dot{\theta} \end{array} \right\} \tag{41.4}$$

$$\begin{aligned} a_{11} &= J_{z'z'} \\ a_{12} &= -J_{z'z'} \cdot \cos\theta - J_{z'x'} \cdot \sin\varphi \cdot \sin\theta - J_{y'z'} \cdot \cos\varphi \cdot \sin\theta \end{aligned}$$

$$a_{13} = -J_{z'x'} \cdot \cos\varphi + J_{y'z'} \cdot \sin\varphi \tag{41.5}$$

$$b_{11} = 0$$

$$b_{12} = \begin{pmatrix} -\frac{1}{2} \cdot \sin 2\varphi \cdot \sin^2\theta \cdot \left(J_{x'x'} + J_{y'y'}\right) + +J_{x'y'} \cdot \cos 2\varphi \cdot \sin^2\theta + \\ +\frac{1}{2} \cdot \sin 2\theta \cdot \left(J_{z'x'} \cdot \cos\varphi - J_{y'z'} \cdot \sin\varphi\right) \end{pmatrix}$$

$$b_{13} = \left(\frac{1}{2} \cdot \left(J_{x'x'} - J_{y'y'}\right) \cdot \sin 2\varphi - J_{x'y'} \cdot \cos 2\varphi\right) \tag{41.6}$$

$$c_{11} = 0;$$

$$c_{12} = 0$$

$$c_{13} = \left(\cos 2\varphi \sin\theta \left(J_{x'x'} + J_{y'y'}\right) - J_{z'z'} \cdot \sin\theta - 2\begin{pmatrix} J_{x'y'} \cdot \sin 2\varphi \sin\theta + \\ +J_{z'x'} \cdot \sin\varphi \cos\theta + \\ +J_{y'z'} \cdot \cos\varphi \cos\theta \end{pmatrix}\right) \tag{41.7}$$

$$a_{21} = \left(J_{z'z'} \cdot \cos\theta - J_{z'x'} \cdot \sin\varphi \cdot \sin\theta - J_{y'z'} \cdot \cos\varphi \cdot \sin\theta\right)$$

$$a_{22} = \begin{pmatrix} J_{x'x'} \cdot \sin^2\varphi \cdot \sin^2\theta + J_{y'y'} \cdot \cos^2\varphi \cdot \sin^2\theta + \\ +J_{z'z'} \cdot \cos^2\theta - J_{x'y'} \cdot \sin 2\varphi \cdot \sin^2\theta \\ -J_{x'z'} \cdot \sin\varphi \cdot \sin 2\theta - J_{y'z'} \cdot \cos\varphi \cdot \sin 2\theta \end{pmatrix}$$

$$a_{23} = \begin{pmatrix} 0,5 \cdot J_{x'x'} \cdot \sin 2\varphi \cdot \sin\theta - \frac{1}{2} \cdot J_{y'y'} \cdot \sin 2\varphi \cdot \sin\theta - \\ -J_{x'y'} \cdot \cos 2\varphi \cdot \sin\theta - J_{z'x'} \cdot \cos\varphi \cdot \cos\theta + \\ +J_{y'z'} \cdot \sin\varphi \cdot \cos\theta \end{pmatrix} \tag{41.8}$$

$$b_{21} = \left(-J_{z'x'} \cdot \cos\varphi + J_{y'z'} \cdot \sin\varphi\right) \cdot \sin\theta$$

$$b_{22} = 0$$

$$b_{23} = \begin{pmatrix} \left(0,5 \cdot J_{x'x'} \cdot \sin 2\varphi - \frac{1}{2} \cdot J_{y'y'} \cdot \sin 2\varphi - J_{x'y'} \cdot \cos 2\varphi\right) \cos\theta + \\ +\left(J_{z'x'} \cdot \cos\varphi - J_{y'z'} \cdot \sin\varphi\right) \cdot \sin\theta \end{pmatrix} \tag{41.9}$$

$$c_{21} = \begin{pmatrix} \left(J_{x'x'} \cdot \sin 2\varphi - J_{y'y'} \cdot \sin 2\varphi - 2 \cdot J_{x'y'} \cdot \cos 2\varphi\right) \sin^2\theta - \\ -\left(J_{z'x'} \cdot \cos\varphi - J_{y'z'} \cdot \sin\varphi\right) \cdot \sin 2\theta \end{pmatrix}$$

$$c_{22} = \begin{pmatrix} \left(J_{x'x'} \cdot \cos 2\varphi - J_{y'y'} \cdot \cos 2\varphi + 2 \cdot J_{x'y'} \cdot \sin 2\varphi\right) \sin\theta - \\ -J_{z'z'} \cdot \sin\theta \end{pmatrix}$$

$$c_{23} = \begin{pmatrix} \left(J_{x'x'} \cdot \sin^2\varphi + J_{y'y'} \cdot \cos^2\varphi - J_{x'y'} \cdot \sin 2\varphi - J_{z'z'}\right) \sin 2\theta - \\ -2 \cdot \left(J_{z'x'} \cdot \sin\varphi + J_{y'z'} \cdot \cos\varphi\right) \cdot \cos 2\theta \end{pmatrix} \tag{41.10}$$

$$a_{31} = J_{z'x'} \cdot \cos\varphi + J_{y'z'} \cdot \sin\varphi$$
$$a_{32} = \begin{bmatrix} 0,5 \cdot \left(J_{x'x'} - J_{y'y'}\right) \cdot \sin 2\varphi \cdot \sin\theta - J_{x'y'} \cdot \cos 2\varphi \cdot \sin\theta - \\ - J_{z'x'} \cdot \cos\varphi \cdot \cos\theta + J_{y'z'} \cdot \sin\varphi \cdot \cos\theta \end{bmatrix}$$
$$a_{33} = J_{x'x'} \cdot \cos^2\varphi + J_{y'y'} \cdot \sin^2\varphi + \frac{1}{2} \cdot J_{x'y'} \cdot \sin 2\varphi \tag{41.11}$$

$$b_{31} = J_{z'x'} \cdot \sin\varphi + J_{y'z'} \cdot \cos\varphi$$
$$b_{32} = \begin{bmatrix} -\left[0,5\left(J_{x'x'} \sin^2\varphi + J_{y'y'} \cos^2\varphi + J_{z'z'} - J_{x'y'} \sin 2\varphi\right)\right] \sin 2\theta + \\ + \left(J_{z'x'} \cdot \sin\varphi + J_{y'z'} \cdot \cos\varphi\right) \cdot \cos 2\theta \end{bmatrix}$$
$$b_{33} = 0 \tag{41.12}$$

$$c_{31} = \begin{bmatrix} \left[\left(J_{x'x'} + J_{y'y'}\right) \cdot \cos 2\varphi + 2 \cdot J_{x'y'} \cdot \sin 2\varphi + J_{z'z'}\right] \cdot \sin\theta + \\ + 2 \cdot \left(J_{z'x'} \cdot \sin\varphi + J_{y'z'} \cdot \cos\varphi\right) \cdot \cos\theta \end{bmatrix}$$
$$c_{32} = \left[\left(-J_{x'x'} + J_{y'y'}\right) \cdot \sin 2\varphi + 2 \cdot J_{x'y'} \cdot \cos 2\varphi\right]$$
$$c_{33} = 0 \tag{41.13}$$

We substitute the equations before $\delta_{\varphi i}$ and $\delta_{\theta i}$ using the notation:

$$\delta_{\varphi i} = \left[(\cos\varphi \cdot \sin\theta) \cdot x'_{ki} + (-\sin\varphi \cdot \sin\theta) \cdot y'_{ki}\right] \tag{41.14}$$

$$\delta_{\theta i} = \left[(\sin\varphi \cos\theta) \cdot x'_{ki} + (\cos\varphi \cos\theta) y'_{ki} + (-\sin\theta) \cdot z'_{ki}\right] \tag{41.15}$$

To facilitate notation, substitution has been done, which looks like as follows:

$$f_{\varphi_{xi}} = \begin{bmatrix} \begin{pmatrix} -\cos\psi \cdot \sin\varphi - \\ -\sin\psi \cdot \cos\varphi \cdot \cos\theta \end{pmatrix} \delta x'_{Pi} + \\ + \begin{pmatrix} -\cos\psi \cdot \cos\varphi + \\ +\sin\psi \cdot \sin\varphi \cdot \cos\theta \end{pmatrix} \delta y'_{Pi} \end{bmatrix};$$
$$f_{\psi_{xi}} = \begin{bmatrix} \begin{pmatrix} -\sin\psi \cdot \cos\varphi - \\ -\cos\psi \cdot \sin\varphi \cdot \cos\theta \end{pmatrix} \delta x_{ki} + \\ + \begin{pmatrix} \sin\psi \cdot \sin\varphi - \\ -\cos\psi \cdot \cos\varphi \cdot \cos\theta \end{pmatrix} \delta y'_{ki} + \\ + (-\cos\psi \cdot \sin\theta) \delta z'_{ki} \end{bmatrix}$$
$$f_{\theta_{xi}} = \begin{bmatrix} (\sin\theta \cdot \sin\psi \cdot \sin\varphi) \delta x'_{ki} + \\ + (\sin\theta \cdot \sin\psi \cdot \cos\varphi) \delta y'_{ki} + \\ + (-\cos\theta \cdot \sin\psi) \delta z'_{ki} \end{bmatrix} \tag{41.16}$$

$$f_{\varphi_{yi}} = \begin{bmatrix} \begin{pmatrix} -\sin\psi \cdot \sin\varphi + \\ +\cos\psi \cdot \sin\varphi \cdot \cos\theta \end{pmatrix} \delta x'_{ki} + \\ + \begin{pmatrix} -\sin\psi \cdot \cos\varphi - \\ -\cos\psi \cdot \sin\varphi \cdot \cos\theta \end{pmatrix} \delta y'_{ki} \end{bmatrix};$$

$$f_{\psi_{yi}} = \begin{bmatrix} \begin{pmatrix} \cos\psi \cdot \cos\varphi - \\ -\sin\psi \cdot \sin\varphi \cdot \cos\theta \end{pmatrix} \delta x'_{ki} + \\ + \begin{pmatrix} -\cos\psi \cdot \sin\varphi - \\ -\sin\psi \cdot \cos\varphi \cdot \cos\theta \end{pmatrix} \delta y'_{ki} + \\ + (\sin\psi \cdot \sin\theta)\delta z'_{ki} \end{bmatrix}$$

$$f_{\theta_y} = \begin{bmatrix} (-\cos\psi \cdot \sin\varphi \cdot \sin\theta)\delta x'_{ki} + \\ + (-\cos\psi \cdot \cos\varphi \cdot \sin\theta)\delta y'_{ki} + \\ + (-\cos\psi \cdot \sin\theta)\delta z'_{ki} \end{bmatrix} \quad (41.17)$$

The relative motion of the wheels, the differential(s) and the engine are characterized by a system of four differential equations derived by the Lagrangian method, which has the form of:

$$[I_\gamma] \cdot [\ddot{\gamma}] = [M_{\gamma i}]; \quad M_{\gamma i} = \{F_{i\tau} \cdot r_i + sign(\dot{\gamma}_i) \cdot [M_{di} - f_i \cdot N_i - M_{si}]\} \quad (41.18)$$

$\vec{F}_{i\tau}$ is tangential component of the tire-road friction force, the positive direction of which is taken backwards, in the more frequent cases of braking or loss of stiffness.

Where μ is friction coefficient depending on slipping speed on the contact spot; $\vec{r}_i$—radius of the wheel; f_i—coefficient of rolling friction; $\vec{N}_i$—normal reaction of the road on wheels; $[I_\gamma]$—a square matrix of coefficients in front the actual angular acceleration of the drive wheels, depending on the moment of inertia of the wheels and the engine; $\dot{\gamma}_i/i = 1 \div 4/$—wheel angular velocity; $[\ddot{\gamma}]$—a matrix-column of the actual angular acceleration of the wheels, two or four of which are propulsive; M_{di}, M_{si}—corresponding engine and brake torque applied to each wheel.

Figure 41.2a shows the dynamic model of an active suspension system. Figure 41.2b shows the dynamic diagram of a driving or sliding wheel.

In the system solving the differential equations of motion, a module has been added for the analysis of the two angles of rotation $\varphi_{x'}$ and $\varphi_{y'}$ which are determined by the kinematic equations of Euler (41.16 and 41.17). They represent the rotation of the unsprung mass around its own coordinate system relative to a parallel moving coordinate system of the fixed coordinate system, invariably connected to its center of mass. Its change above a certain value in the positive and negative direction determines the change in the stiffness of suspension.

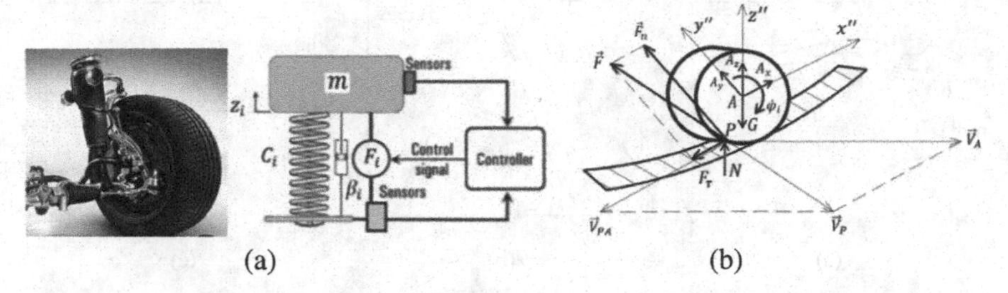

Fig. 41.2 **a** Dynamic model of an active suspension system and **b** drive wheel diagram

41.3 Numerical Experiment of a Spatial Motion Model of an Automobile

Mechanomathematical modeling of vehicle's motion with front-wheel drive, in the presence of a modern active safety system, active suspension, is associated with analysis of the change of angle around its own longitudinal and transverse axis of the car.

Technical data of an automobile are: mass $m = 1180$ kg; length $a = 4{,}31$ m; width $b = 1{,}74$ m; longitudinal base $b = 2{,}65$ m.

Initial linear and angular velocity of the car are as follows:

$$\begin{aligned} &Vx = 100\ \text{km/h};\ Vy = 0\ \text{km/h};\ Vz = 0\ \text{km/h} \\ &\dot{\psi} = 0\ \text{s}^{-1};\ \dot{\theta} = 0\ \text{s}^{-1};\ \dot{\varphi} = 0\ \text{s}^{-1} \end{aligned} \tag{41.19}$$

Elastic constants of springs without active and with active suspension are as follows:

$$\begin{aligned} c_i &= \left[\, 20000\ 20000\ 18000\ 18000 \,\right] (\text{N/cm}) \\ c_{i1} &= \left[\, 100000\ 100000\ 100000\ 100000 \,\right] (\text{N/cm}) \end{aligned} \tag{41.20}$$

Damping factor without active and with active suspension is as follows:

$$\begin{aligned} \beta_i &= \left[\, 5246\ 5246\ 4208\ 4208 \,\right] (\text{N s/cm}) \\ \beta_{i1} &= \left[\, 11731\ 11731\ 9919\ 9919 \,\right] (\text{N s/cm}) \end{aligned} \tag{41.21}$$

Initial linear coordinates of the center of mass and angles of rotation:

$$\begin{aligned} &x_c = 0\ \text{m};\ y_c = -1,5\ \text{m};\ z_c = 0,55\ \text{m} \\ &\psi = 0^\circ;\ \theta = 0^\circ;\ \varphi = 0^\circ \end{aligned} \tag{41.22}$$

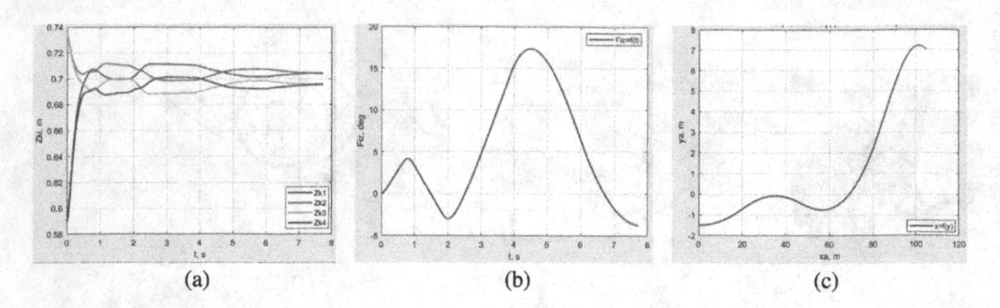

Fig. 41.3 **a** Coordinates of suspension points, **b** changes in the angle around the Oz axis and **b** trajectory of center of mass

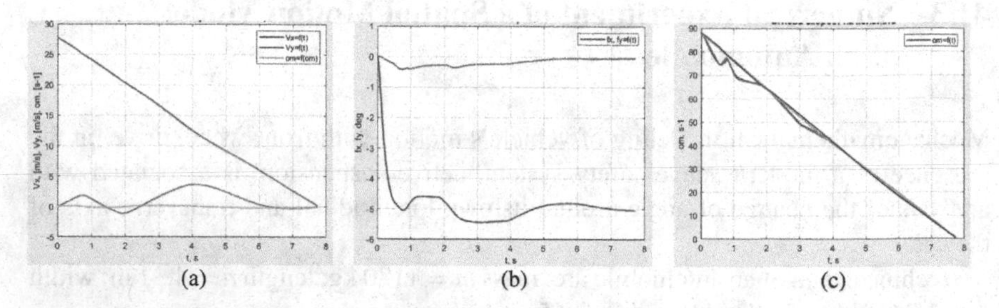

Fig. 41.4 **a** Center of mass velocity and angular velocity, **b** changes in angle $\varphi_{x'}$ *and* $\varphi_{y'}$ and **b** angular velocity of the wheels

41.3.1 Numerical Experiment of Vehicle's Motion Without the Presence of an Active System in Suspension

When active suspension system is not activated, the following graphical dependencies are present (Figs. 41.3 and 41.4):

41.3.2 Numerical Experiment of Vehicle's Motion with the Presence of an Active System in Suspension

When active suspension system is activated, the following graphic dependencies are present (Figs. 41.5 and 41.6).

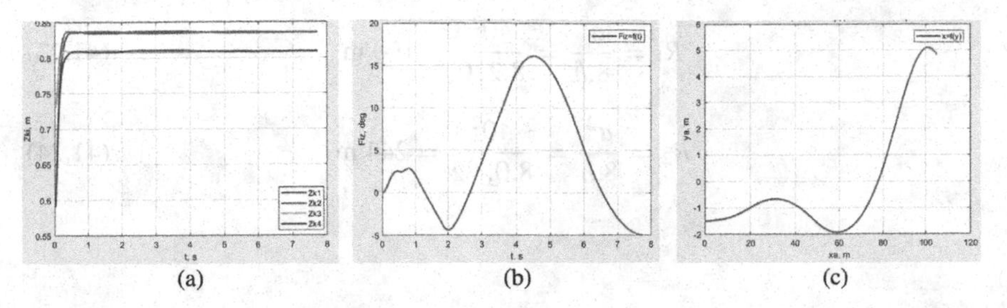

Fig. 41.5 **a** Coordinates of suspension points, **b** changes in the angle around the Oz axis and **b** trajectory of center of mass

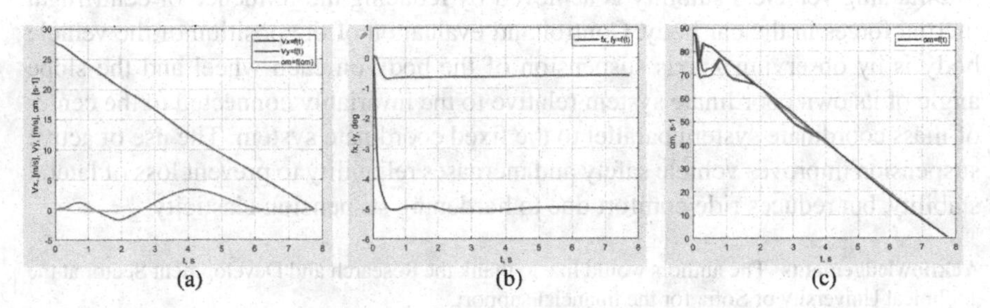

Fig. 41.6 **a** Center of mass velocity and angular velocity, **b** changes in angle $\varphi_{x'}$ and $\varphi_{y'}$ and **b** angular velocity of the wheels

41.3.3 Comparative Analysis of Motion Trajectories and Critical Speed of an Automobile in a Turn. Critical Speed of a Vehicle's Curve

Examination of critical speed of an automobile is based on reduced tire-road friction coefficient of 0.7. The vehicle in motion is in successive curves without any inclination on the roadway (Fig. 41.7).

The radii of the turn with respect to the trajectory of the center of mass are obtained according to the dependence:

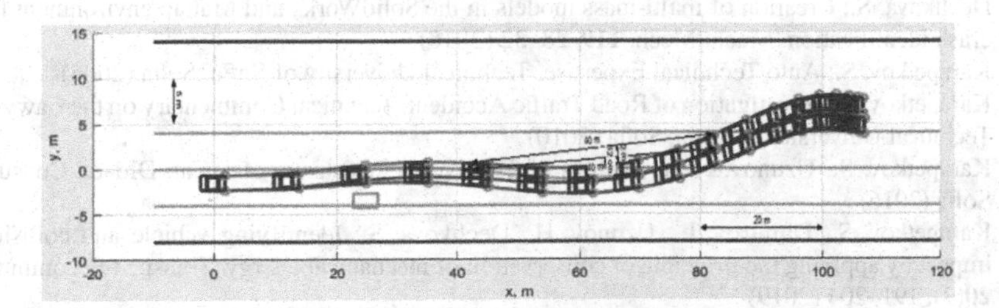

Fig. 41.7 Graphical measurement of the turning radius

$$R_1 = \frac{a^2}{8.h} = \frac{40^2}{8.2,0} = 100 \text{ m} \tag{41.23}$$

$$R_2 = \frac{a^2}{8.h} = \frac{40^2}{8.0,82} = 244 \text{ m} \tag{41.24}$$

41.4 Conclusion

Maintaining vehicle's stability is achieved by reducing the influence of centrifugal inertial forces in the car body. Control and evaluation of the position of the vehicle body is by observing z-axis suspension of the body on each wheel and the slope angle of its own coordinate system relative to the invariably connected to the center of mass coordinate system parallel to the fixed coordinate system. The use of active suspension improves vehicle safety and increases reliability to prevent loss of lateral stability, but reduces ride comfort due to hardening suspension elasticity.

Acknowledgements The authors would like to thank the Research and Development Sector at the Technical University of Sofia for the financial support.

References

1. Daily, J., Shigemura, N., Daily, J.: Fundamentals of Traffic Crash Reconstruction. Institute of Police Technololgy and Managment, Florida (2006)
2. Schmidt, B.F., Haight, W.R., Szabo, T.J., Welcher, J.B.: System-based energy and momentum analysis of collisions. SAE Trans. **107**(6), 120–132 (1998)
3. Sharma, D., Stern, S., Brophy, J.: An overview of NHTSA's crash reconstruction software WinSMASH. In: Proceedings of the 20th International Technical Conference on the Enhanced Safety of Vehicles, ESV 2007, pp. 1–13 (2007)
4. Wach, W.: Analiza deformacji samochodu według standardu CRASH3. Część 2: Pomiar głębokości odkształcenia (analysis of motor vehicle deformation according to the CRASH3 standard. Part 2: measurement of deformation depth). Paragraf na Drodze 12 (2003)
5. Dechkova, S.: Creation of multi-mass models in the SolidWorks and Matlab environment for crash identification. Mach. Mech. **119**, 28–32 (2018)
6. Karapetkov, S.: Auto Technical Expertise. Technical University of Sofia, Sofia (2005)
7. Karapetkov, S.: Investigation of Road Traffic Accident. Technical Commentary on the Lawyer. Technical University of Sofia, Sofia (2010)
8. Karapetkov, S., Uzunov, H.: Dynamics of Transverse Resistance of a Car. Didada Consult, Sofia (2016)
9. Karapetkov, S., Dimitrov, L., Uzunov, H., Dechkova, S.: Identifying vehicle and collision impact by applying the principle of conservation of mechanical energy. Transp. Telecommun. **20**(3), 191–204 (2019)
10. Karapetkov, S., Dimitrov, L., Uzunov, H., Dechkova, S.: Examination of vehicle impact against stationary roadside objects. IOP Conf. Ser. Mater. Sci. Eng **659**, 012063-1–012063-13 (2019)

11. Jiang, T., Grzebieta, R.H., Rechnitzer, G., Richardson, S., Zhao, X.L.: Review of car frontal stiffness equations for estimating vehicle impact velocities. In: Proceedings of the 18th International Technical Conference on the Enhanced Safety of Vehicles Conference, ESV 2003, pp. 1–11. Nagoya (2003)
12. Niehoff, P., Gabler, H.C.: The accuracy of WinSMASH delta-V estimates: the influence of vehicle type, stiffness, and impact mode. Annu. Proc. Assoc. Adv. Automot. Med. **50**, 73–89 (2006)
13. Owsiański, R.: Szacowanie energii deformacji nadwozi kompaktowych samochodów osobowych (estimation of the bodywork deformation energy of compact passenger cars). Paragraf na Drodze 4 (2007)
14. Stronge, W.: Impact Mechanics. Cambridge University Press, Cambridge (2000)

Chapter 42
Mechanical Mathematical Modelling of Two-Vehicle Collisions

Stanimir Karapetkov, Lubomir Dimitrov, Hristo Uzunov, and Silvia Dechkova

Abstract Globally, statistical analysis reports show that the most serious vehicle collisions mainly constitute of two-vehicle accidents. The dynamic and kinematic model of car crashes is particularly challenging when requirements for improved accuracy of the analysis are introduced. For more sophisticated models the more accurate the analysis, the more variables in each phase of motion and impact need to be introduced. Two basic problems must be solved, defined by the final rest position of the cars: whether it is known or must be found by solving Cautchy problem. This article presents a dynamic impact analysis between two vehicles, using data for known final rest position and avoiding the uneasy task of selecting the correct coefficient of restitution. Undeniably, it represents the ratio between post impact and prior to impact relative velocity of the centers of mass of the two vehicles in projection in the direction of the crash pulse. The presented results were obtained by simulation and graphical dependencies of the proposed algorithm for reliability analysis estimation, followed by the deduced discrete positions of post-impact vehicles motion, which confirms the useful application of the proposed algorithm. A comparative study was carried out based on the known treds from first-hand inspection in the field accident.

Keywords Car · Crash · Vehicle collisions · Velocity · Mechanics

42.1 Introduction

The two-car collision was reconstructed as two solids were mounted on elastic supports. The latter were connected to a platform with wheels attached onto it [1–5]. Proper selection of vehicle technical parameters such as mass characteristics, support elasticity, damping properties of shock absorbers and so on determined the accuracy

S. Karapetkov · H. Uzunov · S. Dechkova (✉)
Faculty and College, Technical University of Sofia, Sliven, Bulgaria
e-mail: si_yana@abv.bg

L. Dimitrov
Technical University of Sofia, Kliment Ohridski Blvd., Sofia 1000, Bulgaria

M. Rackov et al. (eds.), *Machine and Industrial Design in Mechanical Engineering*, Mechanisms and Machine Science 109,
https://doi.org/10.1007/978-3-030-88465-9_42

of the dynamic investigation of the post-impact macro motion of the bodies. On the other hand, the problem of impact was solved using a mechanical-mathematical model developed for the purpose. The two vehicles were placed and oriented in the coordinate system as they had been at the point of impact and in position to each other, which completely corresponded to first-hand inspection in the field accident, including vehicle deformation data [6–10].

42.2 Impact Problem

Determine magnitude and direction of the crash pulse, pre-impact velocity of the centers of mass of the two vehicles if post-impact velocity of the centers of mass of the two vehicles are known (Fig. 42.1).

Boundary-value problem of impact: Determine the following initial conditions: post-impact linear and angular velocity of both vehicles if some of the initial conditions are known, such as the position of the two vehicles at the moment of impact and the vehicles final rest position.

The simplest and easiest way to solve the problem when two vehicles are impacted is by applying Momentum Conservation Principle that has the form [11–17]:

$$m_1.\vec{V}_1 + m_2.\vec{V}_2 = m_1.\vec{u}_1 + m_2.\vec{u}_2 \tag{42.1}$$

where m_1 and m_2 are total vehicles mass; V_1 and V_2—velocity of the centers of mass of the vehicles right before impact; u_1 and u_2—velocity of the centers of mass of the vehicles right after impact. The application of the principle has some limitations, assuming that both cars are material points. The second equation of the conservation of momentum principle involves trigonometric function of "sin", in which a small change of the angle causes significant deviations in the final solution of the system.

This ratio is actually the coefficient of restitution, which is as follows:

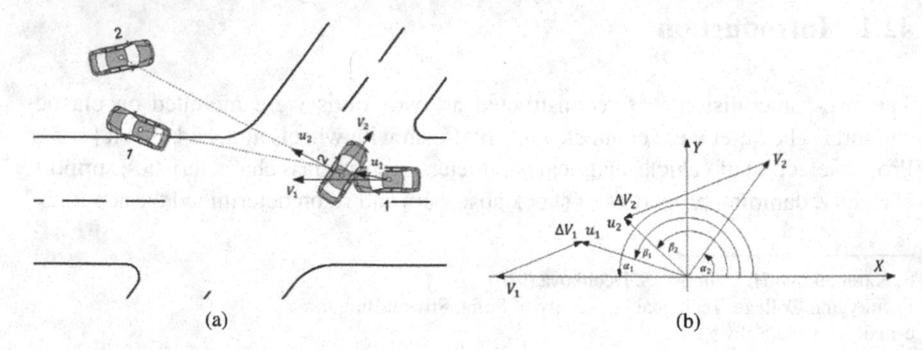

Fig. 42.1 **a** Car accident diagram and **b** vector analysis

$$k = \frac{|\Delta u_n|}{|\Delta V_n|} = \frac{|(\vec{u}_2 - \vec{u}_1).\vec{e}|}{\left|\left(\vec{V}_1 - \vec{V}_2\right).\vec{e}\right|} \tag{42.2}$$

where $\vec{e}$ is a single vector of the crash pulse vector and the "k" is the coefficient of recovery. "Here Δu_n is projection of the relative velocity between the centers of mass of the vehicles after the impact on the crash pulse directrix, and ΔV_n—projection of the relative velocity between the centers of mass of the vehicles before impact on the crash pulse directrix.

It is obvious that there is a necessity to select a particular value for the coefficient of restitution in the range of $0 \leq k \leq 1$, with a possible increase in steps of 0.001.

The impact for each of the vehicles is characterized by the impulse-momentum change theorem for the time of impact, which for each of the vehicles has the form:

$$m.\vec{u} - m.\vec{V} = \vec{S} \tag{42.3}$$

where $\vec{u}$ is the velocity of the center of mass of the post-impact vehicle; $\vec{V}$ is the velocity of the center of mass of prior-to-impact vehicle; $\vec{S}$—crash pulse.

The impulse-momentum change theorem applied to the mechanical system of the car related to the vehicle center of mass during its relative motion around it has the form of:

$$\frac{d\vec{K}_C^r}{dt} = \vec{M}_C^{(e)} \tag{42.4}$$

where $\vec{M}_C^{(e)}$ is the principle of moments of impact forces related to the vehicle center of mass.

After solving the system of Eq. (42.4), the projections of the crash pulse are determined as follows:

$$S_x = \frac{J_{1z}\omega_{z1}\rho_{2x} + J_{2z}\omega_{z2}\rho_{1x}}{\rho_{1x}\rho_{2y} - \rho_{2x}\rho_{1y}};$$
$$S_y = \frac{J_{1z}\omega_{z1}\rho_{2y} + J_{2z}\omega_{z2}\rho_{1y}}{\rho_{1x}\rho_{2y} - \rho_{2x}\rho_{1y}} \tag{42.5}$$

where J_{1z}, J_{2z} are the mass moments of inertia of the vehicles around the central vertical axes perpendicular to the plane of motion; ω_{z1}, ω_{z2}—angular velocities of the vehicles after collision; $\rho_{x_j}, \rho_{y_j}, j = 1, 2$—coordinates of the point of application A_j of the impact force related to a center of mass reference frame, moving translationally.

The magnitude of the projections of the crash pulse is used to obtain the velocities of the centers of mass of the two vehicles before collision according to Eq. (42.3) after their projection on two mutually perpendicular axes:

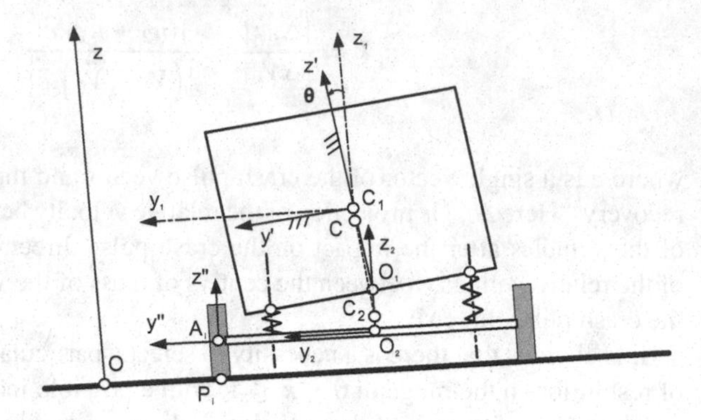

Fig. 42.2 Rear view of the automobile

$$V_{1x} = u_{1x} - \frac{S_x}{m_1};\ V_{1y} = u_{1y} - \frac{S_y}{m_1};\ V_{2x} = u_{2x} + \frac{S_x}{m_2};\ V_{2y} = u_{2y} + \frac{S_y}{m_2} \tag{42.6}$$

In the above equations, the magnitude of ω_{z1}, ω_{z2}, i.e. the angular velocities of the cars after collision and $\vec{u}_j$—the vehicle velocities of the centers of mass after collision should be analyzed.

The differential equations of motion for each vehicle after impact, considered as a multi-mass spatial mechanical system, have the type (Fig. 42.2):

$$m\ddot{x}_C = \sum_{i=1}^{4}[F_{ix}] + mg\sin\xi - W_x\sqrt{\dot{x}_C^2 + \dot{y}_C^2}\,\dot{x}_C \tag{42.7}$$

$$m\ddot{y}_C = \sum_{i=1}^{4}[F_{iy}] + mg\sin\nu - W_y\sqrt{\dot{x}_C^2 + \dot{y}_C^2}\,\dot{y}_C \tag{42.8}$$

$$m\ddot{z}_C = \sum_{i=1}^{4} N_i - \frac{mg}{\sqrt{1 + tg^2\xi + tg^2\nu}} \tag{42.9}$$

where ξ—longitudinal slope of road; ν—transverse slope of road.

After some transformations and projection of Eq. (42.5) on the axes permanently connected to the unsprung mass, the system of differential equations in matrix is obtained:

$$\begin{aligned}\{[J_{C_1}] + [J_{C_3}]\}[\dot{\omega}] = [M_{C_\omega}] + [M_{C_{3\omega}}] + [M_{C_a1}] + [M_{C_a2}] \\ + [M_{C_a3}] + [M_{C_k}] + [M_{C_m}]\end{aligned} \tag{42.10}$$

where $[\dot{\omega}] = [\dot{\omega}_{x'}\dot{\omega}_{y'}\dot{\omega}_{z'}]^T$ is a matrix-column of the derivatives of the angular velocity projections on the coordinate axes permanently connected to the unsprung mass:

$$[M_{C_\omega}] = \begin{bmatrix} M_{F_{x'}} + M_{N_{x'}} + J_{3x'z'}\omega_{3x'}\omega_{3y'} - J_{3x'y'}\omega_{3x'}\omega_{3z'} + \\ (J_{3y'} - J_{3z'})\omega_{3y'}\omega_{3z'} + J_{3y'z'}(\omega_{3y'}^2 - \omega_{3z'}^2) \\ M_{F_{y'}} + M_{N_{y'}} + J_{3x'y'}\omega_{3y'}\omega_{3z'} - J_{3y'z'}\omega_{3x'}\omega_{3y'} + \\ (J_{3z'} - J_{3x'})\omega_{3x'}\omega_{3z'} + J_{3x'z'}(\omega_{3z'}^2 - \omega_{3x'}^2) \\ M_{F_{z'}} + M_{N_{z'}} + J_{3y'z'}\omega_{3x'}\omega_{3z'} - J_{3x'z'}\omega_{3y'}\omega_{3z'} + \\ + (J_{3x'} - J_{3y'})\omega_{3x'}\omega_{3y'} + J_{3x'y'}(\omega_{3x'}^2 - \omega_{3y'}^2) \end{bmatrix} \tag{42.11}$$

$$[M_{Ck}] = -\begin{bmatrix} \begin{bmatrix} \sum_{i=1}^{4} J_{ky''i}\ddot{\gamma}_i \vec{j}_{2i}'' + \vec{\omega}_2 \times \sum_{i=1}^{4} J_{k2i}\ddot{\gamma}_i \vec{j}_{2i}'' + \sum_{i=1}^{4}[J_{kz'i}(\dot{\omega}_2 + \ddot{\vartheta}_{ki}) + \\ + m_{ki}\left|C_2\vec{A}_i\right|^2 \dot{\omega}_2]\vec{k}_2 \end{bmatrix}_{x'} \\ \begin{bmatrix} \sum_{i=1}^{4} J_{ky''i}\ddot{\gamma}_i \vec{j}_{2i}'' + \vec{\omega}_2 \times \sum_{i=1}^{4} J_{k2i}\ddot{\gamma}_i \vec{j}_{2i}'' + \sum_{i=1}^{4}[J_{kz'i}(\dot{\omega}_2 + \ddot{\vartheta}_{ki}) + \\ + m_{ki}\left|C_2\vec{A}_i\right|^2 \dot{\omega}_2]\vec{k}_2 \end{bmatrix}_{y'} \\ \begin{bmatrix} \sum_{i=1}^{4} J_{ky''i}\ddot{\gamma}_i \vec{j}_{2i}'' + \vec{\omega}_2 \times \sum_{i=1}^{4} J_{k2i}\ddot{\gamma}_i \vec{j}_{2i}'' + \sum_{i=1}^{4}[J_{kz'i}(\dot{\omega}_2 + \ddot{\vartheta}_{ki}) + \\ + m_{ki}\left|C_2\vec{A}_i\right|^2 \dot{\omega}_2]\vec{k}_2 \end{bmatrix}_{z'} \end{bmatrix} \tag{42.12}$$

$$M_{Ca1} = \begin{bmatrix} [(\vec{r}_C - \vec{r}_{C_1}) \times (\vec{a}_{C_1} - \vec{a}_C)]_{x'} \\ [(\vec{r}_C - \vec{r}_{C_1}) \times (\vec{a}_{C_1} - \vec{a}_C)]_{y'} \\ [(\vec{r}_C - \vec{r}_{C_1}) \times (\vec{a}_{C_1} - \vec{a}_C)]_{z'} \end{bmatrix} \tag{42.13}$$

$$M_{Ca2} = \begin{bmatrix} [(\vec{r}_C - \vec{r}_{C_2}) \times (\vec{a}_{C_2} - \vec{a}_C)]_{x'} \\ [(\vec{r}_C - \vec{r}_{C_2}) \times (\vec{a}_{C_2} - \vec{a}_C)]_{y'} \\ [(\vec{r}_C - \vec{r}_{C_2}) \times (\vec{a}_{C_2} - \vec{a}_C)]_{z'} \end{bmatrix} \tag{42.14}$$

$$M_{Ca3} = \begin{bmatrix} [(\vec{r}_C - \vec{r}_{C_3}) \times (\vec{a}_{C_3} - \vec{a}_C)]_{x'} \\ [(\vec{r}_C - \vec{r}_{C_3}) \times (\vec{a}_{C_3} - \vec{a}_C)]_{y'} \\ [(\vec{r}_C - \vec{r}_{C_3}) \times (\vec{a}_{C_3} - \vec{a}_C)]_{z'} \end{bmatrix} \tag{42.15}$$

$$[M_{Cm}] = -\begin{bmatrix} \sum_{i=1}^{2}\left[J_{mi} + m_{mi}\left|C_2\vec{C}_{mi}\right|^2\right]\dot{\omega}_2\vec{k}_{2x'} \\ \sum_{i=1}^{2}\left[J_{mi} + m_{mi}\left|C_2\vec{C}_{mi}\right|^2\right]\dot{\omega}_2\vec{k}_{2y'} \\ \sum_{i=1}^{2}\left[J_{mi} + m_{mi}\left|C_2\vec{C}_{mi}\right|^2\right]\dot{\omega}_2\vec{k}_{2z'} \end{bmatrix} \tag{42.16}$$

Here, m is the total mass of the vehicle; $m_{ki}/i = 1 \div 4/$—mass of each of the wheels; $m_{mi}/i = 1 \div 2/$—mass of each of the drives; x_c, y_c, z_c—coordinates of the vehicle center of mass in relation to the fixed coordinate system; ϑ_k—average angle of rotation of the steering wheels around their axles; $\ddot{\vartheta}_k$—angular acceleration; $\dot{\gamma}_i/i = 1 \div 4/$—angles of rotation of the wheels on their own rotary axis; $\dot{\gamma}_i/i = 1 \div 4/$—wheel angular velocity; $\ddot{\gamma}$—angular acceleration of the wheels; $F_i, /i = 1 \div 4/$—friction forces in the wheels; $N_i/i = 1 \div 4/$—normal reactions in the wheels; w—drag coefficient; $\vec{\omega}$—angular velocity of the movable coordinate system $C_1x'y'z'$ permanently connected to the unsprung mass; $\vec{\omega}_2$—angular velocity of the sprung mass and the constantly connected to it movable coordinate system $O_2x_2y_2z_2$;; $M_{F,N'_x}, M_{F,N'_y}, M_{F,N'_z}$—moments of the frictional forces on the wheels and the normal reactions to the permanently connected to the vehicle coordinate axes; $[J_{C1}]$—matrix of the mass inertia of the bodywork related to the coordinate axes, permanently connected to it; $[\omega]$—matrix column of the projections of angular velocity on the same axes determined by Euler's formula; $J_{ky''i}, J_{kz''i}/i = 1 \div 4/$—mass inertia of each wheel relative to its own axis of rotation and its radial axis;—$J_{m_i} - /i = 1 \div 2/$—intrinsic mass moment of inertia of each of the drives relative to its central axis parallel to z_2.

The relative movement of the wheels, differential/s / and the engine is characterized by a system of four differential equations obtained by Lagrange method, which has the type:

$$[I_\gamma][\ddot{\gamma}] = [M_{\gamma i}];\ M_{\gamma i} = \{F_{i\tau}r_i + sign(\dot{\gamma}_i)[M_{di} - f_iN_i - M_{si}]\} \tag{42.17}$$

where μ is friction coefficient depending on slipping speed on the contact spot; $\vec{r}_i$—radius of the wheel; f_i—coefficient of rolling friction; $\vec{F}_{i\tau}$ is tangential component of the tire-road friction force, the positive direction of which is taken backwards, in the more frequent cases of braking or loss of stiffness; M_{di}, M_{si}—corresponding engine and brake torque applied to each wheel.

Example: A head-on collision between Volkswagen Tuareg and Opel Vectra with known geometric dimensions, masses and mass moment of inertia was analysed. Their final rest positions, location impact and the position of treds left on the roadway were known.

Data from the accident was the available treds on the roadway from the vehicles motion at the time of collision and after collision. They are shown in the photographic material in Fig. 42.3a. The two-car collision diagram is shown in Fig. 42.3b.

The developed dynamic deformation model determined the location of the two vehicles at the moment of impact and their final rest position (Figs. 42.4, 42.5, 42.6 and 42.7 and Table 42.1).

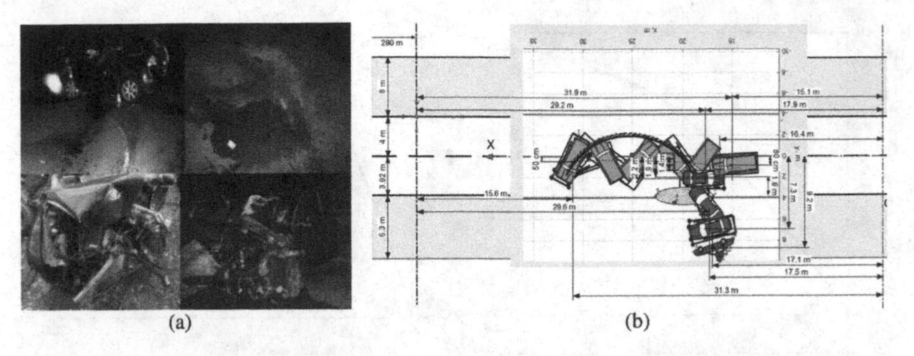

Fig. 42.3 **a** Photographic material, **b** diagram of the two-vehicle crash (comparative analysis)

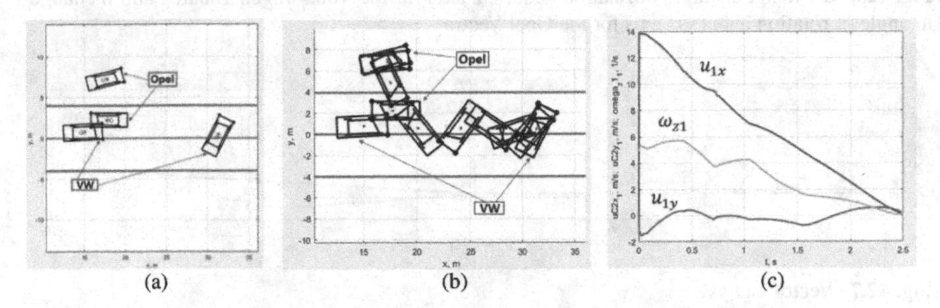

Fig. 42.4 **a** Location of vehicles at the point of impact and final rest positions, **b** discrete positions of the two vehicles motion after collision and **c** projections of velocity centre of mass after impact and angular velocity about Oz axis for Volkswagen Touareg

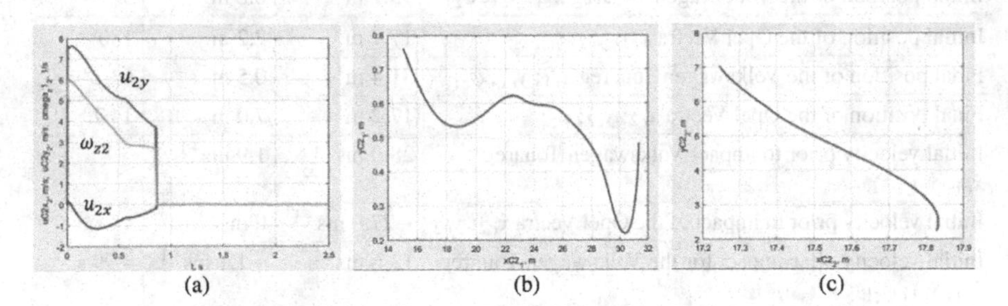

Fig. 42.5 **a** Projections of velocity centre of mass after impact and angular velocity about Oz axis for Opel Vectra, **b** trajectory of the Volkswagen Touareg center of mass and **c** Trajectory of the Opel Vectra center of mass

42.3 Conclusion

A mathematical model was developed to assist computer simulation of two-vehicle collisions. It has the advantage to determine with great accuracy the place of impact,

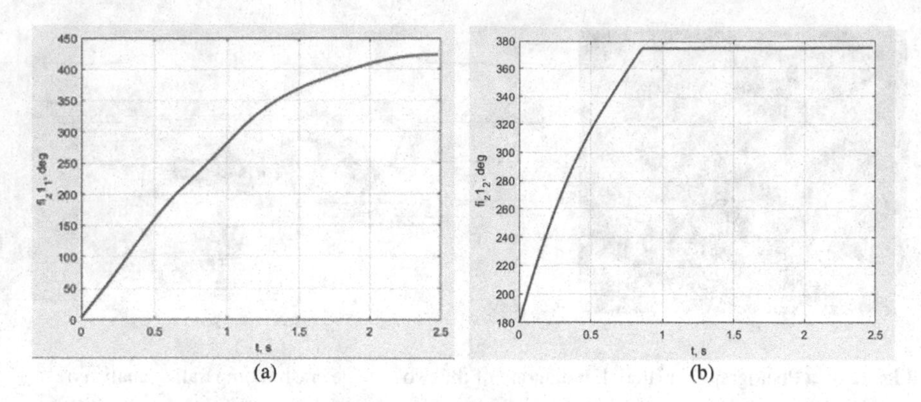

Fig. 42.6 **a** Change in angle of rotation about Oz axis for the Volkswagen Touareg and **b** change in angle of rotation about Oz axis for the Opel Vectra

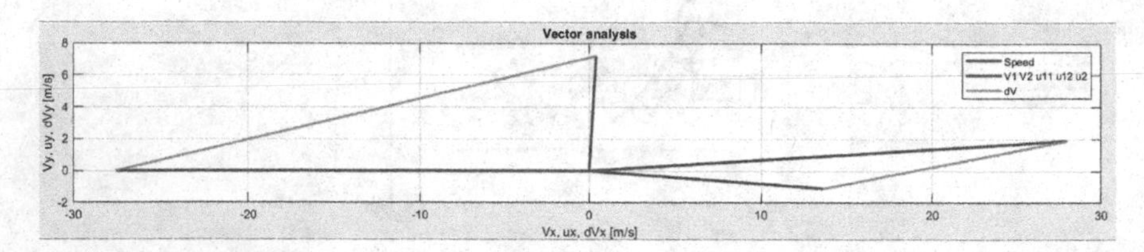

Fig. 42.7 Vector analysis

Table 42.1 Results for the solved impact problem

Kinematic quantities	Results		
Initial position of the Volkswagen Touareg x_{c1}; y_{c1}; φ_{z1}	15.1 m	0.8 m	3°
Initial position of the Opel Vectra x_{c2}; y_{c2}; φ_{z2}	17.7 m	2.2 m	180°
Final position of the Volkswagen Touareg x_{c1}; y_{c1}; φ_{z1}	31.4 m	0.5 m	63.1°
Final position of the Opel Vectra x_{c2}; y_{c2}; φ_{z2}	17.2 m	7.0 m	14.7°
Initial velocity prior to impact Volkswagen Touareg $\dot{x}_{c1}$; $\dot{y}_{c1}$	28.0 ms^{-1}	1.9 ms^{-1}	
Initial velocity prior to impact of the Opel Vectra $\dot{x}_{c2}$; $\dot{y}_{c2}$	−27.6 ms^{-1}	0 ms^{-1}	
Initial velocity after impact for the Volkswagen Touareg $\dot{x}_{c1}$; $\dot{y}_{c1}$; $\dot{\varphi}_{z1}$	13.6 ms^{-1}	−1.1 ms^{-1}	4.99 s^{-1}
Initial velocity after impact for the Opel Vectra $\dot{x}_{c2}$; $\dot{y}_{c2}$; $\dot{\varphi}_{z2}$	0.4 ms^{-1}	7.2 ms^{-1}	4.98 s^{-1}
Crash pulse magnitude S_x; S_y	−3,040 Ns	−6801 Ns	
Coefficient of restitution	k = 0.207		

the position of the two vehicles at the time of impact and the velocities of the centers of mass right before impact, based on the following reliability criteria:

- The trajectories of the vehicles centers of the wheels exactly correspond to the configuration and the position of the tire tracks on the lane.
- The direction of the velocity vectors of the centers of mass of the two vehicles prior to impact correspond completely to the position of the two vehicles at that moment.
- The simulation results demonstrate that the crash pulse vector and its directix correspond completely to the deformations of the vehicles and their yaw rotation after impact.
- The absolute values of the velocity change $|\Delta \overline{V}|$ for each car correlate well with the strain energy.
- It is not necessary to select coefficient of restitution according to the theory of impact, but it is determined on the basis of the presented scientific approach.

Acknowledgements The author/s would like to thank the Research and Development Sector at the Technical University of Sofia for the financial support.

References

1. Tomasch, E., Sinz, W., Hoschopf, H., Kolk, H., Steffan, H.: Prospektive Bewertung der Kollisionsschwere von Le Fahrzeugen unter Berьcksichtigung eines Kollisionsminderungssystems. In: Proceedings of the 10. VDI-Tagung Fahrzeugsicherheit, Sicherheit 2.0, pp. 407–418. Berlin (2015)
2. Tomasch, E., Steffan, H.: ZEDATU (Zentrale Datenbank tödlicher Unfälle in Österreich). A central database of fatalities in Austria. In: Proceedings of the International Conference Expert Symposium on Accident Research, ESAR 2006, pp. 183–185. Hannover (2006)
3. Winner, H., Hakuli, S., Lotz, F., Singer, C.: Handbuch Fahrerassistenzsysteme, Grundlagen. Komponenten und Systeme für aktive Sicherheit und Komfort. Springer Vieweg, Wiesbaden (2015)
4. Zauner, C., Tomasch, E., Sinz, W., Ellersdorfer, C., Steffan, H.: Assessment of the effectiveness of intersection assistance systems at urban and rural accident sites. In: Proceedings of the International Conference Expert Symposium on Accident Research, ESAR 2014, pp. 1–10. Fachverlag NW in der Carl Schuenemann Verlag GmbH (2015)
5. Zhou, J., Lu, J., Peng, H.: Vehicle stabilization in response to exogenous impulsive disturbances to the vehicle body. Int. J. Veh. Auton. Syst. **8**(2/3/4), 242–262 (2010)
6. Tomasch, E., Sinz, W., Hoschopf, H., Kolk, H., Steffan, H.: Bewertungsmethodik von integralen Sicherheitssystemen durch Kombination von Test und Simulation am Beispiel von Fugngerunfllen. In: Proceedings of the 10. VDI-Tagung Fahrzeugsicherheit, Sicherheit 2.0, pp. 157–169. Berlin (2015)
7. Billicsich, S., Tomasch, E., Sinz, W., Eichberger, A., Markovic, G., Magosi, Z.: Potentieller Einfluss von C2X auf die Vermeidung von Motorradunfllen bzw. Reduktion der Verletzungsschwere. In: Proceedings of the 10. VDI-Tagung Fahrzeugsicherheit, Sicherheit 2.0, pp. 383–392. Berlin (2015)
8. Burg, H., Moser, A.: Handbuch Verkehrsunfallrekonstruktion: Unfallaufnahme, Fahrdynamik. Simulation. Vieweg+Teubner Verlag, Wiesbaden (2009)

9. Jung, H.G., Cho, Y.H., Kim, J.: Isrss: integrated side/rear safety system. Int. J. Automot. Technol. **11**, 541–553 (2010)
10. Tomasch, E., Steffan, H., Darok, M.: Retrospective accident investigation using information from court. In: Proceedings of the European Road Transport Research Arena Conference, TRA 2008, poster. Ljubljana (2008)
11. Eidehall, A., Pohl, J., Gustafsson, F., Ekmark, J.: Toward autonomous collision avoidance by steering. IEEE T. Intell. Transp. **8**(1), 84–94 (2007)
12. Karapetkov, S.: Mechanics of solids and impact between bodies in identifying road accidents (2006)
13. Karapetkov, S.: Mechanical and mathematical modelling of vehicle motions in identifying road accidents (2012)
14. Karapetkov, S., Dimitrov, L., Uzunov, Hr., Dechkova, S.: Identifying vehicle and collision impact by applying the principle of conservation of mechanical energy. Transp. Telecommun. **20**(3), 191–204 (2019)
15. Karapetkov, S., Dimitrov, L., Uzunov, Hr., Dechkova, S.: Examination of vehicle impact against stationary roadside objects. IOP Conf. Ser. Mater. Sci. Eng. **659**, 012063–1–012063–13 (2019)
16. Sander, U., Mroz, K., Bostroem, O., Fredriksson, R.: The effect of pre-pretensioning in multiple impact crashes. In: Proceedings of the 21st International Technical Conference on the Enhanced Safety of Vehicles, ESV 2009, pp. 1–11. Stuttgart (2009)
17. Togawa, A., Murakami, D., Saeki, H., Pal, C., Okabe, T.: An insight into multiple impact crash statistics to search for future directions of counter-approaches. In: Proceedings of the 22nd International Technical Conference on the Enhanced Safety of Vehicles, ESV 2011, pp. 1–5. Washington (2011)

Chapter 43
Checking the Validity of the Simulation for a Vehicle Test Collision

Dan-Marius Mustață, Attila-Iuliu Gönczi, Ioana Ionel, and Ramon Mihai Balogh

Abstract Road accident reconstruction is a very complex task, which is more and more based on computer simulation of the vehicles' motion (kinematic and kinetic approach) and of the collision between vehicles, pedestrians and/or different other objects. The research presented in the paper is focused on the validation of a vehicle collision simulation programme, namely Virtual Crash ver. 4.0, based on a real test collision. The test collision refers to test nr. 358 achieved by DSD GmbH in Austria, in 2018, and consisted of an experiment during which a passenger car hits, at high speed, a stationary truck. The paper focuses on the comparison between the crash results and the results obtained from a numerical simulation, by using the Virtual Crash software. The conclusion of the validation is that the positioning errors (the linear and angular) are well under the acceptable values for road accident reconstruction. First of all, it is important to emphasize, that the determined value of the impact speed in the simulation (90,123 km/h) was practically the same as the real speed (90,6 km/h), which is a remarkably good simulated value (the relative error is only −0.53%) and it could be related to the measurement errors of the speed. In a case of the reconstruction of a real accident, the errors could be much higher, because of the errors of the investigation on site and the lack of precise data related to many parameters which influence the reconstruction. The research is part of the MSc thesis of the main author.

Keywords Accident reconstruction software · Validation · Virtual crash

43.1 Introduction

In the automotive industry, one of the highest concerns is vehicle crashworthiness. Rating systems, like Euro NCAP and national control bodies like National Highway

D.-M. Mustață · A.-I. Gönczi · I. Ionel (✉) · R. M. Balogh
Politehnica University Timisoara, Timișoara, Romania
e-mail: ioana.ionel@upt.ro

D.-M. Mustață
e-mail: dan.mustata@student.upt.ro

M. Rackov et al. (eds.), *Machine and Industrial Design in Mechanical Engineering*, Mechanisms and Machine Science 109,
https://doi.org/10.1007/978-3-030-88465-9_43

Traffic Safety Administration in the USA, are regulating all safety aspects, which automotive industry companies have to comply with [1, 2].

A crash test is represented as a destructive test, performed in order to ensure safe design standards in crashworthiness and crash compatibility for vehicles or their components [3]. Experimental crash tests represent a challenge, due to the necessity of correct and well-documented execution. Efforts have been made to be able to assess completely the overall vehicle performance during a collision, without needing to prepare and execute a full-scale experiment, which implies higher costs. Crash tests are performed not only once a vehicle prototype is produced, but already from the first steps of the conception of a vehicle, throughout the entire development, designing and validation process. Virtual crash-test software, using mathematical models (mainly finite element method [4]), are used to model and predicts real vehicle behavior, deformations, appearing during and after a collision, and the effect of the impact on the vehicle occupants.

In this paper, the authors are focusing on reconstructing a real test collision using a vehicle accident simulation software, called Virtual Crash ver. 4.0, created for accident reconstruction purposes and validating the program by comparing the main input and output values. Virtual Crash is a vehicle dynamics and collision simulation tool based mainly on rigid body dynamics and momentum analysis.

43.2 Initial Data

The real test collision was done by DSD GmbH in Austria, in 2018 (DSD test nr. 358) [5], using a passenger car (P05) hitting at high speed a stationary truck (L13). Table 43.1 contains technical and collision data related to each vehicle, subjected to the test collision.

43.3 Analysis and Results

The real test collision was reconstructed with the simulation software Virtual Crash ver. 4. In Figs. 43.1 and 43.2, collision data comparison between the real test collision and the simulated process using the above-mentioned software are presented.

Virtual Crash 4.0 uses, for its principal collision model for vehicle-to-vehicle impacts, a momentum based mathematical model (the Kudlich-Slibar), which is based mainly on momentum conservation (but is much more complex, see [6, 7]). One of the main characteristics of the model is the fact, that the changing of the momentum takes place instantaneously (in reality is a gradual process), which occurs at a given moment after the first contact of the objects entering in collision. The difference between the two moments is the so-called "depth of penetration" which in our case was considered 0.04 s. The default value in the software is 0.03 s. During this interval, the vehicles just penetrate each other, without any change in their dynamic

Table 43.1 General technical and collision data for both vehicles subjected to test [5]

Label	P05	L13
Make	Ford	MAN
Model	Focus Ambiente 1.8 TDDI	19.403 FLLC
Year	2000	1997
Velocity (km/h)	90.6	0
Weight (kg)	1214	10,754
Length (m)	4.15	9.6
Width (m)	1.70	2.5
Height (m)	1.44	3.6
Wheelbase (m)	2.62	5.5
Trackwidth (m)	1.50	2.04
Axle load front (kg)	775	5400
Axle load rear (kg)	439	5354
Dist. COG to front axle (m)	0.95	2.74

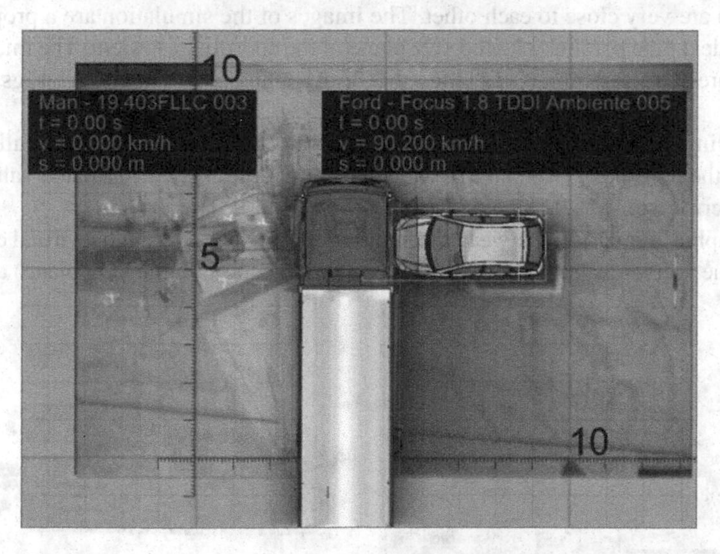

Fig. 43.1 The position of the vehicles at the moment of the first contact (t = 0 s) in the simulation

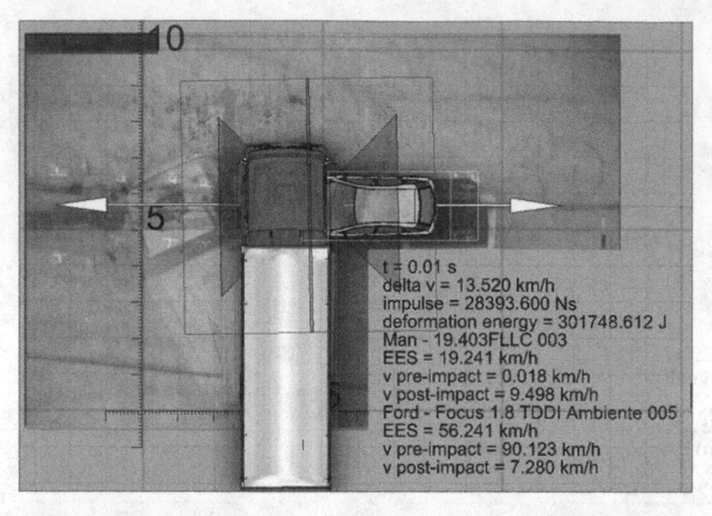

Fig. 43.2 Positions of the vehicles at the first impact (generated automatically by the software at the moment t = 0.04 s) and some of the main parameters of the simulated collision

parameters. In Figs. 43.3, 43.4, 43.5 and 43.6, a comparison between real test collision and software simulation at certain characteristic moments are presented. It is obvious, that excepting the first 0.038 s (the depth of penetration), the simulation and the real collision are very close to each other. The images of the simulation are a projection on a vertical plan parallel with the xOz plane of the coordinate system. The images of the test are extracted frames of a video shot with a high-speed camera (see Figs. 43.3–43.6).

The simulated impact speed in the simulation (90,123 km/h) is practically very close to the real speed (90,6 km/h), which is a remarkably good simulated value (the relative error is only—0.53%).

A graphic comparison between the simulation (modelled through a virtual camera within the software application, which is not a projection on the xOy plane) and the

Fig. 43.3 First contact (at t = 0.000 s in the simulation-left, real test collision-right [5])

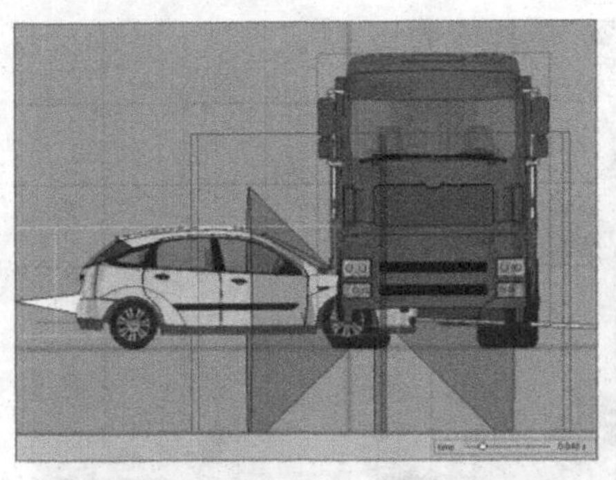

Fig. 43.4 Collision (at t = 0.038 s in the simulation-left, real test collision-right [5])

Fig. 43.5 Positions at time t = 0.175 s (simulation-left, real test collision-right [5])

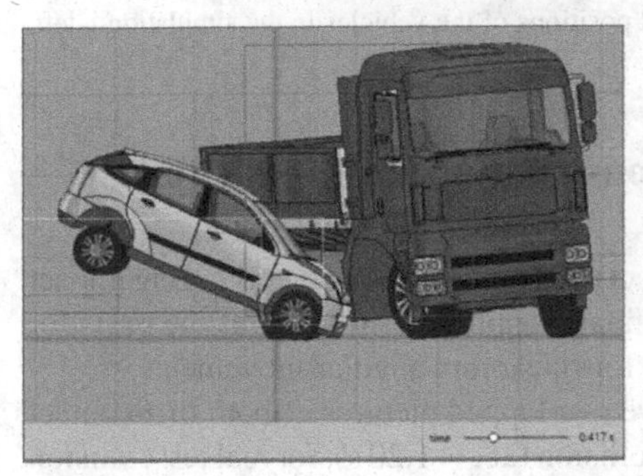

Fig. 43.6 Positions at the moment t = 0.417 s (simulation-left, real test collision-right [5])

real test collision (filmed with a high-speed camera), observed from above, is revealed by Figs. 43.7 and 43.8. Both images are in perspective and not at exactly the same scale.

Fig. 43.7 Positions at t = 0 s, at first contact (simulation—left, real test collision—right [5])

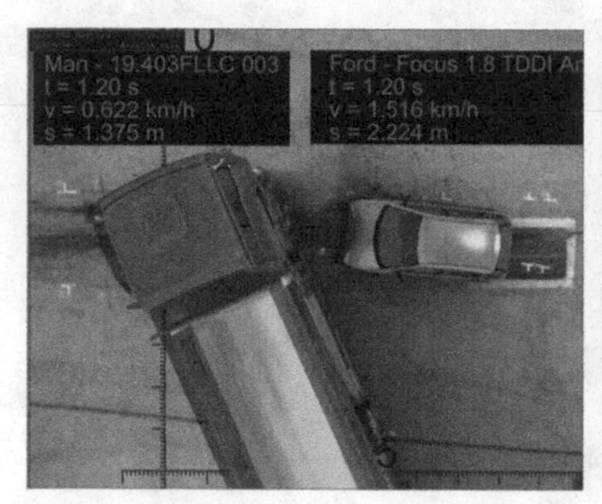

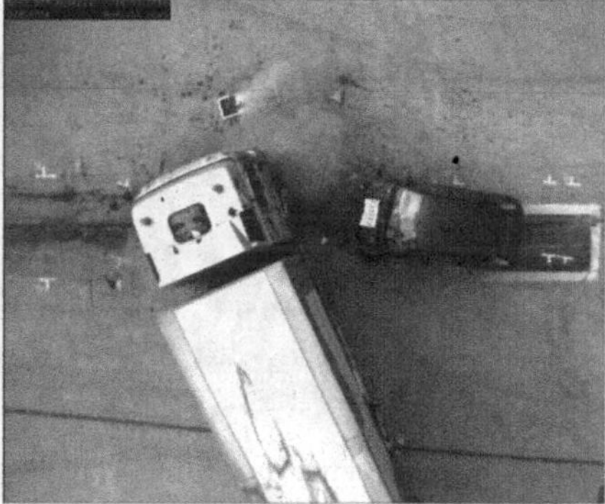

Fig. 43.8 Positions at t = 1.2 s (the almost final positions of the vehicles in the simulation—left, real test collision—right [5])

43.4 Comparison Between Reconstruction and Crash Test

The determination of the positions (at impact) and of the displacements during impact (from the positions at first contact to the final stopping positions) was the basis for error calculations, which were performed for linear and angular movements.

In Figs. 43.9 and 43.10 one can observe overlapped pictures of both first contact (at t = 0 s) and almost final position after collision (at t = 1.20 s), for real test collision and Virtual Crash 4.0 software simulation, which were used for measurement.

The results of the linear measurements and the error calculations are centralized in Tables 43.2 and 43.3, for both vehicles, Ford and MAN.

The angular errors of the last (stopping after impact) position (i.e., the angular difference between the longitudinal axis in the simulation and in the real crash-test) are very small, namely, 1.648° for the passenger car, and just −0.926° for the truck.

The errors between the real crash-test and the simulation (as a result of the reconstruction of the collision with Virtual Crash ver. 4) can be explained by the following

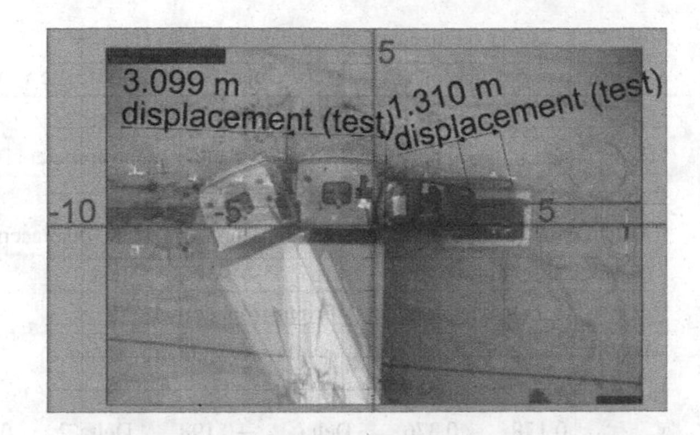

Fig. 43.9 Overlapped pictures of the real test collision [5] at first contact and final position, including displacement direct measurement (Ford, MAN) done in Virtual Crash ver. 4

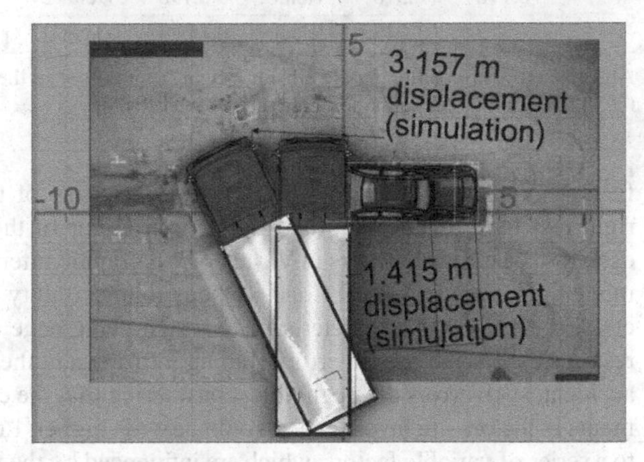

Fig. 43.10 Overlapped images from Virtual Crash 4.0 simulation at first contact and final position, including displacement direct measurement (Ford, MAN) done in Virtual Crash ver. 4

Table 43.2 Linear relative and absolute error results for Ford

Relative error calculation using measurements in the coordinate system							
Ford	Simulation			Ford	Test		
Displacement in meters			1.237	Displacement in meters			1.303
Relative error (%) of displacement of right—rear corner							−5.04%
Relative error calculation using direct measurement of the displacements							
Ford	Simulation		1.415	Ford	Test		1.31
Relative error (%) displacement of right—rear corner							−8.02%
Absolute error calculation for the position of right-rear corner							
Ford	Sim	Test					Distance (m)
x	4.156	4.341	Delta	−0.185	Delta^2	0.034225	0.251
y	1.206	1.037	Delta	0.169	Delta^2	0.028561	

Absolute error of positioning at collision. Back-right corner in the simulation was determined at 0.251 m from the real point measured in the test collision

Table 43.3 Linear relative and absolute error results for MAN

Relative error calculation using measurements in the coordinate system						
MAN	Simulation		MAN	Test		
Displacement in meters		3.169	Displacement in meters			3.092
Relative error (%) of displacement of right-front corner						−2.49%
Relative error calculation using direct measurement of the displacements						
MAN	Simulation	3.157	MAN	Test		3.099
Relative error (%) displacement of right-front corner						−1.87%
Absolute error calculation for the position of right-front corner						
MAN	Sim	Test				Distance (m)
x	0.178	0.376	Delta	−0.198	Delta^2	0.039204	0.238
y	1.877	1.745	Delta	0.132	Delta^2	0.017424	

Absolute error of positioning at collision. Right-front corner in the simulation was determined at 0.238 m from the real point measured in the test collision

factors: original DSD image resolution, coarseness of the position of the corner right-rear lamp of the Ford and right front corner of the MAN (it has a rounded design) measurement errors in Virtual Crash graphic interface and, last but not least, modelling errors in Virtual Crash 4.0, due to uncertainty of some important parameters. It is impossible to determine to which percentage every mentioned factor, is responsible for the overall errors. Taking into account these results, in case of a real accident, such errors are negligible, considering that the errors of the field measurements in the on-site investigation could be even higher. Real accidents are subjected to a series of variable factors, which are influenced by the way the investigation of an accident is done. Investigations are indeed achieved by specialized police officers, but even so, taking the human factor into account, it is almost certain that errors are due to lack of precise data and incomplete measurements within the field. This definitely could lead to much higher errors than the incomplete modelling.

43.5 Conclusion

By comparing the data sets given by the software simulation and the crash-test, there is only a slight difference. For example, the positioning errors (linear and angular) are well under the acceptable values for road accident reconstruction. As for the "reconstructed" value of the collision speed, the relative error is practically zero (−0.53%).

Based on this analysis, it could be sustained, even by this example, that Virtual Crash ver. 4, as a road vehicle collision simulation software, is usable with a high level of confidence for accident reconstruction. Real accidents are subjected to a series of variable factors, which are influenced by the way the investigation of an accident is done and it is almost certain that errors are due to lack of precise data and incomplete measurements within the field.

Acknowledgements The first author expresses his gratitude to the teaching staff of the Politehnica University of Timisoara (Romania) and to the co-authors, especially to Senior Lecturer A. Gönczi for offering his expertise in the field.

References

1. Pawlus, W., Robbersmyr, K.G., Karimi, H.R.: Mathematical modeling and parameters estimation of a car crash using data-based regressive model approach. Appl. Math. Model. **35**(10), 5091–5107 (2011)
2. https://ec.europa.eu/transport/road_safety/specialist/knowledge/safetyratings/safety_ratings_in_use/vehicle_safety_en. Last Accessed 02 Apr 2021
3. https://en.wikipedia.org/wiki/Crash_test. Last Accessed 02 Apr 2021
4. Pawlus, W., Karimi, H.R., Robbersmyr, K.G.: Investigation of vehicle crash modeling techniques: theory and application. Int. J. Adv. Manuf. Technol. **70**, 965–993 (2014)
5. DSD (Dr. Steffan Datentechnik) Homepage. Crash Tests Database. http://www.dsd.at/CrashTests/ShowCrashTest.php?Dir=.%2FCDs%2FDSD+2018%2FTests%2FTest04%2F. Last Accessed 15 Mar 2021
6. Steffan, H., Moser, A.: The collision and trajectory models of PC-CRASH. SAE Tech. Pap. **960886** (1996)
7. Single Body Collision Physics Model. https://www.vcrashusa.com/guide-appendix1. Last Accessed 02 Apr 2021

Chapter 44
Mechanical Metastructure in Structural Engineering: A Short Review

Livija Cveticanin and Sinisa Kraljevic

Abstract In this paper a short review on the mechanical metastructure applied in civil, mechanical and structural engineering is considered. This metastructure represents the macro version of the metamaterial which is a kind of composite. The main property of the mechanical metastructure is to mitigate and suppress vibration in systems. Depending on the configuration it represents vibration isolator or absorber which stops the low-frequency vibrations and forms an oscillation band-gap in a certain frequency interval. Various types of mechanical metastructures with negative effective mass and negative stiffness are presented. In the paper the auxetic structures with negative Poison's coefficient for vibration elimination are also considered. The future investigation in the matter is suggested.

Keywords Mechanical metastructure · Auxetic structure · Vibration isolator · Vibration absorber

44.1 Introduction

One of the most important environmental pollutant is vibration. Generally, vibrations are caused by external periodic forces which often occur in rotors. As is well known the rotating part of the machines has the eccentricity due to inaccuracy during production and mounting, inhomogeneity in material, unsymmetrical wear in working process, etc. The physical and the geometric center of the rotor differ and the centrifugal force acts which causes the transformation of the useful working energy into mechanical vibration. Vibration in our surrounding may cause health problems which may be of local (on hands or legs) or global (for whole body vibration) character. Usually, vibration are accompanied with noise. Noise is the

L. Cveticanin (✉) · S. Kraljevic
Faculty of Technical Sciences, University of Novi Sad, Trg Dositeja Obradovića 6, 21000 Novi Sad, Serbia
e-mail: cveticanin@uns.ac.rs

L. Cveticanin
Obuda University, Nepszinhaz u. 18, Budapest, Hungary

M. Rackov et al. (eds.), *Machine and Industrial Design in Mechanical Engineering*, Mechanisms and Machine Science 109,
https://doi.org/10.1007/978-3-030-88465-9_44

unwilling sound which represents also an environmental pollutant. Namely, part of the vibration energy is transformed into sound wave which gives the noise. Noise may cause disturbance during the work or relaxation but also give very intensive health problems. Because of that, strong restrictions to the noise level in the life and working environment are introduced. Depending on the noise frequency and its density, given in decibel (dB), the influence on the health condition are determined. Namely, according to measuring it is concluded that the noise up to 30 dB gives psychical disturbance, the noise in the interval of 30–65 dB makes problems in vegetative nerve system, noise between from 65 till 90 dB disturbs hearing, noise higher than 120 dB causes physical pain and that of 175 dB causes death. Effect of noise depends also on the frequency spectar. The noise which is in the hearing interval of 100 Hz to 1 kHz is the most dangerous and has to be decreased or eliminated. Problem of vibration and noise in the environment and working space is not new one. Significant number of methods, procedures and also technical devices are developed. Nevertheless, the final solution of the problem is not obtained, yet. In addition to mechanical absorbers very often various types of isolators are applied. According to the theory it is known that the quality of absorption is valuated with transmission lost. This parameter depends on the thickness of the isolator. To increase the noise reduction for 5–6 dB at a certain frequency, it is necessary to double the isolator layer. It seems that for decrease of the noise for 30 dB the isolator thickens is multiplied more than 60 times. In vehicles the requirement for noise reduction is very strong, but the isolator with so high mass is not impossible to be incorporated into the cabine. To fulfill the task there is the requirement for new investigation, including new structures, materials and devices.

Usually, the pre-existing vibrations are treated in two ways: one, by installing of isolator and reduction of vibration amplitude and the second, by using of the dynamic absorber which eliminates vibration at the certain frequency. In both cases for realization of the task masses or mass-spring systems are added. This causes variations in structure properties due to increase of the mass and the change of its rigidity. In addition, this intervention is not costless. Therefore, for the last 20 years, intensive work has been done on the development of new concepts of vibration reduction, but in a way to eliminate the mentioned shortcomings, without changing the functionality as well as the existing physical characteristics of the structure, above all, load-bearing capacity, strength and rigidity.

One of the possibility is to develope the structure based on the model of a composite called metamaterial which is used for elimination of the electromagnetic wave at the certain frequency. Development of electric metamaterial is closely connected with the stealth technology. For this metamaterial it is concluded that the wave absorption is due to negative permeativity, negative permittivity and negative index of reflection.

Using the analogy between the electromagnetic and acoustic wave there was the idea to produce the acoustic metamaterial which would have analog properties: negative mass, negative rigidity and negative Poisson's coefficient. As it is known, the material with negative properties is not available in the nature. Using the principles of composite, the structures with theoretically negative effective characteristics are constructed. The values of mass and stiffness of the metastructure are introduced to be fictively i.e. theoretically negative. According to the theory these types of metastructures are introduced. However, development of new metastructures is quite new and needs more investigation in the future.

Since 2000, research in this area has focused on finding materials that will be able to ensure the elimination of low-frequency oscillations in a certain range. It turned out that there are no such materials in nature, so they resorted to making artificial so-called metamaterials [1]. Meta-materials are composites whose structure is adjusted to meet certain requirements regarding physical properties. Thus, the idea came up to make a material of periodic structure consisting of a series of basic units, each of which contains a vibration absorber. Thus, instead of only one vibration absorber a large number of continually distributed absorbers have to be used to eliminate vibrations. This type of metamaterial is called mechanical or elastic. In the basic honeycomb structure absorbers, made of material which differs from the basic one, are incorporated. The fabrication of such a microscopic order structure proved to be technologically very difficult to do. However, the structure was realized at the macro level and an appropriate metastructure was formed. The paper [2] shows an aluminum chiral plate (dimensions 470 × 91 × 10 mm), in which rubber-coated metal cylinders are placed.

Experimental measurements of the vibration characteristics of the structure with and without inserted absorbers were done. According to different layouts, different frequency responses are indicated, and the existence of a vibration gap i.e. absence of vibration for certain frequencies is evident. A theoretical explanation of this phenomenon is given in [3]. It was concluded that each of the basic units of the metastructure has a negative effective mass. The area of existence of the negative effective mass is directly related to the elimination of oscillations at a certain frequency. The larger the area of the negative effective mass, the larger the area for which this absorber ensures the absence of oscillations [4]. This area has been shown to be larger if the nonlinearity of the absorber is larger [5, 6].

Three types of mechanical metastructures are considered: with negative effective mass, negative stiffness and negative Poison's ratio. The aim of this paper is to show some new types of metastructures for noise and vibration reduction. The paper is divided into 5 Sections. In Sects. 44.2, 44.3 and 44.4 metastructures with negative effective mass, rigidity and Poisson's rate are presented. In Sect. 44.5, the suggested metastructure is connected with energetic harvester which transforms the vibration energy of the structure into electric energy. Two types of harvester are mentioned. Application of the obtained energy is discussed. The paper ends with conclusion.

44.2 Metastructure with Negative Effective Mass

The appearance of elastic or mechanical metamaterials, i.e. metastructures with added absorbers has attracted much attention from researchers [7, 8]. The basic element of the structure is a mass-in-mass unit (see Fig. 44.1).

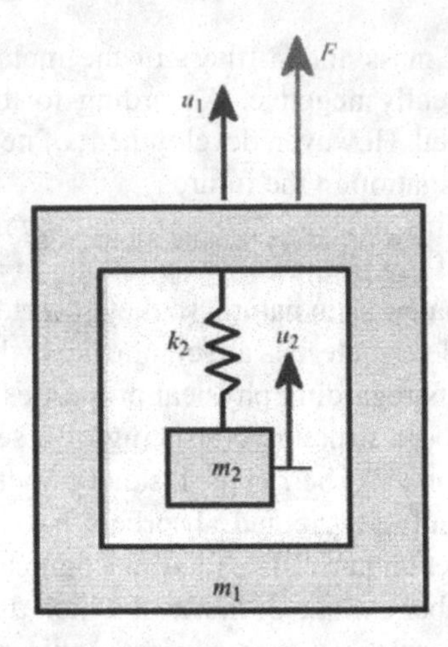

Fig. 44.1 Mass-in-mass unit

Namely, in the basic mass m_1 a spring-mass system is added. The added mass is m_2 and the rigidity of the spring is k_2. On the mass m_1 the external excitation force F acts. In general, displacement of mass m_1 is u_1 and of the added mass is u_2. Investigation in mass-in-mass unit show that for certain excitation frequency and corresponding values m_2 and k_2 the motion of the basic mass m_1 is zero. The added system acts as vibration absorber. Combining the units various beam and plane structures are formed (for example Fig. 44.2).

The absorption property of various metastructures is tested. Metastructures with different basic structures and shape, number and position of the inserted absorbers is investigated.

The disadvantage of these metastructures is the complexity of their fabrication. Because of that new metastructure where the material of basic structure and of absorbers is the same is designed. However, the basic concept of a single unit with

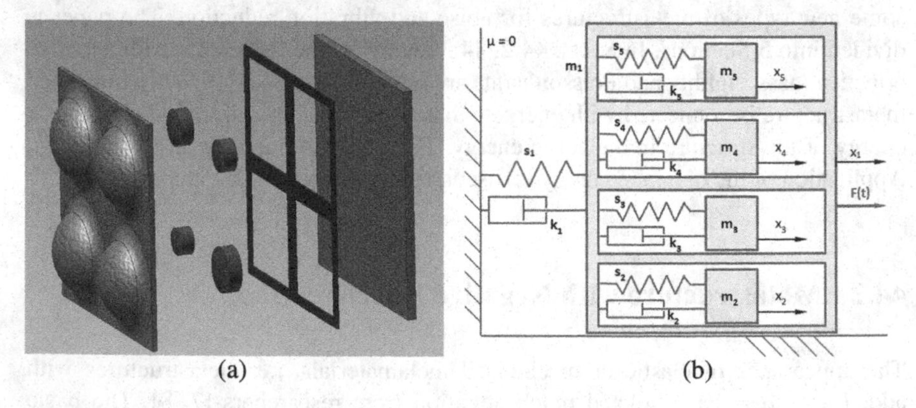

Fig. 44.2 **a** Plane metastructure and **b** model of the metastructure

negative effective mass is not changed [9]. The basic structure is produced on the 3D printer of polyurethane material. The basic units of the structure are interconnected in the form of a 1D rod [10], a 2D plate [11] and a 3D form [12].

44.3 Metastructure with Negative Stiffness

A metastructure that has a negative stiffness can also be used as a vibration isolator. Theoretical investigation show that the basic unit of this vibration isolator is a mass—stiffness system. The basic element is massless but with stiffness coefficient k_1 (Fig. 44.3). Inside the basic mass a mass-spring system is settled. The added mass is m_2 and the spring stiffness is k_2. The external force acts on the basic element. It causes the system to vibrate. However, for certain mass-stiffness coefficient relation, the amplitude of vibration of the basic element is very small (but it is not zero). Negative stiffness is manifested when the direction of force is not parallel to the direction of displacement in the basic structure [13]. For that case, the basic unit has two stable positions.

In addition, due to elasticity of metastructure, the energy of absorption is reversible. In [14] the special type of metastructure with units having quasi-zero dynamic stiffness is considered.

The units are designed in the form of sinusoidal rod or semicircular arches. This metastructure has significantly better vibration isolation capabilities than those exhibited by a linear isolator. The oscillation amplitude is significantly reduced with metastructure. In addition, studies have shown that the vibration characteristic is better if the damping in the material is low and the excitation frequency is high.

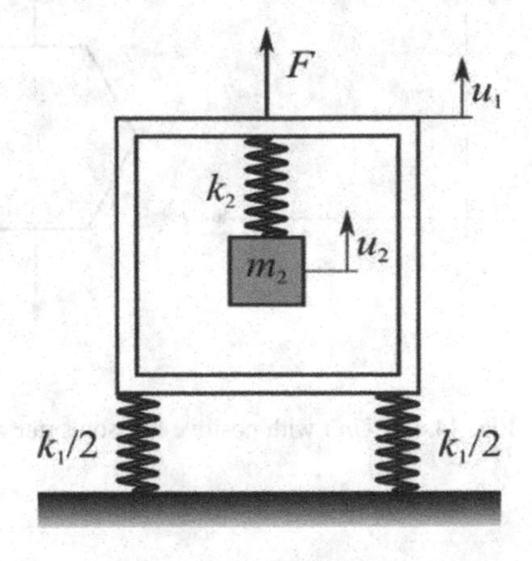

Fig. 44.3 Mass-in-stiffness unit

44.4 Auxenic Structures

Usually, if on the probe a tensile force acts, the length of the probe is increased, while the cross-section is decreased (Fig. 44.4a).

The relation between deformation in the direction of load and deformation in orthogonal direction is called Poisson's rate. It is assumed that this rate is positive However, new concave hexagonal unit is constructed (Fig. 44.4b) for which the character of the Poisson's rate is opposite to the previous one: both deformation in loading direction and also in orthogonal direction are increasing (not only the length but also the cross section).

Structure constructed from these units is called auxetic. The basic feature of the structures is that the dimension in the direction perpendicular to the direction of action of the tensile force increases and at the same time the structure absorbs energy. Basically, these structures are lattice with nodes (polygonal or circular) and elastic bending ligaments. Depending on the spatial arrangement of the nodes and ligaments, there are two basic unit structures [15]: one in which the ligaments are on the same side of node and the other, where the ligaments are on the opposite sides of the node. In [16], a modification of the basic structure was performed in order to reduce the stress concentration in the nodes. The modified S lattice metastructure is designed. Very often, a combination of tetra and anti-tetra units is used in the construction of the metastructure [17]. Depending on the number of ligaments, the basic structure can be four-sided or six-sided [18]. This metastructure is diluted according to the skin structure of lizards and snakes [19].

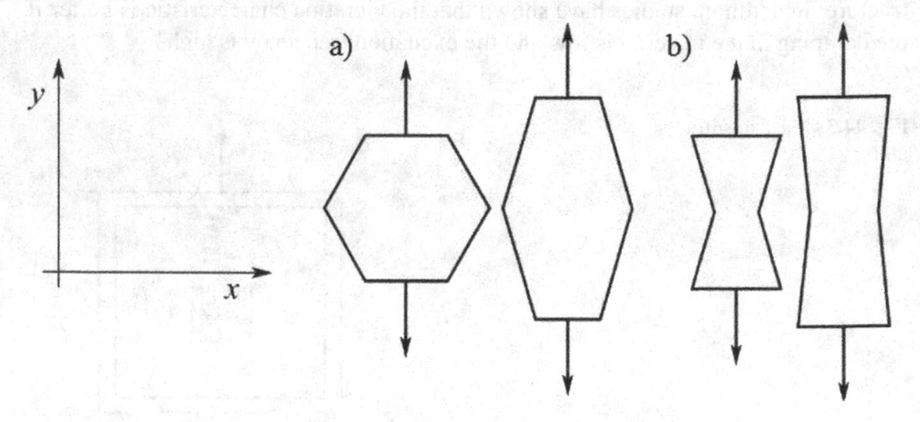

Fig. 44.4 **a** Unit with positive Poisson's rate and **b** unit with negative Poisson's rate

44.5 Conclusion

Although the plate with metastructure has been shown to be very suitable as a vibration absorber or vibration isolator (depending on the structure), tests have also shown a major drawback. Namely, the structure shows anisotropy, i.e. when a force acts in one direction it shows an absorption property, but when a force acts in the other direction, this property does not exist. In addition, the characteristics of the slab with metastructure do not meet some of the requirements that designers usually set before it: they do not have sufficient strength, resistance to temperature changes, thermal conductivity, sufficient rigidity, etc. Hence, there is a need to harmonize the type of material and structure into a single whole in order to ensure: isotropy, suitability for load transfer, easy technical implementation, i.e. production, resistance to temperature changes, but also low cost of fabrication.

Acknowledgements The investigation is the part of the Project of the Department of Technical Mechanics of the Faculty of Technical Sciences, No.54/2021.

1. References

1. Cveticanin, L., Mester, G.: Theory of acoustic metamaterials and metamaterial beams: an overview. Acta Polytech. Hung. **13**(7), 43–62 (2016)
2. Zhu, R., Liu, X.N., Hu, G.K., Sun, C.T., Huang, G.I.: A chiral elastic metamaterial beam for broadband. J. Sound Vib. **333**, 2759–2773 (2014)
3. Milton, G.W.: New metamaterils with macroscopic behavior outside that of continuum elastodynamics. New J. Phys. **359**(9), 1–13 (2007)
4. Cveticanin, L., Zukovic, M., Cveticanin, D.: On the elastic metamaterial with negative effective mass. J. Sound Vib. **436**, 295–309 (2018)
5. Cveticanin, L., Zukovic, M.: Negative effective mass in acoustic metamaterial with nonlinear mass-in-mass subsystems. Commun. Nonlinear Sci. **51**, 89–104 (2014)
6. Cveticanin, L., Zukovic, M., Cveticanin, D.: Influence of nonlineaer subunits on the resonance frequency band gaps of acoustic metamaterial. Nonlinear Dynam. **93**(3), 1341–1354 (2018)
7. Reichl, K.K., Inman, D.J.: Lumped mass model of a 1D metastructure for vibration suppression with an additional mass. J. Sound Vib. **403**, 75–89 (2017)
8. Pierce, C.D., Willey, C.L., Chen, V.W., Hardin, J.O., Berrigan, J.D., Juhl, A.T., Matlack, K.H.: Adaptive elastic metastructures from magneto-active elastomers. Smart Mater. Struct. **29**, 065004-1–065004-11 (2020)
9. Hobeck, J.D., Laurent, C.M.V., Inman, D.J.: 3D printing of metastructures for passive broadband vibration suppression. In: Proceedings of the 20th International Conference on Composite Materials, pp. 1–8. Copenhagen (2015)
10. Wang, T., Sheng, M.P., Qin, Q.H.: Multi-flexural band gaps in an Euler-Bernoulli beam with lateral local resonators. Phys. Lett. A **380**, 525–529 (2016)
11. Peng, H., Pai, P.F., Deng, H.: Acoustic multi-stopband metamaterial plates design for broadband elastic wave absorption and vibration suppression. Int. J. Mech. Sci. **109**, 104–114 (2016)
12. Essink, B.C., Inman, D.J.: Three-dimensional mechanical metamaterial for vibration suppression. In: Dervilis, N. (ed.) Special Topics in Structural Dynamics & Experimental Techniques, Volume 5: Conference Proceedings of the Society for Experimental Mechanics Series, pp. 43–48. Springer, Cham (2020)

13. Zhang, Y., Wang, Q., Tichem, M., vab Keulen, F.: Design and characterization of multi-stable mechanical metastructures with level and tilted stable configurations. Extreme Mech. Lett. **34**, 100593-1–100593-9 (2020)
14. Fan, H., Yang, L., Tian, Y., Wang, Z.: Design of metastructures with quasi-zero dynamic stiffness for vibration isolation. Compos. Struct. **243**, 112244-1–112244-13 (2020)
15. Xia, R., Song, X., Sun, L., Wu, W., Li, C., Cheng, T., Qian, G.: Mechanical properties of 3D isotropic anti-tetrachiral metastructure. Phys. Status Solidi B **255**, 1700343-1–1700343-9 (2018)
16. Meena, K., Singamneni, S.: A new auxetic structure with significantly reduced stress concentration effects. Mater. Design **173**, 107779-1–107779-11 (2019)
17. Li, H., Ma, Y., Wen, W., Wu, W., Lei, H., Fang, D.: In plane mechanical properties of tetrachiral and antitetralchiral hybrid metastructures. J. Appl. Mech. **84**, 081006-1–081006-11 (2017)
18. Mir, M., Ali, M.N., Sami, J., Ansaari, U.: Review of mechanics and applications of auxetic structures. Adv. Mater. Sci. Eng. 753496-1–753496-17 (2014)
19. Santulli, C., Langella, C.: Study and development of of concept of auxetic structures in bio-inspired design. Int. J. Sustain. Design **3**(1), 20–37 (2016)

Chapter 45
Numerical Investigation of 3D Lattice Infill Pattern Cellular Structure for Orthopedic Implant Design

Rashwan Alkentar, Dávid Huri, and Tamás Mankovits

Abstract Due to the significant increase in medical implant replacement surgeries, the demand for the best cellular design has become a strong necessity. With the additive manufacturing methods in place, researchers are producing many types of cellular structures that help to optimize the shape and design of the implants to get the best biomechanical properties. The paper discusses the numerical investigation of a titanium alloy cellular structure applied as an implant using the finite element method. The aim of the research is the establishment of a proper simulation process for compression. Based on the stress–strain curves, the compressive response of three different sets of porosity for the same cellular structure type are compared.

Keywords Cellular structure · Additive manufacturing · Finite element analysis

45.1 Introduction

Along the way of the implant life in the human body, many failures may rise and cause irregularities to the implant function. Researchers have identified the most famous faults of the medical implants to be the dislocation, stress shielding and loosening of the prosthesis planted inside the body [1–3]. Thus, it has been more than important to revise the shape and design of the implant before moving into the manufacturing process. The new feature of the lattice-structured unit cells has provided a new solution for most of the failures mentioned by improving the mechanical and biological properties of the implant structure and being more suitable for bone ingrowth [4–8].

The mechanical and biological properties of the cellular structure used in the medical implants depend on the topology and the porosity of the lattice [9, 10], and that urged the ANSYS Software Company to produce many options to apply some

R. Alkentar · D. Huri
Doctoral School of Informatics, University of Debrecen, Kassai 26, 4028 Debrecen, Hungary

D. Huri · T. Mankovits (✉)
Faculty of Engineering, University of Debrecen, Ótemető 2-4, 4028 Debrecen, Hungary
e-mail: tamas.mankovits@eng.unideb.hu

M. Rackov et al. (eds.), *Machine and Industrial Design in Mechanical Engineering*, Mechanisms and Machine Science 109,
https://doi.org/10.1007/978-3-030-88465-9_45

predefined lattice structures. This feature has the option to manipulate the filling percentage and the dimensions of the strut of the cell to have the required shape.

The design of the unit cells is much more complicated than the normal structure, and it is harder and more expensive to manufacture using conventional methods. Consequently, additive manufacturing (AM) has emerged over the past few years to replace traditional manufacturing. With AM, it is now possible to create intricate geometries and personalized pieces such as the cellular structures under research [11].

All the current researches focus on the titanium alloy Ti6Al4V to be the option for medical implants thanks to its high performance both mechanically and biologically and the good resistance against corrosion [12].

In this paper, a cellular structure is chosen to be numerically analyzed. The finite element analysis includes simulating the compression test to determine the stress–strain curve related. The effect of the structural porosity on the material response is also investigated by analyzing three sets of porosities of the cellular structure.

45.2 Material and Methods

45.2.1 Materials

For the sake of biocompatibility, the titanium Ti6Al4V has been used in the current research as material. This alloy is famous for the high degradability, high strength, corrosion-resistance, low density and wide usage in the biomaterial industry [13–15].

The titanium alloy industrial name used in the paper is Ti64, which corresponds to the standards ASTM F1472 and ASTM F2924. Parts made from such material can be later machined and heat-treated as well. Table 45.1 shows the mechanical and physical properties of the alloy used. The material response was modeled using the isotropic elastic material model. The aim of the investigation was to determine the force–displacement relationship within the elastic zone.

Table 45.1 Properties of the Ti64 [16]

Property	Value
Density (kg/m^3)	4429
Yield strength (MPa)	910
Compressive strength (MPa)	1080
Tensile strength (MPa)	1200
Young's modulus (MPa)	113,800
Poisson's ratio	0.34

45.2.2 Preparation of the Geometry of the Specimens

The sample was modeled with correspondence to the standard ISO 13314 of metals mechanical testing, especially for the compression test. The rectangular specimen has been chosen for the testing process, see in Fig. 45.1. Since the specimen is undergoing compression, and for the sake of making the simulation experiment as precise as possible, extra bulk material was added with the amount of 5 mm on both sides top and bottom, which gives better results. The standard states, for a rectangular specimen, that the width of the specimen should be ten times or more than the diameter of the pore applied. As for the height of the specimen, it should be equal or up to double the width [17].

According to the standard mentioned, the unit cell sample under investigation is of rectangular cross-section with 15 × 15 mm dimensions, and the height of the latticed part is 20 mm as shown in Fig. 45.1. The specimen was produced using the EOS M 290 3D printing machine at the University of Debrecen.

Out of the types of infill options available in SpaceClaim geometric modeling software, the 3D lattice infill option was chosen to be in our experiment. The sample was drawn using the CAD application within the software interface and the infill process was applied thereafter to get the latticed structure according to the predefined infill option mentioned. Table 45.2 lists the chosen parameters of the unit cells for three different sets of porosities.

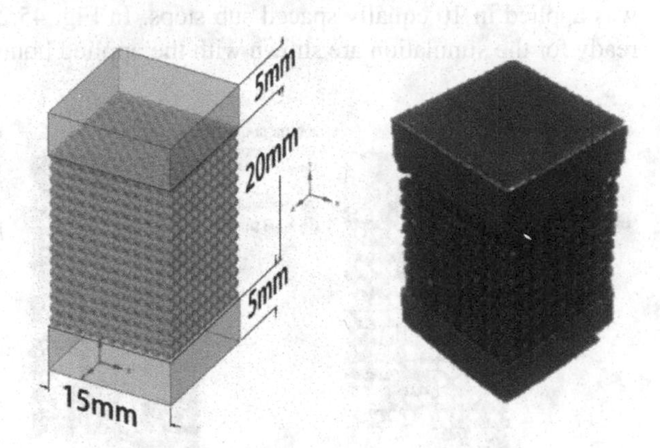

Fig. 45.1 The model of the specimen and the 3D printed specimen

Table 45.2 3D infill option parameters set

3D lattice infill pattern	Porosity (%)	Actual volume of latticed body (mm^3)	Beam (strut) thickness (mm)	Beam (strut) cross section
Case 1	56	2012	0.8	Oval
Case 2	44	2556	0.6	Oval
Case 3	28	3269	0.4	Oval

The porosity of the structure ϕ was also validated using Eq. 45.1.

$$\phi\ (\%) = \frac{V_{bulk} - V_{structure}}{V_{bulk}} \cdot 100 \tag{45.1}$$

where the $V_{structure}$ is the volume of the latticed structure, and V_{bulk} is the bulk volume of the specimen.

45.2.3 Preparation of the Finite Element Model

A three-dimensional finite element analysis has been carried out on the sample using the software. The latticed part was inserted into the mechanical application and tetrahedral elements were applied for the meshing process using patch independent meshing algorithm. Bonded connections were defined between the latticed part and the bulk parts.

In the Y-axis direction, a prescribed displacement of 0.5 mm was applied on the upper surface and fixed support on the lower surface, which puts the sample in the same position as in the real compression experiment. Because of the large displacement, it could happen that the geometry rigidity changes under the deformation. Therefore nonlinear solver was selected and the prescribed displacement load step was applied in 10 equally spaced sub steps. In Fig. 45.2 all three meshed samples ready for the simulation are shown with the applied boundary conditions.

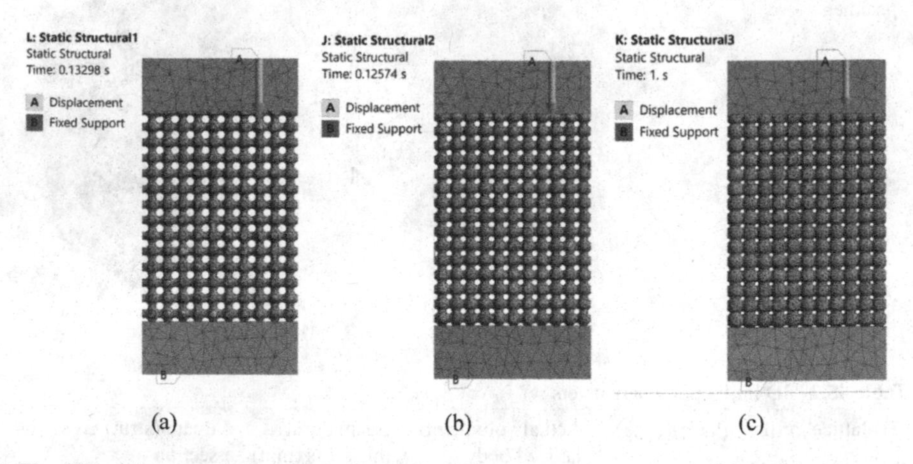

Fig. 45.2 Mesh sets for the samples: **a** Case 1 56%, **b** Case 2 44% and **c** Case 3 28%

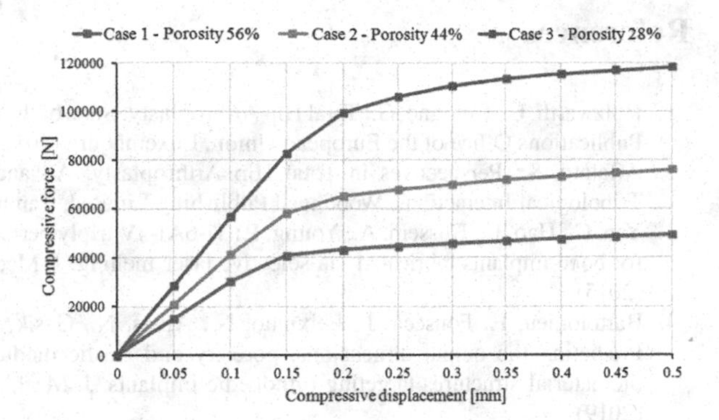

Fig. 45.3 Force–displacement curves under compression

45.3 Results and Discussions

In this paper, the properties of the three samples were numerically analyzed through the simulation method using the above-mentioned computer software. With the application of the 3D latticed infill pattern option on the sample, the calculation of the finite element analysis provides us with the parameters of the unit cell chosen listed in Table 45.2, and that is for three different sets of porosities. The force-displacements curves for each case can be seen in Fig. 45.2.

As seen from the results, the specimens exhibit with some differences among the sets under study as shown in Fig. 45.3. Judging by the slope of the curve for each set, the lower the porosity is, the stiffer the structure becomes.

In general, depending on the type and properties needed for the implants, designers can set the percentage of the porosity. It is worth mentioning that many other factors play an important role in choosing a suitable implant. However, the shape and geometry of the unit cell are of the major roles to be watched for.

45.4 Conclusion

A review on the behavior of the 3D lattice infill pattern was analyzed numerically through the finite element simulation study in the research to investigate the mechanical properties. The paper proposed some of the main numerical properties that might help designers with choosing this infill option when matching with certain bone types under treatment. The main results included the effect of the porosity of the unit cell on the mechanical properties of the structure of the medical implant. In the near future, it is planned to validate the finite element model results with laboratory measurements.

References

1. Holzwarth, U., Cotogno, G.: Total Hip Arthroplasty: State of the Art. Challenges and Prospects. Publications Office of the European Union, Luxembourg (2012)
2. Affatato, S.: Perspectives in Total Hip Arthroplasty: Advances in Biomaterials and their Tribological Interactions. Woodhead Publishing Limited, Cambridge (2014)
3. Yan, C., Hao, L., Hussein, A., Young, P.: Ti-6Al-4V triply periodic minimal surface structures for bone implants fabricated via selective laser melting. J. Mech. Behav. Biomed. **51**, 61–73 (2015)
4. Bartolomeu, F., Fonseca, J., Peixinho, N., Alves, N., Gasik, M., Silva, F.S., Miranda, G.: Predicting the output dimensions, porosity and elastic modulus of additive manufactured biomaterial structures targeting orthopedic implants. J. Mech. Behav. Biomed. **99**, 104–117 (2019)
5. Melo-Fonseca, F., Lima, R., Costa, M.M., Bartolomeu, F., Alves, N., Miranda, A., Gasik, M., Silva, F.S., Silva, N.A., Miranda, G.: Structures 45S5 BAG-Ti6Al4V. The influence of the design on some of the physical and chemical interactions that drive cellular response. Mater. Design **160**, 95–105 (2018)
6. Costa, M.M., Lima, R., Melo-Fonseca, F., Bartolomeu, F., Alves, N., Miranda, A., Gasik, M., Silva, F.S., Silva, N.A., Miranda, G.: Development of β-TCP-Ti6Al4V structures. Driving cellular response by modulating physical and chemical properties. Mater. Sci. Eng. C. **98**, 705–716 (2019)
7. Wang, X., Xu, S., Zhou, S., Xu, W., Leary, M., Choong, P., Qian, M., Brandt, M., Xie, Y.M.: Topological design and additive manufacturing of porous metals for bone scaffolds and orthopaedic implants: a review. Biomaterials **83**, 127–141 (2016)
8. Bartolomeu, F., Dourado, N., Pereira, F., Alves, N., Miranda, G., Silva, F.S.: Additive manufactured porous biomaterials targeting orthopedic implants. A suitable combination of mechanical, physical and topological properties. Mater. Sci. Eng. C. **107**, 110–342 (2020)
9. Gibson, L.J.: Biomechanics of cellular solids. J. Biomech. **38**, 377–399 (2005)
10. Hollister, S.: Porous scaffold design for tissue engineering. Nat. Mater. **4**, 518–524 (2005)
11. Gisario, A., Kazarian, M., Martina, F., Mehrpouya, M.: Metal additive manufacturing in the commercial aviation industry. J. Manuf. Syst. **53**, 124–149 (2019)
12. Yan, X., Li, Q., Yin, S., Chen, Z., Jenkins, R., Chen, C., Wang, J., Ma, W., Bolot, R., Lupoi, R., Ren, Z.: Mechanical and in vitro study of an isotropic Ti6Al4V lattice structure fabricated using selective laser melting. J. Alloy. Compd. **782**, 209–223 (2019)
13. Khan, S.N., Tomin, E., Lane, J.M.: Clinical applications of bone graft substitutes. Orthop. Clin. N. Am. **31**, 389–398 (2000)
14. Robertson, D.M., Pierre, L., Chahal, R.: Preliminary observations of bone ingrowth into porous materials. J. Biomed. Mater. Res. **10**, 335–344 (1976)
15. Ryan, G.E., Pandit, A.S., Apatsidis, D.P.: Porous titanium scaffolds fabricated using a rapid prototyping and powder metallurgy technique. Biomaterials **29**, 3625–3635 (2008)
16. Titanium Alloys—Ti6Al4V Grade 5. https://www.azom.com/article.aspx?ArticleID=1547. Last Accessed 19 Feb 2021
17. ISO 13314.: Mechanical Testing of Metals. Ductility Testing. Compression Test for Porous and Cellular Metals, Geneva (2011)

Chapter 46
Euler Buckling and Minimal Element Length Constraints in Sizing and Shape Optimization of Planar Trusses

Nenad Petrovic, Nenad Marjanovic, and Nenad Kostic

Abstract Structural optimization of trusses is a complex process and requires a realistic representation of the problem in order to achieve applicable results. In addition to using typical constraints such as minimal and maximal allowable stress and minimal displacement, in this paper dynamic constraints for Euler bucking have been used as well as minimal element length constraints for shape optimization. As shape variables can give short, impractical elements or even be in the same location, making the element length effectively 0, the minimal length constraint ensures that solutions can be produced. These constraints were used on a standard test examples with 10 and 47 bars using genetic algorithm optimization. Shape optimization results are compared to results of simultaneous sizing and shape optimization.

Keywords Structural optimization · Truss · Buckling · Dynamic constraints

46.1 Introduction

Researchers in the field of structural optimization have redirected their focus on the benefits of using different optimization methods to find optimal structures using existing constraint models. This approach has led to marginal improvements in the time needed to achieve results, but the problems being solved still do not represent realistic problem requirements. In order to have optimization results directly applicable in practice the mathematical model must represent all factors which are used in conventional design. Specifically in truss design there are numerous factors which still need to be addressed and implemented in the optimization process in order to make results practically applicable. In recent years authors of have made the transition from using the simpler and less accurate fixed buckling constraints to using Euler buckling dynamic constraints or other dynamic buckling constraints.

Authors in [1, 2] used different optimization methods to achieve minimal weight results using Euler buckling constraints with continuous cross-section variables. In

N. Petrovic (✉) · N. Marjanovic · N. Kostic
Faculty of Engineering, University of Kragujevac, 34000 Kragujevac, Serbia
e-mail: npetrovic@kg.ac.rs

M. Rackov et al. (eds.), *Machine and Industrial Design in Mechanical Engineering*, Mechanisms and Machine Science 109,
https://doi.org/10.1007/978-3-030-88465-9_46

[3, 4] researchers compared the effects on results between mathematical models which don't use buckling constraints and ones which use dynamic constraints for buckling and continuous cross-section variables. A key factor in defining sizing optimization problems is cross-section variable definition. In practice, continuous cross-sections are not practically applicable, which is why authors in [5, 6] presented a comparison of optimal results for models which use continuous versus discrete cross-section variables. Most recently in [7] a new approach to sizing optimization was presented which limits the number of different cross-sections which can be used in any given solution further increasing the complexity of the problem, but at the same time bringing results closer to practical use.

Shape constraints are generally given as constraints of node coordinates, as is done in papers [8, 9]. The problem with this approach is the possibility of creating overly short elements which are practically inapplicable. Researchers in [10–12] all achieved impractical shape optimization results with points converging to create elements shorter than 30 cm, even reaching element lengths shorter than 3 cm. These results point to a need to apply additional minimal element length constraints when using shape optimization. This research presents a use of minimal length constraints for shape optimization. The approach is used on shape and sizing shape truss optimization examples most commonly found in literature.

46.2 Minimal Weight Truss Optimization

Sizing optimization views cross section parameters as variables and requires an initial model of the truss. These parameters can be cross section geometry (cross-section type) and/or dimensions. Shape optimization considers node positions as variables. The x, y, and z (in space trusses) coordinates can take any value in a previously set range which is also determined by the precision (minimal used units) of the variables. Shape optimization is rarely used as the only optimization type. It is usually part of a sequential optimization process with sizing optimization, or as in this paper it can also be simultaneously used with sizing optimization. The typical weight minimization optimization problem is mathematically defined as follows:

$$\begin{cases} minW = \sum_{i=1}^{i=n} \rho_i A_i l_i \text{ where } A = \{A_1, \ldots, A_n\}, \\ \text{subjected to} \begin{cases} A_{min} \leq A_i \leq A_{max} & \text{for } i = 1, \ldots, n, \\ \sigma_{min} \leq \sigma_i \leq \sigma_{max} & \text{for } i = 1, \ldots, n, \\ u_{min} \leq u_j \leq u_{max} & \text{for } j = 1, \ldots, k. \end{cases} \end{cases} \tag{46.1}$$

This paper proposes the use of two more constraints in order to achieve optimal results which can be applicable in practice. These constraints are an Euler buckling dynamic constraint and a minimal bar length constraint for shape optimization problems.

46.2.1 Euler Buckling

Long thin elements subjected to compression are susceptible to buckling therefore they are checked for stresses which exceed critical buckling stress. The existence of even a single element exceeding critical values compromises the structures stability and such solutions are discarded. The use of Euler buckling constraints implies that only compressed bars are tested using the following expressions:

$$\sigma_{Ai}^{P} \le \sigma_{ki} \text{ where } \sigma_{Ai}^{P} = \frac{F_{Ai}^{P}}{A_i} \text{ and } \sigma_{ki} = \frac{F_{ki}}{A_i} \tag{46.2}$$

Stress in compressed elements is given as σ_{Ai}^{P}, the critical buckling stress is σ_{ki}, the cross-section area is A_i, calculated axial compression force is F_{Ai}^{P} and F_{ki} is Euler's critical buckling force of the ith element.

$$F_{ki} = \frac{\pi^2 \cdot E_i \cdot I_i}{l_{ki}^2}, \quad \left|F_{Ai}^{P}\right| \le F_{ki} \text{ for } i = 1, \ldots, n \tag{46.3}$$

Module of elasticity is E_i, I_i is the minimal axial moment of inertia for the cross-section of the ith element and l_{ki} is the buckling length of the ith element from the set of 1 to n. Depending on the software used for finite element analysis force or stress constraints can be used. Since sizing constraints vary cross-section areas, and therefore the moments of inertia in each iteration and shape optimization varies both the length (critical length) of bars and in some cases direction of the forces in the bars this constraint is considered as dynamic.

46.2.2 Minimal Element Length Constraint

The minimal element length constraint value for each example is determined by ex-perience or design guidelines given in literature or corresponding standards. The mathematical formulation of this constraint is given as:

$$\begin{gathered} l_i \ge l_{\min} \quad for\ i = 1, \ldots, n \\ l_i = \sqrt{\left(x_b^i - x_a^i\right)^2 + \left(y_b^i - y_a^i\right)^2} \end{gathered} \tag{46.4}$$

The element length l_i is from the 1 to n range which is between nodes a and b with coordinates (x_a^i, y_a^i) and (x_b^i,y_b^i) in that order. Existing node coordinate constraints implicitly define maximal element values therefore they are not necessary, however if there were a need for such a constraint the same method could be used to create it. This could be an interesting constraint for avoiding the use of extensions if bar lengths exceed stock lengths.

46.3 Optimization Method and Examples

The optimization used for this research is genetic algorithm (GA). The algorithm is comprised of three elementary operators: selection, crossover, and mutation. Selection is the process of conveying genetic information through generations. Crossover denotes the operations (process) between two parents, where an exchange of genetic information is conducted, and new generations are produced. The mutation operator creates a random change in the genetic structure of some of the individuals for overcoming early convergence [2]. Algorithm operation is based on survival of the fittest, through evolution which allows for the exchange genetic material. Selection is used as a process of ranking individuals in the population using values from the fitness function, which defines the quality of the individual.

An original software was created in Rhinoceros 7 using Grasshopper, Galapagos and Karamba plugins, which allow for the choice of optimization type and the choice of used constraints. The Galapagos optimization operator uses GA as its optimization method. Constraints are developed in the form of penalty functions in order to avoid unusable solutions.

46.3.1 10 Bar Truss

A common example found in literature for truss optimization is the 10 bar truss (Fig. 46.1). This truss has 10 independent sizing variables (full round cross-sections) and 4 shape variables (x and y coordinates for nodes (3) and (4)). Bar elements are made from Aluminum 6063-T5 whose characteristics are: Young modulus 68,947 MPa, and a density of 2.7 g/cm^3. The applied loads are $F = 444.82$ kN, in nodes 2 and 4, as shown in Fig. 46.1. The model is constrained with to a maximal displacement of ± 0.0508 m of all nodes in all directions, and axial stress of ± 72.3689 MPa for all bars. Discrete variables for cross-section diameters are taken from [5].

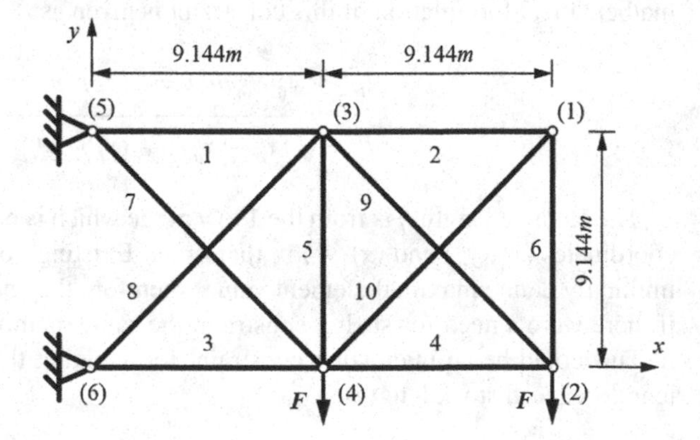

Fig. 46.1 10 bar truss problem

46.3.2 47 Bar Truss

The 47 bar truss problem has 22 nodes placed symmetrically around the y axis, as shown in Fig. 46.2. Cross-section elements are grouped in 27 groups according to the symmetry. This example also uses full round cross-sections, except the material used is construction steel. Material characteristics used are: Young modulus 206,842.719 MPa, and a density of 7.4 g/cm^3. The structure is subjected to three

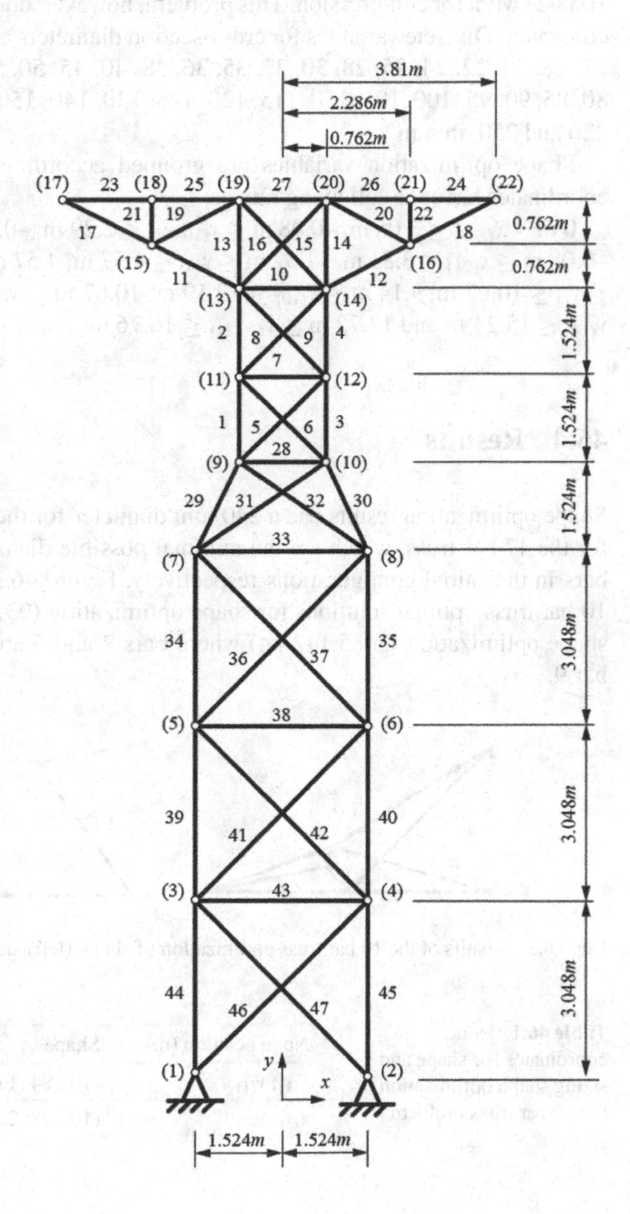

Fig. 46.2 47 bar truss problem

independent load cases (LC1, LC2 and LC3). The first load case (LC1) consists of forces $F = 26.689$ kN in the $+x$ direction and 66.275 kN in the $-y$ direction in nodes 17 and 22. The second load case (LC2) consists of forces F = 26.689 kN in the + x direction and 66.275 kN in the $-y$ direction in node 17. The third load case (LC3) consists of forces $F_1 = 26.689$ kN in the +x direction and force $F_2 = 66.275$ kN in the $-y$ direction of node 22.

The 47 bar truss problem has stress constraints of 137.895 MPa for tension and 103.421 MPa for compression. This problem, however, does not have a displacement constraint. Discrete variables for cross-section diameters are: 6, 8, 12, 12, 14, 15, 16, 17, 18, 20, 22, 24, 25, 28, 30, 32, 35, 36, 38, 40, 45, 50, 55, 56, 60, 63, 65, 70, 75, 80, 85, 90, 95, 100, 105, 110, 115, 120, 125, 130, 140, 150, 160, 170, 180, 190, 200, 220 and 250, in mm.

Shape optimization variables are grouped according to the symmetry. Node coordinates have the following ranges.

$0 \text{ m} \le x_{(2-8)} \le 3.05$ m, $-0.08 \text{ m} \le x_{(10-14)} \le 2.29$ m, $-0.08 \text{ m} \le x_{(18-20)} \le 2.29$ m, $-0.08 \text{ m} \le x_{(21)} \le 3.81$ m, $-1.52 \text{ m} \le y_{(4)} \le 4.57$ m, $4.57 \text{ m} \le y_{(6)} \le 7.62$ m, 7.62 m $\le y_{()} \le 10.67$ m, $9.14 \text{ m} \le y_{(10)} \le 12.19$ m, $10.67 \text{ m} \le y_{(12)} \le 13.72$ m, 12.19 m $\le$ $y_{(14)} \le 15.24$ m and $13.72 \text{ m} \le y_{(20,21)} \le 16.76$ m.

46.4 Results

Shape optimization results use a 240 mm diameter for the 10 bar truss, and 75 mm for the 47 bar truss, which are the minimal possible diameters of the most stressed bars in the initial configurations respectively. Figure 46.3 shows the layout of the 10 bar truss optimal solutions for shape optimization (9371.591 kg) and sizing and shape optimization (3685.142 kg) where bars 2 and 6 are practically parallel with bar 9.

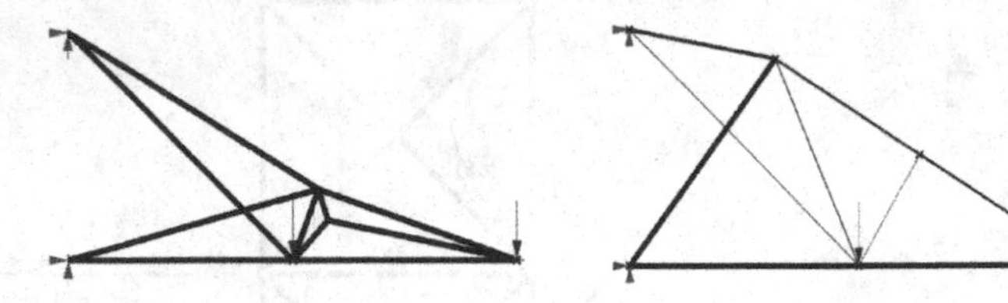

Fig. 46.3 Results of the 10 bar truss optimization of shape (left), and sizing and shape (right)

Table 46.1 Node coordinates for shape and sizing shape optimization of the 10 bar truss problem

Node position (m)	Shape	Sizing and shape
$x_{(1)}$; $y_{(1)}$	(10.584; 1.603)	(11.596; 4.284)
$x_{(3)}$; $y_{(3)}$	(10.178; 2.853)	(5.789; 8.003)

Table 46.2 Node coordinates for shape and sizing shape optimization of the 47 bar truss problem

Node position (m)	Shape	Sizing and shape
$x_{(2)}; y_{(2)}/-x_{(1)}, y_{(1)}$	(0.91; 0.00)	(1.88; 0.00)
$x_{(4)}; y_{(4)}/-x_{(3)}, y_{(3)}$	(1.03; 2.63)	(1.66; 3.11)
$x_{(6)}; y_{(6)}/-x_{(5)}, y_{(5)}$	(0.96; 5.54)	(1.65; 5.92)
$x_{(8)}; y_{(8)}/-x_{(7)}; y_{(7)}$	(0.68; 8.62)	(1.47; 8.55)
$x_{(10)}; y_{(10)}/-x_{(9)}; y_{(9)}$	(0.42; 10.67)	(0.57; 10.43)
$x_{(12)}; y_{(12)}/-x_{(11)}; y_{(11)}$	(0.37; 11.7)	(0.65; 11.91)
$x_{(14)}; y_{(14)}/-x_{(13)}; y_{(13)}$	(0.39; 12.9)	(0.54; 13.31)
$x_{(20)}; y_{(20)}/-x_{(19)}; y_{(19)}$	(0.39; 12.9)	(0.24; 14.62)
$x_{(21)}; y_{(21)}/-x_{(18)}; y_{(18)}$	(1.71; 15.11)	(1.95; 15.15)

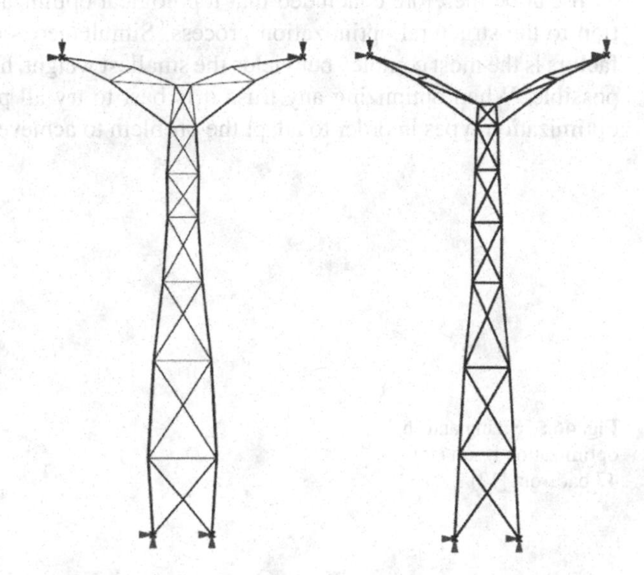

Fig. 46.4 Results of the 47 bar truss optimization of shape (left), and sizing and shape (right)

Tables 46.1 and 46.2 show the node coordinates for the 10 bar and 47 bar truss results for both optimization cases respectively.

Figure 46.4 shows the layout of the 47 bar truss optimal solutions for shape optimization (3204.269 kg) and sizing and shape optimization (1687.102 kg).

All solutions are the best results of at least 10 repeated optimizations of the same problem with the same initial values repeated for each optimization.

46.5 Conclusion

The results of examples presented in this paper show that the addition of a minimal length constraint solves the problem of impractically short elements in shape or sizing and shape optimization. The data however also shows the imperfections in the initial configurations. The 10 bar truss when optimized for sizing and shape gives a resulting structure where even though the nodes do not overlap, the bars are in the same plane. This leads to the conclusion that the 10 bar truss can be expected to not need bar (9) in the initial topology if shape is optimized.

Similarly, compared to results from literature of the 47 bar truss which use sequential optimization of sizing then shape, the solutions presented by this research do not have a convergence of nodes which lead to an impractical solution (Fig. 46.5).

These optimal results are, however, useful for presenting the need for elementary changes in the initial topology. The solution presented in [13], similarly to [10] shows that there is possibly no need for bars 15 and 16, and that nodes (19) and (20) should be joined in the initial model.

It can be therefore concluded that topological optimization is a necessary addition to the structural optimization process. Simultaneous optimization of all three factors is the most complex but yields the smallest weight, however this is not always possible. When optimizing any truss it is best to try all possible combinations of optimization types in order to adapt the problem to achieve the best results.

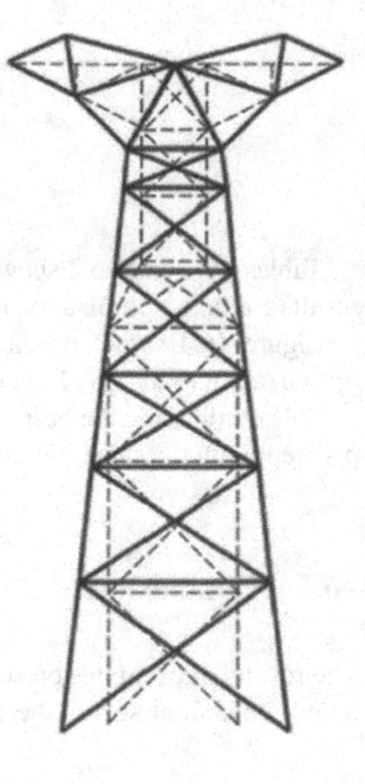

Fig. 46.5 Sizing and shape optimization results for 47 bar from [13]

References

1. Xu, T., Zuo, W., Xu, T., Song, G., Li, R.: An adaptive reanalysis method for genetic algorithm with application to fast truss optimization. Acta Mech. Sinica **26**(2), 225–234 (2009)
2. Petrovic, N., Kostic, N., Marjanovic, N.: Comparison of approaches to 10 bar truss structural optimization with included buckling constraints. Appl. Eng. Lett. **2**(3), 98–103 (2017)
3. Petrovic, N., Marjanovic, N., Kostic, N., Blagojevic, M., Matejic, M., Troha, S.: Effects of introducing dynamic constraints for buckling to truss sizing optimization problems. FME Trans. **46**(1), 117–123 (2018)
4. Petrović, N., Kostić, N., Marjanović, N., Blagojević, M., Matejić, M.: Influence of buckling constraints on truss structural optimization. In: Proceedings of the 14th International Conference on Accomplishments in Mechanical and Industrial Engineering, DEMI 2019, pp. 415–422. Banja Luka (2019)
5. Petrovic, N., Kostic, N., Marjanovic, N.: Discrete variable truss structural optimization using buckling dynamic constraints. Mach. Des. **10**(2), 51–56 (2018)
6. Petrović, N., Kostić, N., Marjanović, N., Marjanović, V.: Influence of using discrete cross-section variables for all types of truss structural optimization with dynamic constraints for buckling. Appl. Eng. Lett. J. Eng. Appl. Sci. **3**(2), 78–83 (2018)
7. Petrović, N., Marjanović, V., Kostić, N., Marjanović, N., Viorel Dragoi, M.: Means and effects of constraining the number of used cross-sections in truss sizing optimization. T. Famena **44**(3), 35–46 (2020)
8. Miguel, L.F.F., Lopez, R.H., Miguel, L.F.F.: Multimodal size, shape, and topology optimisation of truss structures using the firefly algorithm. Adv. Eng. Softw. **56**, 23–37 (2013)
9. Prayogo, D., Gaby, G., Wijaya, B.H., Wong, F.T.: Reliability-based design with size and shape optimization of truss structure using symbiotic organisms search. IOP Conf Ser Earth Environ Sci **506**, 0120471-1–0120471-8 (2020)
10. Gholizadeh, S.: Layout optimization of truss structures by hybridizing cellular automata and particle swarm optimization. Comput. Struct. **125**, 86–99 (2013)
11. Ahrari, A., Atai, A.A.: Fully stressed design evolution strategy for shape and size optimization of truss structures. Comput. Struct. **123**, 58–67 (2013)
12. Xiao, A., Wang, B., Sun, C., Zhang, S., Yang, Z.: Fitness estimation based particle swarm optimization algorithm for layout design of truss structures. Math. Probl. Eng. **2014**, 1–11 (2014)
13. Mortazavi, A., ToğAn, V., NuhoĞLu, A.: Weight minimization of truss structures with sizing and layout variables using integrated particle swarm optimizer. J. Civ. Eng. Manag. **23**(8), 985–1001 (2017)

Chapter 47
Modeling and Simulation of High-Voltage Transmission Line Insulators in a Virtual and Laboratory Environment

Bojan Mitev, Monika Fidanchevska, Marko Naseski, Kristina Miceva, and Atanas Kochov

Abstract In this paper, the physical and mechanical parameters of high-voltage polymer insulators are tested. Already defined forms made according to standard specifications of insulators with a capacity of 12 and 24 kV were analysed. Based on these shapes, virtual 3D models have been made that correspond to the physical ones and optimisation has been performed in order to increase the mechanical characteristics and reduce the volume. The optimized models are then subjected to virtual simulations which resulted in satisfactory and significant performance which was later verified by testing in laboratory conditions with 1:1 physical model made with "rapid manufacturing" processes. These models are made of a composite polymer that consists of epoxy resin and silica and will have a real application in the ever-growing power transmission network.

Keywords High-voltage insulators · Rapid manufacturing · Virtual simulations · Verification of 3D models · Polymers

47.1 Introduction

When it comes to the operating efficiency and operational safety of transmission systems of electrical power, high-voltage insulators are of great importance. Thus, there is no surprise that these components must meet high demands in terms of reliability. In addition to both long rod insulators of conventional design made of porcelain and cap and pin insulators made of glass or porcelain, as were previously used as standard, in the field of insulation technology for high-voltage overhead transmission lines and substations, great importance has been gained by composite insulators [1]. Composite insulators have convincing qualities when designed appropriately, both in terms of construction and in terms of material selection. Non-ceramic insulators offer many benefits as improved damage tolerance, good impact resistance, high mechanical strength-to-weight ratio, flexibility and ease of installation, compared

B. Mitev (✉) · M. Fidanchevska · M. Naseski · K. Miceva · A. Kochov
Faculty of Mechanical Engineering, Ss. Cyril and Methodius University, Skopje, North Macedonia

M. Rackov et al. (eds.), *Machine and Industrial Design in Mechanical Engineering*, Mechanisms and Machine Science 109,
https://doi.org/10.1007/978-3-030-88465-9_47

with their porcelain counterparts. Despite the many benefits, they can fail mechanically in service by rod fracture. Brittle fracture is a failure process of the insulator's mechanical failure modes, and is caused by the stress corrosion cracking of the GRP (glass reinforced polymer) rods [2]. In their article, Kumosa et al. state that a brittle fracture can occur usually either inside the fitting or above the hardware.

Maintaining system integrity requires that the damage limit for any line component be chosen so that irreversible damage to the component does not occur at the maximum design load [3]. The damage limit for composite insulators have been extensively studied because early in their development, it was assumed that mechanical creep was the controlling phenomenon for the lifetime of the insulator.

Static time-load tests on composite insulator dielectric materials suggest that irreversible damage occurs at about 65% of average ultimate strength. Under some circumstances, dynamic loads can result in insulator failure at lower loads, which was not considered by Baker et al. in their study. As they mention, past experience can provide guidance for insulator application in areas where such condition may occur. Damage limits for application with static line load and strength require a consideration of the minimum strength that individual insulators in a lot may have [3].

This paper looks at the mechanical properties of 12 and 24 kV capacity insulators in both a virtual and a laboratory setting. The technical specifications for the insulators were provided by the manufacturer to which several dimension changes were made in order to minimize material usage while maintaining similar properties, Fig. 47.1. The material used in the experiment is a composite from epoxy resin and 30 vol% SiO_2 (71 μm granulation). The elastic modulus is 27,240 MPa and a tensile strength of 73 MPa.

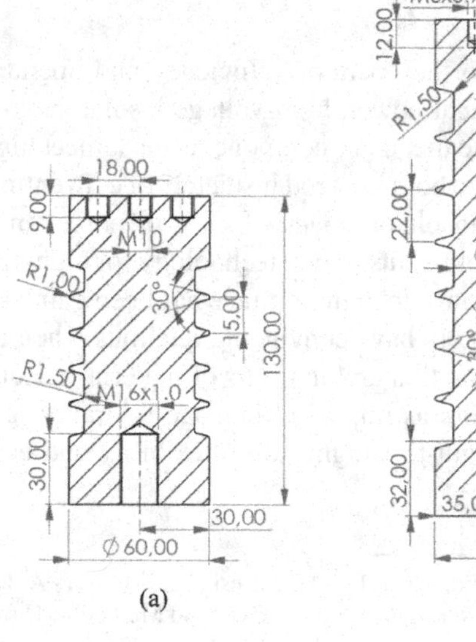

Fig. 47.1 Technical specifications for **a** 12 kV and **b** 24 kV insulator

There is a wide selection of materials that can be used as insulating materials, many of which have great properties and are inexpensive [4]. Silica filled epoxy-based materials are chosen because of their advantages such as low cost, good adhesion to most of the substrate, adjustable curing temperature etc. However, it is reported that the epoxy resin without silica filler cannot meet the requirement for its thermo-mechanical properties [5]. In a study by Wang et al. titled Combined effects of silica filler and its interface in epoxy resin, it is mentioned that the addition of silica filler in the epoxy resin increases Young's modulus of the materials. However, it was noticeable that at 115 °C the Young's modulus of the sample with 28 vol% silica is a bit lower than that of the sample with 14%. According to their research, the material with 14 or 21 vol% filler content has the highest inelastic strain, while the material with 28 vol% silica has the lowest inelastic strain.

47.2 Experimental (Methods)

The mechanical properties testing of the insulators consists of a compression test, tensile test and a three-point bending test, for the virtual static simulations and the laboratory tests.

47.2.1 Virtual Environment

The developed 3D models are analysed with the help of SolidWorks Simulation, a static analysis. The external forces and fixtures of the models in the simulations were made to match those of the actual tests in order to obtain more reliable results. For the static analysis, the simulation procedure consists of 3 steps, namely: applying external forces, determining the restriction of movement of the body and meshing the body. Both 12 and 24 kV have the same simulation configuration, unless specified otherwise.

Compression test simulation

The external force used during the compression simulation is placed on the top face, Fig. 47.2, matching the real-world tests. The bottom face is geometrically fixed.

Tensile test simulation

The virtual tensile test, Fig. 47.3, is performed by adding fixtures in the threaded holes where the clamps from the tensile testing machine clamps on.

Three-point bending test simulation

The last type of stress the model encounters is the stress because of bending, simulated in a virtual environment as a three-point bending test. Figure 47.4 shows the external forces and fixtures for the three-point bending test simulation, with screws.

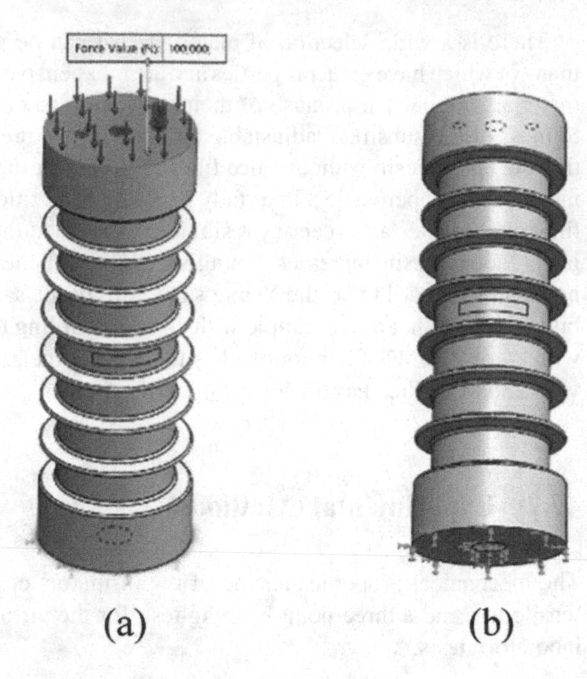

Fig. 47.2 **a** External forces and **b** fixture

47.2.2 Laboratory Environment

The laboratory tests were done in the accredited laboratory for testing of mechanical properties LT-04 at the Faculty of Mechanical Engineering in Skopje on a Shimadzu AGX-V series material testing machine. The testing was comprised of a tensile test, compression test and a three-point bending test. To avoid bending or buckling effects, appropriate techniques were applied to ensure the specimen was clamped and aligned well [5].

Figure 47.5a shows the final prototype insulators. The taller insulators are 24 kV and the shorter are 12 kV. The Shimadzu testing machine used in the laboratory tests is shown in Fig. 47.5b.

47.3 Results

The results from the virtual simulations are shown in Table 47.1. Two different forces were used for each of the simulations, 100 kN, for a comparison between the insulators, and the maximum force from the laboratory tests so a comparison can be made between the laboratory and simulation tests. The results from the laboratory tests are shown in Table 47.2.

The compression test was stopped at 226.151 kN for the 12 kV insulator and 250.230 kN for the 24 kN insulator for safety reasons, as the machine was close to

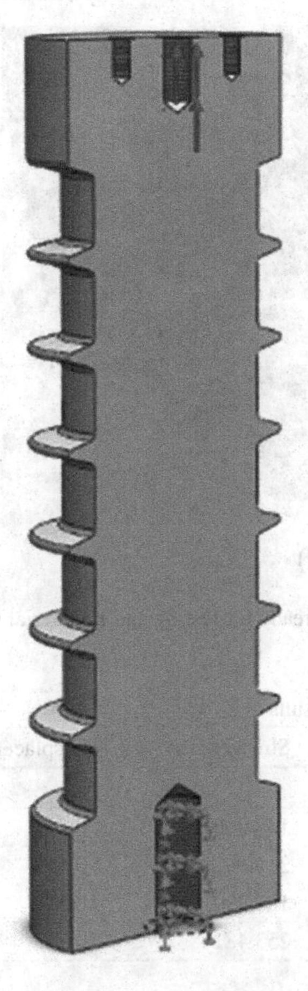

Fig. 47.3 External forces and fixtures for the tensile simulation

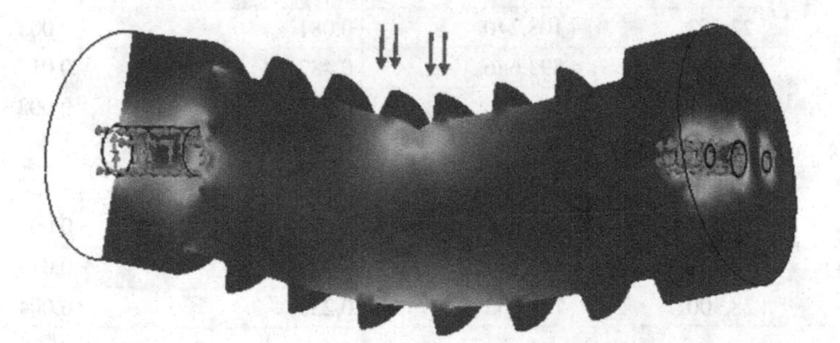

Fig. 47.4 External forces and fixtures for the three-point bending test simulation

(a) (b)

Fig. 47.5 **a** Prototype insulators ready for testing and **b** material testing machine

Table 47.1 Virtual simulation results

	Force (kN)	Stress (MPa)	Displacement (mm)	Strain
Compression				
SMI 12	100	138.006	0.241	0.003
	226.151	312.102	0.545	0.008
SMI 24	100	101.299	0.307	0.003
	250.230	253.481	0.769	0.006
Tensile				
SMI 12	100	479.935	0.361	0.013
	22.553	108.240	0.081	0.003
SMI 24	100	593.646	0.487	0.014
	14.964	88.833	0.073	0.002
Bending				
SMI 12	100	782.031	0.480	0.020
	16.038	125.442	0.077	0.003
SMI 24	100	544.287	0.829	0.015
	28.300	154.033	0.235	0.004

its maximum force. There were no visible cracks. In the tensile test, both insulators broke at the end of the threaded holes, Fig. 47.6a. The three-point bending test finished with the insulators breaking with a burst caused around the end of the threaded hole, Fig. 47.6b.

Table 47.2 Laboratory test results

	Force (kN)	Stress (MPa)	Displacement (mm)
Compression			
SMI 12	226.151	130.351	7.461989
SMI 24	250.230	109.260	5.244969
Tensile			
SMI 12	22.553	13	6.63326
SMI 24	14.964	6.533	4.403313
Bending			
SMI 12	16.038	9.244	5.800042
SMI 24	28.300	12.357	3.38999

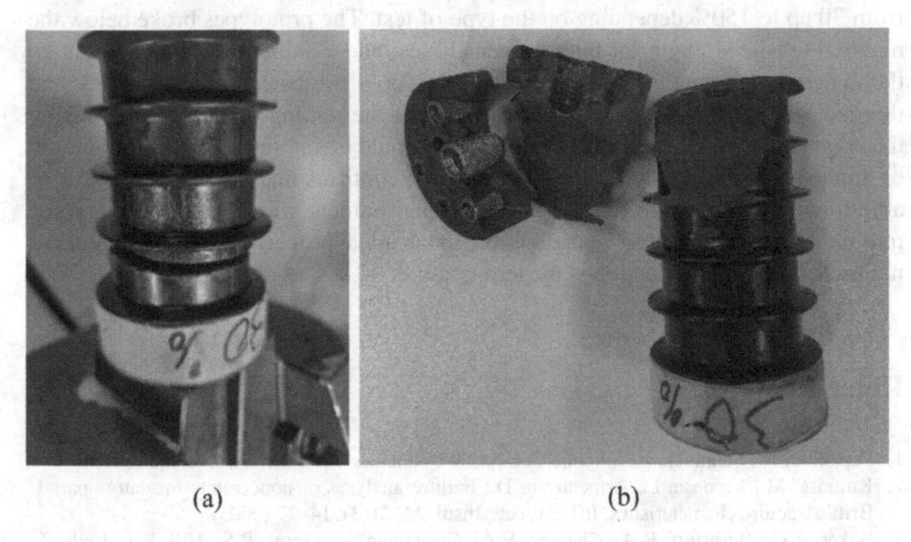

Fig. 47.6 **a** Insulator after the tensile test and **b** Insulator after the three-point bending test

The calculated stress in Table 47.3 is the stress at the point and area of breaking, which will be compared with the corresponding simulation results.

47.4 Conclusion

Finding the perfect material for making affordable high-voltage insulators with good properties is a challenge. The point of the research was to see whether the insulators made from an epoxy and silica (30 vol%) composite have decent mechanical properties for exploitation. The comparison in the results shows that the laboratory test values differ from the simulations. When it comes to the stress difference, it differs

Table 47.3 Stress comparison between laboratory and virtual simulation tests

	Laboratory	Simulation	% difference
Compression	Stress (MPa)	Stress (MPa)	–
SMI 12	130.351	178.041	30.93
SMI 24	109.260	161.086	38.34
Tensile	Stress (MPa)	Stress (MPa)	–
SMI 12	13	53.897	122.269
SMI 24	6.533	36.701	139.557
Bending	Stress (MPa)	Stress (MPa)	–
SMI 12	9.244	46.016	133.087
SMI 24	12.357	93.820	153.448

from 30 up to 150% depending on the type of test. The prototypes broke below the material tensile strength for both the tensile test and bending test, which is due to the material not being injected properly in the mold, which caused air bubbles, and the geometry of the insulators. That means that the mixing process while making the composite and the injection into the mold should be improved. The huge amount of applied force in the laboratory experiments, confirms that the insulators will be able to withhold the forces in a real-world environment where more factors come into play. One of the factors that affect the mechanical properties of the insulators as mentioned in the introduction is the temperature.

References

1. Papailiou, K., Schmuck, F.: Silicone Composite Insulators. Springer, Switzerland (2013)
2. Kumosa, M., Kumosa, L., Armentrout, D.: Failure analyses of nonceramic insulators part 1: Brittle fracture characteristics. IEEE Electr. Insul. M. **21**(3), 14–27 (2005)
3. Baker, A.C., Bernstorf, R.A., Cherney, E.A., Christman, R., Gorur, R.S., Hill, R.J., Lodi, Z., Marra, S., Powell, D.G., Schwalm, A.E., Shaffner, D.H., Stewart, G.A., Varner, J.: High voltage insulators mechanical load limits—part I: overhead line load and strength requirements. T. Power Deliver. **27**(3), 1106–1115 (2012)
4. Naeem, U.J., Hassan, A.D., Kadhim, E.H.: Manufacturing of electrical insulator from composite material. In: Proceedings of the 1st International Conference on Mechanics of Advanced Materials and Equipment, CMME 2018, pp. 1–9. Ahvaz (2018)
5. Wang, H., Bai, Y., Liu, S., Wu, J., Wong, C.P.: Combined effects of silica filler and its interface in epoxy resin. Acta Mater. **50**(17), 4369–4377 (2002)

Chapter 48
Application of the Finite Element Method for Analysis of Piston Characteristics

Dalibor Feher, Jovan Dorić, and Nebojša Nikolić

Abstract Within this paper an application of modern software for prediction of the piston's mechanical characteristics in reference to its material is presented. Those predictions were used to optimize the piston design in order to achieve the lowest mass within an acceptable range of the piston load. Using the software for 3D modeling, the base model of the piston was initially created on the basis of data previously defined by thermal calculation of an engine. On the model made in this way, a structural analysis was performed using the finite element method with the aim of obtaining a picture of its stress state. After that, the piston design was optimized in accordance with the stress state, with the aim of creating a piston that is minimally loaded with low material consumption. A stress analysis was performed for three different materials to compare their properties.

Keywords Internal combustion engine · Piston · FEA · Optimization

48.1 Introduction

Mechanization in agriculture is predominantly driven by internal combustion engines due to their robustness and autonomy. However, due to legislations, which increasingly require a reduction in emissions, manufacturers face the problem that development of the internal combustion engines in some cases can't achieve this reduction quickly enough to be in line with changes in regulations. In order to accelerate the development of the engines, modern computer methods are widely used. The great advantage of those methods, in addition to saving time, is also the saving of financial resources. By applying the software, results that are close to realistic can be obtained, which skips a few steps in the production chain and development itself [1, 2].

The parameter that indirectly affects the reduction of emissions of harmful gases and the fuel efficiency, is the mass of the engine components. Reducing the mass of the engine components also significantly affects the increase in the specific power of the

D. Feher (✉) · J. Dorić · N. Nikolić
Faculty of Technical Sciences, University of Novi Sad, Novi Sad, Serbia
e-mail: daliborfeher@uns.ac.rs

M. Rackov et al. (eds.), *Machine and Industrial Design in Mechanical Engineering*, Mechanisms and Machine Science 109,
https://doi.org/10.1007/978-3-030-88465-9_48

engine. The reduction of the mass of the engine components can be achieved in two way, the first of which is use of lighter materials and the second one is optimization of the structure with the aim of consuming a smaller amount of material [3].

The optimization method with the aim of reducing the mass was also applied by Aisha Muhammad, where he showed that by applying the finite element method, a reduction in the mass of connecting rod up to 60% can be achieved. He performed the analysis by loading a small end of the connecting rod in only one direction, varying the forces from 100 to 500 N. Accordingly to that, he carried out the optimization of the connecting rod with aim of reducing the mass [4].

Rayapati Subbarao used the method of structural analysis of the piston, where he performed an analysis on three different materials Alcoa Deltalloy 4032-T651 Aluminium, ATI Allegheny Ludlum Stainless Steel Type 201L Annealed (UNS S20103) and NIMONIC Alloy 81. The two boundary conditions set by R. Subbarao are such that the piston crown is loaded with a pressure of 5 MPa, and frictionless support is placed on the pin holes of piston. In the analysis he conducted, it was concluded that Alcoa Deltalloy 4032-T651 Aluminium proved to be the most deformable compared to the other two materials [5].

It can be noticed that there is a room for improvement of the previously mentioned analyzes. The authors performed simulations where they loaded the elements of piston mechanism separately, in only one direction, and did not take into account the influence of the piston assembly on the connecting rod and vice versa. Engine parts are loaded with much more complexity during their exploitation. In that case, it would be expedient to initially perform the calculation of the pressure change above the piston crown and according to that pressure to define the forces acting on all the components of the piston mechanism. By applying such determined forces during simulation of load, results that are significantly closer to the real loads, can be achieved.

48.2 Methodology

48.2.1 Piston Modeling

The development of the basic three-dimensional model of the piston (Fig. 48.1) according to the adopted dimensions is done with the use of the Autodesk Inventor software. The dimensions (Fig. 48.1) of the piston were adopted according to recommendations obtained on the basis of experimental testing of the existing engine designs which operates in similar conditions [6, 7].

In order to make the results of the analysis as relevant as possible, beside piston model, models of the other components of the piston assembly are also made. In this way, their effect on the piston load is also taken into account. Appropriate materials are assigned to the created model. Following the example of Subbaro and Gupta [5], the materials used for the simulation were Alcoa Deltalloy 4032-T651 Aluminium,

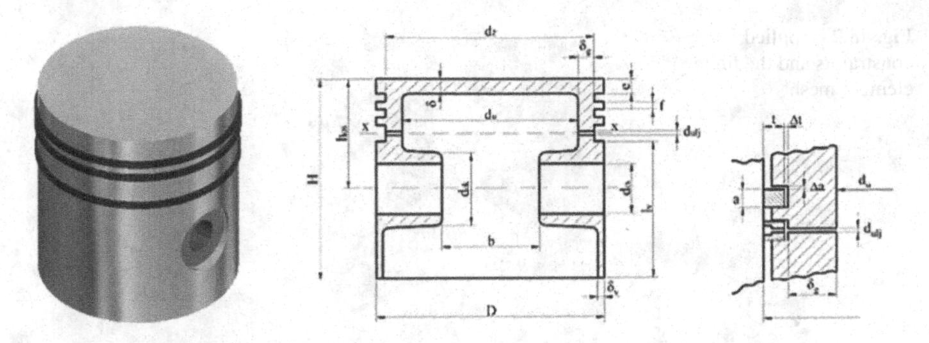

Fig. 48.1 Initial model of the piston assembly and dimensions of the piston [7]

ATI Allegheny Ludlum Stainless Steel Type 201L Annealed (UNS S20103) and NIMONIC Alloy 81.

48.2.2 Finite Element Analysis

The load simulation of the piston model is performed by analyzing the effect of the gas force, i.e. the effect of other members of the piston assembly on the piston. The load modeling is performed for the position of the crankshaft at which the maximum gas pressure is achieved. The stress picture of the piston is analyzed and then the optimization and analysis of the optimized piston model are performed.

Before conducting the simulation and analysis, the first step is to create a finite element mesh of appropriate characteristics. The characteristics of the mesh are chosen so that the simulation achieves a sufficiently high accuracy without creating the simulation too demanding and time-consuming. For that purpose, first-order tetrahedral elements were chosen and the average size of element is 5% of the largest dimension of the model. It is known that first-order tetrahedral elements tend to overestimate the stress level, but for the purpose of comparing different cases in order to see each other characteristics, quick and useful results can be obtained.

The simulation is performed for the whole piston assembly in order to notice the effect of other components on the piston. It is necessary to model a connecting rod whose role is to simulate a reaction in the direction of its axis. As the connecting rod has a role of creating the reaction, it is not necessary to perform its calculation. It is modeled as a rigid body and its axis in relation to the cylinder axis create the angle of 1.49°. The angle between axes is necessary because the maximum pressure above the piston head is achieved at the position of the crankshaft of 365°. The applied constraints are "Frictionless Constraint" and "Pin Constraint" (Fig. 48.2). In this way, the reaction of the cylinder is simulated, also, the reactions of the wrist pin and the connecting rod are taken into account.

The piston load is caused by the pressure acting on the piston crown and the inertial force (Fig. 48.2). The maximum pressure above the piston is 12.569 MPa

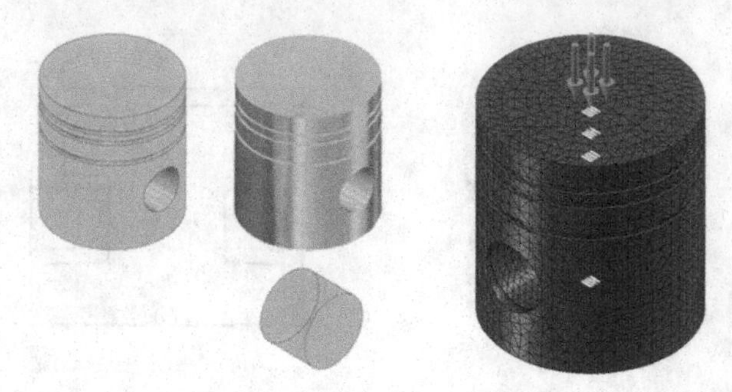

Fig. 48.2 Applied constraints and the finite element mesh

and the inertial force is defined using the "Gravity" function. The acceleration value is defined as the piston acceleration at the moment of maximum gas pressure and its calculated value is 9872 m/s^2. The acceleration of the piston is opposite in regards to the action of the gas force.

Validation of the model and boundary conditions is performed by comparing the normal force occurring at the cylindrical constraint "Frictionless Constraint" and the analytically obtained normal force within the dynamic calculation of the piston mechanism. The deviation of the forces being compared is 7.5%. Based on this, it can be established that the boundary conditions are set precisely enough and the simulation results can be considered valid.

48.2.3 Optimization

Optimization process is used with the aim of achieving the most favorable stress state with possible reduction of the piston mass in reference to the initial model of the piston. Due to the complexity of the model itself and the differences in the parameters that vary for the examined three different cases, it was difficult to define specific parameters that vary. Therefore, optimization process was performed using manual trial-and-error method in the same CAD software used to design the initial piston model. Some of the parameters that varied in all cases are wall thickness, radius between wall and piston crown as well as piston crown thickness. It is also important to mention that main dimensions of piston, like the diameter of the piston crown or piston pin hole, were not varied. Material was removed from the unloaded areas of the piston, while certain changes were introduced at the areas of high stress, in the form of adding material and local geometric adjustments (Fig. 48.3). The target of optimization was to achieve minimal mass of the pistons, but also to maintain the sufficient safety factor for all of the pistons. This was performed for the pistons made of different materials, and then such an optimized piston models (Fig. 48.4) were used for comparison of their mechanical characteristics.

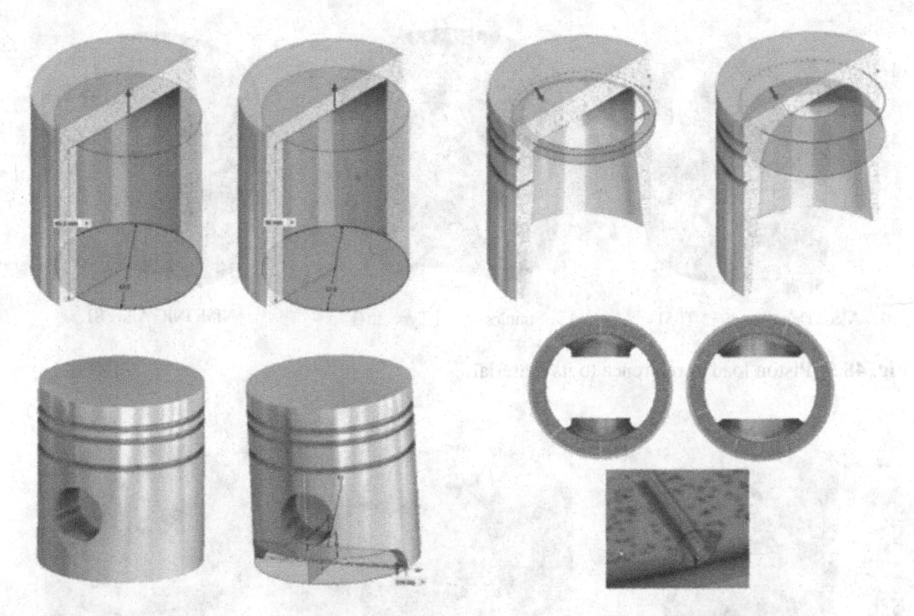

Fig. 48.3 Some steps of the optimization of the aluminium piston

Fig. 48.4 Optimized models of the piston according to their materials

48.3 Results and Discussion

Optimization successfully reduced the piston mass of the all three initial models. After the simulation over the optimized designs and the analysis of the stress states presented in Fig. 48.5. It was noticed that the zones of increased stress are located on the piston crown, inner face of the piston crown, at the cross section of the oil piston ring groove and the piston pin hole. From the aspect of Von Mises stress, the piston made of Alcoa Deltalloy 4032-T651 Aluminium behaves slightly better than other two. Also, the piston made of this material has the lower mass compared to the other two, but also the largest deformation are observed. Based on the deformations (Fig. 48.6), it can be said that the most critically loaded piston in this case is the one made of aluminium, but also the aluminium piston is the one that has the most

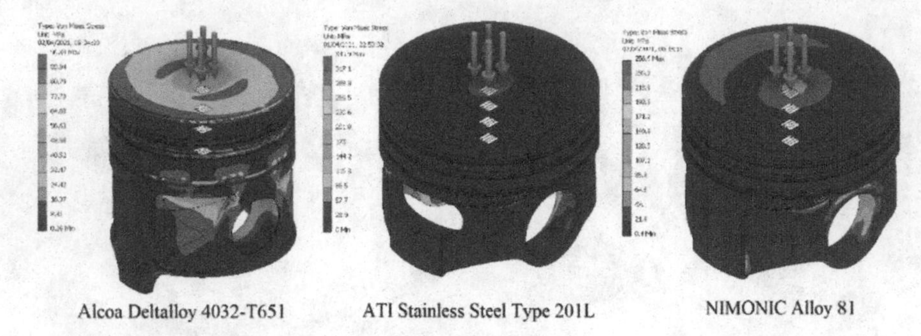

Fig. 48.5 Piston load in reference to its material

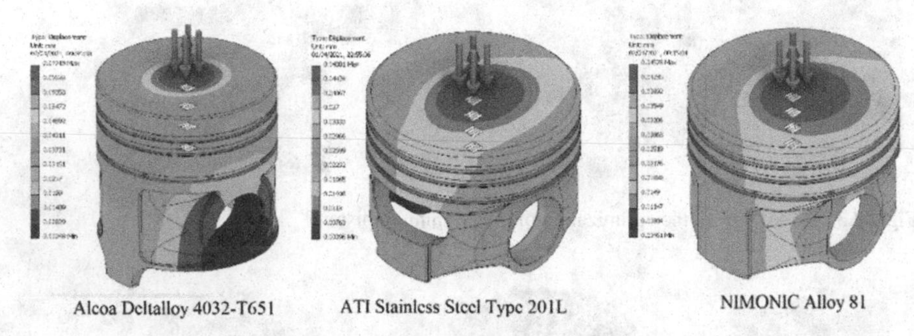

Fig. 48.6 Piston deformation in reference to its material

symmetrical deformation due to the application of a larger amount of material. Steel and nickel alloy pistons are suitable for more robust constructions due to their superior strength compared to aluminium. The biggest disadvantage of those two material is significantly larger mass. Specific comparative values of characteristic parameters are presented in the Table 48.1.

Table 48.1 Maximum values of the characteristic parameters

Material	Von Mises stress (MPa)	Deformation (mm)	Mass (kg)
Alcoa Deltalloy 4032-T651 Aluminium	96.8	0.0721	0.703
ATI Stainless Steel Type 201L	345.9	0.048	1.254
NIMONIC Alloy 81	256.6	0.0458	1.413

48.4 Conclusion

It can be concluded that this approach to piston load modeling provides a clearer picture of stresses that can be expected during exploitation because several factors are taken into account—gas pressure force, influence of the piston assembly components on the piston and normal force caused by the connecting rod in certain position. Also, the optimization procedure itself in this case provided a reduction in the mass. Such a model can be applied to the analysis of piston properties in correlation with piston material. Alcoa Deltalloy 4032-T651 Aluminium proved to be a suitable material if the goal of piston design is lower mass, which is indirectly related to the reduction of harmful exhaust emissions and increased specific power. On the other hand, ATI Stainless Steel Type 201L and NIMONIC Alloy 81 are a better choice if there is a need for a more robust construction which is expected to have high reliability and long service life as would be the case with agricultural machinery.

Acknowledgements This research has been supported by the Ministry of Education, Science and Technological Development through project no. 451-03-9/2021-14/200156: "Innovative scientific and artistic research from the FTS activity domain".

References

1. Basshuysen, R., Schäfer, F.: Internal Combustion Engine Handbook: Basics, Components, Systems, and Perspectives. Vieweg Verlag, Wiesbaden, Germany (2002)
2. Raut, R.D., Mishra, S.: Stress optimization of S.I. engine piston. Int. J. Sci. Res. **4**(9), 1159–1165 (2015)
3. Dorić, J.: Theory of Internal Combustion Engines. Faculty of technical sciences, Novi Sad (2015) (in Serbian)
4. Muhammad, A., Ali, M.A.H., Shanono, I.H.: Design optimization of a diesel connecting rod. Mater. Today **22**(4), 1600–1609 (2020)
5. Subbarao, R., Gupta, S.V.: Thermal and structural analyses of an internal combustion engine piston with suitable different super alloys. Mater. Today **22**(4), 2950–2956 (2020)
6. Heywood, J.B.: Internal Combustion Engine Fundamentals. McGraw-Hill Education, New York (2018)
7. Filerl, E.K.: Internal Combustion Engines: Motor Mechanics, Calculation and Design of the Reciprocating Engines. Friedr Vieweg & Sohn Verlag, Wiesbaden (2006)

Chapter 49
Numerical and Analytical Analysis Methods for Radial Response of Flexible Ring Dampers

Mykola Tkachuk, Andriy Grabovskiy, and Anton Tkachuk

Abstract Radial stiffness of flexible ring dampers used in rotor supports is analyzed in this paper. Two major approaches are proposed for this purpose. The first is the conventional finite element modeling of this contact mechanics problem. The second is an analytical method that can be used as alternative to the costly numerical computations. This method is based on the Kalker's principle of minimum complementary energy. A special variational formulation is developed in the closed form using Euler–Bernoulli beam approximation for the elastic ring and a simplified model of normal contact at the ring flanges. It has been shown that the surface tolerances of the parts have a substantial effect on the radial response of the flexible ring that may become nonlinear. The tight fit of the ring on both sides makes it much stiffer, while the loose fit results in free motion of the rotor and much weaker damping of its motion. Both methods produced results that are in excellent agreement for the considered cases.

Keywords Flexible ring damper · Rotor vibrations · Complementary energy

49.1 Introduction

Rotary machines are prone to vibrations due to various factors of their design and operation [1]. The arrangement and characteristics of rotor supports play the key role on its dynamic stability. In particular this applies to high-velocity impellers in turbocharged two-stroke internal combustion engines [2]. Due to the compactness and high performance of this component it is a challenging task to detune this rotor with a heavy disk from critical rotational velocities. Passive and active measures can be taken to achieve this goal [3, 4]. The main task is to control the stiffness of the bearings and other rotor supports. This characteristics often turns out to be nonlinear.

M. Tkachuk (✉) · A. Grabovskiy
National Technical University "Kharkiv Polytechnic Institute", vul. Kyrpychova 2, Kharkiv 61002, Ukraine
e-mail: m.tkachuk@tmm-sapr.org

A. Tkachuk
Fakulteten för hälsa natur-och teknikvetenskap, Karlstads Universitet, Universitetsgatan 2, 65188 Karlstad, Sweden

M. Rackov et al. (eds.), *Machine and Industrial Design in Mechanical Engineering*, Mechanisms and Machine Science 109,
https://doi.org/10.1007/978-3-030-88465-9_49

Besides the methods of nonlinear rotor dynamics [5] that describe the vibrations in these systems special methods are required to model the radial response of the supports themselves as a part of the greater analysis.

Finite element analysis is a utilitarian tool to perform such analysis for arbitrary designs of the rotor supports [6]. Nonetheless it might not be readily accessible in everyday engineering practice where simple and quick solutions are highly appreciated. That is why an alternative analysis method is proposed in this paper. It is based on a well-established variational principle of contact mechanics proposed by J.J. Kalker [7]. This formulation has previously proved to be an efficient mathematical basis for numerical methods of contact analysis [8]. Furthermore, there is a very efficient application of this approach to a special design of flexible ring dampers used as intermediate elastic supports of rotors. Its radial response is found from a solution of a quadratic constrained minimization problem each part of which is computed analytically.

49.2 Flexible Ring Damper for Moderation of Rotor Vibrations

Flexible ring dampers find place in many rotary machines as cheap and efficient solution for moderation of unwanted vibrations. They provide the required compliance in the bearing supports which alter eigen frequencies of rotor. It is often possible to detune several lowest modes from the operational domain of rotational velocities. These elastic rings are easy to manufacture, they are extremely compact and thus can be fitted into an existing structure once any problems with vibrations are encountered even at the latest stages of the design process, since they do not require excessive changes to the rotor, the chosen bearings and the housing.

The ring is a simple thin cylindrical shell with intermeshed cut-outs on the inner and the outer surface. The remaining narrow flanges rest on the outer ring of the bearing from inner side and the stator housing on the outer side. A stopping ring may be added in order to prevent the rotational sliding of the elastic damper. The portions of the ring between flanges are thinner than the nominal thickness of the ring due to the gap introduced by the cut-outs. Thus they are not supported from any side and can deform freely as a curvilinear beam sections. This is the source of flexibility of this structural element.

Note that the contact configuration plays an essential part in the overall behavior of the ring. If contact status remains unchanged throughout the loading then the reaction force will be linear function of the displacement. Thus the effective response of the elastic ring can be determined by a single scalar parameter, the radial stiffness. Nevertheless, the assumption of constant contact configuration in reality is often violated. Thus the elastic response of the flexible ring can be expected to be in general nonlinear. The corresponding design cases and their practical implications are examined in the subsequent section of the paper.

In order to determine the elastic properties of the ring the appropriate analysis tools are required first. The considered contact problem is solved by two numerical methods. The first approach is to perform finite element analysis. However, a completely new semi-analytical solution method is developed as an alternative to this costly numerical procedure.

49.3 Finite Element Model of the Elastic Ring Damper

The elastic ring damper has relatively simple geometry that can be fully parameterized in any modern CAD software. The built-in geometrical pre-processor of FEA package ANSYS© Workbench™ was chosen to perform this task instead. The model of the elastic ring is built for any arbitrary number of flanges, cut-out arc angles and depth automatically.

It can be further assumed that the flexible ring undergoes plain stress deformations. Apparently, there are no forces or constraints acting in the transverse direction along the axis of the rotor. In particular, friction can be neglected since this part of rotary machines is usually well lubricated. This consideration reduces the dimensionality of the contact problem and the computational cost of a single analysis. The outer ring of the bearing and the rotor support are treated as rigid solids since they are much bulkier compared to the slender ring damper. The radial displacement of the rotor with the bearing is applied as a boundary condition along one of the symmetry planes. This choice of loading direction makes it possible to consider only a half of the model. Furthermore, it eliminates the issue of ring rotation since the symmetry boundary conditions naturally restrict it. Obviously, the response of the flexible ring varies with the direction of loading. It is chosen to leave this aspect beyond the scope of the current study, though.

Frictionless contact pairs were introduced at inner and outer surfaces of the ring including the faces of the flanges and the cut-outs. When ring dimensions fit perfectly the gap between the bearings and the housing the flanges will be initially in full contact with the corresponding surfaces. Including the rest of the ring surface into contact model makes it possible to detect when the gap introduced by the cut-outs gets closed, which marks a rapid change in the overall flexibility of the damper.

It is also possible to include such important factor as tolerances into the finite element analysis. Oversize or undersize fits can be assigned from both inner and outer side independently through the offset property of the corresponding contact pairs. Positive value of this parameter means a tight fit while negative value introduces an initial gap between the parts.

49.4 A Semi-analytical Variational Method

It is possible to reduce the dimensionality of the problem even further down. The elastic ring can be considered as a curved beam of variable cross-section. Contact problems for beams can be tackled by displacement based finite element method without any particular difficulties [9]. Nevertheless, an alternative semi-analytical method is proposed instead. In the considered special case there is no other structural elements besides a single looped beam, no framework-like connectivity, the flexible ring has a simple circular form. The system of forces acting on the ring damper is well defined. Thus static indeterminacy of this structure can be resolved easily. There is no need to determine the continuous displacement field in the deformed beam. The internal forces and reactions that come in form of contact tractions may be determined instead. This can be done by means of a variational principle which enables to handle the inequality constraints of unilateral normal contact in a straight-forward way.

A new formulation of the well-known principle of minimum complementary energy proposed by J. J. Kalker for normal contact of two deformable solids is developed for this special case. The complementary energy functional takes the form:

$$\Phi = \int_0^{2\pi} R\left(\frac{T^2(\phi)}{2EA(\phi)} + \frac{M^2(\phi)}{2EI_z(\phi)}\right)d(\phi) + \left[\sum_i P_i^+ g_i^+ - \sum_j P_j^- g_j^-\right] \tag{49.1}$$

The integral term in (49.1) is the internal complementary energy of the deformed ring as an Euler–Bernoulli beam in terms of the tangential force $T(\phi)$ and the bending momentum $M(\phi)$. $A(\phi)$ and $I_z(\phi)$ are the variable cross-section area and the moment of inertia, correspondingly. P_i^+ and P_j^- are the reaction forces at the contact points on inner and outer side of the ring. The total gap values g_i^+ and g_j^- at these points is formed by the initial offset δ^+ and δ^- on both sides as well as the normal projection of the rotor displacement vector with horizontal component U and vertical component V that counters the gap from the internal side of the ring:

$$g_i^+ = \delta^+ - (U\cos(\phi_i) + V\sin(\phi_i)), \quad g_j^- = -\delta^- \tag{49.2}$$

Note that the contact in this formulation was intentionally restricted to the fixed locations and is assumed to be pointwise. The working hypothesis is that the contact occurs exclusively on the edges of the ring flanges. This statement is supported in particular by the finite element simulations as discussed below. As a consequence the entire ring can be divided into a discrete set of segments between the well-defined potential contact points. Once the tangent force T, shear force Q and the bending moment M are known for any given internal point of the segment their distribution can be determined for the entire element through equilibrium conditions. Without any unknown forces acting or constraints imposed along any portion of the flexible ring the problem is statically determinate. The internal forces though will experience jumps or kinks at each end of the segment where contact reactions are applied. Altogether it is sufficient to know the internal forces T_0, Q_0 and M_0 at any arbitrary cross-section ϕ_0 on the ring and the contact reactions P_i^+ and P_j^- in order to determine the stressed state of the entire ring. This means that the complementary energy (49.1) can be analytically computed as a function of these force variables:

$$\Phi = \Phi\Big(T_0, Q_0, M_0; P_i^+; P_j^-\Big) \tag{49.3}$$

This task is performed by means of symbolic computing environment Maple. The internal forces depend linearly on the listed variables. Hence the complementary energy functional (49.3) turns out to be quadratic.

Naturally, the reaction forces and cannot be arbitrary. First of all they need to be statically admissible, that is to satisfy the equilibrium conditions:

$$\begin{cases} \sum_i P_i^+ \cos(\phi_i) - \sum_j P_j^- \cos(\phi_j) = 0 \\ \sum_i P_i^+ \sin(\phi_i) - \sum_j P_j^- \sin(\phi_j) = 0 \end{cases} \tag{49.4}$$

that only concern horizontal and vertical components of the resulting force since all the reactions are central and produce zero momentum relative the axis of the rotor. Besides these two equality constraints the unilateral nature of the contact restricts the unknown contact reactions to positive values only:

$$P_i^+ > 0, \; P_j^- > 0 \tag{49.5}$$

Ultimately the variational principle is formulated in the form of a constrained minimization problem for complementary energy Φ defined in the closed form (49.3) subject to a set of equality and inequality constraints (49.4) and (49.5). The response of the elastic ring to any vertical or horizontal displacement of the rotor is recovered from the determined contact forces on inner or outer side of the ring:

$$\begin{cases} F_U = -\sum_i P_i^+ \cos(\phi_i) = \sum_j P_j^- \cos(\phi_j) \\ F_V = -\sum_i P_i^+ \sin(\phi_i) = \sum_j P_j^- \sin(\phi_j) \end{cases} \tag{49.6}$$

The proposed analysis method does not explicitly operate with the elastic displacements of the flexible ring or introduces any differential equations that need to be solved. No higher-order discretization of this displacement field is required correspondingly. Instead a discretized formulation of the problem in terms of a few force variables is naturally obtained from a well-established variational principle and a simple assumption regarding the arrangement of contact in the closed form.

49.5 Radial Stiffness of the Flexible Ring Damper

When the rotor undergoes certain radial displacement relative to the housing it will push onto the inner flanges of the flexible ring that are facing the direction of motion. As a consequence the curved sections of the ring squeezed between the rotor and the housing will get bent and flatten out. This elastic response will create a reaction force opposing the movement of the rotor that will perform the damping function of the ring. The force to displacement curves for the vertical loading computed analytically by the proposed variational method based on the principle of minimum complementary energy (pMCE) and obtained from the finite-element analysis (FEA) are shown for three different cases in Fig. 49.1

When the all elements are assembled with zero tolerances this response will be linear as can be seen in Fig. 49.1. The contact will be instantly established at the edges of the flanges located at the bottom half of the ring towards which the rotor gets pressed. This contact set will remain intact throughout most part of the loading. The constant radial stiffness of the flexible ring computed analytically equals 18.1 kN/mm. This value is in excellent agreement with the finite element analysis which gives the estimate of 18.5 kN/mm for the initial stage of the loading. The growing discrepancy between the results for larger vertical displacements should be attributed to the simplifying assumptions that have been taken for the developed analytical method. The displacements on the left and the right end of the loaded half of the ring will be much closer to tangential direction rather than the normal. Meanwhile the proposed variational formulation is based on the model of normal contact. The contact detection in finite element analysis is able to identify the change of normal as the edges of the flanges are sliding along the rigid bodies of the rotor and the housing. More drastic change of the behavior comes in the form of abrupt stiffening of the radial response at the displacement that equals approximately the depth of the cut-outs. When this gap gets closed due to the approach of the rotor the ring will also get in contact at the side opposite to the flange facing the direction of radial motion. Once this portion of the ring gets squeezed between the rotor and the housing it will become as stiff as the two bulk solids. Obviously the proposed analytical method completely disregards this scenario.

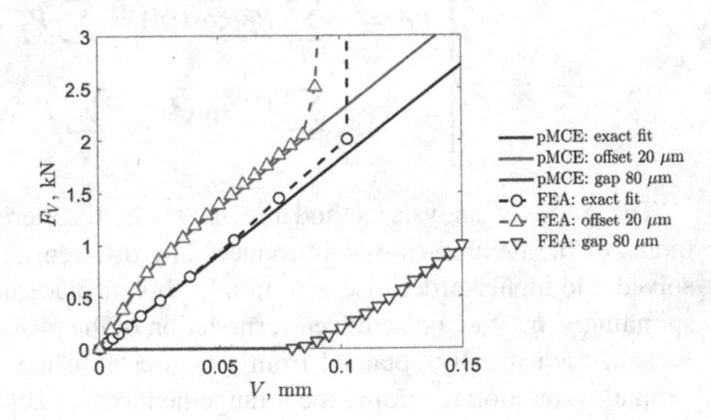

Fig. 49.1 Reaction of the flexible ring to the vertical displacement of the rotor

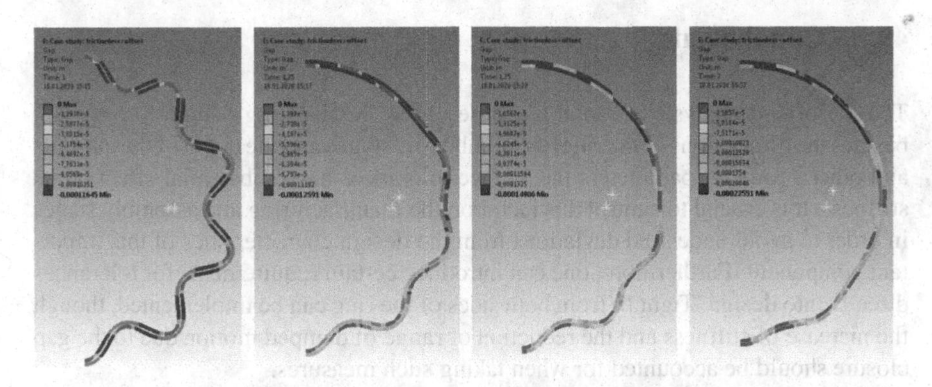

Fig. 49.2 Gap closing/opening at inner and outer side of the flexible ring with tight fit on both sides at different stages of vertical loading

In the other case when the ring is fitted tightly both to the rotor and the housing the radial response becomes quantitatively and qualitatively different. Even for the relatively small positive tolerance of 10 μm on both sides which gives the total offset of 20 μm the flexible ring displays significantly stiffer behavior. The picture of ring deformations and the change of gap status shown in Fig. 49.2 explain this difference. As can be seen, the tight fit creates a pre-stress state and initial deformations at the beginning of the loading for the neutral location of the rotor. The ring becomes uniformly compressed and its flanges are pressed inwards and outwards with non-zero forces. As a result the flanges at the top haft of the ring remain engaged into contact for some time.

There are more active contact constraints in the beginning of the loading compared to the case of exact fit. Together with the pre-stress this causes much stiffer elastic response of the ring. As the flanges in the top half of the ring get out of contact the stiffness reduces as can be seen from the flattening of the loading curve in Fig. 49.1. Ultimately it gets nearly parallel to the linear response of the perfectly fitted ring. The upturn of the reaction force occurs sooner in this case since the surface offset consumes some part of the initial gap. This stiffening though is more gradual compared to the previously considered case.

When the fit is loose some free motion of the rotor relative to the housing occurs. Its magnitude equals the sum of tolerances on both sides. In the considered case the rotor can travel 80 μm until it encounters any resistance: 40 μm before it touches the ring and another 40 μm they would move together towards the housing. The elastic response after this threshold is much softer when compared to the other two cases. The stiffness will grow slowly as more and more flanges will consecutively get into contact. The analysis window excludes the stiffness upturn due to the gap elimination.

49.6 Conclusions

The performed analysis of radial response of the flexible ring damper proves that besides the dimensions of the ring, the number of flanges and the depth of the cut-outs and other geometry parameters the surface tolerances have substantial effect on its stiffness. It is crucial to control this factor on the manufacturing and assembly stages in order to avoid undesired deviations from the design characteristics of this important component. Furthermore, one can introduce certain requirements for tolerances directly into design. Tight fit from both sides of the ring can be implemented, though the increase of stiffness and the reduction of range of damped motion due to the gap closure should be accounted for when taking such measures.

In order to predict reliably these crucial characteristics thorough analysis need to be performed. Finite element analysis can be suggested for the experienced users due to its versatility. One can model a broad variety of damper designs and expand analysis to the multiphysics including for instance fluid dynamics of the lubricant [6]. However, simple and analytical method developed in this paper has its own merits for many engineers. For the considered type of flexible ring dampers it displays accuracy that is not inferior to the finite element analysis. It can be easily implemented in any symbolic computing environment or even Microsoft Excel which makes it much more accessible for common use by engineers.

References

1. Kelson, A.S., Cymanskii, H.P., Yakovlev, B.H.: Dynamics of Rotor-Bearing Systems. Nauka, Moskow (1982)
2. Tkachuk, M.M., Grabovskiy, A., Tkachuk, M.A., Zarubina, A., Lipeyko, A.: Analysis of elastic supports and rotor flexibility for dynamics of a cantilever impeller. J. Phys.: Conf. Ser. **1741**, 012043-1–012043-10 (2021)
3. Martynenko, G.: Resonance mode detuning in rotor systems employing active and passive magnetic bearings with controlled stiffness. Int. J. Automot. Mech. Eng. **13**(2), 3293–3308 (2016)
4. Avramov, K., Shulzhenko, M., Borysiuk, O., Pierre, C.: Influence of periodic excitation on self-sustained vibrations of one disk rotors in arbitrary length journals bearings. Int. J. Nonlin. Mech. **77**, 274–280 (2015)
5. Avramov, K.V., Mikhlin, Y.V.: Review of applications of nonlinear normal modes for vibrating mechanical systems. Appl. Mech. Rev. **65**(2), 020801-1–020801-20 (2013)
6. Zhang, Y., He, L., Yang, J., Wan, F., Gao, J.: Vibration control of an unbalanced single-side cantilevered rotor system with a novel integral squeeze film bearing damper. Appl. Sci. **9**(20), 4371-1–4371-18 (2019)
7. Kalker, J.J.: Variational and non-variational theory of frictionless adhesive contact between elastic bodies. Wear **119**(1), 63–76 (1987)
8. Tkachuk, M.: A Numerical method for axisymmetric adhesive contact based on Kalker's variational principle. East.-Eur. J. Enterp. Technol. **3**(7), 34–41 (2018)
9. Atroshenko, O., Tkachuk, M.A., Martynenko, O., Tkachuk, M.M., Saverska, M., Hrechka, I., Khovanskyi, S.: The study of multicomponent loading effect on thinwalled structures with bolted connections. East.-Eur. J. Enterp. Technol. **1**(7), 15–25 (2019)

Chapter 50
Developing a Methodology for Design of Patient-Specific Plate Type Implants and Defining the Relative Deformation of the Implant

Yavor Sofronov and Krasimira Dimova

Abstract Nowadays, one of the common fields where Engineering collaborates with Medicine is the field of Implantology where the cooperation between engineers and surgeons results in creation of personalized implants. The aim of those innovations is to create a methodology from the beginning of the process, where the CT (computer tomography) scan data of the injured area is received to the end of the process where the implants are inserted. After the Virtual Prototyping the Physical prototyping could be accomplished with additive technology—the technology of 3D printing. In some clinical cases, the additive technology via 3D printing is very suitable for creating patient-specific implants, because of the variety of the materials. In the whole process are involved group of researchers including engineers and neurosurgeons. The research is based on introduced design process solution which consists of six steps and few proposed software. The research focuses on the design of a patient-specific implant and the ability to define in advance its relative deformations via FEA (Finite Element Analysis). It is important to note that after the production of the implant and before its insertion the implant should cover the Standard Test Methods for Spinal Implant Constructs in a Vertebrectomy Model.

Keywords Patient-specific implant · Spine · FEM analysis

50.1 Introduction

Biotechnologies are involved in the rush progress of the technologies, especially the design, manufacturing and integration of personalized implants. The design of the implant is becoming more and more complex with improving the Additive Manufacturing technologies. The process of the implants implementation tends to ease because of the developing patient-specific implant which requires less treatment during the operation [1]. Plate type spinal implant are usually attached with screws

Y. Sofronov · K. Dimova (✉)
FIT-Faculty of Industrial Technology, Technical University of Sofia, Sofia, Bulgaria
e-mail: krdimova@tu-sofia.bg

Laboratory "CAD/CAM/CAE in Industry", Technical University of Sofia, Sofia, Bulgaria

M. Rackov et al. (eds.), *Machine and Industrial Design in Mechanical Engineering*, Mechanisms and Machine Science 109,
https://doi.org/10.1007/978-3-030-88465-9_50

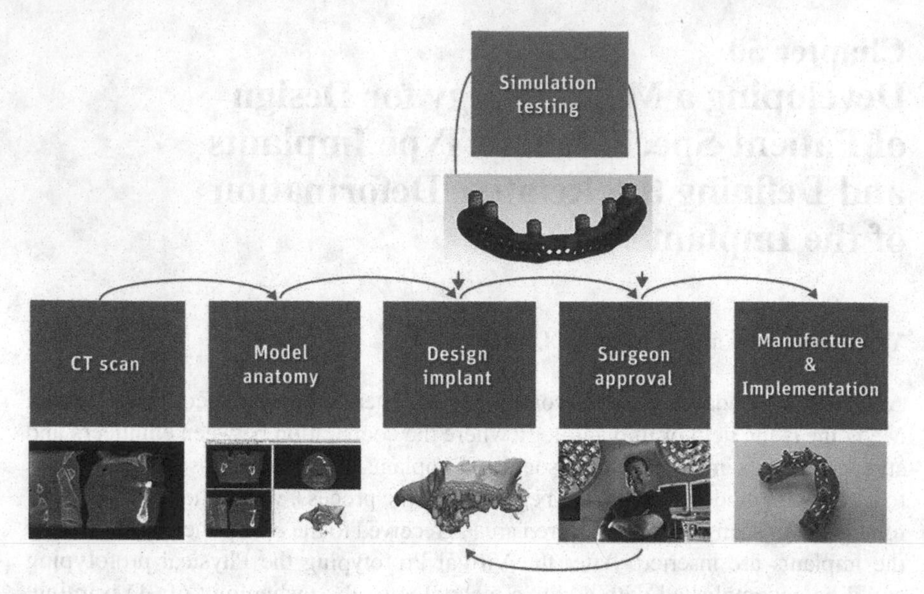

Fig. 50.1 Design process solution [3]

to the vertebrae. They assist in stabilizing the spine. This type of fusion is a procedure in which a surgeon uses spinal implants and screws to ensure the bones fusion. They are still flexible enough to allow the spine to bend [2]. Based on the origin design process, which is shown on Fig. 50.1, are made further analysis in the field of spinal personalized implants. The innovation is connected with the opportunity to define the relative deformation of plate type implants in order not to affect the natural movements of human spine. Based on the fact that the most movable part of the human spine is a cervical spine the plate is designed to answer the human anatomy without limiting the spine movements and as it would be of healthy spine.

50.2 Materials and Methods

The main aim of a plate type implant with screws is to create a pressure between one piece of the vertebrae to another. This results in stability which leads to healing the injured bones. When inserted the patient-specific implant limits the natural movements of a human spine. The purpose of this scientific research is to introduce a method to control the relative movements of the implant. The cervical plate type implant was used for the research. In some clinical cases appear cervical instability or nerve pressure. It can result from trauma or spinal reconstruction [3]. Depending on the procedure and the number of spinal levels involved, one or more plates could be implanted. The plate is held in place by screws and adjusted to the vertebrae [4]. The initial design and the materials are chosen considering the fact that deformation

Table 50.1 Material properties [5]

Density (g/cm^3)	Young Modulus (GPa)	Poisson's ratio	Yield strength (MPa)	Elongation (%)
4.42	114	0.35	930	5–18

should be defined in advance. The plate type spinal design is compatible with the loads occurring in the spinal cord, with its convex middle part, which ensures elasticity. The implant is inserted with four surgical screws. For the aims of the study is used FEA (Finite Element Analysis) via ANSYS software.

The most common materials in Implantology are stainless steel, plastics, titanium and titanium alloys. For the research it is chosen titanium alloy Ti6Al4V, which is the most used because of it is corrosion resistant, wear resistant, non-toxic, biocompatible properties [4]. The material's mechanical properties are the following (Table 50.1) [5].

The methods used in the research are based on Virtual Prototyping [6]. The model of the implant is designed in CAD software, then it is imported in ANSYS, where it's applied load on the implant and the main part of the research is held (Fig. 50.2a). Because of the symmetric structure for the analysis was used half model (Fig. 50.2b). This plate type implant is inserted with four screws, each two on one vertebrae. The vertebrae are conditionally defined as two parallelepipeds (Fig. 50.4). After that are defined the materials and the contact areas. The material of the implant and the screws is Ti6Al4V. The contacts between screws' head and the implant are defined as No separation, the contact between the implant and the vertebrae is defined as bonded. The screws have pretention defined in advance for that reason the analysis is non-linear. On the implant are applied 10 N force. This results in excessive deformation of the implant, total deformations of 2.2 mm and Equivalent Stress of 660 MPa which

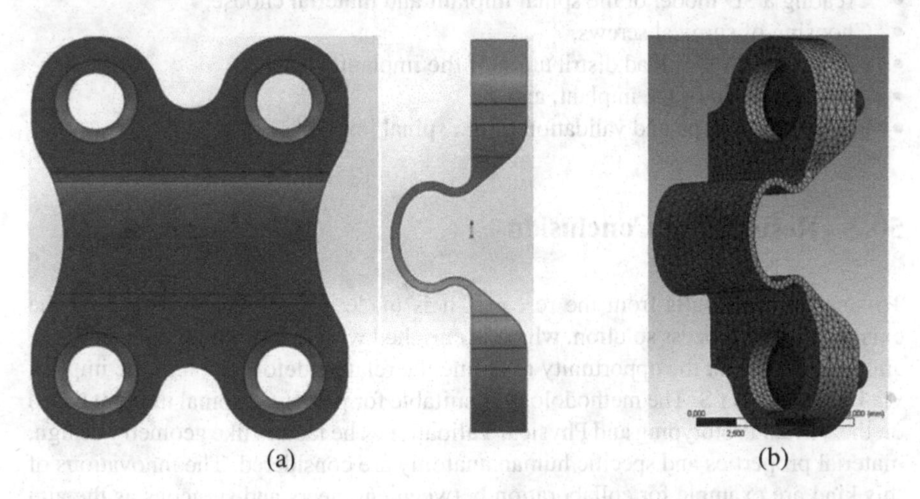

Fig. 50.2 **a** 3D model of the implant and **b** meshed half model of the implant for ANSYS analysis

Table of Schematic B2: Optimization , Candidate Points

	A	B		C	D	E	F
1	Reference	Name		P1 - Rzsmall (mm)	P4 - Rysmall (mm)	P7 - Directional Deformation Maximum (mm)	
2						Parameter Value	Variation from Reference
3	○	Candidate Point 1	DP 30	1,8356	2,805	-1,0017	-1,58 %
4	●	Candidate Point 2	DP 24	1,7345	2,8305	-0,98612	0,00 %
5	○	Candidate Point 3	DP 19	1,9272	2,6804	-0,97983	0,64 %
6	○	Candidate Point 4	DP 14	1,9203	2,6674	-1,0255	-4,00 %
7	○	Candidate Point 5	DP 20	1,827	2,7745	-0,97099	1,53 %
*		*New Custom Candidate Point*		*1,75*	*2,75*		

Fig. 50.3 Candidate points

are under Tensile Yield Strength of the material. The design should be changed for reaching the set deformation of 1 mm. In Direct Optimization module in ANSYS target deformation of 1 mm is defined and are selected parameters to be varied. As result are generated five candidate points shown on Fig. 50.3 and the one with less deviation is chosen. The initial design should be changed referring the optimization data. The radius of the middle part was changed referring to the optimization [7]. In addition, Von-Mises Stress value is 392.5 MPa which is below Tensile Yield Strength for the material [8]. When calculated the geometry of the model is changed referring to optimization data.

Based on the design process solution a methodology about the design of a patient-specific plate type spinal implant and defining its relative deformations is created [3, 9, 10].

The methodology (Fig. 50.5) includes the following steps:

- Patient's data and generating a 3D model of patient's spine,
- Creating a 3D model of the spinal implant and material choose,
- Choosing of surgical screws,
- FEA in ANSYS of load distribution of the implant,
- Personalization of the implant, and
- Physical prototype and validation of the spinal implant.

50.3 Results and Conclusion

To sum up the results from the research it is made a methodology based on the existing design process solution, which is enriched with further simulation analysis, and improved with the opportunity to define the relative deformation of the implant via FEA in ANSYS. The methodology is suitable for plate type spinal implant based on the Virtual Prototyping and Physical Validation. The factors like geometry design, material properties and specific human anatomy are considered. The innovations of this kind are example for collaboration between engineers and surgeons as the aim to improve the quality of the treatment and to reduce the time for integration of the

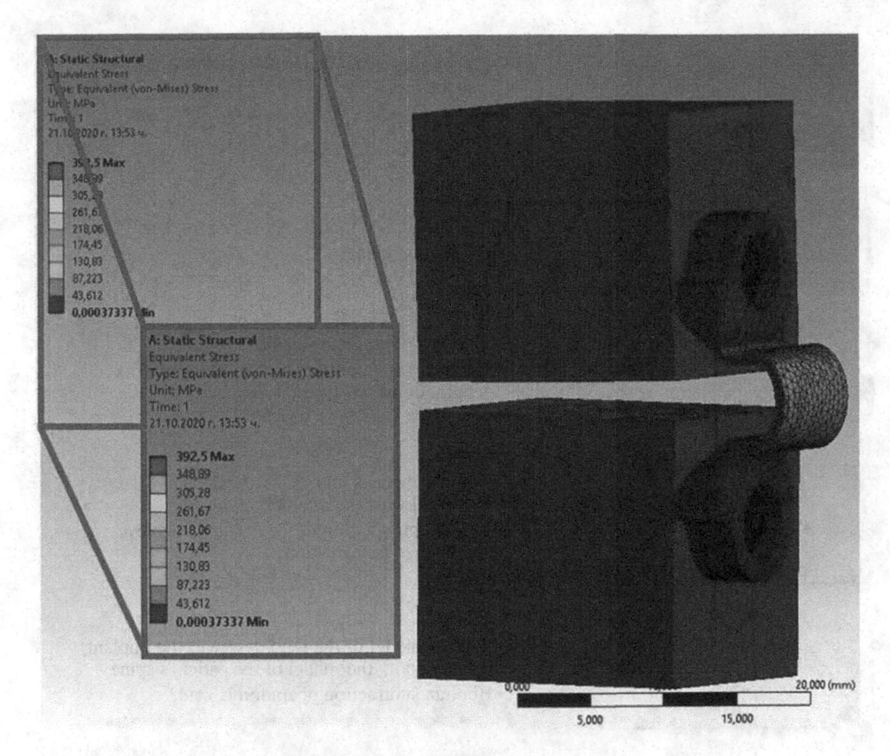

Fig. 50.4 Von-Mises Stress distribution of the implant

implant. Moreover, applying the methodology will improve the design of these types of implants because the implant will ensure the defined relative deformation as it would be of healthy spine. The essence of the research aims to create improve spinal personal implants where the relative movements could be set in advance. Based on the methods and materials it was made a step forward in the field of innovations connected with human health and mechanical engineering. Furthermore, the innovation will find a direct application in modern neurosurgery where they will provide access to that kind of innovations to broad group of patients.

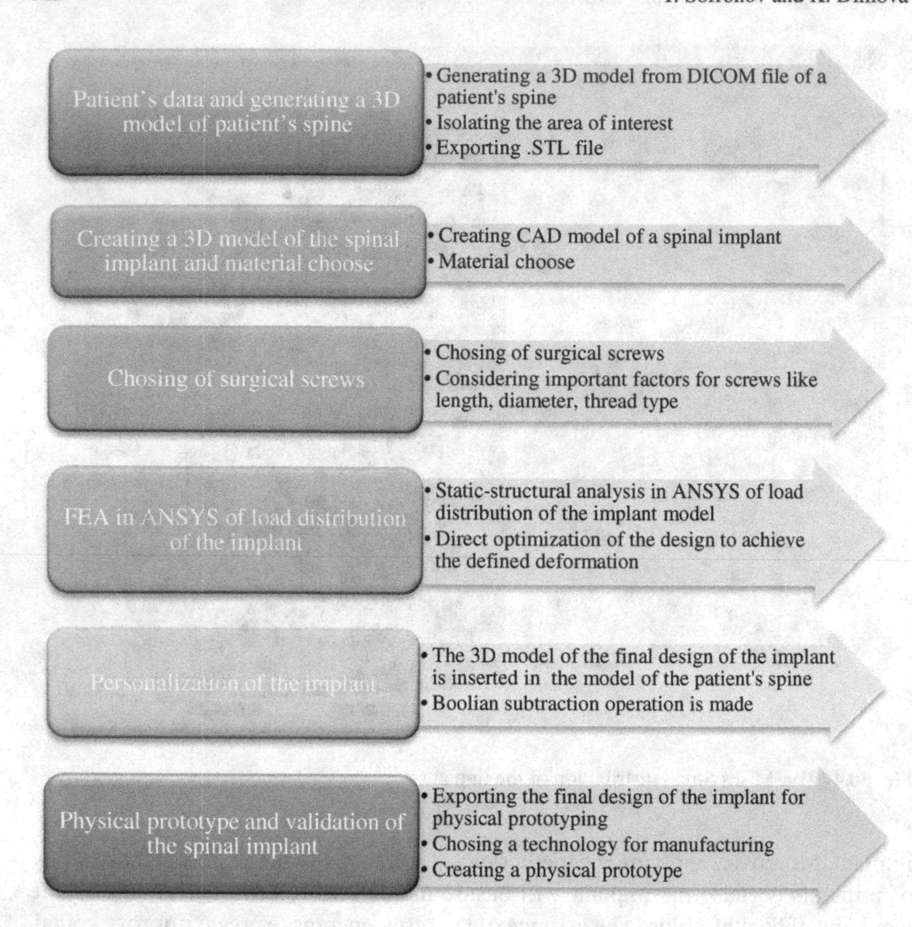

Fig. 50.5 Methodology for design of patient-specific plate type implants

Acknowledgements This study is performed by the support of project DN-17-23 "Developing an approach for bone reconstruction and implant manufacturing through virtual engineering tools" of National Science Fund, Ministry of Education and Science, Bulgaria.

References

1. Lowe, T.G.: Cervical plates: provide neck stability using spinal instrumentation. https://www.spineuniverse.com/treatments/surgery/cervical-plates-provide-neck-stability-using-spinal-instrumentation (2016). Last accessed 01 Dec 2020
2. Dombrowski, J.: Spinal implants: types, usage and options. https://www.lumitex.com/blog/spinal-implants (2018). Last accessed 01 Nov 2020

3. Fink, S., Atkinson, A.: Personalized implants restore smiles. https://www.ansys.com/about-ansys/advantage-magazine/volume-xii-issue-3-2018/personalized-implants-restore-smiles (2018). Last accessed 01 Nov 2020
4. De Viteri, V.S., Fuentes, E.: Titanium and titanium alloys as biomaterials. In: Gegner, J. (ed) Tribology: Fundamentals and Advancements, pp. 155–181. IntechOpen (2013)
5. AZoM become a member search, Search menu properties this article has property data, Titanium Alloys—Ti6Al4V Grade 5. www.azom.com. Last accessed 01 Oct 2021
6. Todorov, G., Kamberov, K.: Virtual Engineering (2017)
7. Wang, H., Zhou, C., Yu, Y., Wang, C., Tsai, T.-Y., Han, C., Li, G., Cha, T.: Quantifying the ranges of relative motions of the intervertebral discs and facet joints in the normal cervical spine. J. Biomech. **112**, 110023-1–110023-8 (2020)
8. Kamberov, K., Todorov, G., Sofronov, Y., Nikolov, N.: Methodology for designing, manufacturing and integration of personalized spinal implants for surgical treatment of the cervical spine. AIP Conf. Proc. **2333**, 110009-1–110009-5 (2021)
9. Ortho Baltic implants, Process chain for manufacturing patient-specific implants. http://balticimplants.eu/patient-specific-medical-devices/process/. Last accessed 1 Oct 2020
10. Todorov, G., Nikolov, N., Sofronov, Y., Gabrovski, N., Laleva, M., Gavrilov, T.: Computer aided design of customized implants based on CT-scan data and virtual prototypes. In: Poulkov, V. (ed) Future Access Enablers for Ubiquitous and Intelligent Infrastructures: FABULOUS 2019. LNICS-SITE, vol. 283, pp. 339–3346. Springer, Cham (2019)

Chapter 51
Assessment of Aircraft Cylinder Assembly Lifetime at Elevated Temperature

Nikola Vučetić, Ranko Antunović, and Lazar Šarović

Abstract After 1560 h of operation the crack was noticed on the aircraft cylinder head. Aircraft piston engine cylinder assemblies are component exposed to fatigue load during its operation. In this paper an assessment of aircraft cylinder assembly lifetime determined numerically has been shown. The mentioned research was preceded by numerical thermal and structural analysis of cylinder assembly. Also, the influence of material porosity on the cylinder assembly lifetime is considered. Based on the conducted research the potential cause of the crack appearance on the cylinder head was determined.

Keywords Crack · Cylinder head · Fatigue analysis

51.1 Introduction

In this paper the fatigue analysis of the cylinder assembly at elevated temperature was performed using Ansys Workbench. The cylinder assembly failed due to the appearance of a crack on the cylinder head. Failure of aircraft cylinder assemblies is accompanied by the occurrence of fatigue as a consequence of the existence of multi-axial thermomechanical loads during operation [1]. Due to the long-term action of periodically changing loads, gradual destruction of the material occurs, especially at elevated temperatures [2, 3]. The number of cycles that the cylinder assembly can perform under the action of the applied load is obtained.

Within the research, the influence of material porosity on the lifetime of the cylinder assembly is shown. It is fact that porosity significantly shortens the lifetime of the structure [4–8].

N. Vučetić (✉) · R. Antunović · L. Šarović
Faculty of Mechanical Engineering, University of East Sarajevo, 71123 East Sarajevo, Bosnia and Herzegovina

M. Rackov et al. (eds.), *Machine and Industrial Design in Mechanical Engineering*, Mechanisms and Machine Science 109,
https://doi.org/10.1007/978-3-030-88465-9_51

51.2 Model and Boundary Conditions

The parts of an aircraft cylinder air-cooled assembly are the cylinder body and the cylinder head, Fig. 51.1 The cylinder body is made of AISI 4140 steel. Based on the source [9] the properties of the material for the cylinder body are defined. The internal surfaces of the cylinder body are sanded and honed, while there are deep cooling ribs on the outer side. The material of the cylinder head is aluminum alloy 242.0. The material properties necessary for structural FEM analysis were determined in [10].

The cylinder head and the cylinder body are connected in the way that the cylinder head is heated to a temperature of 3500 °C and so heated it is placed on the cylinder body. By cooling of the cylinder head a rigid connection between threads on the inner side of the cylinder head and the outer side of the cylinder body is achieved. The connection between cylinder body and cylinder head is defined as "Bonded", Fig. 51.2a.

Fig. 51.1 Model of the cylinder assembly

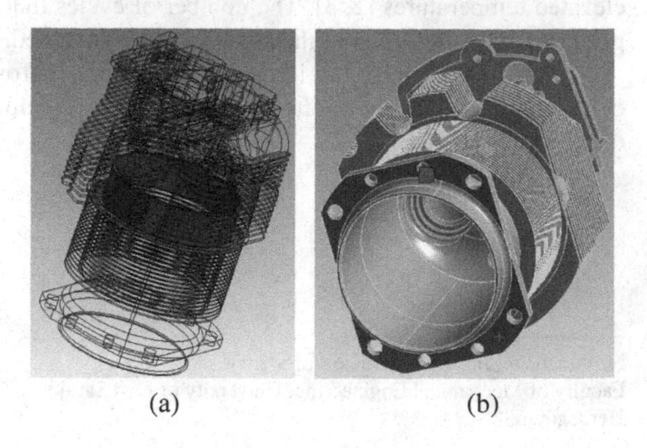

Fig. 51.2 **a** Connection between cylinder body and cylinder head; **b** the cylinder body to the engine housing connection

The connection between the cylinder body and the engine housing is made using 8 screws. The fixed connection is defined on the contact surface, Fig. 51.2b. The mesh of tetrahedral finite elements with midside nodes was generated on the 3D model of the Lycoming IO-360-B1F aircraft engine cylinder assembly. The finite element mesh consists of 145,604 elements and 277,367 nodes. The minimum finite element size was 0.062 mm and the maximum was 12.3 mm. The cylinder head workload is presented by a mean effective pressure of 998 kPa [10].

51.3 Fatigue Analysis of Cylinder Assembly

Within this chapter, the fatigue analysis of the cylinder assembly at elevated temperature was performed, that is the fatigue analysis due to the combined thermomechanical load to which the assembly was exposed during operation. In previous research, a structural analysis of the cylinder assembly was performed and the distribution of the stress state of the cylinder assembly was determined [10].

One cycle of operation of a cylindrical piston assembly involves the intake of a mixture of air and fuel, compression, explosion and exhaust after combustion [11]. The lowest stress values occur during the first stroke (Fig. 51.3a), while the highest stress values occur during the ignition of the mixture (Fig. 51.3b).

If we take into account that the whole cycle of 4 strokes lasts 0.044 s, it means that each of the stroke lasts 0.011 s and that the stress states change in a negligibly short time interval, so that the stress state obtained on the basis of the average effective pressure at operating temperature is considered relevant [10]. The lifetime of the cylinder assembly at elevated temperature is shown in Figs. 51.4 and 51.5.

Fatigue analysis of the cylinder assembly revealed that the lifetime of the material in the area of crack formation ranges from 2.80×10^8 to 2.98×10^8 cycles. Thus, this confirms the fact of the life of the cylinder head which is estimated at 3600 h of operation [12], which corresponds to the number of cycles of 2.92×10^8, taking into account that one cycle implies one engine stroke, or two full crankshaft revolutions that last 0.044 s [10].

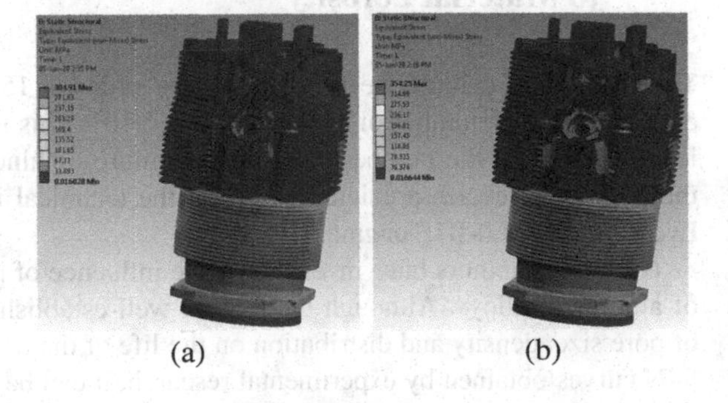

Fig. 51.3 Stress state of the cylinder assembly: **a** first stroke and **b** third stroke

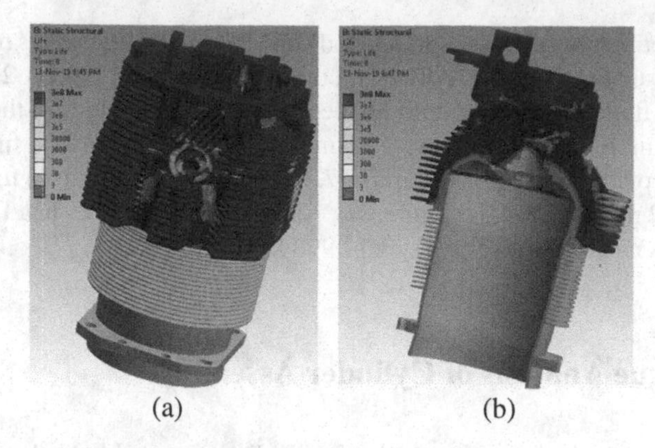

Fig. 51.4 The lifetime of the cylinder assembly at elevated temperature

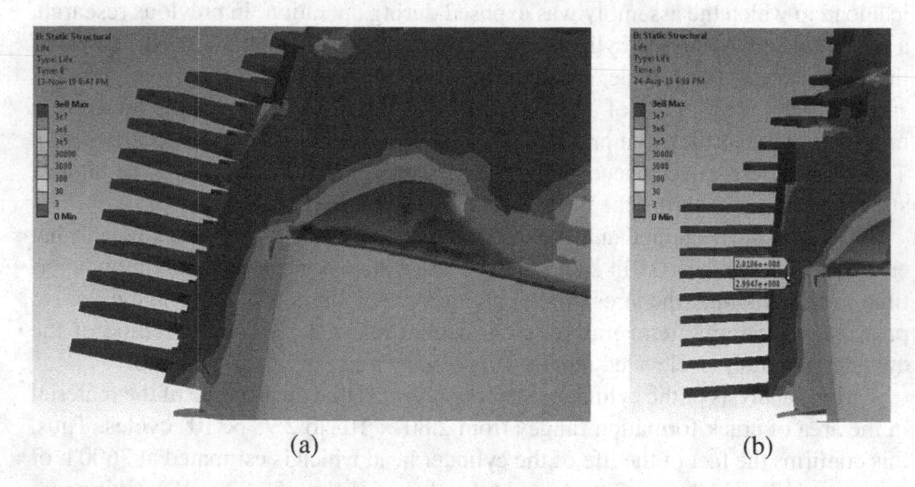

Fig. 51.5 The lifetime of the cylinder assembly crack location at elevated temperature

51.4 Fatigue Analysis of Cylinder Assembly Due to Material Porosity

The problem of cylinder head failure was recorded at 1560 operating hours, which corresponds to a number of cycles of 1.26×10^8. This number of cycles is significantly less than the number of cycles to failure obtained numerically, as well as the number of cycles predicted based on the technical instructions for use of the Lycoming IO-360-B1F engine [12].

Numerous authors have investigated the influence of porosity on the fatigue life of aluminum alloys. Although there is no well-established model of the influence of pore size, density and distribution on the life of the material, based on numerous S–N curves obtained by experimental research, it can be concluded that in the case of material porosity, the lifetime, or the number of cycles to failure, decreases by an average of 40% [4–8].

Taking into account the existence of porosity of the cylinder head material proven in previous research [10], the S–N curve in the material properties settings was modified and the fatigue analysis of the cylinder assembly was repeated. The modified S–N curve is shown in Fig. 51.6. The lifetime of the cylinder assembly is shown in Fig. 51.7.

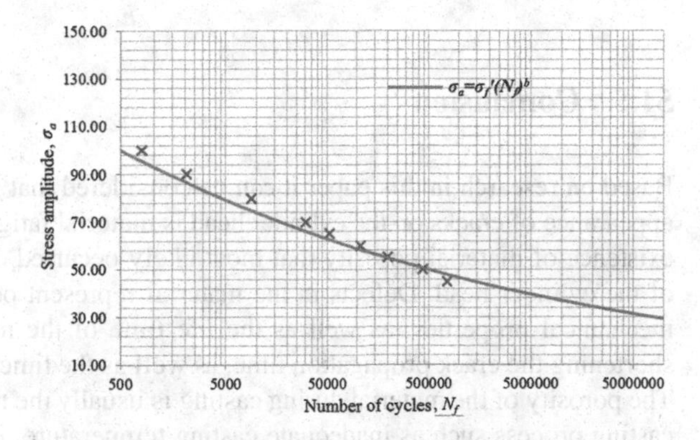

Fig. 51.6 The modified S–N curve due to the existence of porosity

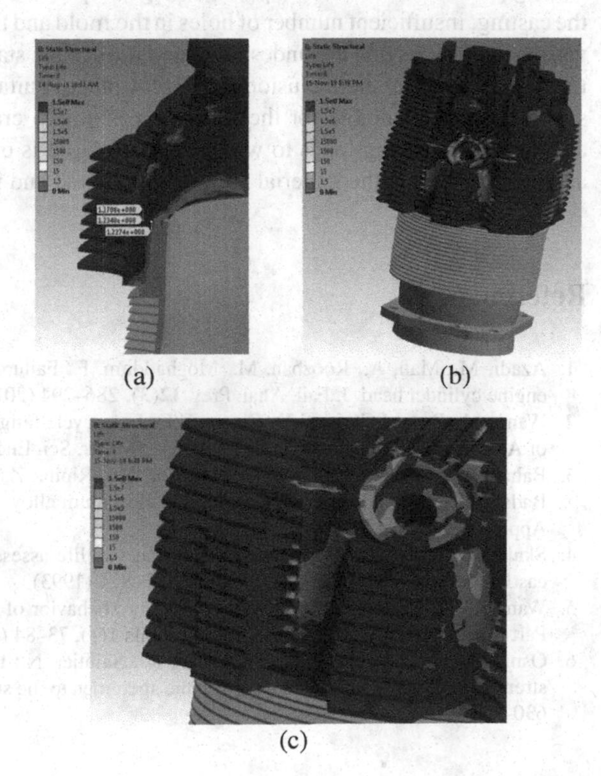

Fig. 51.7 Cylinder assembly lifetime in case of material porosity

Based on the obtained results ranging from 1.22×10^8 to 1.27×10^8 cycles to failure, it can be concluded that the lifetime of the cylinder head has been confirmed, which corresponds to the number of cycles to failure that occurred after 1560 h of engine operation.

.

Thus, it can be concluded that the dominant cause of the appearance of cracks on the cylinder head is material fatigue caused by the previous existence of material porosity that most likely occurred during the casting process of the cylinder head.

51.5 Conclusion

Based on research in this paper it can be considered that the dominant cause of the appearance of cracks on the cylinder head is material fatigue caused by the previous existence of material porosity that most likely occurred during the casting process of the cylinder head. Defects in the material represent potential cracks reduce the mechanical properties, as well as the life time of the material due to fatigue by shortening the crack propagation time, as well as the time required for its initiation. The porosity of the material during casting is usually the result of a poorly designed casting process such as inadequate casting temperature, inadequate cooling rate of the casting, insufficient number of holes in the mold and the like. Metal alloys contain various inclusions that are undesirable and reduce the static load-bearing capacity of the structure. Also, the inclusions represent places suitable for stress concentration sources that are suitable for the formation of micro cracks. During very complex and variable working loads to which such material is exposed during exploitation, adjacent cavities in the material are interconnected and initial cracks are formed.

References

1. Azadi, M., Mafi, A., Roozban, M., Moghaddam, F.: Failure analysis of a cracked gasoline engine cylinder head. J. Fail. Anal. Prev. **12**(3), 286–294 (2012)
2. Wang, M., Pang, J.C., Li, S.X., Zhang, Z.F.: Low-cycle fatigue properties and life prediction of Al–Si piston alloy at elevated temperature. Mater. Sci. Eng. **704**, 480–492 (2017)
3. Bahaideen, F.B., Saleem, A.M., Hussain, K., Ripin, Z.M., Ahmad, Z.A., Samad, Z., Badarulzaman, N.A.: Fatigue behaviour of aluminum alloy at elevated temperature. Modern Appl. Sci. **3**, 52–61 (2009)
4. Skallerud, B., Iveland, T., Harkegard, G.: Fatigue life assessment of aluminium alloys with casting defects. Eng. Fract. Mech. **44**(6), 857–874 (1993)
5. Wang, Q.G., Apelian, D., Lados, D.A.: Fatigue behavior of A356-T6 aluminum cast alloys. Part I. Effect of casting defects. J. Light Metals **1**(1), 73–84 (2001)
6. Osmond, P., Le, V.-D., Morel, F., Bellett, D., Saintier, N.: Effect of porosity on the fatigue strength of cast aluminium alloys: from the specimen to the structure. Procedia Engineer. **213**, 630–643 (2018)

7. Mayer, H., Papakyriacou, M., Zettl, B., Stanzl-Tschegg, S.E.: Influence of porosity on the fatigue limit of die cast magnesium and aluminium alloys. Int. J. Fatigue **25**(3), 245–256 (2003)
8. Fintova, S., Konecna, R., Nicoletto, G.: Microstructure, defects and fatigue behavior of cast AlSi7Mg alloy. Acta Metall. Slovaca **19**(3), 223–231 (2013)
9. MakeItFrom.com.: https://www.makeitfrom.com/material-properties/Normalized-4140-Cr-Mo-Steel. Last accessed 26 Apr 2019
10. Vučetić, N.: Razvoj metodologije za procjenu integriteta vazdušno hlađenog avionskog klipnog motora izloženog visokocikličnom mehaničkom i termičkom opterećenju. PhD Thesis, Faculty of Engineering, University of Kragujevac (2020)
11. Heywood, J.B.: Internal Combustion Engine Fundamentals. McGraw-Hill, NY (1988)
12. Operator's manual Lycoming O-360, HO-360, IO-360, AIO-360, HIO-360 & TIO-360 series aircraft engines. Part No. 60297-12, Rev. No. 60297-12-5. Williamsport (2005)

Chapter 52
Dynamic Simulation of a Planetary Gearbox with Double Satellite

Imre Zsolt Miklos, Cristina Carmen Miklos, and Carmen Inge Alic

Abstract Double Satellite planetary gearbox is part of the complex mechanisms with mobile axis. These gearboxes are mainly used in the mining and metallurgical industries, when to operate equipment using hydraulic motors with speeds lower than electric motors, torques transmitted respectively higher gear ratios. This paper presents elements of kinematics and design of the planetary gearbox with double satellite, respectively 3D modeling and its dynamic analysis using the Dynamic Simulation module from the Autodesk Inventor Professional application. Dynamic simulation involves transforming the 3D model of the gear into a mechanism, modeling the kinematic torques, defining the parameters of the driving movement and external loads, respectively visualizing in graphical form the kinematic and kinetostatic results for the different components of the mechanism, in real operating conditions, during a kinematic cycle. At the same time, based on the obtained results, the analysis by the finite element method of the component elements is presented at the time step in which their stress are maximum.

Keywords Planetary mechanism · Degree of mobility · Central gear · Satellite · Mechanism analysis

52.1 Introduction

The planetary gearboxes are mechanical transmissions, having in their component gears with moving axis, being characterized by the fact that they ensure high transmission ratios in the conditions of a small size. In industry, we can find planetary gearboxes with different numbers of central gears, having single or double satellites [1].

I. Z. Miklos (✉) · C. C. Miklos (✉) · C. I. Alic
Faculty of Engineering Hunedoara, Politehnica University Timişoara, Revoluţiei Str. 5, 331128 Hunedoara, Romania
e-mail: imre.miklos@upt.ro

C. C. Miklos
e-mail: cristina.miklos@upt.ro

M. Rackov et al. (eds.), *Machine and Industrial Design in Mechanical Engineering*, Mechanisms and Machine Science 109,
https://doi.org/10.1007/978-3-030-88465-9_52

The use of drive motors for specific equipment from metallurgical or mining industry, requires in certain situations, conception and design of special gearboxes with parameters that are not found in the standard series. Such cases are found when using hydraulic motors to drive, with speeds lower than electric motors, high transmitted torques, or high gear ratios. These gearboxes are those with double satellite, they are part of the category of complex transmissions, they result from the serial connection of one or more single satellite transmissions [1, 2].

Below are two constructive variants, studied, of planetary gearboxes with double satellite.

- The model P-C-II (Fig. 52.1a), is made with two internal cylindrical gears, the driving element being the satellite arm like an crank shaft, H on which is mounted the double satellite 2–3. The central gear 1 is fixed, ensuring the planetary gearbox a single degree of mobility, and the central wheel 4 is connected to the driven shaft of the gearbox. For this planetary gearbox model, two dimensional variants have been made, which are currently being tested. They are characterized by constructive simplicity and very wide kinematic possibilities. The disadvantage is that due to the eccentricity of the satellite arm, high inertia forces appear, respectively the manufacturing costs are high [2].
- The model N-C-EI-I (Fig. 52.1b), is made with three cylindrical gears, one external and two internal, the driving element being the central gear 1. The 2–3 double satellite is mounted on the H satellite arm. Similarly the central gear 4 is fixed and the central wheel 4 is connected to the driven shaft of the gearbox. This type of planetary gearbox was made in a wide variety of types, sizes and construction variants, respectively 9 sizes with 6 transmission ratios. The N-C-EI-I model

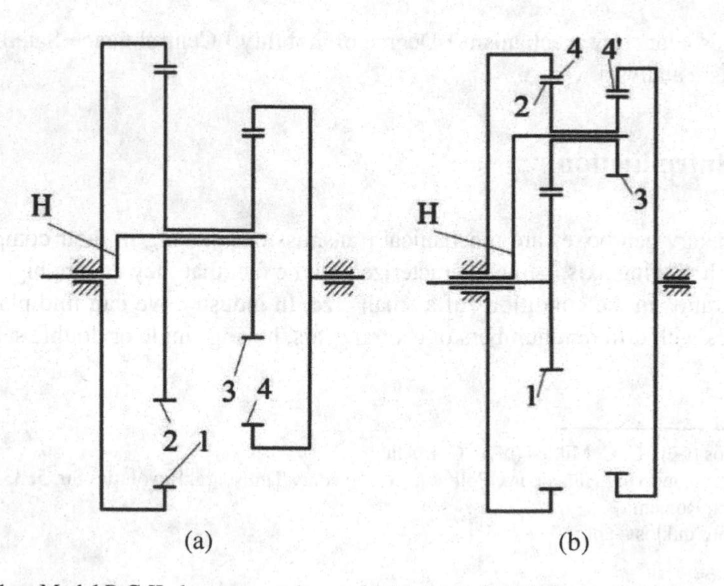

Fig. 52.1 **a** Model P-C-II planetary gearbox and **b** Model N-C-EI-I planetary gearbox

have advantages such as: safety in working, noise-free operation, high reliability, coaxiality between the driving and driven shafts, high efficiency, and respectively easy installation [3].

52.2 Kinematics and Synthesis of Planetary Mechanisms

The total transmission ratios, depending on the number of gears teeth, for the two models of the studied planetary gearboxes, are presented in the relations below [3].

- P-C-II model:

$$i_{H4}^{1} = \frac{1}{1 - i_{41}^{h}} = \left(1 - \frac{z_3 z_1}{z_4 z_2}\right)^{-1} \quad (52.1)$$

- N-C-EI-I model:

$$i_{14'}^{4} = \frac{z_2 z_{4'}}{z_1 z_4} \cdot \frac{z_1 + z_4}{z_2 + z_3} \quad (52.2)$$

An essential problem in the planetary mechanisms design is the correct choice of tooth numbers, in order to ensure the coaxiality condition. When this condition is accomplished, without gear unit correction, the transmission ratios of the planetary gearboxes can be defined with *z*, *a*, *b* arbitrarily chosen parameters, which are simultaneously integers or simultaneously odd multiples of 0.5. The definition of the tooth numbers in this form allows that, at the synthesis of the planetary mechanisms, they be determined directly, resulting at the same time an information on the specific losses due to the gear, from the relation of the sum or difference of these tooth numbers. When determining the number of teeth, it is necessary, when there is a transmission crotch of the movement, to verify the fulfillment of the mounting and neighborhood conditions [3].

The specific losses due to gearing can be very small, by the appropriate choice of parameters *a* and *b*. These losses can also be influenced by adopting negative values for parameter *a*, $a < 0$ [3].

The relations between the transmission ratios and the parameters *z*, *a*, *b* for the two constructive variants of the planetary gearboxes, respectively the calculation relations of the gears tooth numbers are presented in Table 52.1.

52.3 3D Model of the Planetary Gearbox with Double Satellite, N-C-EI-I Model

The 3D model of the planetary gearbox assembly, with double satellite, in the constructive variant N-C-EI-I was made in the Autodesk Inventor Professional

Table 52.1 Calculation relations of tooth numbers

Gearbox model	P-C-II	N-C-EI-I
$z,\ i^4_{14'}$	$z = \sqrt{i^1_{H4}(b^2 - a^2) + a^2}$ $z \gg a;\ z \gg b;\ \lvert a \rvert < b;$	$i^4_{14'} = \frac{2(z^2-a^2)(2z+b-a)}{(b-a)(3z+b-2a)(z+b)}$ $a < z < b$
z_1	$z_1 = z + b$	$z_1 = z + b$
z_2	$z_2 = z - a$	$z_2 = z - a$
z_3	$z_3 = z - b$	$z_3 = -z + b$
z_4	$z_4 = z + a$	$z_4 = z + a$
$z_{4'}$	–	$z_4 = 3z + b - 2a$

program (Fig. 52.2), with the specification of the component elements, according to its kinematic scheme (Fig. 52.1b).

The wheels of the planetary gearbox with double satellite have been modeled as individual components of the gears of which they are part with the help of the Design Accelerator module of the Autodesk Inventor Professional program [4].

The "assembly" of the wheels as spur gears, respectively of the other components of the planetary gearbox, in order to build the kinematic chain of the mechanism was done by properly defining the assembly constraints, between them, but also to the housing, considered fixed element. Gear mobility was achieved by introducing mo-tion type constraints, specifying the corresponding transmission ratios [4].

The coaxiality condition between the two internal gears, with 0 unit correction, was ensured by adopting modules with different values.

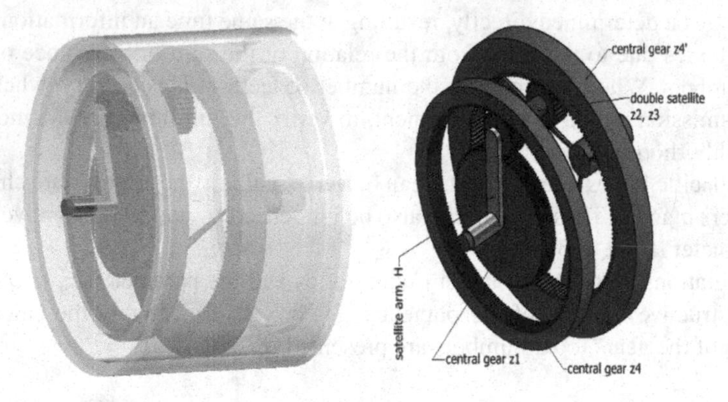

Fig. 52.2 3D model of the planetary gearbox

52.4 Dynamic Simulation of the Planetary Gearbox with Double Satellite, N-C-EI-I Model

The dynamic study of the planetary gear was performed with the Dynamic Simulation module, of the Autodesk Inventor Professional application, on the principle of the theory of multibody mechanical systems (MBS). In this situation the equations of motion are self-formulated by the software based on the geometric-elastic model of the mechanism and the restrictions in the movement of the elements [5].

Thus, real virtual prototypes can be created in order to obtain products that functionally meet the market requirements, by accurately modeling both the planetary gearbox components and its real operating conditions, which allows fast testing of many geometric—constructive variants, in order to optimize the mechanism [6].

In order to achieve the planetary gear dynamic study, a dynamic model was created, by defining the kinematic couplings, the driving motion, respectively the external loads, taking into account the mass of the component elements [7].

Mostly the kinematic couplings were modeled by self-converting the assembly constraints of the planetary gearbox components. In the case of the external gear $z_1 - z_2$, respectively internal gear $z_2 - z_4$, (see Fig. 52.1b), being a double gearing of the gearwheel z_2, the rolling joints (IV class) were modeled manually by indicating the corresponding pitch circles (Fig. 52.3).

The driving (input) motion of the planetary gearbox was modeled as a rotational motion in the kinematic coupling: fixed element—central wheel z_1, it was defined as a linear variable angular velocity, from 0 to the maximum valuc 360 deg/s (see Fig. 52.5).

Similarly, the external load was modeled as a torque, linear variable from 0 to the maximum value, being applied to the driven element by the kinematic coupling: fixed element—central wheel $z_{4'}$. At the same time, the masses of the mobile components was taken into account.

Following the simulation of the planetary gear, defined over a time of 5 s and 250 steps, it was found a properly work, by distinguishing the movements of the components. As a results of the simulation, in the defined loading conditions, it is possible to visualize, in graphical form, the variations of some kinematic and kinetostatic parameters. For example, the variations of angular velocities in the kinematic couplings corresponding to the satellite arm—double satellite and the central wheel

Fig. 52.3 Rolling joint manual modeling

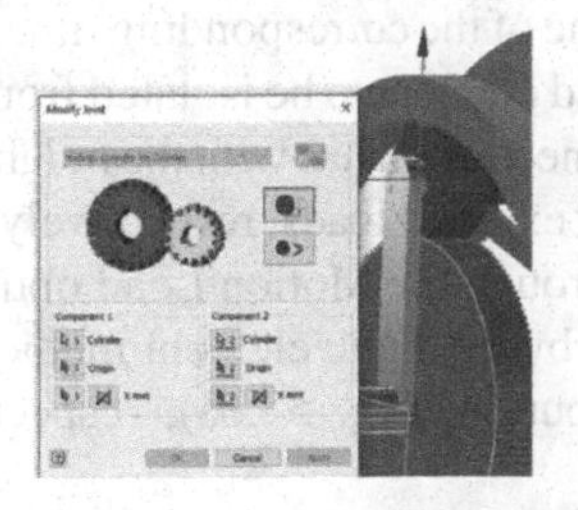

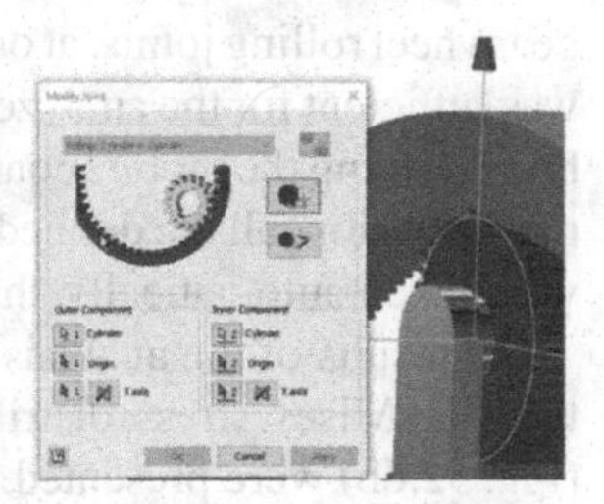

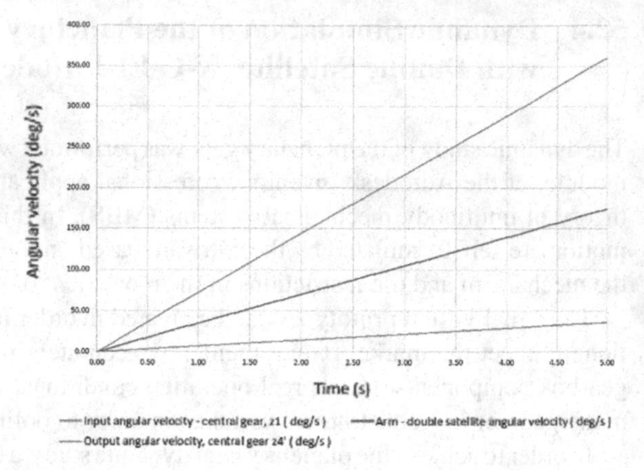

Fig. 52.4 Angular velocities variation in kinematic couplings

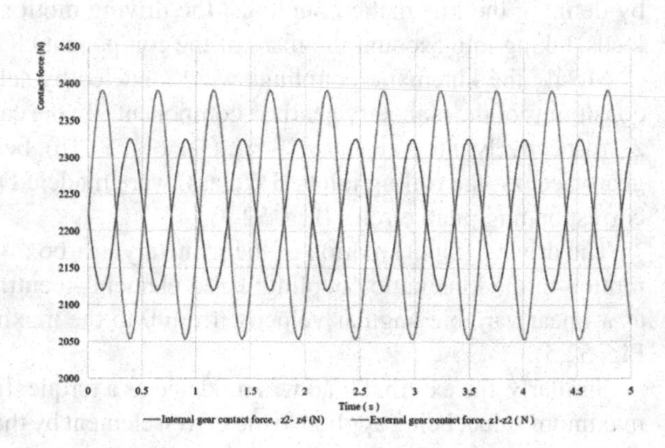

Fig. 52.5 Contact force variation at the gear, z_2

$z_{4'}$ (driven element)—fixed element (Fig. 52.4), respectively contact force variations in rolling joints, corresponding to the gear z_2 (Fig. 52.5) are presented.

Because the z_2 gearwheel (component of the double satellite) can be considered as the most stressed element due to the double gearing, with the gearwheels z_1 and z_4, an analysis by the finite element method was required, using the Stress Analysis module of Autodesk Inventor program.

Since that the finite element analysis was performed on the basis of the results obtained from the dynamic simulation, respectively maximum contact forces in the z_2 gearwheel rolling joints, at one of the corresponding time steps, t = 4.7 s (Fig. 52.5), it was sufficient for the analyzed element to be isolated from the rest of the mechanism, by specifying the mobile connections it has with the neighboring elements. Following this operation, all the defined external loads, respectively the resulting internal loads, were taken automatically, through the Motion Load option [8, 9].

As results of the analysis by the finite element method of the studied gearwheel, the von Mises stress distribution (Fig. 52.6a), respectively nodal displacements (Fig. 52.6b) were presented.

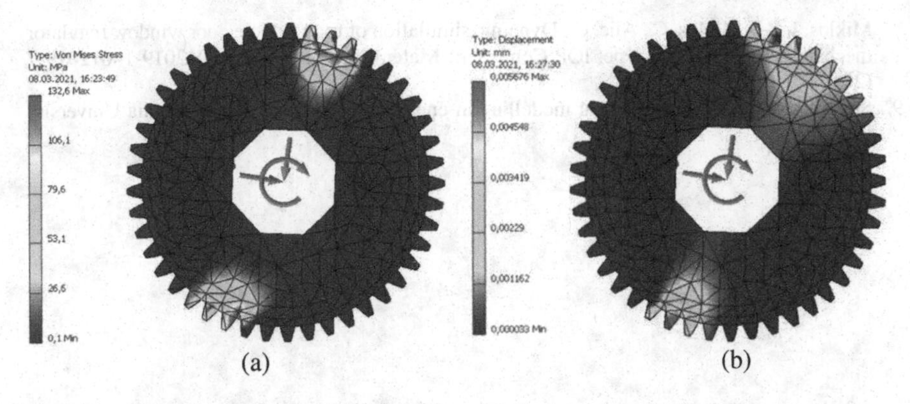

Fig. 52.6 **a** Von Mises stress distribution and **b** nodal displacements distribution

52.5 Conclusion

Dynamic simulation is a component of computer-aided design and is a method of analyzing a product in real operating conditions. It can be compared to a virtual prototyping based on a 3D model, but without the expenses necessary to create a real prototype. Following the dynamic simulation, results in the form of graphical variations of kinematic and kinetostatic sizes were obtained, based on which it was possible analyze by the finite element method of the planetary gear components. In the case of the z_2 gearwheel, the maximum values of tension and displacement are in the gear areas with the conjugate gearwheels, they being within the allowable limits for the chosen material, 132.6 MPa, respectively 0.0056 mm.

References

1. Miklos, I., Miklos, I. Zs., Cioată, V.: Proiectarea asistată de calculator a unui reductor planetar cu dublu satelit. In: Proceedings of the 7th National Symposium, PRASIC 2002, pp. 283–286. Brașov (2002)
2. Miloiu, Gh., Dudiță, F.: Transmisii mecanice moderne. Editura Tehnică, București (1971)
3. Vasu, T.A., Bularda, Gh.: Transmisii planetare cu roți dințate. Editura Tehnică, București (1970)
4. Dolga, L., Ravencu, M., Maci, C.A.: Parametric and feature based modelling with aplications in Catia and Inventor. Editura Politehnica, Timișoara (2004)
5. Alexandru, C.: Aspecte privind analiza mecanismelor ştergătoarelor de parbriz considerate sisteme multicorp. AUO-FMTE **V**(XV), 592–593 (2006)
6. Rusu-Anghel, S., Lihaciu, I., Rusu-Anghel, N.: State feedback control of a robotic arm. Ann. Fac. Eng. Hunedoara—Int. J. Eng. **13**(1), 75–82 (2015)
7. Autodesk Knowledge Network.: https://knowledge.autodesk.com/support/inventor. Last accessed 09 March 2021

8. Miklos, I. Zs., Miklos, C., Alic, C.: Dynamic simulation of road vehicle door window regulator mechanism of cross arm type. IOP Conf. Ser.: Mater. Sci. Eng. **163**(1), 012019-1–012019-7 (2017)
9. Spyrakos, C. Ch.: Finite element modelling in engineering practice. West Virginia University Press (1994)

Chapter 53
Design and Simulation of a Multi-pole, Multi-layer, Double-Sided Magnetorheological Brake

Aleksandar Poznić and Boris Stojić

Abstract Materials, whose rheological properties change reversibly under the influence of an external force, belong to a group called smart materials. This group includes fluids, greases, gels, polymers, etc. Materials whose rheological properties change under the influence of an external magnetic field are called magnetorheological materials. Magnetorheological grease e.g. is a type of material whose rheological properties also changes due to an external magnetic field influence. The main disadvantages of any magnetorheological system e.g. brakes and clutches, are insufficient torque and/or settling effect. The specific design of the device addresses the torque issue. Dealing with the magnetorheological material's settling effect, with the usage of magnetorheological greases, has been suggested. The main challenges in any torque base device are maintenance, response time, and braking torque. Increasing the magnetorheological device's torque potential by varying its unique design and the number of active surfaces in contact with magnetorheological material was the key aspect of this research. This paper is a special part of ongoing research that aims to increase the total magnetorheological brake torque. Research partially relies on the results gained from previous experiments and numerous simulations carried out using commercial finite element method software—COMSOL Multiphysics, AC/DC module. Materials' magnetic properties, required for the simulation process, were previously obtained by the experimental measurements or were obtained from the manufacturer and applied to the simulations. Post-processing was utilized to calculate the magnetic attributes across the models' specific cross-sectional areas. The proposed magnetorheological brake design shows great potential.

Keywords Multi-pole · Multi-layer · Magnetorheological brake · Grease · Simulation

---

A. Poznić (✉) · B. Stojić
Faculty of Technical Sciences, University of Novi Sad, Trg Dositeja Obradovića 6, 21000 Novi Sad, Serbia
e-mail: alpoznic@uns.ac.rs

M. Rackov et al. (eds.), *Machine and Industrial Design in Mechanical Engineering*, Mechanisms and Machine Science 109,
https://doi.org/10.1007/978-3-030-88465-9_53

53.1 Introduction

Magnetorheological—MR brake is a type of electromechanical brake that consists of a stator, rotor, working medium, and one or more excitation coils. Magnetorheological grease—MRG, is the working medium of the MR brake—MRB and is contained between active surfaces of the stator and the rotor. When excited, by the control current, each coil generates a magnetic field through MRB's body. Affected by the magnetic field, the MRG's viscosity changes [1, 2], thus generating the braking torque of the MRB. There are a number of MRB types [3]. Regardless of their construction differences, the direction of the magnetic flux density vector is their common feature. The magnetic flux density, magnetic flux, direction needs to be as perpendicular as possible to MRG's flow direction i.e. MRG's active surfaces and to form a closed magnetic loop.

From the magnetic and construction point of view, typical MRB is composed of MRG, as (ferromagnetic) working medium, nonmagnetic and magnetic materials. Magnetic properties of a MRGs' can easily be obtained from their manufacturers. Nonmagnetic materials, such as aluminum, have known magnetic properties. On the other hand, magnetic properties of magnetic material such as steel, usually are not known or are not available, and need to be determined by experimental measurements. The most important material's magnetic property, in this case, is the initial magnetization curve, which has a highly nonlinear characteristic [4].

The main drawback with any MRB's type and its application is still insufficient overall braking torque. Still, there are several ways to increase it. One of them is to use a working medium with better yield characteristics and to reduce the corresponding gap size inside the MRB. The second is to increase the applied magnetic flux acting on the working medium. The last one is to enhance the size of the MRB's active surface area by multiplying the number of its layers.

The primary goal of this work was to present progress on the Authors new multilayer MRB's design. New design utilizes a MRG as a working medium in FEM simulation. Layers are equidistant and double-sided. Particular stator and rotor element thicknesses and their influence on the magnetic flux distribution were analyzed using a commercial Finite Element Method—FEM software. The important aspect of the work was aspiration to form a uniform magnetic flux distribution to reduce unnecessary concentration points in MRB's ferromagnetic construction. Magnetic flux distribution was even more important in layers of the model. The intention was to distribute magnetic flux in such a manner so that all layers contribute equally to the overall braking torque value. Obtained magnetic flux values then can easily be converted into braking torque values for each layer.

53.2 New Design

The authors of this paper have focused on a hybrid MRB design, which combines known Disk [3] and Multi-pole [5] MRB type design. The design is a variation of the two aforementioned.

53.2.1 Innovation

In Fig. 53.1a, c, cross-section of MRB was presented, with emphasis on main, magnetically conductive elements, relative to one another. The same figure, but (b) and (d) illustrate these elements in pair and full shape i.e. on both sides of the brake. This, completely new design, relies partially on a previous design variant of the T-shape rotor MRB [6]. However, this variation of the T-rotor design was utilized to streamline the magnetic flux through magnetically conductive parts. Magnetic flux was generated by excitation coils, Fig. 53.2a, which are placed on both sides of the brake Fig. 53.2b. Overall magnetic flux spreads through the MRB, but was separated into two by magnetically nonconductive disk, Fig. 53.2c, d, and contained with magnetically nonconductive housing, Fig. 53.2e, f. This reduced magnetic flux dissipation and focused it on the active surface areas. More details in follow up text.

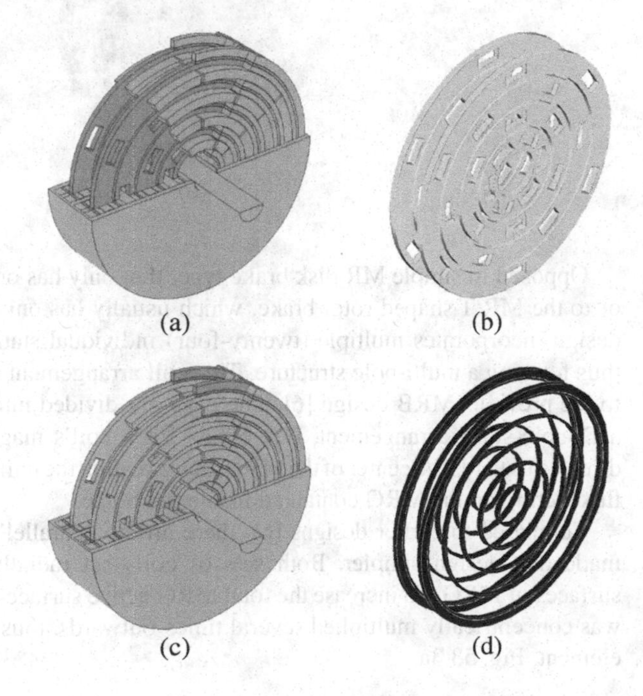

Fig. 53.1 Illustrations of the new multi-pole, multi-layer, double-sided magnetorheological brake, **a** segmental cross-section with emphasis on cores, **b** cores, **c** segmental cross-section with emphasis on magnetically conductive rings, **d** magnetically conductive rings

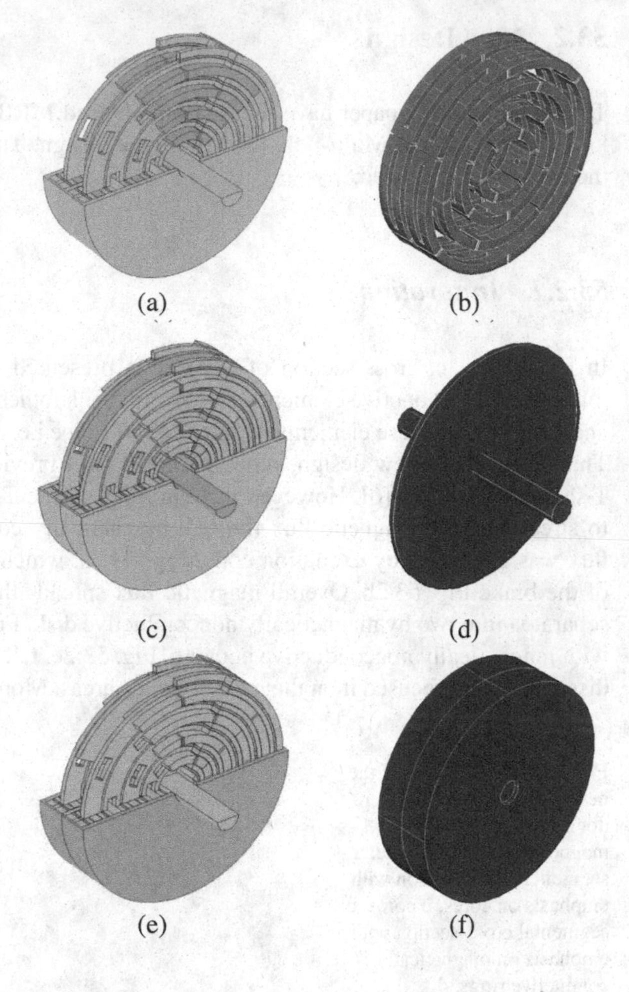

Fig. 53.2 Illustrations of the new multi-pole, multi-layer, double-sided magnetorheological brake, **a** segmental cross-section with emphasis on coils, **b** coils, **c** segmental cross-section with emphasis on nonmagnetic disk and shaft, **d** nonmagnetic disk and shaft, **e** segmental cross-section with emphasis on nonmagnetic housing, **f** nonmagnetic housing

Opposed to simple MR disk brake type, that only has one excitation coil (coils), or to the MR T-shaped rotor brake, which usually has only two separate coils, this design incorporates multiple (twenty-four) individual stationary coils, Fig. 53.2b, thus forming a multi-pole structure. This coil arrangement is to some extent similar to the previous MRB design [6]. The coils are divided into two sets, thus forming a Double-sided arrangement, Fig. 53.2b. Each coil's magnetic flux vector $-\vec{B}$ is directed towards the center of the MRB, thus creating the uniformly oriented magnetic flux, acting on the MRG contained inside the brake.

Compared to earlier designs [6], there are no "parallel" excitation coils, which made this design simpler. Both sets of coils act radially on the coaxial active surfaces of MRG. To increase the total MRG active surface area, the T-rotor element was concentrically multiplied several times outwards, thus forming a multi-T-rotor element, Fig. 53.3a.

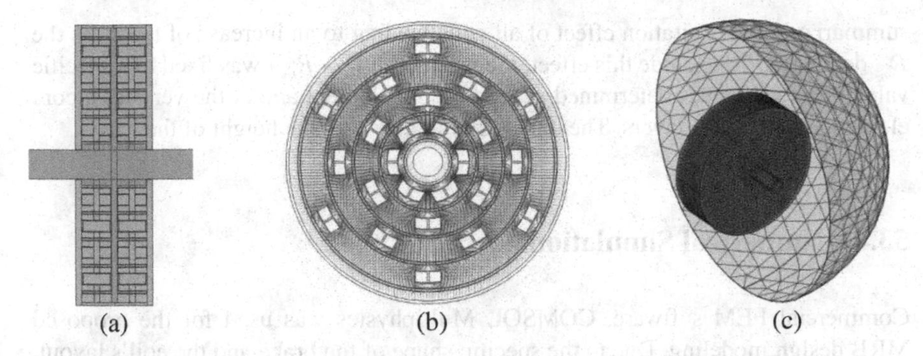

Fig. 53.3 Illustrations of **a** T-rotor element, **b** one side magnetic flux spreading, **c** the mesh

Proposed MRB multi-T-rotor assembly, i.e. shaft and multi-T element is comprised of nonmagnetic and magnetic elements. The nonmagnetic shaft also features a nonmagnetic disk, designated as multi-T-element inner support, Fig. 53.2c, d. The nonmagnetic disk magnetically separates the body of the MRB and in combination with nonmagnetic housing, Fig. 53.2f, effectively containing two magnetic fluxes each to its own side, Fig. 53.3b. Magnetic fluxes act uniformly onto separate but geometrically equal sides of the MRB active surfaces, i.e. MRG layers.

53.3 Mathematical Model

In the case of the multi-T-rotor brake design, the torque generating properties can be described by the same analytical model used for the MR Drum brake model with the results adjustment for additional MRG layers and their specific radiuses and heights for both sides. The maximum field-induced torque, for MR drum brake, is given by:

$$T_\tau = \sum_{1}^{k} 2 \cdot R_{O_k} \cdot \left(2 \cdot \pi \cdot R_{O_k} \cdot l_k\right) \cdot \tau_y = \sum_{1}^{k} 4 \cdot \pi \cdot \tau_y \cdot \left(l_k \cdot R_{O_k}^2\right) \tag{53.1}$$

Similarly, the maximum viscous torque is:

$$T_\eta = \sum_{1}^{k} 4 \cdot \pi \cdot \eta \cdot \frac{\dot{\theta}}{g} \cdot \left(R_{O_k}^3 \cdot l_k\right) \tag{53.2}$$

where k is the number of MRG layers, R_{O_k} is a radius of a specific MRG layer, l_k is the MRG layer's height, τ_y is the yield stress developed in response to the applied magnetic field, η is the viscosity of the MRG with no applied magnetic field, $\dot{\theta}$ is the angular velocity of the rotor and g is the thickness of the MRG gap. It was noticeable that the overall intensity of the $\vec{B}$ increases as it progresses toward inner MRG layers, summarizing the excitation effect of all coils leading to an increase of the τ_y as the R_{o_k} decreases. To exclude this effect, the product of $(l_k \cdot R_{o_k}^2)$ was fixed to a specific value. This value was determined at the far core element and at the very least core element of the MRG layers. The only variable here was the height of the MRG.

53.4 Numerical Simulation

Commercial FEM software, COMSOL Multiphysics was used for the proposed MRB design modeling. Due to the specific shape of the brake and the coil's layout, COMSOL's 3D space dimension option was utilized. The magnetic field was considered to be static, so the Stationary Study was used. To reduce computation power required the model could be cut along the shaft's axis and then split into two across the disk's middle line.

The entire model was surrounded by a sphere-shaped air boundary, several times the volume of the model. Appropriate material *nodes* and *boundary* conditions were assigned to every element of the model. In this specific simulation, materials such as nonmagnetic air and aluminum were selected from *COMSOL*'s database, but nonlinear magnetic materials, such as construction steel and MRG, were defined using previously obtained data. These data have been loaded to the *COMSOL* as separate files. Note—the influence of the elements like ball bearings was not taken into account due to their composition and volume share in the overall construction. In the *Magnetic Fields* subsection, additional *Ampère's Laws* were added.

Model's mesh, Fig. 53.3c, was generated using the *User-controlled mash*. The MRG's layers were meshed using the *Free tetrahedral* with custom element size. The minimum tetrahedral element size was 0.05 mm. Also, special attention was placed on the curvatures and the narrow regions of MRG segments of the brake. The curvature radii were multiplied by the Curvature factor parameter which in return gives the maximum allowed element size along a specific boundary. The Resolution of narrow regions parameter controls the number of the element created in narrow regions. These parameters greatly improved the mash quality of the models, which is now at the threshold of 0.1, which is considered a satisfactory mesh. The solver was stationary but nonlinear.

To determine the overall magnetic flux intensity in a specific MRG layer FEM simulation was carried out. A median magnetic flux value, for each MRG layer, was determined along predetermined circular lines. A *1D Plot Group* line graphs were used to depicture magnetic flux magnitude changes along these circular lines. Circular lines were positioned at the very center of the MRG layers.

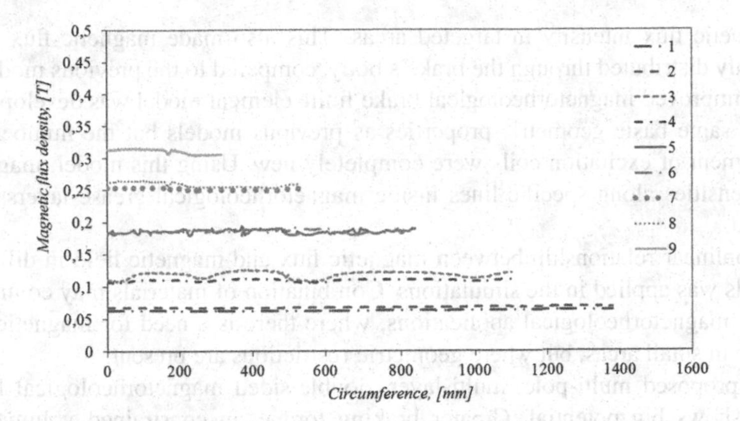

Fig. 53.4 Change of magnetic flux intensity for each magnetorheological grease layer

53.5 Results and Discussion

53.5.1 Magnetic Flux Distribution

The magnetic flux distribution pattern within the proposed MRB was studied and part of the results was presented in this paper. Average magnetic flux intensity changes in particular MRG layers are graphically presented in Fig. 53.4. These values were determined along single circular lines for each MRG layer. Minimum and maximum values of magnetic flux, along these lines, ranged from around 0.06 T up to 0.32 T across all MRG layers. Good results repeatability is achieved especially in the outer MRG layers area. Inconsistencies in magnetic flux results are due to exciting coils periodical arrangement, the specific shape of the MRG layers, and coarse mesh in boundary areas between MRG layers and the rest of MRB construction. Future work will include much finer mesh but it will be at the expense of computational requirements.

Linearity in magnetic flux change through MR layers is noticeable. This may lead to consistency in the field-induced torque values among MRG layers. This was predicted and considered in an early stage of the MRB design.

53.6 Conclusion

This study represents a part of a continued effort in magnetorheological brake design improvement. A completely new magnetorheological brake design was presented. By combining magnetic and nonmagnetic materials, in a specific manner, a new space may have been opened for magnetorheological technology. The use of nonmagnetic material in combination with magnetic ones of a specific shape leads to an increase in magnetic flux intensity in targeted areas. This also made magnetic flux more uniformly distributed through the brake's body, compared to the previous models.

The improved magnetorheological brake finite element model was developed. It had the same basic geometric properties as previous models but the number and arrangement of excitation coils were completely new. Using this model, magnetic flux intensities along specific lines inside magnetorheological grease layers were obtained.

A nonlinear relationship between magnetic flux and magnetic field in different materials was applied in the simulations. Combination of materials may contribute to other magnetorheological applications, where there is a need for magnetic flux increase in small areas, but where geometric restrictions are present.

The proposed multi-pole, multi-layer, double-sided magnetorheological brake design shows big potential. Greater braking torque, in constrained volume and weight, is now achievable, (2390 Nm/0.0153 m^3; 2390 Nm/125 kg).[1] Future work should be focused on scaling the magnetic flux intensity through magnetorheological grease layers which was not the case now.

Acknowledgements This paper has been supported by the Ministry of Education, Science and Technological Development through the project no. 451-03-68/2020-14/200156: "Innovative scientific and artistic research from the FTS domain".

References

1. Raju, A., Choi, S., Ferdaus, M.: A state of art on magneto-rheological materials and their potential applications. J. Intel. Mat. Syst. Str. **29**(10), 2051–2095 (2018)
2. Mohamad, N., Mazlan, S.A., Ubaidillah, Choi, S.-B., Nordin, M.F.M.: The field-dependent rheological properties of magnetorheological grease based on carbonyl-iron-particles. Smart Mater. Struct. **25**(9), 095043-1–095043-10 (2016)
3. Avraam, M.T.: MR-fluid brake design and its application to a portable muscular rehabilitation device. Ph.D. Thesis, Active Structures Laboratory Department of Mechanical Engineering and Robotics, Bruxelles (2009)
4. Poznić, A., Miloradović, D., Zelić, A.: Determination of magnetic characteristics of some steels suitable for magnetorheological brake construction. In: Proceedings of the 3rd International Conference and Workshop Mechatronics in Practice and Education, MECHEDU 2015, pp 130–133. Subotica (2015)
5. Yaojung, S., Quang-Anh, N.: Development of a multi-pole magnetorheological brake. Smart Mater. Struct. **22**(6), 065008-1–065008-13 (2013)
6. Poznić, A., Stojić, B.: A contribution to the development of automotive magnetorheological brake. In: Proceedings of the 8th International Congress Motor Vehicles and Motors 2020, pp 1–8. Kragujevac (2020)

[1] Rounded FEM simulation values.

Chapter 54
Comparative Analysis of Rotation Welding Positioners Based on Friction Forces

Marija Matejic, Lozica Ivanovic, and Milos Matejic

Abstract In this paper a two solution of rotation welding positioners are presented. The presented welding positioner can be used for a various dimension of workpieces. First of the presented solutions is a conventional one. As the conventional solution it can be purchased at market as machine. A concept of second solution is presented in one of the authors previous papers. This paper presents way of functioning for both solutions. Also, in this paper is presented a friction forces determination, and its dependence of the workpiece position. A detailed mathematical model for friction forces are presented and for one size, the biggest one allowed to be on device, the reaction and friction forces are calculated. The conducted analysis is used for comparison between friction forces in conventional and new concept design. The paper concludes with clearly pointed advantages and disadvantages of the proposed new concept related to the conventional solution. At the very end the future research plans and directions are given.

Keywords Friction force · Welding positioner · Ship winches

54.1 Introduction

Welding presents a process of jointing different workpieces made occasionally of same material. Mainly the welded materials are metals and their alloys. In last decade welding expands in term that it can be applied for non-metal materials as thermoplastics etc. In modern machine industry the welding is one of the essential manufacturing processes. In the steel designed machines, such as processing equipment, transporters, mines equipment and shipping equipment the must use processes are MIG/MAG welding processes [1, 2].

M. Matejic
Faculty of Technical Sciences, University of Pristina With Temporary Settled in Kosovska Mitrovica, Knjaza Milosa 1, 38220 Kosovska Mitrovica, Serbia

L. Ivanovic · M. Matejic (✉)
Faculty of Engineering, University of Kragujevac, Sestre Janjic 6, 34000 Kragujevac, Serbia
e-mail: mmatejic@kg.ac.rs

M. Rackov et al. (eds.), *Machine and Industrial Design in Mechanical Engineering*, Mechanisms and Machine Science 109,
https://doi.org/10.1007/978-3-030-88465-9_54

For jointing two or more metal workpieces using welding techniques the essential need is heat [3]. But if the heat is occurred more than is necessary on places, it can be counterproductive. The non-uniform heat, especially while welding a big length can cause workpiece failure during the exploitation. For that purposes, welding automation, especially for bigger workpieces is required. The welding positioners are very helpful in avoiding unwanted effects during the welding processes. Welding positioners keeps in place the elements while being welded and providing relative, progressive and smooth movement between the welding head and joint to be welded [4]. In the present times, where the mass production becomes default, it is often required to automate the manufacturing processes that were conventionally conducted manually. While using the rotation positioner, the welding gun is fixed and the workpiece rotate continuously. The workpiece is rotating by using of electric motor and worm and worm wheel arrangement [5]. The research described in this paper is related to the problems of friction force that occurs between the workpiece and the rotating workpieces. Torque transmission is done by friction contact with rubber covered wheels and metal workpieces. Previous investigations on this topic of friction forces that occur between the contact elements in rotary positioners for welding has been insufficiently researched. Jeremić et al. [6] analyzed the impact of the construction element position of ship winch drum on the effects of torque transmission by friction in the mechanization welding process. In their study, a new design of a rotary positioner for welding was proposed and an analysis of the friction forces that occur between the workpiece and the rotating elements was performed. Research is based on determining the friction force between rubber covered wheels and metal in static condition. Static condition is considered because of the small rotational speeds.

Static friction coefficient depends of many parameters, especially from the contact surface, normal force, and temperature of the atmosphere in which contact occurs, surface absorption, quality of contact surface materials [7–11]. Persson and Volokitin [12] have developed a model which considers the thermal fluctuations influence on the depending of little contact patches (stress domains) at the rubber-substrate interface. The theory of those authors predicts that the macroscopic shear stress velocity dependence has a bell-shaped form. That paper shows that the low-velocity side exhibits has the same temperature dependence as the bulk viscoelastic modulus, in qualitative agreement with experimental data. Shanahan et al. [13] investigated the mechanism of adhesion that occurs in the contact pair between rubber and hard metal rolling bodies. If the contact time and pressure reach sufficient values, high adhesion can be apparent even at room temperature. The influence that accompanies hysteresis adhesive separation is based on obtained results and that energy was determined, which is dissipated during rolling, refers not only to, but also to the losses caused by loading with large cylinder.

In order to optimize the design of welding positioner it was performed large number of theoretical considerations and preliminary idea was done for a detailed review and analysis of the literature that investigates this issue. Research is based on determining the transmission of rotating movement from the welding positioner to the winch drum. The rotating movement is conducted by friction between rubber covered wheel and metal workpiece. In the paper is presented two solutions: conventional and new one. Both are analyzed from friction forces aspects. The paper concludes with their comparative analyses and future direction of research in this field.

54.2 Conceptual Design of Rotation Welding Positioner

The costs is a very important aspect in implementation of any automated systems into the manufacturing process. Due to this statement, the design of the production items should be analysed before the automation step in production is even considered. This paper is related to automation process of drums production for ship winches in factory *RAPP Zastava*. The CAD model of the considered winch drum is shown in Fig. 54.1.

These drums has a very strict quality control requirements, because they are the main part of the winch. Usually, the drums are carrying very expensive equipment for fishing, recording ocean life or for ocean research purposes. Due to given reasons the production must be without any fault, especially in welding joints. The strict rules in winch production justifies the certain level of automation process. The problem is that, the Factory production line has a large number of different winches, which leads to large number of winch drums for various purposes. The overriding of that problem is a modular welding automation equipment.

At market there are numerous solutions for welding automation process. One of the solutions, called welding rotary positioner or turning rolls, is shown in Fig. 54.2. The shown solution is realized by two electric gear motors for the workpiece driving and one motor with screw shaft for positioning in order to receive variety of workpiece dimensions. The principle of operation of the electric motor which drives workpieces requires ensuring of movement synchronization. A movement synchronization with enabling of rotating speed variation is a serious theoretical and practical problem.

The movement problem with two driving units becomes even harder when the movement is transmitted by friction. One of the problems of providing synchronized movement is the possible difference of phases of propulsion motors. The described

Fig. 54.1 Drum of the ship winch

Fig. 54.2 Rotating positioner [14]

problems lead to the significant cost increase in building and purchasing devices which is given in Fig. 54.2. The more acceptable solution can be that driving unit should be powered with one single gear motor. The concept of the presented idea is given in Fig. 54.3.

All partial functions of the presented conceptual solution executes the main function which is achieving of almost constant rotation speed of the workpiece which is winch drum. The presented idea of welding positioner consists of machine parts, sub-assemblies, subgroups and groups linked to a functional unit. A pair of flexible coupling with shaft transmits the torque and power from the engine and one power wheel another power wheel which would be behind the shown projection in Fig. 54.3. Driven wheels rotating function performed under the influence of rotation of the winch drum. Frames design would rely frames which should be used as distance pieces [6].

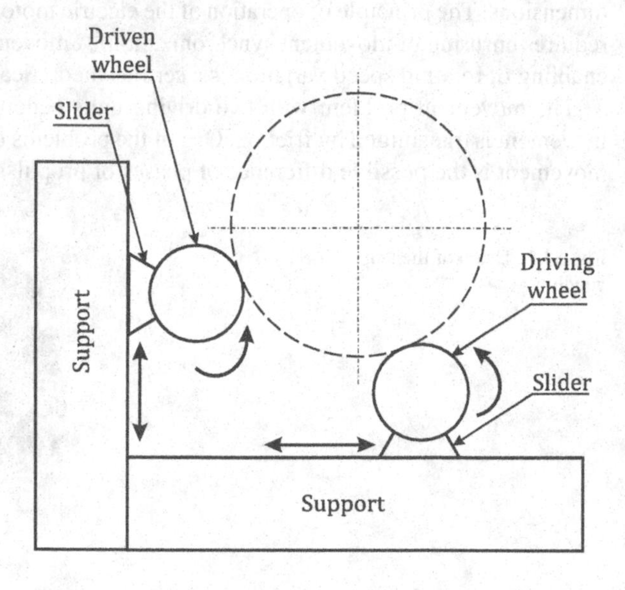

Fig. 54.3 New conceptual design of the welding rotary positioner

54.3 Analysis of Transmission Forces on Conventional Concept

The conventional welding positioner has equal, or almost equal friction forces in the contact of the drum pipe and driving units. The force analysis for conventional welding positioner is shown in Fig. 54.4.

The Fig. 54.4 shows workpiece position variation as well. The normal forces N_1 and N_2 are equal because the conventional positioner is a symmetric solution. Solving the equations for normal forces is given by expressions (54.1):

$$N_1 = N_2 = \frac{mg}{2\cos\varphi + \mu_1 \cos(90^\circ - \varphi) + \mu_2 \cos(90^\circ - \varphi)}$$
$$N_1 = N_2 = \frac{mg}{2\cos\varphi + \sin\varphi(\mu_1 + \mu_2)} \tag{54.1}$$

where are: N_1 and N_2 normal reaction forces (N), m workpiece mass (kg), μ_1 and μ_1 friction coefficients between the workpiece and driving wheel 1 and 2 respectively.

Friction forces which are occurring on convectional welding positioner solution are given as Eqs. (54.2) and (54.3):

$$F_{t1} = N\mu_1 \tag{54.2}$$

$$F_{t2} = N\mu_2 \tag{54.3}$$

where are: F_{t1} friction force on driving wheel 1 (N) and F_{t2} friction force on driving wheel 2 (N).

The analysis of friction forces took into consideration specific example of the winch drum with diameter $D = 1000$ mm and length $l = 2000$ mm, which loads the

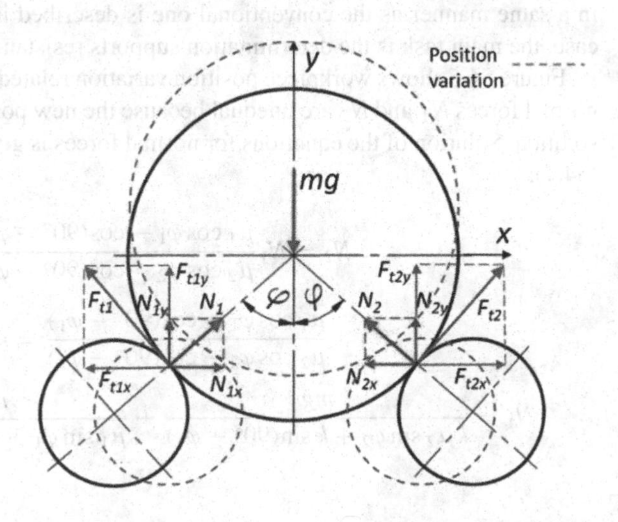

Fig. 54.4 The friction forces analysis for conventional rotary welding positioner concept

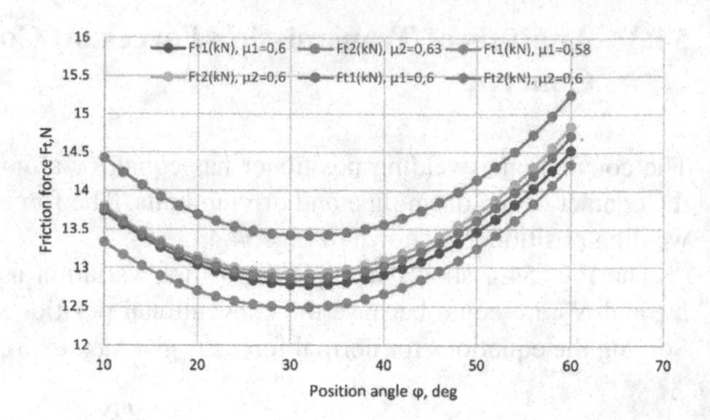

Fig. 54.5 Results of friction forces analysis for various positions and friction coefficients

pair of wheels shown in Fig. 54.4 with load of $mg = 50{,}000$ N. The different positions must be used into considerations as well as different friction coefficients because of the wear of driving wheels. The wheels theoretically should have a uniform wear, which is not a case in practice. Figure 54.5 shows the diagram with different angle (φ) and different pairs of friction coefficient.

The driving wheels are covered with rubber. Friction coefficient between rubber and steel goes from $\mu = 0.58$ to $\mu = 0.63$, [15]. Figure 54.5 shows 3 pair of forces with friction coefficients $\mu_1 = 0.60$ and $\mu_2 = 0.60$, $\mu_1 = 0.6$ and $\mu_2 = 0.63$ and $\mu_1 = 0.58$ and $\mu_2 = 0.60$.

54.4 Analysis of Transmission Forces on New Concept

The new concept of welding positioner has unequal, friction forces in the contact of the drum pipe, driving unit and driven unit. The force analysis for the new concept of welding positioner is shown in Fig. 54.6. The design is considered as planar problem in a same manner as the conventional one is described in previous chapter. In this case, the main task is the determination supports resistance and friction forces.

Figure 54.6 shows workpiece position variation related to the marked angles. The normal forces N_1 and N_2 are unequal because the new positioner is not a symmetric solution. Solution of the equations for normal forces is given by expressions (54.4)–(54.7):

$$N_2 = N_1 \frac{\mu_1 \cos\varphi_1 - \cos(90^\circ - \varphi_1)}{\mu_2 \cos\varphi_2 - \cos(90^\circ - \varphi_2)} \tag{54.4}$$

$$\frac{\mu_1 \cos\varphi_1 - \cos(90^\circ - \varphi_1)}{\mu_2 \cos\varphi_2 - \cos(90^\circ - \varphi_2)} = k \tag{54.5}$$

$$N_1 = \frac{mg}{k\mu_2 \sin\varphi_2 + k\sin(90^\circ - \varphi_2)} + \frac{mg}{\mu_1 \sin\varphi_1 + \sin(90^\circ - \varphi_2)} \tag{54.6}$$

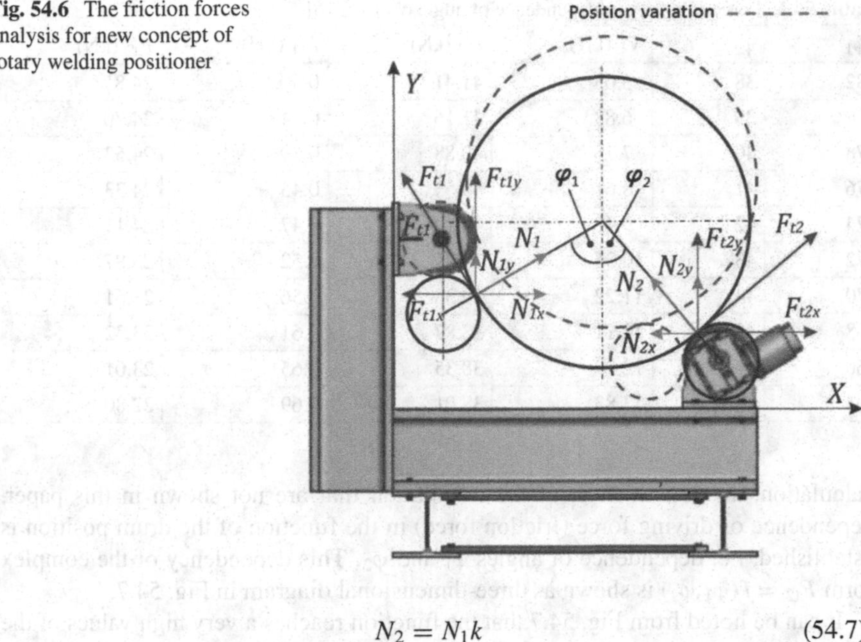

Fig. 54.6 The friction forces analysis for new concept of rotary welding positioner

$$N_2 = N_1 k \tag{54.7}$$

where are: N_1 and N_2 normal reaction forces (N), *m* workpiece mass (kg), μ_1 and μ_1 friction coefficients between the workpiece and driving wheel 1 and 2 respectively and *k* is unitless coefficient introduced for easier equation manipulation and simulation.

The problem is solved by introducing the normal force reaction on wheels N_1 and N_2, in the same way like it is done for conventional design concept. The analyses is based on the effect of the friction forces that appear between the drum and wheels. The friction coefficient for driving wheel has a value $\mu_2 = 0.60$, and the value of coefficient of rolling friction between the driven wheels and drums $\mu_1 = 0.05$, because it is unloaded and it serves just to hold the workpiece in process of welding [6]. The friction forces are calculated with same expressions as on conventional concept. The expressions for friction forces are given by (54.8) and (54.9):

$$F_{t1} = N_1 \mu_1 \tag{54.8}$$

$$F_{t2} = N_2 \mu_2 \tag{54.9}$$

The force intensity in dependence of angle changes, i.e. depending on the position of the drum, and with the same diameter and weight of the drum, is calculated given in Table 54.1.

Based on analysis of obtained values of the normal forces and frictional force from Table 54.1, it can be noted that for settings in any position, the highest load has a force N_2, i.e. maximum load is on the drive wheel. Based on comparative analysis

Table 54.1 Force intensity in dependence of angle changes [6]

φ_1	φ_2	N_1 (kN)	N_2 (kN)	F_{t1} (kN)	F_{t2} (kN)
82	38	6.03	41.41	0.30	24.85
80	39	6.88	41.16	0.34	24.70
78	40	7.75	40.88	0.39	24.53
76	41	8.61	40.55	0.43	24.33
74	42	9.48	40.19	0.47	24.11
72	43	10.35	39.79	0.52	23.87
70	44	11.22	39.35	0.56	23.61
68	45	12.11	38.87	0.61	23.32
66	46	12.99	38.35	0.65	23.01
65	47	13.83	38.01	0.69	22.80

calculation, as well as most of the calculation that are not shown in this paper, dependence of driving force (friction force) in the function of the drum position is established, i.e. dependence of angles φ_1 and φ_2. This dependency of the complex form $F_{t2} = f(\varphi_1,\varphi_2)$ is shown as three-dimensional diagram in Fig. 54.7.

It can be noted from Fig. 54.7 that the function reaches a very high values of the force intensity around angles φ_1 and φ_2 which are close to 90°. It is obvious that this is a so-called wedge effect, which occurs at extremely high intensity of tangential force. Ostensibly, this "extreme function" can be good solution from the driving

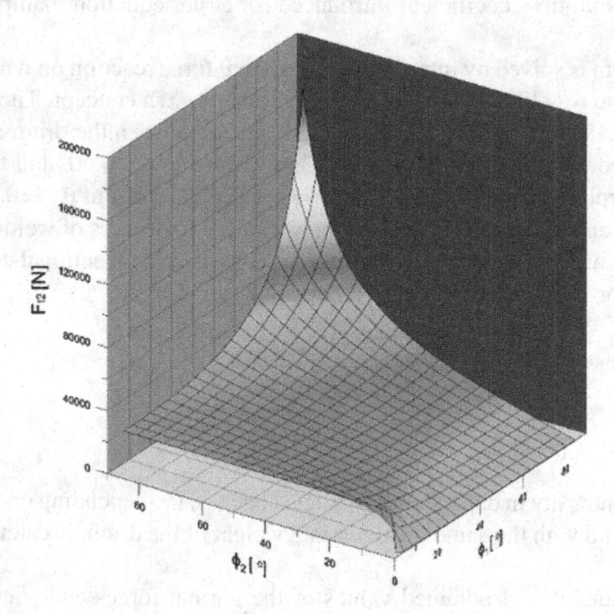

Fig. 54.7 Friction force diagram for driving wheel in dependence of position angles [6]

forces aspect. However, this maximum force can cause unallowable stress on vital welding positioner parts. For the explained reason, the authors find that the angles of φ_1 and φ_2 should be chosen in a range with much lower values, with which will not cause intensive wear and stability of device will be kept.

54.5 Conclusion

From the literature sources related to the considered problem, it can be concluded that the area of presented research is very complex. This paper presents two concepts of welding positioners. First one is the conventional and it can be purchased on market. Second one is developed in order to improve the winch drum manufacturing process by introducing the new approach to this problem. Theoretical considerations was made before the design is fully developed. The proposed solution does not require synchronization device for movement of two drive motors, which have the conventional one. The introduction of new design of welding positioner into the production process has many advantages as follows:

- Improving the quality of welded joints is achieved thereby enabling semiautomatic welding. During rotation of the positioner, rotation is continuous and it is not conditioned by equal wear of the driving wheels as on the conventional positioner. This is achieved by selecting one wheel per plane to be a driving one.
- Friction force in new concept is increased by 44%, which ensures that rotating will be smooth and constant.
- For the first concept reaction forces are equal, while at new concept reaction force on driving wheel is multiple times bigger than reaction on the driven wheel.
- Reduced the preparatory time during welding. This is the same in both concepts.
- Increase in profit resulting from reduced preparation time. Plus in profit for the new concept is that new concept is significantly lower in price.
- Technical features of the device are such that it is possible to perform the procedure on the positioner gas cutting, as well as welding. This is the same for both concepts.

The conventional concept has only one advantage related to the new concept. That advantage is a bit easer usage in term when different drum size is inserted on positioner. This is negligible when the series of similar drums is performed, but it can be a problem if the different sizes are required during a single day. A very important advantage of this design is that the new concept of welding positioner can be made with a lot of low cost elements. With closer look at new design, it can be noted that new design do not require a particularly high accuracy and precision manufacturing. It was tended to meet function during design of device, to be easy for manufacture, to be easy for manipulation, to have lowest cost possible and to satisfy safety requirements.

The next step of this research will be manufacturing and practical testing of the new device concept. The device prototype is planned to be with much smaller scale than it is described in this paper and it will be used just to proving the presented concept.

Acknowledgements This research was funded by the Ministry of Education and Science of the Republic of Serbia under the contracts TR33015 and TR35033.

References

1. Miletić, I., Ilić, A., Nikolić, R., Ulewicz, R., Ivanović, L., Sczygiol, N.: Analysis of selected properties of welded joints of the HSLA steels. Materials **13**(6), 1–12 (2020)
2. Ilić, A., Miletić, I., Nikolić, R., Marjanović, V., Ulewicz, R., Stojanović, B., Ivanović, L.: Analysis of influence of the welding procedure on impact toughness of welded joints of the high-strength low-alloyed steels. Appl. Sci. Basel **10**(7), 1–14 (2020)
3. Khanna, O.P.: Welding Technology. Dhanapat Rai Publication (1999)
4. Totala, N.B., Bhutada, S.S., Katruwar, N.R., Rai, R.R., Dhumke, K.N.: Design, manufacturing and testing of circular welding positioner. IJERD **10**(2), 8–15 (2014)
5. Badgujar, P., Gunjal, S., Hatte, P., Chopade, R., Bangar, V.: Welding rotator with proximity sensor. Int. J. Sci. Technol. Manag. **4**(3), 65–75 (2015)
6. Jeremić, M., Bogdanović, B., Tadić, B., Miloradović, D.: Analyzing the influence of the construction element position on torque transmission by friction. Tribol. Ind. **36**(3), 300–307 (2004)
7. Zum Gahr, K. Voelker, K.: Friction and wear of SiC fiber-reinforced borosilicate glass mated to steel. Wear 225–229(Part 2), 885–895 (1999)
8. Blau, P.: The significance and use of the friction coefficient. Tribol. Int. **34**(9), 585–591 (2001)
9. Ivkovic, B., Djurdjanovic, M., Stamenkovic, D.: The influence of the contact surface roughness on the static friction coefficient. Tribol. Ind. **22**(3), 41–44 (2000)
10. Muller, U., Hauert, R.: Investigations of the coefficient of static friction diamond-like carbon films. Surf. Coat. Tech. **174–175**(1), 421–426 (2003)
11. Polyakov, B., Vlassov, S., Dorogin, L., Kulis, P., Kink, I., Lohmus, R.: The effect of substrate roughness on the static friction of CuO nanowires. Surf. Sci. **606**(17–18), 1393–1399 (2012)
12. Persson, B., Volokitin, A.: Rubber friction on smooth surfaces. Eur. Phys. J. E **21**(1), 69–80 (2006)
13. Shanahan, M., Zaghzi, N., Schultz, J., Carré, A.: Hard rubber/metal adhesion assessment using a heavy cylinder rolling test. Adhesion **12**, 223–238 (1988)
14. https://www.ljwelding.com/, last accessed 2021/03/17
15. Fuller, D.D.: Theory and Practice of Lubrication for Engineers. John Wiley and Sons, New York (1984)

Chapter 55
Wear Analysis of IC Engine Crankshaft Bearings Depending on the Connecting Rod Length

Nebojša Nikolić, Jovan Dorić, and Dalibor Feher

Abstract Considering the crankshaft of an internal combustion (IC) engine to be one of the most loaded components, forces acting on its bearings are very important to be known. Gas forces dominantly affect load of the bearings but inertia forces are not negligible, too. Influence of the connecting rod length on wear of the crankshaft main bearings is studied and analyzed in this paper. To that end, a previously developed computer software was used for theoretical wear diagrams of all main bearings to be obtained. The procedure has been conducted on a six-cylinder engine, for two different values of the connecting rod length. Based on the obtained results a short analysis has been done in order for some general conclusions to be reached.

Keywords IC engine · Crankshaft bearings · Wear · Connecting rod length

55.1 Introduction

In majority of internal combustion engines, forces acting between a crankshaft and its bearings are the most influential factor considering wear of the bearings. The wear phenomenon is very difficult to understand and describe, especially if all the influencing factors (the geometry and temperature of the contact, the physical and chemical properties of the contact materials and so on) are to be taken into account.

Recently, many papers have been published that show the great importance of wear in mechanical systems [1–6]. In these papers [1–6], the authors developed models for the wear depth calculation, in order to predict the dimensions of the worn-out surface after a period of time or after a number of operating cycles. However, when it comes to the IC engine crankshaft bearings, it is not so important to predict the depth of worn-out material, but to recognize most jeopardized bearings in terms of wear. Also, it is important to find the area on the bearing surface that is most exposed to wear. For that reason, Nikolic et al. [7] proposed an algorithm for constructing a theoretical wear diagram, which can be used to find the bearing of a crankshaft

N. Nikolić (✉) · J. Dorić · D. Feher
Faculty of Technical Sciences, University of Novi Sad, Novi Sad, Serbia
e-mail: nebnik@uns.ac.rs

M. Rackov et al. (eds.), *Machine and Industrial Design in Mechanical Engineering*, Mechanisms and Machine Science 109,
https://doi.org/10.1007/978-3-030-88465-9_55

and the area on the bearing surface that are most exposed to wear. The theoretical wear diagram is named after its shape that looks like a profile of a worn-out bearing [8]. The algorithm is based on the assumption that the journal-bearing contact is not lubricated. This assumption is valid to the theoretical case, when journal and bearing are in direct contact, but it can also refer to the critical operating modes of IC engines, when lubrication is missing or insufficient. During the research described in reference [7], a computer software named "Wear Diagrams" was developed, which enables the theoretical wear diagrams of all the crankshaft bearings of an engine to be generated automatically, for given input data. The possibilities offered by this software have been used to investigate how the connecting rod length affects the wear of the crankshaft bearings, and the results of the research are shown in this paper.

55.2 A Model of Bearing Wear Calculation

The software named "Wear Diagrams" is an upgrade of the previously developed "Polar Diagrams" software, and the two are presented in references [9] and [10]. The "Wear Diagrams" software is based on the procedure for constructing theoretical wear diagrams of IC engine main bearings, which is described in reference [7]. An interested reader can there find a detailed presentation of the developed algorithm and a complete mathematical model as well. According to the model, a crankshaft was considered to be a statically indeterminate beam. Furthermore, the load distribution on the journal-bearing contact surface was assumed to be elliptical.

Figure 55.1 shows a simplified drawing of the crankshaft of an IC engine with (n + 1) main bearings and the forces acting on them. The OXY coordinate system is stationary and the OX1Y1 coordinate system rotates together with the crankshaft at an angular velocity ω. The position of the OX1Y1 in relation to the OXY coordinate system is defined by the angle φ. All the forces in Fig. 55.1 depend on that angle.

Fc_i(i = 1, ..., n) denotes the force originating from the crank "i", the action of which is distributed to all the main bearings and not only to the adjacent ones. This force is calculated taking into account the gas and inertia forces, which are dominant in comparison to all the other forces acting in the crank train. Gas forces can be determined either by measuring the pressure in the combustion chamber or by mathematical modelling using one of the known methods. Inertia forces are easily determined using Newton's second law of motion, assuming that the angular velocity of the crankshaft is constant. Hence, the forces Fc_i can be treated as input data.

Reference [7] shows that each force $Fb_j(\phi)$ acting on the main bearing b_j (j = 1,..., n + 1) can be decomposed into the X and Y directions as follows:

$$\begin{aligned} FbX_j(\varphi) &= FbX1_j(\varphi) \cdot \cos\varphi + FbY1_j(\varphi) \cdot \sin\varphi \\ FbY_j(\varphi) &= -FbX1_j(\varphi) \cdot \sin\varphi + FbY1_j(\varphi) \cdot \cos\varphi \end{aligned} \tag{55.1}$$

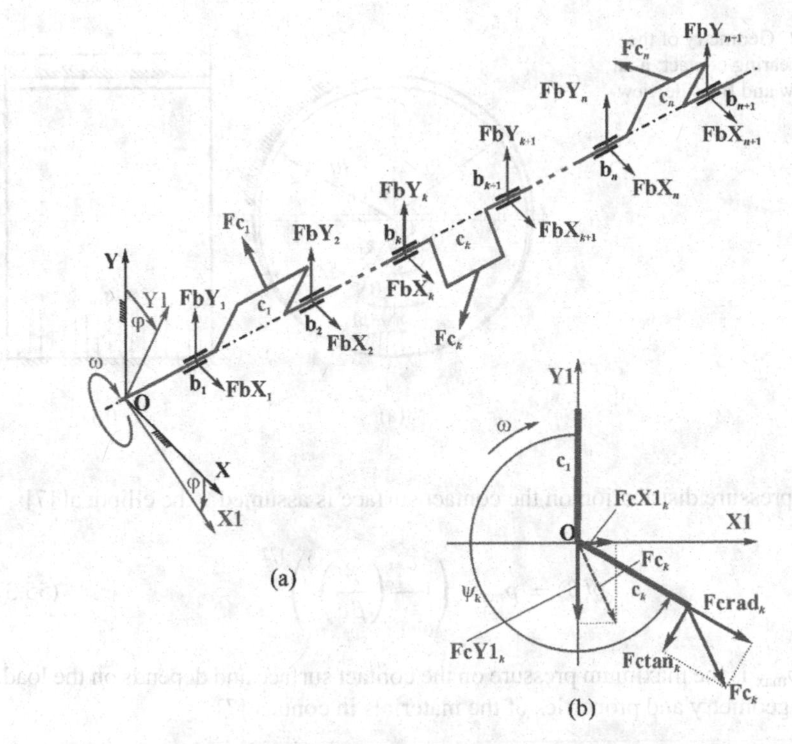

Fig. 55.1 Forces affecting the bearings of a crankshaft: **a** an isometric view and **b** a side view

where $FbX1_j$ and $FbY1_j$ are the components of the force $Fb_j(\varphi)$ in X1 and Y1 directions:

$$FbX1_j(\varphi) = \sum_{i=1}^{n} \left(-\rho_{i,j} \cdot Fcrad_i(\varphi_i) \cdot \sin\psi_i + \rho_{i,j} \cdot Fctan_i(\varphi_i) \cdot \cos\psi_i\right)$$

$$FbY1_j(\varphi) = \sum_{i=1}^{n} \left(\rho_{i,j} \cdot Fcrad_i(\varphi_i) \cdot \cos\psi_i + \rho_{i,j} \cdot Fctan_i(\varphi_i) \cdot \sin\psi_i\right) \quad (55.2)$$

$Fcrad_i$ and $Fctan_i$ represent the radial and tangential components of the crank force Fc_i, $\rho_{i,j}$ are the influence coefficients [7], ψ_k is the counterclockwise angle between cranks c_1 and c_k (Fig. 55.1b) and φ_i is the angle defining the position of crank c_i with respect to the beginning of a corresponding engine cycle. Taking into account that the engine cycle in cylinder i advances by an angle θ_i in relation to the engine cycle in cylinder 1, the following relation between the angles φ_i and φ applies: $\varphi_i = \varphi + \theta_i$. The force $Fb_j(\varphi)$, defined by Eq. (55.1), is transferred from the journal to the bearing, bringing these two elements into contact, as shown in Fig. 55.2.

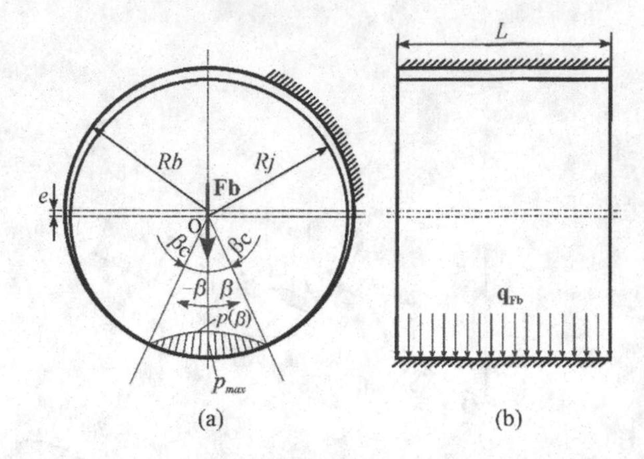

Fig. 55.2 Geometry of the journal-bearing contact: **a** an axial view and **b** a side view

The pressure distribution on the contact surface is assumed to be elliptical [7]:

$$p(\beta) = p_{\max} \cdot \left(1 - \left(\frac{\beta}{\beta c}\right)^2\right)^{1/2} \tag{55.3}$$

where $p_{\max}$ is the maximum pressure on the contact surface and depends on the load, contact geometry and properties of the materials in contact [7]:

$$p_{\max} = 0.55 \cdot \frac{Fb}{L \cdot Rb} \cdot \left(\frac{1}{\beta c} + 0.35\right) \tag{55.4}$$

Assuming that the bearing wear is proportional to the pressure $p(\beta)$ and looping the variable φ from the beginning to the end of the engine cycle, it is possible to calculate the cumulative conditional wear depth ΔR as a dimensionless quantity, which is described in reference [7]. Based on that calculation, the computer software "Wear Diagrams" was developed, which was used in this research. A detailed description of the software is given in reference [9] and its user interface is shown here just for illustration (Fig. 55.3).

55.3 Results and Discussion

To determine whether the connecting rod length L affects wear of the crankshaft bearings, the theoretical wear diagrams were generated taking two extreme values of the connecting rod length. The values were chosen so that the connecting rod ratio was close to the lower and upper limits of the recommended range (0.22 and 0.33). This means that the values of 238 and 159 mm were taken for the maximum and the minimum length of the connecting rod, respectively. Having in mind all the above-mentioned, a total of 14 theoretical wear diagrams were obtained (7 crankshaft

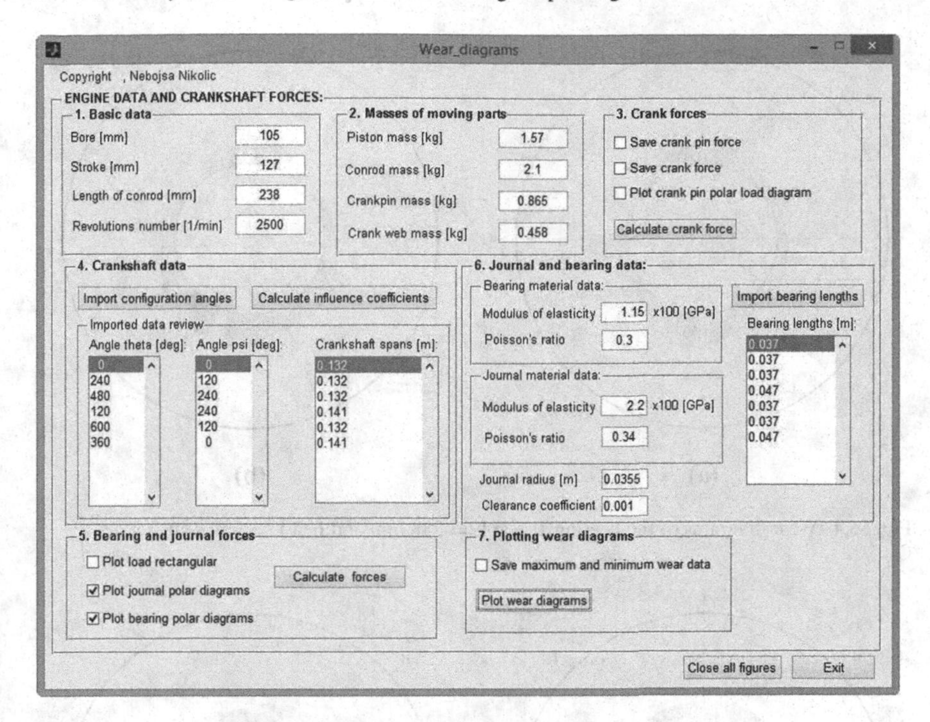

Fig. 55.3 User interface of the "wear diagrams" software

bearings multiplied by two connecting rod lengths chosen). Taking into account the symmetry of the crankshaft, the wear diagrams of the bearings 1 and 7 are almost identical, which also applies for the following pairs of bearings: 2–6 and 3–5. For that reason, only the wear diagrams for the bearings 1, 2, 3 and 4 are shown in Figs. 55.4, 55.5, 55.6 and 55.7, respectively.

Observing the pairs of diagrams in Figs. 55.4, 55.5, 55.6 and 55.7, one can conclude that there is no significant influence of the connecting rod length on the wear profile of the main crankshaft bearings. Namely, when the length of the connecting rod is changed, the wear profiles of the main bearings remain almost unchanged, with the exception of some small differences in wear intensity. The wear intensity is expressed by the dimensionless quantity ΔR named cumulative conditional wear depth, as mentioned earlier. In Figs. 55.4, 55.5, 55.6 and 55.7 the maximum and the minimum values of ΔR are denoted with red and green colors, respectively.

Table 55.1 shows the maximum values of ΔR and the angles $\alpha(\Delta R_{\max})$ that define the location on the bearing surface where ΔR reaches its maximum. The differences of maximum wear intensities at the two considered cases, expressed as a percentage, are also given in Table 55.1. It can be noticed that the use of a shorter connecting rod leads to a reduction in the maximum wear depth to some extent (between 2 and 9%). This reduction can be considered as small, given that two extreme cases have been observed. It is interesting that connecting rod shortening has the least impact

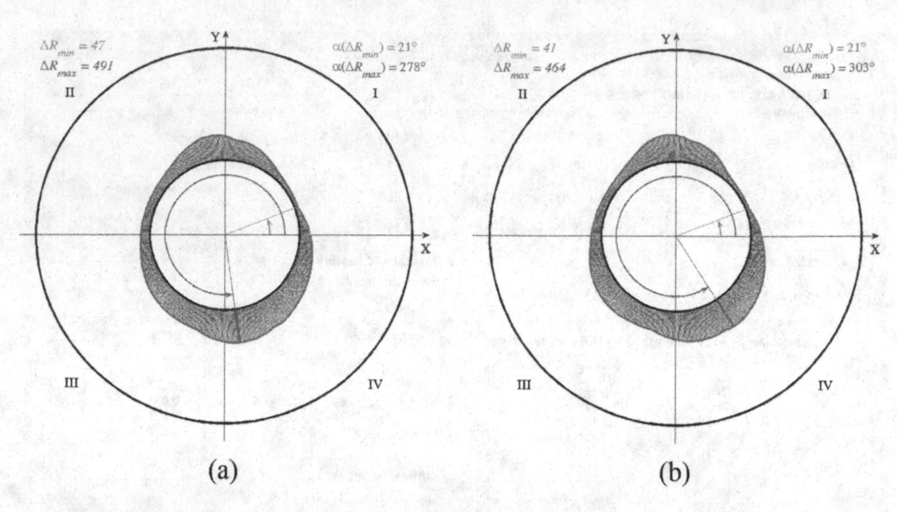

Fig. 55.4 Wear diagrams of the bearing 1: **a** at L = 238 mm and **b** at L = 159 mm

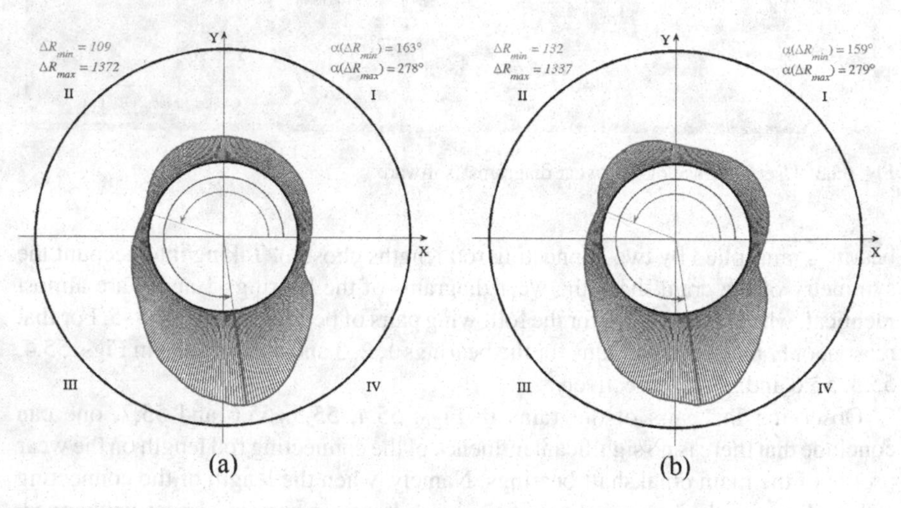

Fig. 55.5 Wear diagrams of the bearing 2: **a** at L = 238 mm and **b** at L = 159 mm

on the maximum wear depth of the most loaded bearings. However, this cannot be considered a general rule before a research is conducted on a larger sample of engines.

55.4 Conclusion

The main contribution of the paper is that it presents an investigation of how the connecting rod length affects the wear of the main crankshaft bearings. As far as

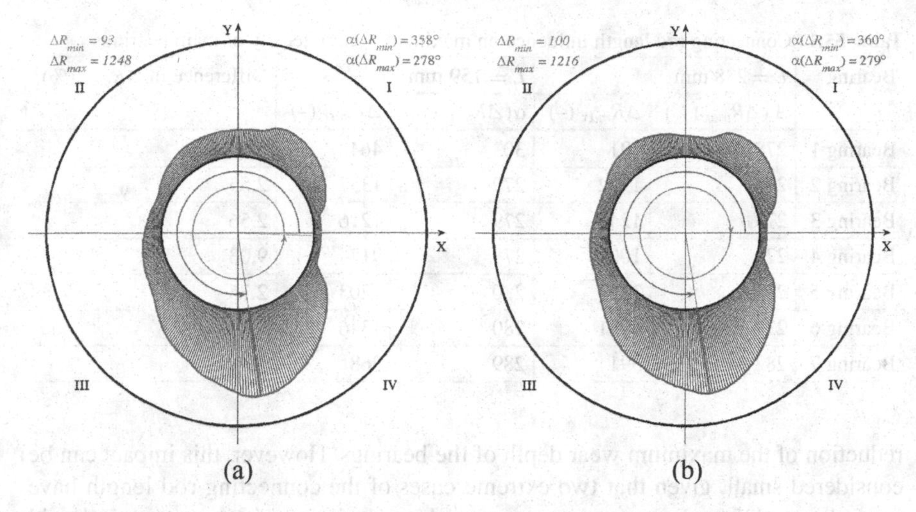

Fig. 55.6 Wear diagrams of the bearing 3: **a** at L = 238 mm and **b** at L = 159 mm

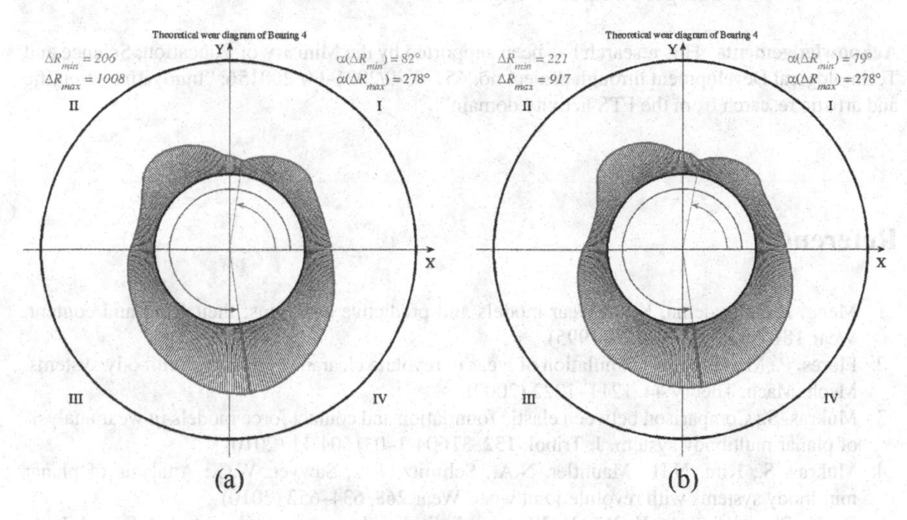

Fig. 55.7 Wear diagrams of the bearing 4: **a** at L = 238 mm and **b** at L = 159 mm

the authors are aware, such a research is not published in the available literature, or at least it is not conducted in this way, automatically, using the software developed by authors themselves. It is known that the connecting rod length has an influence on the wear of the piston-cylinder assembly of an IC engine. However, the question arises, whether the change in length of the connecting rod affects the wear of the main crankshaft bearings. To answer this question, a study was conducted on a 6-cylinder engine, which is described in this paper. It has been shown that by changing the connecting rod length, the wear profiles of the crankshaft bearings change almost negligibly. Also, it was noticed that shortening of the connecting rod leads to a

Table 55.1 Connecting rod length influence on maximum wear intensity of main bearings

Bearing	$L = 238$ mm		$L = 159$ mm		Difference in ΔR_{max} (%)
	A (ΔR_{max}) (°)	ΔR_{max} (–)	$\alpha(\Delta R_{max})$ (°)	ΔR_{max} (–)	
Bearing 1	278	491	303	464	5,50
Bearing 2	278	1372	279	1337	2,55
Bearing 3	278	1248	279	1216	2,56
Bearing 4	278	1008	278	917	9,03
Bearing 5	278	1237	280	1203	2,75
Bearing 6	277	1381	280	1346	2,53
Bearing 7	282	391	289	368	5,88

reduction of the maximum wear depth of the bearings. However, this impact can be considered small, given that two extreme cases of the connecting rod length have been observed. In order to reach more general conclusions, a similar research should be conducted on a larger sample of engines.

Acknowledgements This research has been supported by the Ministry of Education, Science and Technological Development through project no. 451-03-9/2021-14/ 200156: "Innovative scientific and artistic research from the FTS activity domain".

References

1. Meng, H.C., Ludema, K.C.: Wear models and predictive equations: their form and content. Wear **181–183**(2), 443–457 (1995)
2. Flores, P.: Modeling and simulation of wear in revolute clearance joints in multibody systems. Mech. Mach. Theory **44**, 1211–1222 (2009)
3. Mukras, S.: Comparison between elastic foundation and contact force models in wear analysis of planar multibody system. J. Tribol. **132**, 31604-1–031604-11 (2010)
4. Mukras, S., Kim, N.H., Mauntler, N.A., Schmitz, T.L., Sawyer, W.G.: Analysis of planar multibody systems with revolute joint wear. Wear **268**, 634–652 (2010)
5. Su, Y., Chen, W., Tong, Y., Xie, Y.: Wear prediction of clearance joint by integrating multibody kinematics with finite-element method. P. I. Mech. Eng. J.-J. Eng. **224**, 815–823 (2010)
6. Dorić, J., Klinar, J.: Influence of conrod length on wear of IC engine crankshaft bearing. Mach. Des. **7**(4), 125–128 (2015)
7. Nikolic, N., Torovic, T., Antonic, Z.: A procedure for constructing a theoretical wear diagram of IC engine crankshaft main bearings. Mech. Mach. Theory **58**, 120–136 (2012)
8. Lukanin, V.N., Shatrov, M.G.: Internal Combustion Engines, Dynamics and Design. Vysshaya Shkola, Moscow (2007)
9. Nikolić, N., Dorić, J., Jocanović, M.: A computer program for the visualization of IC engine crankshaft main bearings load. In: Proceedings of the 9th International Symposium on Machine and Industrial Design in Mechanical Engineering, KOD 2016, pp. 89–92. Balatonfured (2016)
10. Nikolić, N., Dorić, J., Stojić, B.: A computer program for IC engine crankshaft main bearings wear diagrams. IOP Conf. Ser.: Mater. Sci. Eng. **393**, 012069-1–012069-8 (2018)

Chapter 56
Application of Techniques and Systems for Additive Manufacturing in Rapid Tooling

Marko Popovic, Vesna Mandic, and Marko Delic

Abstract The paper presents the integrated application of additive manufacturing and reverse engineering technologies for the rapid tooling for two-component plastic casting, as a faster and cheaper approach than injection moulding, particularly for small-scale production of plastic parts or spare parts when CAD models and technical documentation are not available. An optical scanner based on white structured light was used for 3D digitalization of the selected plastic gas handle. Based on point cloud, a master CAD model was prepared for the design of casting tool cavities. The design of the master model was verified through its additive manufacturing using FDM technology. After repeated 3D digitization of the printed master model and comparison with the original part, the master CAD model was redesigned. The process of two-component casting in tools obtained by additive FDM manufacturing was realized successfully and verified by comparing the cast and original gas handle.

Keywords Rapid tooling · Additive manufacturing · FDM · 3D digitalization · Two-component plastic casting

56.1 Introduction

The integration of virtual engineering (VE) technologies is the best support for concurrent engineering approach because they can be of great use to engineers in making decisions and establishing control over the product development process and its production. Although they belong to relatively new technologies, they are characterized by a large number of advantages, but the key ones are reflected in the reduction of time and costs of product development [1]. One of the most significant VE technologies is additive manufacturing (AM) which is defined as the process of 3D fabrication of a part directly from a CAD model, usually layer by layer, as

M. Popovic (✉)
Gorenje MDM, Kosovska 4, 34000 Kragujevac, Serbia
e-mail: marko.popovic@gorenje-mdm.com

V. Mandic · M. Delic
Faculty of Engineering, University of Kragujevac, Sestre Janjic 6, 34000 Kragujevac, Serbia

M. Rackov et al. (eds.), *Machine and Industrial Design in Mechanical Engineering*, Mechanisms and Machine Science 109,
https://doi.org/10.1007/978-3-030-88465-9_56

opposed to the method of subtractive manufacturing. AM technologies have been developed to such an extent that, in addition to rapid prototyping (RP), they also enable rapid tooling (RT). Tools made in this way enable effective application for the small-scale production of products or spare parts, especially in situations where the production of tools by conventional procedures would be extremely expensive [2].

Today, a lot of papers can be found in the literature, which through practical examples show the advantages of applying additive production technologies for fast tool making. 3D RP technology was applied to make a master model of the implant, which was later used to form a silicone rubber mold for vacuum casting [3]. Also, authors showed that the RT technology can be used very successfully in the process of precision casting of metal implants, where the tool manufacturing time is significantly shorter, and the production costs are twice lower than conventional methods. Len-Cabezas et al. [4] in their work used Stereolithography (SLA), Selective Laser Sintering (SLS) and PolyJet technology to prototype plastic injection moulding tools. The obtained tool prototypes have been successfully used for injection moulding in the production of small series using conventional polymeric materials such as polypropylene and ABS plastic [4]. The possibility of making tools for two-component plastic casting with PolyJet additive technology is presented through a practical example in [5], and it is noted that this method of production could become an optimal solution for small batches if the main limiting factors related to the curing process are overcome. Dongaonkar et al. [6] based on the results of 3D scanning developed a 3D CAD model which was used as input for rapid prototyping by Fused Deposition Modelling (FDM) technology, and finally concluded that the obtained prototype can be used as a master model for the casting process [6]. Four case studies of the RT concept in casting technology were presented, where it was concluded that Reverse Engineering (RE) supported by 3D digitization enables faster metrological control of shape and geometry, by calculating deviations between CAD models and 3D scanning results, i.e. the obtained cloud points [7]. Also, the integration of RE, RT and CAE (Computer Aided Engineering) technology can reduce delivery time and associated costs in the industry which contributes to the improvement of the competitive position in the market. The case study presented in [8] demonstrates the advantages and possibilities of integration of VE technologies. It has been shown that the application of VE technologies can successfully realize different phases of the product life cycle, while achieving savings in terms of time and costs of product development.

The aim of this paper is the application of additive production techniques and systems for rapid prototyping of master model and rapid tooling for two-component plastic casting. For the production of tools for two-component casting of the selected part (plastic gas handle), FDM technology was applied. In order to obtain tool cavities that correspond to the geometry of the handle, it was necessary to perform its 3D scanning to obtain point cloud from which a 3D master CAD model was modelled. In the end, the procedure of two-component casting using produced tools was realized.

56.2 Design of Tools for Two-Component Plastic Casting

For the additive manufacturing of tools for two-component casting of the selected part of the gas handle, shown in Fig. 56.1a, the reverse engineering procedure was applied to obtain a 3D CAD model of the handle. The first phase of reverse engineering is 3D digitization, i.e. the digital transformation of the collected data on the coordinates of points from the surface of the scanned object. The precision of 3D digitization is very important in the process of reverse engineering because it directly affects the quality of the resulting CAD model [9]. The accuracy of the applied David SLS-2 3D scanner is 0.1% of the scan size, or 0.06 mm. It takes a few seconds to get one scan, giving more than 1,200,000 points.

Adequate object preparation is necessary for successful 3D scanning using optical methods. Reflection can negatively affect the quality of the obtained results, so a thin layer of matting spray is applied to the scanning object as in Fig. 56.1b. Before scanning the object, the David SLS-2 3D scanner was calibrated in order to precisely collect point cloud data. The scanning object, just before the start of point cloud data collection, is shown in Fig. 56.1c. Figure 56.2a shows one scan of the handle in the scanner software editor. To get the entire point cloud, it was necessary to scan the object in different positions and views. The position of the lever was changed by manual manipulation, taking into account that the obtained scans must have common areas due to their correct positioning and interconnection. An overview of all obtained scans is given in Fig. 56.2b [10].

All noises and objects from the handle environment that were mistakenly "caught" during the scanning process, had to be removed so as not to impair the quality of

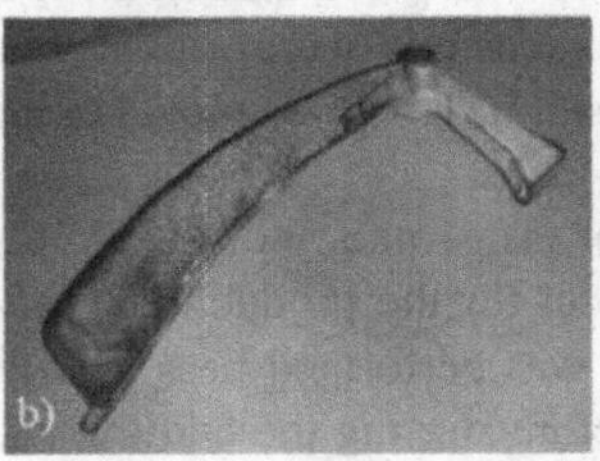
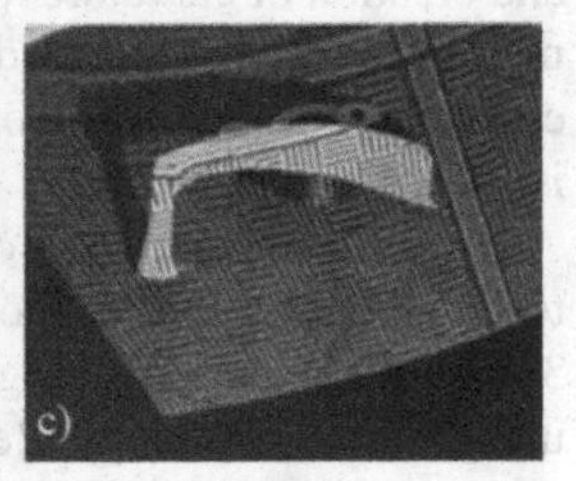

Fig. 56.1 **a** Gas handle, **b** matting gas handle and **c** 3D scanning of the handle

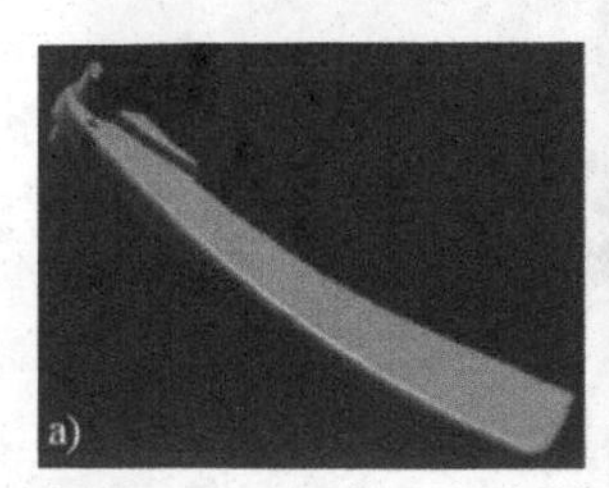

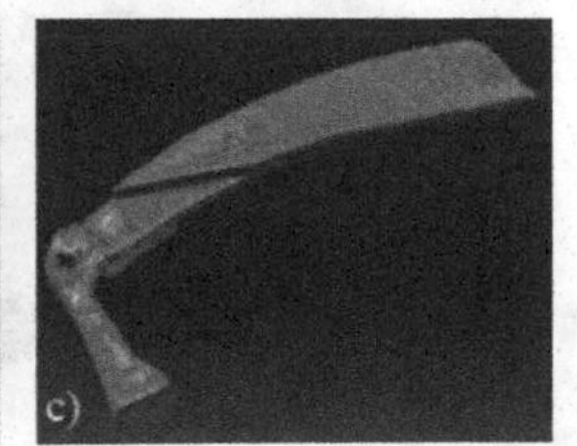

Fig. 56.2 Software processing of the obtained 3D scan results

the 3D CAD modelling. The Fusion option in 3D scanner software was used with set values for resolution and sharpness to merge all scans and form a closed CAD and STL model of the handle. By Checking the Close Holes option small holes on the resulting model was closed. The final appearance of the handle point cloud, after eliminating the noises and merging all the necessary scans, is shown in Fig. 56.2c [10].

All noises and objects from the handle environment that were mistakenly "caught" during the scanning process, had to be removed so as not to impair the quality of the 3D CAD modelling. The *Fusion* option in 3D scanner software was used with set values for resolution and sharpness to merge all scans and form a closed CAD and STL model of the handle. By Checking the *Close Holes* option small holes on the resulting model was closed. The final appearance of the handle point cloud, after eliminating the noises and merging all the necessary scans, is shown in Fig. 56.2c [10].

Due to the observed irregularities, which were caused by the inability of the optical scanner to "see" inaccessible areas and very small details, certain corrections of the 3D model were made in the CATIA software, so that the digital model had precise dimensions and closed shape. Figure 56.3a shows the observed shortcomings on the STL model of the handle, and the corrections included the reconstruction of the cylindrical supports (1), the gap between the supports and the limiter (2) and the wedge (3). Prior to the model reconstruction process, the point cloud was imported into the CATIA software using the *Digitalized Shape Editor* module. After importing the point cloud, the surface reconstruction was performed by a combination of the *Quick Surface Reconstruction* and the *Generative Shape Design* modules, using a number of useful tools. Reconstruction of critical areas, and then chamfering of edges and creation of curvature radii was performed in the module *Part Design.* Using the measuring equipment, the dimensions of the critical areas of the original handle were checked, as well as their position. The CAD model of the handle after reconstruction is shown in Fig. 56.3b.

To verify the accuracy of the geometry of the 3D CAD model of the gas handle to be used as a master model for the production of two-component casting tools, 3D additive manufacturing was performed on a MarkForged Onyx Pro 3D printer using FDM technology. Before starting the additive manufacturing process, it was necessary to set up the 3D printer, i.e. to level the printing table, which is shown in Fig. 56.4a. The STL model of the gas handle (Fig. 56.4b) was imported into the *Eiger*

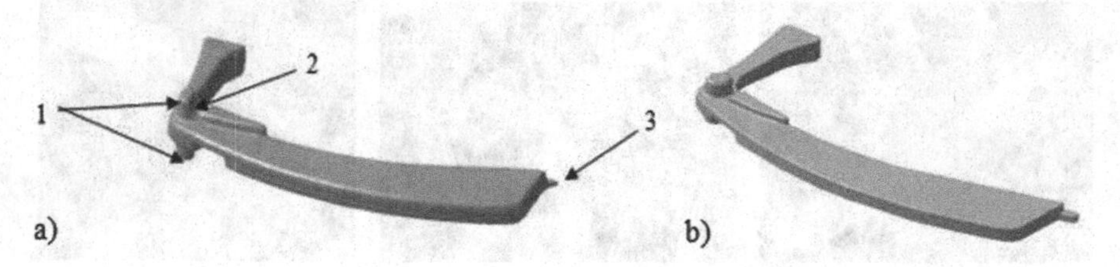

Fig. 56.3 **a** Observed irregularities on STL model and **b** CAD model after reconstruction

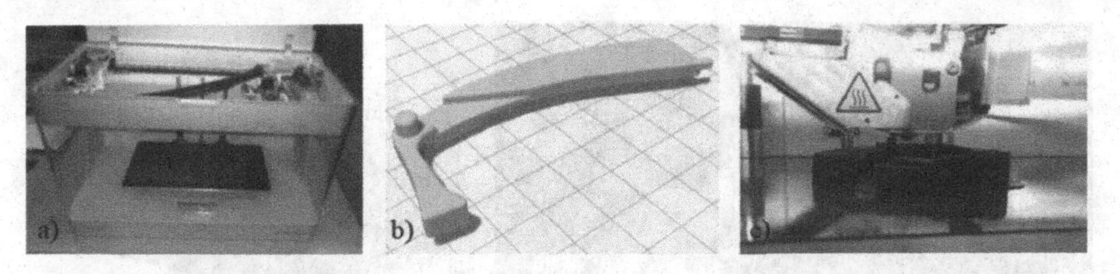

Fig. 56.4 Stages of AM process: **a** 3D printer setup, **b** STL model, **c** printed master model

virtual software library. After positioning the model on the table, the parameters of the additive manufacturing process were set. A value of 0.1 mm was selected for the *Layer Height*, and then the *Use Supports* option was activated. *Onyx* was chosen as the base material, with the *Solid Fill* option for filling the model with plastic filament. The reinforcement material was *Fiberglass*, applied with *Concentric Fiber* and *All Walls* options for 2 concentric reinforcements along all model walls.

After the completion of the additive manufacturing process of the master model, the table was removed from the 3D printer, and the printed handle from the table with increased caution in order not to damage the master model. The last step involved the removal of the support material, as well as the subsequent treatment of the model surfaces in order to remove traces of the support material and the "step effect" on the sloping surfaces. The final appearance of the master model of the handle after removal of the support material and surface treatment is shown in Fig. 56.4c.

In order to finally check the accuracy of the master model, a 3D scanning of the printed master model was performed. The point cloud (STL) thus obtained was compared with the point cloud obtained by 3D scanning of the original handle, using Geomagic Studio software. In Fig. 56.5 a comparison of the printed master model of the handle and the reconstructed CAD model is shown. It can be noticed that the deviation values of the master model in relation to the CAD model mostly range from -0.061 until $+0.061$ (green fields), which is considered to be extremely good results. However, higher values of deviations in the range of $-0.254 \div 0.254$ can be also considered acceptable since they are primarily a consequence of the subsequent manual processing of the surfaces of the printed handle and the rest of the support material.

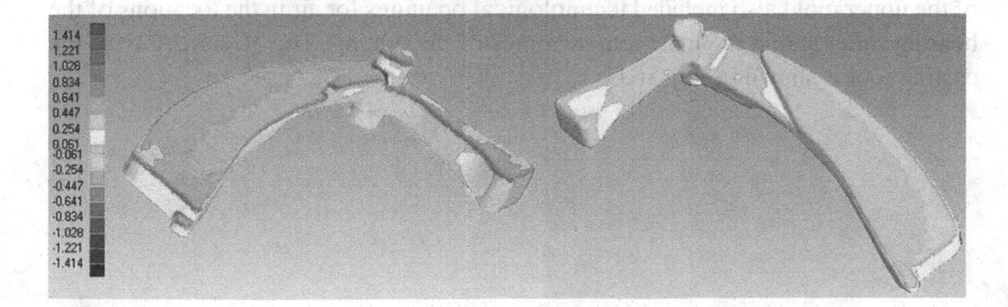

Fig. 56.5 Comparison of the master model obtained by additive manufacturing and CAD model

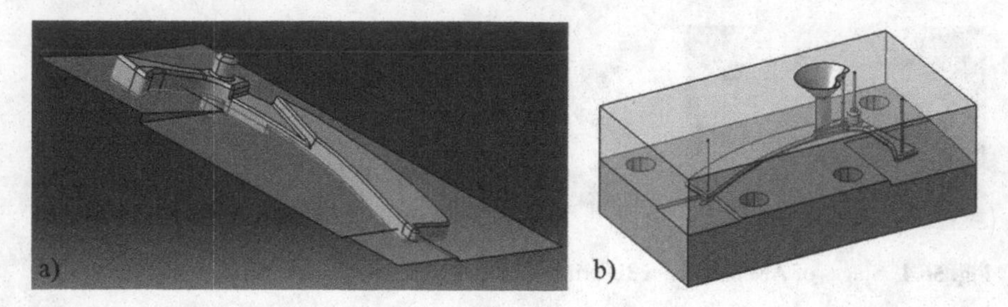

Fig. 56.6 **a** Splitting plane and **b** 3D assembly of two-component casting tool

The two-component casting tool design process included the following steps:

- determination of the split line,
- determination of overall dimensions of mold plates,
- modelling of mold cavities,
- modelling of guides, and
- selection of the appropriate location and shape of inflow channel and air vents, and their modelling.

The two-component casting tool was modelled in CATIA software. After determining the split line, a complex splitting plane was designed as shown in Fig. 56.6a. A very important step in modelling the tool was to emboss the CAD model, i.e. the geometry of the master model of the handle into the upper and lower mold plate, after which the obtained mold cavities corresponding to the geometry of the plastic casting of the handle.

The choice of the position and shape of the inflow channel is important for the casting process. An inflow channel of elliptical cross-section measuring 6 × 4.2 mm was chosen. In addition, at the very top of the upper mold, the channel is conically widened to facilitate the casting of liquid material into the tool.

Four guide holes, 10 mm in diameter and 10 mm deep, were modelled on the lower mold. The holes are conical in shape with a slope of 3° for easier separation of mold plates. The guides modelled on the upper mold, 10 mm in diameter and 9.5 mm high, are conical in shape with the same slope. The distance between the axes of the guide elements and the contour edges of the tool is 10 mm. The design of the upper mold also included technological openings for air in the locations of the bearing roller and in the most remote zones of the casting. The 3D assembly of the casting tool is shown in Fig. 56.6b.

Fig. 56.7 **a** Casting molds during its additive manufacturing, **b** after removing and **c** final

56.3 Additive Manufacturing of the Tool for Two-Component Plastic Casting

Additive manufacturing of molds for two-component casting of the handle was performed on a 3D printer MarkForged Onyx Pro. The manufacturing process included all phases as in the manufacturing of master model of the handle by FDM technology. The basic material is Onyx, but unlike the printing of the master model, no fiberglass reinforcement was applied, as well as the complete filling of the tool prototype with plastic filament. The *Triangular Fill* option was chosen, which enables the creation of a ribbed structure inside the printed molds whose volume is 37% of the total volume. In order to obtain the finest possible tool surfaces, a layer thickness of 0.1 mm was chosen. According to the Eiger software estimation, it took 16 h 38 min, 103.98 cm^3 of Onyx material to make the gas handle molds, and the estimated cost of the material in the US is 24.57$.

The MarkForged 3D printer has the ability to print multiple parts on a table, so both molds are printed at the same time, as shown in Fig. 56.7a. The appearance of the two-component casting tool after the completion of the AM process and the removal of the printing table from the printer is shown in Fig. 56.7b. After removing the support material, manually treating the sloping surfaces and checking the gap in the split plane, the tool was ready for casting (Fig. 56.7c).

56.4 Two-Component Plastic Part Casting Process

The tool made by additive manufacturing technology was used during the realization of the two-component casting process, consisting of several main steps.

The first step was the preparation of the casting molds. It was necessary to inspect the molds in detail and remove all impurities that could cause various defects on the obtained castings. After that, a thin layer of separating material was applied on mould surfaces which will come into contact with the casting material, in order to facilitate the process of removing the casting from the tool.

The material used in the two-component casting process was NEUKADUR AF Neu blue, i.e. the so-called surface resin. The main feature of this material is that it has

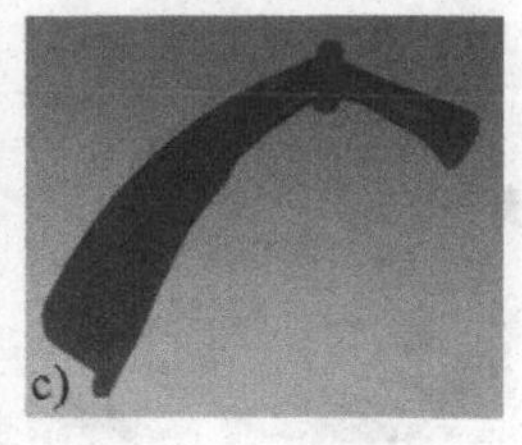

Fig. 56.8 **a** Injection of resin into the inflow channel, **b** appearance of the casting after separation of the upper and lower molds and **c** appearance of the casting after removal from the tool

great hardness and abrasion resistance. It is also characterized by improved resistance to chemicals and heat by using BWS hardeners. The casting resin was mixed with 10% of BWS hardener to give a total of 22 ml of a homogenized two-component mixture. In the next step, the mixture was injected through a inflow channel system made in the upper mold, as shown in Fig. 56.8a. After the completion of the hardening process of the mixture in the tool, which lasted about 18 h, the upper and lower molds were separated (Fig. 56.8b). The appearance of the casting, gas handle, after removal from the molds is shown in Fig. 56.8c.

56.5 Conclusions

Due to the significant advantages that characterize it, the AM technology is increasingly used, both for the rapid prototyping, and for the rapid tooling. Reverse engineering, which is defined as the opposite process to the classical process of designing and modelling of product components and tools, can be very useful in situations where there are no technical drawings, documentation or CAD models, because a 3D CAD model can be created based on 3D scanning and point cloud data.

Two-component casting in tools made by additive manufacturing process, as an alternative technology to injection molding, offers high quality plastic products with minimal costs and minimal production time, so it is suitable for small series and production of spare parts. Also, this can be useful for making functional parts that do not have excessive operating loads and special aesthetic requirements. With the application of the described methodology in the work, a plastic casting was obtained which dimensionally and functionally corresponds to the original part of the handle, so it can be used for installation in a functional assembly.

To obtain more precise parts for complex assemblies, it is necessary to apply the presented AM methodology by PolyJet technology of printing, which enables obtaining smooth surfaces of mold cavity. Optimization of the position and number of air vents can be achieved by subsequently making vents on existing printed molds. As a subject of future research, the behaviors of different casting materials in a mold made with FDM technology can be examined. This would help in choosing the most suitable mixtures of materials for performing the casting process in such molds.

Acknowledgements The paper includes research conducted within the project TR34002, funded by the Ministry of Education, Science and Technological Development of the Republic of Serbia.

References

1. Mandić, V.: Virtual engineering (in Serbian). University of Kragujevac, Mechanical Engineering Faculty (2007)
2. Gebhardt, A.: Understanding additive manufacturing. Hanser Publishers, Munich (2011)
3. Rajić, A.: Application of additive manufacturing technologies in the process of precise casting of orthopedic implants (in Serbian). University of Novi Sad, Novi Sad (2015)
4. León-Cabezas, M.A., Martínez-García, A., Varela-Gandía, F.J.: Innovative advances in additive manufactured moulds for short plastic injection series. Procedia Manuf. **13**, 732–737 (2017)
5. Gavrilovic Z., Mandic, V., Urosevic, V.: Application of poly-jet technology in rapid tooling. In: Proceedings of the International Conference Modernization of Universities through Strengthening of Knowledge Transfer, Research and Innovation, WBCInno 2015, pp. 101–104. Novi Sad (2015)
6. Dongaonkar, A.V., Metkar, R.M.: Reconstruction of damaged parts by integration reverse engineering (RE) and rapid prototyping (RP). In: Kumar, L., Pandey, P., Wimpenny, D. (eds.) 3D Printing and Additive Manufacturing Technologies, pp. 159–171. Springer, Singapore (2019)
7. Ferreira, J.C., Alves, N.F.: Integration of reverse engineering and rapid tooling in foundry technology. J. Mater. Process. Tech. **142**, 374–382 (2003)
8. Mandić, V., Ćosić, P.: Integrated product and process development in collaborative virtual engineering environment. Teh. Vjesn. **18**, 369–378 (2011)
9. Wang, W.: Reverse Engineering. Technology of Reinvention. CRC Press, London (2010)
10. Popovic, M.: Application of techniques and systems for additive manufacturing in rapid tooling (in Serbian). University of Kragujevac, MSc Theses. Faculty of Engineering (2020)

Part V
Mechatronics and Robotics: Design, Kinematics and Dynamics

Chapter 57
L-CaPaMan Design and Performance Analysis

Alexander Titov and **Marco Ceccarelli**

Abstract This paper presents a solution for low-cost lightweight design of L-CaPaMan by using market components and 3D printing manufactured parts. The aim of this new solution is to provide a parallel manipulator prototype for research and formation activities, such as for performance evaluation and education of parallel manipulators. The proposed CAD design is used in dynamic simulation whose results are discussed in terms of advantages and limitations of the designed solution.

Keywords Robotics · Parallel manipulators · Performance analysis · Simulation · CaPaMan

57.1 Introduction

Parallel manipulators are closed-loop mechanisms whose moving platform connected to the base by several independent kinematic chains [1, 2]. Ability of reaching high accelerations and high accuracy in positioning defines its applications in wide variety of fields, such as manufacturing and additive technology [3], geology (earthquake simulations [4]), surgery (assistant systems [5]), aerospace (flight simulators [6], telescope positioning [7]) and the other fields. The first design of CaPaMan with a prototype was reported in 1997 [4]. The improvements in prototypes were difficult and expensive in manufacturing, as reported in [8]. A 3D printed prototype has partially solved the problem with low-cost components [8]. This paper describes the results of CaPaMan design improvements. The prismatic joint has been replaced with a novel mechanism solution. The motion has been represented and verified in CAD simulation and checked on a 3D printed model, whose results are discussed for a performance characterization.

A. Titov (✉) · M. Ceccarelli
LARM2 Laboratory of Robot Mechatronics, Department of Industrial Engineering, University of Rome Tor Vergata, Rome, Italy
e-mail: aleksandr.titov@students.uniroma2.eu
URL: https://phdindustrialengineering.uniroma2.it/

M. Ceccarelli
e-mail: marco.ceccarelli@uniroma2.eu

M. Rackov et al. (eds.), *Machine and Industrial Design in Mechanical Engineering*, Mechanisms and Machine Science 109,
https://doi.org/10.1007/978-3-030-88465-9_57

57.2 Problems and Requirements

The first generation of CaPaMan, as shown in Fig. 57.1a, has translational joints in its structure [4], and the following improvements [8], as shown in Fig. 57.1b, keep this scheme of the parallel mechanism.

The translational joint had been designed by using manufactured commercial prismatic parts. The disadvantage of this solution is the contact area between parts of a joint is large, so that friction and resistance forces can give negative effects to the mechanism motion. Referring to Fig. 57.1, the following problems can be identified: high cost of parts manufacturing (cutting from metal blank); large weight; friction in a prismatic joint (large area of the contact). Requirements for a new design are summarized in the Fig. 57.2.

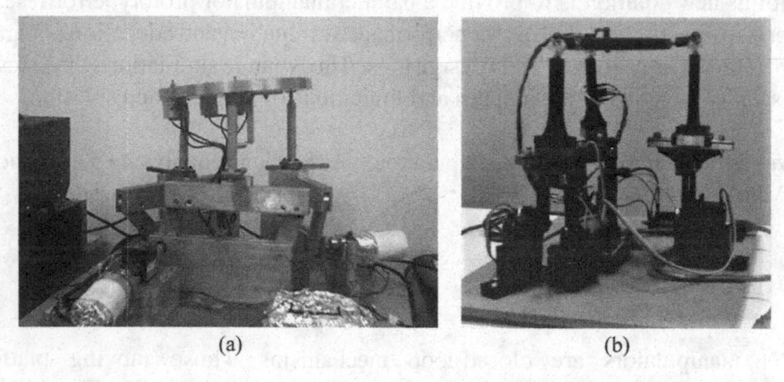

Fig. 57.1 CaPaMan prototypes: **a** original design in 1997 [4] and **b** 3D printed solution in 2017 [8]

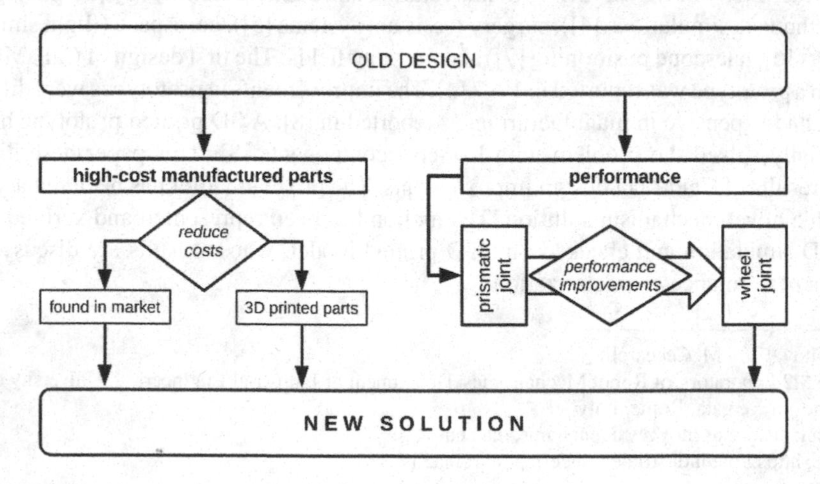

Fig. 57.2 Requirements for a new solution of L-CaPaMan design

57.3 A CAD Design

In the new solution prismatic joint has been replaced with a wheel-based construction. PLA plastic has been used for 3D printing of the structure part. The scheme of mechanism is shown in Fig. 57.3a, and the dimensions are listed in the Table 57.1. Dimensions of the fixed platform in Table 57.1 are a × b, $\mathbf{T_{1i}}$ is a torque of the motor, l_{ki} (k is a number of the part, i is a manipulator leg number) are dimensions of the links. The CAD design scheme of a new L-CaPaMan solution is presented in Fig. 57.3b.

The translational link with two moving parts in a prismatic joint has been redesigned as 3D printed bodies equipped by market ball bearings, as shown in Fig. 57.4. To avoid unexpected movements, a suitable surface complicated-form road has been designed with bearing roads, as in Fig. 57.4b. Three points of contact are used in the translational part to ensure stability of motion.

The design of a construction allows changing the dimensions of parts without rebuilding other components. The mechanical design is developed with a tolerance of 0.1 mm among the moving parts.

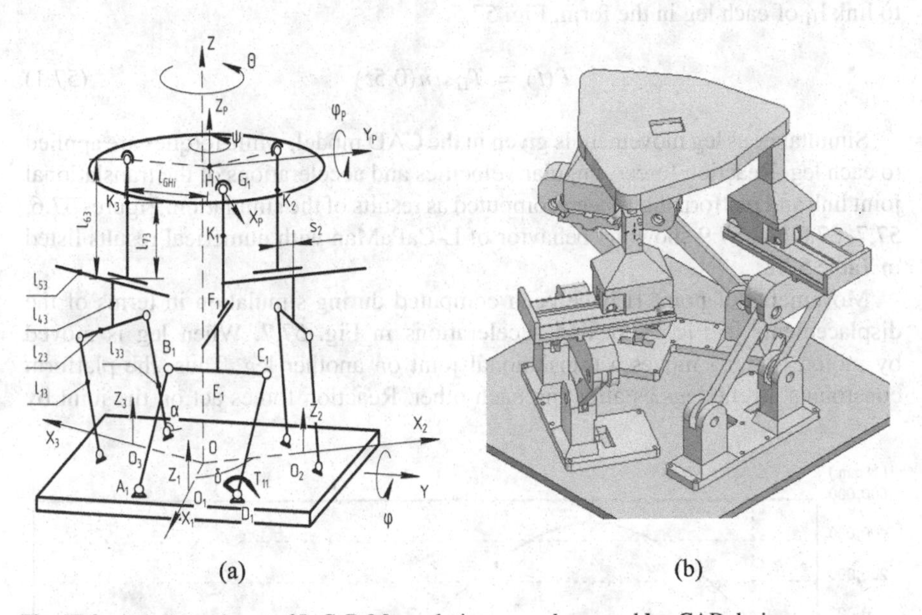

Fig. 57.3 A design scheme of L-CaPaMan solution: **a** a scheme and **b** a CAD design

Table 57.1 Design parameters of prototype in Fig. 57.3

	a × b	l_{1i}	l_{2i}	l_{4i}	l_{5i}	l_{6i}	l_{7i}	l_{GHi}
mm	200 × 200	70	60	25	90	70	50	60

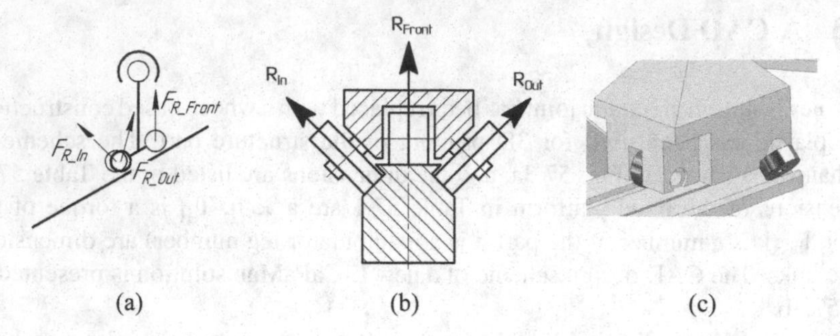

Fig. 57.4 The new prismatic joint in Fig. 57.4: **a** a scheme, **b** a mechanical design, **c** a CAD design

57.4 Performance Analysis via Simulation

The model in Fig. 57.3b has been simulated in Autodesk Inventor (Educational Version) [9] with the scheme of a leg in Figs. 57.3 and 57.4. The wheels are indicated in Fig. 57.3c. An input torque with maximum value $T_{1i} = 500$ Nmm has been applied to link l_{1i} of each leg in the form, Fig. 57.5.

$$T(t) = T_{1i} sin(0.5t) \tag{57.1}$$

Simultaneous leg movement is given in the CAD model, while torques are applied to each leg. Reaction forces, angular velocities and accelerations of the translational joint link and platform have been computed as results of the simulation. Figures 57.6, 57.7, 57.8 and 57.9 show the behavior of L-CaPaMan with numerical results listed in Table 57.2.

Movements of point H have been computed during simulation in terms of the displacements in Fig. 57.6, and accelerations in Fig. 57.7. When leg is moved by motor, this leg moves a translational joint on another leg. Thus, the platform constraints all the legs as affecting each other. Reaction forces act on the joint by

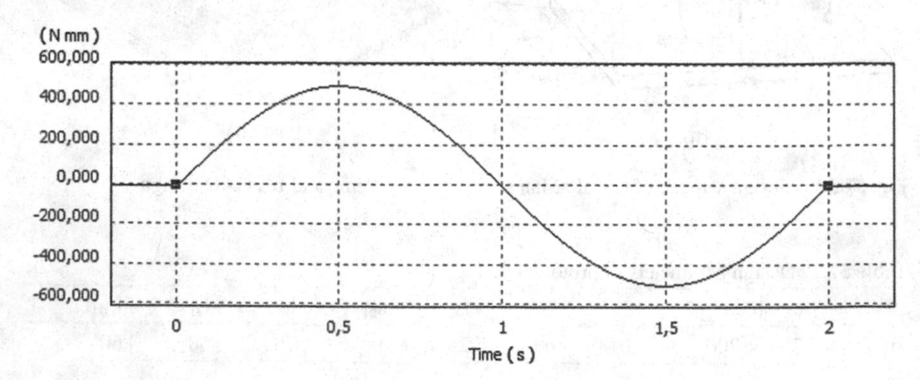

Fig. 57.5 Input torque on leg cranks for a simulated operation

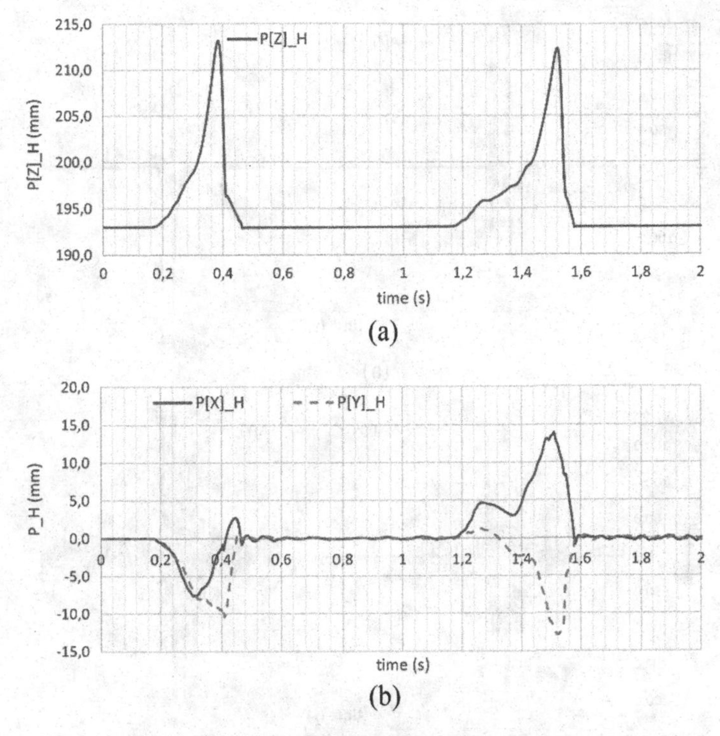

Fig. 57.6 Computed results of the point H displacements along: **a** Z axis, **b** X and Y axis

angle 60°, causing high specific reaction forces on wheels. The largest values of reaction forces act on wheel, which is on opposite side of motion of another leg. Due to this, platform moves not only along Z direction, but also along X and Y.

The accelerations in Fig. 57.7 are computed with peaks of 800 m/s^2. Rotations of legs in a simulation are strongly constrained; when the leg reaches extreme angle (±45°), the kick occurs, and in this moment, accelerations reach extremely high level.

One of the disadvantages of the new construction has been detected during a simulation in the fact that small surface of a contact makes the joint flexible and gives oscillations when moving, so we have undesirable high reaction forces and moments and high accelerations—vibrations in a joint. These negative effects can be reduced by improving construction qualities of parts by reducing gaps. To improve quality of a simulation friction in parts could also be added.

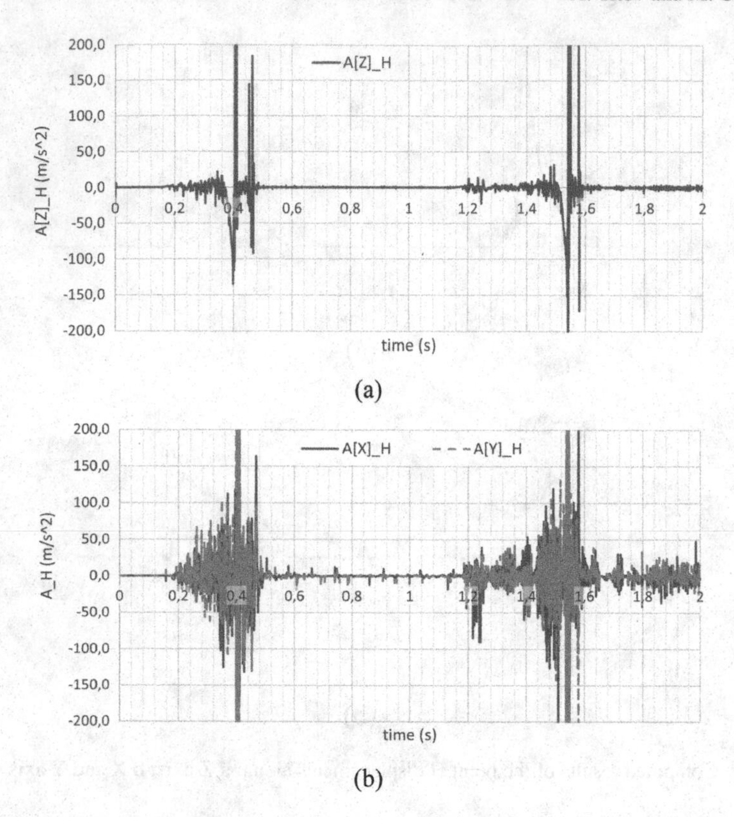

Fig. 57.7 Computed results of the point H accelerations along: **a** Z axis, **b** X and Y axis

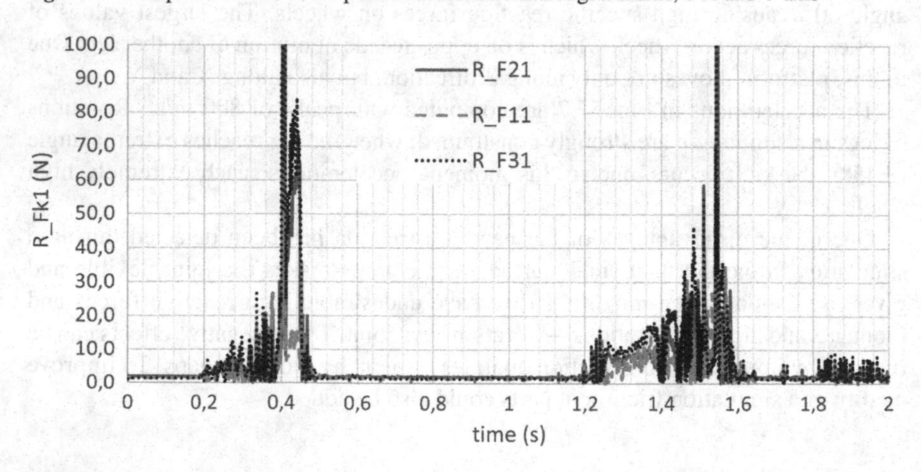

Fig. 57.8 Computed reaction forces in translation joint wheels of leg 1

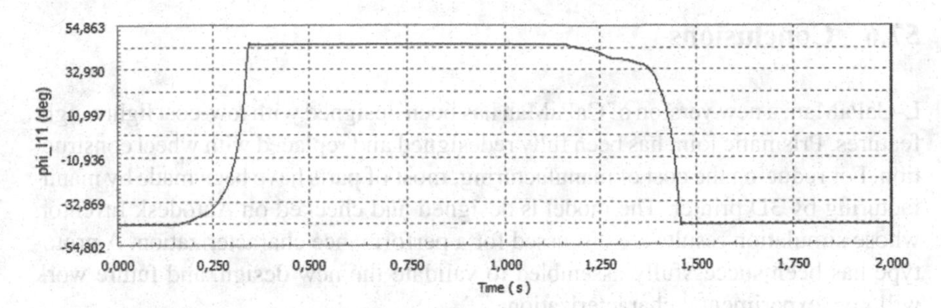

Fig. 57.9 Computed angular position of crank l_{11}, degrees

Table 57.2 Computed reaction forces and torques for the wheels in the prismatic joint, Fig. 57.3

Wheel	t_1 (s)	R_{1max} (N)	T_{1max} (Nmm)	t_2 (s)	R_{2max} (N)	T_{2max} (Nmm)
Inner	0.15	266.1	1745.8	1.14	286.9	366.2
Outer	0.15	223.6	928.1	1.15	354.2	2555.7
Central	0.15	80.4	29.5	1.15	93.3	15.8

57.5 A Build Prototype

The prototype has been built and assembled for checking the results. MG996R servo motors are used for moving legs, and signals for motion are generated by Arduino Mega 2560 and Arduino Toolkit. The prototype is shown in Fig. 57.10.

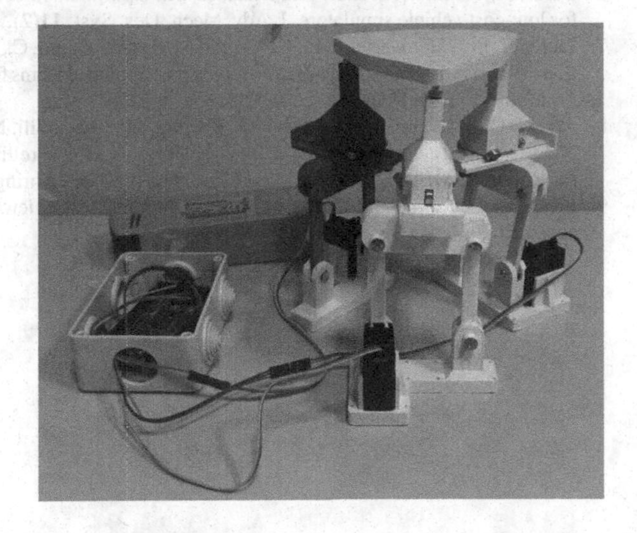

Fig. 57.10 The L-CaPaMan prototype at LARM2 in Rome

57.6 Conclusions

L-CaPaMan, a new version of CaPaMan has been designed with low-cost lightweight features. Prismatic joint has been fully redesigned and replaced with wheel construction. For reducing the cost of manufacturing, most of parts have been made by manufacturing by 3D printer. The model is designed and checked on Autodesk Inventor, whose simulation results are discussed for a performance characterization. A prototype has been successfully assembled to validate the new design, and future work will give experimental characterization.

References

1. Ceccarelli, M.: Fundamentals of Mechanics of Robotic Manipulation. Kluwer Academic Publishers, Dordrecht (2004)
2. Tsai, L.W.: Robot Analysis: The Mechanics of Serial and Parallel Manipulators. Wiley, New-York (1999)
3. Song, X., Pan, Y., Chen, Y.: Development of a low-cost parallel kinematic machine for multidirectional additive manufacturing. J. Manuf. Sci. Eng. **137**(2), 021005-1–021005-13 (2015)
4. Ceccarelli, M.: Historical development of CaPaMan, Cassino parallel manipulator. In: Viadero, F., Ceccarelli, M. (eds.) New Trends in Mechanism and Machine Science. MMS, vol. 7, pp. 749–757. Springer, Dordrecht (2013)
5. Saracino, A., Oude-Vrielink, T.J.C., Menciassi, A., Sinibaldi, E., Mylonas, G.P.: Haptic intracorporeal palpation using a cable-driven parallel robot: A user study. IEEE T. Bio.-Med. Eng. **67**(12), 3452–3463 (2020)
6. Casas, S., Coma, I., Portales, C., Fernandez, M.: Optimization of 3-DOF parallel motion devices for low-cost vehicle simulators. J. Adv. Mech. Des. Syst. **11**(2), 17–23 (2017)
7. Jauregui, J.C., Hernandes, E.E., Ceccarelli, M., Lopez-Cajun, C., Garcıa, A.: Kinematic calibration of precise 6-DOF Stewart platform-type positioning systems for radio telescope applications. Front. Mech. Eng. **8**(3), 252–260 (2013)
8. Arslan, O., Karaahmet, S.B., Selvi, Ö., Cafolla, D., Ceccarelli, M.: Redesign and construction of a low-cost CaPaMan prototype. In: Gasparetto, A., Ceccarelli, M. (eds.) Mechanism Design for Robotics: MEDER 2018. MMS, vol. 66, pp. 158–165. Springer, Cham (2019)
9. Inventor user's manual 2021, https://help.autodesk.com/view/INVNTOR/2021/ENU/. Last accessed 2021/03/03

Chapter 58
Walking Robot with Modified Jansen Linkage

Lucian Alexeev, Andreea Dobra, and Erwin Lovasz

Abstract The paper deals with a walking robot that has as legs a modified version of the Jansen mechanism, four of them on each side. The proposed walking robot uses two actuators for the driving cranks that move opposite pairs of legs. The modified Jansen linkage has all the rotational joints on the mobile platform collinear and it ensures a better linearity of the coupler point in contact with the ground, offering a higher stability for the walking robot.

Keywords Jansen linkage · Legged walking robot · Kinematic analysis

58.1 Introduction

The legged walking robots are an important type of mobile robots, which are increasingly being developed in last decades. The research topics aim to increase the: stability of the mobile platform during the motion, movement on rough and uneven terrain, omnidirectional movement, length and height of the step to climb over obstacles, higher speed of motion and higher energetical efficiency to increase the autonomy of the walking robot.

For this goal were developed several mechanism structures for the legs, for example: pantograph 2D linkage [1–3], Mammal type leg [4], serial linkage [5, 6], 5-link belt mechanism [7] with 2 DOF mechanism and Klann linkage [8], Jansen linkage [9], Strider linkage [10], etc. using 1 DOF mechanism.

The Jansen linkage created by Jansen [10] reproduces very well the movement of a large terrestrial animal's foot. Since 2007 many studies have been done on this mechanism. Komoda et al. [11] proposes a parametric change from the precise circle to the ellipses of additional cyclic motion that generate different locomotive orbits, including upward movement, stepping in the same place and back. Nansai et al.

L. Alexeev · A. Dobra · E. Lovasz (✉)
Politehnica University of Timişoara, Bv. Mihai Viteazul 1, 300222 Timişoara, Romania
e-mail: erwin.lovasz@upt.ro

L. Alexeev
e-mail: lucian.alexeev@student.upt.ro

M. Rackov et al. (eds.), *Machine and Industrial Design in Mechanical Engineering*, Mechanisms and Machine Science 109,
https://doi.org/10.1007/978-3-030-88465-9_58

[12] set a theoretical basis for further investigation, optimization and extension of the Theo Jansen mechanism by analyzing the dynamics of a four-legged robot that uses that mechanism. In [13] Pop et al. developed an experimental test bench in order to analyze the trajectory shape described by Jansen linkage during a walking sequence. Lovasz et al. [7] followed by Mohsenizadeh and Zhou [14] developed a theoretical method for kinematic analysis using loop closing equations and simulated in MathCAD [7] and MATLAB/SolidWorks [14]. Wang and Hou studied in [15] the Jansen linkage with various changes in the position and dimensions of the elements. The implementation of the Jansen mechanism in the realization of an eight-legged stepping robot was studied by Vujošević et al. [16]. Chaterjee and Kanungo [17] analyzed the robot's movement by using the image processing method.

The paper propose a study of a modified Jansen linkages, where all the rotational joints on the mobile platform of the walking robot are collinear.

58.2 Modified Jansen Linkage Used for Walking Robot

The Jansen linkage proposed by Theo Jansen is an 8-link planar mechanism (Fig. 58.1) with 2 ternary and 6 binary links. The drive element is a crank (2), which transmit the motion through 2 four bar linkage (A_0ABB_0 and A_0AEB_0) to the

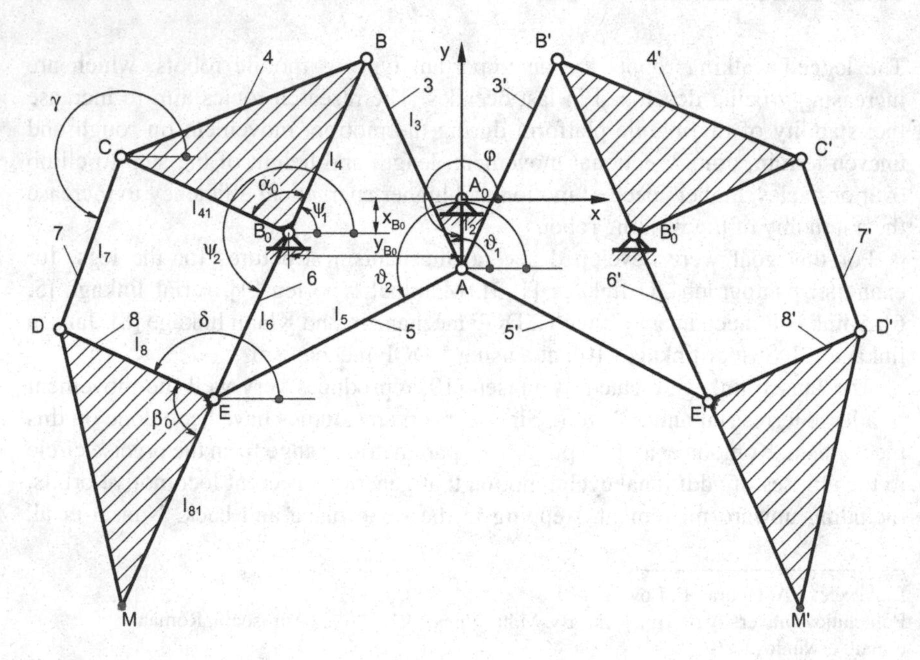

Fig. 58.1 Walking robot with opposite Jansen linkage legs

rockers (4) and (6). The movement of the two cranks generates through a five bar linkage (B_0CDEB_0) the path of the characteristic point M belonging to link (8).

The mathematical model for geometrical analysis was developed in [7] for a whole walking cycle. The parametric coordinates of the characteristic point M were established using the three closed loops equations:

$$\begin{cases} x_M(\varphi) = x_{B_0} + l_6 \cdot \cos\psi_2(\varphi) + l_{81} \cdot \cos(\delta(\varphi) + \beta_0) \\ y_M(\varphi) = y_{B_0} + l_6 \cdot \sin\psi_2(\varphi) + l_{81} \cdot \sin(\delta(\varphi) + \beta_0) \end{cases} \tag{58.1}$$

where:

$$\psi_i(\varphi) = 2 \cdot atan \frac{B_i(\varphi) - \sqrt{A_i(\varphi)^2 + B_i(\varphi)^2 - C_i(\varphi)^2}}{A_i(\varphi) - C_i(\varphi)} \tag{58.2}$$

$$\begin{aligned} &A_1(\varphi) = 2l_4x_{B_0} - 2l_2l_4 \cdot \cos\varphi, \quad B_1(\varphi) = 2l_4y_{B_0} + 2l_2l_4 \cdot \sin\varphi \\ &C_1(\varphi) = x^2B_0 + y^2B_0 + l_2^2 - l_3^2 + l_4^2 - 2l_2x_{B_0} \cdot \cos\varphi - 2l_2y_{B_0} \cdot \sin\varphi \\ &A_2(\varphi) = 2l_6x_{B_0} - 2l_2l_6 \cdot \cos\varphi, \quad B_2(\varphi) = 2l_6y_{B_0} + 2l_2l_6 \cdot \sin\varphi \\ &C_2(\varphi) = x_{B_0}^2 + y_{B_0}^2 + l_2^2 - l_5^2 + l_6^2 - 2l_2x_{B_0} \cdot \cos\varphi - 2l_2y_{B_0} \cdot \sin\varphi \end{aligned} \tag{58.3}$$

$$\delta(\varphi) = 2 \cdot atan \frac{B_3(\varphi) - \sqrt{A_3(\varphi)^2 + B_3(\varphi)^2 - C_3(\varphi)^2}}{A_3(\varphi) - C_3(\varphi)} \tag{58.4}$$

where:

$$\begin{aligned} &A_3(\varphi) = 2l_6l_8 \cdot \cos\psi_2(\varphi) - 2l_{41}l_8 \cdot \cos(\psi_1(\varphi) + \alpha_0) \\ &B_3(\varphi) = -2l_6l_8 \cdot \sin\psi_2(\varphi) + 2l_{41}l_8 \cdot \sin(\psi_1(\varphi) + \alpha_0) \\ &C_3(\varphi) = l_{41}^2 + l_6^2 - l_7^2 + l_8^2 + 2l_6l_{41} \cdot \cos(\psi_1(\varphi) + \alpha_0) \end{aligned} \tag{58.5}$$

The walking robot using opposite modified Jansen linkage legs is shown in Fig. 58.2. The modified Jansen linkage contains all rotation joints on the mobile platform (A_0, B_0, B'_0) collinear and the length between them higher. As a result of these changes, the other links will undergo dimensional changes in order to ensure a better linearity of the coupler point M in contact with the ground.

In Table 58.1 are given the geometrical parameters for the Jansen linkage versus the proposed modified Jansen linkage. For the two variants of Jansen linkage were computed the full path of the coupler point M during a full cycle Fig. 58.3.

The analysis of the paths generated by Jansen and modified Jansen linkages are presented in Table 58.2. The modified Jansen linkage shows a higher step height H, but a shorter step length L. The height variation between the return points of the flat curve domain is quite similar by both Jansen linkage versions.

In Figs. 58.4 and 58.5 were represented the paths of the coupler point M for Jansen and modified Jansen linkages with their leg and opposite leg respectively the complementary Jansen and modified Jansen linkage, which has the same crank

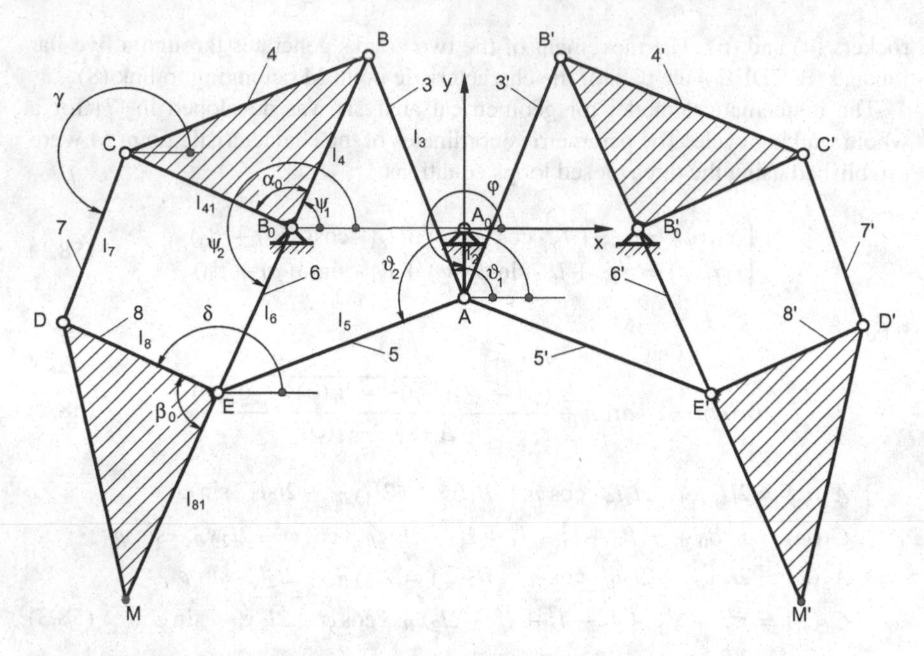

Fig. 58.2 Walking robot with opposite modified Jansen linkage legs

Table 58.1 Geometrical parameters of the Jansen versus modified Jansen linkage

Length/Angle	Jansen linkage (mm)	Modif. Jansen linkage	Length/Angle	Jansen linkage	Modif. Jansen linkage
A_0A/l_2	15.0	15.0	DE/l_8	36.7 mm	36.7 mm
AB/l_3	50.0	59.0	EM/l_{81}	49.0 mm	49.0 mm
B_0B/l_4	41.5	40.0	x_{A0}	− 38.0 mm	− 45.0 mm
B_0C/l_{41}	40.1	41.5	y_{A0}	− 7.5 mm	0.0 mm
AE/l_5	61.9	63.0	α_0	90.0°	86.4°
B_0El_6	39.3	39.3	β_0	90.0°	99.1°
CD/l_7	39.4	43.0			

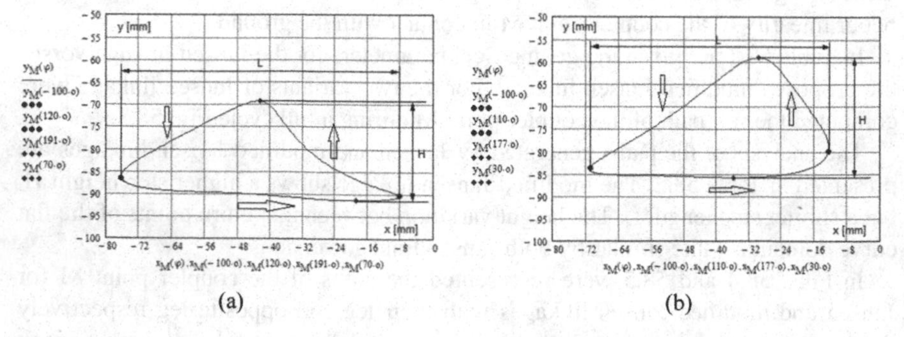

Fig. 58.3 Full path of the coupler point M of Jensen **a** modified Jansen **b** linkage

Table 58.2 Path parameters of the Jansen versus modified Jansen linkage

Parameter	Jansen linkage (mm)	Angular range φ	Modif. Jansen linkage (mm)	Angular range φ
Step length L	68.08	– 100° ÷ 120°	58.49	– 100° ÷ 110°
Step height H	22.08	– 100° ÷ 120°	26.73	– 100° ÷ 110°
Height variation flat curve domain	3.64	70° ÷ 191°	3.61	– 100° ÷ 110°
Step length on ground	63.13	– 70° ÷ 120°	50.19	– 100° ÷ 80°
Height variation step on ground	1.61	– 70° ÷ 120°	0.43	– 100° ÷ 80°

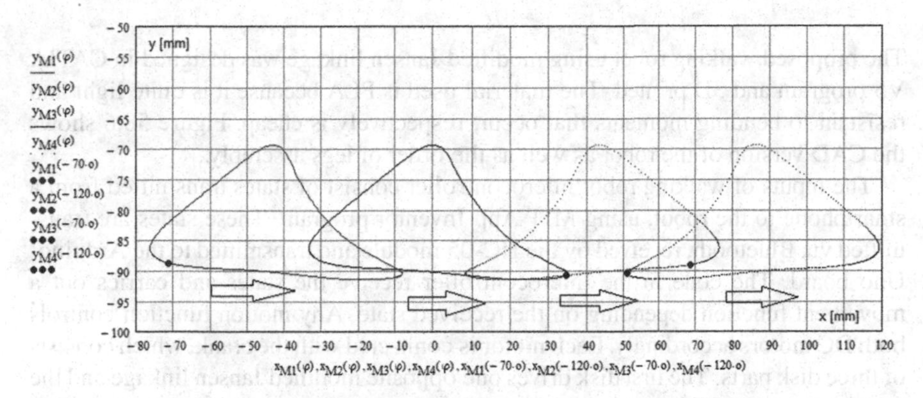

Fig. 58.4 Path of the coupler point M of Jensen linkage with their leg and opposite leg (red) respectively the complementary Jansen linkage (blue)

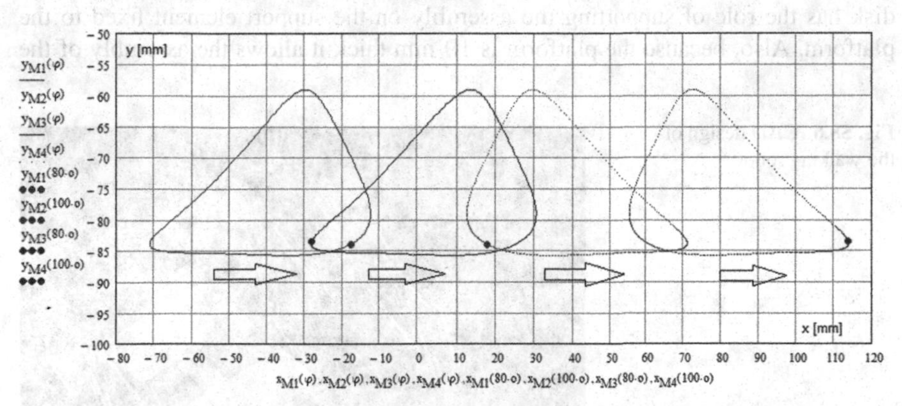

Fig. 58.5 Path of the coupler point M of modified Jensen linkage with their leg and opposite leg (red) respectively the complementary modified Jansen linkage (blue)

rotated with 180° and works on the same side of the platform. By analyzing the steps from the Jansen and complementary Jansen linkage can be observed that the beginning of the step on the legs is not superposed with the end of the step on the flat curve domain. That means the step length on the ground is shorter as the full length of the flat domain by the both versions of Jansen linkage, but the height variation of the modified Jansen linkages is like 25% lower as by the Jansen linkage. This advantage is very important for the waking robot, offering a higher stability during the motion for the mobile platform.

58.3 Experimental Prototype of the Walking Robot

The proposed walking robot using modified Jansen linkage was designed in CATIA V5 program and 3D printed. The material used is PLA because it is quite light and resistant to bending moments that occur, respectively is cheap. Figure 58.6 shows the CAD version of the robot as well as the order of legs assembly.

The inputs of walking robot microcontroller consist of states transmitted from a smartphone to the robot, using MIT App Inventor program. These states are transmitted via Bluetooth, received by the HC-05 module and transmitted to the Arduino-Uno board. The code in the microcontroller receive the states and carries out a movement function depending on the received state. Any motion function controls both DC motors accordingly. Each motor is connected with the crank, which consists of three disk parts. The first disk drives one opposite modified Jansen linkage and the second disk drives another opposite modified Jansen linkage, with the crank rotated with 180°. Both of them are on the same side of the walking platform. The third disk has the role of supporting the assembly on the support element fixed to the platform. Also, because the platform is 10 mm thick it allows the assembly of the

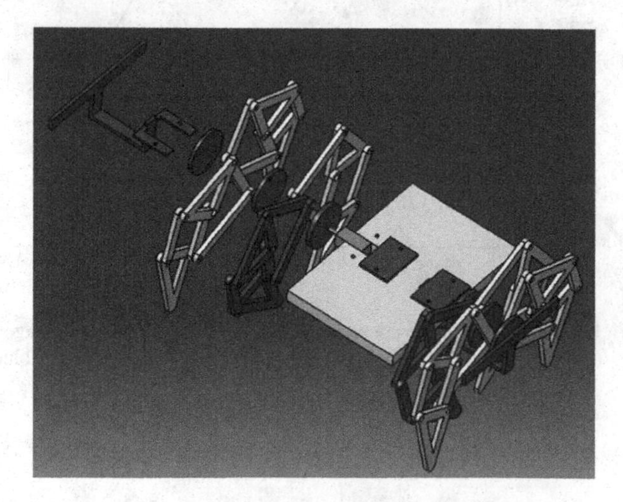

Fig. 58.6 CAD design of the walking robot

Fig. 58.7 Experimental prototype of the walking robot

motors in the slots located inside the platform. The motors used are DC motors with micro-gearbox, having a transmission ratio of 1:30. The final version of the walking robot can be seen in Fig. 58.7.

58.4 Conclusions

The paper proposed a modified Jansen linkage by redesigning the original Jansen linkage with collinear rotational joints on the mobile platform of the robot for all legs. The design of the mobile platform uses four legs on each side, first pair actuated by the same crank, and the second pair having the same crank rotated with 180°.

The Jansen and modified Jansen linkage were analyzed regarding the generated path of the coupler point M. The study shows that the modified Jansen linkage allows a higher step height, but a shorter step length. Also by analyzing the step order results that the step length on the ground is shorter by both Jansen linkages, but the height variation of the modified Jansen linkage is very low. That means a better linearity of the path in contact with the ground and a higher stability of the mobile platform of the robot.

Further researches will test the walking robot prototype and studied study their kinematic and dynamic behavior.

References

1. Waldron, K.J., Kinzel, G.L.: Kinematics, Dynamics, and Design of Machinery. Wiley (2003)
2. Simionescu, P.A., Talpasanu, I.: Kinematics of the eccentric RPRPR chain with applications to robotics, materials handling and manipulation. Int. J. Mech. Robot. Syst. **2**(3/4), 314–340 (2015)
3. Liang, C., Ceccarelli, M., Takeda, Y.: Operation analysis of a Chebyshev-Pantograph leg mechanism for a single DOF biped robot. Front. Mech. Eng. **7**(4), 357–370 (2012)
4. Song, S.M., Waldron, K.J.: Machine that Walk: The Adaptive Suspension Vehicle. MIT Press Cambridge (1989)
5. Preumont, A., Alexandre, P., Doroftei, I., Goffin, F.: A conceptual walking vehicle for planetary exploration. Mechatronics **7**(3), 287–296 (1997)
6. Doroftei, I., Preumont, A.: Development of an autonomous micro walking robot with articulated body. In: Proceedings of the 2nd International Conference on Climbing and Walking Robots, CLAWAR 1999, pp. 497–507. Wiley (1999)
7. Lovasz, E.-C., Pop, C., Pop, F., Dolga, V.: Novel solution for leg motion with 5 link belt mechanism. Int. J. Appl. Mech. Eng. **19**(4), 699–708 (2014)
8. Klann, J.C.: Walking Device. US Patent, No. US 6.260.862 B1 (2001)
9. Jansen, T.: The Great Pretender. Nai010 Publishers, Roterdam (2007)
10. DIY Walkers.com. https://www.diywalkers.com/strider-linkage-plans.html. Last accessed 2021/03/20
11. Komoda, K., Wagatsuma, H.: A proposal of the extended mechanism for Theo Jansen linkage to modify the walking elliptic orbit and a study of cyclic base function. In: Proceedings of the 7th Annual Dynamic Walking Conference, IHMC pp. 2–4. Pensacola (2012)
12. Nansai, S., Rajesh Elrab, M., Iwasea, M.: Dynamic analysis and modeling of Jansen mechanism. Procedia Eng. **64**, 1562–1571 (2013)
13. Pop, F., Dolga, V., Pop, C.: Analysis of the trajectory shape described by a robotic leg during a walking sequence. Appl. Mech. Mater. **658**, 684–689 (2014)
14. Mohsenizadeh, M., Zhou, J.: Kinematic analysis and simulation of Theo Jansen mechanism. In: Proceedings of the 15th Annual Early Career Technical Conference, ECTC 2015, pp. 98–106. Birmingham (2015)
15. Wang, C.Y., Hou, J.H.: Analysis and applications of Theo Jansen's linkage mechanism. In: Proceedings of the 23rd International Conference of the Association for Computer-Aided Architectural Design Research in Asia, CAADRIA 2018, pp. 359–368. Hong Kong (2018)
16. Vujošević, V., Mumović, M., Tomović, A., Tomović, R.: Robot based on walking Jansen mechanism. IOP Conf. Ser.: Mater. Sci. Eng. **393**, 012109-1–012109-1 (2018)
17. Chaterjee, S., Kanungo, P.: Experimental investigation of foot trajectory for a Jansen mechanism based walker. In: Proceedings of the International Conference on Engineering and Technology, ICET 2016, pp. 1–4. Coimabatore (2016)

Chapter 59
Sensor System and Control of a 3D Printed Bionic Hand

Toni Duspara and Hasan Smajić

Abstract In previously published article on this topic, design of a low-budget 3D printed prosthetic hand was discussed. A functioning prototype, which has independent movement for all fingers was manufactured for under 30€. In this paper, different means of hand control will be explained. Arduino microcontroller that has been linked to micro servomotors will transmit instructions for motor movement. Idea is to involve sensor system into this project, where microcontroller would calculate displacement of each finger depending on the sensor state. Two types of sensor system were tested. First system contained specially designed rings equipped with custom stretch sensors. Using this means of control, it is possible to replicate movement of a human hand onto the bionic model. Rings transmit stretch state of each finger to the microcontroller. Another means of control involve using "AI" and Machine learning. This system includes a camera linked to Google's open-source library called "Teachable machine". Camera takes a photo sequence of human hand in various positions. Later on, software trains the model and develops an algorithm which recognizes position of each part of the hand, and transfers it to the model. Next phase would involve using neurotransmission sensors, where hand would be controlled using brainwaves as signals that are transformed in movement.

Keywords Bionic hand · 3D printer · Machine learning · AI · Mechatronics · Actuator

59.1 Introduction

In an earlier published article titled "Development and Manufacturing of a controlled 3D printed bionic hand", a movable prothesis was made [1]. The average amputee pays tenths of thousands of Euros for a new custom-built arm/hand. With the advancement of technology through time, manufacturing processes became cheaper and more

T. Duspara (✉) · H. Smajić
Faculty of Vehicle Systems and Production, TH Köln, Betzdorfer Str. 2, 50679 Cologne, Germany

H. Smajić
e-mail: hasan.smajic@th-koeln.de

M. Rackov et al. (eds.), *Machine and Industrial Design in Mechanical Engineering*, Mechanisms and Machine Science 109,
https://doi.org/10.1007/978-3-030-88465-9_59

reachable. Conclusion of a previously written article was the fact that a functional prosthetic hand prototype can be built for under 30€. Even though this prototype does not have all the functions and proficiency as the full priced prosthetic hand, but it can replicate all the movements as the real device. All the fingers are capable of moving individually, sideways and with the work on the new version, gripping function could be perfected. Main essence of this article is based on control system of said hand. Different ways of actuator control are discussed. Movement signals are sent from Arduino microcontroller towards servomotors and thusly affecting the position state of each finger.

59.2 Smart Control Based on Standard Microcontroller

Control system requires a microcontroller that moves the servo motors. Arduino nano has 13 digital I/O pins. For control of the actuators, each servo requires one PWM output pin. Usually, Arduino can supply the voltage to actuate these micro servos, but because there are seven servos attached to this assembly, Arduino voltage output cannot provide enough current to move all of them simultaneously. For this project a 20 W additional power source is used. This is more than sufficient to provide continuous and simulations movement of all servo motors. Using pulse width modulation, signals for exact position of each servo is transmitted. Depending on output voltage of desired PWM pin, axle position of appropriate servomotor is set. When output pin is provided with 5 V, servo axle is placed in 180° position. Accordingly, 0 V on said pin results in 0° position of same axle. Servo levers are designed in a way that 180° lever position pulls the finger fully open, whereas 0° lever position pulls the finger fully closed. Servo connection cables are pulled between motors mounted on the arm, and later on to the electronics housing. Electronics housing holds inside an Arduino controller with soldered connectors for servo motors and power source. All the components and connectors are soldered onto a PCB board. This housing is used only for testing. In final design, (shown on Fig. 59.1), all electronic components are placed onto the arm assembly.

To connect all the servos to Arduino, a custom PCB board was designed. On one end of the board there is a standard voltage connector. The positive lead of voltage connector is soldered to the VIN pin of Arduino. The same pin is connected to all of the middle pins of servo connector (red wire). Other pin of power supply connector is soldered to the Arduino GND pin. With this, all the devices including servos and Arduino are powered with the same power source. Third servo lead is the signal lead and it is linked with PWM pins on Arduino. Used pins are D2 to D8 for the eight servomotors as it is shown on figure below.

Figure 59.2 shows wiring schematics for entire assembly and PCB board layout for Arduino and connectors. Using Arduino IDE, control system can be programmed and transferred to the device in order to achieve desired movement.

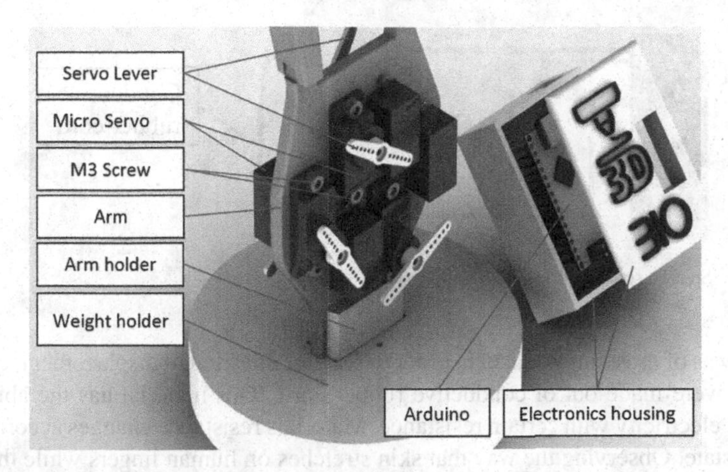

Fig. 59.1 Actuator assembly

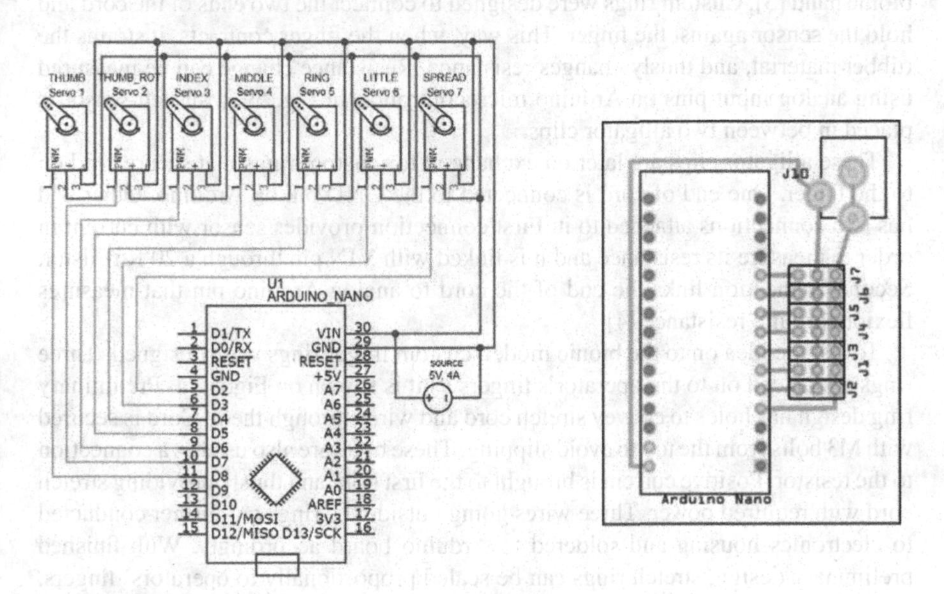

Fig. 59.2 Wiring schematics

59.3 Stretch Sensors

First tested means of control that could be applied on this assembly was based on movement replication. Idea is to build a sensor system that would work as a glove. This glove would be placed on to operators' hand, and depending on displacement of his fingers, movement would be replicated in real time onto the bionic model [2]. There are many ways to detect movement in space, but in this case, flexion of

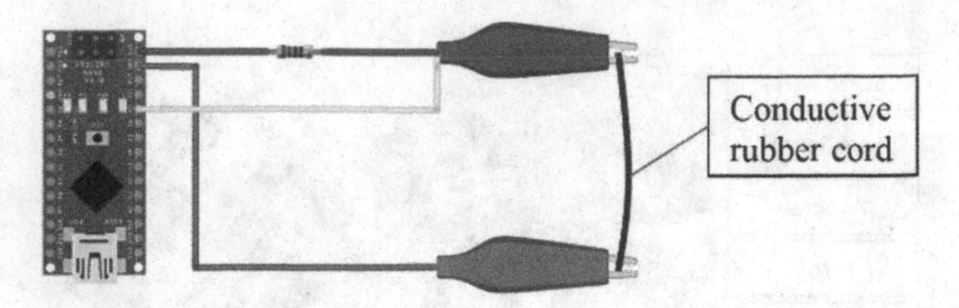

Fig. 59.3 Stretch sensor wiring

fingers was of most importance. In order to register this kind of displacement, custom sensors were made out of conductive rubber cord. This material has the ability to conduct electricity with certain resistance. Materials resistance changes according to stretch state. Observing the way that skin stretches on human fingers while they are flexing, lead to an idea that the same principle can be replicated in order to control bionic hand [3]. Custom rings were designed to connect the two ends of the cord and hold the sensor against the finger. This way, when the finger contracts, it strains the rubber material, and thusly changes resistance. Resistance change can be measured using analog input pins on Arduino microcontroller. In Fig. 59.3, stretch sensor is placed in between two alligator clips.

These alligator clips are later on exchanged for custom rings that secure cord on to the finger. One end of cord is connected to the GND pin on Arduino. Other end has two connections attached to it. First connection provides sensor with current in order to measure its resistance and it is linked with VIN pin through a 20 K resistor. Second connection links the end of the cord to analog Arduino pin that measures flexion through resistance [4].

To test the idea on to the bionic model, custom finger rings were designed. Three rings are placed on to the operator's fingers as it is shown on Fig. 59.4. Preliminary ring design has holes to convey stretch cord and wires through them. Cord is secured with M3 bolts from the top to avoid slipping. These bolts are also used as a connection to the resistor. Positive current is brought to the first bolt, and thusly providing stretch cord with required power. Three wires going outside the rings are further conducted to electronics housing and soldered to Arduino board accordingly. With finished preliminary design, stretch rings can be scaled proportionally to operators' fingers. To scale the rings, diameter of each finger must be measured in three places. More

Fig. 59.4 Rings placement

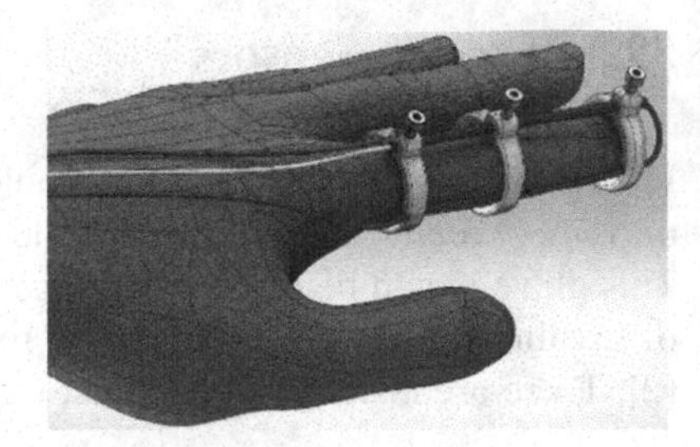

places that hold the cord fixed results in higher measuring precision. For basic testing, three rings were made for each finger. Rings are manufactured using rapid prototyping technology.

In order to control the bionic model with this sensor system, measured values must be converted into servomotor impulses. Analog pin is provided with voltage that changes its value according to flexion state. Arduino reads this value in range from 0 to 1023 digits which correspond to 0–5 V values. While observing analog readings, programmer marks value change in fully open and fully closed positions. These values changed in the range from 43 to 65. Servomotor positions are typed in a form of degrees, so analog read values 43–65 must be mapped to servo angles 0°–180°. An example code for a single finger is shown below.

```
#include <Servo.h>
int Angle = 0;
Servo Finger;
void setup () {
Serial.begin (9600);
Finger.attach (3); }
void loop () {
Angle = analogRead (A7);
Angle = map (Angle, 43, 65, 0, 180);
delay (10);
Finger.write (val);
delay(5); }
```

Code is uploaded to Arduino board using an USB Serial connection. After the upload is done, USB cable can be disconnected, and power cable plugged in. This test showed promising results.[1]

59.4 Machine Learning Recognition Approach

Second explored means of control involves machine learning. This type of system is also based on movement replication but without physical contact to the operator. Device used for movement recognition is a simple camera that records sequence of frames [5].

[1] Video preview of test: https://youtu.be/slh-K-N4hdc.

59.4.1 Teachable Machine as Learning Transfer WebApp

Google has developed a web-based tool called "Teachable machine" that creates machine learning models fast and easy. Teachable machine is a project from Google creative lab that incorporates transfer learning [6]. Transfer learning (TL) is a research problem in machine learning (ML) that focuses on stocking knowledge gained through solving one task later on using the same method, solving a different but related task [7]. Transfer learning is explained through tasks and domains. They define the transfer learning principle [8]. A domain $\mathcal{D}$ consists of: feature space χ and a marginal probability distribution $P(\mathrm{X})$, where $\mathrm{X} = \{x_1, \ldots, x_n\} \in \chi$. Given a specific domain $\mathcal{D} = \{\chi, \mathrm{P(X)}\}$, a task consists of two components: a label space $\mathcal{Y}$ and an objective predictive function $f: \chi \rightarrow \mathcal{Y}$. The function f is used to predict the corresponding label $f(x)$ of a new instance x. This task, denoted by $\tau = \{\mathcal{Y}, f(x)\}$, is learned from the training data consisting of pairs $\{x_i, y_i\}$, where $x_i \in \mathrm{X}$ and $y_i \in \mathcal{Y}$. Given a source domain $\mathcal{D}_S$ and learning task τ_s, a target domain $\mathcal{D}_T$ and learning task τ_T, where $\mathcal{D}_S \neq \mathcal{D}_T$, or $\tau_s \neq \tau_T$, transfer learning aims to help improve the learning of the target predictive function $f_T(\cdot)$ in $\mathcal{D}_T$ using the knowledge in $\mathcal{D}_S$ and τ_s [9].

For the control of the arm, different classes are created. Classes represent frame sequences using whom model is trained. For preliminary testing, two classes were created—Hand open and Hand closed. Sequence of 150–200 frames are imported into class. Number of imported frames can affect model quality (More frames implicates better quality). Into first class, operator imports a photo array of his open hand in different angles. Same principle is used for generating the second class. After naming the classes, model can be trained. Initiation of training process starts with preparing training data, and later on developing an algorithm using machine learning. Now web tool opens a camera live preview where model can be tested. By placing the hand in front of the camera and recreating some of the positions that were earlier recorded, algorithm should recognize the position and link it with class. For example, when someone places open hand in front of the camera, software outputs a class that was recognized and percentage of certainty. As it is shown on Fig. 59.5, after placing an open hand in front of the camera, software outputs with 100% certainty, that an open hand is placed in front of the camera.

These outputs can be used as binary variables transmitted to Arduino board. When a binary variable "Open" changes its state to high, Arduino gives a command to servomotors to place the bionic model into a fully open position. When variable "Close" changes its state from 0 to 1, controller signals the motors to close the hand. Arduino code can be set up in a way that only at adequate percentage of certainty changes the variable state from 0 to 1. While testing, algorithm was coded so only at 90% or higher certainty, variable changes its state. This sensor system provides control for the bionic model without physical contact with operator. Both of the mentioned means of control replicate the movement of human hand onto the bionic model. These sensor systems are used to establish capabilities of model and test its movement and response. Further work on this project involves using neurotransmission sensors. Usage of this technology could help develop this model

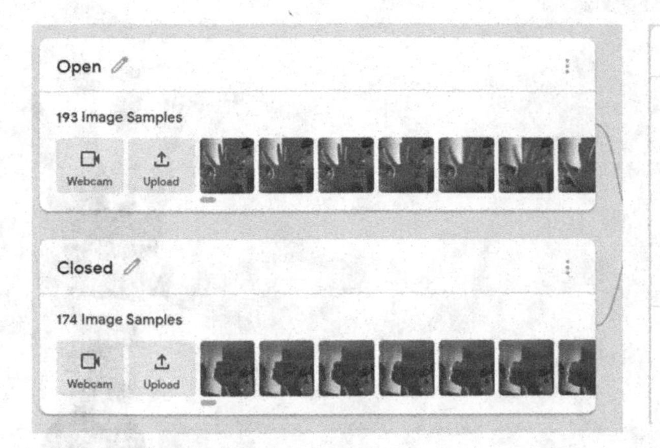

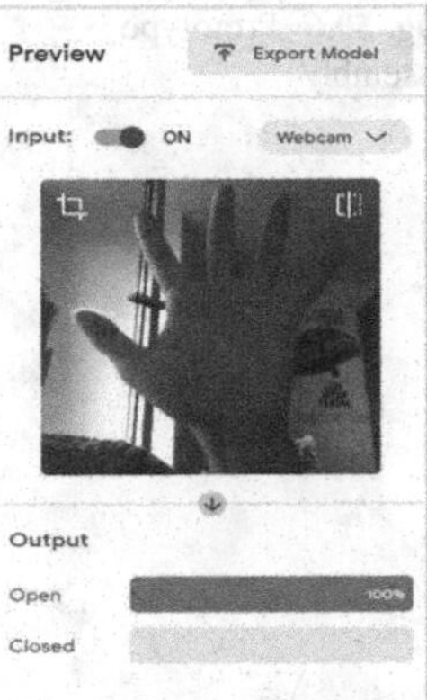

Fig. 59.5 Teachable machine preview

into a functioning hand prosthesis. Sensor system in that case wouldn't work on movement replication techniques. Operator could control this bionic model using only his brainwave activity.

59.5 Advantages in the Field

Prosthetic hands and limbs exist for a long period of time. In the new age, due to technical advancements of sensor and actuator systems, and the low-cost factor of materials and production, manufacturing technologies became more reachable [10]. Actuation part of this article shows an example of the low-cost advancement factor. Sensor systems, involving two different means of implementation, demonstrates the contactless and on-skin control. The contactless control is easier to implement, but there are certain limitations involving the actual position detection. Actual finger position can be registered only at certain angles. Tests conducted showed the immediate detection whether the finger is open or closed but not the actual position. Using the "On-skin" control via tension strings transmits the actual position of each and every finger. Further experimenting with different cameras and filters could improve recall of the contactless control.

59.6 Conclusion

During the work on this project, a final prototype was built. The average amputee pays 30.000$ for a new custom-built arm/hand. With the advancement of technology

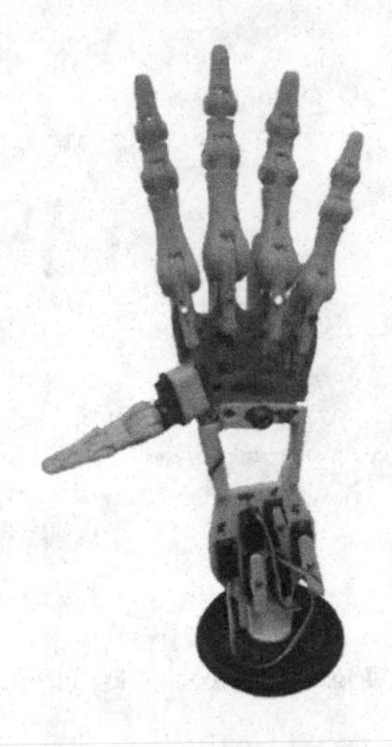
Fig. 59.6 Prototype assembly

through time, manufacturing processes became cheaper and more accessible. Technical innovation of this project was the fact that a functional prosthetic hand prototype was built for under 50$. Figure 59.6 shows the fully assembled picture of prototype built during this project.

This prototype does not have all the functions and capabilities as the full priced custom prosthetic hand, but it can replicate all the movements as the real device. All the fingers are capable of moving individually, sideways and with the work on the new version, gripping function could be perfected. Two means of control were tested and explained. A custom-made sensor system was developed. This system uses conductive rubber cords to register flexion of human fingers and replicates that movement on bionic model. Second system involves artificial intelligence and machine learning. Using a simple camera, a web-based tool trains an algorithm that recognizes human hand in front of the camera, and transmits its state onto the bionic model. Further work on this project would involve testing with different kinds of materials to improve the working stability. Second part of research would involve exploring of different sensor systems. Next phase would involve using neurotransmission sensors, where arm would be controlled using brainwaves as signals that are transformed in movement.

References

1. Thingiverse Homepage, www.thingiverse.com/thing:2838239. Last accessed 2020/06/01
2. Shi, W.T., Lyu, Z.J., Tang, S.T., Chia, T.L., Yang, C.Y.: A bionic hand controlled by hand gesture recognition based on surface EMG signals: a preliminary study. Biocybern. Biomed. Eng. **38**(1), 126–135 (2018)
3. Bronzino, J.D., Peterson, D.R.: Respiratory models and control. In: Bronzino, J.D. (ed.) Biomedical Engineering Fundamentals, pp. 197–216. CRC Press (2006)

4. West, J., Ventura, D., Warnick, S.: Spring Research Presentation: A Theoretical Foundation for Inductive Transfer. Brigham Young University, College of Physical and Mathematical Sciences (2007)
5. Michie, D., Spiegelhalter, D.J., Taylor, C.C.: Machine Learning, Neural and Statistical Classification. Ellis Horwood, Upper Saddle River (1995)
6. Teachabele Machine Homepage, https://teachablemachine.withgoogle.com/train/image. Last accessed 2020/09/09
7. Carney, M., Webster, B., Alvarado, I., Phillips, K., Howell, N., Griffith, J., Chen, A.: Teachable machine: Approachable web-based tool for exploring machine learning classification. In: Proceedings of the Conference on Human Factors in Computing Systems, CHI 2020, pp. 1–8. Honolulu (2020)
8. Lin, Y.P., Jung, T.P. Improving EEG-based emotion classification using conditional transfer learning. Front. Hum. Neurosci. **11**, 334-1–334-11 (2017)
9. Olivas, E.S., Guerrero, J.D.M., Martinez-Sober, M., Magdalena-Benedito, J.R., Serrano, L.: Handbook of Research on Machine Learning Applications and Trends: Algorithms, Methods, and Techniques. IGI Global (2009)
10. Clement, R.G.E., Bugler, K.E., Oliver, C.W.: Bionic prosthetic hands: a review of present technology and future aspirations. Surgeon **9**(6), 336–340 (2011)

Chapter 60
Sensors in Self-Driving Car

Livija Cveticanin and Ivona Ninkov

Abstract In this paper a short review on sensors applied in self-driving car are presented. The self-driving car is represents a cyber-physical system where the motion of the vehicle is automotive. To realize the driving the perceptional, orientation, communication and control for the car has to be developed. The process of perception and orientation is connected with a system of sensors. Various types of sensors are suggested: radar, lidar, ultrasonic sensor, video-, thermal-, and infrared camera. For navigation the Global positioning system and the Inertial measurement unit are usually applied. In the paper all of these sensors are considered. Advantages and disadvantages for all of them are discussed. It is concluded that depending on the type and number of embedded sensors the self-driving car is more or less efficient.

Keywords Radar · Lidar · Ultrasonic sensor · GPS · Inertial measurement unit

60.1 Introduction

Human beings have always been forced to move and travel shorter or longer distances in order to provide themselves with basic necessities of life, and, above all, with food. In the beginning he moved on foot, but soon there was a need to provide himself with a means of transport that would allow him to get to a certain place quickly, but also the ability to cover longer distances. The development of human beings and their way of life was accompanied by technical achievements. Development of civilization came to the emergence of various means of transport. Certainly one of the most significant is the emergence of personal vehicle for transport i.e. cars. The primary function of cars is to transport people from one to another location efficiently and in safe manner. The first cars appears on the roads in 1886 and was designed by Karl Benz

---

L. Cveticanin (✉)
Faculty of Technical Sciences, University of Novi Sad, Trg Dositeja Obradovića 6, 21000 Novi Sad, Serbia
e-mail: cveticanin@uns.ac.rs

L. Cveticanin · I. Ninkov
Obuda University, Nepszinhaz u. 18, Budapest, Hungary

M. Rackov et al. (eds.), *Machine and Industrial Design in Mechanical Engineering*, Mechanisms and Machine Science 109,
https://doi.org/10.1007/978-3-030-88465-9_60

in Germany. Since that time many modification and improvement are done on the motor and also on the car body. So, the speed of the car is increased up to 442 km/h (Hennesey Venom GT) and the motor is made to be extremely efficient with minimal fuel consumption and small carbon emission. Cars are extremely comfortable with decreased vibration and noise. However, for car motion the human driver is necessary. However, with the growing numbers of cars on the roads, the comfort in driving is disturbed. The safety on the road is decreased. By 2030 it is predicted that 70% of the world's population will live in big cities. In addition, it is taught that in following ten years the global car fleet would be doubled from currently 1.4 billion. These data directed the researcher to new investigation toward cars which would be more appropriate for modern passengers and give the idea of realization of the futuristic idea of un-manned car. Namely, the result which is expected in future is to have cars without human drivers which would move along roads safely without accidents giving the possibility to passengers to enjoy the time which spend in the vehicle. In addition, the automated cars are expected to reduce the fuel consumption, to reduce the CO_2 emissions, to optimize traffic flow, and so on. The challenge is how to fulfil the task.

Nowadays, we are in the era of the new technical revolution. As it is known, the first industrial revolution was in period of 1760–1870 when according to new inventions in steam and water power there was the transformation from the hand production to production with machines. This era is known as early mechanization period. The significant improvement in textile and iron industry, mining, agriculture etc. is achieved. The period of implementation of new technologies lasted for a long time. After discovering of electricity and its application the Second revolution called Technical revolution follows which lasts for almost hundred years. Factories became larger and were adopted for massive production. During this period there was the improvement in communication and expansion of railroads. The third so called Digital Revolution, connected with development of automation and digitalization, was in the period of 1962 to 2011. During this short time computers and specially supercomputers significantly improved the production process. However, the main result of this period was the improvement of communication and information (IT) technologies. Intensive networking established strong communication. Introduction of the Internet gave the worldwide connection in the population. The Fourth industrial revolution with the high-tech strategies of computerization of manufacturing, called Industry 4.0, started just ten year ago. The main aim of this revolution is to obtain automatic production without human intervention, i.e. to develop new methods and system of communication between physical systems and computers, i.e. the cyber-physical systems (CPS) technology. This system includes automation which gives the system the possibility of self-optimization, self-configuration, self-diagnosis, cognition. It requires advances in communication but also connections. It requires the improvements in some areas like: Robotics, Nanotechnology, Quantum computing, Biotechnology, Artificial intelligence, the Internet of things IoT, the Industrial internet of things IIoT, Fifth generation of wireless technology, etc. It is expected that using the CPS and additional technologies the realization of the project of the self-driving car would be fulfilled.

60.2 What is a Cyber-Physical System

There are a significant number of definition what a cyber-physical system (CPS) [1] is. In general, CPS represent an assembly of physical and a highly sophistic artificial computing system which is able to make perception, collect data, plan, realize and control the process. Integrated physics and logics is applied for the function of interacting components of human, physical analog and digital type which are the constitutive elements of CPS. Namely, computer-based algorithms controls or monitors the realization of the process in the real physical part of CPS. Physical and software parts of CPS are deeply coupled. The main advantage is that due to operation in different modes of space and time, it is possible to produce quite new behavior which may satisfy the new prescribed involvements. For realization the CPS requires mesh of various approaches: theory of cybernetics, mechatronics, designs and process science.

As is already mentioned, CPS is an integration of computation and physical process [2]. Devices which construct a CPS include sensors (for detection of physical values) and from simple hardware to high-end work computers for data managing and control to the most complex hardware for overall of the system. The working of the system is available due to a range of reliable, unreliable and compromised networks which move information and commands from one to another parts of working system. Due to connection system two types of CPS are developed: one, fully contained without outside connection, and second, with Internet connection.

The first type of CPS has its own independent system from collecting data up to making self-decision. Then the CPS is an isolated system focused of effective, reliable, accurate, real time and secure data transmission and control. The examples for these systems are those in smart bomb in flight to target, Mars rover operating between massages from Earth, in original vehicle in the first Defense Advanced Research Projects Agency (DARPA) challenge, etc.

The second type of CPS is usually connected with Internet. As Internet is a global system of interconnected computer networks, i.e. it is the network of networks which consists of all networks, it carries enormous range of information resources and services which are useful for making decision in CPS. Connection with Cloud servers, which are accessed over the Internet, many software and databases are available to CPS. Usually, Internet of Things (IoT) is viewed as the Internet of CPS which gives the interaction between cyber world and physical world.

The usual question is: What is Internet of Things (IoT) and what is the difference to CDS?

The Internet of Things (IoT) is the mesh of things, which are physical objects supplied with systems like sensors, software, etc., available for connecting and exchange data with other systems and devices by using the Internet. IoT focuses on effective source sharing and management, interface among different networks, massive-scale data and storage, data mining, data aggregation and information extraction, high quality of network, etc. Sensors in IoT provide the application of the obtained data.

The similarity between CPS and IoT is due to their primary architecture. Physical and computational elements are, however, more coupled in CPS than in IoT. Due to its complexity CPS gives addition means for realization of required tasks than IoT.

The control the process is in technics often called 'embedded system'. It contains a computer processor and memory, and peripheral devices for input and output. However, the main accent of the embedded systems is on the elements of computation, and not on the connection between physical and computational components. It is appropriate for control physical operations of machines, electrical and electronic devices as the computation is in real-time. Such systems are the part of the most CPS connected with IoT. The most important system of this type is the self-driving car.

60.3 Self-Driving Car (SDC)—Definition

A self-driving car, also called an autonomous vehicle (AV or auto) is a driverless car, or robot-car which percepts the environment and drives with little or no human help.

Society of Automotive Engineers SAE made the classification—J3016 201609 in 2016 in autonomous cars by introducing 6 levels [3].

Level 0 (No automation): Automated system issues warnings and may momentarily intervene but has no sustained vehicle control.

Level 1 ("hands on"—Driver assistance): The driver and the automated system share control. At this stage the control of steering is done by driver, while the automated system controls the power of the engine and regulates the speed of the car, and also controls the power of brake during speed variation. Steering during parking is automated and the speed is controlled manual. The car driver has to be ready to control the whole driving process at any moment.

Level 2 ("hands off"—Partial automation): Steering, braking and accelerating of vehicle is under full control of the automated system. However, the driver has to be active in the monitoring of process and to act at the moment when there is the lack in the automated system.

Level 3 ("eyes off"—Conditional automation): The attention of driver must not be directed to driving for the whole time as the vehicle will act immediately in some situations like emergency braking. However, the driver has to be ready to act in some real time interval.

Level 4 ("mind off"—High automation): This level represents the improvement in safety to level 3. The driver need not to give attention to driving process. Such cars is allowed to travel on roads in limited spatial areas or under special circumstances. Outside of these areas the driver has to retake control.

Level 5 ("steering wheel optional"—Full automation): No human intervention is required at all. It is a CPS that can travel to predominant destinations without human intervention of the driver.

The realization of the full automation of the 5th level for SDC requires two general systems (electronic devices and computers) which have to replace the activity of human brain in driving and these are divided into [4]:

- Perceptual systems,
- Positioning system of vehicle,
- Path planning and navigation systems for global and local rout planning, and
- Control and Decision making systems.

To percept and sense the environment the vehicle must be supplied with a system of various sensors which would detect the position of vehicle among other objects (see Fig. 60.1). Based on these information the path of motion is planned and the navigation paths are identified. Decision making computer systems are supplied with artificial intelligence AI and deep learning technologies which are used for control of driving [5], too.

For perception of the environment, the SDC is supplied with the external and internal sensor network system. In addition, it is found that the IoT assists in the integration of communications in vehicular systems. The first step in autonomous driving is to establish the so called vehicle-to-everything communication (V2X), i.e. the connection with other objects in environment and road infrastructure [6]. V2X is the automatic connectivity which use computer vision, localization and intelligent communication techniques. It consists of following main components:

- vehicle to vehicle communication (V2V)—very often is the WiFi connection,
- vehicle to infrastructure communication (V2I)—traffic infrastructure, road signs, lane marking, traffic lights,
- vehicle to networks communication (V2N)—using mobiles, tablets, navigation systems,
- vehicle to grids communication (V2G)—uses and produces electrical energy stored in car batteries by communication with electricity grids,
- vehicle to pedestrian communications (V2P)—using wireless smart-phones, and
- vehicle to device communication (V2D)—all other communications.

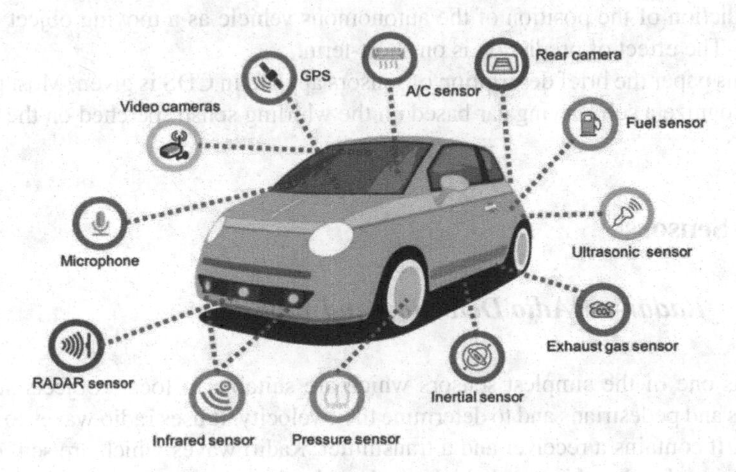

Fig. 60.1 Scheme of sensors in self-driving car

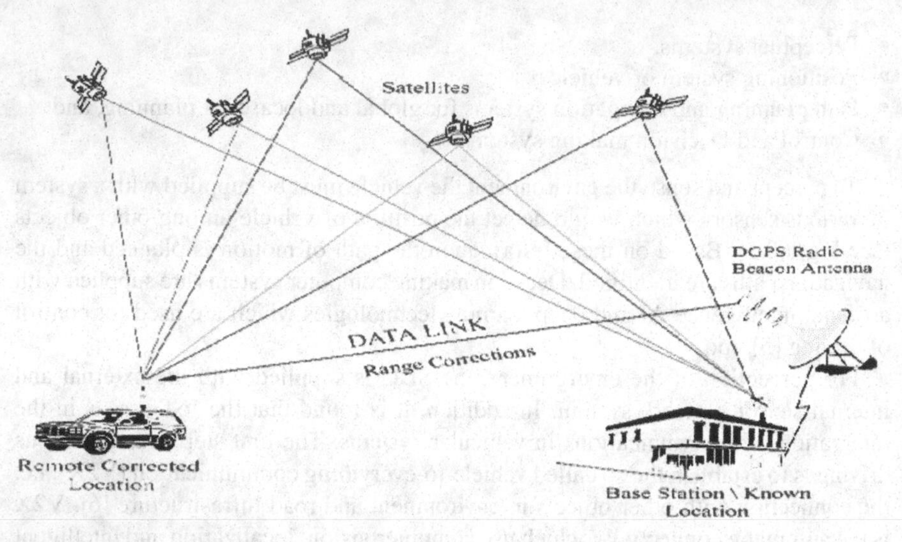

Fig. 60.2 Scheme of a GPS system

It can be concluded that this interaction between vehicles and inside SDC enables safety drive and internal control of the vehicle functionality, but also control of the traffic and management of transport on road.

For positioning of the vehicle complex systems are necessary. For navigation local and global route plans are done and the certain system performs the navigation operations (Fig. 60.2). Control of drive is usually of internal character. The system calculates and applies control actions to maintain the local trajectory of movement and the target states of vehicle.

In addition to hardware, various software programs are implemented in SDC: for prediction of the position of the autonomous vehicle as a moving object on the ground. The effect of prediction is on short-term.

In this paper the brief description of sensors applied in CDS is given. Most people will recognize a self-driving car based on the whirling sensor perched on the roof.

60.4 Sensors

60.4.1 Radar—RAdio Detection and Ranging

Radar is one of the simplest sensors which are suitable to locate objects such as vehicles and pedestrians and to determine their velocity. It uses radio waves to detect objects. It contains a receiver and a transmitter. Radio waves which are sent out by a transmitter hit the object or vehicle, and get back to the receiver. Radar available the detection of the distance of the object or vehicle, its direction and velocity. These data are necessary for control of speed, braking and control of safety systems as the response to sudden changes in traffic changes. Depending on the application the long range radars, medium range radars and short range radars are applied. Long range radars are used for measuring the distance to and speed of other vehicles. Medium range radars are used for detecting objects within a wider field of view e.g. for cross traffic alert systems. Short range radars are used for sensing in the vicinity of the car, e.g. for parking aid or obstacle detection. The radar fail to give accurate results if more objects are at the same distance and moving with different velocities. Radar resolve the velocities of the different objects but it needs time. In addition, radar does not give the exact size and shape of an object.

In SDC three types of radars are used: the co called 'short-range radar' (SRR) which is convenient for detecting of an object or vehicle on the distance of 1–20 m, 'medium-range radar' (MRR) for objects on the distance from 1 to 60 m and 'long-range radar' (LRR) for distances up to 250 m. Usually, these radars are extended with sensors for lane-change assistance (LCA) and detection of blind-spot (BSD).

60.4.2 Lidar—LIght Detection and Ranging

Lidar scanning is the latest development in surveying technology [7]. It is one of the more crucial and argumentative sensors that is used on self-driving cars used for surround view, detection of objects and to create detailed maps that self-driving cars need to get around. Lidar's function is similar to that of the radar. It helps autonomous vehicle to see other objects like cars, pedestrians and cyclists. Instead of radio waves to scan environment, lidar uses laser light pulses.

Lidar system contains four key elements: transmitter, receiver, optical analyzing system and a computer. The transmitter pulses laser pulses, i.e. laser light, while the receiver receives the reflection of the object i.e. intercept the reflected light pulses. Hundreds of thousands of laser pulses are transmitted and received every second. Input data are analyzed with optical analyzing system. The computer has to be powerful and able to give the real-life visualization. Images may be with two or three dimension. Namely, the onboard computer records each laser's reflection point, and forms an updating "point cloud" which is animated into 2D or 3D representation or picture. So, lidar literally map surroundings at the speed of light. Its versatility in direct air and in the vacuum of space allows lidar to operate on a short-wave, near-infrared optical signal—resulting in a much finer scan accuracy than longer waves, such as microwaves, could allow. However, the disadvantage of the lidar is that it cannot work normally in bad weather (rain, snow, dust). In addition, it is an expansive device.

60.4.3 Ultrasonic Sensor

Ultrasonic sensors are used to detect nearby objects. In addition, these sensors are helpful in parking of the vehicle. The ultrasonic sensor sends out short ultrasonic impulses. The ultra sound waves are reflected by obstacles and the echo signals are received and processed. Unlike the Lidar, ultrasonic sensors are able to work in bad weather conditions (including fog) and in low light night time situation, too. Ultrasonic sensors are able to see through objects (it is not the case in Lidar). Some newer versions have resolutions and object recognition capabilities comparable to Lidar. Ultrasonic sensor is relatively cheap and inexpensive. However, ultrasonic sensor has not the resolution to detect small objects or multiple objects moving at fast speeds. It has shorter field of view and accuracy compared to Lidar. Ultrasonic sensor cannot see color.

60.4.4 Video, Thermal and Far Infra-Red Cameras

Video cameras are sensors that continuously record the view in front of the car. Cameras can automatically send pictures and video on the screen. To obtain good quality of video, the resolution of camera need to be high. Usually, thermal sensing camera are applied, that passively collects heat signatures from nearby objects, and converts them into a video. Computer vision algorithms detect and classify the objects. Nowadays, the far-infrared (FIR) cameras are applied [8]. These sensors deliver reliable, accurate detection in real time and in any environmental condition. While radar and lidar are sensors which emit and receive signals, a FIR camera collects them. Namely, by determination of the thermal energy, which is radiated from the object, the data set is obtained which identifies the object in front of the vehicle. FIR cameras achieve better sensing than other sensors due to the fact that infrared wavelength is far longer than of the visible light. Unlike other sensing options, thermal sensors do not require any light to accurately detect, segment and classify objects and pedestrians. They give complete detection of the road and its surroundings in all-weather conditions. FIR can work to detect a car's surroundings without ever upsetting the sensors of other vehicles i.e. without interference. Lidar and radar, which are installed and acting on a SDC, may cause interference with other vehicles which are passing away.

60.5 Navigation

60.5.1 Global Positioning System (GPS)

One of the global navigation satellite systems (GNSS) is the so called Global Positioning System (GPS). It is a satellite-based radio-navigation system that provides geo-location and time information to a GPS receiver anywhere on or near the Earth. The time and three position coordinates of receiver are computed due to data from four or more GPS satellites whose locations are known with great precision. The GPS operates independently of any reception (telephonic or internet) and need not a user to transmit any data. GPS gives the positioning information for all users at any time. For calculation of the position and time parameter of the GPS on the SDC, the receiver of GPS need data from minimum 4 GPS satellites. Each satellite have to carry an accurate record of its position and time, and have to be able to transmit those data to the receiver. It is worth to be mentioned that the velocity of radio wave is constant and does not dependent on the satellite velocity. It causes the time delay between emitted and received signal. The time is proportional to the distance from the satellite to the receiver. Based on this procedure the calculated data are accurate enough. However, GPS fails if the line of sight is obstructed. Obstacles such as mountains and buildings block but also tunnels etc. give relatively weak or no GPS signals.

60.5.2 Inertial Measurement Unit (IMU)

To overcome the environmental problem in navigation with GPS, the Inertial Measurement Unit (IMU) for global positioning is designed. IMU is a kind of Micro-Electro-Mechanical System (MEMS). The unit determines the vehicle's acceleration, heading angle and relative position. The device is the combination of an accelerometer, gyroscope and magnetometer. The accelerometer measures acceleration in three directions and gives the information of the specific force. The gyroscope gives the rotation rate giving three independent angles in the space. Magnetometer is used as a heading reference giving the orientation of the body. The IMU is an autonomous precise navigation system which applies the principles of gravity and inertia and not of the external environment. IMU is a reliable data source for precise navigation of SDC. Namely, the IMU available the determination of the position of the vehicle mostly accurate, by applying of the algorithm for comparison of location.

Recently, GPS devices with enabled IMU are produced. When GPS-signals are unavailable, for example in tunnels, inside buildings, or when electronic interference is present, IMU takes over the navigation function. Otherwise, the GPS performs navigation.

60.6 Conclusion

In this paper a short review on sensors which are already applied in the self-driving car is presented. The attention is given to radar, lidar, ultrasonic sensor and video, thermal and far infra-red cameras as sensors which are widely used in the automotive vehicles. For navigation the GPS and inertial measurement unit are also considered. The advantages and disadvantages of each sensor is reported. It can be concluded that improvement the properties and elimination the lacks further investigation in sensors for self-driving car is necessary.

Acknowledgements The investigation is the part of the Project of the Department of Technical Mechanics of the Faculty of Technical Sciences, No.54/2021.

References

1. Putnik, G.D., Ferreira, L., Lopes, N., Putnik, Z.: What is a cyber-physical system: definitions and models spectrum. FME Trans. **47**, 663–674 (2019)
2. Madden, J.: Security analysis of a cyber-physical system: a car example. MSc Thesis. Missouri University of Science and Technology (2013)
3. Aria, M.: A survey of self-driving urban vehicles development. IOP Conf. Ser.: Mater. Sci. Eng. **662**, 042006-1–042006-6 (2019)
4. Vdovin, D.S., Khrenov, I.O.: Systems of the self-driving vehicle. IOP Conf. Ser.: Mater. Sci. Eng. **534**, 012016-1–012016-6 (2019)
5. Grigorescu, S., Trasnea, B., Cocias, T., Macesanu, G.: A survey of deep learning techniques for autonomous driving. J. Field Robot. **37**(3), 362–386 (2019)
6. Gwak, J., Jung, J., Oh, R.D., Park, M., Rakhimov, M.A.K., Ahn, J.: A review of intelligent self-driving vehicle software research. KSII Trans. Internet Inf. **13**(11), 5299–5320 (2019)
7. Chang, Y.P., Liu, C.N., Pei, Z., Lee, S.M., Lai, Y.K., Han, P., Shih, H.K., Cheng, W.H.: New scheme of LiDAR-embedded smart laser headlight for autonomous vehicles. Opt. Express **27**(20), A1481–A1489 (2019)
8. Thakur, R.: Infrared sensors for autonomous vehicles. In: Srivastava, R. (ed.) Recent Development in Optoelectronic Devices, pp. 81–96. IntechOpen (2018)

Chapter 61
High Speed Spindle Simulation Using Multibody Siemens NX MCD

Hasan Smajić, Milos Knezev, Aleksander Stekolschik, and Aleksandar Zivkovic

Abstract The digital product model is the key in setting up an end-to-end process from design to production: it works right from technical specifications to the finished product. Previously, the use of CAD systems was limited only to the creation of (2D) 3D models for linking and creating drawings. Now the main goal of introducing and using high-end CAD systems is to provide all product development parties through the necessary engineering tools and product information to resolve their problems. This problem is generally resolved not only by the CAD system in which the data is created, but also by the life cycle management system (PLM). In machine tools industry has been required growing tendencies aimed to development of digital twins, what allows designers to be more accurate in prediction of spindle lifetime. The number of sensor available to measure loads during spindle rotation is limited, but virtual model comparing with prototype cope better with that issue. This paper presents the first step of digital twin development of high speed spindle. Modeling 3D parts and assembly, defining simulation model has been developed in Siemens NX PLM software and dynamic results has been presented.

Keywords Mechatronic concept design simulation · Siemens NX · High speed spindle—HSS

61.1 Introduction

Multibody simulation has grown rapidly with the computer and software development, during the last decade, as the market for complex mechanical systems has an exciting trend. Designers face daily challenges related to the analysis of complex

H. Smajić · A. Stekolschik
Faculty of Vehicle Systems and Production, Technology Artz Science TH Cologne, 50968 Cologne, Germany

M. Knezev (✉) · A. Zivkovic
Faculty of Technical Science, University of Novi Sad, 21000 Novi Sad, Serbia
e-mail: knezev@uns.ac.rs

M. Rackov et al. (eds.), *Machine and Industrial Design in Mechanical Engineering*, Mechanisms and Machine Science 109,
https://doi.org/10.1007/978-3-030-88465-9_61

mechatronic systems. Dynamic simulation of multibody systems has several significant advantages related to lower cost and design time, but increased quality. As is known, early time finding deficiency of product, decrease unnecessary expenses [1, 2].

In other hand, regarding to the need of higher productivity but keeping or improving quality, came demand for development high speed spindles. Using high speed spindles can increase production, lower costs, and improve overall efficiency.

From all mentioned above, came an idea for developing digital twin. Most of the papers doesn't take into account bearing simulation, but consider bearings as a simple cylindrical joint and sliding joints. This paper presents first step of digital twin development i.e. multibody simulation model of spindle with accent on bearings. It gathers modeling all parts and assembly basis which can be created multibody simulation. Simulation in PLM software as is Siemens NX allows using various sensors to get required results.

61.2 PLM Software

The problem of choosing a software product arises because of the introduction of computer aided design systems in industry, educational and research processes at the university. The CAD market offers a wide range of software products focusing on resolving global or local problems. These products can be classified into groups:

- Lower level: Mainly two-dimensional programs focusing on the creation of design and technological documentation. These programs are generally not connected by a single data structure.
- Middle-range level: Specialized products (e.g. plant design software) working with a single source data structure.
- High-end level: Multifunctional integrated systems with a single source data structure and a variety of problem-oriented applications. Modern high-end engineering software development centers on the effective and fast integration. And the automation of different methods of engineering design and production along with the development of new functionality.

NX is an interactive high-end 3D-based engineering software system for engineering computer-aided design, manufacturing, and calculation of mechanical, electrical, and mechatronic products. NX is a three-dimensional modeling system for creating products of any degree of complexity (starting from general mechanical engineering to complex automotive and aerospace products). Considerable changes in computing performances and wider assortment of software tools for engineers i.e. software for LCM—Life Cycle Management enabled to use CAD, CAM, and CAE systems or modules to automate their design and production processes. NX provides key capabilities for these phases in the product creation process:

- NX CAD (computer-aided design): Integrated solutions for conceptual design, 3D modeling, and documentation. The design process is supported by various modeling tools and is flexible regarding to changes. Some modeling options are wireframe, surface, parametric solid, or the so-called direct modeling. Drawings can be linked with the 3D CAD model.
- NX CAE (computer-aided engineering): Multi-discipline simulation for mechanics, motion, thermal, flow, and multi-physics applications. It supports the development process by tools and methods for simulation and calculation based on a previously created 3D-CAD model.
- NX CAM (computer-aided manufacturing): Manufacturing solutions for tooling, machining, and quality inspection. It supports the manufacturing process in the planning and execution phase, including programming of manufacturing machines and the production process itself.

61.3 Mechatronic Concept Design

The MCD (mechatronic concept designer) application in NX is designed to simulate the behavior of mechanical and mechatronic systems, both in the early and late stages of product development. The user is provided with a virtual modeling environment in which can describe the product, being developed by using the physical characteristics of its components and set the boundary conditions, acting forces, and environmental parameters.

Moreover, the behavior of the system can be simulated in real time by using virtual sensors, signals, and interfaces to third-party products. The Mechatronics Concept Designer enables the combination of various disciplines like mechanical engineering, electrical engineering, and automation technology. Engineering designers and systems planners can provide a virtual definition of the entire engineering process from the rough concept to mechanics, electronics, and software. Errors can be recognized very early—this contributes to a low change cost and time of delivery [3, 4].

The 3D CAD model can be extended within MCD for:

- Mechanical connections,
- Springs and dampers,
- Collision definition,
- Position and velocity control,
- Conveying surfaces,
- Sensors and operators,
- Sequences and signals.

Another functionality of MCD is especially interesting for the Institute for Production in a series of research projects and education, emphasizing its multidisciplinary focus. The main focus and the critical part of the plant construction is the commissioning. The difficult error correction, which can only take place after the complete physical assembly of the production system, takes up 60% of the time.

The MCD environment contains the so-called virtual commissioning functionality. It allows simulations combining the physical and virtual components of a mechatronic product. In particular, a real digital controller can be combined with a virtual digital model of the mechanism or plant. The virtual machine model in Mechatronics Concept Designer can be used to test the real control technology; diverse automation interfaces can be connected. MCD behaves with the controller like the real machine or plant. The commissioning time on the real machine can be reduced considerably. The delivered software quality increases due to tests on the digital twin instead of on the real system without any risk in advance. Software issues in the industrial product lead to issues and even product recalls. The virtual commissioning of the 100% original software code is, therefore, the top priority with the use of tools like MCD. The typical process using MCD can be:

- Creation of 3D-CAD models of the product or plant (static product model),
- Creation of systems engineering model (e.g. function structure of the project, customer and legal requirements, re-use of knowledge),
- Definition of the product concept model (e.g. basic physical properties of the product concept, geometry and physical properties, allocation of components to functions, sensors and movements),
- Product completion to the detailed model (e.g. 3D design completion, use of electrical and electronic components, motor drives),
- Performing virtual commissioning (detailed movement analysis, signal tests, simulation execution, result exports to third-parties' simulation tools).

Different concepts of the product can be defined and the variants virtually compared without creating physical prototypes. So, the results can be achieved rather quickly. However, the quality of the results must be checked in detail because it is directly dependent on the virtual model properties. For example, some essential factors for realistic commissioning are the quality of the material properties definition or signal description.

61.4 High Speed Spindle—HSS

Spindle is one of the key parts of every machine tool. As a fact that industry requires high productivity, high accuracy with reducing a production costs. Although using the high speed spindle is reasonable solution. This paper is aimed to development of mechatronic concept design model of HSS so modeling of parts will not be taken into account. To be known CAD model was obtained by digitalization of physical HSS model.

Machine tools able to work with high speed and high precision conditions have become a major goal in machining industries in past years. The heart of these kind

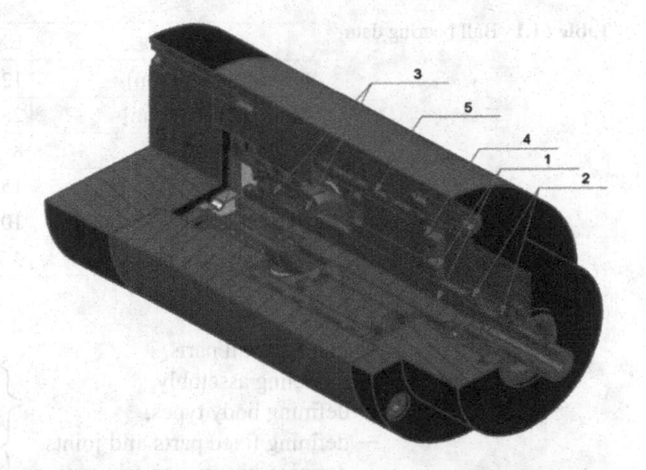

Fig. 61.1 The cross section of high speed spindle 3D model

of machine tools is high speed spindle, which is able to provide rotations of tens to hundreds thousands numbers of revolutions per minute [5–7].

Using HSS have a purpose to make machine tools capable to achieve excellent performance, and they are well-suited for grinding, milling, drilling or even for some special applications. The schematic of HSS elements is shown on Fig. 61.1. This HSS is actually the motor spindle and the shaft is integrated with rotor and it can be considered as one part (1). The spindle has been mounted with two pairs of high precise angular contact ball bearing in front side of spindle (2), are SNFA EX12 and the rear bearings are SHFA EX10 (3). This bearing system has most significant influence on spindle thermal, dynamic, etc. behavior. The number 4 is Stator, and beside ball bearings it is heat source, so it requires active cooling with water, the coolant flows through cooling jackets placed around stator. On the other hand bearings are cooled and lubricated with compressed air mist.

If would make a comparison, between conventional spindles and motorized spindles. Than can be said that motorized spindles are equipped with an embedded-in motor, so power transmission devices as gears and belts are obviated. This design provides better exploitation conditions, i.e. vibrations are lowed, rotational balance is increased, and control of rotational accelerations and decelerations are more precise. It soon became obvious that the high speed operation and the very high heat dissipation of built-in motors push other spindle components to their limits [3, 8]. The bearings dimensions are given in Table 61.1.

61.5 Mechatronic Concept Design Model

Developing the mechatronic concept design model, of high speed spindle includes following procedures:

Table 61.1 Ball bearing data

	EX12	EX10
Inner diameter [mm]	12	10
Outer diameter [mm]	28	26
Width [mm]	8	8
Contact angle [°]	15	15
Number of balls	10	10
Ball diameter [mm]	4.76	4.37

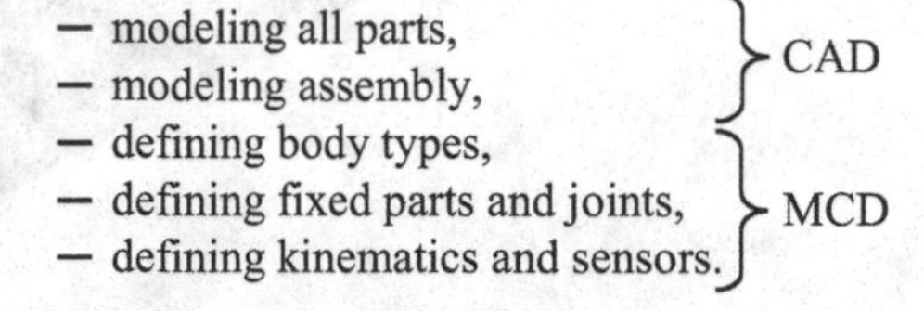

– modeling all parts,
– modeling assembly,
CAD
– defining body types,
– defining fixed parts and joints,
– defining kinematics and sensors.
MCD

Within developing MCD model of high speed spindle, non-movable parts are connected with housing, so they are fixed. List of fixed joints is shown on Fig. 61.2. Fixed joint connects a motion body to a fixed position (such as ground), or to another joint. The default position is the center of gravity. Two joints that are connected as fixed move together as one body. A fixed joint allows zero degrees of freedom.

Rotor/shaft is connected with stator by hinge joint. Hinge Joint represents the boundary condition for creating a rotational joint between two models/bodies. It

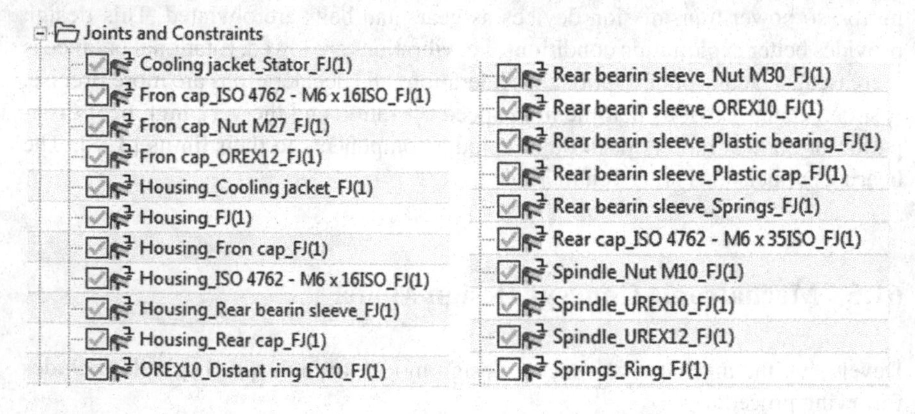

Fig. 61.2 List of fixed joints applied between housing and non-movable parts

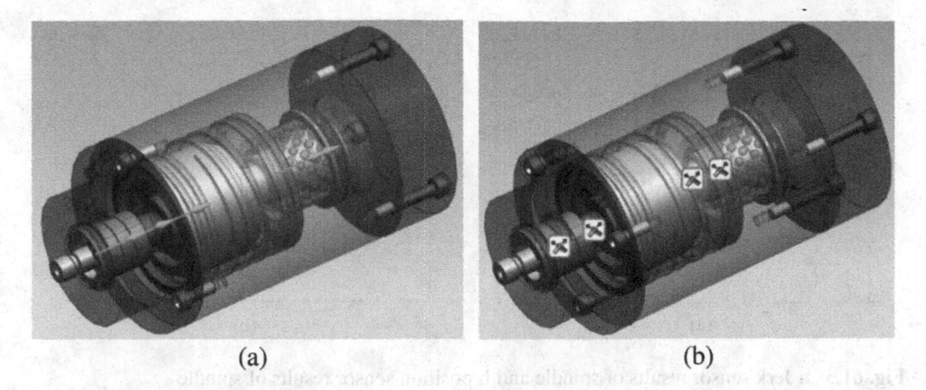

(a) (b)

Fig. 61.3 **a** Hinge joints and **b** prevent collision constrains

represents movement drive, during the simulation. Hinge joints, but not driving are defined between all four rolling bearings and shaft (Fig. 3a). While bearing are not in focus, their stiffness are defined by material parts. During the rotational between movable and fixed parts, which are in contact, to avoid parts penetration, have been used Prevent collision constraints (Fig. 3b).

61.6 Results and Discussion

Figure 4a shows velocity increasing from minimum to maximum, and the diagram is linear because acceleration is constant. On the next diagram shown on Fig. 4b, can be see how acceleration characteristic at the beginning is with constant trend, with small oscillations, but when spindle achieve maximum speed, acceleration oscillate around 0. Figure 5a represents results of jerk sensors. Shaft position change sensor, which results are shown on Fig. 5b shows the limit where position passes from nonlinear to linear. It is related with acceleration, at the same time (point) on acceleration diagram

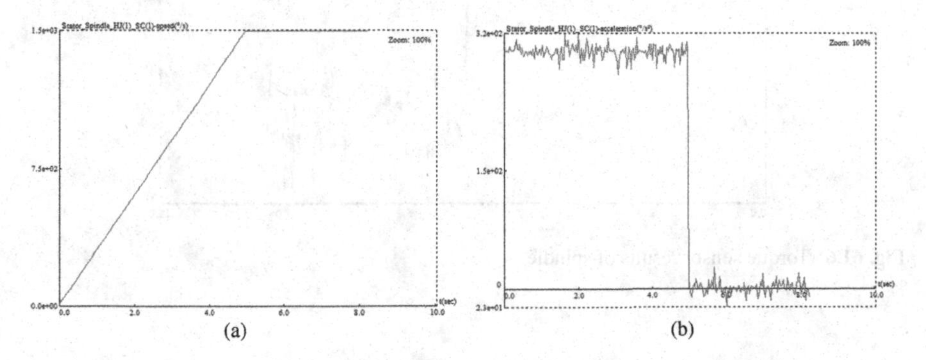

(a) (b)

Fig. 61.4 **a** Velocity sensor results of spindle and **b** acceleration sensor results of spindle

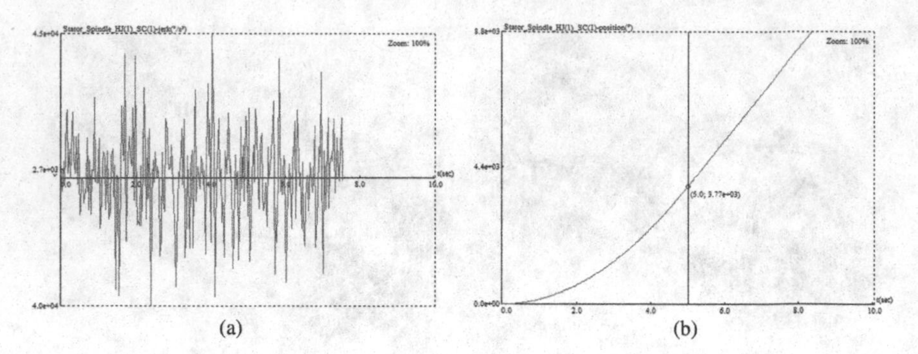

Fig. 61.5 **a** Jerk sensor results of spindle and **b** position sensor results of spindle

value drops around 0 o/s^2. During acceleration torque increases and after reaching maximum speed sharply drops little bit above 0 (Fig. 61.6).

Multibody simulation gives some advantages as follows. Virtual model simulation, enables to engineers relatively easy test of virtual prototype. Those mechatronic systems can be very complex, this tool makes testing possible in a short time. Comparing with the time required for building the physical prototype, thus automatically draws reducing costs. Software as is Siemens NX provides flexibility. Parametric software as Sienens NX, allows for change the any part dimension, and will automatically change it in assembly as well as in simulation module and quick simulate again, without a limitation of iteration with aim to obtain optimal solution.

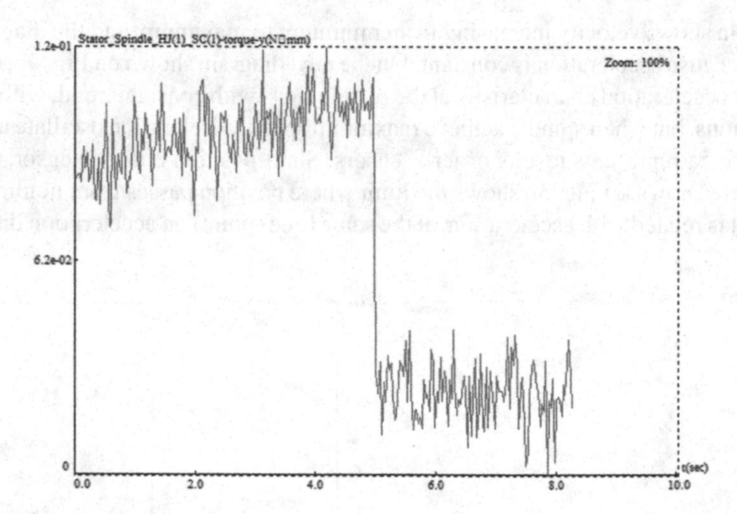

Fig. 61.6 Torque sensor results of spindle

61.7 Conclusion

A stepwise approach for first step to build digital twin of High Speed Spindle has been presented in this paper with accent on mechatronic concept design model. On first look motion of high speed spindle is very simple, but in fact it is totally opposite. For this spindle complexity is mainly related to high precision and high speed angular contact ball bearings. While in mechatronic concept design module, all parts are meshed, and during high speed rotation there is a critical point of speed where bearing falls apart. It is a problem which cannot be found in literature, so it is a great challenge and contribution for Industry 4.0 and digital twin development. Future plans for research and main challenge is to overcome it, because than will not be limitations regarding to speed. This paper presents the results of kinematic parameters of HSS spindle from mechatronic concept design model.

Acknowledgements This research (paper) has been supported by the Ministry of Education, Science and Technological Development through the project no. 451-03-68/2020-14/200156: "Innovative scientific and artistic research from the FTS (activity) domain".

References

1. Wünsch, G.: Methoden für die virtuelle Inbetriebnahme automatisierter Produktionssysteme. Herbert Utz Verlag, München (2007)
2. Patrick, A., Simons, S.: Virtuelle Inbetriebnahme mit NX 11 Mechatronics Concept Designer. In: Proceedings of the Fachkonferenz, Das Forum für Fachleute der Automatisierungstechnik aus Hochschulen und Wirtschaft, AALE 2017, pp. 215–224. Wildau (2017)
3. VDI 363-1: Virtuelle Inbetriebnahme—Modellarten und Glossar. Beuth Verlag (2016)
4. Zäh, M., Wünsch, G., Hensel, T., Lindworsky, A.: Feldstudie—Virtuelle Inbetriebnahme. Werkstattstechnik online (2006)
5. Zahedi, A., Movahhedy, M.R.: Thermo-mechanical modeling of high speed spindles. Sci. Iran. **19**(2), 282–293
6. Osinde, N.O., Byiringiro, J.B., Gichane, M.M., Smajic, H.: Process modelling of geothermal drilling system using digital twin for real-time monitoring and control. Designs **3**(3), 45-1–45-12 (2019)
7. Robert Lacour, F.-F.: Modellbildung für die physikbasierte Virtuelle Inbetriebnahme materialflussintensiver Produktionsanlagen. Herbert Utz Verlag, München (2011)
8. Bossmanns, B., Tu, J.F.: A thermal model for high speed motorized spindles. Int. J. Mach. Tools Manuf. **39**(9), 1345–1366 (1999)

Part VI
Quality Engineering and Management/Economical Aspects of Construction of Technological Equipment and Systems

Chapter 62
Influence of Maintenance Practice on MTBF of Industrial and Mobile Hydraulic Failures: A West Balkan Study

Marko Orošnjak, Milan Delić, and Sandra Ramos

Abstract The article investigates the influence of maintenance practice on the MTBF (Mean-Time-Between-Failures) of hydraulic system failures. Firstly, the paper starts by challenging the argument that contamination is at least 70% responsible for hydraulic system failures. Secondly, independent maintenance variables that potentially influence MTBF are synthesised and investigated via Person's correlation factor. Although some predictors (variables) show good prediction properties, however, show discrepancy while being subjected to different maintenance policies. Eight selected predictors were subjected to Stepwise Multiple Regression (SMR) for selecting the most appropriate solution. Finally, four main predictors ($p < 0.05$) are selected: Machine Age (MA), Filter Replacement Time (FRT), Failure Analysis Personnel (FAP) and Maintenance Policy (MP) applied. The results show that the suggested model shows good prediction properties ($R^2 = 83.51$) in estimating the MTBF of hydraulic machines.

Keywords Contamination control · Filter management · Mean-time-between-failures · Hydraulic failures · Stepwise multiple regression · Maintenance policy

62.1 Introduction

It has been an ever-present notion that the most common cause of failures in hydraulic systems is contamination [1]. Divided opinions of engineers and scientists stating that between 70 and 90% [2–4] of failures attributes to contamination (solid particles [5], air and water contamination [6], temperatures [7]) resonate with the need for more up-to-date evidence on the matter. Such statements suggest that hydraulic

M. Orošnjak (✉) · M. Delić
Faculty of Technical Sciences, University of Novi Sad, Trg Dositeja Obradovića 6, 21000 Novi Sad, Serbia
e-mail: orosnjak@uns.ac.rs

S. Ramos
Instituto Politécnico Do Porto, Instituto Superior de Engenharia, R. Dr. António Bernardino de Almeida, 431, Porto, Portugal

M. Rackov et al. (eds.), *Machine and Industrial Design in Mechanical Engineering*, Mechanisms and Machine Science 109,
https://doi.org/10.1007/978-3-030-88465-9_62

system reliability [8] is strongly associated with the fluid condition [9]. Consequently, this affects the actuator's response rate and precision, thus the product quality [10]. Therefore, one would expect that MTBF (Mean-Time-Between-Failures) depends on Maintenance Decision-Making (MDM) at the operational and tactical levels, thus encompassing the importance of the Maintenance Analysis Program (MAP), for instance, oil condition monitoring (OCM) [11] or Prognostics and Health Management (PHM) [12] in reducing MTBF. At the strategic level, there is a lack of evidence that compares the value of MTBF subjected to different MP (Maintenance Policy). However, assuming that innovative policies improve MTBF, studies highlight the cases where the most advance MP does not always show the best performance. For instance, Sellitto [13] investigated the influence of opportunistic, corrective and preventive maintenance policy performance. The study shows that corrective maintenance outperforms opportunistic and partial corrective. Vineyard et al. [14] study the influence of five MPs on flexible manufacturing system (FMS), showing that the opportunistic maintenance policy outperforms preventive and corrective policies. In a recent study, Paprocka et al. [15] pointed out that frequent maintenance actions reduce overall maintenance performance, questioning companies' ability to adapt CBM considering the trade-off between reducing stoppages and increasing profit.

To follow up on the maintenance practice perturbations and the influence on MTBF, specifically in an oil hydraulics sphere, we first challenge the argument that contamination is at least 70% ($P \geq 0.70$) responsible for the hydraulic system' failures (***hypothesis—H_1***). From our practical experience, pipes and hoses' bursting is the most common component failure. However, if the contamination is still a relatively high cause of failure, we can presume that filter replacement time (FRT) must have a strong correlation ($r > 0.5$, $p < 0.01$) with MTBF (***H_2***) given the sample size. Since our focus group comprises the companies utilising hydraulic systems, the study eligibility criteria consider filter replacement practice regardless of the policy. Therefore, we set the final hypothesis as (***H_3***): "Maintenance policy does not play a significant role as a predictor in regression modelling" (with $p > 0.05$).

This study aims to provide a clear understanding of the maintenance practice effect on hydraulic MTBF's with a specific regression model considering predictors of interest. Three additional research objectives are defined for accomplishing the study aim: (1) develop a questionnaire-based instrument for collection of empirical evidence for a defined time-span; (2) determine critical predictors using stepwise multiple regression (SMR); (3) validate the model through hypothesis testing. The rest of the study is as follows. Section 62.2 explains the general research model and methodology used for setting the research context. Section 62.3 depicts the questionnaire meta-data, failure-related and maintenance-related empirical evidence on the West Balkan territory, and the research findings. The final section encapsulates the research findings and proposes future research agenda.

62.2 Methodology

The questionnaire-based survey is set for the region of West Balkan territory. The survey is disseminated to the companies explicitly utilising hydraulic mobile and industrial machines for servicing and manufacturing purposes. The questionnaire instrument validation is developed through three stages (Fig. 62.1): (1) survey design—literature review, determining variables, and expert judgment (panel 1); (2) survey simulation—setting region, period, and focus group to determine if the survey is understandable and if variables can be measured (panel 2); (3) survey analysis—meta-data, data sorting and filtering in respect to eligibility criteria, and evaluation of empirical evidence collected.

Additionally, companies willing to share the data could bypass the survey if all the data available is appropriate. As stated, MTBF served as a function for investigating latent maintenance variables affecting the result. Aside from machine age (MA), the maintenance variables collected for estimating the influence on MTBF include maintenance personnel per machine (MPPM), Maintenance Department Team (MDT), Failure Analysis Personnel (FAP), Filter Replacement Time (FRT), Condition Monitoring Sensors (CMS), Oil Replacement Time (ORT), MP and MAP.

Based on the data, we presume that the systems are in a somewhat mean exploitation stage of the bath-tub curve and not in the early (infant) nor wear-out stage since models show good prediction properties as a linear function. Therefore, with statistically significant confidence, we can confirm that the MTBF can follow the proposed model's linearity if, and only if, the fundamental assumptions are confirmed—linearity, multivariate normality, absence of multicollinearity and homoscedasticity. The multivariate normality of regression analysis assumes that the residuals are normally distributed. Multicollinearity assumes that independent variables are not highly correlated, which can be confirmed using Pearson's correlation (<0.80).

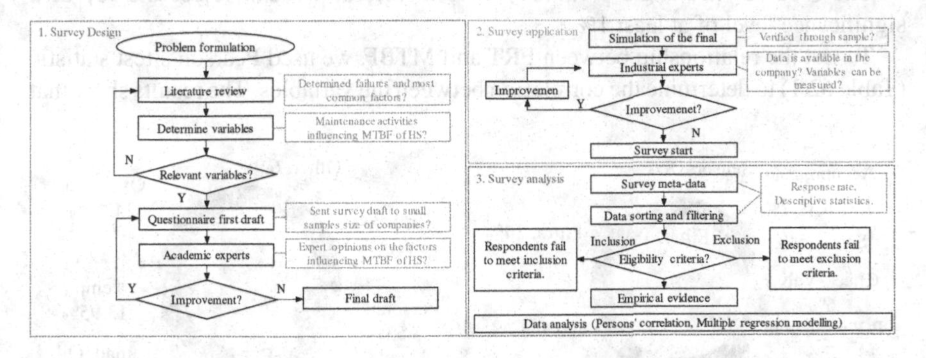

Fig. 62.1 Survey design and general research model

Homoscedasticity assumes that the error variance is similar across independent variables. The plot shows if the points are equally distributed around the dependent variable. Finally, to determine the maximum number of predictors used for the regression model, we used a rule of thumb that a sample should be between 5 and 15 per variable (predictor).

62.3 Results and Discussion

The survey results show a 37% response rate from 220 companies in the evaluation realised between May 2018 and September 2019 on West Balkan's territory. However, after analysing the data and concerning eligibility criteria, 18 respondent applications were flagged as ineligible. In the final 63 survey applications (1153 hydraulic machines) were eligible for the analysis of which 31 companies were from the manufacturing sector, while 32 companies were from construction, open- and closed-mining sectors.

Given the results, the evidence suggests that the most common failures are hoses and pipes (Fig. 62.2). Thus, to test the claim that the most common cause of failure is contamination ($H_1, P \geq 0.70$), we used a z-test statistic for one sample proportion. The data shows that contamination is responsible for 37.98% of failures as a sum of contamination due to air, water, temperature, and solid particles (Fig. 62.2).

Thus, the test statistic shows the following results:

$$Z = \frac{\hat{P} - P}{\sqrt{\frac{P \cdot q}{n}}} = \frac{0.3798 - 0.7}{\sqrt{\frac{0.7 \cdot 0.3}{1153}}} = -23.73 \tag{62.1}$$

The *z* score shows the *p-value* <0.01; therefore, we can reject the H_1 at a significance level of at least 1%.

To test the relationship between FRT and MTBF, we used Pearson's test statistic (Table 62.1) to determine the correlation between the variables. The results show that

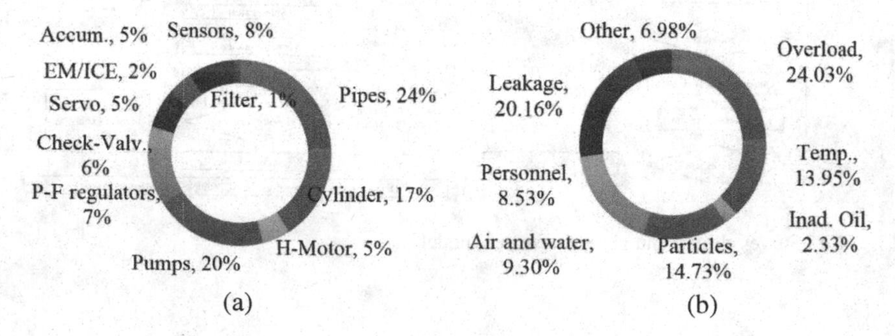

Fig. 62.2 Most common component failures (**a**) and the most common causes of failures (**b**)

Table 62.1 Pearson's correlation matrix of potential factors influencing MTBF

	MTBF	MP	MAP	MPPM	MDT	FAP	MA	FRT
MTBF								
MP	0.40***							
MAP	0.26**	0.41***						
MPPM	– 0.06	– 0.05	0.18					
MDT	0.18*	0.18*	0.11	– 0.01				
FAP	0.48***	0.37***	0.13	– 0.06	0.36***			
MA	– 0.83***	– 0.21*	– 0.14	0.18	– 0.02	– 0.36***		
FRT	– 0.65***	– 0.20	– 0.16	– 0.03	– 0.24*	– 0.45***	0.58***	
CMS	0.13	0.04	0.21*	0.11	– 0.10	0.13	– 0.25**	– 0.21*

Note MTBF Mean Time Between Failures; *MP* Maintenance Policy; *MAP* Maintenance Analysis Program; *MPPM* Maintenance Personnel Per Machine; *MDT* Maintenance Department Team; *FAP* Failure Analysis Personnel; *MA* Machine Age; *FRT* Filter Replacement Time; *CMS* Condition Monitoring Sensor. *p*-value < 0.01***, *p*-value < 0.05**; *p*-value < 0.1*

variables MA, FRT, FAP and MP show a strong correlation ($p < 0.01$) with resulting MTBF function, while MAP and MDT show lower tendency with $p < 0.5$ and <0.1, respectively. Thus, the value of –0.65 ($p < 0.01$) shows a high negative correlation of FRT and MTBF, by which case H_2 is proven and accepted as valid.

Displayed results show a low *p*-value of MDT, MPPM and MAP, and after running an SMR in MINITAB and using ANOVA, the coefficients show the *p-value* > 0.05 and are excluded from the modelling. Besides, we ran a Grubbs test for outliers and used a scatter plot to check for linearity. The scatter plot shows good linearity, and the outlier test ($G = 1.91$) shows the absence of outliers. The second assumption is to check multivariate normality, i.e. errors of observed and predicted values are normally distributed (Fig. 62.3—Normal Probability Plot). The third assumption is to check the absence of multicollinearity, which is proven with no coefficients between independent variables $r > 0.80$ (Table 62.1). The final assumption is to check for heteroscedasticity in the data. The expansion in the model's linearity suggests heteroscedasticity, and data, in this case, is homoscedastic (Fig. 62.3—Versus Fits; Versus Order).

After elaborating and confirming linearity assumptions, a general MLR analysis model is formulated as:

$$y = \beta_0 + \beta_1 x_1 + \beta_2 x_2 + \cdots + \beta_n x_n + \varepsilon \quad (62.2)$$

Additionally, for multiple non-linear regression, we used exponential regression as:

$$y = \exp(\beta_0 + \beta_1 x_1 + \beta_2 x_2 + \cdots + \beta_n x_n + \varepsilon) \quad (62.3)$$

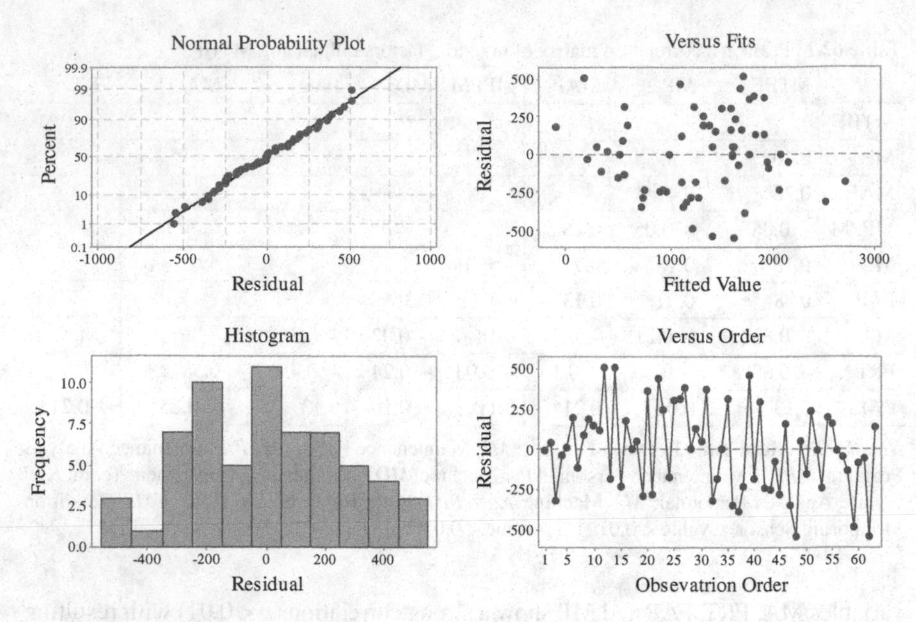

Fig. 62.3 MTBF residuals testing the predicted and observed value

The y is the dependent variable, x_i are independent variables, β_i are maintenance parameters/coefficients, and ε is the error ~$N(\mu, \sigma^2)$. $MTBF_{NLR}$ model is an exponential function model, and $MTBF_{LR}$ is a multi-linear model with the function of MTBF modelled from coefficients in Table 62.1. Since FAP and MP variables are qualitative expressions, the predictors are categorical (dummy) variables (0, 1). The resulting models are represented in the following equations.

$$MTBF_{NLR} = e^{8.641 - 0.1119 \cdot MA - 0.000132 \cdot FRT + MP + FAP} \quad (62.4)$$

$$MTBF_{LR} = 3112 - 0.1052 \cdot FRT - 100.2 \cdot MA + MP + FAP \quad (62.5)$$

Both models show good R^2 and R^2_{adj} values; however, prediction properties of R^2 show a relatively high deviation with the exponential case (Table 62.2). Finally, we used ANOVA analysis (Table 62.3), showing that MP ($p < 0.05$) rejects the hypothesis ($\boldsymbol{H_3}$) and shows that the influence of MP as a predictor is significant in the case of $MTBF_{LR}$. In exponential case, SMR excludes candidate predictor MP

Table 62.2 Resulting R^2 values for MLR and MNLR optimised models

Model	S	R^2 (%)	R^2_{adj} (%)	R^2_{pred} (%)
$MTBF_{NLR}$	0.329	81.45	77.88	69.73
$MTBF_{LR}$	287.294	83.51	80.34	75.17

Table 62.3 ANOVA results of coefficients in optimised MLR and MNLR model

Source	DF	Adj SS	Adj MS	*F*-value	*P*-value
Multiple non-linear regression model ($MTBF_{NLR}$)					
Regression	10	24.7028	2.4703	22.83	0.000
FRT	1	0.5450	0.5450	5.04	0.029
MA	1	7.9028	7.9028	73.05	0.000
MP	4	0.8808	0.2202	2.04	0.103
FAP	4	1.2177	0.3044	2.81	0.034
Error	52	5.6258	0.1082		
Total	62	30.3285			
Multiple linear regression model ($MTBF_{LR}$)					
Regression	10	21,733,903	2,173,390	26.33	0.000
FRT	1	348,134	348,134	4.22	0.045
MA	1	6,338,796	6,338,796	76.80	0.000
MP	4	1,648,447	412,112	4.99	0.002
FAP	4	1,191,207	297,802	3.61	0.011
Error	52	4,291,981	82,538		
Total	62	26,025,884			

because the *p-value* > 0.05. The results show that the MP' coefficients of individual policies significantly affect ($p < 0.05$) in improving the R^2 of the $MTBF_{LR}$ model (Table 62.4).

For more in-depth analysis and discussion of MP and FAP coefficients, the data is given in Table 62.4. There are individually five categorical variables of both MPs and FAP. The linear model's function shows that the best policy is CBM, while other policies show degradation in MTBF given in coefficients. The worst policy for maintaining hydraulic machinery is DM. We suspect that engineers in redesigning or modifying the equipment did not consider the MDT and expertise of FAP in later removal or reducing the equipment failures. The FBM also shows a result of a 574 h reduction in the MTBF indicator. The PM shows a slight reduction of 467 h in the MTBF indicator, whereas OM shows the lowest 400 h reduction. Indeed, CBM shows the best performance compared to other policies, suggesting that condition monitoring in predictive analytics and reducing stoppages plays an important role in increasing the availability of hydraulic machines.

Considering the FAP, the failure analysis is divided into five categories. The best performance on the MTBF indicator is if a specialist performs failure analysis. Engineer show no effect on MTBF improvement nor reduction. Outsourcing failure analysis shows a reduction of 46 h in MTBF, while if there is no failure analysis personnel, i.e. if parts are replaced in the "as-good-as-new" approach, the 156 h reduction MTBF is observed. Finally, if failure analysis performs a technician on the "as-bad-as-old" basis, the evidence suggests the highest MTBF reduction by 368 h. Overall, it can be said that accurate analytics plays one of the most crucial roles in

Table 62.4 MTBFLR model-independent continuous and categorical coefficients

Term	Coefficient	SE Coeff	*T*-value	*P*-value	VIF
Constant	3112	176	17.66	0.000	
MA	– 100.2	11.4	−8.76	0.000	1.66
FRT	– 0.1052	0.0512	−2.05	0.045	1.84
MP					
CBM	0	0	*	*	*
DM	– 800	209	−3.83	0.000	1.51
FBM	– 574	156	−3.68	0.001	2.68
PM	- 467	131	−3.57	0.001	3.07
OM	- 400	194	−2.06	0.045	1.71
FAP					
Engineer	0	0	*	*	*
None	– 156	134	−1.16	0.251	2.12
Outsource	– 46	114	−0.41	0.686	2.19
Specialist	145	157	0.92	0.361	1.62
Technician	– 368	128	−2.88	0.006	2.15

Note Maintenance Policies—*CBM* Condition-Based Maintenance; *DM* Design-Out Maintenance; *FBM* Failure Based Maintenance; *OM* Opportunity Maintenance; *PM* Preventive Maintenance. *VIF* Variance Inflation Factor. *SE Coeff.* Standard Error Coefficient

reducing the MTBF of a hydraulic system. Inappropriate failure analysis by unspecialised personnel leads to the escalation of MTBF, or in the cases of a replacement of the parts on “an as-good-as-new” basis leads to more financial investments.

62.4 Conclusion

The lack of recent studies on the causes of hydraulic failures provoked the authors to conduct the study. Current findings suggest that contamination is still the primary cause of failures at a much lower proportion. However, although system overload and personnel mistakes are considered the causes of failure, an intrinsic relationship with contamination is present. This is because it is impossible to isolate and provide an exact reproduction of particular conditions of failure; even though the study included 63 companies utilising 1153 machines, the research may be impeded by the lack of complete details surrounding failure causes in practice. The fact that FRT plays a significant role in MTBF reduction stress this implicit causality. Finally, the model highlights that coefficients (MP and FAP) are statistically significant predictors in improving the R^2_{adj} value. Further research includes an in-depth exploration of the causality between hydraulic MTBF and MP on all MDM levels.

References

1. Espinosa-Garza, G., De Jesus Loera-Hernandez, I.: Improvement of productivity in hydraulic systems with servomechanisms. Procedia Manuf. **41**, 779–786 (2019)
2. Ng, F., Harding, J.A., Glass, J.: Improving hydraulic excavator performance through in line hydraulic oil contamination monitoring. Mech. Syst. Signal Process. **83**, 176–193 (2017)
3. Tič, V., Edler, J., Lovrec, D.: Operation and accuracy of particle counters for on-line condition monitoring of hydraulic oils. Ann. Fac. Eng. Hunedoara 425–428 (2012)
4. Mariusz, D., Hassan, M., Joanna, D.: Simulation of particle erosion in a hydraulic valve. Terotechnology **5**, 17–24 (2018)
5. Karanović, V., Jocanović, M., Baloš, S., Knežević, D., Mačužić, I.: Impact of contaminated fluid on the working performances of hydraulic directional control valves. Stroj. Vestnik/J. Mech. Eng. **65**, 139–147 (2019)
6. Baker, M.: Most Common Causes of Hydraulic Systems Failure. https://yorkpmh.com/resources/common-hydraulic-system-problems
7. MACHydraulics: Effects of Temperature on Hydraulic Systems. https://mac-hyd.com/blog/hydraulic-system-temperatures
8. Orošnjak, M., Jocanović, M., Karanović, V.: Quality analysis of hydraulic systems in function of reliability theory. In: Proceedings of the 27th DAAAM International Symposium on Intelligent Manufacturing and Automation, pp. 569–577. DAAAM International, Vienna (2016)
9. Jocanovic, M., Agarski, B., Karanovic, V., Orosnjak, M., Ilic Micunovic, M., Ostojic, G., Stankovski, S.: LCA/LCC Model for evaluation of pump units in water distribution systems. Symmetry-Basel **11**(9), 1181-1–1181-21 (2019)
10. Runje, B., Horvatic Novak, A., Keran, Z.: Impact of the quality of measurement results on conformity assessment. In: Proceedings of the 29th DAAAM International Symposium on Intelligent Manufacturing and Automation, pp. 51–55. DAAAM International, Vienna (2018)
11. Karanović, V.V, Jocanović, M.T., Wakiru, J.M., Orošnjak, M.D.: Benefits of lubricant oil analysis for maintenance decision support: a case study. IOP Conf. Ser.: Mater. Sci. Eng. **393**, 012013-1–012013-8 (2018)
12. Orošnjak, M., Jocanović, M., Karanović, V.: Applying contamination control for improved prognostics and health management of hydraulic systems. In: Ball, A., Gelman, L., Rao, B. (eds.) Advances in Asset Management and Condition Monitoring: Smart Innovation, Systems and Technologies, vol. 166, pp. 583–596. Springer, Cham (2020)
13. Sellitto, M.A.: Analysis of maintenance policies supported by simulation in a flexible manufacturing cell. Ingeniare **28**(2), 293–303 (2020)
14. Vineyard, M., Amoako-Gyampah, K., Meredith, J.R.: An evaluation of maintenance policies for flexible manufacturing systems. Int. J. Oper. Prod. Manag. **20**, 409–426 (2000)
15. Paprocka, I., Kempa, W.M., Skołud, B.: Predictive maintenance scheduling with reliability characteristics depending on the phase of the machine life cycle. Eng. Optim. **53**, 165–183 (2021)

Chapter 63
Automated MIG Welding Application: An Industrial Case Study

Miguel Ángel Zamarripa Muñoz, Pedro Agustín Ojeda Escoto, and Gerardo Brianza Gordillo

Abstract Manufacture companies continually seek to improve processes trying to save money, in this case, developed equipment has the specific objective to avoid an industrial manipulator purchase for weld application. Gas metal arc welding (GMAW) or as called metal inert gas (MIG) is the selected process for welding a boom arm with irregular geometry. Most versatile method for welding in manufacture industry is through an industrial manipulator, not all companies are economy capable to get this kind of technology. There is cheaper technology like the welding carriage, but it isn't capable to follow trajectory changes. This paper shows low cost equipment (case study) to apply automated and continuously MIG welding on irregular geometry boom arm. Developed equipment allows to get a continuously weld bead and is adaptive to the boom geometry. Boom arm has straight and curve sections along itself and weld should be applied continuously (no overlapping bead) for product esthetics. Weldment device design was developed under criteria of design for manufacturing (DFM) methodology, DFM practices leads to more competitive products because it directly addresses cost. Finally, weld penetration, porosity and deformations on arm were the parameters checked after welding having good results.

Keywords Weld · Equipment · Boom

63.1 Introduction

Welding is a metallurgical fusion process, where parts to be joined are brought together such that heating and solidification results into permanent joint. It is used in

---

M. Á. Z. Muñoz (✉) · P. A. O. Escoto · G. B. Gordillo
Universidad Tecnológica de Aguascalientes, 20200 La Cantera, México
e-mail: miguel.zamarripa@utags.edu.mx

P. A. O. Escoto
e-mail: pedro.ojeda@utags.edu.mx

G. B. Gordillo
e-mail: gerardo.brianza@utags.edu.mx

M. Rackov et al. (eds.), *Machine and Industrial Design in Mechanical Engineering*, Mechanisms and Machine Science 109,
https://doi.org/10.1007/978-3-030-88465-9_63

every large or small industry. In metal products, it is one of the most important fabricating and repairing process. The process is economical, efficient and dependable as a means of metal joint. MIG (Metal Inert Gas) welding and in the USA known as GMAW (Gas Metal Arc Welding), is now a widely process used for welding a variety of materials, ferrous and non-ferrous. In gas shielded arc welding both the arc and the molten weld pool are shielded from the atmosphere by a stream of gas. The arc may be produced between a continuously feed wire and the work pieces [1–3].

There are many methods for welding, robotic arc welding process requires complicated control and sensing techniques applied to various process parameters. These measures improve repeatability and enhance the quality of welds. Weld penetration depth is one of the critical weld profile parameters that have significant influence on fatigue life and structural integrity.

This situation can be observed for load carrying welds, where the stress concentration factors at the weld root and weld toe decrease with increasing weld penetration depth [4].

MIG welding robots are attractive to manufacturers due their versatility but large investment is required. Welding carriage are capable to automates welding and cutting operations, it can increase speed and improve efficiency in production plants, but this technology has limitations to apply weld bead on irregular geometries. Another option is welding manually, but the amount of heat in the welded part is higher than automated welding and error probability is present. When long weld bead is applied on a part, heat amount is a critical factor to consider, this heating causes deformations in the part.

In order to reduce the cost of the project and trying to keep automated welding to guarantee functionality of the part, an equipment capable for applying the weld bead continuously on a boom arm was developed and is described in this paper. A coordinate table with driving device and torch support was developed for this welding application. The arm to be welded is built on ASTM A36 steel.

63.2 Theoretical Framework

Welding is a manufacturing process in all industries, small or large. In metal products, it is one of the most important fabricating and repairing process. The process is economical, efficient and dependable as a means of metal joint. The process finds its applications underwater, space and in air. Why welding is used—Because it is:

- Suitable for small thicknesses to a third of a meter, and
- Versatile, it can be applied to a wide range of component sizes and shapes.

MIG may be operated in semiautomatic, machine, or automatic modes. All commercially important applicable metals such as carbon steel, high-strength, low-alloy steel, and stainless steel, aluminum, copper, titanium, and nickel alloys can be welded in all positions with this process by choosing the correct electrode, shielding gas and welding variables (see Fig. 63.1).

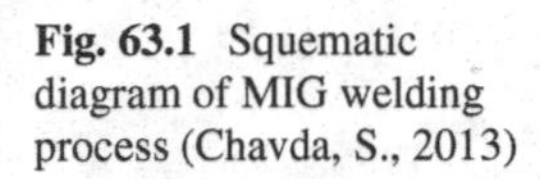

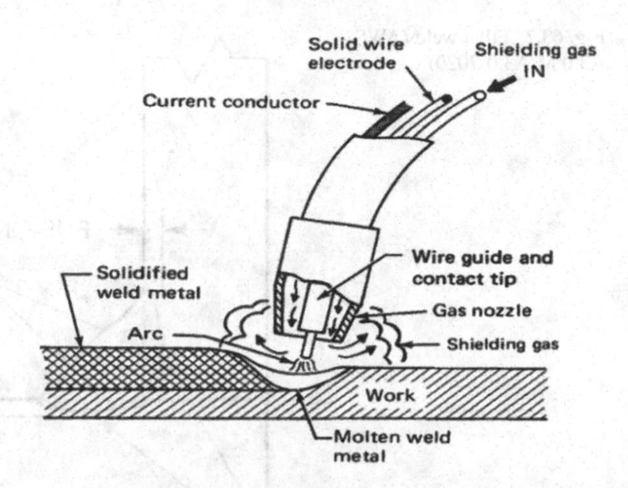

Fig. 63.1 Squematic diagram of MIG welding process (Chavda, S., 2013)

The welding parameters are selected by operator based on a handbook or experience. But, this does not guarantee that the selected welding parameters can produce the optimal weld pool geometry for that particular welding machine and environment [5–7].

In design process, the goal is to develop satisfactory product at the lowest cost. Welding is a vital approach in design because it is arguably the best joining process [8].

The arc may be produced between the work and a continuously fed wire. Continuous welding with coiled wire helps high welding speed and high metal depositions rate. Generally, the filler wire is connected to the positive polarity of DC source becoming one of the electrodes. The welded part is connected to the negative polarity. The power source could be constant voltage DC power source, with electrode positive it yields a smooth metal transfer with least spatter for the entire current range and stable arc [9].

In order to validate weld quality, terms of "fusion" and "fusion depth (penetration)" should be clarified. The American Welding Society (AWS) defines fusion as "The melting together of filler metal and base metal (substrate), or of base metal only which results in coalescence" [10]. Fusion occurs once you have atomic bonding of the metals. On the other hand, penetration, or properly termed fusion depth, is defined as "The distance that fusion extends into the base metal or previous bead from the surface melted during welding". To obtain the correct weld strength, all welding requires complete fusion to occur between pieces of metal and filler metal but not all joints require a large fusion depth or deep penetration [10]. Figure 63.2 shows fusion depth on a fillet weld.

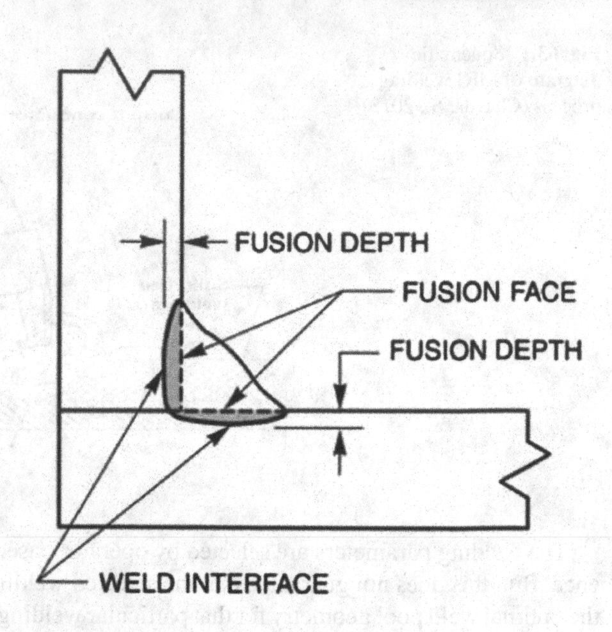

Fig. 63.2 Fillet weld (AWS A3.0 M/A3.0-2020)

63.2.1 *Design for Manufacturing (DFM)*

The general rules of DFM consist of designing assembly with a minimum number of parts, standard parts, modular design, and multi-functional parts, making parts standard for multiple products, maximum surface roughness and tolerance, avoiding secondary processes, using materials that are easy to manufacture, minimizing the handling of parts, and setting the guidelines of design and shape. These general rules focus on the cost and manufacturability of the process, which lead to uniform/standardized products [11].

The objective to use DFM applied to one particular process is design products easy to maintain, reliable, in less time and keep it simple [12]. To reach the goal, there are some principles that design team should keep in mind:

- Minimize number of components,
- Use modular design,
- Use standard commercially available components,
- Design multifunctional parts,
- Design for ease part fabrication,
- Avoid separate parts,
- Eliminate or reduce adjustment required,
- Use wide tolerances,
- Minimize the number of operations,
- Avoid secondary operations,
- Redesign components to eliminate process steps,

- Minimize operations that do not add value, and
- Design for the process.

63.3 Methodology

Weldment device design focuses on DFM method in order to reduce time and cost of fabrication (see Fig. 63.3). Modular design, minimizing number of components, using wide tolerances, designing for easy part fabrication, number of components reduced design for easy assembly, using standard commercially available components were some of the considerations for project development.

Assembly configuration allows to have multi-functional parts on driven device and rail supports of platform, generation of left and right hand parts is not required avoiding bending parts. Wide tolerances on platform allows to save manufacturing time. Motor, pulleys, bearings and springs were some of the purchased parts on assembly, all of them are standard components and easy to get them.

On the other hand, about welding, characteristics and advantages provided by MIG welding were considered [10].

- It can be used in many kind of materials (Carbon steel, Stainless steel, Aluminum, etc.),
- Continuous electrode, it increases productivity, no waste time changing electrode; Welding speed is higher than coated electrode,
- Welding can be applied in any position,
- Long weld beads, they can be applied with no overlapping, and
- Slag weld removing is not required.

There are fundamental parameters that should be checked to get a good welding. Control and adjustment of these parameters helps to get quality on welds. These variables are: Electric tension, speed wire feed, polarity and protection gas.

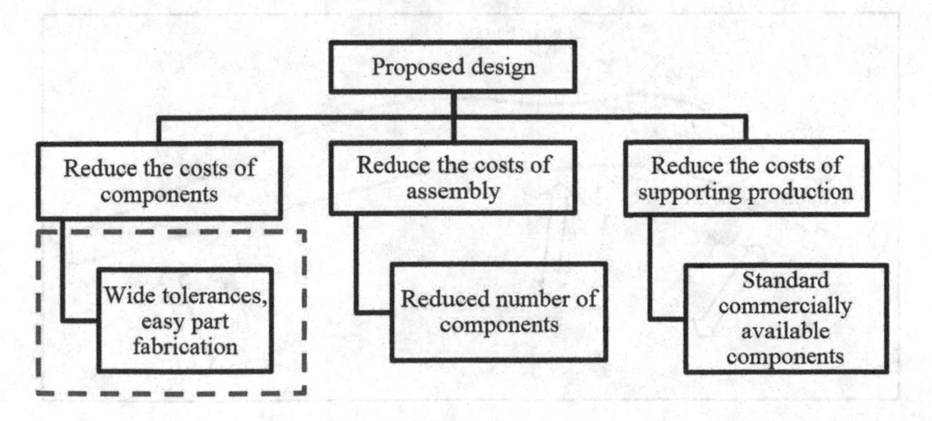

Fig. 63.3 Using DFM on weldment device (own production)

Often, the coordinate tables have motors in the axes to provide the movement, in this equipment axes X, Y and Z are free. Driven device (cyan) is hanging from the rail (green) though the yellow part. Conceptual design is shown in Fig. 63.4.

Developed idea was based on the use of a coordinate table, under this method, driving device has contact with the welded part and it is able to copy the arm geometry, welding torch is located through support arm installed on the driven device, in this way, torch follows the arm geometry for welding.

Mechanism has three wheels in contact with part, one of them has a compression spring guaranteeing contact of three wheels anytime (Fig. 63.5), torch is located on the welding edge through positioning arm. Electronic speed variator on driven wheel provides speed adjustment. Part cross section changes along arm (Fig. 63.6), mechanism is adaptive for this geometry and device runs along arm (Fig. 63.5).

Welded part requires 4 long weld beads and the heat on part is a critical factor. Once the two weld beads on the top of the piece are applied (location #1 and #2, Fig. 63.6), the piece is turned for the opposite side for welding location #3 and #4

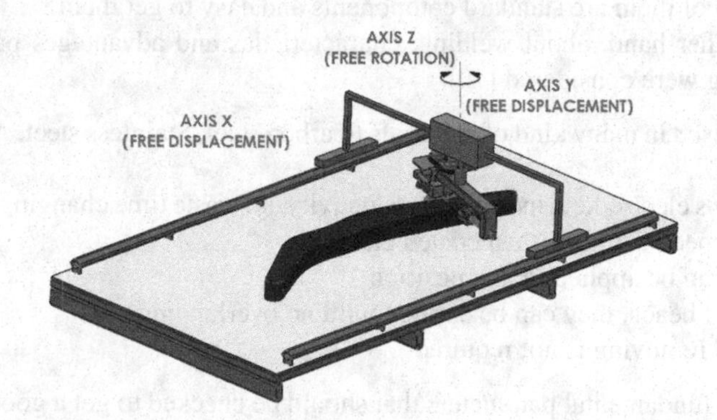

Fig. 63.4 Conceptual design, degrees of freedom (own production)

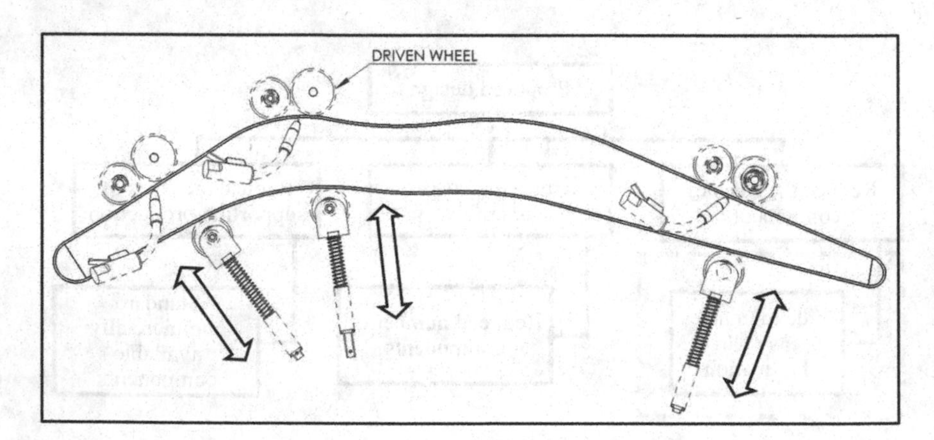

Fig. 63.5 Adaptive mechanism, different locations along arm (own production)

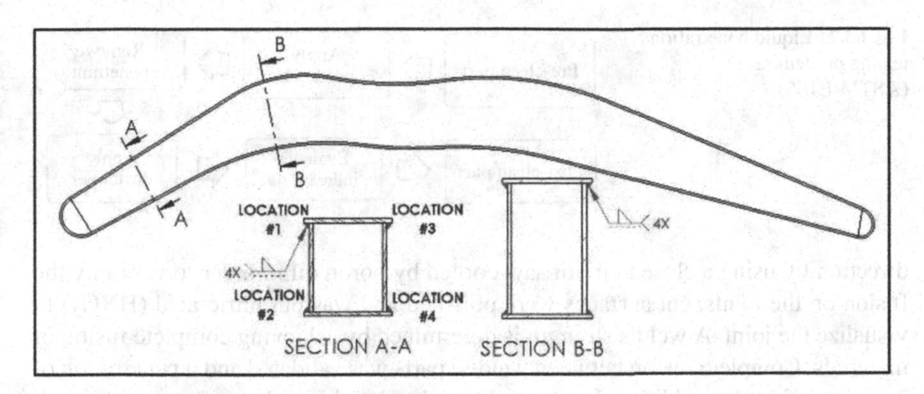

Fig. 63.6 Arm cross section (own production)

Table 63.1 Final MIG welding parameters (own production)

Material thickness (in)	Wire diameter (in)	Feed speed (in/min)	Amperes	Volts	Welding speed (in/min)
3/16	1/32	285	280	34	38

(Fig. 63.6). Induced heat on part is considerably reduced increasing the welding speed (see Table 63.1). Weld bead size on welded part is 1/4 inch and the length of each weld bead is 80 in.

63.4 Results

In order to guarantee weld quality, all welds should be reviewed. In some cases, the revision involves nothing more than a visual examination by the welder. But, a good-looking weld does not always guarantee internal quality, so it is important to make some form of nondestructive testing (NDT). Many different types of NDT methods exist, the most commonly used ones being ultrasonic testing, magnetic particle testing, liquid penetrant inspection, radiographic testing and eddy current testing. Liquid penetration testing is an inspection for examine, interpretation, evaluate and indication surface open defect of the good surface condition tested metallic and non-metallic material that coated penetrant with a visible or fluorescent dye for prevent product, component and structure failure caused performance of erosion, wear, fatigue crack, shrinkage crack, shrinkage porosity, corrosion and creep [13]. This type of testing is limited to the detection of surface-breaking discontinuities or discontinuities that are open to the surface where the penetrant has been applied (Fig. 63.7).

Liquid penetration inspection was applied to the arm welds and there were no discontinuities detected. Having finished NDT, arm was cut perpendicular to welding

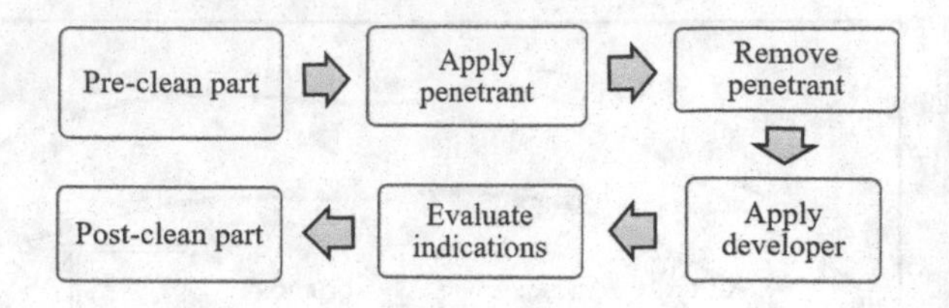

Fig. 63.7 Liquid penetration testing procedure (ASTM-E165)

direction by using a closed circuit saw cooled by boron oil in order to warranty the fusion on the joints, cut surfaces were polished and was put nitric acid (HNO_3) to visualize the joint. A weld's strength is determined by achieving complete fusion of materials. Complete fusion between welded parts was validated and a penetration of 2 mm was found as additional parameter. Table 63.1 shows the final parameters for welding.

Methodology of DFM brought great benefits on project, wide tolerances and easy part fabrication provided the greatest benefits on project, saving time and money. Reducing number of components helps to save time on drawings generation and assembly analysis. Using standard commercially components provides the possibility to have spare parts available if necessary.

63.5 Conclusion

Saving time and cost is an essential consideration in design. In this paper, obtained results shows a practical automated MIG welding application on manufacturing industry. For company, this project represents significant savings avoiding an industrial manipulator purchase. One of the best practices in design is "keep it simple", and the intention of this paper is show it as a case study. Using this equipment, company saved about 15,000 USD avoiding industrial manipulator purchase.

References

1. Zaidi, A., Madavi, K.: Improvement of welding penetration in MIG welding. Int. J. Sci. Res. Sci. **4**(5), 1198–1203 (2018)
2. Thakre, D., Dhore, D., Mehar, C., Mungmode, N., Khobragde, J., Mustkim, M., Ghosh, S.: A review on defect minimization in MIG welding. Int. J. Res. Appl. Sci. Eng. Tech. **8**(9), 93–96 (2020)
3. Kikani, P.: Analysis of process parameters for dissimilar metal welding using MIG welding: a review. JoMME **6**(2), 1–4 (2016)
4. Mansour, R., Zhu, J., Edgren, M., Barsoum, Z.: A probabilistic model of weld penetration depth based on process parameters. Int. J. Adv. Manuf. Tech. **105**, 499–514 (2019)
5. Chavda, S., Patel, T.: A review on parametric optimization of MIG welding for medium carbon steel using FEA-DOE hybrid modeling. Int. J. Sci. Res. Dev. **1**(9), 1785–1788 (2013)
6. Salunke, S., Mali, M., Yande, O., Kulkarni, V.: Automatic welding machine for exhaust pipes using MIG welding. Int. Res. J. Eng. Tech. **7**(6), 5791–5794 (2020)

7. Mustafa, F., Rao'f, M.: Automatic welding machine for pipeline using MIG welding process. Int. Res. J. Eng. Tech. **3**(12), 1448–1454 (2016)
8. Butola, R., Meena, S., Kumar, J.: Effect of welding parameter on micro hardness of synergic MIG welding of 304L austenitic stainless steel. Int. J. Mech. Eng. Tech. **4**(3), 337–343 (2013)
9. Chavda, S., Desai, J., Patel, T.: A review on optimization of MIG welding parameters using Taguchi's DOE method. Int. J. Eng. Man. Res. **4**(1), 16–21 (2014)
10. Standard Welding Terms and Definitions. Approved American National Standard, AWS A3.0M/A3.0-2020
11. Chu, W.-S., Kim, M.-S., Jang, K.-H., Song, J.-H., Rodrigue, H., Chun, D.-M., Cho, Y.T., Ko, S.H., Cho, K.-J., Cha, S.W., Min, S., Jeong, S.H., Jeong, H., Lee, C.-M., Chu, C.N., Ahn, S.-H.: From design for manufacturing (DFM) to manufacturing for design (MFD) via hybrid manufacturing and smart factory: a review and perspective of paradigm shift. Int. J. Pr. Eng. Man.-G.T. **3**(2), 209–222 (2016)
12. Sánchez, C., Cortés, C.: Concepts of design for manufacturing (DFM) of lost wax parts. Rev. Ing. Invest. **25**(3), 49–60 (2005)
13. Suhaila, Y., Rafidah, A., Nurul, A., Azrina, A., Shaiful, I., Fairul, A., Mustaffa, I.: Development time in liquid penetration testing for metal butt joint. Appl. Mech. Mater. **465–466**, 1109–1113 (2014)

Part VII
Ecology, Ergonomics, Energy Efficiency, Environment Protection, Safety and Health at Work

Chapter 64
Application of Ergonomic Principles in Solving Workplaces in Industrial Enterprises in the Slovak Republic

Petra Marková and Vanessa Prajová

Abstract The basic idea of proper workplace design is that the workplace needs to be adapted to human needs and not to be adapted to the needs of technology. At the same time, it is necessary to start from the idea that there is no reason for a person to be in pain at work. Findings from various studies point to the significant impact of work and working conditions on the occurrence of difficulties of the musculoskeletal system of employees. These difficulties occur in high times in workplaces with older equipment but also in workplaces with modern equipment. By respecting the basic ergonomic principles in the solution of workplaces, we can effectively prevent the occurrence of cumulative risk factors affecting employees and causing a decrease in work performance, painful syndromes, and the overall well-being of employees. It turns out that, in general, there is little correct information among employees about the possibilities of individual adaptation of work and working conditions at workplaces. Employees do not know the causes of their problems, which are often related to the areas of anatomy and physiology. The issue of work positions is not clearly characterized in the commonly available literature. The paper examines whether in Slovak industrial companies there are respected principles of ergonomics in workplace solutions. It also describes the consequences of non-compliance with these principles on the difficulties and health of exposed workers.

Keywords Workplace · Ergonomics · Employees

64.1 Introduction

The aim of ergonomics is to maintain health, i.e. physical, mental, and social satisfaction of a person, to create conditions for optimal human activity, as well as to create a feeling of well-being in the workplace [1].

P. Marková (✉) · V. Prajová
Faculty of Materials Science and Technology in Trnava, Slovak University of Technology in Bratislava, Jána Bottu č. 2781/25, 917 24, Trnava, Slovakia
e-mail: petra.markova@stuba.sk

M. Rackov et al. (eds.), *Machine and Industrial Design in Mechanical Engineering*, Mechanisms and Machine Science 109,
https://doi.org/10.1007/978-3-030-88465-9_64

The importance of ergonomics lies mainly in the provision of materials and proposals for the design, construction, and rationalization of relationships in the system man-machine-environment [1].

The subject of ergonomics research is, in the first place, the human being. It follows that the ergonomic principles of designing workplaces and the working environment are aimed at improving human participation in the work process [1].

Creating a workspace that meets the demands and needs of a person in all respects is a demanding process that requires not only technical but also ergonomic knowledge. The better the workspace is adapted to the intended work of a person, the higher the culture and productivity of his work [2].

The basic aspect in assessing and creating new workplaces is a person with his physical and mental capabilities and abilities. Therefore, when designing a workplace correctly, it is necessary to carefully evaluate the factors influencing the creation of the workspace, use the knowledge from their analysis, and analyze secondary factors that may or may not affect the workplace [2].

The main principle for achieving the greatest possible working comfort when designing a suitable workplace is to constantly evaluate the possible presence of adverse effects and to implement sufficient measures in timely measures. It is, therefore, necessary to eliminate harmful, disruptive, and annoying effects. The legislation stipulates that the employer must create suitable and healthy working conditions for his employees [3].

When designing a workplace, it is necessary to pay attention to the factors that affect the occurrence of difficulties in the human musculoskeletal system, and especially [2]:

- Choosing a suitable working position,
- Determining the optimal working height,
- Determination of optimal visual conditions at work,
- The optimal solution for work seats,
- Optimal handling space,
- Use of the principles of economics of work movements,
- Suitable deployment notification and controls,
- Functional consistency,
- Principle of optimal distribution,
- Principle degree of importance,
- Principle of sequential use.

It is appropriate to state the characteristics of the workplace in connection with any regulation of activity at the workplace. These data should be part of the labor standards and should be revised before any intended change of workplace [4]. If there is a change at this level, it should always follow the new ergonomic assessment of the workplace [5].

The method of performing work is based on the possibilities given by the workplace. Defining labor standards is a process of finding the most suitable known method of work. The standard created should take into account occupational safety and ergonomics. The standard determines the level of work performance and well-being in the workplace [2].

64.2 Material and Methods

In preparing the paper, we use the material that was collected in industrial enterprises in Slovakia Assoc. Prof. Hatiar and the team of refugees cooperating with him. Data were collected through a special “Nordic Questionnaire”. The questionnaire was used to monitor and evaluate the impact of work and working conditions on employees in Slovak companies [6, 7]. A significant part of the data was obtained within the work of the occupational health service Probenefit [8]. Furthermore, the research material was obtained through diploma and dissertation theses of students of the Faculty of Materials Science and Technology in the study fields “Industrial Management” and “Personnel work in industrial companies”. The obtained data were processed by epidemiological methods of a retrospective cohort and a cohort study on the incidence and intensity of difficulties and diseases of the musculoskeletal system as indicators of workplace deficiencies and working conditions in terms of ergonomics [8]. In addition to the epidemiological analysis, the results were confronted with physical observation of the evaluated workplaces, interviews with the employees concerned, evaluation of video recordings of work cycles in specific operating conditions. Another basis for the elaboration of the article was the analysis in 51 industrial enterprises in Slovakia through the Evaluation Form consisting of critical points or critical situations in terms of ergonomics in the company, where one of the evaluated areas is the issue of workplace solutions [9]. We also confronted the results of surveys in industrial companies in Slovakia with the outputs of the European Working Conditions Survey (EWCS). The results of research by foreign authors were also an important source for evaluation.

64.3 Results

In the data from the European Working Conditions Survey (EWCS), we focused on manual workers in Slovakia with regard to their qualifications, which they need for work performance. We assessed what percentages of employees perform work in a sitting position, their work demanding compliance with quality standards works, whether their work performs work activities that are forced to perform in painful or tiring working postures. We also investigated whether, for a group of manual workers, job rotation is applied to reduce work monotony and the ability to vary the working position at work. The results of the survey show, as we can see in Tables 64.1 and 64.2, that the possibility of job rotation as a tool to improve working conditions at the workplace is not sufficiently used for manual workers.

It is clear from the results of the survey that manual workers in Slovak industrial companies perform their work mainly standing up and at the same time their work is demanding to comply with quality standards. When it comes to working in unnatural, painful, and tiring working positions, the situation is average. For more than half of low-skilled manual workers do not occur and less than half of highly qualified manual

Table 64.1 Selected factors of working conditions for manual workers in the Slovak Republic according to the EWCS of working conditions [10]

The monitored factor of the European working conditions survey	Job position—manual worker					
	(Almost) all-time %		(Almost) never %		Between 1/4 and 3/4 time %	
	Competence		Competence		Competence	
	Low	High	Low	High	Low	High
Sitting at work	22	14	**52**	**67**	26	20
Work by high-quality standard	**64**	88	36	12		
Work in painful and tiring postures	8	**14**	**53**	39	39	46

Table 64.2 Evaluation of the possibility of work rotation for manual workers in the Slovak Republic according to the EWCS [10]

Factor	Job position—manual worker	Autonomous rotation of fixed tasks in %	Autonomous multitasking in %	Management-driven job rotation in %	No job rotation in %
Possibility of job rotation	Low competence	1	2	11	**54**
	High competence	0	4	32	**61**

workers were experience painful and tiring working positions 1/4 to 3/4 of the work changes.

Another source for assessing respect ergonomic principles in resolving workplace, the survey, which was realized in 51 industrial companies in Slovakia. Companies were selected at random, regardless of their size and the industry type in which they operate. However, most of them are companies from the engineering and automotive industry. In the survey, the workplaces of manual workers were assessed. The survey shows that most companies adhere to the monitored parameters, even if there are shortcomings, as we can see in Table 64.3.

The most significant shortcomings appear to be that more than 40% of the monitored companies do not consistently apply the placement of the working height at the level of the elbow of employees, which is the most suitable for long-term optimal work performance. From the point of view of respecting the physical dimensions of the employees, workplaces are adapted more to the higher employees. Also, up to 35% of companies have reserves in the setting of optimal reach levels for frequently used materials, controls, and tools that the employee uses at work. In the monitored companies, there were shortcomings in the possibilities of changing work positions during work, and employees also do not have the opportunity to rest during a short work break in a different working position than the one in which they also work. Although shortcomings were identified in less than half of the enterprises surveyed,

Table 64.3 Factors of the solution of workplaces in industrial companies of the Slovak Republic

Factor in solving workplaces	Companies without shortcomings (%)	Companies with deficiencies in workplace solutions (%)
Working height for each employee at elbow height	58.8	**41.2**
Adaptation of the workplace for employees of lower figures	64.7	**35.3**
Adapting workplaces for employees higher figures	74.5	**25.5**
Frequently used materials, tools, and controls in the workplace within easy reach	64.7	**35.3**
Multi-purpose, stable work surface for every workplace	90.2	9.8
Comfortable standing on both legs and near the front edge of the handling plane of the workplace for standing work	94.1	5.9
Possibility to change working position during work (sitting—standing)	64.7	**35.3**
Possibility of short breaks at work in standing and sitting position by changing working position	58.8	**41.2**
Possibility of individually adjustable chair with backrest for sitting work	70.6	**29.4**

there are still industrial companies that not consistently apply ergonomic principles in workplace solutions. Consequently, there may be the occurrence of musculoskeletal difficulties of company employees, and by long-term exposure also the emergence of painful syndromes in employees. Which results in a decrease in their work performance, followed by quality and volume of production.

Table 64.4 shows the development of localization of intensive musculoskeletal difficulties in total of 4419 employees in industrial companies in the Slovak Republic in the last years, which were processed in the health service Probenefit, both overall and also depending on the prevailing working position sitting and standing. Based on this research it was found that in a ratio of about 3:1 the work activities performed by the posture prevail in standing in comparison with the occurrence of the work performed by sitting down [8].

The results of research carried out in Slovak industrial companies can be classified according to severity as overall intense musculoskeletal difficulties, which are most projected in the area of crosses, palms and hands, the area of the neck, back, ankles and feet, followed by knees and shoulders.

Table 64.4 Comparison of occurrence of localization and intensity of musculoskeletal difficulties in sitting and standing work [8]

Localization of musculoskeletal disorders difficulties	Working postures								Significance of differences
	Overall (n = 4419)		Work in sitting (n = 1213)			Work in standing (n = 3206)			
	Freq	%	Freq	%	95% conf. limit	Freq	%	95% conf. limit	
Neck	2499	**56.6**	918	**75.7**	**73.29–78.11**	1581	49.3	47.57–51.03	***
Shoulders	1556	**35.2**	463	**38.2**	**35.79–40.61**	1093	34.1	32.46–35.74	***
Back	2067	46.8	682	**56.2**	**53.79–58.61**	1385	43.2	41.56–44.84	***
Elbows	781	17.7	146	12.0	10.18–3.82	635	19.8	18.42–21.18	***
Crosses	2734	**61.9**	410	33.8	31.14–36.46	2324	**72.5**	**70.95–74.05**	***
Palm/Hands	2718	**61.5**	657	54.2	51.40–57.00	2061	**64.3**	**62.64–65.96**	***
Hips/Thighs	680	15.4	109	9.0	7.39–10.61	571	17.8	16.49–19.11	***
Knees	1568	**35.5**	167	13.8	11.86–15.74	1401	**43.7**	40.67–46.73	***
Ankles/Feet	1697	**38.4**	280	23.1	20.73–25.47	1417	**44.2**	41.17–47.23	***

When comparing the research results in Table 64.4 in terms of the predominant working position, there are statistically significant differences in musculoskeletal difficulties projected into the body parts. When working while sitting, these are mainly:

- Neck—the difficulties related to insufficient vision conditions at work, suggest work when turning the head will lead to deterioration of work performance,
- Shoulders and upper back—the difficulties indicate a high holding of the elbow at work, the handling plane of the operator is too high and does not correspond to the range and weight of the objects he is handling.

When assessing the research results in Table 64.4 for standing work, it is possible to determine the order of intensive musculoskeletal difficulties projected into body parts as follows:

- Crosses—difficulties indicate the occurrence of rotations and deep forward bends, long reach distances in forced working positions and high working pace,
- Palms and hands—difficulties indicate the handling of too small or, on the contrary, larger and heavier objects at a high pace and often in a forced working position,
- Knees and ankles—when working while standing, this means a lack of opportunities to relax while sitting,
- Elbows, hips and thighs—are not a problem when working in a standing position, but the results are statistically significant compared to working in a sitting position.

64.4 Discussion

In Slovakia, work-related problems and disorders of the musculoskeletal system, together with respiratory diseases, are among the most common causes of doctor visits and incapacity for work [8]. It should be noted that the incidence of work-related musculoskeletal difficulties and illnesses is also gradually showing a negative impact on the psyche [11]. This constitutes a complication in the secondary and tertiary prevention of the above-mentioned health and mental impacts on the employee. There are a number of factors that are at risk for musculoskeletal disorders and diseases, the negative impact of which is already proven by long-term epidemiological studies [6, 12].

In industrial practice, it often happens that the dimensions and design of workplaces are not in accordance with the physical dimensions of the adult population in Slovakia. This is due to the fact that many foreign-owned companies come with technology from the parent company, and therefore workplace solutions can be designed with a population of other body dimensions in mind [13].

From the surveys carried out, it is clear that there are shortcomings in industrial companies in Slovakia in respecting ergonomic principles in solving the workplace. Employees often do not have the opportunity to change their working position, there are shortcomings in the arrangement of work items and the workplace itself. From the results of ergonomic analyzes in companies in Slovakia, it can be stated that in the field of work organization, the most gentle procedures for the implementation of new automated and automatic technology to protect the health of employees are not currently used. At the same time, it should be noted that in Western Europe, labor standards for the same type of work are much lower than in Eastern European branches.

Companies are trying to address ergonomic shortcomings in the rationalization of work in a purely virtual way, which, however, is not sufficient to determine the real impact of the proposed measures.

64.5 Conclusion

Improperly designed workplace solutions limit the ability to adhere to ergonomic principles when performing work. This has an impact on the health of workers working in such workplaces. It is necessary to evaluate the possible risks and adverse effects on employees during use but also changes in the workplace. At the same time, it is necessary to inform exposed employees about the risks and pay attention to the implementation of corrective measures for workplaces that show shortcomings in terms of ergonomics. Various tools are available for this, from simple checklists to virtual reality tools, the use of ergonomic consulting services, or the creation of a team at the company level. Any path that leads to increased performance and comfort in the workplace is the path in the right direction.

Acknowledgements This publication has been written thanks to support of the research project KEGA 013TUKE-4/2019 "Modern educational tools and methods for shaping creativity and increasing the practical skills and habits of graduates of technical departments of universities".

References

1. Szombathyová, E.: The use of some ergonomic principles in the design of workplaces. Transfér inovácií **7**, 134–135 (2004)
2. Krišťak, J.: Ergonomic layout of the workplace. https://www.ipaslovakia.sk/clanok/ergonomicke-usporiadanie-pracoviska. Last accessed 2021/03/12
3. Král, M.: Methods and Techniques Used in Ergonomics. Occupational Safety Research Institute, Praha (2001)
4. Miller, A., Bures, M., Simon, M.: Proactive approach during design and optimization of production system. In: Proceedings of the Annals of DAAAM and 22nd International DAAAM Symposium, pp. 559–560. DAAAM International, Vienna (2011)
5. Vavruška, J.: Ergonomics way to MOCAP. In: Proceedings of the Conference Applied Ergonomics, pp. 2–9. Faculty of Mechanical Engineering, Praha (2019)
6. Hatiar, K., Kobetičová, L., Hájnik, B.: Ergonomics and preventive ergonomic programs (4): ergonomic analysis with modified questionnaire: Nordic questionnaire. Bezpečná práca **35**(4), 20–28 (2004)

7. Kuorinka, B., Jonsson, B., Kilbom, A., Vinterberg, H., Biering-Sørensen, F., Andersson, G., Jørgensen, K.: Standardized Nordic questionnaires for the analysis of musculoskeletal symptoms. Appl. Ergon. **18**(3), 233–237 (1987)
8. Hatiar, K., Bršiak, V.: Industrial revolution and prevention of work-related difficulties and disorders of the musculoskeletal system. In: Jurkovičová, J., Štefániková, Z. (eds) Proceedings of Living Conditions and Health 2017, pp. 195–211. Public Health Authority of the Slovak Republic, Slovak Society of Hygienists SLS, Institute of Hygiene, Faculty of Medicine, Comenius University, Bratislava (2018)
9. ILO and IEA Ergonomic checkpoints: Practical and Easy-to-Implement Solutions for Improving Safety, Health and Working Conditions. International Labour Office, Geneva (2010)
10. Sixth European Working Conditions Survey. https://www.eurofound.europa.eu/sk/surveys/european-working-conditions-surveys/sixth-european-working-conditions-survey-2015. Last accessed 2021/03/15
11. Barr, A.E., Barbe, M.F.: Pathophysiological tissue changes associated with repetitive movement: a review of the evidence. Phys. Ther. **82**(2), 173–187 (2002)
12. NIOSH: Muskuloskeletal disorders and workplace factors: a critical review of epidemiologic evidence for work—related muskuloskeletal disorders of the neck, upper extremity and low back. In: Bernard, B. (ed.) DHHS (NIOSH), pp. 97–141. Cincinnati (1997)
13. Marková, P., Prajová, V., Homokyová, M., Horváthová, M.: Human factor in industry 4.0 in point of view ergonomics in Slovak Republic. In: Proceedings of the 30th International DAAAM Symposium Intelligent Manufacturing and Automation, pp. 284–289. DAAAM International, Vienna (2019)

Chapter 65
Impact of Long-Term Sitting on Productivity, Burnout, and Health: Employees' View on Workplace Ergonomics

Mihalj Bakator, Eleonora Desnica, Milan Nikolić, and Ivan Palinkaš

Abstract The main purpose of this research is to determine the impact of chair ergonomics, posture, and long-term sitting on employee productivity. Furthermore, the paper addresses the relationship between workplace ergonomics, prolonged and continuous sitting, employee burnout, and employee health. A structured survey was used to evaluate employees in manufacturing and service industries. The obtained data was analyzed with reliability statistics, regression analysis, and correlation analysis. The results are interesting, and contradict the majority of the existing findings in this domain. Even though there are limitations, this current study addresses work environment ergonomics in a concise way, providing significant and unexpected results.

Keywords Workplace ergonomics · Employee productivity · Sedentary jobs

65.1 Introduction

Sedentary lifestyle is the new norm and prolonged sitting can cause lumbar discomfort and pain [1], which negatively affect overall health, work ability and quality of life [2]. In addition, prolonged sitting is a major risk factor for musculoskeletal disorder development. Sitting and standing without freedom to change position is a source of health problems in the modern-day workplace [3]. Breaking up sedentary time has a positive effect on long-term sitting discomfort [4].

Based on these research findings, it is evident that long-term sitting is a big health risk that managers have to regulate. The existing literature in this domain provides mainly objective data which includes various measurement instruments of the spine and other metrics regarding sitting or standing positions and levels of discomfort. Now, objective measurements are certainly significant and are the cornerstone of workplace ergonomics research. However, subjective research, or

M. Bakator (✉) · E. Desnica · M. Nikolić · I. Palinkaš
Technical Faculty "Mihajlo Pupin" in Zrenjanin, University of Novi Sad, Novi Sad, Republic of Serbia
e-mail: mihalj.bakator@uns.ac.rs

M. Rackov et al. (eds.), *Machine and Industrial Design in Mechanical Engineering*, Mechanisms and Machine Science 109,
https://doi.org/10.1007/978-3-030-88465-9_65

more precisely, employees' views on workplace ergonomics are also an important metric. With objective measurements, subjective bias is inevitable. With subjective research, such bias is expected. Such subjective studies on workplace ergonomics are scarce. This may pose a problem, as subjective observations can provide also valuable insight on how employees perceive workplace ergonomics.

This paper aims to fill this gap by analyzing the influence of long-term continuous sitting on employee productivity. In addition, employee health, and employee burnout are analyzed. The paper begins by approaching, and analyzing the existing literature. Scientific articles in the domain of workplace ergonomics, long-term sitting, employee health, employee burnout, and employee productivity are analyzed. Further, descriptive statistics is conducted in order to determine the average values, and standard deviation values of the measured items. Next, the internal consistency of the measured items is analyzed. In addition, regression analysis, and correlation analysis are used to investigate the relationships between the measured variables.

65.2 Theoretical Background

Ergonomics as an interdisciplinary field is present in a large number of domains and it can be related to human performance [5]. When it comes to workplace ergonomics, long-term continuous sitting increases employee discomfort [6]. Prolonged sitting has a negative impact on mortality, and chronic diseases [7]. Complementary to those findings Wilks et al. [8], argued that height modifiable desks resulted in improved productivity opposed to fixed desks.

Further, in 2016 a complex, and in-depth research [9] of prolonged sitting, and discomfort was conducted. In their research, they noted that prolonged sitting results in chair movement to compensate the occurring, and growing discomfort. I addition, it was also discussed that the longer the continuous sitting, the higher the number of in chair movements, thus indicating a higher level of discomfort. A major influential factor on prolonged sitting discomfort, and productivity is the chair design [10]. Undoubtedly, there is a negative influence of continuous sitting on health, and productivity. Workers in offices, in the automobile industry, and electric cables industry, are at high risk of developing the mentioned negative effects of prolonged sitting. Continuous sitting caused hand/wrist, upper back, and lower back pain, as well as a substantial decrease in productivity. It also increases the number of errors at assembly lines [11]. However, changing positions from sitting to standing in predetermined periods of time, reduced the negative effects.

Furthermore, Despres [12] examined the negative effects of sedentary behavior, and noted that there is a higher risk of chronic metabolic, and cardiovascular disease in workers that sit for longer continuous periods of time, in opposite of workers who have a more active jobs.

In this research, employee burnout correlated to continuous sitting is investigated. Employee burnout manifests itself as emotional exhaustion, depression episodes, impaired personal sense of accomplishment, and self-achievement [13]. It was found that the literature on employee burnout, and long-term continuous sitting relationship is very scarce. Therefore, in this paper the correlation between continuous sitting, and employee burnout will be researched.

Literature that addresses various forms of prolonged sitting [14], and office ergonomics [15], are more focused on health issues of workers, rather than productivity, and employee burnout with correlation to continuous sitting [16]. Therefore, in this research the relationships between employee burnout, employee health and productivity is also examined.

65.3 Methodology

65.3.1 Research Framework

On Fig. 65.1, the structural framework of the research is presented. The researched relations between the dependent variable (Employee productivity), and independent variables (employee burnout, time spent sitting in continuity, employee health) are presented. In addition, the effect of time spent sitting in continuity on employee burnout and employee health is observed. Finally, the effect of employee burnout on employee health is also addressed.

Based on the above mentioned literature analysis, and the goal of this paper, the following hypotheses are analyzed:

- H_1: Employee burnout is in a negative relation with employee productivity.
- H_2: Time spent sitting in continuity is in a negative relation with to employee burnout.
- H_3: Employee health is in a positive relation with employee productivity.

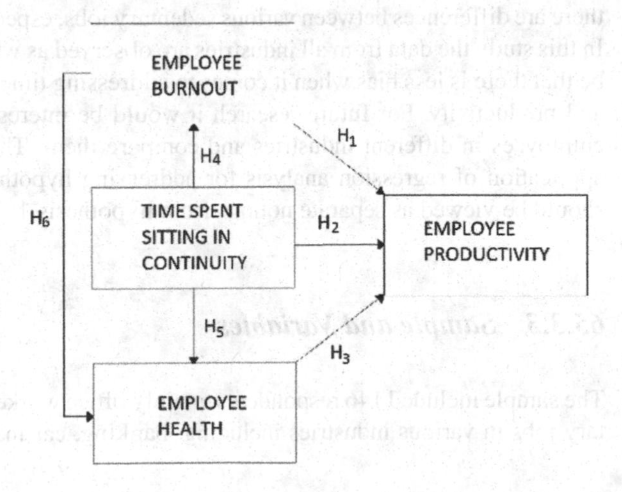

Fig. 65.1 Research framework

- H_4: Time spent sitting continuity is in a statistically significant relation with employee burnout.
- H_5: Time spent sitting continuity is in a statistically significant relation with employee health.
- H_6: Employee burnout is in a statistically significant relation with employee health.

The effects of time spent sitting in continuity and other factors in hypotheses 4, 5, and 6, are proposed as a *statistically significant relation*, rather than *positive relation* due to the assumptions derived from the existing body of literature.

65.3.2 Research Phases and Data Analysis

The **first phase** of the research included the development of a concise survey based on similar studies in this domain. Afterwards, the surveys were distributed online to workers who have sedentary jobs in various industries. The obtained data was stored in Excel spreadsheets.

Next, in the **second phase** of the research, the obtained data was analyzed. First, to ensure internal consistency of the items, a reliability test was conducted to determine the Cronbach's alpha values. Further, descriptive statistics was used to present the standard deviation, and mean of the items according to the measured dimensions. Next, regression analysis was conducted to describe the influential relationship between the dependent and independent variables. Additionally, the influences between the independent variables are observed through separate regression analyses. Finally, a correlation analysis was used to show the correlation between the measured dimensions.

The third phase of the research included the analysis and discussion of the obtained results. As every research approach has limitations, this one is no exception. First, the there are differences between various sedentary jobs, especially in various industries. In this study the data from all industries are observed as whole. A positive side could be that there is less bias when it comes to addressing time spent sitting in continuity and productivity. For future research it would be interesting to address data from employees in different industries and compare them. The second limitation is the application of regression analysis for addressing hypotheses 4–6. These relations should be viewed as separate notions from hypothesis 1–3.

65.3.3 Sample and Variables

The sample included 146 respondents, mainly office workers, and people with sedentary jobs in various industries including: banking, car manufacturing, shoemakers, etc. The survey was designed according to similar research in this domain [17] and through the creative process from the authors of this paper [18].

Next, the survey included 32 items from which 25 were seven-point Likert scale items. (1—Totally Disagree; 7—Totally Agree). The investigated variables were: time sitting in continuity, employee health, employee burnout, and employee productivity. The noted variables were measured via self-assessment of the respondent.

65.4 Results

The results of the descriptive statistics are given in Table 65.1.

Furthermore, the regression analysis is conducted. Time sitting in continuity, employee burnout, and employee health are independent variables, while employee productivity is the dependent variable. Table 65.2 shows the results of regression analysis.

Further, in order to test the hypothesis H_4 and H_5, TSC is noted as the independent variable, while EBO and EHT are viewed as a dependent variable. The results are presented in Table 65.3.

Next, in order to test the hypothesis H_6, EBO is noted as the independent variable, while EHT is viewed as a dependent variable. The results are presented in Table 65.4.

Next, the results of the correlation analysis are presented in Table 65.5.

In the discussion section, the obtained results are analyzed.

Table 65.1 Descriptive statistics

Variable	Min	Max	Mean	Std. Dev
Time spent sitting in continuity (TSC)	1	7	4.39	2.09
Employee burnout (EBO)	1	7	3.95	2.26
Employee health (EHT)	1	7	4.74	2.09
Employee productivity (EP)	1	7	4.24	2.29

Table 65.2 Regression analysis (for hypotheses H_1, H_2, and H_3)

Dependent Var	Independent Var	β	R^2	*p*-value
Employee productivity (EP)	Time spent sitting in continuity (TSC)	1.020	0.008	<0.0001
	Employee burnout (EBO)	0.256	0.108	<0.0001
	Employee health (EHT)	0.876	0.221	<0.0001

Table 65.3 Regression analysis (for hypotheses H_4, and H_5)

Independent Var	Dependent Var	β	R^2	*p*-value
Time spent sitting in continuity (TSC)	Employee burnout (EBO)	1.020	0.188	<0.0001
	Employee health (EHT)	0.922	−0.111	<0.0001

Table 65.4 Regression analysis (for hypothesis H_6)

Independent Var	Dependent Var	β	R^2	*p*-value
Employee burnout (EBO)	Employee health (EHT)	0.844	−0.106	<0.0001

Table 65.5 Correlation analysis

Variables	TSC	EBO	EHT	EP
Time spent sitting in continuity (TSC)	1.000			
Employee burnout (EBO)	−0.040	1.000		
Employee health (EHT)	0.015	0.338	1.000	
Employee productivity (EP)	−0.036	0.667	0.356	1.000

65.5 Discussion

In this research paper the influence of prolonged sitting in continuity on employee productivity was analyzed. In addition, employee health, and employee burnout were measured. The regression analysis results indicate that there is no relation between time spent sitting in continuity (TSC) and employee productivity (EP) ($R^2 = 0.008$). Further, there is insignificant relation between employee burnout (EBO) and employee productivity (EP) ($R^2 = 0.108$). Weak relation between is noted between employee health (EHT) and employee productivity (EP) ($R^2 = 0.221$).

Furthermore, the results of the regression analysis where TSC was noted as the independent variable, while ETH and EBO were observed as dependent variables (two separate regression calculations were made), indicate that there is a statistically significant influence of TSC on ETH and EBO, however this influence is weak as the regression values for EBO, and ETH are $R^2 = 0.188$ and $R^2 = -0.111$. The situation is similar with the influence of EBO on ETH, which is statistically significant, but weak ($R^2 = -0.106$).

The correlation analysis results indicate interesting results. Employee burnout is positively correlated with employee productivity (EP), with a correlation coefficient of 0.667. Positive correlation was also noted between employee health (ETH) and employee productivity (EP) with a correlation coefficient of 0.356. Now, the proposed hypotheses are addressed:

- H_1: Employee burnout is in a negative relation with employee productivity. **did not gain support**.

- H_2: Time spent sitting in continuity is in a negative relation with to employee burnout. **did not gain support**.
- H_3: Employee health is in a positive relation with employee productivity. **failed to reject**.
- H_4: Time spent sitting continuity is in a statistically significant relation with employee burnout. **is supported**.
- H_5: Time spent sitting continuity is in a statistically significant relation with employee health. **is supported**.
- H_6: Employee burnout is in a statistically significant relation with employee health. **is supported**.

The hypotheses were constructed in accordance with a logical and analytical approach and are based on the theoretical background. It is rather interesting how the research came up with such non-expected findings. Nevertheless, the study was able to demonstrate how some of the measured factors are in relationship with each other. This is indeed significant for future research.

65.6 Conclusion

In conclusion, the results are unexpected, as two of the three analyzed hypotheses did not gain support. This paper significantly contributes to the vast literature on workplace ergonomics, and sedentary job risks. Even though, the results are not complementary with the suggested hypothesis, the research itself provides a significant insight into this complex domain. There is room for improvement. Further, it is necessary to address the type of work the workers are doing, because highly dynamic, stressful work may have a stronger influence on employee productivity, burnout and health.

The findings may be the result of different non-measured factors, including specific sedentary job type, organization size, type, and even organization culture. It sure will be interesting to address, and broaden this research in the future. Even though, there are some relations noted between variables, they don't indicate and don't support the research hypotheses. Therefore, it is necessary to address, and isolate major influential factors in workplace environments, in order to surely determine the negative impact of long-term sitting.

To summarize, it is recommended to define other factors which may affect employee productivity, employee burnout, and employee health. This way, the impact of long-term sitting can be more precisely defined. It is quite interesting, that the findings in this paper negated the possible negative influence of long-term sitting on employee health. Therefore, a future research in this domain is suggested.

References

1. Agarwal, S., Steinmaus, C., Harris-Adamson, C.: Sit-stand workstations and impact on low back discomfort: a systematic review and meta-analysis. Ergonomics **61**(4), 538–552 (2017)
2. Sowah, D., Boyko, R., Antle, D., Miller, L., Zakhary, M., Straube, S.: Occupational interventions for the prevention of back pain: overview of systematic reviews. J. Safety. Res. **66**, 39–59 (2018)
3. Tissot, F., Messing, K., Stock, S.: Studying the relationship between low back pain and working postures among those who stand and those who sit most of the working day. Ergonomics **52**(11), 1402–1418 (2009)
4. Owen, N., Healy, G.N., Matthews, C.E., Dunstan, D.W.: Too much sitting: the population-health science of sedentary behavior. Exerc. Sport Sci. Rev. **38**(3), 105–113 (2010)
5. Gligorović, B., Desnica, E., Palinkaš, I.: The importance of ergonomics in schools—secondary technical school students' opinion on the comfort of furniture in the classroom for computer aided design. IOP Conf. Ser.: Mater. Sci. Eng. **393**, 012111-1–012111-7 (2018)
6. Callaghan, J.P., Gregory, D.E., Durkin, J.L.: Do NIRS measures relate to subjective low back discomfort during sedentary tasks? Int. J. Ind. Ergonom. **40**(2), 165–170 (2010)
7. Patel, A.V., Bernstein, L., Deka, A., Feigelson, H.S., Campbell, P.T., Gapstur, S.M., Thun, M.J.: Leisure time spent sitting in relation to total mortality in a prospective cohort of US adults. Am. J. Epidemiol. **172**(4), 419–429 (2010)
8. Wilks, S., Mortimer, M., Nylén, P.: The introduction of sit–stand worktables: aspects of attitudes, compliance and satisfaction. Appl. Ergon. **37**(3), 359–365 (2006)
9. Cascioli, V., Liu, Z., Heusch, A., McCarthy, P.W.: A methodology using in-chair movements as an objective measure of discomfort for the purpose of statistically distinguishing between similar seat surfaces. Appl. Ergon. **54**, 100–109 (2016)
10. Zemp, R., Taylor, W.R., Lorenzetti, S.: Are pressure measurements effective in the assessment of office chair comfort/discomfort? A review. Appl. Ergon. **48**, 273–282 (2015)
11. Falck, A.C., Örtengren, R., Högberg, D.: The impact of poor assembly ergonomics on product quality: a cost–benefit analysis in car manufacturing. Hum. Factor. Ergon. Man. **20**(1), 24–41 (2010)
12. Despres, J.P.: Physical activity, sedentary behaviours, and cardiovascular health: when will cardiorespiratory fitness become a vital sign? Can. J. Cardiol. **32**(4), 505–513 (2016)
13. Brewer, E.W., Shapard, L.: Employee burnout: a meta-analysis of the relationship between age or years of experience. Hum. Resour. Dev. Rev. **3**(2), 102–123 (2016)
14. Wennberg, P., Boraxbekk, C.-J., Wheeler, M., Howard, B., Dempsey, P.C., Lambert, G.: Acute effects of breaking up prolonged sitting on fatigue and cognition: a pilot study. BMJ Open **6**(2), e009630-1–e009630-9 (2016)
15. Haynes, S., Williams, K.: Impact of seating posture on user comfort and typing performance for people with chronic low back pain. Int. J. Ind. Ergonom. **38**(1), 35–46 (2008)
16. Groenesteijn, L., Vink, P., de Looze, M., Krause, F.: Effects of differences in office chair controls, seat and backrest angle design in relation to tasks. Appl. Ergon. **40**(3), 362–370 (2009)
17. Choi, B., Schnall, P.L., Yang, H., Dobson, M., Landsbergis, P., Israel, L., Baker, D.: Sedentary work, low physical job demand, and obesity in US workers. Am. J. Ind. Med. **53**(11), 1088–1101 (2010)
18. Helander, M.G.: Forget about ergonomics in chair design? Focus on aesthetics and comfort! Ergonomics **46**(13–14), 1306–1319 (2003)

Chapter 66
Geometrical Considerations of Air-Conditioner Vent Arrangement in a Farm Tractor Cab

Dragan Ružić and Tanasije Jojić

Abstract The forced convection is the main mode of farm tractor operator's body heat exchange in hot as well as in cold conditions. Tractor cabs have an air distribution system with air vents, whose position is important in terms of proper air distribution inside the cab and around the operator's body. In this paper a method for determination of potential positions of the air-conditioner vents inside the farm tractor cab using CAD software CATIA is presented. The criterions are based on the operator's body geometry, the air jet characteristics, air flow demands as well as cab interior geometry. The resulting intersections of geometric requirements show the way of arrangement of the vents that would be in compliance with defined constraints.

Keywords Farm tractor cab · Ventilation · Thermal comfort · CAD

66.1 Introduction

A modern farm tractor cab is designed to protect an operator and to provide comfortable environment, since the conditions inside farm tractors affect the health, performance and comfort of the operator. A system for tractor cab air conditioning is significant energy consumer, therefore influencing overall tractor economy and/or range, like in any other vehicle [1, 2]. Since the forced convection is the main mode of the operator's body heat exchange, apart from energy consumption the conditioned air distribution strongly affects an operator's thermal sensation and comfort. Tractor cabs have an air distribution system with air vents (outlets of the ventilation system) placed in front of the operator or at the ceiling. The vents are mostly of circular cross-section. The air velocity, direction of the air jet and the air temperature should be under control of an operator.

Airflow from the HVAC system, characterized by spatially distributed local air velocities and air temperatures, influences local and overall microclimate conditions, consequently the heat loss from the operator's body. The heat loss should be such as

D. Ružić (✉) · T. Jojić
Faculty of Technical Sciences, University of Novi Sad, Novi Sad, Serbia
e-mail: ruzic@uns.ac.rs

M. Rackov et al. (eds.), *Machine and Industrial Design in Mechanical Engineering*, Mechanisms and Machine Science 109,
https://doi.org/10.1007/978-3-030-88465-9_66

to provide a feeling of thermal comfort. Additionally, the fresh air must be supplied into the operator's breathing zone, but without dry-eye discomfort.

An analysis of the influence of different air distribution designs on cooling the operator's body under the same other conditions using CFD technique is presented in research published by Ružić et al. [3]. Among different cases, the largest heat loss from the whole body was obtained in the system with the nozzles on the dashboard. However, the results show that setting of direction have more significance for total body heat loss than positioning of the nozzles (e.g. on the dashboard or ceiling). In the study done by Oh et al. [4], thermal comfort was simulated using CFD software through flow analysis. The vent arrangements were based on the actual location of the air conditioner vents in the tractor cab: the ceiling, dashboard and side pillar. They concluded that the lowest kinetic energy loss in the flow field was achieved when the vents are positioned on the dashboard. In the papers [5, 6] a comparison of three different arrangements of the air distribution system in the farm tractor cab is presented. The research was done on the virtual models of the cab and the operator, using the CFD software. The results show that design with four vents or design with a perforated opening on the ceiling have better performances than the design with three vents in terms of convective heat loss distribution and control.

This paper presents a method for determination of potential positions of the air-conditioner vents inside the farm tractor cab in relation to the operator's body. Geometrical constrains based on air jet characteristics, air flow around human body demands, anthropometry, visibility and cab interior geometry were considered.

66.2 Characteristics of Air Jet

In a quiet uniform environment, around the human body an upward free convection flow exists, whose characteristics depends on body posture, surrounding air temperature, etc. At comfortable room air temperature a flow from the front of the body with a mean velocity as low as 0.1 m/s will disturb the free convection flow and above 0.2 m/s will penetrate this layer [7]. Standard ASHRAE 55-2009 suggests elevated air velocities up to 1.2 m/s, to increase the maximum temperature for acceptable comfort, if the air speed is under the control of an exposed person. These values could be applied to a lightly clothed operator (0.5–0.7 Clo) with metabolic rate between 1.0 and 1.3 Met [8, 9]. Exposure to the direct air flow from the vents is used when intensive heat exchange is necessary, like during the heat-up or cool-down regimes.

Air can be discharged from circular or rectangular vents with grilles or relatively large perforated panels. The circular vents have less turbulence and less non-uniformity than rectangular vents [7]. In this study the air jet was treated as an isothermal. The velocity distribution in a non-isothermal jet is similar to that in an isothermal jet [7].

The length of the jet potential core region is around 4–6 opening diameters where the maximum velocity (temperature in the case of a non-isothermal jet) of

Table 66.1 Distances from vent opening to target surface based on jet behaviour and vent diameter

Vent diameter, D_V (m)	0.04	0.12
Jet core (m)	0.24	0.48–0.72
Jet transition (m)	0.24–1.00	0.72–3.00

the airstream remains practically unchanged [10]. The body surface is usually at a greater distance than the potential core. The highest velocity is along jet centerline and this will be taken as a reference target velocity. In transition region (up to 25 opening diameters [10]) the air velocity (m/s) in a jet at distance x (m) is proportional to $x^{-0.5}$ [10].

In the developed zone the air velocity (m/s) and temperature at distance x (m) from circular opening of area A_V (m^2) are proportional to x^{-1} and depending on the type of opening [10, 11]. At distance r (m) from the centreline air velocity is decreasing, in dependence of r, x and, of course, type of opening [10, 11].

Chosen characteristic dimensions of typical vents (D_V) are in 40–120 mm range. Calculated values of some characteristic distances between the vent and the body surface that will be considered in this research are given in Table 66.1.

66.3 Hand Reach Range

Air jet direction is changed by adjusting the vent by hand. This control is considered to be not critical and the adjustment can be done by the hand and arm movements with comfortable shifts of the upper body with slight rotation or with changes of posture from the seated position, while the body remains in contact with the seat.

Hand reach range in this study is taken from the standard for commercial vehicle driver's place (standard ISO 16121-3 [12]). The hand reach ranges are considered as two forward-facing hemispheres of 750 mm radius. The centers of the hemispheres are in shoulder points.

66.4 Design Recommendations and Constrains

Summary of air jet characteristics, anthropometrical parameters of the tractor operator and recommendations for efficient air distribution system and personalized ventilation, that are to be used in geometrical consideration of vent arrangement, are as follows:

- Vent direction has more significance to total body heat loss than the positioning of the vents [2]: primary position is based on normal projection on target surface because of high heat transfer coefficient [13].
- The boundary conditions and/or need for local heat loss from the body are not uniform [5, 6, 14]: more symmetrically arranged vents are necessary.

- Large round vent cross-section is preferable because of an air flow with a uniform velocity profile and low turbulence [6].
- The minimum target surface air velocity is 0.3 m/s [7], except the face zone where air velocity should be 0.3 m/s maximum [15].
- Operator's breathing zone should be in the core zone of air flow to be supplied with clean fresh laminar air flow [7].
- Body surface should be within the transition zone in order to accurate control air velocity (and temperature as well).
- Since the vent direction is manually adjustable (even in systems with automatic control), the vents should be within the hand reach.

Due to the complexity of human body geometry, these requirements would be too difficult to define analytically. For this reason, 3D geometrical model built in software CATIA was used to generate spatial surfaces suitable for vent positioning. Chosen values of offset from the operator's body surfaces are:

- 0.24 m ($D_V = 0.04$ m) to 0.48 m ($D_V = 0.12$ m) for supplying of breathing zone,
- 0.48–0.75 m for body exposure to air jet transition zone ($D_V = 0.04$–0.12 m).

A minimum internal clearance inside the cab is according to standard ISO 4252 [16]. Boundaries of internal space of a typical farm tractor cab are also presented in simplified shape [17].

Besides airflow characteristics and structural requirements for positioning of vents, forward view field must be considered too. Operator's field of vision in farm tractor is defined according to UN ECE regulation R71, where allowed sizes of obstructions are given [18]. In this research, only a part of the field that corresponds to the proposed direct area of vision is taken as a constraint where should not be any obstructions. According to the regulation, the field of vision in lateral direction must also have a minimum of obstructions.

66.5 Results and Discussion

Model of the sitting operator, as well as entire geometrical analysis, was made in CAD software CATIA. Dimensions of the operator's body model were retrieved from the database of CAD program CATIA, for 50-percentile male [19]. Surfaces suitable for vent location are generated using CAD as an offset from the body segments: trunk as a largest surface, head—breathing zone, upper legs, and lower legs. Only potential vent locations that are in front of an operator are considered. Some of the cab boundaries and vent location surfaces are shown only for right hand side (RHS), in order to simplify graphical presentation.

Geometrical constraints derived from minimum and standard operator's space inside the cab as well right hand reach envelope are shown in Fig. 66.1. Spatial surfaces for potential vent location derived as an offset from chosen body surfaces

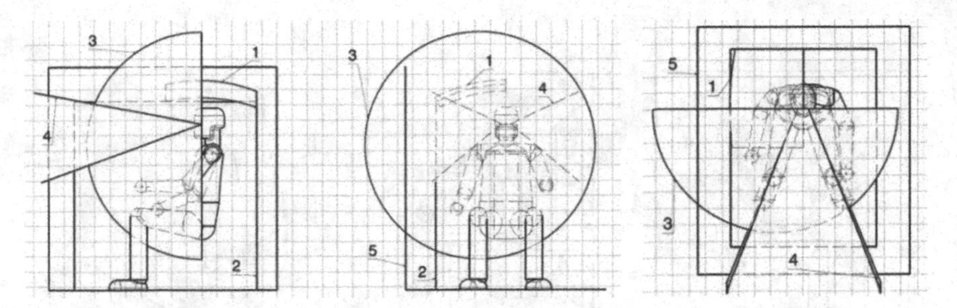

Fig. 66.1 Boundary conditions: 1, 2—minimum space inside the cab (RHS only), 3—right hand reach range, 4—direct area of vision limits, 5—typical tractor cab interior boundaries (RHS only)

are shown in Fig. 66.2. The intersection of both surfaces is presented in Fig. 66.3. In the background of the drawings is a 100 × 100 mm grid.

Implementing the resulting surfaces in a typical cab design, it can be noted that ceiling, side pillar and instrument panels are suitable for vent positioning. The front pillar would be also a potential place for vents directed to the operator's trunk.

Vents that are directed towards head and breathing zone, as the most sensitive to draught, must be within the hand reach. However, some vents could be outside of chosen hand reach (directed to lower and upper leg, for example) because the sensitivity of those body segments is not critical.

The most suitable positions for breathing zone (supply of fresh air) and trunk (largest area for heat transfer) are in front of those body parts. However, those positions would be within the direct field of view and inside the minimum operator's space.

The generated surfaces depending on operator's body size and shape as well as the limits (the cab dimensions, hand reach etc.) are not definite. Combinations of vent diameter and the air flow rate enable a wider range of potential distance.

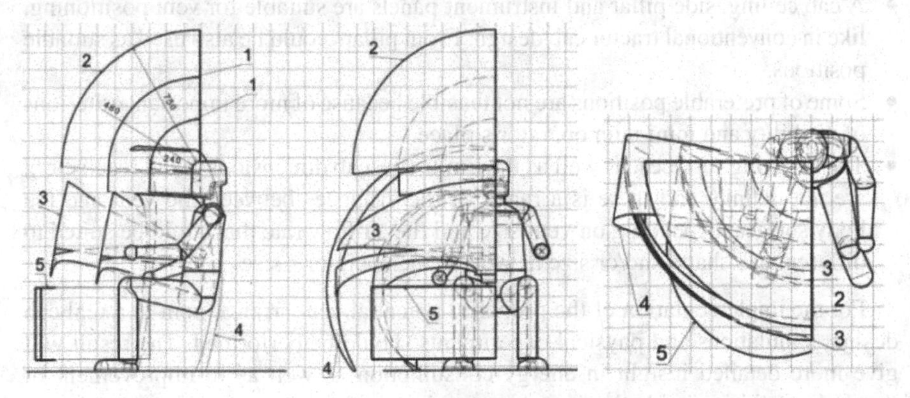

Fig. 66.2 Potential vent positioning based on operator's body surface. Positions for vents directed to 1—breathing zone, 2—head, 3—trunk, 4—upper leg, 5—lower leg

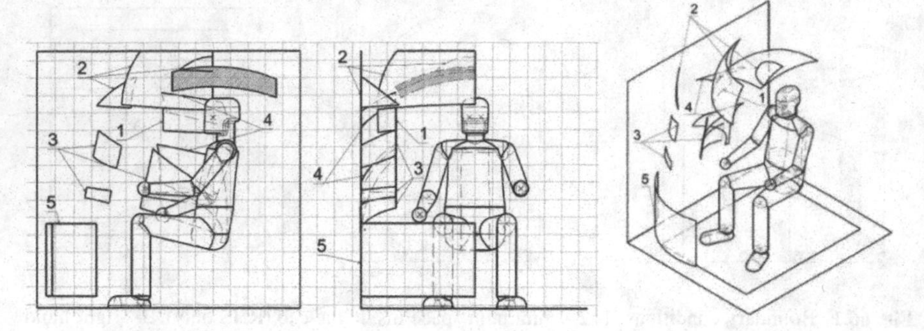

Fig. 66.3 Resulting surfaces for potential vent positioning (RHS only). The hatched surface is operator's minimum space above the head. Positions for vents directed to 1—breathing zone, 2—head, 3—trunk, 4—upper leg, 5—lower leg

According to air jet theory, vent diameter should be proportional to the distance between the vent and the target surface. The final proposal of vent arrangement should be checked under various regimes and boundary conditions using simulations and physical experiments.

66.6 Conclusions

A method for determination of potential positions of the air-conditioner vents inside the farm tractor cab is presented in this paper. Constrains are based on air jet characteristics, air flow around human body demands, anthropometry, visibility and cab interior geometry. The entire geometrical analysis was done in CAD software CATIA. Results of the geometrical analysis of the potential vent positions are as follows:

- A cab ceiling, side pillar and instrument panels are suitable for vent positioning, like in conventional tractor cab design. Front pillars could be also used as suitable positions.
- Some of preferable positions are not feasible because of more important criterions of visibility and minimum operator's place.
- The resulting surfaces, as well as the limits (the cab dimensions, hand reach etc.), are not definite and there is a range of the distances between the vent and the body surface, depending on vent size and the air flow rate that could be suited to different cab shape and/or size as well as the operator's size.

For the implementation of the proposed potential vent arrangement in a real cab design, simulations and physical experiments should be performed. The result will give more detailed insight in energy consumption as well as in improvement of thermal conditions inside the farm tractor cab.

Acknowledgements This paper has been supported by the Ministry of Education, Science and Technological Development through project no. 451-03-9/2021-14/200156: "Innovative scientific and artistic research from the FTS activity domain".

References

1. Leighton, D., Ruth, J.: Increasing EDV range through intelligent cabin air handling strategies. Annual Progress Report, National Renewable Energy Laboratory USA (2016)
2. Ružić, D.: Advanced methods to reduce energy consumption for farm tractor cab air-conditioning. In: Proceedings of International Congress Motor Vehicles and Motors, MVM 2020, pp. 239–245. Kragujevac (2020)
3. Ružić, D., Časnji, F., Poznić, A.: Efficiency assessment of different cab air distribution system layouts. In: Proceedings of 15th Symposium on Thermal Science and Engineering of Serbia, SimTerm 2011, pp. 819–828. Sokobanja (2011)
4. Oh, J., Choi, K., Son, G., Park, Y., Kang, Y., Kim, Y.: Flow analysis inside tractor cabin for determining air conditioner vent location. Comput. Electron. Agr. **169**, 1–12 (2020)
5. Ružić, D., Simikić, M.: The influence of air distribution system design and setting in agricultural tractor cab on the operator's thermal sensation. In: Proceedings of 19th Conference on Thermal Science and Engineering of Serbia, SimTerm 2019, pp. 948–959. Sokobanja (2019)
6. Ružić, D.: Comparison of different air vents arrangements in a tractor cab using the CFD techniques. In: Proceedings of 8th International Symposium Machine and Industrial Design in Mechanical Engineering, KOD 2014, pp. 97–104. Novi Sad (2014)
7. Melikov, A.: Personalized ventilation. Indoor Air **14**(7), 157–167 (2004)
8. ASHRAE Standard 55P. Thermal environmental conditions for occupancy. Third public review, American Society of Heating, Refrigerating and Air Conditioning Engineers, Atlanta (2003)
9. Ružić, D.: Improvement of thermal comfort in a passenger car by localized air distribution. In: Proceedings of the 11th International Symposium Interdisciplinary Regional Research, ISIRR 2010, p. 210. Szeged (2010)
10. ASHRAE Fundamentals Handbook: Chapter 31: Space Air Diffusion. American Society of Heating, Refrigerating and Air Conditioning Engineers, Atlanta (1997)
11. Stoecker, W.F., Jones, J.W.: Refrigeration and Air Conditioning. McGraw-Hill Inc., New York (1982)
12. ISO 16121-3: Road vehicles: ergonomic requirements for the driver's workplace in line-service buses. ISO (2005)
13. Incropera, F.P., DeWitt, D.P.: Fundamentals of Heat and Mass Transfer. Wiley, New York (1981)
14. Ružić, D., Časnji, F.: Design of vehicle cabin air distribution system based on human thermal sensation. In: Proceedings of 7th International Symposium about Machine and Industrial Design in Mechanical Engineering, KOD 2012, pp. 255–260. Novi Sad (2012)
15. ASABE/ISO 14269-2: Tractors and self-propelled machines for agriculture and forestry: operator enclosure environment—part 2: heating, ventilation and air-conditioning test method and performance. ISO (1997)
16. ISO 4252: Agricultural tractors—operator's workplace, access and exit—dimensions. ISO (2007)
17. Ružić, D., Časnji, F.: Thermal interaction between a human body and a vehicle cabin, In: Kazi, S.N. (ed.) Heat Transfer Phenomena and Applications, pp. 295–318. IntechOpen (2012)

18. UN ECE regulation R 71: Uniform provisions concerning the approval of agricultural tractors with regard to the driver's field of vision (1987)
19. Ružić, D., Simikić, M.: Geometry characteristics of human body model suitable for simulation of thermal comfort in an agricultural vehicle. In: Proceedings of 9th International Symposium about Machine and Industrial Design in Mechanical Engineering, KOD 2016, pp. 131–136. Novi Sad (2016)

Chapter 67
Energy Efficiency of Food Cooling and Freezing Plants in Serbia

Marko Mančić, Dragoljub Živković, Milena Rajić, Milena Mančić, Milan Đorđević, Bojana Vukadinović, and Milan Banić

Abstract The food industry is one of the largest users of refrigeration technology. Refrigeration is a process by which heat is transferred from one place to the environment, with application of heat or mechanical work. The majority of refrigeration plants designed for cooling and freezing of fruits and vegetables and account for more than 50% of all electricity used at on these sites. Cooling and freezing facilities typically found in Serbia, range from small applications, standalone refrigerators with hydrofluorocarbons as refrigerants to large walk-in cold rooms with ammonia freezing cycles. Freezing of raw goods can be performed in small freezing chambers for small applications, or continual freezing tunnels typical for large installations. In a commercial context even, a small reduction in refrigeration energy use can offer significant cost savings. In this paper, small and large freezing and cooling plants in Serbia are analyzed, typical energy efficiency measures applicable in the sector are presented, and national energy efficiency indicators for this sector are presented based on the performance of the analyzed plants.

Keywords Energy efficiency indicators · Refrigeration · Cooling and freezing plants

67.1 Introduction

The food industry is one of the largest users of refrigeration technology. In the food sector in general, refrigeration plants electricity consumption can account for around

M. Mančić (✉) · D. Živković · M. Rajić · M. Banić
Faculty of Mechanical Engineering, University of Niš, 18000 Niš, Serbia
e-mail: marko.mancic@masfak.ni.ac.rs

M. Mančić
Faculty of Occupational Health and Safety, University of Niš, 18000 Niš, Serbia

M. Đorđević
Faculty of Technical Sciences, University of Priština, Kosovska Mitrovica, Serbia

B. Vukadinović
Faculty of Technology and Metallurgy of the University of Belgrade, 11000 Belgrade, Serbia

M. Rackov et al. (eds.), *Machine and Industrial Design in Mechanical Engineering*, Mechanisms and Machine Science 109,
https://doi.org/10.1007/978-3-030-88465-9_67

50% of the electricity consumption [1]. Specialized freezing and cooling facilities, used primarily to freeze and store raw goods, such as agricultural products, electricity consumption, originating primarily from powering the cooling facilities can account for 66% up to 96% of the total energy consumption of the plant. This ratio is directly affected by the type of production, processes, equipment and business of a particular plant. Simply put, refrigeration is a process by which heat is moved from one location to another [1]. Typical coolants applied in the cooling and freezing facilities in Serbia for large scale installations are ammonia (R717), and hydrocarbons, whereas for smaller installations typically hydrocarbons are applied. Piston compressors are mostly applied, where newer installations are fitted with variable speed control drives.

In a commercial context even, a small reduction in refrigeration energy use can offer significant cost savings [1]. Energy efficiency of the cooling processes with respect to building energy consumption is partially considered in the national legislative [2, 3], where some attention is brought to coefficient of performance values (COP) [2], but the buildings and facilities for cooling are not identified as a separate consumer type with regulated nor recommended specific energy consumption values per unit square or volume area [3]. Energy efficiency of cooling facilities and possible improvements are analyzed in the scientific research literature [4–13]. The effect of optimal refrigerant charge and energy efficiency on a small scale unit is analyzed in [4]. Economics of cooling appliances energy label classification is analyzed in [5], while a model for assessment of energy efficiency of commercial systems is proposed in [6]. Effect of refrigerant and operating conditions is found to be able to improve energy efficiency of industrial refrigeration for 16.3% up to 27.2% [7]. Energy performance indicators is proposed as tool to tackle the problem of energy efficiency of commercial refrigeration applications in [8], while coefficient of performance (COP) calculated as part of energetic analysis is used for energy efficiency assessment in [9]. Impact of control and operation on energy efficiency of vapor compression cooling systems, which are most commonly applied is tackled in [10], while application of cold storages and phase change materials is analyzed in [11]. Finally, exergy analysis in addition to energy balance is performed in [12]. Application of photovoltaics for cooling and freezing facilities could provide significant reduction of electricity consumption from the grid [13]. In addition to pollution and prevention aspects, energy efficiency measures and energy efficiency indicators for food industry in general, but also cooling and freezing facilities is given in the best available techniques reference documents published by the European Union [14–17].

In this paper, an attempt is made to provide national energy efficiency indicators for the cooling and freezing facilities in Serbia, based on the energy balances of the sample installations, which include installations ranging from small to large. Indicators in general are significant for benchmarking with resource management and resource efficiency assessments [17]. Installations may be controlled manually, by integrated technology or by a centralized refrigeration plant. The optimization of performance and improvement of energy efficiency is always installation specific, however, typical solutions may be found in literature identified as general. Best available techniques related to energy efficiency [16, 17]. Some of these solutions have been found applicable for the analyzed installations in Serbia. In general, the available and applicable options heat recuperation must have been pre-examined so that the level of non-recoverable heat is reduced, prior to dissipation of heat from the industrial process into the environment. Improved energy efficiency is a result of an integrated approach reducing the environmental impact of industrial cooling systems at the same time with cost cutting and energy efficiency improvement. Selection between wet, dry and wet/dry cooling to meet process and site requirements should aim at the highest overall energy efficiency [1].

67.2 Energy Efficiency of Food Cooling and Freezing Plants in Serbia

Refrigeration is a mechanical process which enables the temperature in a given space to be reduced [1]. The refrigeration cycle of mechanical refrigeration takes advantage of the mechanical work used by the compressor which draws from the evaporator and thus increases coolant pressure and temperature. Pressure in the evaporator is then reduced, causing the liquid particles to evaporate. The refrigerant extracts heat from the warmer objects in the insulated refrigerator cabinet, thus providing the cooling effect. The replacement of the liquid refrigerant is controlled by an expansion valve, restricting flow of the liquid refrigerant in the liquid line, changing the high-pressure, sub-cooled liquid refrigerant to low-pressure, low-temperature liquid, which continues the cooling cycle by absorbing heat. The refrigerant low-pressure vapour is then sucked from the evaporator by the compressor, and then compressed by the compressor to a high-pressure vapor. The high pressure coolant vapor is then forced into the condenser, where it condenses to a liquid under high pressure and rejects heat to the condenser. The heat from the condenser can be extracted by air, water or other fluids. The condensed liquid coolant is typically led into a liquid coolant storage tank, led to the expansion valve by pressure created by the compressor, enabling the realization of a complete cooling cycle.

67.2.1 The Case-Study Cooling and Freezing Facilities in Serbia

Typical coolants applied in the cooling and freezing facilities in Serbia for large scale installations are ammonia (R717), whereas for smaller installations typically hydrocarbons are applied. Piston compressors are mostly applied, where newer installations are fitted with variable speed control drives. Installations specialized for freezing and cooling of raw materials were selected, with a broad range of freezing and storage capacity. For the purposes of this study, the installations are classified in the following manner: (1) installations with power rating greater than 1 MW are considered large scale, (2) installations with power rating ranging from 100 kW to 1 MW are considered small scale, (3) installation with power rating less than 100 kW are is considered micro scale.

A refrigeration system is always used for freezing purposes and for cooling demands of cold storages. Large scale installations (see Table 67.1) use reciprocating or rotary screw compressors with ammonia as used as coolant. The evaporators responsible for keeping temperatures in ranges of: –30 to –40 °C in freezers, continuous or batch, while cold rooms are typically maintained at –20 to –25 °C. In addition, with respect to process technological requirements, and additional cooling temperature is maintained by evaporators at –5 to –10 °C in spaces for packaging and product treatment and manipulation. The ammonia refrigeration systems use double or multi stage cooling, with "booster" vapor compressors, thus improving the cooling cycle efficiency. This also enables reduced energy consumption in periods when freezing tunnels are not used, since the compressors responsible for near – 40 °C may be shut down. Smaller scale systems use R404a as coolant, reciprocating compressors and batch freezing chambers evaporators reach maintain temperatures up to –28 to –30 °C.

Generally, operations of washing, classification, pit removal and cutting are performed prior to the freezing process. Large installations with hot processing are equipped with steam boilers, where heat is applied to produce products or byproducts and in some cases pasteurization. Installations without heat processing, use heat only for space heating purposes. This has great impact on the energy balance of the installation.

Analyzed installations are used primarily to freeze and store raw goods (fruits or vegetables), where electricity consumption used to power the vapor compressors is the most significant factor in the energy balance accounting for 66% up to 96% of the total energy consumption of the plant. This ratio is directly affected by the type of production, processes, equipment and business of a particular plant. The installations which have some kind of additional product treatment (such as hot processing) of the raw material show lower share of the electricity and hence cooling energy in the energy balance. The capacities, temperature levels and applied coolants are presented in Table 67.1. Large scale installations typically use ammonia (R717) as a coolant. With the reduction of power, smaller scale systems use R404a coolant, but there are also mixed solutions with both systems applied, as a result of additional capacity

Table 67.1 Cooling and freezing installation technical and consumption data

No.	Power rating	Temperature levels	Hot processing yes/no	Refrigerant
Installation 1	4787 kW	– 10 °C/–38 °C	Yes	R717
Installation 2	340 kW	– 10 °C/–30 °C	No	R717
Installation 3	1030 kW	– 10 °C/–32 °C/–40 °C	Yes	R717
Installation 4	140.2 kW	– 5 °C/–30 °C/–35 °C	No	R404a
Installation 5	1167 kW	– 5 °C/–25 °C/–40 °C	No	R717, R404a
Installation 6	18.7	– 5 °C/–28 °C	Yes	R404a

Table 67.2 Ratio of electricity in the installation energy balance

No.	Share of electricity in total energy consumption (%)	Share of electricity in total energy costs (%)	Heat medium used	Energy source used for heating
Installation 1	51	46.1	Steam	Natural gas
Installation 2	96	92	N/A	Wood, fuel oil
Installation 3	66	76	Steam	Wood, fruit pits
Installation 4	96	96	N/A	N/A
Installation 5	92	100	Hot water	Fruit pits
Installation 6	84	63	N/A	Fire wood

increase. Three of the analyzed 6 installations performed some kind of hot treatment of their products, while 3 installations perform freezing and storage of goods.

Most of the large scale installations covered by this study were relatively old installations (originally built in the 70 s, or 80 s), which have modernized their equipment to date to smaller or larger extent. Small scale and micro installations were mostly relatively new, built according to new standards in the last decade. Despite the age of the design of large installations, insulation thickness of the cooling chambers does not differ much the new installations, and is of 20 cm or more. The micro installation, on the other hand side, is of panel construction with only 10–12 cm insulation thickness found in the panels.

Since the greatest electricity consumers in these installations are in fact the vapor compressors, and additional cooling and freezing equipment to some extent, such as evaporator fans and defrosting heaters, condenser fans and pumps, cooling chamber floor heaters, cooling chamber door heaters, the cooling and freezing systems in the analyzed installations can be considered responsible for most of the electricity consumption.

The share of electricity in total energy consumption is higher than 50% in all of the analyzed installations, and goes up to 96%, whereas the cost for electricity range from 46% up to 100% (see Table 67.2). Installations 3 and 5 take advantage of fruit pits to fire their boilers, and thus reduce heat production costs, where installation 5 generates enough fruit pits for to meet its entire heat demands. Installation 2 and 4 use heat only for space heating and washing purposes, while installation 6 uses wood fired stoves for hot processing purposes with many manual operations. Large scale facilities, such as installation 1, have better energy management practices, and based on their monitoring data, cooling and freezing facilities utilize 71% of their total electricity consumption (see Table 67.3). Other installations have no monitoring of electricity consumption per sector or product, however, based on the bottom-up approach, this ratio is estimated in the range from 85% for installation 3, to 95% or more for installations 2, 4 and 5.

Table 67.3 Specific electricity consumption in the installation energy balance

No.	Total electricity consumption kWh per t of product	Heat consumption per kWh per t of product	Steam consumption kg steam per t of product	Total energy consumption kWh per t of product
Installation 1	397.73	337.29	596.7	735.03
Installation 2	243	N/A	N/A	243
Installation 3	324.89	1711	2983.53	2035.89
Installation 4	444.7	N/A	N/A	444.7
Installation 5	800.25	N/A	N/A	800.25*
Installation 6	841.5	196.95	N/A	1038.45

*Total energy consumption does not account for heat, since the installation does not meter quantity of combusted fruit pits

67.2.2 Energy Efficiency Indicators

Energy efficiency indicators represent specific consumption, reflect the level of energy efficiency of an installation and enable comparison of energy efficiency of similar installations. The indicators are calculated for the analyzed installations as electricity and heat consumption per t of product, and as total energy consumption per t of product. For the installations with steam boilers. In addition, indicators are given as kg of steam per t of product, whereas effects of pressure and temperature of steam (i.e. enthalpy) are neglected with such choice of the indicator. All calculated values represent values based on annual energy and production values.

Benchmark values for electricity and heat can be acquired in literature [14], and are presented in Table 67.4. Benchmark values are result of performance of mostly large installations from EU, where energy consumption benchmark indicator values

Table 67.4 Benchmark values of energy efficiency indicators [14]

No.	Electricity consumption kWh per t of product	Heat consumption per t of product
Electricity (cooling)		
Fruit sorting	0–20	kWh_e/t frozen vegetable
Deep freezing –30 to –40 °C	80–280	kWh_e/t of frozen vegetables
Washing	up to 28	kWh_e/t
Product storage	20–65	kWh_e/m^3 of storage space/year
Total electricity	100–373	kWh_e/t
Hot processing		
Electricity	90–125	kWh/t
Thermal energy	2300–2800	(kg steam/t)

may be a result of detailed monitoring of energy consumption per process (see Table 67.4).

By comparison of values in Tables 67.3 and 67.4, it can be observed that, due to lack of monitoring data in Serbian installations, we can compare the indicators for total electricity consumption, whereas thermal energy can be compared just for the two installations with steam based hot processing.

Confronting the benchmark values to the calculated energy indicator values (see Table 67.5), we conclude that the Serbian installations consume 2.4–8.4 times more electricity compared to the lowest energy consumption of the benchmark range. Electricity consumption of the large installations is within ±10% of the highest benchmark range value, except for the installation 5, where consumption is 2.1 times higher. Electricity consumption of small scale installations is within –30/+20% range of the highest range benchmark value Micro scale installation uses 2.3 times more electricity than the highest benchmark range value, and can be considered as the least energy efficient installation based on this indicator. Confronting the steam consumption, we can conclude that Installation 1 has 3–5 times lower steam consumption compared to the benchmark range, where as there is 20–30% potential for energy saving in installation 3.

The benchmark value account for the entire food, drink and milk sector, whereas the indicator values and installations analyzed in this study correspond to operations where cooling and freezing is dominant operation and as such a dominant contributor to energy consumption, as shown in Table 67.2.

Table 67.5 Comparison of the installation electricity consumption from the benchmark value

No.	Electricity consumption kWh per t of product	Deviation from benchmark best value	Deviation from benchmark highest range value	Steam consumption kg steam per t of product	Deviation from benchmark	Deviation from benchmark best value
Installation 1	397.73	4.0	1.1	596.7	0.3	0.2
Installation 2	243	2.4	0.7	N/A	N/A	N/A
Installation 3	324.89	3.2	0.9	2983.53	1.3	1.2
Installation 4	444.7	4.4	1.2	N/A	N/A	N/A
Installation 5	800.25	8.0	2.1	N/A	N/A	N/A
Installation 6	841.5	8.4	2.3	N/A	N/A	N/A

67.3 Conclusion

In this paper, six cooling and freezing facilities were analyzed. The sample of these six installations was chosen to best represent the energy consumption of cooling and freezing, thus enabling consumption levels for Serbian installations to be defined. Based on the results of the assessment of the sample of 6 installations, which include micro to large scale installations from Serbia, the following energy efficiency indicators, i.e. benchmarks can be calculated:

- Total electricity consumption of cooling and freezing installations in Serbia ranges from 243 to 841.25 kWh per t of frozen or stored product, with an average value of 463.5 kWh/t of product,
- Total heat consumption for cooling and freezing ranges from 196.95 to 1711 kWh per t of product, with an average value of 748.4 kWh per ton product, and
- Total energy consumption ranges from 243 to 2035.89 kWh per t of product, with the average value of 882.9 kWh per t of product.

Acknowledgements This research was financially supported by the Ministry of Education, Science and Technological Development of the Republic of Serbia (Contract No.#451-03-9/2021-14/200109).

References

1. Fresner, J., Stamenić, M., Krenn, C., Tanasić, N., Mančić, M.: Energetska efikasnost u sektoru prehrambene industrije u postupku ishodovanja integrisane dozvole. Faculty of Technology and Metallurgy of the University of Belgrade (2020)
2. Pravilnik o energetskoj efikasnosti zgrada. Službeni glasnik RS, br. 61 (2011)
3. Pravilnik o uslovima, sadržini i načinu izdavanja sertifikata o energetskim svojstvima zgrada. Službeni glasnik RS, br. 69 (2012)
4. Li, Z., Jiang, H., Chen, X., Liang, K.: Optimal refrigerant charge and energy efficiency of an oil-free refrigeration system using R134a. Appl. Therm. Eng. **164**, 114473-1–114473-8 (2020)
5. Goeschl, T.: Cold case: the forensic economics of energy efficiency labels for domestic refrigeration appliances. Energ. Econ. (Suppl. 1), 104468-1–104468-12 (2019)
6. Minetto, S., Rossetti, A., Marinetti, S.: Seasonal energy efficiency ratio for remote condensing units in commercial refrigeration systems. Int. J. Refrig. **85**, 85–96 (2018)
7. Oh, J.S., Binns, M., Park, S., Kim, J.K.: Improving the energy efficiency of industrial refrigeration systems. Energy **112**, 826–835 (2016)
8. Acha, S., Du, Y., Shah, N.: Enhancing energy efficiency in supermarket refrigeration systems through a robust energy performance indicator. Int. J. Refrig. **64**, 40–50 (2016)
9. Gazda, W., Kozioł, J.: The estimation of energy efficiency for hybrid refrigeration system. Appl. Energ. **101**, 49–57 (2013)
10. Yin, X., Wang, X., Li, S., Cai, W.: Energy-efficiency-oriented cascade control for vapor compression refrigeration cycle systems. Energy **116**(Part 1), 1006–1019 (2016)
11. Selvnes, H., Allouche, Y., Manescu, R.I., Hafner, A.: Review on cold thermal energy storage applied to refrigeration systems using phase change materials. Therm. Sci. Eng. Prog. **22**, 100807-1–100807-26 (2021)

12. Qin, Y., Li, N., Zhang, H., Liu, B.: Energy and exergy performance evaluation of a three-stage auto-cascade refrigeration system using low-GWP alternative refrigerants. Int. J. Refrig **126**, 66–75 (2021)
13. Mančić, M., Živković, D., Laković Paunović, M., Đorđević, M., Vukadinović, B., Rajić, M.: Application of rooftop photovoltaics in cooling and freezing facilities. In: Proceedings of the 19th International Conference on Thermal Science and Engineering of Serbia, SimTerm 2019, pp. 808–819. Sokobanja (2019)
14. Best Available Techniques Reference Document for the Food, Drink and Milk Industries. Industrial Emissions Directive 2010/75/EU (Integrated Pollution Prevention and Control), EUR 29978 EN. Publications Office of the European Union, Luxembourg (2019)
15. Reference Document on Best Available Techniques on Energy Efficiency. Publications Office of the European Union (2009)
16. Reference Document on the application of Best Available Techniques to Industrial Cooling Systems. Publications Office of the European Union (2001)
17. Milovanović, M.B., Antić, D.S., Rajić, M.N., Milosavljević, P.M., Pavlović, A., Fragassa, C.: Wood resource management using an endocrine NARX neural network. Eur. J. Wood Wood Prod. **76**(2), 687–697 (2018)

[illegible] Qiu Y [illegible] Zhang [illegible] Liu [illegible] performance evaluation of [illegible] stage [illegible] photocatalytic [illegible] system [illegible] Int J [illegible] 120 [illegible] (2017)

[illegible] M [illegible] M [illegible] A [illegible] Rajn [illegible] M [illegible] application of [illegible] photovoltaic [illegible] in the [illegible] of the [illegible] 9th International [illegible] Scientific [illegible] Conference [illegible] (2019) [illegible]

[illegible] Best Available Techniques [illegible] Reference Document for [illegible] Industrial [illegible] Integrated Pollution Prevention and Control [illegible] Office of the European Union [illegible] (2016) [illegible]

[illegible] Reference Document [illegible] Available Techniques [illegible] Energy Efficiency [illegible] Publications Office [illegible] (2009)

[illegible] Reference Document on the application of [illegible] Best Available Techniques to Industrial Cooling Systems [illegible] (2001)

[illegible] M [illegible] R [illegible] PM [illegible] A [illegible] Wood [illegible] (2016)

Chapter 68
Preliminary Research Concerning the Pollution Generated by a Diesel Engine in Order to Higher Its Performance

Cosmin Constantin Suciu, Daniel Ostoia, Nicolae Stelian Lontis, Ion Vetres, Sorin Vlad Igret, and Ioana Ionel

Abstract The article studies the impact on both the pollution and climate change generated by the vehicles themselves, covering results generated by their engines, and starting from the end user's demands up to the final results. The first part refers to statistics and covers the differences of market needs, in terms of types of vehicles and their number, based on several countries across Europe and the impact they have on the environment. The analysis consists in measuring and interpretation of main characteristics of a Diesel engine (classified euro 3) equipped vehicle, still very common in the East European countries, with both an opacimeter and a gas analyzer. The purpose is to obtain base values for comparison with further improvements to be proposed and achieved on a Diesel engine, and get a better understanding on the resulted influences towards pollution.

Keywords Pollution · Climate change · Diesel · Engines · Improvements

68.1 Introduction

An important sector regarding the terms of demand in primary materials in the industry sectors is represented by the passenger vehicles, representing between 0.26 and 0.67 per person in the EU, and an estimative 0.5 vehicles person in the EU fleet, according to the EUROSTAT, in 2020.

The mass of material represented by these vehicles cover the interval between 450 and 750 kg per person. This number lowers in Easter Europe, in Romania being 260 kg. Higher values are to be found in Western and Central Europe, for example

C. C. Suciu (✉) · D. Ostoia · N. S. Lontis · I. Vetres · I. Ionel (✉)
Faculty of Mechanical Engineering, Politehnica University Timisoara, Timisoara, Romania
e-mail: cosmin.suciu@student.upt.ro

I. Ionel
e-mail: ioana.ionel@upt.ro

S. V. Igret
Faculty of Engineering, Aurel Vlaicu University, Arad, Romania

M. Rackov et al. (eds.), *Machine and Industrial Design in Mechanical Engineering*, Mechanisms and Machine Science 109,
https://doi.org/10.1007/978-3-030-88465-9_68

in Luxemburg the value reaches 1,000 kg per person. These numbers are influenced by the number of vehicles and also their mass [1]. Taking this fact into account, one must think about the impact on the environment of the fact that it is expected and normal to consider and conclude that it takes a lot more energy to put in motion a vehicle of higher mass and the number of vehicles of that kind in a region [2].

In order not to suffer the worst consequences of climate change, a rapid reduce in the global energy system is necessary. Even with the urge to reduce the global greenhouse gas emissions, it still remains at unsustainably high levels. About 60% of the emissions today are originated in the functionality of economic developer and heavy industry, together with the current energy infrastructure, with an expected raise up to nearly 100% by 2050 if no action will be taken [3].

In principle, the pollutant, which is part of both local and regional pollution problems, reacts as either a radiative forcing agent or changes the distribution of radiative forcing agents with the risk of producing a linkage between air quality and climate change issues. Figure 68.1 presents some of the potential linkages [4].

An important parameter related to the assessment of the environment is the fuel used by the vehicles. Since there are more types of fuel, they can be classified by their energy content and density. The amount of direct CO_2 emissions is related to the amount of carbon contained in the fuel, while the sulfur content is related to the SO_2 emissions. In Table 68.1 are presented the properties of several fuels [5]. Note that the used fuel in this research is diesel.

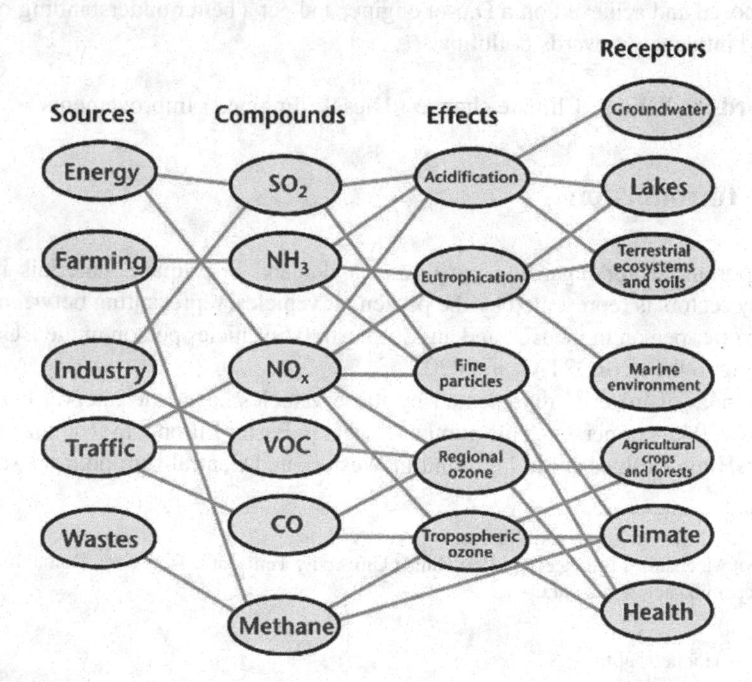

Fig. 68.1 Examples of some of the potential linkages between regional air quality and climate change issues

Table 68.1 Fuel properties [4]

	Energy content [kJ/kg]	Density [g/L]	CO_2 emissions factor [kg CO_2/Liter]	Sulphur content [ppm]	SO_2 emission factor [g SO_2/L]
Petrol	42,715	755	2.212	50	7.55 E−4
Diesel	43,274	850	2.697	50	8.50 E−4
CNG (G20)	49,578	717	2.712	0	0
LPG	45,114	550	1.549	15	1.65 E−4
Bio-diesel (RME)	37,700	880	2.470	100	1.76 E−3

Based on the fuel properties presented in Table 68.1, one factor that might be a cause for a fuel economy of the diesel fuel is its mass energy content. Diesel contains about 13% more energy [expressed in kJ/kg] than petrol. This means that a reduced quantity requirement of Diesel is needed to obtain the same output results, in comparison to petrol. However, diesel fuel is denser than petrol and despite that, it can be considered a quality fuel, analyzing only its calorific value.

The mechanical major contributor to the higher efficiency of the diesel engine is certainly also based on its higher compression ratio, compared to the Otto (petrol) engine. An important factor for the higher efficiency of the diesel engine can be also explained by its capability of running on leaner mixture. Compared to the petrol one, the compression-ignited engines, where the fuel is injected into highly compressed air towards the end of the compression stroke, do not require a knock sensor since the fuel in the combustion chamber will ignite in the areas of greater oxygen density. Even so, in order to keep the pollutants levels low and burn the fuel as efficient as possible, the diesel engine, as well as the petrol one, work with stoichiometric air/fuel mixtures.

Indirect emissions linked to the amount of fuel used by the vehicles can be calculated by knowing the energy content and density while the direct emissions are directly determined by the vehicle itself. The type approval test all and each vehicle sold on the European market gives information regarding the regulated emissions: carbon monoxide (CO), hydrocarbons (HC), nitrogen oxides (NO_x) and, in the specific case of diesel vehicles, particulate matter (PM) and also unregulated emissions: carbon dioxide (CO_2), Sulphur dioxide (SO_2), nitrous oxide (N_2O) and methane (CH_4). While the N_20 emissions mostly depend on the type of technology, both the carbon dioxide and sulfur dioxide can be found in the table presented, and calculated [6, 7].

It is known that diesel engines' due to their high thermal and fuel efficiency produce emissions low in carbon dioxide. The problem consists in these two pollutants: nitrogen oxide (NO_x) and particulates. These two are exchanged between each other in many aspects of the engine design. Higher temperatures reduce the emission of soot but increase the levels of nitric oxide (NO). To lower the amount of NO produces, the combustion chamber's temperature must be lowered but this increases the soot formation. The answer consists in a better stoichiometric report to lower the emissions. The NO resulted oxides are converted to NO_x which, combined with hydrocarbons or volatile organic compounds in the presence of sunlight, result in low level ozone leading to smog formation [8].

This article presents the analyze results of the outcome of the engine, in terms of emission values, characterized by factory parameters measure, with both an Otto gas analyzer and a Diesel gas analyzer. This complementary research method was adopted in order to achieve best results and accommodate the scientific results into an original conclusion.

Analyzing the current tendencies and direction by improvement of air quality and climate change, manufacturers focus to orientate to environment-friendly vehicles. Based on many studies, solutions for compression ignition engines to be kept on roads for a while more include bio-fuel, multifuel and also hybridization with addition of electrical motors.

68.2 Engine and Fuel System Specifications

The vehicle chosen for these measurements is equipped with a diesel-cycle 1,896 cm^3, 4 cylinders. It has 2 valves per cylinder, and a turbocharged with intercooler compression ignition traverse alignment engine.

This has a bore of 79.5 mm with a stroke of 95.5 mm, with a compression ratio of 19.5, being equipped with a VP37 Bosch injection pump and a Garrett GT1749V variable geometry turbocharger. It measures an output of 89 bhp, at 3,750 rpm, with 210 Nm of torque, at 1900 rpm.

Being a compression ignition type of engine, it features direct injection from 4 mechanical injectors with 0.184 μm nozzles, characterized by 5 holes and a 10 mm VE injection pump from Bosch, the VP37. For the 2 stage injectors, the pre-injection takes place at 220 bar and the second stage takes place when the pressure reaches the pressure of 300 bar.

68.3 Research Method

The measurements were achieved by using the Capelec 3,200 stand-alone opacimeter, presented in Fig. 68.2. It offers the possibility of a multi-gas analysis and a roadworthiness testing. ISO 11614, NFR 10025-2016 and CEM approvals and certifications, meaning that it can also be used for the technical inspection, are attested.

The machine is also upgraded with a Capelec gas analyzer module, thus it can offer data from both Otto and Diesel engines.

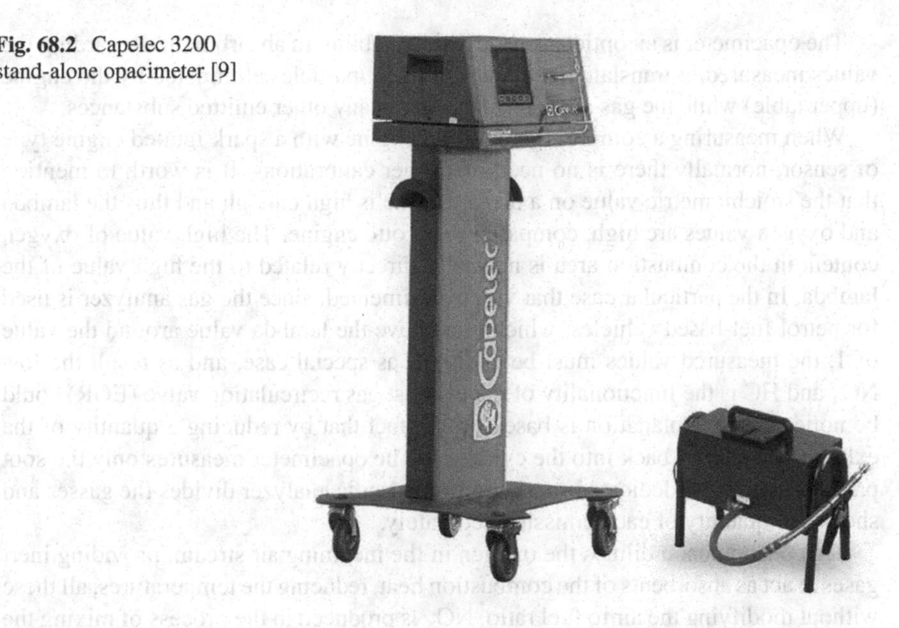

Fig. 68.2 Capelec 3200 stand-alone opacimeter [9]

Measurements have been achieved in a certified and authorized periodical inspection institution by the Romanian Automotive Register through the Product Certification Body (RAR-OCP) with SR EN ISO/CEI 17065:2013 certification. They consist in a measurement at idle and another one at 3,000 rpm status, with both analyzers.

68.4 Measurements and Results

Comparing the two different working regimes presented in Table 68.2, one concludes that there is a 363% increase of the light absorbent coefficient (K) value, when the engine is running at 3000 rpm. Considering that the environment where the sensor is positioned is a fix volume, if the temperature is constant, more particles will fit in this volume as the pressure rises. However, if the pressure is constant instead of the temperature, the number of particles inside the volume will be inverse proportional to the temperature.

Table 68.2 Opacimeter measurement results at different engine speeds

Engine speed [rpm]	Absorbent coefficient K [–]	Engine oil temperature [°C]
903	0.033	105
3,000	0.120	105

The opacimeter is an optical sensor with the ability to absorb light. Regarding the values measured, it translates as quantity of soot/particles eliminated by the engine (upper table) while the gas analyzer measures many other emitted substances.

When measuring a compression ignition engine with a spark ignited engine type of sensor, normally there is no need of further calibrations. It is worth to mention that the stoichiometric value on a diesel engine is high enough and thus the lambda and oxygen values are high, compared to an otto engine. The high value of oxygen content in the combustion area is normally directly related to the high value of the lambda. In the particular case that was experimented, since the gas analyzer is used for petrol fuel-based vehicles, which must have the lambda value around the value of 1, the measured values must be analyzed as special case, and as result the low NO_x and HC is the functionality of the exhaust gas recirculation valve (EGR) could be noticed. The explanation is based on the fact that by reducing a quantity of the exhaust gas, it goes back into the cylinders. The opacimeter measures only the soot particles, while the dedicated spark ignition specific analyzer divides the gasses and shows the quantity of each emission separately.

This phenomenon dilutes the oxygen in the incoming air stream, providing inert gases to act as absorbents of the combustion heat, reducing the temperatures, all these without modifying the air to fuel ratio. NO_x is produced in the process of mixing the atmospheric nitrogen and oxygen with the high temperature inside the combustion chamber, occurring when the cylinder is a peak pressure. Re-introduced gases from the EGR (exhaust gas recirculation) contain concentrations of NO_x and CO, which inhibits or lowers the total net production of these species, and not only. Comparing the air excess ratio (lambda value) and the oxygen values between the two situations from Table 68.3, one concludes that even though the lambda value droping by 8.9% in the 2nd regime, the measured oxygen quantity drops by 1.4%. This result means that the mixture is very lean, at higher constant rotations per minute, compared to the idle engine speed regime, but it also determines higher CO_2 values.

For both tests, run comparative, the engine measured parameters are presented in Table 68.4.

Table 68.3 Comparison of measurement results, for different engine speeds

Measured parameter	Regime 1	Regime 2
Engine speed [rpm]	903	3,000
CO [% vol.]	0.01	0.01
CO_2 [% vol.]	1.20	1.30
O_2 [% vol.]	19.28	19.01
HC [ppm vol.]	0	0
NOx [ppmv]	0	0
Air excess ratio lambda [–]	12.19	11.19
COcorr [% vol.]	0.12	0.11
Engine oil temperature [ºC]	106	105

Table 68.4 Measured parameters at different engine speeds

Measured parameter	Value first test	Value second test
Engine speed [rpm]	903	3,000
Intake air pressure [mbar]	1,020	1,091.4
Injection quantity [mg/str]	2.8	4.4
Start of injection [°BTDC]	0.4	0.2
Coolant temperature [°C]	90.9	91.8

Since there will be a continuation on methods of reducing the pollution levels of the diesel engines, one intends to approach the topic of multi fuel and fuel blends in further articles, in the near future.

68.5 Conclusion

The paper is a first step in the research concerning the identification of modalities for Diesel engine to be turned into more friendly and economic solutions, against the general trend on the market to renounce on their utility. The reason therefore is that presently a large number of engines are still in function, and they must be retrofitted either for reducing the pollution, either for reducing the fuel consumption, that leads to a reduced pollution exhaust.

The measurements have been done in a static regime, with no load. The results lead to following conclusions: (i) explanations concerning the trend on diesel engines are reasonable and proved by the results and (ii) reducing the emissions' concentration is possible, as a priority and also from the perspective of pollution control, knowing that diesel engines are a higher source of high emissions of CO, CO_2 and NO_x. Due to the high reliability and low fuel consumption, in the future a great use of for hybrid propulsion solutions or even range-extender, even better than a gasoline engine is very probable. A few manufacturers already launched on the market. Even if on average the costs of manufacturing a diesel engine are higher compared to an internal combustion spark ignited ones by around 15%, it compensates with the greater fuel efficiency, lower fuel price and torque driving characteristics.

A continuation of the research is expected, covering more ways of lowering the pollution level of the diesel engines and offering motivations concerning technical reasons sustaining why they should still be kept on the market, in the next future, of course under strict regulations concerning the pollution level.

Acknowledgements The first author expresses his gratitude to the teaching staff of the Politehnica University of Timisoara (Romania) and University of Aurel Vlaicu (Romania), where he studied. This paper marks the beginning of a journey into researching for finding ways to keep the diesel engine alive. The research is part of the PhD program of the first author at the Politehnica University of Timisoara. The authors acknowledge also the Romanian Automotive Register for the substantial support.

References

1. Bobba, S., Marques Dos Santos, F., Maury, T., Tecchio, P., Mėhn, D., Weiland, F., Pekar, F., Mathieux, F., Ardente, F., Sustainable use of Materials through Automotive Remanufacturing to boost resource efficiency in the road Transport system (SMART), EUR 30567 EN. Publications Office of the European Union, Luxembourg (2021)
2. Giavazzi, F., Buttini, P., Perego, C.: Polluting emissions caused by transport. In: Beccari, M., Romano, U. (eds.) Encyclopedia of Hydrocarbons, pp. 717–729. Istituto della Enciclopedia Italiana Giovanni Treccan (2007)
3. IEA, Energy Technology Perspectives. IEA, Paris (2020) https://www.iea.org/reports/energy-technology-perspectives-2020
4. Air quality expert group, Air Quality and Climate Change: A UK Perspective (2007)
5. Timmermans, J.-M., Matheys, J., Van Mierlo, J., Lataire, P.: Environmental rating of vehicles with different fuels and drive trains: a univocal and applicable methodology. Eur. J. Transp. Infrast. **6**(4), 313–334 (2006)
6. Vardoulakis, S.: Human exposure: indoor and outdoor. In: Hester, R.E., Harrison, R.M. (eds.) Air Quality in Urban Environments, pp. 85–99. Royal Society of Chemistry, UK (2009)
7. Vallero, D.A.: Fundamentals of Air Pollution, Forth Academic Press is an imprint of Elsevier, Burlington (2007)
8. Pignon, J.: Diesel engines: design and emissions. Platin. Met. Rev. **49**(3), 119–121 (2005)
9. https://www.capelec.fr/en/equipements/emission-testers/cap3201-o. Accessed 2021/04/02

Chapter 69
A Bayesian Analysis of CO_2 Emissions and National Industrial Production in Romania

Carmelia Mariana Bălănică Dragomir, Geanina Podaru, and Cristian Munteniță

Abstract Policies to reduce carbon dioxide emissions are frequently deemed individually, relating to electrical power and energy, residential heating, transport, and industrial sector. This may result in specific actions being considered a priority in the erroneous sectors, and omits others. The policymaker's study should focus on greenhouse gas reduction objectives, notwithstanding the incertitude concerning what aims are suitable and the proper period. Starting with 1993 Romania implemented the European statistical data collection procedure based on annual questionnaires. The article focuses on assessing the CO_2 emissions and the linkage with national industrial production. By analysing the CO_2 emissions and the production of main industrial goods two important conclusions are outlined: (1) the trend of the total production is an important factor to generate the emissions; (2) the modify in industrial structure has the distinct share to the emissions in several periods, in 1993–1998 and 2002–2012, the higher production implies over the same period changes in CO_2 trend of the emissions.

Keywords Climate change · Sustainable industrialization · Bayesian analysis

69.1 Introduction

Inclusive and sustainable industrialization is essential to achieve sustainable development. Readaptation and using the green and additional eco-friendly methods should have fulfil the requirements of the nowadays—without comprising the demands of further generations [1]. Greening manufacture and production systems has a crucial aspect to change to green economies and to accomplish a viable progress [2].

The green industry involves powerful and concerted economic drivers that engender operation and profit, stimulate international commerce and capacitate efficacious usage of resources [3].

C. M. B. Dragomir (✉) · G. Podaru · C. Munteniță
"Dunărea de Jos" University of Galati, Galati, Romania
e-mail: cdragomir@ugal.ro

M. Rackov et al. (eds.), *Machine and Industrial Design in Mechanical Engineering*, Mechanisms and Machine Science 109,
https://doi.org/10.1007/978-3-030-88465-9_69

In the last 200 years, with a constant population increase, an accelerating industrial development, urban sprawl and global energy consumption has been precipitously expanded and ecological conditions is progressively deteriorating. The air pollution, particularly the greenhouse gases and carbon dioxide (CO_2) pollution, becomes a significant general issue to adapt to climate change [4].

CO_2 emissions are absolutely dependent on the production and utilization of energy and consequently the pattern of the relationship among CO_2 emissions and economic progress has relevant repercussion for the designation of a proper common economic and environmental strategy [5, 6].

According to the 2018 IPCC several current worldwide trends have led any persons to incorrectly deduce that industry is no more a principal sector of the economy. A commonly idea is that fabrication's significance has been decreasing through the last decades, according to the beginning of the "post-industrial" period. The observational proofs used to support this demand is generally relied on the nominal value added produced in industries as a proportion of nominal gross domestic product (GDP). Apparently, both at the general level and between particular country groups, the value of industrial production has diminished proportional to other sectors, insinuating a process of deindustrialization. It is extremely important that CO_2 emissions and the use of materials increased in industry during the period 1995 until 2014 [7]. The industrial sector includes a large variety of subsectors containing production (e.g., steel, construction materials, chemicals), mining and metallurgy.

Greenhouse gas emissions were roughly halved among 1990 and 2007 in every Baltic Member States, with the highest reduce registered in Latvia (–54.7%). Furthermore, there were substantial decreased in Romania, Slovakia and Bulgaria and Poland [8].

Between 1990 and 2018, CO_2 emissions were considerably reduced by Latvia (82%), Luxembourg (82%), Lithuania (80%), Czechia (79%), Bulgaria (76%), Romania (75%) and Estonia (73%) comparatively with the level of CO_2 emissions in 1990. Only Austria, Cyprus, Ireland and Spain account emission increases [9].

The relationship within economic increase and environmental contamination is assessed one of the most significant empirical correlation currently, having as one of its major supposition that in a country's growth process, as production of main industrial goods environmental quality firstly deteriorates to a specific point, and then the environmental feature improves [10, 11].

In this paper we analyzed the empirical connection among carbon dioxide (CO_2) emissions and production of main industrial goods in Romania over the period 1993–2019. This empirical interdependence, widely known in the literature as the Environmental Kuznets Curve (EKC) hypothesis, indicates that the link between these variables, in process of time, pursues an inverse U-shape, and increased in economic development would be followed by betterment in environmental characteristic. Bayesian statistics is a major issue nowadays in several domains in which statistics is applied.

The Bayesian method has multiple advantages like offering more correct and eloquent assumptions, answering complex issues directly and precisely, by relying on entire accessible data, and being especially appropriate for decision-making [12].

69.2 Data and Methods

The datasets were collected by the National Institute of Statistics, Romania starting from 1993 until 2019. The 27 years of values recorded for both industrial productions, finished products and CO_2 emissions have been the basis of this article.

The information was collected on the basis of questionnaires and was carried out in accordance with the recommendations and rules of Eurostat, with the main purpose of providing to the internal and external users complete, current and reliable information regarding the industrial physical production realized in our country, comparable with those realized by the member states of the European Union. The information was collected through personalized questionnaires, as pre-printed with the identification data of the company, with the codes and with the names of the products that they make during the reference year. The reference period of the research is the previous calendar year.

The statistical research PRODROM A (Industrial products and services), has as main objective the collection of data and information of industrial physical production, according to the Nomenclature of industrial products and services PRODROM, integrated in the national classifications system and in direct correspondence with the Classification of Activities in the National Economy. CANE Rev 2), respectively with the Classification of Products and Services associated with Activities [13].

Industrial production is the direct result of the processes of extraction from nature of raw materials and existing materials, of subsequent processing of industrial products or the restoration of their initial technical and qualitative parameters.

The National Institute of Statistics of Romania has implemented the methodology of Air Emission Accounts (AEA) developed by Eurostat and data have been reported since the reference year 2008, in accordance with the requirements and the required reporting format—imposed by division of economic activities. Greenhouse gases (GHGs) are gases of natural and anthropogenic origin, which absorb and emit radiation with wavelengths specific to the spectrum of infrared radiation emitted by the earth's surface, atmosphere and clouds. The Kyoto Protocol regulates the main greenhouse gases: dioxide carbon dioxide (CO_2), methane (CH_4), nitrous oxide (N_2O) and three groups of fluorinated gases: hydrofluorocarbons (HFCs), perfluorocarbons (PFCs) and sulfur hexafluoride (SF_6).

The Bayes Factor Inference on Pairwise Correlations were calculated based on formula (69.1):

$$P\left(\frac{A}{B}\right) = \frac{P(B/A)P(A)}{P(B)} \tag{69.1}$$

where

P(A/B)	Probability of A given B,
P(B/A)	Probability of B given A,
P(A)	Probability of A,
P(B)	Probability of B.

P(A\B) and P(B\A) are known as conditional probabilities, which is the probability of one event (A or B) occurring given another event (A or B) has already occurred.

Prior Distributions: The first step in a Bayesian analysis is to determine what is known as the Prior Distribution. In Bayesian statistics, the variance of our Prior Distribution is commonly noted to as precision; the higher the precision, the more confident it is. Distributions with upper accuracy will be more peaked, with a smaller variance and conversely [14, 15]. The final step in a Bayesian study is to observe what is known as the Posterior Distribution using Bayes' Theorem. The Posterior Distribution is usually obtained by Markov Chain Monte Carlo Methods using statistical software [16].

In this study we used IBM-SPSS Version 26, a powerful tool for advanced statistical analysis.

69.3 Results and Discussion

In order to quantify the production of main industrial products we selected from national inventory the main industrial goods and we represented in Fig. 69.1, both depending on the type of product and depending on the year of manufacture. The main products analysed are: raw steel, electric crude steel, finished heavy steel laminates for pipes and forging, steel pipes, cold drawn steel wires, non-alloy steel, finished hot rolled steel sheets, medium and lights steel laminates, seamless steel pipes and laminates of aluminium and aluminium alloy. The industrial goods used in this article are in fact the first ten products manufactured in Romania during 1993–2019.

The trend is obviously decreasing, clearly observing the downward trend in the production of industrial goods. If in 1993 the national production was 18,167,993 tones it decreased in 2019 to 13,487,649 tones. The difference of 4,680,344 tones is

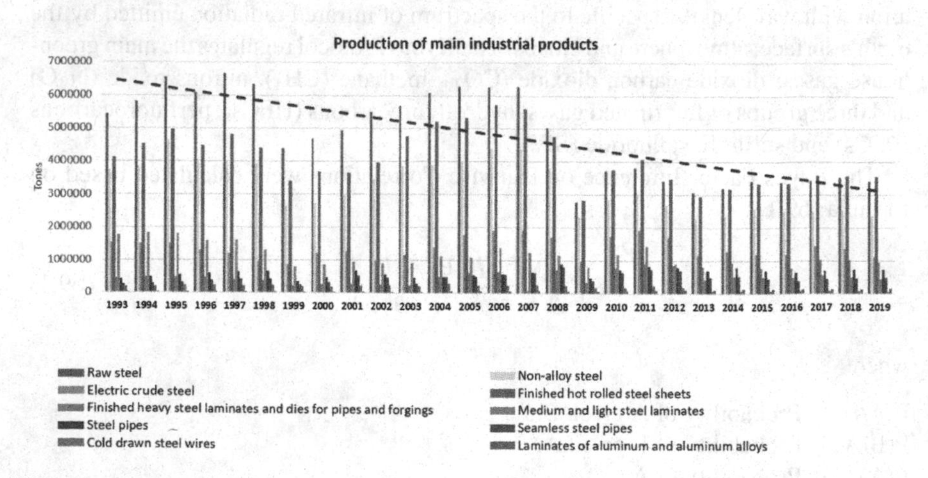

Fig. 69.1 Trend of national production of main industrial products between 1993 and 2019

25.76% of the national production. From the analyzed period of 27 years, one can obviously observe the upward trend starting with 2000 until 2007, a decrease in 2009 and then an attempt to recover from 2010 to 2012. The peak of industrial production was registered in 2006, more precisely 21,978,943 tones. In the last five years there is a period of stagnation of production, around 13,608,441 tones.

To observe the type of industrial goods produced in the twenty-seven years, we calculated the percentage. Approximately 75% of the total production is represented by 3 types of products: raw steel—29%, non-alloy steel—20% and 25% finished hot rolled steel sheets. The lowest percentages are 0.41% for laminates of aluminium and aluminium alloy and 1.06%—seamless steel pipes. From this analysis it is clear which are the most frequently requested products on the Romanian market.

From the comparison between the national production and the CO_2 emissions, in the period 1993–2019, it is clear that the two analyzed elements have the same trend. If between 1993 and 1998 we have a period of maintenance, followed in 1999 by a sharp decline and then a slow increase until 2006. In 2009 both the amount of industrial goods produced and CO_2 emissions have a minimum point, followed by an increase in 2011 and then both elements continue relatively linearly. At the same time, Fig. 69.2 clearly highlights the ten-year economic cycle of the economy.

Using IBM-SPSS, we analyzed the Bayes Factor Inference on Pairwise Correlations and Posterior Distribution Characterization for Pairwise Correlations. The pattern of the Bayesian inference about Pearson correlation coefficient permits to outline Bayesian inference by appraising Bayes factors and describing posterior distributions. Production of main industrial goods and CO_2 emissions are the two elements analyzed in our study.

The Pearson correlation calculated is 0.717 and it indicate a indicate a strong positive linear relationship between the two parameters, while the Bayes Factor is 0.001. Bayes Factors range from 0 to infinity; values less than 1 support the null hypothesis as being more likely than the alternate hypothesis. 27 represents the

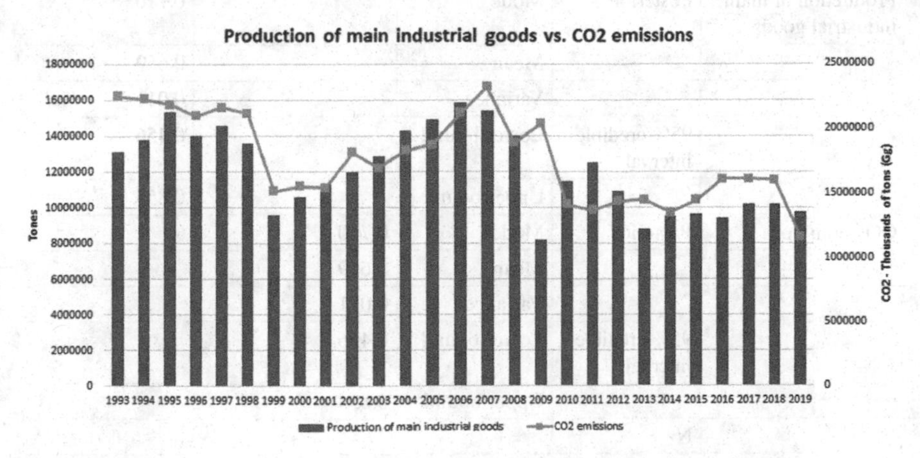

Fig. 69.2 Comparison between national production and CO_2 emissions for 1993–2019

number of cases, more exactly the number of years from 1993 to 2019 (see Table 69.1).

For the Posterior Distribution Characterization for Pairwise Correlations, we used 95% Credible Interval. The IBM SPSS analysis indicate 0.710 the coefficient for mode, 0.669 the coefficient for mean, while the variance for analyzed data is 0.011 (Table 69.2).

Function of log likelihood values can be used to compare the fit of models, in our case, the production of main industrial goods and CO_2 emissions. The log likelihood function indicates an evident connection between production of industrial goods and CO_2 emissions. The two factors are computed using a likelihood function, which

Table 69.1 Bayes factor inference on pairwise correlations

		Production of main industrial goods	CO_2 emissions
Production of main industrial goods	Pearson correlation	1	0.717
	Bayes factor		0.001
	N	27	27
CO_2 emissions	Pearson correlation	0.717	1
	Bayes factor	0.001	
	N	27	27

Bayes factor: Null versus alternative hypothesis

Table 69.2 Posterior distribution characterization for pairwise correlations

			Production of main industrial goods	CO_2 emissions
Production of main industrial goods	Posterior	Mode		0.710
		Mean		0.669
		Variance		0.011
	95% credible interval	Lower bound		0.456
		Upper bound		0.855
CO_2 emissions	Posterior	Mode	0.710	
		Mean	0.669	
		Variance	0.011	
	95% credible interval	Lower bound	0.456	
		Upper bound	0.855	
	N		27	

The analyses assume reference priors (c = 0)

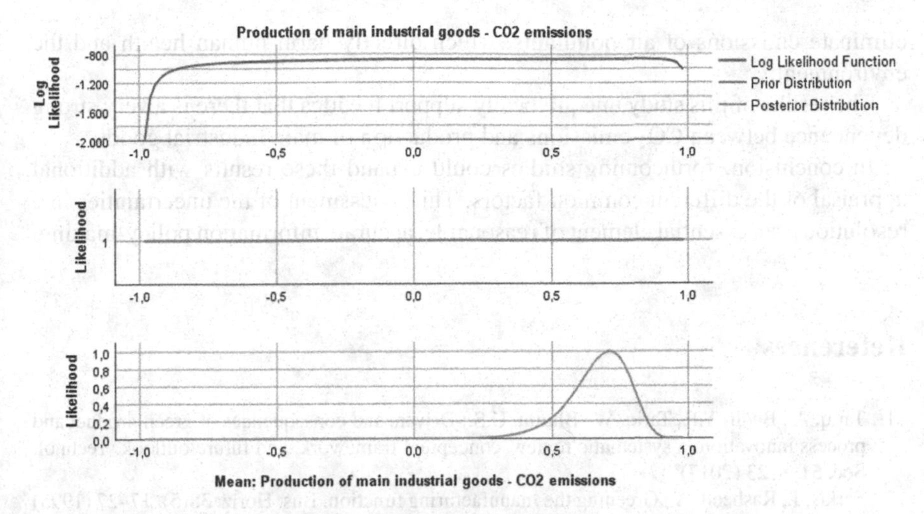

Fig. 69.3 Plots for pair 1: production of main industrial goods—CO_2 emissions

indicates the most likely values for the unknown parameters given our data. The prior distribution is linear in our case, while posterior distribution has a peak between 0.5 and 1.0, around the 0.7 value (see Fig. 69.3).

The Bayesian analysis indicates the most likely difference between mean production of main industrial goods is 0.669 that can outline the null is a more probable explanation for the data than the alternate.

69.4 Conclusion

In this paper we applied a Bayesian method to test the empirical connection between production of main industrial goods—CO_2 emissions for Romania over the period 1993–2019. The Pearson correlation indicated a indicate a strong positive linear relationship between the two parameters, the result being confirmed by the Function of log likelihood values. Bayesian Inference is becoming increasingly popular and may be used in a range of statistical tests.

Using the 27 years dataset collected by the National Institute of Statistics we observed the trend of Romanian production of main industrial goods analyzing 10 types of products. The tendency of national production is a downward one and it is overlapping extremely well with the trend of CO_2 emissions. Certainly, other elements influence the amount of CO_2 in the atmosphere, but the industry is on the first place. Or at least it was in the 27 years of analyzed period.

Our major contribution to the literature is the analysis of industrial production in close connection with CO_2 emissions. By collecting and evaluating this data, technical measures can be taken to protect the environment, prevent, reduce and eliminate emissions of air pollutants, which directly harm human health and the environment.

The results of this study unequivocally support the idea that there is a very strong dependence between CO_2 emissions and production of main industrial goods.

In conclusion, forthcoming studies could expand these results with additional appraisal of the different common factors. This assessment of the uncertainties in a resolution is an essential element of reasonable, accurate information policy-making.

References

1. Tariq, A., Badir, Y.F., Tariq, W., Bhutta, U.S.: Drivers and consequences of green product and process innovation: a systematic review, conceptual framework, and future outlook. Technol. Soc. **51**, 8–23 (2017)
2. Sarkis, J., Rasheed, A.: Greening the manufacturing function. Bus. Horiz. **38**(5), 17–27 (1995)
3. Winter, S., Lasch, R.: Environmental and social criteria in supplier evaluation e lessons from the fashion and apparel industry. J. Clean. Prod. **139**, 175–190 (2016)
4. Sun, J.W.: Changes in energy consumption and energy intensity: a complete decomposition model. Energy Econ. **20**(1), 85–100 (1998)
5. Rypdal, K., Winiwarter, W.: Uncertainties in greenhouse gas emission inventories: evaluation, comparability and implications. Proc. Environ. Sci. Policy **4**(2), 107–116 (2001)
6. Orabi, W., Zhu, Y., Ozcan-Deniz, G.: Minimizing greenhouse gas emissions from construction activities and processes. In: Proceedings of the Construction Research Congress, ASCE 2012, pp. 1859–1868. Reston (2012)
7. IPCC Report Demand for Manufacturing: Driving Inclusive and Sustainable Industrial Development 2018. Last accessed 2021/04/02
8. EEA, Annual European Community greenhouse gas inventory 1990–2007 and inventory report 2009. http://www.eea.europa.eu. Last accessed 2021/04/04
9. EEA, Annual European Community greenhouse gas inventory 1990–2018 and inventory report 2020. http://www.eea.europa.eu. Last accessed 2021/04/04
10. Jardon, A., Kuik, O., Tol, R.S.J.: Economic growth and carbon dioxide emissions: an analysis of Latin America and the Caribbean. Atmósfera **30**(2), 87–100 (2017)
11. Kruschke, J.K.: Bayesian estimation supersedes the t test. J. Exp. Psychol. Gen. **142**(2), 573–603 (2012)
12. O'Hagan, A.: Bayesian statistics: principles and benefits. Frontis **3**, 31–45 (2004)
13. National Institute of Statistics. http://statistici.insse.ro. Last accessed 2021/04/02
14. Hoeting, J.A., Madigan, D., Raftery, A.E., Volinsky, C.T.: Bayesian model averaging: a tutorial. Stat. Sci. **14**(4), 382–401 (1999)
15. Raftery, A.E., Zheng, Y.: Discussion: performance of Bayesian model averaging. J. Am. Stat. Assoc. **98**, 931–938 (2003)
16. Touran, A., Wiser, E.P.: Monte Carlo technique with correlated random variables. J. Constr. Eng. Manage. **118**, 258–272 (1992)

Chapter 70
Environmental Impact of Minimal Order Quantity Constraint in (*R, s, S*) Inventory Policy

Jasmina Žic and Samir Žic

Abstract Inventory management and logistics, recognised as key components of supply chain management, significantly impact the environment. In this paper, the correlation between stochastic market demand, inventory control working under periodic review policy, and environmental influence of product replenishment by road transportation are considered. By conducting extensive numerical simulation research of the supply chain echelon model, we study the impact of minimum order quantity constraint on (*R, s, S*) inventory system and its environmental performance measured by greenhouse gas emissions using EN 16258 methodology. The simulation design includes one level of average daily market demand and standard deviation, one level of targeted fill rate, eleven levels of lead time, one level of work time, six levels of minimum order quantity and five categories of delivery vehicles. Results of extensive numerical simulations indicate the sensitivity regarding logistics and environmental performance of the supply chain depending on the minimum order quantity and lead time change.

Keywords Supply chain management · Minimum order quantity · Greenhouse gas emission

70.1 Introduction

The companies operating in supply chains nowadays face the necessity to adjust their operations to meet the legislative and market requirements for environmental impact reduction, simultaneously increasing or at least maintaining profitability. Integration of environmental thinking into traditional supply chain management is being referred to as Green Supply Chain Management (GSCM), and its focus is on the influence

J. Žic
Faculty of Mechanical Engineering and Naval Architecture, University of Zagreb, 10002 Zagreb, Croatia

S. Žic (✉)
Faculty of Engineering, University of Rijeka, 51000 Rijeka, Croatia
e-mail: samir.zic@riteh.hr

M. Rackov et al. (eds.), *Machine and Industrial Design in Mechanical Engineering*, Mechanisms and Machine Science 109,
https://doi.org/10.1007/978-3-030-88465-9_70

and relationships between supply chain management and the natural environment. GSCM is driven mainly by increasing customer awareness and regulatory obligations resulting from the raw materials' resources depletion, growing problems with pollution and waste, intense degradation of the environment, etc. [1].

This paper explores inventory management as one of the key segments of supply chain management and its environmental impact, specifically greenhouse gas (GHG) emissions related to inventory replenishments. As an indicator of the environmental performance of supply chains, emissions are highly sensitive to the mode and frequency of deliveries and generally increase with frequent replenishments, particularly with globalisation and longer-distance trade [2]. Relevant literature recognises efficient inventory management and operational adjustments as valuable tools for emissions' reduction without investing in new technologies [3–5]. This paper aims at providing valuable insights in that direction by exploring the influence of operational constraints such as minimum order quantity (MOQ) and lead time on output variables.

MOQ is defined by the manufacturer or the supplier, representing the lowest quantity of goods that can be ordered, measured either by product units or monetary value. It is recognised as one of the common ways to achieve economies of scale in distribution and production [6]. MOQ at the supplier's side reflects on the processes of handling and executing the orders, such as invoicing, bookkeeping, manhandling, shipping, and related costs [7]. It is not unusual that workflows, processes and production batches are also adjusted to MOQ size in order to achieve better process efficiency. However, large MOQs bring significant challenges to the efficient management of supply chains, as they oblige retailers to choose between ordering either none or many units, which outcomes in reduced flexibility in responding to the market demand and increased inventory costs [6]. Although common in real-world settings, the MOQ constraint related to (R, s, S) inventory control policy is rarely studied in the SCM literature and particularly, GSCM field. The work of [8] studies the periodic review inventory model to indicate the inventory policies that minimise the costs, taking into account the MOQ parameter. Reference [6] consider the impact of MOQ on stochastic inventory system performance, analysing demand variability. The work of [9] studies the cost performance of periodic review inventory system with stochastic demand considering MOQ constraint. The authors provide expressions to calculate near-optimal policy parameters. A single MOQ constraint is included in the study of [10] examining economic, environmental and inventory management performance of a supply chain model operating under (R, s, S) policy.

The rest of the paper is organised as follows. Section 70.2 presents the setup of the supply chain model and calculations' procedure. In Sect. 70.3, results and discussion are provided. Section 70.4 brings conclusions.

70.2 Experimental Work

In combination with a large set of numerical simulation scenarios, modelling and simulation method are recognised as the most suitable approach for examining a wide range of influential parameters in complex systems [11]. In this paper, we use numerical simulations to study the complex interdependencies between stochastic market demand, inventory control under (*R*, *s*, *S*) inventory control policy, and consequently, environmental influence of product replenishments by road transportation.

70.2.1 Supply Chain Model Setup

We study a single product single echelon supply chain, as displayed in Fig. 70.1. The distribution centre (DC) observes market demand and regulates its inventory levels with the supplier's replenishments.

Inventory in DC is controlled by periodic review or (*R*, *s*, *S*) policy, with a review period of one day. The targeted fill rate is set to 99.9%. Safety stock, partial deliveries or backlogging are not considered. The experimental design includes one level of average daily market demand and standard deviation of demand, one level of targeted fill rate, eleven levels of lead time, one level of work time and six levels of MOQ, resulting in a total of 1650 simulation experiments.

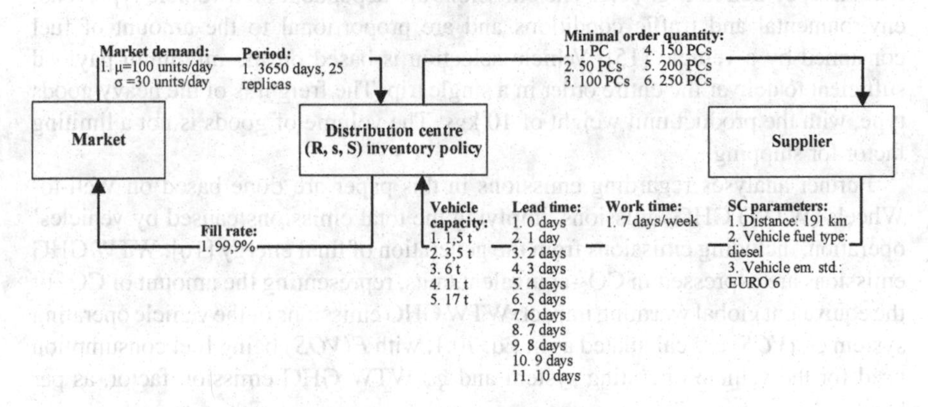

Fig. 70.1 Supply chain model scheme

70.2.2 Numerical Simulations and Environmental Performance

There are no simple procedures or algorithms available to set the optimal inventory levels—reorder point s and order-up-to level S in practice. Therefore, in most ERP environments they are chosen manually, without algorithmic determination [12]. This research uses the two-dimensional brute force search algorithm to determine the lowest combinations of s and S, which satisfy the targeted fill rate and the rest of the model constraints: work time, lead times, and MOQ.

We observe the total period of 3650 days to calculate inventory and logistic parameters: inventory levels s and S, average inventory level (AIL), number of orders, size and number of replenishment deliveries (N_d). These information represent the basis for calculating GHG emissions from deliveries on the relation supplier—DC, using EN 16258 methodology. GHG emissions are under the influence of the number and size of shipments required for satisfying the predetermined fill rate of market demand. The rest of the logistic parameters, such as delivery distance, vehicle type and capacity, fuel type, vehicle emission standard, are specified in Fig. 70.1.

Methodology for calculation and declaration of energy consumption and GHG emissions of transport services EN 16258 [13] is used to determine the levels of emissions released in road freight transportation by using Ecotransit software [14]. Deliveries are organised on the 191 km long route between cities of Frankfurt and Cologne, with vehicles of 5 different payload capacities and EURO6 emission standard, as specified in Fig. 70.1. The emissions are dependent on a vehicle type, load, environmental and traffic conditions and are proportional to the amount of fuel consumed by a vehicle [15]. Vehicle selection is based on the minimum payload sufficient to deliver the entire order in a single trip. The freight is of the heavy goods type, with the product unit weight of 10 kgs. The volume of goods is not a limiting factor for shipping.

Further analyses regarding emissions in this paper are done based on Well-to-Wheels (WTW) GHG emissions, implying the total emissions caused by vehicles' operation, including emissions from the generation of final energy [16]. WTW GHG emissions are expressed in CO_2-equivalent units, representing the amount of CO_2 of the equivalent global warming impact. WTW GHG emissions of the vehicle operating system $G_w(VOS)$ are calculated as in Eq. 70.1, with $F(VOS)$ being fuel consumption used for the vehicle operating system and g_w WTW GHG emission factor, as per [13].

$$G_w(VOS) = F(VOS) \cdot g_w \tag{70.1}$$

Total WTW GHG emissions G_w emitted during the observed period are calculated according to Eq. 70.2.

$$G_w = G_w(VOS) \cdot N_d \tag{70.2}$$

70.3 Results and Discussion

The experimental design described in Subsect. 70.2.1 resulted in 1650 simulation experiments. Results regarding delivery size, the number of deliveries and related WTW GHG emissions are grouped for further analysis, based on lead times and MOQs, and presented in Figs. 70.2, 70.3 and 70.4.

Study results show that, with the increase of lead time length, regardless of the MOQ change, average delivery size grows and the number of deliveries decreases. For MOQ1 and MOQ50, the average delivery size is more than 11 times larger in conditions of the longest lead time (LT10) compared to the shortest one (LT0). In

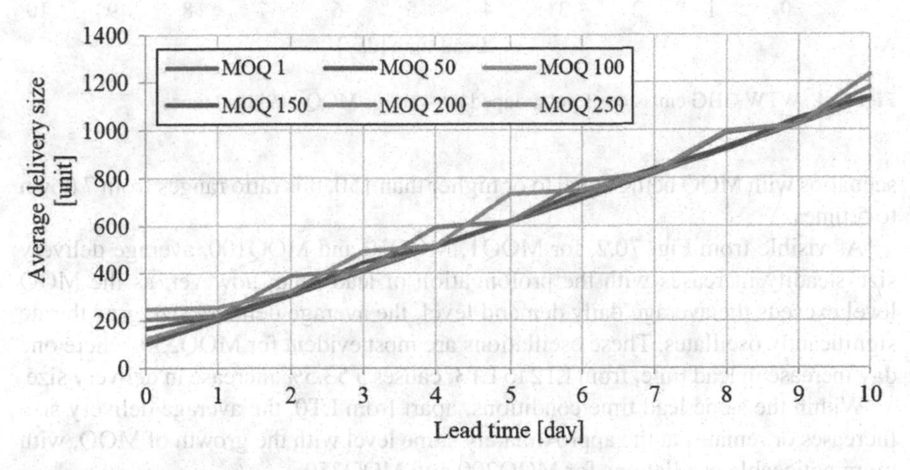

Fig. 70.2 Average delivery size grouped per various MOQs and lead times

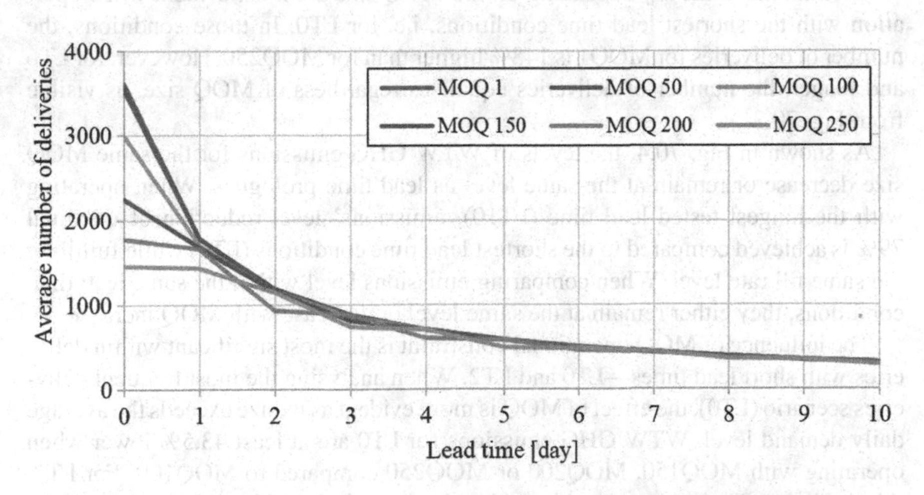

Fig. 70.3 The average number of deliveries grouped per various MOQs and lead times

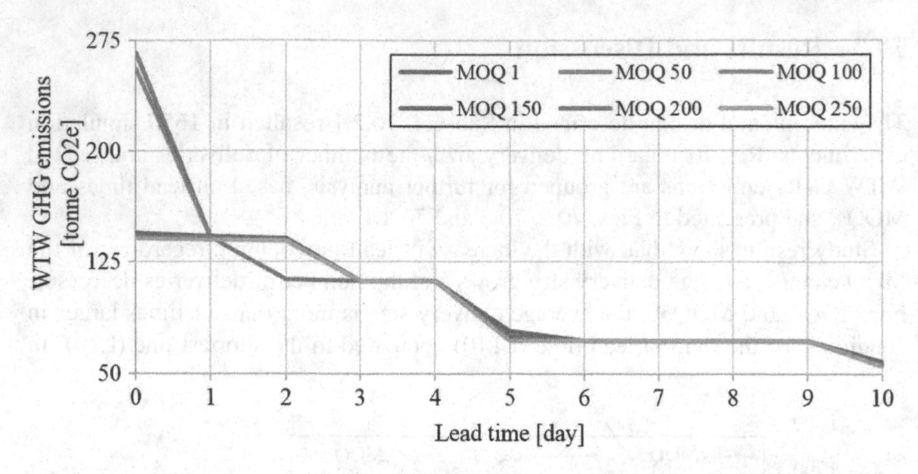

Fig. 70.4 WTW GHG emissions level grouped per various MOQs and lead times

scenarios with MOQ being equal to or higher than 150, this ratio ranges from 7 down to 5 times.

As visible from Fig. 70.2, for MOQ1, MOQ50 and MOQ100, average delivery size steadily increases with the prolongation of lead time; however, as the MOQ level exceeds the average daily demand level, the average delivery size growth rate significantly oscillates. These oscillations are most evident for MOQ250, where one day increase in lead time, from LT2 to LT3, causes a 53.5% increase in delivery size.

Within the same lead time conditions, apart from LT0, the average delivery size increases or remains at the approximately same level with the growth of MOQ, with more noticeable oscillations for MOQ200 and MOQ250.

Oscillations in the N_d depending on the MOQ size are most evident when operation with the shortest lead time conditions, i.e. for LT0. In those conditions, the number of deliveries for MOQ1 is 148% higher than for MOQ250. However, for LT6 and longer, the number of deliveries equalises regardless of MOQ size, as visible from Fig. 70.3.

As shown in Fig. 70.4, the levels of WTW GHG emissions for the same MOQ size decrease or remain at the same level as lead time prologues. When operating with the longest tested lead time (LT10), emissions' level reduction of maximal 79% is achieved compared to the shortest lead time conditions (LT0) while fulfilling the same fill rate level. When comparing emissions level within the same lead time conditions, they either remain at the same level or decrease with MOQ increase.

The influence of MOQ operational constraint is the most significant within deliveries with short lead times—LT0 and LT2. When analysing the most frequent deliveries scenario (LT0), the effect of MOQ is more evident as its size exceeds the average daily demand level. WTW GHG emissions for LT0 are at least 43.5% lower when operating with MOQ150, MOQ200 or MOQ250 compared to MOQ100. For LT2, this reduction of emissions level is evident between MOQ150 and MOQ200, and it totals 19.5%.

70.4 Conclusions

In this paper, we study the interdependencies between stochastic market demand, inventory control under (*R, s, S*) policy, and the effect that certain operational parameters have on logistics and environmental influence of supply chains. Specifically, we analysed how MOQ constraint, together with lead time length influences the number and size of deliveries related to inventory replenishments needed for fulfilling targeted fill rate and the levels of WTW GHG emissions from these transportations.

Results imply that with the prolongation of lead time, for all tested MOQs, average delivery size grows and the number of deliveries decreases. Additionally, as the MOQ level exceeds the average daily demand, the average delivery size growth rate significantly oscillates. The most significant oscillations in the number of deliveries depending on MOQ change are evident during the shortest lead time. As the lead time prolongs, this effect minimises, and there are no considerable differences in the number of deliveries after the lead time of 6 days between tested MOQs. Lastly, it is evident that the levels of WTW GHG emissions significantly decrease with the increase of lead time. The influence of MOQ size on emissions' levels is most evident during the shortest lead time, and it means up to 43.5% lower levels when operating with MOQ higher than average daily demand.

Study results based on extensive numerical simulation of supply chain operating under stochastic, normally distributed market demand showed a strong influence of minimum order quantity and lead time change on supply chain logistic and environmental performance. Future research resulting from this paper will examine the abovementioned relationships and outcomes with additional supply chain variables and complexities.

References

1. Srivastara, S.K.: Green supply-chain management: a state-of-the-art literature review. Int. J. Manag. Rev. **9**(1), 53–80 (2007)
2. Venkat, K., Wakeland, W.: Is lean necessarily green? In: Proceedings of the 50th Annual Meeting of the International Society for the Systems Sciences, ISSS 2006, pp. 1–16. Curran Associates, Inc. (2013)
3. Benjaafar, S., Li, Y., Daskin, M.: Carbon footprint and the management of supply chains: insights from simple models. IEEE T. Autom. Sci. Eng. **10**(1), 99–116 (2013)
4. Marcilio, G.P., de Assis Range, J.J., de Souza, C.L.M., Shimoda, E., da Silva, F.F., Peixoto, T.A.: Analysis of greenhouse gas emissions in the road freight transportation using simulation. J. Clean. Prod. **170**, 298–309 (2017)
5. Rout, C., Paul, A., Kumar, R.S., Chakraborty, D.: Cooperative sustainable supply chain for deteriorating item and imperfect production under different carbon emission regulations. J. Clean. Prod. **272**, 122170-1–122170-16 (2020)
6. Zhou, B., Zhao, Y., Katehakis, M.N.: Effective control policies for stochastic inventory systems with a minimum order quantity and linear costs. Int. J. Prod. Econ. **106**(2), 523–531 (2007)
7. Lokad Homepage: https://www.lokad.com. Last accessed 10 Mar 2021
8. Zhao, Y., Katehakis, M.N.: On the structure of optimal ordering policies for stochastic inventory systems with minimum order quantity. Probab. Eng. Inf. Sci. **20**, 257–270 (2006)

9. Kiesmüller, G.P., de Kok, A.G., Dabia, S.: Single item inventory control under periodic review and a minimum order quantity. Int. J. Prod. Econ. **133**(1), 280–285 (2011)
10. Žic, J., Žic, S.: Multi-criteria decision making in supply chain management based on inventory levels, environmental impact and costs. Adv. Prod. Eng. Manag. **15**(2), 151–163 (2020)
11. Banks, J.: Handbook of Simulation: Principles, Methodology, Advances, Applications, and Practice. Wiley, New York (1998)
12. Sani, B., Kingsman, B.G.: Selecting the best periodic inventory control and demand forecasting methods for low demand items. J. Oper. Res. Soc. **48**(7), 700–713 (1997)
13. European Committee for Standardisation: Methodology for calculation and declaration of energy consumption and GHG emissions of transport services (freight and passengers), EN 16258. CEN, Brussels (2012)
14. EcoTransIT World Homepage: https://www.ecotransit.org. Last accessed 15 Jan 2021
15. Demir, E., Bektaş, T., Laporte, G.: A comparative analysis of several vehicle emission models for road freight transportation. Transport. Res. D.-Tr. E. **16**(5), 347–357 (2011)
16. Ifeu, INFRAS, IVE: EcoTransIT World—Methodology and Data, update 2019. Berne, Hannover, Heidelberg (2019)

Chapter 71
Cabin Safety Prediction in a Selected Group of Mobile Working Machines

Adam Vincze, Ladislav Gulan, Roman Ižold, and Andrej Korec

Abstract Protective structures are safety constructions fitted to self-propelled earth moving machinery providing reasonable accident protection for the vehicle operator in the driving position. Verification and certification of the proposed solutions is an important step in the development process or the innovation process of new constructions of mobile working machines. These processes are regulated by valid regulations and standards. They are run in standard conditions, also in the form of destructive tests on fully functional modules. In case of deficiencies, these processes must be rerun on new modules, resulting in additional financial cost. The paper deals with simulation of the required tests with attention to time saving and reducing financial cost. The simulation aims to streamline the design process of the machine or its modules. It has been proven that it is possible to accurately simulate destructive tests of cabins for working machines. The results of a simulation and real destruction tests of the cabin were compared by deformation graphs.

Keywords Mobile working machines · Loader · Safety · Tests · Simulation

71.1 Introduction

Parameter verification in mobile working machine modules is a part of the innovation process run in order to increase their safety, comfort or design changes [1]. One of the tests required for product certification is the standardized conditions test of cabin modules [2, 3]. These tests are costly because for their implementation it is necessary to create more than one functional prototype of the cabin, since cabins are subjected to deformational tests. In case of negative results, the design needs to be improved and the critical points need to be stiffened. Repeating the tests on a new functional model represents considerable financial cost and often a time delay in the design and implementation of a new or innovated mobile working machine.

A. Vincze (✉) · L. Gulan · R. Ižold · A. Korec
Faculty of Mechanical Engineering, Slovak University of Technology in Bratislava, Nám. slobody 17, 81231 Bratislava, Slovakia
e-mail: adam.vincze@stuba.sk

M. Rackov et al. (eds.), *Machine and Industrial Design in Mechanical Engineering*, Mechanisms and Machine Science 109,
https://doi.org/10.1007/978-3-030-88465-9_71

Therefore, in order to optimize the design processes, manufacturers are looking for new possibilities and methods to simulate the important safety tests and gradually approach the final solution [4, 5]. These methods can be realised using CAD software and computing programs, such as the finite element method [6].

71.1.1 Roll-Over Protective Structures—ROPS

They include all elements and parts which are a permanent part of the cabin frame, excluding demountable parts such as cabin door or window, glass structures.

The procedure of destruction tests for various types of machinery, auxiliary test equipment, such as the structures for distributing load stress, or the evaluation and individual assessment of the performed test is specified in the standard. The standard also defines the required load depending on the category, use and weight of the mobile working machine [2, 3].

71.1.2 Falling Object Protective Structures—FOPS

FOPS provide two categories of acceptability, taking into account the usability of a specific mobile working machine.

Category I is applied to the protection of the driver against falling bricks, concrete blocks, hand tools, tools used for road maintenance, tools for landscaping and other service activities at the work station. It is assumed that protection against the intrusion of the above objects will be provided for category I in the event of a round object falling from a height sufficient to generate energy 1365 J. The spherical object is made of sufficiently strong steel or ductile iron, weighing 45 kg and its diameter not exceeding 250 mm. Furthermore, the standard strictly prescribes the procedure of the destructive test itself, as well as the location of the DLV (Deflection Limiting Volume) [7] under the tested protective structure and the exact location of the falling object depending on the machine category.

Category II provides protection for the driver against heavy objects, for example trees or stones. Cylindrical or spherical objects with maximal diameter of 400 mm, capable of achieving energy of 11,600 J on impact are used for category II tests. Their impact surface must be circular with a diameter of 200 mm, based on planar surface.

Basic cabin dimensions are represented by length (1566 mm), width (823 mm) and height (1435 mm), Fig. 71.1. Sheet metal plates and special cabin profiles are designed with wall thickness of 4 mm steel.

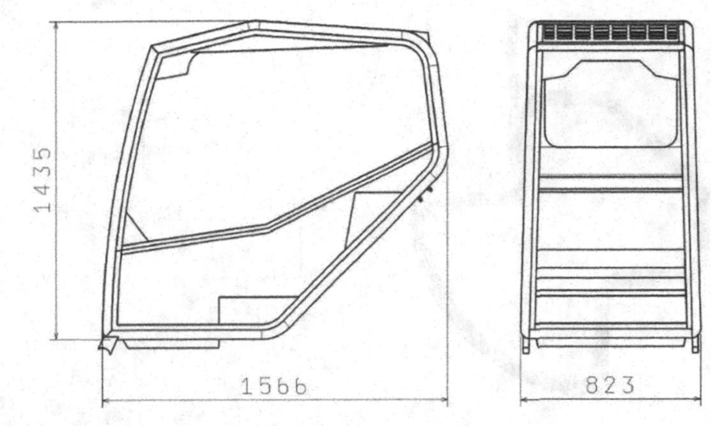

Fig. 71.1 Basic cabin dimensions

71.2 Simulation of Cabin Safety Tests

The process of verifying the resistance of the skid steer loader cabin by simulation with respect to the safety requirements of the FOPS and ROPS regulations can be divided into two main steps. The first step consists of the preparation of 3D data supplied by the cabin manufacturer. During the second step, the strength calculation can be performed by the finite element analysis.

71.2.1 Preparation of 3D Data

The preparation of 3D data from the cabin manufacturer involves the modification of the modular structure so that the shape and properties of the structure are preserved. The preparation simplifies and creates continuous surfaces in places intended for technological tasks—e.g. welds. Since the construction of the cabin of a mobile working machine consists of special cabin profiles (butterfly profiles) and sheets with a constant cross-section and thickness, the original volumetric model can be simplified to a surface model of the cabin. During this modification, it is necessary to create median surfaces of the original profiles from the volumetric parts of the cabin profiles and sheets of a certain thickness. In the interest of a trouble-free mesh creation, but with an effort to maintain the rigidity of the original model, the removal of small radiuses on the cabin profiles is also an important step, Fig. 71.2a, b.

The prepared surface model of the cabin is used as an input to the simulation, where it is necessary to define the boundaries and conditions including loads determined by the standard for safety tests of cabins.

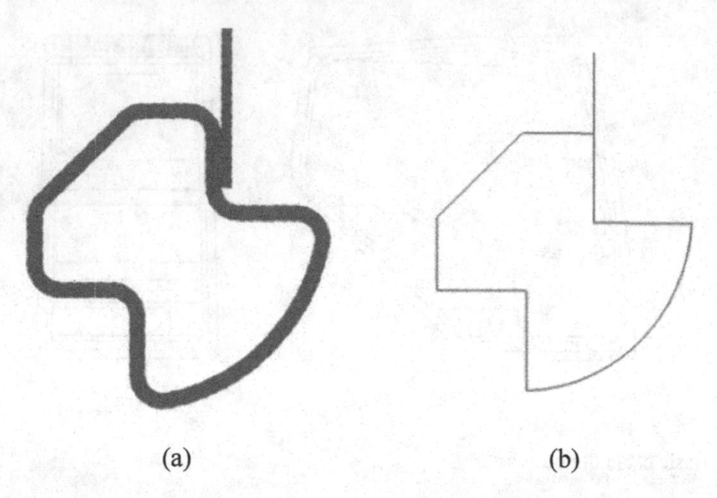

Fig. 71.2 **a** Cross-section of a volumetric model and **b** Cross-section of a surface model

71.2.2 Simulation of the Safety Tests

Simulation was run in Ansys computing software, where a mathematical model was created. Shell linear elements were used for mathematical model creation. Cabin mounting brackets of the frame were discretized by linear volume elements (solid). During the mesh generation, a flat four-node element was used for all cabin profiles, side and rear sheet metal plates. A volumetric ten-node element was generated for the solid attachments of the cabin to the frame, Fig. 71.3.

71.2.3 Boundary Conditions of the Calculation

The cabin frame was fixed (all degrees of freedom were removed) on surfaces that are in contact with the rigid frame of skid steer loader as throughout normal working conditions. In this context, it is necessary to mention the use of "silent blocks" in the places of attachment of the cabin to the frame of the working machine. This fixation was not considered during the mathematical calculation.

71.2.4 Loads

In case of the application of a lateral load, the area representing the plate of the loading press during the real test was modeled. This plate was loaded with the force required to achieve the deformation energy prescribed by the standard for skid steer loaders of a given weight category. After reaching this energy, the plate was gradually

Fig. 71.3 Mesh elements generated on the mobile working machine cabin

relieved until zero load. Cabin deformed in previous step is used as an input for a vertical load. Cabin is vertically loaded on surfaces which are in contact with the load plate of the test press.

71.3 Comparison of Simulation and Test Results

Results from the lateral force loading are taken into account for the purpose of comparison. Based on the mathematical simulation and the results of the experimental destructive tests on the cabin prototype, several conclusions can be drawn.

It is important to mention the fact that the cab subjected to the mathematical calculation was fastened with fixed couplings at the points of attachment to the frame of the working machine. During the real tests, the cabin was attached to the machine frame by means of elastic parts (silent blocks), as well as during its working operation. In addition to the loader frame, the experimental equipment safety test of the cabin also includes work equipment, Fig. 71.4a, b.

The resulting graphs of the dependence of the cab structure deformation on the lateral loading force can be used as comparable outputs. In the case of mathematical calculation, the graph is generated from the calculation software, in the experimental test the values of force and deformation are recorded during loading, Figs. 71.5 and 71.6.

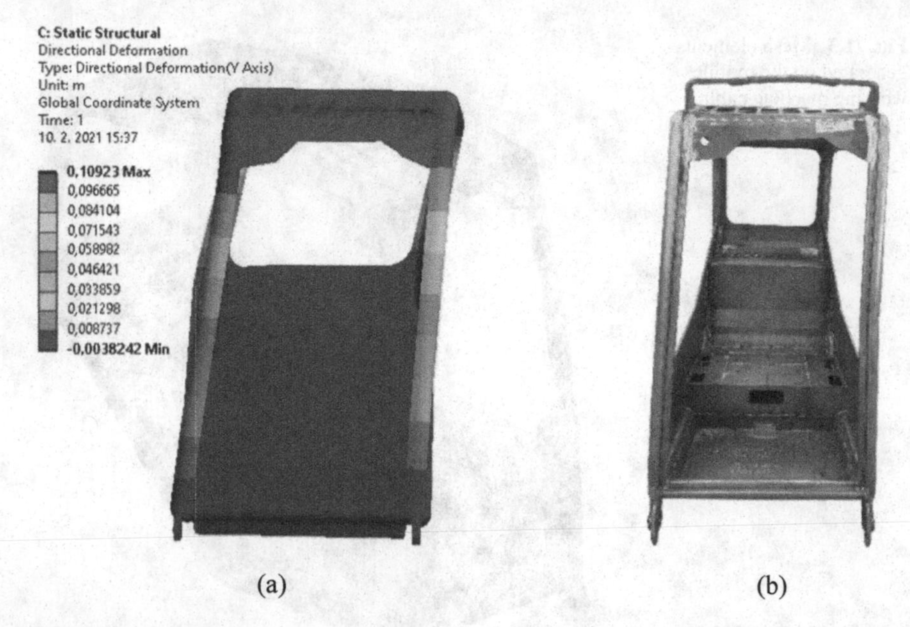

Fig. 71.4 **a** Deformation of the cabin structure at maximal simulated load and **b** Deformation of the structure after the maximal real test load

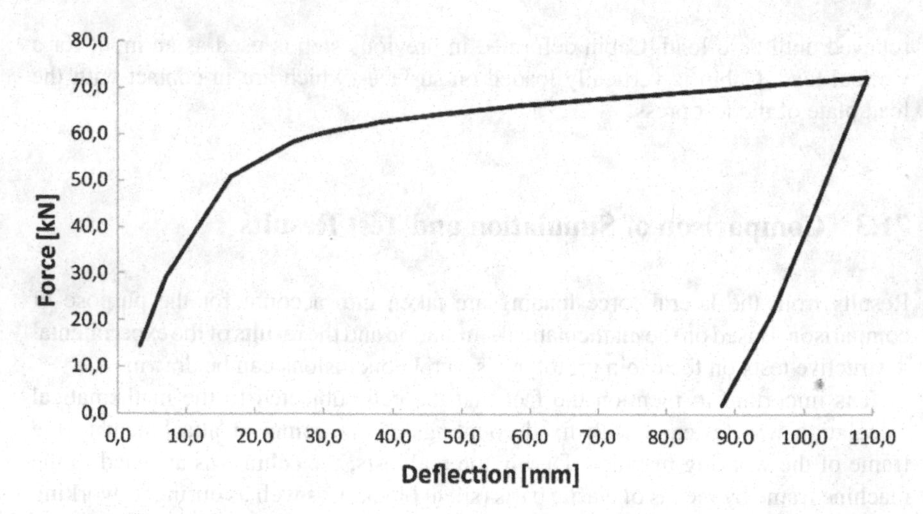

Fig. 71.5 Dependence of cabin deformation on the loading force at lateral simulated load

71.4 Conclusion

Based on the dependence of the cabin structure deformation with respect to the increase of the lateral force from the computational simulation, the maximum deformation of the cabin in the simulation is approximately 110 mm, based on the graph of

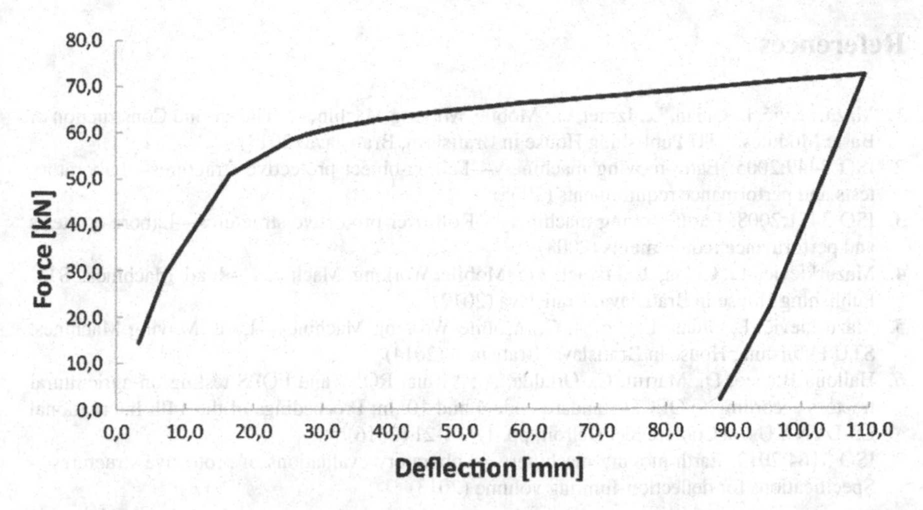

Fig. 71.6 Dependence of cabin deformation on the loading force at the real lateral load

measured values the maximum deformation of the cabin is approximately 160 mm. After evaluating the slow-motion video from the real load test, it was possible to conclude that a higher value of deformation may arise due to the attachment of the cab to the frame of the working machine by means of flexible parts.

The mathematical simulation also verified other test loads prescribed by the standard for ROPS type roll-over protective structures, both vertical and longitudinal loads. Protective structures protecting against falling objects of FOPS category I were also verified and the calculation was performed as a dynamic analysis of a falling spherical body, through explicit time integration. Due to the high time complexity of the calculation, only the first impact of the spherical body was monitored, further reflections were no longer simulated.

Based on a comparison of the load results of the cab of a mobile work machine, either by mathematical simulation or a real load test, it can be concluded that appropriately selected calculation parameters and boundary conditions can confidently verify the safety of mobile work machines cabin with respect to applicable regulations and standards by mathematical simulation.

Such predictive method greatly simplifies and speeds up the design of innovative mobile work machine modules already during the design phase, and allows design changes to be adapted in time, even before the prototype production phase. A reduction in the time required for the development, production and testing of the working machine module is achieved. The financial costs are also reduced due to the fact that the 3D data is tested by simulation, and it is not necessary to subject several pieces of manufactured cabins to a real test.

References

1. Mazurkievič, I., Gulan, L., Izrael, G.: Mobile Working Machines—Theory and Construction of Basic Modules. STU Publishing House in Bratislava, Bratislava (2013)
2. ISO 3449:2005: Earth-moving machinery—Falling-object protective structures—Laboratory tests and performance requirements (2005)
3. ISO 3471:2008: Earth-moving machinery—Roll-over protective structures—Laboratory tests and performance requirements (2008)
4. Mazurkievič, I., Gulan, L., Izrael, G.: Mobile Working Machines—Road Machines. STU Publishing House in Bratislava, Bratislava (2012)
5. Mazurkievič, I., Gulan, L., Izrael, G.: Mobile Working Machines—Earth Moving Machines. STU Publishing House in Bratislava, Bratislava (2014)
6. Hailoua Blanco, D., Martin, C., Ortalda, A.: Virtual ROPS and FOPS testing on agricultural tractors according to OECD standard code 4 and 10. In: Proceedings of the 14th International LS-DYNA Users Conference, Detroit, pp. 1-1–1-21 (2016)
7. ISO 3164:2013: Earth-moving machinery—Laboratory evaluations of protective structures—Specifications for deflection-limiting volume (2013)

Chapter 72
Determination of Zipline Braking Distance

Tanasije Jojić, Jovan Vladić, and Radomir Đokić

Abstract This paper defines the determination of zipline passengers braking distance. Considering that there is currently no legal regulation in Serbia which defines devices for zipline braking and arresting, the determination of safe braking distance is based on some foreign standards. The first part of the paper gives a short description for determination of required kinematic parameters. This is followed by a description of the human body tolerances, and finally, an example of braking distance determination for passengers who are traveling along concrete zipline which was built on Fruška Gora is given.

Keywords Zipline · Braking · Velocity

72.1 Introduction

The analysis of passenger's kinematic parameters consists of two parts. First part includes static analysis which is based on catenary theory, while the second part takes into account inertial forces, air and movement resistance, tightening force, the position of a passenger during lowering, etc. The theoretical analysis is detailed described in [1], and the impact of anchorage type and tension rope force in [2]. The impact of other influential sizes is presented in [3] and [4]. An excerpt from some foreign standards is given in [5], as well as an overview of existing or patented solutions of arresting systems.

T. Jojić (✉) · J. Vladić · R. Đokić
Faculty of Technical Sciences, University of Novi Sad, Novi Sad, Serbia
e-mail: tanasijejojic@uns.ac.rs

M. Rackov et al. (eds.), *Machine and Industrial Design in Mechanical Engineering*, Mechanisms and Machine Science 109,
https://doi.org/10.1007/978-3-030-88465-9_72

72.2 Human Body Tolerances

In the NASA study titled Human Tolerance to Impact Velocities, can be found a diagram shown in Fig. 72.1 that shows the chances of survival when hitting a hard flat surface at different velocities [6]. There are three zones on the diagram:

- zone of certain survival,
- zone of marginal survival, and
- fatal zone.

The "zone of certain survival" still carries significant potential for serious injury, so do not mistake that zone as an acceptable, safe, or desirable outcome.

Knowing that the final speed of the free fall from a certain height is calculated as:

$$v = \sqrt{2 \cdot g \cdot h} \tag{72.1}$$

The diagram given in Fig. 72.2 shows a comparison of arrival velocities to their equivalent free fall distances. These values can be used to assess ziplines arrival speeds and to get an understanding of how far a patron is "falling" when they arrive at a terminal platform. For example rider traveling at 30 m/s has the same velocity as someone falling from 45 m.

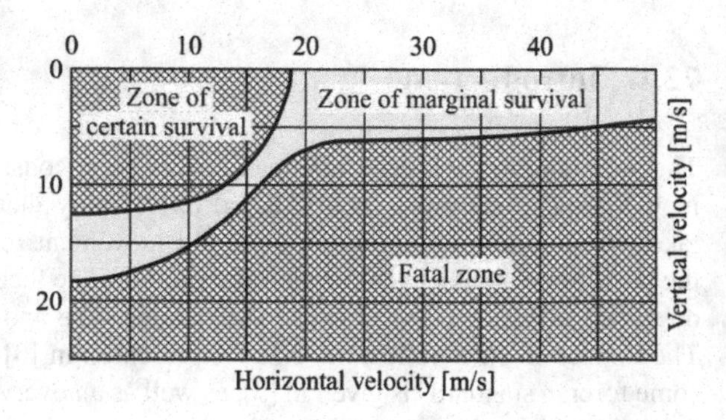

Fig. 72.1 Different survival zones

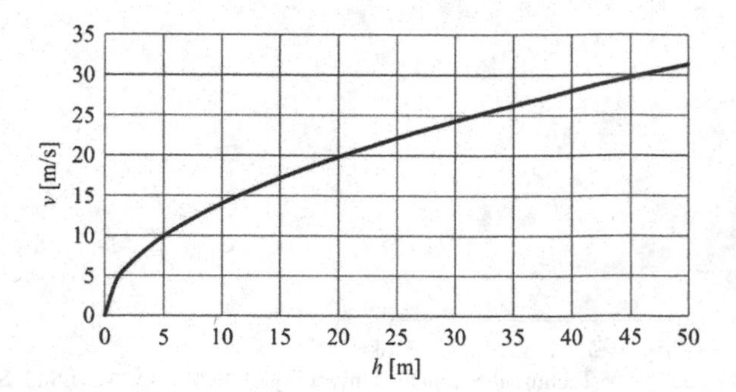

Fig. 72.2 Comparison of arrival velocities to their equivalent free-fall distances

The diagram shown in Fig. 72.1 gives information about the possibility of survival when the body moving at a certain speed stops instantly. However, since there is no instantaneous stop at regular usage of zipline, it is much more practical to observe the acceleration or deceleration of passenger.

The acceleration or deceleration is manifested by the load on the passenger's body, so the value of the acceleration will not be observed, but the force that the acceleration, i.e. deceleration, caused. The force is usually not expressed by its intensity but by the relationship to the weight of the passengers, the so-called G-force.

Since the human body does not receive a certain G-force in all directions equally [7], a coordinate system has been introduced as in Fig. 72.3. Figure 72.4 represents the

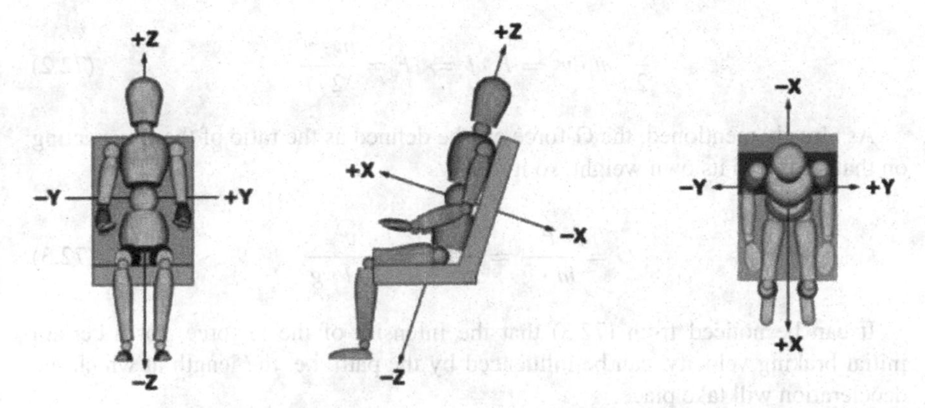

Fig. 72.3 Passenger's coordinate system

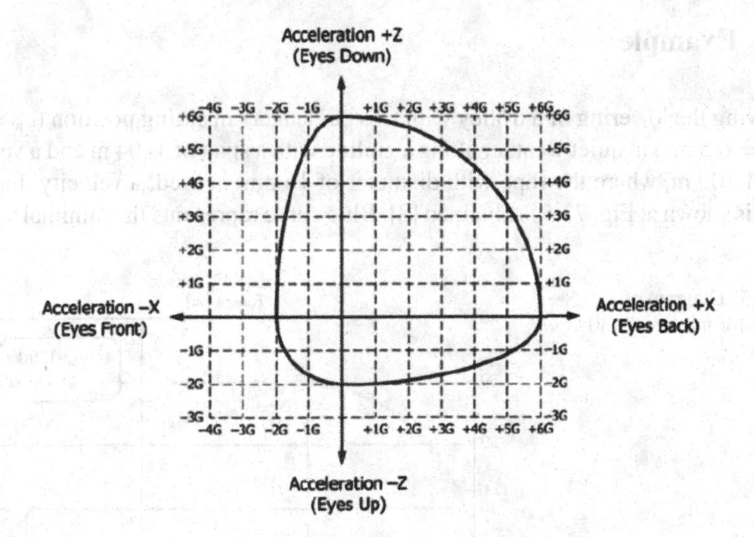

Fig. 72.4 Allowable combined magnitude of X and Z accelerations

allowable combined magnitude of X and Z accelerations for the mentioned coordinate system.

According to the ASTM F2291 standard [6], the maximum value of G-force during braking for sitting position is allowed in + X direction and has intensity of 6 g.

However, since the passenger is connected to the trolley by belts which allow swinging in the direction of movement, a braking force of 6 g will cause an upward swing and the passenger may hit the zipline cable. It has been empirically determined that the braking force should not exceed the intensity of 2.5 g.

From the equation of work it follows that the force in the case of stopping a body which was moving at a velocity v on the length l is:

$$\frac{1}{2} \cdot m \cdot v^2 = F \cdot l \Rightarrow F = \frac{m \cdot v^2}{2 \cdot l} \tag{72.2}$$

As already mentioned, the G-force can be defined as the ratio of the force acting on that body and its own weight, so it follows:

$$G = \frac{F}{m \cdot g} = \frac{\frac{m \cdot v^2}{2 \cdot l}}{m \cdot g} = \frac{v^2}{2 \cdot l \cdot g} \tag{72.3}$$

It can be noticed from (72.3) that the intensity of the G-force, for a certain initial braking velocity, can be influenced by the path, i.e. the length at which the deceleration will take place.

72.3 Example

Observing the lowering of a trolley with two passengers in sitting position ($c_w = 0.5$ and $A = 0.5$ m^2) in quiet weather along a zipline with a span of 1404 m and a vertical drop of 109 m, where the rope with diameter of 16 mm is used, a velocity diagram which is shown at Fig. 72.5 is obtained [8]. Blue curve represents the minimal weight

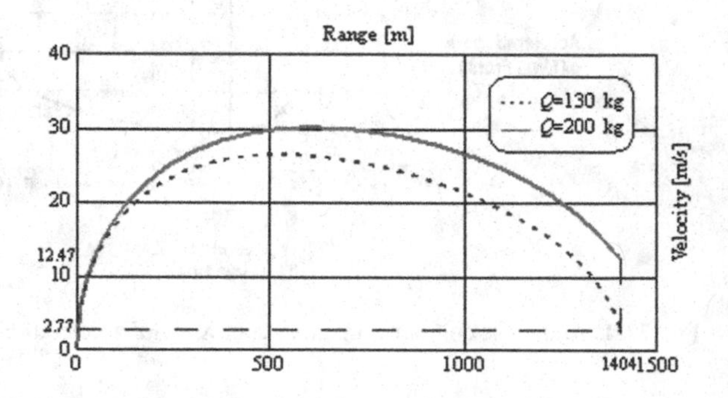

Fig. 72.5 Diagram of velocity for masses of 130 kg and 200 kg

of two people (e.g., mother and child), while red curve represents the maximal weight that can occur (e.g., two men). Since the passengers are sitting one behind the other, the surfaces exposed to the air do not change significantly.

Based on Eq. (72.3), it follows that if we want to stay in the range up to 6 g, the braking distance should be:

$$l_{Q=130\,\mathrm{kg}} = \frac{v^2_{Q=130\,\mathrm{kg}}}{2 \cdot G \cdot g} = \frac{2.77^2}{2 \cdot 6 \cdot 9.81} = 0.06\ \mathrm{m} \tag{72.4}$$

$$l_{Q=200\,\mathrm{kg}} = \frac{v^2_{Q=200\,\mathrm{kg}}}{2 \cdot G \cdot g} = \frac{12.47^2}{2 \cdot 6 \cdot 9.81} = 1.32\ \mathrm{m} \tag{72.5}$$

while if we want to stay in the range up to 2.5 g, the braking distance should be:

$$l_{Q=130\,\mathrm{kg}} = \frac{v^2_{Q=130\,\mathrm{kg}}}{2 \cdot G \cdot g} = \frac{2.77^2}{2 \cdot 2.5 \cdot 9.81} = 0.15\ \mathrm{m} \tag{72.6}$$

$$l_{Q=200\,\mathrm{kg}} = \frac{v^2_{Q=200\,\mathrm{kg}}}{2 \cdot G \cdot g} = \frac{12.47^2}{2 \cdot 2.5 \cdot 9.81} = 3.17\,\mathrm{m} \tag{72.7}$$

Based on the above-calculated parameters, it can be concluded that the braking distance has to be slightly larger than 3 m.

However, considering that the usage of ziplines is allowed even in slightly windy weather, for the case of tailwind with an intensity of 6.5 m/s, the velocity diagram for a trolley with passengers of 200 kg would look as shown in Fig. 72.6.

The braking distance for that case should be at least:

$$l^{wind}_{Q=200\,\mathrm{kg}} = \frac{v^2_{Q=200\,\mathrm{kg}}}{2 \cdot G \cdot g} = \frac{21.12^2}{2 \cdot 6 \cdot 9.81} = 3.79\ \mathrm{m} \tag{72.8}$$

but recommended:

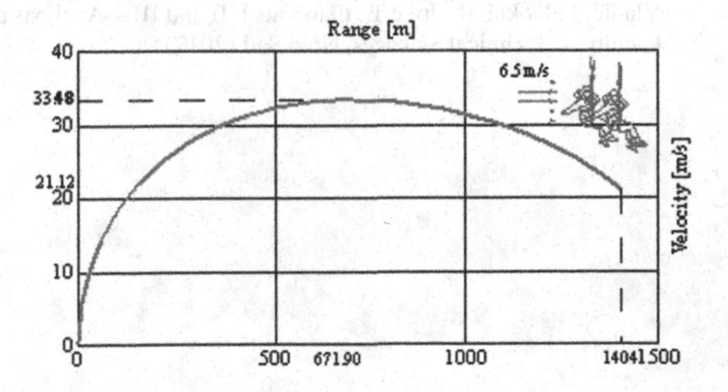

Fig. 72.6 Diagram of velocity for a mass of 200 kg and tailwind

$$l_{Q=200\,\mathrm{kg}}^{wind} = \frac{v_{Q=200\,\mathrm{kg}}^2}{2 \cdot G \cdot g} = \frac{21.12^2}{2 \cdot 2.5 \cdot 9.81} = 9.09\,\mathrm{m} \tag{72.9}$$

72.4 Conclusion

Considering that ziplines are a relatively new system, there are still no appropriate regulations for their construction and usage. For ziplines with small inclination angle or so-called "from three to three" ziplines, on which high velocities cannot be achieved, there was no greater danger of injuring. However, as a large number of high-range ziplines have recently been built on which it is possible to achieve high velocity, the topic of braking has become much more interesting. Within this paper, the importance of the adequate selection of braking or arresting devices is pointed out, as well as a computational example.

References

1. Vladić, J., Đokić, R., Jojić, T.: Theoretical analysis and determination of zipline movement parameters. Tehnika **68**(3), 405–412 (2019)
2. Jojić, T., Vladić, J., Đokić, R.: Anchorage type and tension rope force impact on zipline's kinematic characteristics. Mach. Des. **11**(4), 149–154 (2019)
3. Đokić, R., Vladić, J., Jojić, T.: Zipline computational model forming and impact of influential sizes. In: Proceedings of the Seventh International Conference Transport and Logistics, TIL 2019, pp. 71–74. University of Niš, Niš (2019)
4. Jojić, T., Vladić, J., Đokić, R.: Zipline design issues and analysis of the influencing parameters on passenger's velocity. In: Proceedings of the 5th International Conference Mechanical Engineering in XXI Century, MASING 2020, pp. 129–133. University of Niš, Niš (2020)
5. Vladić, J., Jojić, T., Đokić, R.: Condition analysis and basis for selection of zipline arresting devices. IMK–14–Res Develop Heavy Mach **26**(4), 89–94 (2020)
6. Braking Dynamics, Head Rush Technologies. White Paper, USA (2017)
7. Cargill, R.S.: Amusement Rides: How Much Thrill Is too Much? JP Research, Fort Washington, USA
8. Vladić, J., Đokić, R., Jojić T.: Elaborats I, II and III—Analysis of the Zipline System in Vrdnik. Faculty of Technical Sciences, Novi Sad (2017)

Chapter 73
Risk and Safety of Cylindrical Tank Exposed to Fire

Mirko Đelosević and Goran Tepić

Abstract This study presents a completely new methodology for analysis of fragmentation due to the explosion of cylindrical tanks. The epistemic uncertainty of kinematic parameters is eliminated by the introduction of initial acceleration, which depends on the type of tank material and the temperature effect. The most probable range of the fragment due to the explosion of the tank is between 600 and 650 m. The ranges of the fragments are shown with credible statistical distributions. Relevant factors for assessing fragmentation hazards include the trajectory of the fragment, the height and distance of the target from the tank. Fragments of pronounced aerodynamics do not pose a danger to targets up to 15 m high and at distances up to 50 m. The presented paper proposes an advanced methodological concept for the reliable analysis of the fragmentation of different process equipment in order to adequately manage the process risk.

Keywords Risk · Explosion · Cylindrical tank

73.1 Introduction

The most common causes of accidents in the process industry are related to explosions. The explosion of the tank is accompanied by the effect of fragmentation, which is a serious hazard for neighboring facilities. The first fragmentation models were applied for risk assessment purposes in nuclear facilities [1, 2]. Almost 80% of accidents in the process industry are related to the fragmentation effect, where the number of generated fragments is one-digit [3]. About 60% of the fragments created by the explosion of the tank cover an area of ±30° relative to the center of the tank [4]. Some recent studies are based on the results of these studies [5, 6]. The entropy model is the most common in the literature for estimating the number of generated fragments [5]. The explosion of the tank within the industrial plants are

M. Đelosević (✉) · G. Tepić
Faculty of Technical Sciences, University of Novi Sad, 21101 Novi Sad, Serbia
e-mail: djelosevic.m@uns.ac.rs

M. Rackov et al. (eds.), *Machine and Industrial Design in Mechanical Engineering*, Mechanisms and Machine Science 109,
https://doi.org/10.1007/978-3-030-88465-9_73

usually accompanied by a BLEVE effect [7, 8]. Adequate fragmentation risk assessment is not possible without analysis of fragmentation mechanics. In the literature, this aspect of fragmentation analysis is considered through a simplified model [9]. The main disadvantages of the simplified fragmentation model are that the literature does not state the limitations of its application [10]. The simplified model implies the definition of the initial velocity, the estimation of which is mainly of an orientational character. Experimental determination of the initial velocity of the fragments is presented in [11]. The aim of this paper is to assessment the kinematic parameters fragments and risk due to the explosion of cylindrical tanks caused by the BLEVE effect.

73.2 Critical Zone of the Tank

The critical zones of the cylindrical tank are estimated according to (73.1) and (73.2). The cylindrical tank considered in Fig. 73.1 has three characteristic zones (A–A, B–B) and (C–C). The fracture along the critical zone A–A takes place at the tank with a torispherical end caps, while the elliptical end caps influences the tank fracture in the B–B cross-section (Fig. 73.1a). The fracture along the cross-section C–C occurs exclusively in the tanks with spherical end caps (Fig. 73.1b).

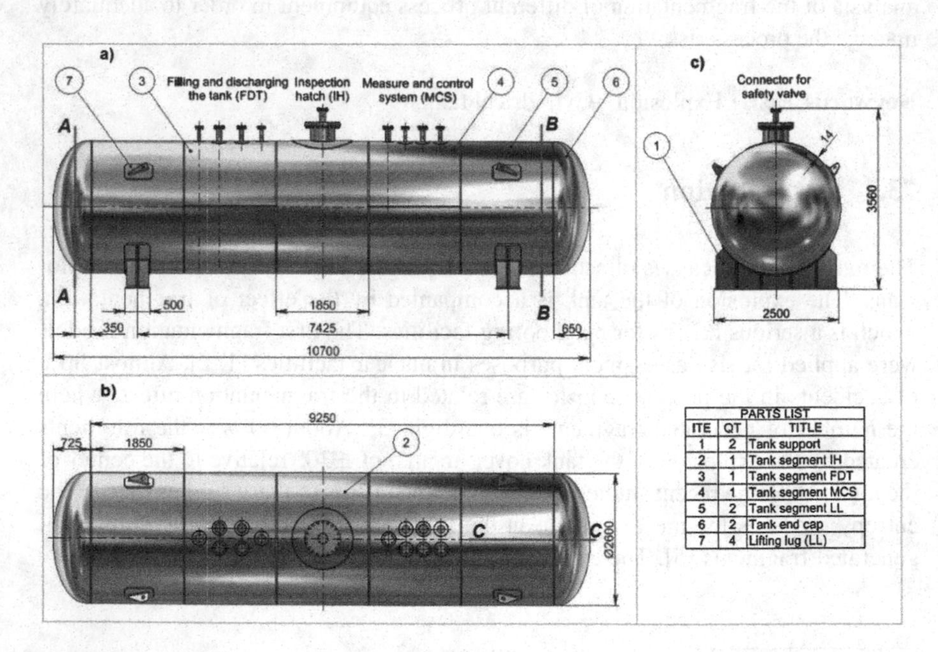

ITE	QT	TITLE
1	2	Tank support
2	1	Tank segment IH
3	1	Tank segment FDT
4	1	Tank segment MCS
5	2	Tank segment LL
6	2	Tank end cap
7	4	Lifting lug (LL)

Fig. 73.1 Construction and dimensions of cylindrical tank according to DIN 28,013

$$\sigma_x = \left[1 + \frac{3}{2}\frac{1}{\sqrt{3(1-\nu^2)}}\left(\frac{D}{2h}\right)^2 e^{-\lambda x}\sin(\lambda x)\right]\cdot\left(\frac{Dp}{4\delta}\right) = 46.5p \tag{73.1}$$

$$\sigma_\theta = \left[1 + \frac{1}{4}\left(\frac{3\nu\sin(\lambda x)}{\sqrt{3(1-\nu^2)}} - \cos(\lambda x)\right)\left(\frac{D}{2h}\right)^2 e^{-\lambda x}\right]\cdot\left(\frac{Dp}{2\delta}\right) = 92.5p \tag{73.2}$$

$$\sigma_{cr} = \sqrt{\sigma_x^2 + \sigma_\theta^2 - \sigma_x\sigma_\theta + \frac{3}{2}(\sigma_x - \sigma_\theta)^2} = 98\cdot p_{cr} \tag{73.3}$$

where

$$\lambda = \sqrt[4]{\frac{12(1-\nu^2)}{D^2\delta^2}} \tag{73.4}$$

A tank fracture occurs when the critical stress (σ_{cr}) reaches the tensile strength of the material ($f_m = 470$ MPa). The pressure that leads to a tank fracture is $p_{cr} =$ 4.8 MPa. Maximum operating pressure according to theoretical concept and standard EN 13,445–3 does not exceed 2.1 MPa. The actual working pressure for the tank from Fig. 73.1 whose wall thickness is 14 mm does not exceed the value of 1.7 MPa.

73.3 Fragmentation Mechanics

Fragmentation mechanics enables the definition of kinematic and dynamic parameters of fragments necessary for risk assessment. These parameters depend on the shape and mass of the generated fragments. The dynamics of the flight of the fragment which has mass m_{fr} and velocity v_{fr} is described with the ordinary differential equation (Fig. 73.2).

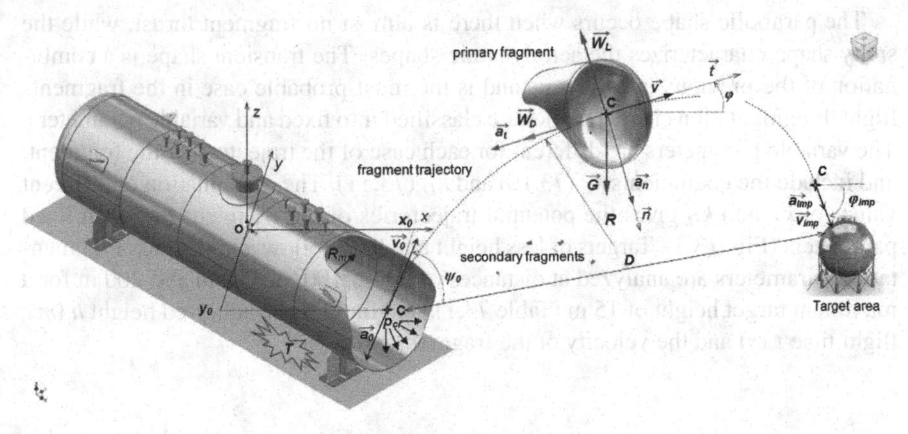

Fig. 73.2 Fragmentation of horizontal cylindrical tank

$$m_{fr} \cdot \left(\frac{dv_{fr}}{dt}\right)_x + \left(\frac{1}{2}\rho_v C_D A_D v_{fr}\right) \cdot (v_{fr} \cos\varphi) + \left(\frac{1}{2}\rho_v C_L A_L v_{fr}\right) \cdot (v_{fr} \sin\varphi) = 0 \quad (73.5)$$

$$m_{fr}\left(\frac{dv_{fr}}{dt}\right)_y = 0 \quad (73.6)$$

$$m_{fr} \cdot \left(\frac{dv_{fr}}{dt}\right)_z + \left(\frac{1}{2}\rho_v C_D A_D v_{fr}\right) \cdot (v_{fr} \sin\varphi) - \left(\frac{1}{2}\rho_v C_L A_L v_{fr}\right) \cdot (v_{fr} \cos\varphi) + m_{fr} g = 0 \quad (73.7)$$

For each of the generated fragments, the initial conditions of the following form should be included:

$$(x_{fr})_0 = x_0 \wedge (z_{fr})_0 = z_0 \quad (73.8)$$

$$(v_{x,fr})_0 = v_{xo} \wedge (v_{z,fr})_0 = v_{zo} \quad (73.9)$$

73.4 Results

Dynamic analysis of fragments flight indicates three characteristic trajectory shapes:

- parabolic,
- spiked and
- transient.

The parabolic shape occurs when there is almost no fragment thrust, while the spiky shape characterizes the aerodynamic shapes. The transient shape is a combination of the previous two variants and is the most probable case in the fragments flight. Fragmentation characteristics are classified into fixed and variable parameters. The variable parameters are different for each case of the trajectory of the fragment, and include the coefficients k_L (73.10) and k_D (73.11). The combination of different values of k_L and k_D gives the potential trajectories of the fragment for given fixed parameters (Fig. 73.3). Targets of less height and base follow a lower risk. Fragmentation parameters are analyzed at distances of 50 m, 100 m, 150 m and 200 m for a maximum target height of 15 m (Table 73.1), and include the achieved height *h* (*m*), flight time *t* (*s*) and the velocity of the fragment *v* (*m/s*).

Fig. 73.3 Trajectory shapes and fragment range (m_{fr} = 200 kg, $\psi_0 = 35°$)

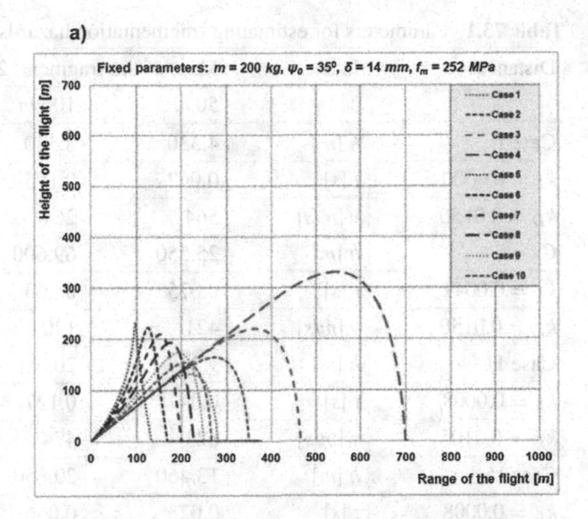

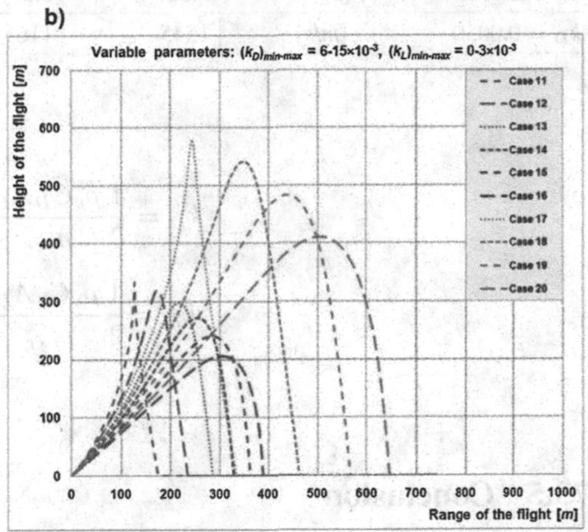

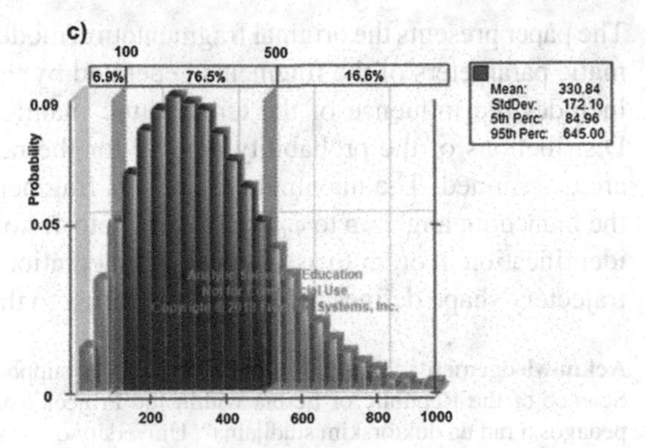

Table 73.1 Parameters for estimating fragmentation hazards with the height of the object

Distance	Index	Mass of the fragment: 200 kg			
		50 *m*	100 *m*	150 *m*	200 *m*
Case 1	h [m]	4.330	8.580	12.400	14.300
$k_L = 0.0000$	t [s]	0.062	0.195	0.47	1.077
$k_D = 0.0150$	v [m/s]	564	264	124	59
Case 6	h [m]	25.550	69.600	153.27	83.740
$k_L = 0.0041$	t [s]	0.075	0.303	1.602	9.885
$k_D = 0.0150$	v [m/s]	421	120	14	6
Case 11	h [m]	9.340	20.610	34.130	49.540
$k_L = 0.0008$	t [s]	0.047	0.127	0.268	0.515
$k_D = 0.0105$	v [m/s]	823	474	272	155
Case 15	h [m]	13.460	29.650	47.760	68.550
$k_L = 0.0008$	t [s]	0.028	0.066	0.115	0.181
$k_D = 0.0050$	v [m/s]	1545	1165	880	658

$$k_D = \frac{1}{2}\frac{\rho_v C_D A_D}{m_{fr}} \tag{73.10}$$

$$k_L = \frac{1}{2}\frac{\rho_v C_L A_L}{m_{fr}} \tag{73.11}$$

73.5 Conclusion

The paper presents the original fragmentation model for the identification of the kinematic parameters of the fragments generated by the explosion of a tank. The model includes the influence of the temperature manifested through the BLEVE effect. Distributions of the probability density for the range of fragment of mass 200 kg are determined. The maximum range of a fragment of mass 200 kg corresponds to the launching angle up to $\varphi_0 = 35°$. Trajectories of fragments are a basis for hazard identification in order to assessment fragmentation risks in the process industry. The trajectory shape defines the ratio of the thrust to the air resistance.

Acknowledgements This work has been partially supported by the Ministry of Education and Science of the Republic of Serbia within the Project No. 34014 and by the project "Naučni i pedagoški rad na doktorskim studijama", University of Novi Sad, faculty of Technical Sciences.

References

1. Moore, C.V.: The design of barricades for hazardous pressure systems. Nucl. Eng. Des. **5**(1), 81–97 (1967)
2. Baker, W.E., Cox, P.A., Westine, P.S., Kulesz, J.J., Strehlow, R.A.: Explosion Hazards and Evaluation. Elsevier, Amsterdam (1983)
3. Holden, P.L., Reeves, A.B.: Fragment hazards from failures of pressurised liquefied gas vessels. IchemE Symp. Series **93**, 205–220 (1985)
4. Holden, P.L.: Assessment of Missile Hazards: Review of Incident Experience Relevant to Major Hazard Plant. Culcheth, Warrington (1988)
5. Mébarki, A., Mercier, F., Nguyen, Q.B., Saada, R.A.: Structural fragments and explosions in industrial facilities. Part I: Probabilistic description of the source terms. J. Loss Prev. Process Ind. **22**, 408–416 (2009)
6. Mébarki, A., Nguyen, Q.B., Mercier, F.: Structural fragments and explosions in industrial facilities. Part II: Projectile trajectory and probability of impact. J. Loss Prev. Process Ind. **22**, 417–425 (2009)
7. Eckhoff, R.K.: Boiling liquid expanding vapor explosions (BLEVEs): A brief review. J. Loss Prev. Process Ind. **32**, 30–43 (2014)
8. Zhang, J., Laboureur, D., Liu, Y., Mannan, M.S.: Lessons learned from a supercritical pressure BLEVE in Nihon Dempa Kogyo Crystal Inc. J. Loss Prev. Process Ind. **41**, 315–322 (2016)
9. Mannan, S.: Lees' Loss Prevention in the Process Industries. Elsevier, Oxford (2012)
10. Gubinelli, G., Zanelli, S., Cozzani, V.: A simplified model for the assessment of the impact probability of fragments. J. Hazard. Mater. A **116**, 175–187 (2004)
11. Baum, M.R.: The velocity of large missiles resulting from axial rupture of gas pressurised cylindrical vessels. J. Loss Prev. Process Ind. **14**, 199–203 (2001)

Part VIII
Academic and Mechanism Education and History of MMS

Chapter 74
Education 4.0 for Industry 4.0

Biljana Marković and Aleksija Đurić

Abstract Industry 4.0 has raised many challenges for the industry as a whole, but also for the education system that generates the necessary knowledge for new expectations and challenges in the industry. Companies that embark on these challenges must be able not only to effectively manage information about their product, throughout the entire life cycle, but also to count on highly qualified professionals who will be able to do so. Therefore, mechanical engineers, who carry out production processes, as well as engineering students at universities, should develop new skills and meet new market demands. This means that the education system must do everything in time to enable the transformation of the curriculum in order to approach the requirements of Industry 4.0. Therefore, Industry 4.0 requires Education 4.0 as a prerequisite for realization.

Keywords Education · Industry 4.0 · Competences

74.1 Introduction

The epochs of industrial development are often called the "industrial revolution". The fourth of such industrial revolutions (Industry 4.0) has led to incredibly rapid changes resulting from the extremely rapid development of science and technology. Changes in production technologies related to automation and digitization are known as Industry 4.0. While Industry 3.0 focused on automating certain business processes, Industry 4.0 focuses on the digital transformation of enterprises. The fourth industrial revolution (Industry 4.0) arose in the correlation between the existing traditional industry with innovations in the field of the Internet, i.e. in the field of information and communication technologies (ICT). This involves digitizing all physical assets and creating new digital ecosystems, including value chain partners [1]. Data generation, analysis and communication enable the performance promised by Industry 4.0, as it networks a wide range of new technologies in order to create value. The focus of

B. Marković (✉) · A. Đurić
University of East Sarajevo, Faculty of Mechanical Engineering, East Sarajevo, Republika Srpska, Bosnia and Herzegovina
e-mail: biljana.markovic@ues.rs.ba

M. Rackov et al. (eds.), *Machine and Industrial Design in Mechanical Engineering*, Mechanisms and Machine Science 109,
https://doi.org/10.1007/978-3-030-88465-9_74

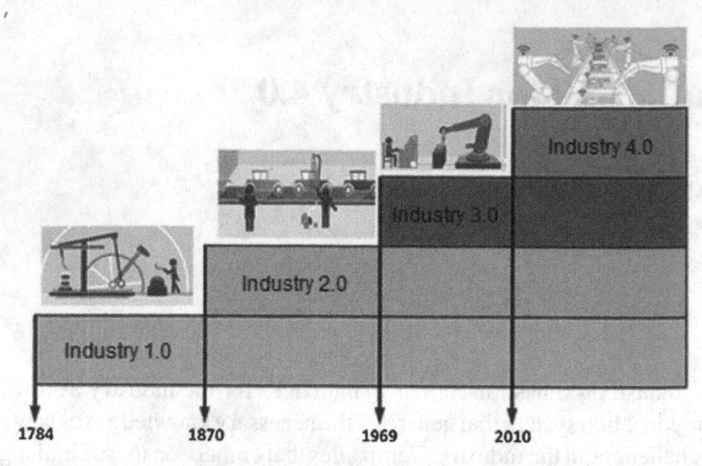

Fig. 74.1 From Industry 1.0 to Industry 4.0 [1]

these changes is mostly related to process models, methods, IT tools and information models in the development of smart products and services, i.e. smart factories. Based on the literature data [1], the following components of Industry 4.0 can be identified (Fig. 74.1):

- Cyber-Physical Systems CPS,
- Internet of Things IoT,
- Internet of Services (IoS) service, and
- Smart Factory.

74.1.1 Industry 4.0 Market Needs for New Professional Skills

Each era of industrial evolution (from 1.0 to 4.0) is innovative and requires new skills from the involved professionals. Technological development has shattered old models of work, requiring even more skilled professionals, especially in process of product development [2]. Industry 4.0 has changed something else, introduced new professions that did not exist in previous industrial revolutions. Thus, the skills of professionals have not changed according to the needs of the industry, but new professions have been introduced for which skills must be acquired, and therefore the way they learn, which has influenced teaching and learning methodologies (BRICS Working Group for Skills Development, 2016). Following the models of developed countries of the world, it would be logical for companies to review what knowledge and skills are acquired at universities and suggest methods on how the next generations should learn, in order to meet their needs in a sharp and fast market competition.

The first step in trying to catch up with world ideas in that sense, within B&H, or the closest environment (Serbia), is the so-called dual education, a level of secondary education that enables "learning through work", in order to achieve the principles of accessibility, relevance, lifelong learning, the right to choose, equal opportunities, partnerships, professionalism and ethics, quality assurance. Legislation that is defined in this sense, in order to establish rules of conduct for employers and students, tries to systematically address and quantify the scope of teaching and the scope of practice, which would optimally enable the acquisition of competencies needed by industry at the moment.

What is happening with university education?

Will the need for human resources, in the environment of Industry 4.0, by the elimination of certain jobs, due to the transformation and automation, digitalization and application of information technologies that are constantly advancing, lead to the disappearance of the need for human labor or will it increase or lead to other, new, different jobs, for people of the new generation, educated according to the principles of education 4.0?

What postulates, what requirements and challenges are thus placed before the new university role in achieving the application of the principles of Industry 4.0 in our framework?

It should be noted here that this challenge, for building a digital culture and properly training people, is equally applicable in developed and underdeveloped countries, even for companies considered to be technologically advanced, as well as for companies from different industrial sectors. This is something that can be expected, because the implementation of Industry 4.0 has serious implications on the organizational structure of the company, the way it works and the applied delivery models.

Although training or retraining of the workforce is one of the biggest challenges facing Industry 4.0, this topic has not received much attention at the level of vocational education, except in the beginnings of dual education, and least in reshaping the curriculum offered at the level of higher education, within our framework.

74.1.2 Results of World Research

Similar trends are reported in the EU Skills Panorama 2014 [3]. According to Eurostat data, in 2013, over 32 million employees worked in the manufacturing sector across the EU. Although employment in the entire EU manufacturing sector is expected to decline by 4% by 2025, employment in high-tech manufacturing sectors is expected to create more than 2 million jobs.

The implementation of Industry 4.0 is expected to face a number of challenges. Since Industry 4.0 continues to the previous generation of the revolution, but also the generation of caught workers of all profiles, who are profiled in it, according to research by Price water house coopers [4] the most important challenges to face are:

- Lack of digital culture and training (50% of respondents),
- Lack of a clear vision of digital operations and support/leadership of top management (40%), and
- Unclear economic benefits and digital investments (38%).

Relevant available research, presented through different methodologies to predict the skills needs required to implement Industry 4.0 provide a description of the general environment, forecasting, skills needs analysis and implementation of educational structures and programs related to the stages of development of new technologies. It can be concluded that the demand for research and development and design, IT and data integration, production robotics and automation, logistics, sales and services, while the demand for occupations for production and quality control is decreasing. However, changing the required technical skills will not be the only change. Staff must adapt to new forms of organizational structures in terms of processes and personnel issues and the new human role in production processes. Further, the digital transformation of the enterprise implied by the Industry 4.0 initiative requires new delivery approaches for software development and application [5].

These changes in the labor market will dramatically affect the labor market, more specifically the skills required and the way companies recruit their staff. The World Economic Forum [6] suggests that several major changes are needed in the way businesses view and manage their employees, both immediately and in the longer term, i.e. the period of transition to the required skills. Mentioned skills are: coordination, ability to negotiate, problem solving, flexibility, ability to make decisions, human resource management, critical thinking, emotional intelligence, creativity and judgment.

The fact that the skills required for Industry 4.0 are numerous and diverse has been recognized in various studies [7, 8], where the identified competencies are grouped into 4 categories, namely:

- Technical competencies such as superior knowledge, process understanding, technical skills, etc.,
- Methodological competencies include the following: ability to solve problems and conflicts, creativity, ability to make decisions, research and analytical skills, orientation towards efficiency,
- Social competencies such as intercultural skills, language skills, skills communication, networking skills, ability to work in a team, ability to compromise and cooperate, ability to transfer knowledge and leadership skills, and
- Personal competencies that include flexibility, tolerance for ambiguity, motivation to learn, ability to work under pressure, sustainable mindset, and compliance.

However, the papers of authors from the earlier period [9], through the analysis of the assessment of competencies of mechanical engineers (Star of competencies), show 5 fields of competencies, which are necessary for successful work in real production capacities of developed countries and highly technologically developed industries. aircraft…).

74.2 Research Results in U B&H, Industry 4.0

74.2.1 Benchmarking of Manufacturing Industry in B&H, Regarding to Croatia

Investigating the place and level of current development of production capacities in B&H, in relation to the needs and requirements of Industry 4.0, through 9 fields, Table 74.1, colleagues from the Faculty of Mechanical Engineering in Sarajevo conducted a research study, modeled on research conducted in Croatia, using the same methodology.

The research methodology applied is identical to the methodology applied in the research of Veža and others which aimed assessing the level of advancement of manufacturing companies in Croatia [10]. Replicating the same methodology for survey in B&H, the same way as it was implemented in Croatia, also aimed to enable benchmarking of Croatian and B&H production companies. Besides, the developed methodology evaluates the real situation in companies based on the field research approach (valuating also the level of business process development of companies), which is a better-quality approach than research and assessment of the state of the industry based only on desk research approaches with certain available indicators, such as presented with the paper of Atika and Unlu [11].

The average B&H manufacturing company, in terms of the level of development in relation to Industry 4.0, is still at the second industrial revolution with a quantified level of 2.19. The research was done in 2019 and included a sample of 47 manufacturing companies (350 contacted companies), that responded to survey, of different sizes and from different industries, located throughout B&H, (in Croatia during 2015, reached level of 2.15, using the same research method). The methodology used in this paper is a very good starting point for a realistic assessment of

Table 74.1 Evaluated processes

No.	Area to which the question relates
1	Product development
2	Degree of automation
3	Work orders management
4	Product traceability through production
5	Input materials inventory and work in progress (WIP) inventory management
6	Finished goods inventory management
7	Quality management
8	Product Lifecycle Management—PLM
9	Green and lean production

the level of development of industrial companies. The methodology could be further refined with on processes that are not well enough covered, such as supply chain management.

However, the previously evaluated results do not show the causes of this situation in factories in B&H (2.19), in relation to Industry 4.0, which are directly related to the necessary knowledge of employees, whether managers at all levels of management or technical staff, engineers or operators. More precisely, the research shows the level of knowledge and use of 9 areas, Table 74.1, without going into the causes and reasons for lagging behind developed countries, from the point of view of the necessary competencies of the relevant staff.

The first results of the research related to the competencies required for Industry 4.0 acquired by students of the Faculty of Mechanical Engineering of public universities in B&H were conducted in March 2021. The focus of the research was on the study programs of the first cycle of studies related to the scientific field of Mechanical Construction and Product Development. The research is based exclusively on information available on the website of the Faculties of Mechanical Engineering in B&H. The aim of the research was to determine the degree of representation of subjects in the mentioned study programs dealing with the development of competencies in students that are important for industry 4.0, namely: knowledge of product development methods, ability to use 3D modeling software, knowledge of algorithms and programming (to develop a program), knowledge of reversible engineering methods (3D scanning and 3D printing), knowledge of mechatronic systems, knowledge of the basics of entrepreneurship and the acquisition of soft skills. In B&H, students of five mechanical faculties of public universities offer a study program of the first cycle of studies in the field of mechanical constructions and product development, namely: Faculty of Mechanical Engineering, University of Sarajevo, Faculty of Mechanical Engineering, University of East Sarajevo, Faculty of Mechanical Engineering, University of Zenica, Faculty of Mechanical Engineering "Džemal Bijedić" Mostar and the Faculty of Mechanical Engineering, Computing and Electrical Engineering, University of Mostar. Table 74.2 shows the results of the previously mentioned research. For each of the mentioned faculties, a subjective analysis determined the project of subjects that enable the acquisition of the previously mentioned competencies. For subjects that have in their syllabuses over 60% of the material related to these competencies, the number 1 is assigned, and for subjects that have from 30 to 60% of the material, the number 0.5 is assigned. The representation of relevant subjects was determined by dividing the total number of subjects, by the number of those gravitating to Industry 4.0, multiplied by 100%.

The results show that there is a very small representation of subjects at the faculties of mechanical engineering in B&H, which enable the acquisition of competencies required by Industry 4.0. Taking into account the complex system of education in B&H as well as the legal framework, in the future it will be very difficult to adjust the content of study programs to the needs of Industry4.0 in order to acquire the necessary competencies. The large number of basic subjects required for the education of mechanical engineers is a limiting factor, too. It is recommended that the existing curricula (20% in the Republic of Srpska without the approval of the Ministry, with

Table 74.2 Degree determination of representation of subjects that enable the earning of competencies required by Industry 4.0 in the study programs at public Faculties of Mechanical Engineering in B&H

Faculty	Faculty of Mechanical Engineering, University of Sarajevo	Faculty of Mechanical Engineering, University of East Sarajevo	Faculty of Mechanical Engineering, University of Zenica	Faculty of Mechanical Engineering, University of "Dzemal Bijedic" Mostar	FSRE, University of Mostar
Study program Name	Mechanical constructions	Machine constructions and product development	Product Engineering Design	Product Design	Constructing and developing products
Year licensing	2012	2017	2015	2020	–
Study duration	3 years	4 years	4 years	3 years	3 years
ECTS points	180	240	240	180	180
Number of subjects in the study program	36	46	56	32	37
Number of subjects that enable the earning of competencies required by Industry 4.0	4.5	8	9	3.5	6
Representation of subjects that enable the earning of competencies required by Industry 4.0	12.5%	17.4%	16.1%	10.9%	16.2%

the approval of the NNV faculty) be changed within the legal framework, and that coordination be done between the contents of relevant subjects, i.e. a course that would involve a student starting certain project assignments in one course and ending in another.

74.3 Education 4.0

74.3.1 *"Good Practice" Examples*

In research studies and examples of practice in developed countries, which have brought the level of education of future mechanical engineers, through established educational models (Karslruhe model. KaLep, Germany), [9] closer to real practical conditions, it is clear that it is necessary to introduce or increase the level of study, so-called soft skills, in relation to the representation of the study of hard engineering skills (hard skills).

In the Netherlands, for example, models of so-called "makerspace" are used. Makerspaces are prototype and digital production environments that integrate machines, devices and enable creativity, stimulating joint innovations and new business development [12]. In these environments, participants are expected to use these devices independently, encouraging learning and knowledge sharing, providing dynamic interactions among participants in research, education, development and production. In addition to providing access to technologies and tools, they also encourage the exchange of knowledge and the creation of synergies, focusing on creativity and innovation [12].

A similar educational concept (the example of KaLeP) was introduced at the Faculty of Mechanical Engineering in East Sarajevo, through the subject Integrated Product Development, up 2008, which very quickly showed all the benefits. Through the course of the mentioned subject, mechanical engineers, for a period of 4 months, acquired soft skills, which enabled them to obtain significant comparative advantages over their colleagues who were not educated according to a similar program when getting a job or during the realization of practical projects. The motivation of students to work in this way has resulted in numerous awards, and some examples of innovation that the student has realized are shown in Fig. 74.2.

74.3.2 *Results of the Survey of Required Competencies for Industry 4.0*

During March and April 2021, a survey of the current competencies of mechanical engineers required for Industry 4.0 was conducted. 44 mechanical engineers participated in the survey, of which 75% are from the real sector, and 25% are professors of professional subjects of mechanical technical schools. The survey took into account the geographical representation, the level of development of the company, the work experience of engineers, so that 19% of engineers with more than 10 years of experience participated in the survey, 35.7% have between 5 and 10 years length of service, while 45.2% of respondents have less than 5 years length of service. Table 74.3 gives the results of the research based on the methodology presented in paper of Ferro dos Santos and others [12]. The results of the survey show that the average grade

Smart pot School year 2017/18 Awards: 1. First place at "Sarajevo unlimited" student competition 2. The best student innovation in Republic of Srpska in 2018 Year	Machine for cleaning the intermediate tube space of thermal power plant boilers School year 2016/17 Awards: 1. The best realized technological innovation in Republic of Srpska in 2018 Year	Machine for simultaneously squeezing and cutting apple fruit School year 2013/14

Fig. 74.2 Papers of students of the Faculty of Mechanical Engineering, University of East Sarajevo realized through the subject Integral Product Development

Table 74.3 Answers from the interviewed individuals

Competences	Real sector engineers	Technical school engineers
Communication skills	3.2	2.8
Teamwork	3.5	3.5
Long life lerning	3.2	3.4
Professionalism	3.5	3.4
Ability to solve problems and make decisions	3.3	3.2
Technical competencies	3.1	2.9
Knowledge based on science and engineering principles	3.1	2.8
Knowledge of contemporary issues	3.2	2.9
Engineering systems approach	2.9	2.6
Competence in specific engineering disciplines	2.5	2.4

of surveyed engineers in the real sector is slightly better than the score obtained by surveying engineers/professors working in mechanical engineering schools, which is to be expected, because they solve real engineering problems every day and are not obliged to respect and monitor items.

74.4 Conclusions

The results obtained by researching current competencies related to Industry 4.0, which are possessed by non-beginner mechanical engineers, are within expectations, given the current development of the economy in B&H and the equipment of industrial capacities with necessary equipment, machines and devices, as well as modern IT support. Also, the results are appropriate and in line with the results of preliminary research study programs at the faculties of mechanical engineering in B&H, which do not offer sufficient knowledge in the ability to acquire the necessary competencies required for Industry 4.0. Thus, higher education that is not adapted to the expected requirements, as a prerequisite for entering the third and then the fourth industrial revolution, does not provide adequate staff that without preparation and additional education can initiate changes in the economy and automatically adapt to Industry 4.0. Therefore, in front of the education system in B&H, as well as in the immediate environment, there is much room for transformation and development of education methodology, based on examples of good practice of successful countries, which can be applied in various industrial areas.

Also, it would be very interesting to repeat the same research with an interval of one or more years, what would enable monitoring of actual progress of the industry development and more effective guiding of such progress, underlining the needed higher education enhancements in that sense.

References

1. Miltenović, V., Antić, D.: Inženjering pametnih proizvoda i usluga. University of Niš (2020)
2. Sallati, C., de Andrade Bertazzi, J., Schützer, K.: Profesional skills in the product development process: The contribution of learning environments to professional skills in the Industry 4.0 scenario. Proc. CIRP **84**, 203–208 (2019)
3. Price water house coopers, Global Industry 4.0 Survey (2016). https://www.pwc.com/gx/en/industries/industries-4.0/landing-page/industry-4.0-building-your-digital-enterprise- april-2016.pdf
4. European Commission. EU Skills Panorama. Advanced manufacturing Analytical Highlight. ICF GHK and Cedefop (2014)
5. Fitsilis, P., Tsoutsa, P., Gerogiannis, V.: Industry 4.0: Required personnel competences. Int. Sci. J. Ind. 4.0 **3**(3), 130–133 (2018)
6. World Economic Forum, The Future of Jobs Employment, Skills and Workforce Strategy for the Fourth Industrial Revolution (2016). http://reports.weforum.org/future-of-jobs-2016/
7. Hecklau, F., Galeitzke, M., Flachs, S., Kohl, H.: Holistic approach for human resource management in Industry 4.0. Proc. CIRP **54**, 1–6 (2016)
8. Leinweber, S.: Etappe 3: Kompetenzmanagement. In: Meifert, M. (ed.) Strategische Personalentwicklung, pp. 145–178. Springer, Berlin, Heidelberg (2010)
9. Marković, B.: Metodološki pristup upravljanju ljudskim resursima u procesu razvoja proizvoda. PhD Thesis, Faculty of Mechanical Engineering, University of Niš (2008)
10. Veza, I., Mladineo, M., Peko, I.: Analysis of the current state of Croatian manufacturing industry with regard to Industry 4.0. In: Proceedings of the 15th International Scientific Conference on Production Engineering, CIM 2015, pp. 1–6. Croatian Association of Production Engineering, Zagreb (2015)

11. Atika, H., Ünlüa, F.: The Measurement of Industry 4.0 performance through Industry 4.0 index: An empirical investigation for Turkey and European countries. Procedia Comput. Sci. **158**, 852–860 (2019)
12. Ferro dos Santos, E., Benneworth, P.: Makerspace for skills development in the Industry 4.0 era. Braz. J. Oper. Prod. Manag. **16**, 303–315 (2019)

Chapter 75
Cyber Physical Systems in Manufacturing Engineers Education

Zivana Jakovljevic and Dusan Nedeljkovic

Abstract Implementation of Industry 4.0 concept in manufacturing environment requires the education of a new generation of engineers capable to address all the challenges that this industrial (r)evolution brings about. Cyber Physical Systems (CPS) represent technological basis of Industry 4.0 and the education of engineers in this highly interdisciplinary area is a paramount for successful implementation of Industry 4.0. In this paper we analyze expected levels of CPS implementation in manufacturing environment, opportunities that they offer with respect to manufacturing customization and high product variety, as well as the changes, challenges and threats that these systems introduce. Based on this analysis, the paper presents the most important topics that should be covered in manufacturing engineers' education to provide them with necessary skills and competences, making them capable to effectively design and implement CPS based solutions and Industry 4.0 concept at factory shop-floor.

Keywords Cyber Physical Systems · Internet of Things · Industry 4.0

75.1 Introduction

Implementation of Cyber Physical Systems (CPS) and Internet of Things (IoT) in manufacturing environment is significantly changing the way we manufacture. It is expected that these changes will have the effect that the introduction of steam engine, electricity, division of labor, electronics, etc. had in the past, i.e., that they lead to new industrial (r)evolution known as Industry 4.0 [1]. Namely, contemporary market conditions induced by globalization and fluctuating demand, force companies to embrace mass customization production paradigm [2] in which products are manufactured according to the customer needs. This kind of production leads to significant decrease of lot sizes up to the level of one-off products and, to survive in such conditions, companies have to employ highly adaptable manufacturing systems.

Z. Jakovljevic (✉) · D. Nedeljkovic
Faculty of Mechanical Engineering, University of Belgrade, Kraljice Marije 16, 11000 Belgrade, Serbia
e-mail: zjakovljevic@mas.bg.ac.rs

M. Rackov et al. (eds.), *Machine and Industrial Design in Mechanical Engineering*, Mechanisms and Machine Science 109,
https://doi.org/10.1007/978-3-030-88465-9_75

Recent class of manufacturing systems—Reconfigurable Manufacturing Systems (RMS) represents the main process enabler for mass customization paradigm [3]. They are characterized by high adaptability to different products that is achieved through fast and economically effective changes in RMS functionality, structure and capacity [4]. Required adaptability is provided through RMS functional and/or physical reconfiguration that can be rapid and effective only if:

- Information about current status of the production process at shop-floor is available in real-time, and
- A very high modularity of the system at hardware and software level is achieved.

Furthermore, for fast response to customer needs the status of the processes at shop-floor should be shared with suppliers, users and other business stakeholders. In other words, the integration of physical systems and their cyber representation through interaction in real time is necessary, i.e., manufacturing systems should be designed in a form of CPS.

Successful implementation of Industry 4.0 and digitalization of all business processes requires highly skilled workforce with new job roles [5], and one of the main pillars of Industry 4.0 is the education of engineers capable to design and implement CPS at factory shop-floor. In this paper we analyze some important aspects that should be addressed during education of manufacturing engineers to provide them with knowledge and skills necessary for CPS design and implementation.

The reminder of the paper is structured as follows. In Sect. 75.2 we analyze the role and different levels of CPS implementation in manufacturing. Section 75.3 addresses the distribution of control tasks to CPS within RMS, and verification and security related issues introduced during this process. In Sect. 75.4 we provide some topics that should be addressed in the education of manufacturing engineers capable to implement CPS in manufacturing environment. Finally, Sect. 75.5 gives concluding remarks.

75.2 CPS in Manufacturing

CPS are systems in which physical processes and computation are integrated through real-time interaction, and the behavior of the system is defined by its physical and cyber part [6]. One example of CPS is smart sensors and actuators that have integrated computation and communication capabilities. Their development is enabled by recent advances in embedded systems and Information and Communication Technologies. Instead of delivering raw data, smart sensors are capable to retrieve suitable information from the system [7]. Figure 75.1 represents an example of smart sensor for detection of abrupt changes in machining process that is based on vibrations measurement. In this sensor, accelerometer as sensing element (ADXL311 Analog Devices dual axis accelerometer) is augmented with wireless node based on low power Atmel Atmega16 microcontroller and Microchip MRF24J40 IEEE 802.15.4

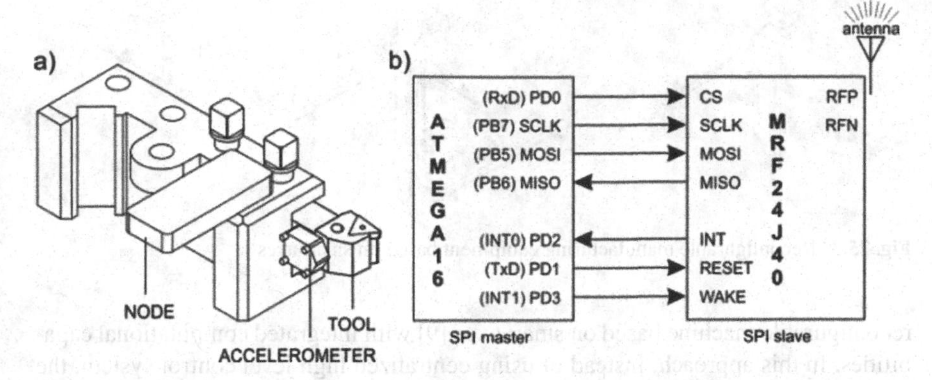

Fig. 75.1 Smart accelerometer: **a** position on cutting tool and **b** node configuration

compatible transceiver [8]. An information machine based on discrete wavelet transform and fuzzy c-means clustering models the behavior of the machining process in normal conditions and in the presence of abrupt changes; it is implemented in microcontroller and used for the detection of abrupt change during machining; the occurrence of this change is transmitted to the higher level control system using wireless transceiver.

Another example of CPS is a smart pneumatic cylinder with integrated proximity sensors for limit positions detection and dual control valve that is augmented with wireless node based on ARM Cortex-M3 MCU and Microchip MRF24J40 IEEE 802.15.4 compatible transceiver (Fig. 75.2). This smart cylinder is capable of performing certain tasks in coordination with other smart actuators and sensors within its network without the need for high level control system as will be elaborated in Sect. 75.3.

Smart sensors and actuators, such as those we have presented, are applied in various manufacturing resources for example machine tools (condition monitoring, machine axes, etc.), robots (grippers, vision sensors, force sensors, etc.), measurement equipment (measuring heads, laser sensors, etc.). These devices promote reconfigurability of manufacturing resources, since, in addition to traditionally present modularity of mechanical subsystems, they inherently provide the modularity of software and control hardware components. Figure 75.3 presents an example of the

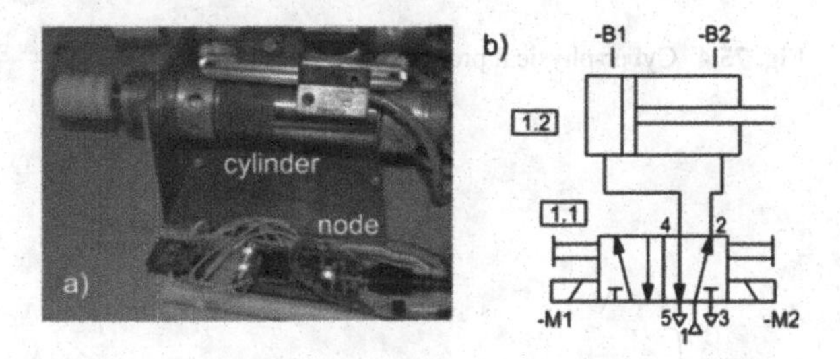

Fig. 75.2 Smart pneumatic cylinder: **a** photo and **b** pneumatic circuit

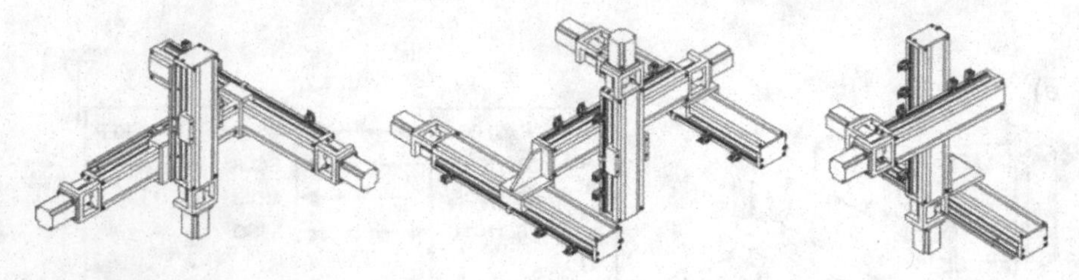

Fig. 75.3 Reconfigurable manufacturing equipment based on smart axes

reconfigurable machine based on smart axes [9] with integrated computational capabilities. In this approach, instead of using centralized high level control system, the performance of the reconfigurable machine is achieved through interaction of axes' local controllers.

Within RMS the CPS are configured as systems of systems where the systems of lower complexity (smart sensors, actuators…) create more complex systems (manipulators, machining systems…). The highest level of CPS implementation in manufacturing represents Cyber-Physical Production System (CPPS) [10]. Within CPPS the whole manufacturing system and its cyber representation are integrated through intensive real-time exchange of data/information (Fig. 75.4). For CPPS the high permeability of data between physical world and cyber model is crucial; it enables reliable virtualization of manufacturing system as well as the feedback from cyber world to the physical system in real-time. It should be noted that cyber part of manufacturing system can be in the form of digital twin with virtual representation or another kind of cyber representation. In addition to smart sensors and actuators that are necessary for feedback, the realization of CPPS requires [10] a reliable network of smart resources with common semantics, and methods for big data analysis and information retrieval.

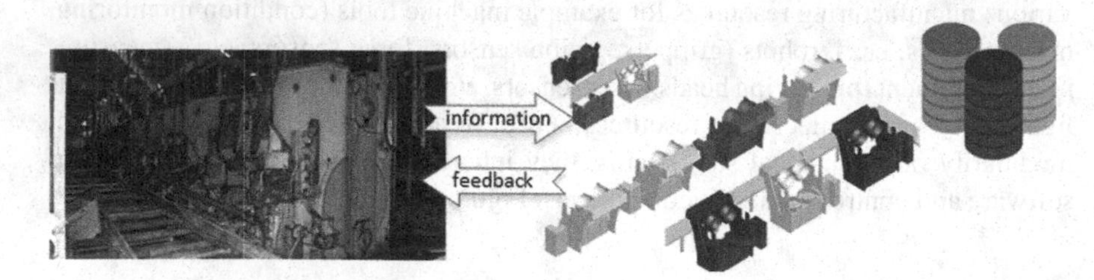

Fig. 75.4 Cyber-physical production system

75.3 Distribution of Control Tasks to CPS

Implementation of smart devices with integrated computational and communication capabilities at factory shop-floor will have significant influence on the manufacturing processes control. Traditionally, in manufacturing environment sensors and actuators are connected to high level control devices which centrally control the behavior of the system [11]. On the other hand, smart devices enable the distribution of control tasks to their local controllers, where each device performs given subtask, and the functionality of the system as a whole is obtained through smart devices' intensive communication and interoperability. It is expected that within Industry 4.0 traditional automation hierarchy standardized by IEC 62264 will give the way to distributed control systems. Nevertheless, all functions of the hierarchy will remain, but they will be distributed over system elements [12, 13].

IEC 61499 standard enables modeling distributed control of manufacturing resources based on smart modular equipment. Using this standard each smart device is represented by its own object—function block that models its functionality, and the behavior of the system as a whole is modeled by function blocks' interconnection. As an example, in Fig. 75.5 we provide the high level model (called application in IEC 61499 formalism) of a manipulator; the modeling is carried out in 4diac software [14]. The manipulator has three degrees of freedom in configuration TRT carried out using smart pneumatic cylinders *A*, *C* and *B* respectively with functionalities such as in cylinder from Fig. 75.2 and a smart gripper denoted *D* augmented with control valve and local controller. Each smart device is modeled using two function blocks—one modeling the cyber and the other modeling the physical part of the device. These parts are integrated through close real-time interaction modeled by interconnection of input and output signals and events represented by solid blue and red lines in Fig. 75.5.

IEC 61499 represents a very useful framework for modeling and simulation of distributed control systems. Nevertheless, it assumes that the behavior and tasks allocated to devices are known and that they are created by system designer; the standard itself does not provide systematic method for the design of system parts behavior.

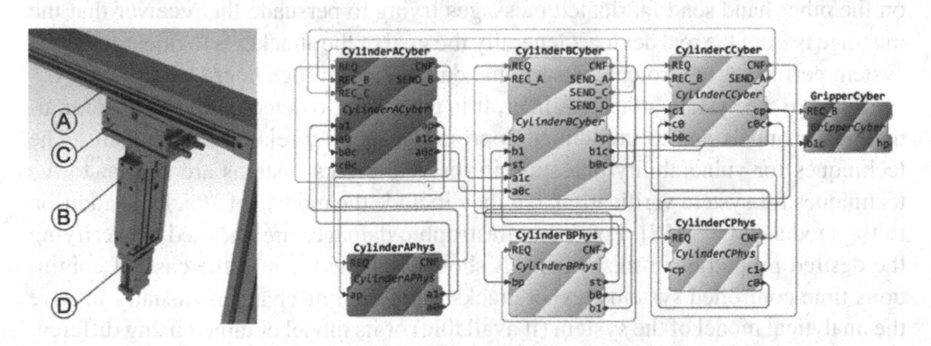

Fig. 75.5 Modular pneumatic manipulator and IEC 61499 application that models its behavior

Distribution of control tasks to smart devices can be very complex and error prone. To ensure safe and desired performance of the system, before deploying distributed tasks to real-world devices, it should be verified that the requested behavior of the system is maintained after the distribution of control tasks to local controllers. As in the case of centralized control, different formal methods (that will be mentioned in the sequel) can be used for this purpose. The starting point in verification is the generation of the system model using the selected method whose behavior can be verified by formally specifying the desired properties of the system and checking their fulfillment using software tools that are available for different formal methods. The most frequently used techniques for specification of the desired CPS properties are Linear Temporal Logic—LTL [6] and Computation Tree Logic—CTL [15].

Intensive communication between the devices raises the issue of their interoperability. The first 4 layers of OSI (Open Systems Interconnection) communication model are well standardized. On the other hand for the upper layers of communication OPC-UA (Open Platform Communication—Unified Architecture) offers the most promising solution with respect to Industry 4.0 framework [1]. OPC-UA represents an IEC (International Electrotechnical Commission) standard for data exchange between devices in industry. Communication according to this standard is platform independent and can be used with different communication media (wired or wireless) [16]. The main goal of this standard is to ensure the interoperability of devices through exchange of information along with its semantic meta-model [17]. One of the main characteristics of OPC-UA is scalability in the sense that it enables networking of different devices starting from sensors, actuators, through programmable logic controllers (PLCs), embedded systems up to the Manufacturing Execution System (MES) and Enterprise Resource Planning (ERP). Essentially, it enables the communication inter and intra all levels of automation pyramid.

The ubiquitous communication opens up space for different cyberattacks and raises security related issues. Generally, there exist a number of different attacks that can be classified into denial of service and deception attacks. Denial of service attacks prevent the data/information from reaching the intended device although sent by transmiter; these kinds of attacks are similar to different communication related issues and can be easily misdiagnosed as connection problems. The deception attacks on the other hand send fabricated messages trying to persuade the receiver that the message is from the real device. Generally the goal of the attacker is to downgrade the system performance through catastrophic damage or through degraded efficiency.

To make the system attack resilient, it is necessary to detect vulnerable communication channels and to implement appropriate security related mechanisms. The techniques for vulnerability detection in discrete events systems are similar to the techniques for system verification, but they include the models of attacks in addition to the model of CPS [18]. Possible catastrophic damages are checked by verifying the desired properties of the system. It should be noted that in the case of continuous time controlled systems, cyberattacks protection mechanisms usually involve the analytical model of the system (if available) or its model obtained using different regression techniques including machine learning [19].

75.4 Physical and Cyber Subsystem Co-Design

As can be observed from examples presented in previous sections, in CPS the behavior of the system is defined by close interaction of cyber and physical worlds, and CPS represent the intersection and not the union of these worlds [6]. Thus, effective CPS require the co-design of cyber and physical part of the system that necessitates highly interdisciplinary knowledge.

During the development of CPS three phases can be distinguished [6]:

- Generation of the system model,
- Design of the system and
- Verification of the system performance through its extensive analysis.

In CPS the design of physical part of the system (e.g., mechanical subsystem) should be closely interconnected with the design of its cyber part, i.e., the control hardware and software. Manufacturing engineers traditionally obtain competencies in the design of mechanical subsystems, as well as in the design and implementation of sensing and actuation elements. Nevertheless, to carry out successful co-design of physical and cyber part of CPS, future engineers should get competencies in different embedded platforms such as microcontrollers, PLCs, DSP (Digital Signal Processor) processors. The students should be familiar with the architecture, capabilities, performances, and programming of such devices.

Since the information retrieval from raw data represents one of the main pillars for successful virtualization of real-world factories within CPPS, the competencies in signal processing and information retrieval are necessary. In this context different techniques for stationary (Fourier Transform, FIR filtering, etc.) and non-stationary (wavelet transform, Hilbert-Huang transform, etc.) signal processing should be studied. Furthermore, significant competencies in Machine Learning and Artificial Intelligence are needed to enable effective information retrieval from raw data.

The design of the cyber representation of the system is highly correlated with the system model. In general, CPS are modeled using traditionally employed techniques for corresponding systems. Discrete Event Systems (DES) are modeled using Finite State Machines or Control Interpreted Petri Nets and the derived GRAFCET and SFC (Sequential Flow Charts) concepts that are widely spread in control engineering practice. For continuous time systems, generation of system model in the form of Differential and Algebraic Equations is most frequently met. Furthermore, for complex and non-linear systems different machine learning based models are very useful. Hybrid models that contain both, DES and continuous time elements can be generated using a number of different techniques among which time automata can be singled out.

As shown in previous sections, to enable the manufacturing systems reconfigurability, it is expected that strict automation hierarchy will be replaced by truly distributed control systems. Thus, in addition to IEC 61131–3 PLC programming languages next generation manufacturing engineers should be familiar with IEC 61499 standard for modeling distributed control systems. Furthermore, the distributed control systems complexity usually makes them less transparent then equivalent centralized control systems thus making their verification time consuming and error prone. Considering this, to facilitate distributed control systems verification, different formal methods (finite state automata, time automata, Petri Nets, Timed Petri Nets to name a few) should be introduced to manufacturing engineers' education, as well as the methods for desired CPS properties specification such as LTL and CTL. In addition, practical engineering techniques for distribution of control tasks to smart devices (CPS) such as one presented in [20] should be studied; although at the moment these techniques are scarce, it is expected that in the following years a number of practically and readily applicable methods will be developed.

CPS represent a basis for Industrial Internet of Things (IIoT), and to successfully implement IIoT manufacturing engineers should have competencies in Industrial Internet Architecture as well as in different wired and wireless networks. In industrial internet, interoperability of different devices is one of the most important aspects. In this context, to ensure the interoperability of multivendor components/systems within factory network engineers should obtain the knowledge and skills in OPC-UA. Finally, within Industry 4.0 cybersecurity becomes one of the most important issues. Different kinds of attacks, as well as the methods for their detection/prevention and for the design of attack resilient systems should be a part of the curricula for the education of manufacturing engineers for Industry 4.0.

75.5 Conclusion

In this paper we have analyzed the most important aspects of CPS implementation within Industry 4.0 manufacturing facility. The most significant opportunities, challenges and threats that this implementation brings about are considered, as well as a number of different scientific and engineering methods/techniques that have been employed for successful CPS design and implementation.

Based on the performed analysis, the crucial skills and competencies as well as topics for education of manufacturing engineers capable to readily and effectively employ CPS in real-world are identified. These topics cover the most important elements necessary for (i) CPS design, development and implementation, i.e., co-design of physical and cyber parts of CPS, (ii) CPS modeling and distribution of control tasks to CPS formed as systems of systems realized through IoT, and (iii) systems verification and security related issues. Our future work will be devoted to implementation of the presented findings at the M.Sc. level education at Faculty of Mechanical Engineering in Belgrade where new courses covering these topics have been established. Furthermore, since Industry 4.0 and CPS represent a very active R&D area, new results will be constantly followed and introduced into corresponding curricula.

Acknowledgements This research was supported by the Science Fund of the Republic of Serbia, grant No. 6523109, AI-MISSION 4.0.

References

1. Kagermann, H., Wahlster, W., Helbig, J.: Recommendations for Implementing the Strategic Initiative INDUSTRIE 4.0. Forschungsunion, Acatech (2013). http://www.acatech.de. Last accessed 2021/02/04
2. ElMaraghy, H.: Smart changeable manufacturing systems. Proc. Manuf. **28**, 3–9 (2019)
3. Jovane, F., Koren, Y., Boër, C.R.: Present and future of flexible automation: Towards new paradigms. CIRP Ann. Manuf. Techn. **52**(2), 543–560 (2003)
4. Koren, Y., Gu, X., Guo, W.: Reconfigurable manufacturing systems: Principles, design, and future trends. Frontiers of Mech. Eng. **13**(2), 121–136 (2018)
5. The Future of Jobs Report 2018, World Economic Forum Centre for the New Economy and Society. http://www3.weforum.org/. Last accessed 2021/02/04
6. Lee, E.A., Seshia, S.A.: Introduction to Embedded Systems: A Cyber-Physical Systems Approach. MIT Press (2017)
7. Berger, C., Hees, A., Braunreuther, S., Reinhart, G.: Characterization of cyber-physical sensor systems. Proc. CIRP **41**, 638–643 (2016)
8. Jakovljevic, Z., Petrovic, M., Mitrovic, S., Miljkovic, Z.: Intelligent Sensing Systems—Status of Research at KaProm. In: Ni, J., Majstorovic, V., Djurdjanovic, D. (eds.) Proceedings of 3rd International Conference on the Industry 4.0 Model for Advanced Manufacturing: AMP 2018. LNME, pp. 18–36. Springer, Cham (2018)
9. Lesi, V., Jakovljevic, Z., Pajic, M.: Towards Plug-n-Play numerical control for Reconfigurable Manufacturing Systems. In: IEEE 21st International Conference on Emerging Technologies and Factory Automation, ETFA2016, pp. 1–8. IEEE Press (2016)
10. Jakovljevic, Z., Majstorovic, V., Stojadinovic, S., Zivkovic, S., Gligorijevic, N., Pajic, M.: Cyber-Physical Manufacturing Systems (CPMS). In: Majstorovic, V., Jakovljevic, Z. (eds.) Proceedings of 5th International Conference on Advanced Manufacturing Engineering and Technologies, NEWTECH 2017. LNME, pp. 199–214. Springer, Cham (2017).
11. Williams, T.J.: The Purdue enterprise reference architecture. Comput. Ind. **24**(2–3), 141–158 (1994)
12. Monostori, L.: Cyber-physical production systems: Roots, expectations and R&D challenges. Proc. CIRP **17**, 9–13 (2014)
13. Jakovljevic, Z., Mitrovic, S., Pajic, M.: Cyber Physical Production Systems—An IEC 61499 Perspective. In: Majstorovic V., Jakovljevic Z. (eds) Proceedings of 5th International Conference on Advanced Manufacturing Engineering and Technologies, NEWTECH 2017. LNME, pp. 27–39. Springer, Cham (2017).
14. 4diac, https://www.eclipse.org/4diac/. Last accessed 2021/02/04
15. Lesi, V., Jakovljevic, Z., Pajic, M.: Reliable industrial IoT-based distributed automation. In: Proceedings of the International Conference on Internet of Things Design and Implementation, IoTDI 2019, pp. 94–105. ACM, New York (2019)
16. OPC UA Online Reference, OPC foundation. https://reference.opcfoundation.org/v104/Core/docs/Part1/. Last accessed 2021/02/04
17. Leitner, S., Mahnke, W.: OPC UA—Service-oriented architecture for industrial applications. GI Softwaretechnik-Trends **26**(4) (2006)
18. Jakovljevic, Z., Lesi, V., Pajic, M.: Attacks on distributed sequential control in manufacturing automation. IEEE T. Ind. Inform. **17**(2), 775–786 (2021)
19. Nedeljković, D., Jakovljević, Z., Miljković, Z.: The detection of sensor signal attacks in industrial control systems. FME Trans. **48**(1), 7–12 (2020)
20. Jakovljevic, Z., Lesi, V., Mitrovic, S., Pajic, M.: Distributing sequential control for manufacturing automation systems. IEEE T. Contr. Syst. T. **28**(4), 1586–1594 (2020)

Chapter 76
Teaching Methodology for Designing Smart Products

Tatjana Kandikjan, Ile Mircheski, and Elena Angeleska

Abstract This paper aims to explain the teaching methodology used for the course "New Product Development" at the Faculty of Mechanical Engineering in Skopje, Republic of North Macedonia, as a method that promotes project-based learning and design exploration as the most effective learning tools for designers and engineers. The main steps and the learning path strategy of the smart product development process, as practiced with the Industrial Design students, are provided and several project examples are elaborated in order to illustrate the advantage of the applied teaching methodology. The paper also includes an explanation of the state-of-the-art smart products from an aspect of their development for users' acceptance, which is what the methodology is based on. The goal of this paper is to provide an overview of the external and internal learning sources used as a part of the methodology, as well as the special topics thought to support the project development process, in order to inspire other designers, professors and students to apply similar tools for smart product development. Such tools help increase their, as well as target users' understanding of the technological aspect of smart items, and in addition, grow the general willingness to use such products, removing the skepticism and unfamiliarity barriers.

Keywords Smart products · Industrial design · User-centered design · Design exploration · Design teaching methodology

76.1 Introduction

Technological advancements are the basis for the development of new ingenious products. The fast-emerging technologies are inevitably becoming a part of our everyday lives aiming to make living easier and more comfortable. As a result, we are surrounded by smart systems and connected products being a part of the "really

T. Kandikjan (✉) · I. Mircheski · E. Angeleska
Faculty of Mechanical Engineering, University Ss Cyril and Methodius, Skopje, Republic of North Macedonia
e-mail: tatjana.kandikjan@mf.ukim.edu.mk

M. Rackov et al. (eds.), *Machine and Industrial Design in Mechanical Engineering*, Mechanisms and Machine Science 109,
https://doi.org/10.1007/978-3-030-88465-9_76

new products (RNPs)" and the "Internet of Things (IoT)" era. RNPs are highly innovative items that allow the users to experience new functionalities [1] and the IoT is the integration of such inventions with information and communication technologies and sensors [2]. IoT is not only machine-to-machine communication, but a way the Internet connects individuals to computing devices [3]. This brings a new challenge for designers who seek the best methods to design smart products and systems in order to bring the technology closer to consumers and enable them to enjoy the benefits it offers. The design process needs to mature and evolve from focusing on a single device to thinking beyond that one product and designing for connected systems [4].

However, all these smart items are not always well accepted among the users due to multiple reasons among which the leading one is using technology-centered approaches for developing products, instead of using a human-centered approach [5]. In this sense, the slow adoption process can happen as a result of: experienced barriers and difficulties to understand the new features [1]; issues with manual programming of the smart connected systems (trigger-action programming; if this-then that) [6]; the invisibility, as a requested quality for interface design, which causes uncertainties when using such seamless, immaterial, invisible technologies (codes, wireless connections, microprocessors, the "cloud") by end-users who are typically used to experience products through their materiality [7] etc. In addition, all the mentioned points can also cause a confusion for designers as well, not only users. The main reason for this is the fact that the complexity of smart products requires the work of a multidisciplinary team, not only one individual [8].

Therefore, the teaching methodology in design and development of new products should prepare the next generations of designers to cope with the revolutionary changes in technology, rapid change of customer needs and the consequent creation of new innovative products. The success of the new products will depend on how the users react when interacting with smart and connected products, their preferences, the flow of the technology adoption process and suitable methods to make smart products more fun and useable. The solution to this problem might be in promoting a teaching process for new product development based on human-centric methods, multidisciplinary strategies and project-based learning. This paper elaborates the practical application of such a teaching methodology for new, smart product development.

Understanding the challenges connected with the design of smart items is the stepping stone to developing successful RNPs. Therefore, Sect. 76.2 includes an overview of the latest trends of implementing smart technologies in product design, from an aspect of p2u communication (with the goal to promote human-centric design). Following the dynamics of ever-changing technology and design trends is an essential part of the teaching process. The drawn conclusions are incorporated in the core of the teaching methodology presented in Sect. 76.3, which is based on combining theoretical knowledge and practical skills gained through experience from working on projects. Students are trained to apply a wide range of changing and evolving knowledge databases, as well as creative thinking strategies and collaborations with students form other study programs.

76.2 Smart Technologies and Their Implementation in Product Design

This section includes an overview of the implementation of smart technologies in design of everyday products focusing on the product-user communication of smart products. Conclusions are drawn as recommendations for industrial design of new smart products that are incorporated into the teaching methodology presented in this paper.

76.2.1 Communication with Smart Technologies

There are [3, 9] four stages of users' acceptance of technological devices: pre-adoption, adoption, adaptation, and use/retention. In these stages, several processes take place: expectation, trust building, behavior change, and minimal use, with an optional stage of routinized use. In order to avoid minimization of use and routinized use, the suggestion is to create products that constantly invite the exploration of their options and can respond to changing conditions. Doing so, IoT devices expose their usefulness, promote new behaviors and additionally increase the trust in the technology. In order to increase the trust in the technology even more, trends for explaining the dematerialization (which is desired for smart products) and finding new ways for p2u communication have emerged. Researches aim to reveal and mediate the invisible materials of technologies used in smart products in order to help individuals learn and understand the high-tech functionality [7, 10]. Examples try to capture the essence of using smart consumer technology as a material. This can be done by designing smart products to be physically inviting and to actively participate in communications with the user, as illustrated by the example shown on Fig. 76.1.

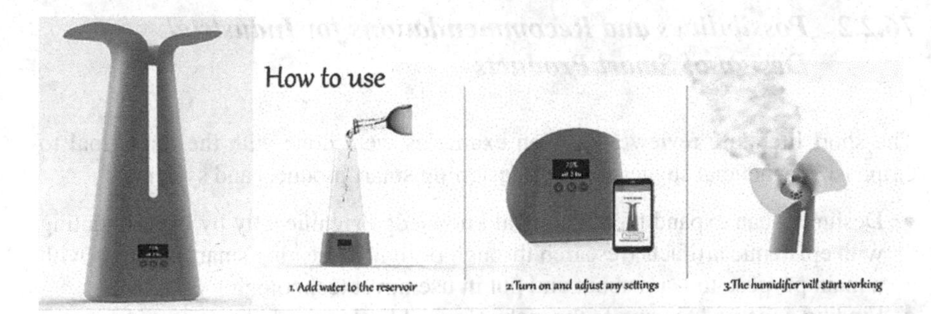

Fig. 76.1 Conceptual design of an air humidifier with basic smart functionality, and with shape inspired form a traditional glass water-pitcher (student project)

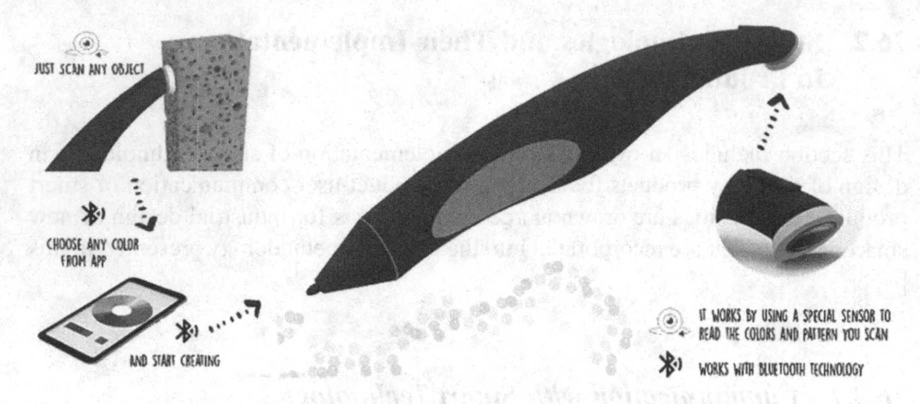

Fig. 76.2 "Pigmentum"—Concept of a device that can scan color and pattern from objects, shaped as a leech to express a simple association to its functionality (student project)

When exploring more methods for understanding the functionalities and innovations, particularly of RNPs, researchers suggest that shaping the products using metaphors can help to reduce the cognitive load of users and allow them to instantly understand how the product works [1]. The product illustrated in Fig. 76.2 is an example of a successfully applied metaphoric design—the smart pen that "sucks colors and patterns from objects" resembles a leech, explaining how it works through its' shape. Metaphorically designed products stimulate interaction with the technology through their physical form.

The mentioned examples illustrate means of human—smart product interaction. Researchers go deeper into these forms of interaction and suggest defining the products' form and functionalities based on the relationships they aim to establish with the users [11]. This is one more trend for designing smart products for which the relationship is the goal for designing an artifact.

76.2.2 Possibilities and Recommendations for Industrial Design of Smart Products

The short literature review and given examples were done with the main goal to emphasize beneficial suggestions for designing smart products and systems:

- Designers can expand their technical knowledge significantly by experimenting with epistemic artifacts (research through design)—creating smart products with a main purpose to learn about and put in use novel technologies,
- Finding a method to express the technological background of smart products in a material way can help bring them closer to end-users,

- Physical engagement and interaction with palpable computing items helps to enhance the learning and cognitive development and through such interaction, a circle of input and feedback is created which can influence users' behavior.

In addition, it is important to emphasize three main points for the development of smart products, as listed by Andreas G. Mysen [8]: user-centered design (understanding users and incorporating them in frequent prototype evaluations), product-user interaction (simplifying the communication between item and user) and user experience (people are emotion-driven creatures and therefore products need to be designed to meet their emotional expectations [12]). All these points are used as an input in the process of developing concepts for smart products which is a part of the teaching methodology for smart product development explained in the next section of this paper.

76.3 Teaching Methodology for Smart Product Development

76.3.1 Goal and Learning Strategy

The main goal of the methodology used for the course "New Product Development" (NPD) at the Faculty of Mechanical Engineering, industrial design studies, is to promote project-based learning as the most effective learning tool for industrial design engineers. The NPD course introduces students to: recognizing opportunities, ethnographic characteristics, usability and aesthetic aspects, development of scenarios, conceptualization, prototyping and improvement, evaluation of product value for the customers, product branding, product/market strategy and improvement of the concepts.

In the past few years, the attention of the practical part of the course was shifted towards training the students to explore and conceptualize smart products. In Fig. 76.3, the learning path strategy of the NPD course is given. The key points of learning NPD are continuous acquisition of knowledge during the process, as well as post-project learning. This is especially important in the development of smart products, which are highly innovative, and whose development is based on the integration of several disciplines. The project-based learning includes using epistemic artifacts, expressing technology in a material way, designing smart products that encourage physical engagement, and generating loops of product-designer and product-user interactions. These points are an essential part of human-centric design of smart items, as described.

Students are guided to use a variety of learning resources. In addition to the teaching materials for the NPD course, selected case-studies of smart products are analyzed, and student projects from previous generations are also selected for reviewing. Students and teacher also use various external learning resources. For

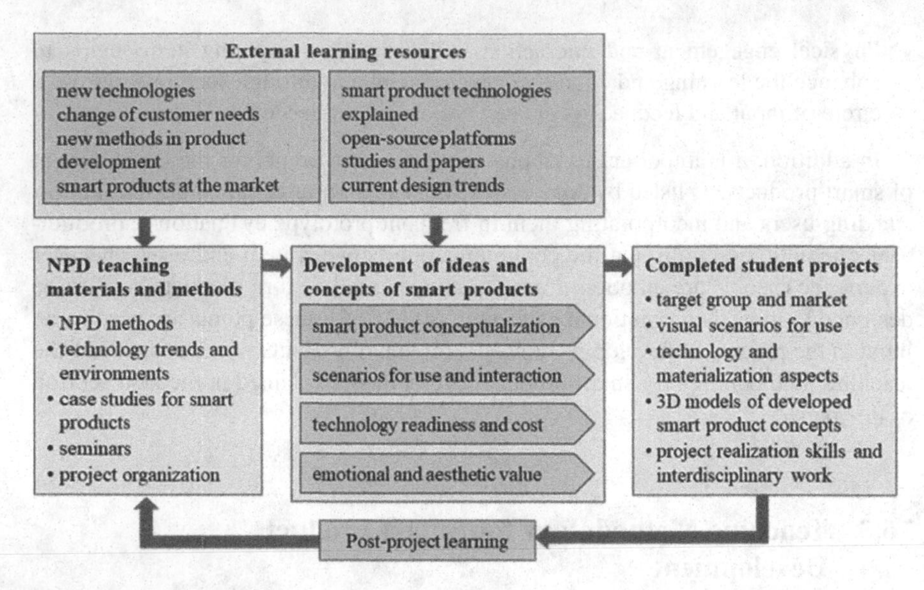

Fig. 76.3 Learning path strategy of the NPD course

example, students are introduced to various ubiquitous technologies using TU Delft's learning platform: WikID, the Industrial Design Engineering Wiki. Similarly, for following current design trends, students are connected through Internet to the latest events, products, papers, summaries and open-source platforms (for example, AWOL Trends).

The need for an industrial designer to participate in an NPD team can vary considerably, depending on the type of the product developed. Therefore, we limit the spectra of student projects to smart products in which: the form is as relevant for customer satisfaction as the implemented technologies; used technologies are widespread and do not require a special environment. The proposed project should be simple enough for the student to be able to develop a virtual prototype concept over a period of 15 weeks.

Special challenge for the industrial design students is to design products that should be perceived as both acting smart and looking smart, through expressing visual identity aspects that can distinguish smart products from other products on the market. The students are encouraged to work on projects they are passionate about, and to use an interdisciplinary approach for collaborating with other undergraduate students from different studies: mechanical engineering, information science and computer engineering.

The design process for smart products is considered as a specialization of the NPD process. The special topics thought to support the project development process for smart products, as practiced with the students in industrial design, include: understanding the purpose of ubiquitous technologies; establishing consultation/teamwork with students from other specialties; design for communication and interaction;

components/modules needed to implement the technology; creating graphic design of web-based applications for product control; and finalizing the visual/textual communication to explain the functionality of the product to the customer.

Post-project learning is necessary to gather and organize the acquired knowledge, which is further used for the next generations of students. Groups of information that are accumulated annually for updating internal learning resources are analyses of innovative smart products; new technologies introduced in consumer products; and completed student projects, presented at the industrial design studies website.

76.3.2 Examples—Student Projects

Students follow the main steps of the methodology in order to complete a different design project chosen by them. For some of those projects, prototypes are created as "epistemic artifacts" which serve the purpose to learn about the implementation of technology and as models for further testing and evaluation. The other part of the projects are concept projects, only presented by posters. For all the projects, students turn in a final report that includes: description of the identified need for the product; analysis of SET factors (Social, Economy and Technology); analysis of other similar products on the market; description of the characteristics of the target group; description of the full idea for the product (inspiration, technology, style, sketches, keywords, descriptions of the environment it should be used in, its' p2u and p2p relationships); the emotions to be triggered when using the product; the materials its composed of; its' practical, sign and symbolic functions; a detailed scenario for packing, transporting, selling, using, storing, maintaining, and other aspects involved in the life cycle of the product; and finally, 3D models of the developed product and technical drawings.

The following section includes several examples of student projects—the first one being a real physical prototype, and the others, concepts illustrated by posters.

The first example is a smart, kinetic installation, which was placed in the hallway of the Faculty of Mechanical Engineering—Skopje (Fig. 76.4). This project was realized by a team of five students from industrial design, computers and information technology, and mechanical engineering. The kinetic installation uses rotating components made of recycled material (cardboard rolls) cut in the same size and differently colored on both sides. As a person approaches the installation, the cylindrical components rotate according to his/her movement and reveal their black side creating a reflection of the viewers' silhouette. To make the movement possible, a "Kinect" camera was installed with a proximity sensor and a three-camera system along with a processor. Software was developed, integrated and tested. The "Arduino" platform was used as an open-source, electronics-prototyping platform and the signal and electrical installations that connect the control units and the motors was planned. The structure of the whole installation, on which the rotating elements are placed, was created from plywood.

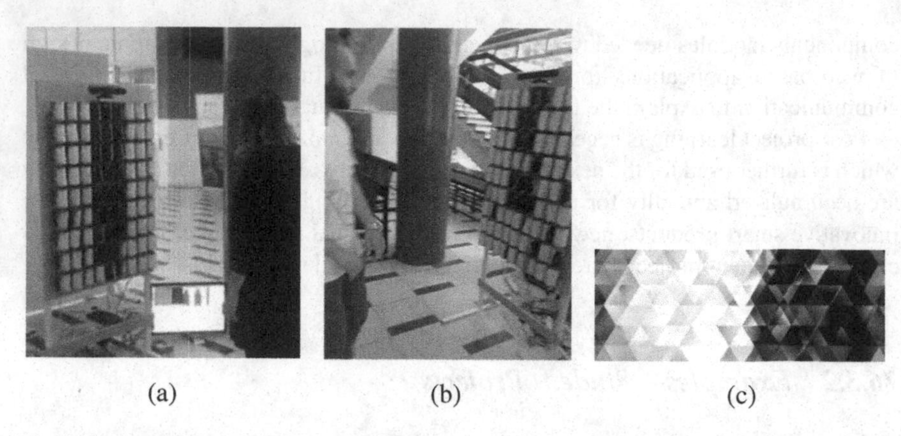

(a) (b) (c)

Fig. 76.4 Smart mirror. Function: Advertisement, amusement, education. Technology: Kinect camera, processor, software, Arduino prototyping shields, servomotors, multiplexers. Components: Back panel, motor holders, cardboard rolls and decorative paper (**c**). Cylinders rotate half-turn as they detect a student approaching, simulating her shadow (**a**). Cylinders rotate following the dancing movements of a student (**b**)

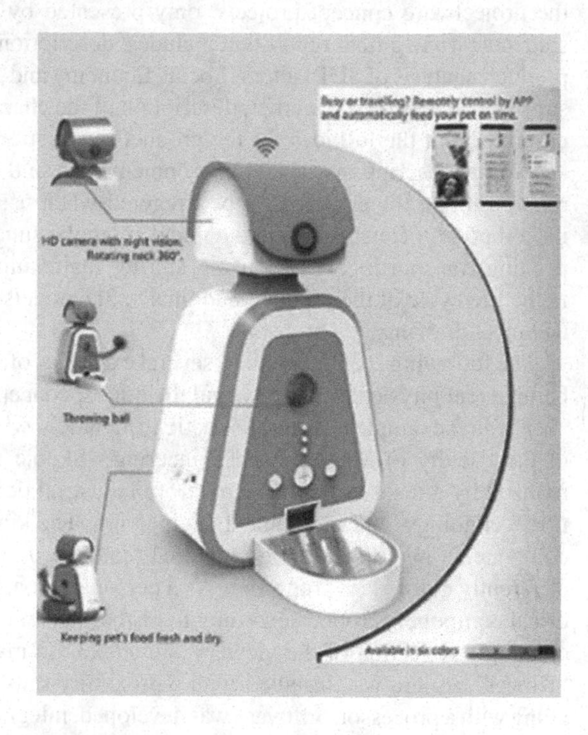

Fig. 76.5 Smart pet-feeder. Function: Food dispenser, pet-monitoring, sound-playing and ball-throwing. Technology: Wi-Fi, app-controlled, HD camera. Components: Case, rotating "head", food-storing container, feeding-bowl, HD camera, voice-recorder, RFID system

Fig. 76.6 Stress-relief nebulizer. Function: Nebulizer offering breathing exercises, emotional data recording. Technology: Bluetooth connected, app-controlled. Components: Case, display, compressor, electronics, microcontroller and sensors, light, battery, USB cable

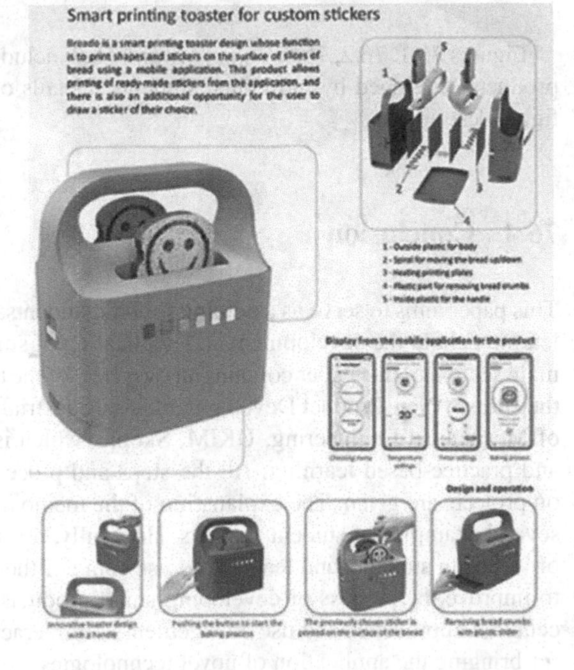

Fig. 76.7 Smart toaster. Function: Toaster producing icons on the slices of bread. Technology: Wi-Fi, app-controlled for choosing icons, temperature and time of toasting. Components: Case, removable lid for crumbs, button, lights, springs, heating plates

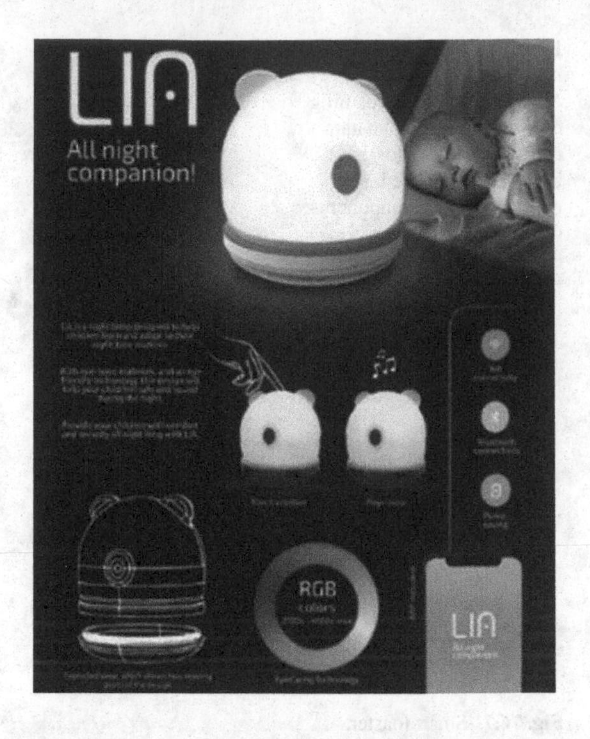

Fig. 76.8 Smart night light. Function: Night light with 3 functions—light, music and baby/child monitoring. Technology: Wi-Fi and Bluetooth, app-controlled, touch-activated, monitoring system. Components: Microphone, speaker, LED lights, electronics, battery

Figures 76.1, 76.2, 76.5, 76.6, 76.7 and 76.8 include few more successful concept products developed by individual students. Details of the projects are given in the figure descriptions.

76.4 Conclusion

This paper aims to serve as a helping tool for students, as well as professors, who are learning about the development and application of smart products and systems. The main section of this paper contains an overview of the teaching methodology used for the course "New Product Development" for Industrial Design Studies, at the Faculty of Mechanical Engineering, UKIM, Skopje, which is based on design exploration and practice-based learning. All the steps and procedures followed when working on projects are given. The explanation of the methodology is enriched by analyzing several examples of student projects. Hopefully, this paper meets the main purpose of inspiring students and teachers to use some of the segments of the methodology to improve the process of developing smart products. The goal is to practice user-centered approaches and use engagement and interaction as extremely useful ways for bringing the application of novel technologies closer to designers and users.

References

1. Cheng, P., Mugge, R., De Bont, C.: Smart home system is like a mother: The potential and risks of product metaphors to influence consumers' comprehension of really new products (RNPs). Int. J. Des. **13**(3), 1–19 (2019)
2. Sayar, D., Er, Ö.: The antecedents of successful IoT service and system design: Cases from the manufacturing industry. Int. J. Des. **12**(1), 67–78 (2018)
3. Cho, M., Lee, S., Lee, K.-P.: How do people adapt to use of an IoT air purifier? From low expectation to minimal use. Int. J. Des. **13**(3), 21–38 (2019)
4. Funk, M., Eggen, B., Hsu, Y.-J.: Designing for systems of smart things. Int. J. Des. **12**(1), 1–5 (2018)
5. Coskun, A., Kaner, G., Bostan, I.: Is smart home a necessity or a fantasy for the mainstream user? A study on users' expectations of smart household appliances. Int. J. Des. **12**(1), 7–20 (2018)
6. Funk, M., Chen, L.-L., Yang, S.-W., Chen, Y.-K.: Addressing the need to capture scenarios, intentions and preferences: Interactive intentional programming in the smart home. Int. J. Des. **12**(1), 53–66 (2018)
7. Arnall, T.: Exploring 'immaterials': Mediating design's invisible materials. Int. J. Des. **8**(2), 101–117 (2014)
8. Greftegreff Mysen, A.: Smart products: An introduction for design students. Smart Design, Edition 2017–2018 (2018)
9. Montalván, J., Shin, H., Cuéllar, F., Lee, K.: Adaptation profiles in first-time robot users: Towards understanding adaptation patterns and their implications for design. Int. J. Des. **11**(1), 1–19 (2017)
10. Nordby, K.: Conceptual designing and technology: Short-range RFID as design material. Int. J. Des. **4**(1), 29–44 (2010)
11. Ghajargar, M., Wiberg, M., Stolterman, E.: Designing IoT systems that support reflective thinking: A relational approach. Int. J. Des. **12**(1), 21–35 (2018)
12. Matthiessen, N.: Interactive design and the human experience: What can industrial design teach us? In: Marcus, A. (eds.) Design, User Experience, and Usability. Design Philosophy, Methods, and Tools, DUXU 2013. LNCS, vol. 8012, pp. 100–106. Springer, Berlin, Heidelberg (2013)

Printed in the United States
by Baker & Taylor Publisher Services

Printed in the United States
by Baker & Taylor Publisher Services